New Frontiers in Quantum Electrodynamics and Quantum Optics

NATO ASI Series

Advanced Science Institutes Series

A series presenting the results of activities sponsored by the NATO Science Committee, which aims at the dissemination of advanced scientific and technological knowledge, with a view to strengthening links between scientific communities.

The series is published by an international board of publishers in conjunction with the NATO Scientific Affairs Division

A	**Life Sciences**	Plenum Publishing Corporation
B	**Physics**	New York and London
C	**Mathematical and Physical Sciences**	Kluwer Academic Publishers Dordrecht, Boston, and London
D	**Behavioral and Social Sciences**	
E	**Applied Sciences**	
F	**Computer and Systems Sciences**	Springer-Verlag
G	**Ecological Sciences**	Berlin, Heidelberg, New York, London,
H	**Cell Biology**	Paris, and Tokyo

Recent Volumes in this Series

Volume 227—Dynamics of Polyatomic Van der Waals Complexes
edited by Nadine Halberstadt and Kenneth C. Janda

Volume 228—Hadrons and Hadronic Matter
edited by D. Vautherin, F. Lenz, and J. W. Negele

Volume 229—Geometry and Thermodynamics: Common Problems of Quasi-Crystals, Liquid Crystals, and Incommensurate Systems
edited by J.-C. Tolédano

Volume 230—Quantum Mechanics in Curved Space-Time
edited by Jürgen Audretsch and Venzo de Sabbata

Volume 231—Electronic Properties of Multilayers and Low-Dimensional Semiconductor Structures
edited by J. M. Chamberlain, L. Eaves, and J.-C. Portal

Volume 232—New Frontiers in Quantum Electrodynamics and Quantum Optics
edited by A. O. Barut

Volume 233—Radiative Corrections: Results and Perspectives
edited by N. Dombey and F. Boudjema

Volume 234—Constructive Quantum Field Theory II
edited by G. Velo and A. S. Wightman

Series B: Physics

New Frontiers in Quantum Electrodynamics and Quantum Optics

Edited by

A. O. Barut

University of Colorado at Boulder
Boulder, Colorado

Plenum Press
New York and London
Published in cooperation with NATO Scientific Affairs Division

Proceedings of a NATO Advanced Study Institute on
New Frontiers in Quantum Electrodynamics and Quantum Optics,
held August 14-26, 1989,
in Istanbul, Turkey

Library of Congress Cataloging in Publication Data

NATO Advanced Study Institute on New Frontiers in Quantum Electrodynamics and Quantum Optics (1989: Istanbul, Turkey)
New frontiers in quantum electrodynamics and quantum optics / edited by A. O. Barut.
p. cm.—(NATO ASI series. Series B, Physics; v. 232)
"Proceedings of a NATO Advanced Study Institute on New Frontiers in Quantum Electrodynamics and Quantum Optics, held August 14-26, 1989, in Istanbul, Turkey"—T.p. verso.
"Published in cooperation with NATO Scientific Affairs Division."
Includes bibliographical references and index.
ISBN 0-306-43669-8
1. Quantum electrodynamics—Congresses. 2. Quantum optics—Congresses. 3. Quantum theory—Congresses. I. Barut, A. O. (Asim Orhan), date. II. North Atlantic Treaty Organization. Scientific Affairs Division. III. Title. IV. Series.
QC679.N34 1989 90-7932
537.6′7—dc20 CIP

A Division of Plenum Publishing Corporation
233 Spring Street, New York, N.Y. 10013

Printed in the United States of America

PREFACE

This volume continues a tradition to review, about every five years, the progess achieved in the borderline areas of quantumelectrodynamics, quantumoptics and quantum theory of measurement. Two such titles have appeared,[1,2] and two other volumes with very related contents.[3,4] It is devoted, again, to highlight the most interesting and fundamental experimental and theoretical results of recent years in the tests and various formulations of quantumelectrodynamic and quantumoptical processes.

There are three Sections in this book. The first Section is devoted to Quantum Optics. It begins with a review of crucial experiments in quantum optics, which sets the stage for other contributions on the coherence, interference and squeezing of light. The second Section deals with Quantumelectrodynamics. The experimental tests of QED are reviewed in three papers dealing with positronium, H-atom and high energy physics tests, respectively. Quantumelectrodynamics, although it is the most accurate of all physical theories, is not entirely satisfactory, because of the infinities, the uncertainty in the convergence of the perturbation series, and practical complexity of higher order calculations and of renormalizations. That is why more and more accurate tests of the theory are required. The experiment is in many places more accurate than the theory and a number of higher order processes which are measured, e.g. in positronium, are not yet fully calculated. There seems to be also a major difference between theory and experiment in the decay rate of orthopositronium into three photons. That is also the reason for new theoretical approaches, discussed in this Section, to overcome these problems; two in the famework of standard QED, and one new approach, the selffield QED, in which the nonlinear selfinteraction of the electron is taken fully into account from the beginning instead of a quantized radiation field.

Finally, Section III collects contributions closely related to the first two beginning with the phenomena of quantum coherence in

---

1 *"Foundations of Radiation Theory and Quantumelectrodynamics"* A.O. Barut, editor, Plenum Press 1980.

2 *"Quantumelectrodynamics and Quantumoptics"* A.O. Barut, editor, Plenum Press 1984.

3 *Quantumoptics, Experimental Gravitation and Measurement Theory"* P. Meystre and M.O. Scully, editors, Plenum Press 1984.

4 *"Frontiers of Nonequilibrium Statistical Mechanics"* G.T. Moore and M.O. Scully, editors, Plenum Press 1986.

atom optics and neutron optics. Quantum coherence in all its manifestations leads inevitably to the quantum theory of measurement and nonlocal effects in QED. Some of these problems are discussed here from several different points of view. And there are many more interesting contributions. The topics included in this Volume are a testimony to the richness and significance of this field of physics all the way from its very foundation to many of its applications. I hope that it will be a useful review and reference book.

The contributions collected here were originally presented at the NATO Advanced Study Institute held in Istanbul, August 14–26, 1989. I should like to thank the NATO Advanced Study Institute Progams and its Director, Dr. Luis V. da Cunha for making this meeting, hence this Volume, possible, and to contributors and participants for their enthusiasm. It was a lively and fruitful meeting.

A.O. Barut
Boulder, Colorado
March 1989

CONTENTS

I. QUANTUM OPTICS

II. QUANTUM ELECTRODYNAMICS

QUANTUM EFFECTS IN THE ATOM-RADIATION INTERACTION

Herbert Walther

Sektion Physik, Universität München
and Max-Planck-Institut für Quantenoptik
Garching, FRG

In this contribution quantum-effects in the radiation-atom interaction will be reviewed. It is well known that most of the phenomena involving laser light can be described by semi-classical methods, however, phenomena such as laser noise, the spectra and photon statistics of resonance fluorescence and the interaction of atoms in the single-atom maser require a quantization of the radiation field. In this paper we will describe results on the latter two effects. We will start with the discussion of resonance fluorescence.

I) Resonance Fluorescence

The resonant interaction of laser light with atomic systems has received considerable theoretical and experimental attention over the past decade. Until the advent of the laser, light sources for spectroscopy consisted of ordinary spectral lamps excited by DC or RF discharges, and produced light having a very broad spectral width and, hence, very short correlation time, and a relatively low intensity. For such fields both the experimental and theoretical results are in general well understood. However, the development of the laser made available light sources which are sufficiently intense that an atomic (or molecular) transition can be very easily saturated. In addition, the lasers are highly monochromatic having a coherence time much greater than typical natural lifetimes of excited atomic states, and

New Frontiers in Quantum Electrodynamics and Quantum Optics
Edited by A. O. Barut, Plenum Press, New York, 1990

finally, tunable, making it possible to selectively excite particular atomic transitions.

The theoretical analysis of this physical situation requires the use of techniques more general than those found adequate in the case of thermal fields. In the latter case the weakness of the atom-field interaction meant that perturbative techniques were generally sufficient. These techniques were based on the assumption that the initial state of the atomic system was essentially unchanged by the interaction. However, as saturation can be easily achieved with an intense laser field, more general nonperturbative methods are required. Furthermore, for a highly coherent field, one cannot consider successive photon emission and absorption processes as being independent as it is now possible for an atomic system to undergo many such processes during the correlation time of the laser field, and hence phase memory effects cannot be neglected.

The simplest such system is also one which has attracted an enormous amount of interest: the problem of theoretically and experimentally determining the spectrum of the fluorescent light radiated by a two-level atom driven by an intense monochromatic field. This is the situation that gives rise to the AC Stark effect in which, for sufficiently strong fields, it is found that the spectrum of the scattered light splits into three peaks: a central peak, centered on the driving field frequency with a width $\gamma/2$ (γ^{-1} being the Einstein A coefficient) and having a height three times that of two symmetrically placed sidebands, each of width $3\gamma/4$ and displaced from the central peak by the Rabi frequency. In addition, there is a delta-function (coherent) contribution, also positioned at the driving frequency. In the limit of strong driving fields, the energy carried by this last contribution is negligible in relation to the three-peak contribution. This result was first predicted by Mollow [1] and has now been very well confirmed experimentally (for a review see Ref.[2]).

However, it is not only the spectral property of the fluorescent light that has come under investigation. Examination of the intensity correlation of the scattered field in the basic two-level atom has also attracted much attention since fluorescent light exhibits interesting statistical properties, especially when there is only a single atom at a time interacting with the laser beam. Under these conditions, the phenomenon of photon-antibunching can be observed [2,3]. The single-atom condition cannot easily be fulfilled if experiments are made on neutral atoms, whereas the new techniques of laser spectroscopy of single ions in a radio-frequency trap are very suitable for this purpose, as has been demonstrated by Diedrich and Walther [4].

The Paul-trap used in the experiments is shown in Fig. 1. Mounted inside a stainless-steel ultrahigh vacuum chamber, its ring diameter of 5 mm and its pole cap separation of 3.54 mm are much larger than for other single-ion radio-frequency traps. The single ion is closely confined in the centre of the trap by photon recoil cooling. The large size of the trap affords a large solid angle for detecting the fluorescence radiation which is transmitted through a molybdenum mesh covering a conical bore in the upper pole cap. The experiment has been performed with $^{24}Mg^{+}$ ions produced by an atomic beam ionized in the centre of the trap by an electron beam. The resonance transition has the wavelength λ = 280 nm.

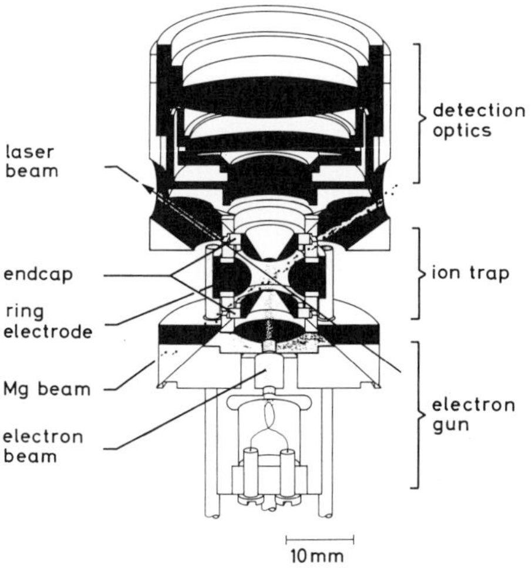

Fig. 1. Scheme of the Paul trap used for the experiments (for details see Ref.[4]).

To investigate the photon statistics, the second-order correlation function (intensity correlation) was measured in a Hanbury-Brown and Twiss Experiment. The intensity correlation is proportional to the probability of detecting a second photon at a time τ after the first. For thermal and non-coherent light this probability has a maximum for $\tau = 0$ and decreases for larger τ. The behaviour is called bunching. The intensity correlation for coherent light is independent of τ. Quantized fields may show additional behaviour: the intensity correlation can have a minimum at $\tau = 0$, a behaviour known as antibunching. Such a field is produced by the single stored ion in the following way: after a photon is emitted, the trapped ion returns to the ground state; before the next photon can be emitted, the ion has to be excited again. This happens through Rabi nutation in the external laser field. On the average, a time of half a Rabi period has to elapse until another photon can be observed. The probability of two photons being emitted in a short time interval is therefore zero.

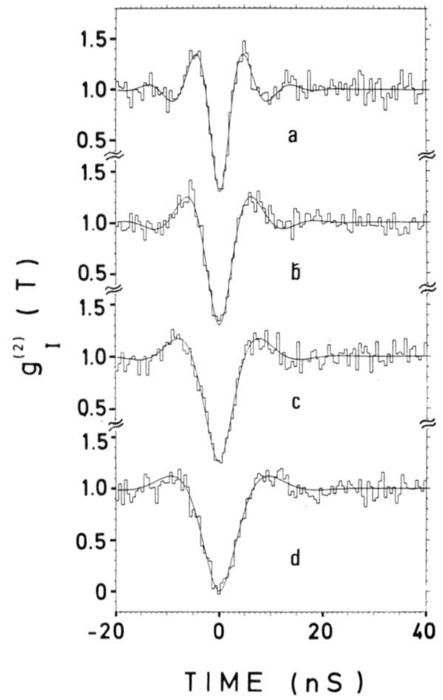

Fig. 2. Results for the intensity correlation. Antibunching for a single ion for different laser intensities, increasing from bottom to top (for details see Ref.[4]).

The results for the intensity correlation of a single stored ion are shown in Fig. 2. Plotted is $g_I^{(2)}(\tau)$, which is defined by: $g_I^{(2)}(\tau) = \langle I(t)I(t+\tau)\rangle / \langle I(t)\rangle^2$.

Owing to a time delay in one of the signal channels the intensity correlation $g_I^{(2)}(\tau)$ could also be measured for negative τ. The laser intensity decreases from a to d and therefore the average time interval in which a second photon follows a first increases.

Previous experiments to investigate antibunching in resonance fluorescence have been performed with laser-excited collimated atomic beams. The initial results obtained by Kimble et al.[3] showed for the second-order correlation function $g^{(2)}(\tau)$ a positive slope which is characteristic for photon antibunching. However, $g^{(2)}(o)$ was larger than $g^{(2)}(\tau)$ for $\tau \rightarrow \infty$, due to number fluctuations in the atomic beam and to the finite interaction time of the atoms. The experiment of Diedrich et al.[4] (Fig. 2) with a trapped ion does not show this restriction. (See also Ref. [2] for other results.)

The fluorescent light of a single stored ion has another interesting quantum property: the fluctuations of the photon number recorded in a small time interval δt is narrower than that expected for a Poisson distribution, i.e. the variance is smaller than the mean value of the photon number and we again find sub-Poissonian statistics. The reason is that the single ion can only emit one photon at a time and fluctuations only occur because of the finite detection probability. Antibunching and sub-Poissonian statistics are often associated, although they are distinct properties and need not necessarily be simultaneously observed [5] , as is the case in the experiment by Diedrich et al.[4]. Although there is evidence of antibunching in the older atomic beam experiments [2,3] the photon counts were not sub-Poissonian as a result of fluctuations in the number of atoms. In the experiment by Short and Mandel [5] this effect was excluded by use of a special trigger scheme for the single-atom event. In the experiment by Diedrich et al. these precautions were not necessary since there are no fluctuations in the atomic number.

II) The One-Atom Maser

The simplest and most fundamental system for studying radiation-matter coupling is a single two-level atom interacting with a single mode of an electromagnetic field in a cavity. This problem received a great deal of attention shortly after the maser was invented. At that time, however, the problem was of purely academic interest: the matrix elements describing the radiation-atom interaction are usually too small, so that the field of a single photon is not sufficient to lead to an atom-field evolution time shorter than the other characteristic times of the system, such as the excited state lifetime, the time of flight of the atom through the cavity and the cavity mode damping time. It was therefore not possible to test experimentally the fundamental theories of radiation-matter interaction. These theories predict, however, some interesting and basic effects. These include the (a) modification of the spontaneous emission rate of a single atom in a resonant cavity, (b) the oscillatory energy exchange between a single atom and the cavity mode, and (c) the disappearance and quantum revival of optical nutation induced in a single atom by a resonant field.

The situation concerning the experimental testing of these basic effects has drastically changed in the last few years since frequency-tunable lasers now allow population of highly excited atomic states characterized by a high main quantum number n of the valence electron. These states are generally called Rydberg states since their energy levels can be described by the simple Rydberg formula. The highly excited atoms are very suitable for observing the quantum effects in radiation-atom coupling for three reasons. Firstly, these states are very strongly coupled to the radiation field (the induced transition rates between neighbouring levels scale as n^4). Secondly, these transitions are in the millimetre wave region, which allows low-order mode cavities that are still sufficiently large to ensure rather long interaction times. Finally, Rydberg states have relatively long lifetimes with respect to spontaneous decay (for reviews see Refs.[6] and [7]).

The strong coupling of Rydberg states to resonant radiation to neighbouring levels can be understood in terms of the correspondence principle: with increasing n the classical evolution frequency of the highly excited electron becomes identical with the transition frequency to the neighbouring level; the atom therefore corresponds to a large dipole oscillating with the resonance frequency. (The dipole moment is very large since the atomic radius scales as n^2.)

In order to understand the modification of the spontaneous emission rate in an external cavity, we have to remember that in quantum electrodynamics this rate is determined by the density of modes of the electromagnetic field at the atomic transition frequency ω_0. The vacuum density of modes per unit volume depends on the square of the frequency. If the atom is not in free space, but in a resonant cavity instead, the continuum of modes is changed into a spectrum of discrete modes with one of them being in resonance with the atom. Since there is energy dissipation within the cavity, a photon radiated at a well-defined frequency will be smeared out over the full spectral width $\Delta\omega_c$

of the resonant mode. The full width at half maximum $\Delta\omega_c$ is related to the cavity quality factor $Q = \omega_c/\Delta\omega_c$.

The spontaneous decay rate of the atom in the cavity γ_c is enhanced in relation to that in free space γ_f by a factor given by the ratio of the corresponding mode densities (V_c is the volume of the cavity):

$$\gamma_c/\gamma_f = \rho_c(\omega_0)/\rho_f(\omega_0) = Q\lambda_0^3/4\pi^2 V_c.$$

For low-order cavities in the microwave region one has $V_c \approx \lambda_0^3$; the spontaneous emission rate is thus roughly increased by a factor of Q in a resonant cavity; conversely, the decay rate decreases when the cavity is mistuned. In this case the atom cannot emit a photon, since the cavity is not able to accept it, and therefore the energy has to stay with the atom.

Recently, quite a few experiments have been conducted with Rydberg atoms to demonstrate the enhancement and inhibition of spontaneous decay in external cavities or cavity-like structures (for the most recent experiment see Ref.[8]).

There are also more subtle effects due to the change of the mode density: radiation corrections such as the Lamb shift and the anomalous magnetic dipole moment of the electron are also modified with respect to the free space value if they are calculated under the boundary conditions of a cavity. The change is just of the order of magnitude of present experimental accuracy. Roughly speaking, one can say that these effects are determined by a change of virtual transitions and not by real transitions as in the case of spontaneous decay (for details see e.g. Ref.[9]).

In the following, attention is focused on discussing the one-atom maser in which the idealized case of a two-level atom interacting with a single mode of a radiation field is realized; the theory of this system was treated by Jaynes and Cummings many years ago [10]. We concentrate on the dynamics of the atom-field interaction predicted by this theory. Some of the features are explicitly a consequence of the quantum nature of the electromagnetic field: the statistical and discrete nature of the photon field leads to new dynamic characteristics such as collapse and revivals in the Rabi nutation.

The experimental setup is shown on Fig. 3. The atoms are injected into a superconducting cavity in the upper state of the maser transition. This was the 63 $p_{3/2}$ Rydberg level, populated by excitation with frequency-doubled light of a dye laser. The atoms are monitored by using field ionization. This detection can be performed state-selectively by choosing the proper field strength [11].

In most of the experiments the transition 63 $p_{3/2}$ - 61 $d_{5/2}$ with a frequency of 21.456 GHz was investigated. When the cavity is tuned in resonance to this transition, the number of atoms in the upper state decreases owing to enhanced spontaneous emission. The tuning of the cavity is performed by squeezing the cavity with piezoelectric elements. The flux of atoms is very low so that the average number of atoms in the cavity at a time is usually much less than unity. The interaction time of the atoms with the cavity field can be varied by means of a Fizeau velocity selector. In this way, the dynamics of the energy exchange between the atom and cavity field can be investigated (Fig. 3).

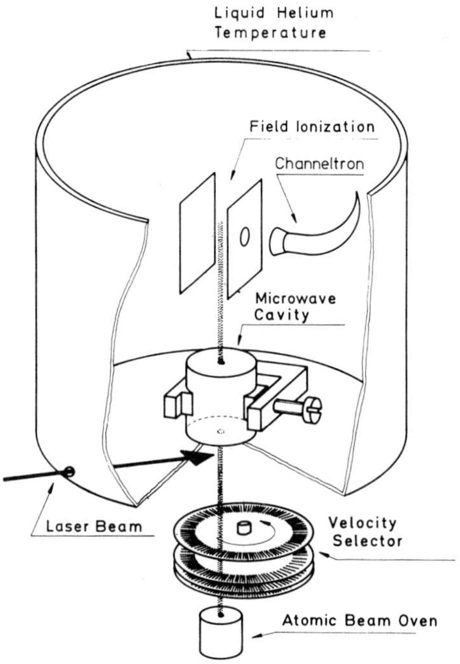

Fig. 3. Scheme of the single-atom maser for measuring quantum collapse and revivals (see Ref.[13]).

With very low atomic-beam flux, the cavity contains essentially thermal photons only, whose number varies randomly obeying Bose-Einstein statistics. When the velocity of the atoms is changed, the probability of the atom being in the excited state $P_e(t)$ after interaction in the cavity varies with the interaction time in an apparently random way. At higher atomic-beam fluxes the atoms deposit energy in the cavity and the maser reaches the threshold so that the number of photons stored in the cavity increases and their statistics changes. For the case of a coherent field the probability distribution is a Poissonian which results in a dephasing of the Rabi oscillations, and therefore the envelope of $P_e(t)$ collapses; after the collapse $P_e(t)$ starts oscillating again in a very complex way. These recurrences occur periodically, the time intervals being proportional to $n^{1/2}$. Both collapse and revivals in the coherent state are pure quantum features without any classical counterpart [12].

The inversion collapses and revives also in the case of a thermal Bose-Einstein field. The spread in the photon number for the Bose-Einstein field is far larger than for a coherent state, and the collapse time is much shorter. In addition, the revivals overlap completely and interfere, producing a very irregular time evolution. On the other hand, a classical thermal field represented by an exponential distribution of the intensity shows collapse, but no revivals. From this it follows that revivals are pure quantum features of the thermal radiation field, whereas the collapse is less clear-cut as a quantum effect [12].

The above-mentioned effects have been demonstrated experimentally. The experimental results clearly show the collapse and revival predicted by the Jaynes-Cummings model. Figure 4 shows a series of measurements obtained with the single-atom maser [13]. Plotted is the probability $P_e(t)$ of finding the atom in the upper maser level for increasing atomic flux N. The strong variation of $P_e(t)$ for interaction times between 50 and 80 μs disappears for larger N and a revival shows up for $N = 3000\ s^{-1}$ for interaction times larger than 140 μs. The average photon number in the cavity varies between 2.5 and 5, about 2 photons being due to the black-body field in the cavity corresponding to a temperature of 2.5 K.

There is another aspect of the single-atom maser which is very interesting, the non-classical statistics of the photons in the

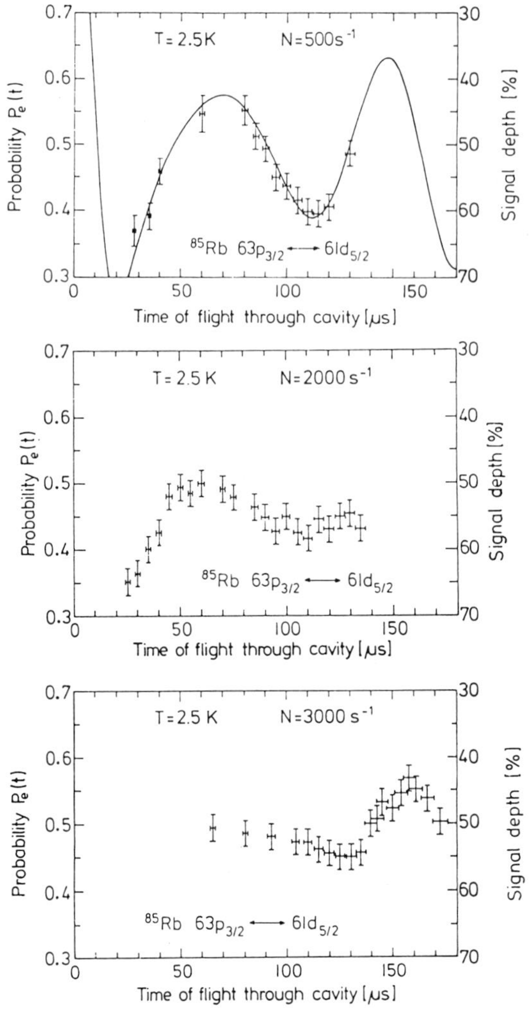

Fig. 4. Quantum collapse and revival in the one-atom maser. Plotted is the probability $P_e(t)$ of finding the atom in the upper maser level for different fluxes N of the atomic beam (see Ref.[13]).

cavity. This problem is briefly discussed in the following. There are two approaches to the quantum theory of the one-atom maser. Filipowicz et al.[14] use a microscopic approach to describe the device. On the other hand, Lugiato et al.[15] show that the standard macroscopic quantum laser theory leads to the same steady-state photon number distribution. The special features of the micromaser were not emphasized in the standard laser theory because the broadening due to spontaneous decay obscured the Rabi cycling of the atoms. When similar averages in the microscopic theory associated with inhomogeneous broadening are performed, equivalent results are obtained.

Both theoretical approaches predict that the statistics of the photons in the maser cavity depends on the interaction time of the atoms in the cavity. It is found that the distribution is mostly sub-Poissonian. In the case of a high Q value of the order of $Q \approx 5 \cdot 10^{10}$ the radiation field is expected to be in a state with a fixed photon number, i.e. a Fock state.

To achieve such a state experimentally, two conditions have to be met. The first concerns the temperature; thermal photons have to be suppressed because they not only induce statistical decay but also result in superposition of number states. We can eliminate thermal photons by cooling the cavity to a low enough temperature. The second condition is that photons stored in the cavity for the duration of the experiment should not be lost, i.e. one needs a cavity in which losses can be neglected for this time, which means that the Q value has to be higher than 10^{10}. Both conditions can now be realized in the Rydberg maser experiment [16].

The results obtained for the single-atom maser also give a new insight into the statistical properties of masers and lasers. It was demonstrated that the photon statistics in the one-atom maser is in general sub-Poissonian or corresponds to a Fock state, in contrast to the usual masers and lasers, where coherent radiation, i.e. Poissonian statistics, is observed. The reason for this is that there is much stronger damping present in the cavities of the usual masers and lasers than in the micromaser; in addition, the atomic or molecular transitions also usually show stronger damping than the Rydberg states, this being especially the case for laser systems. Furthermore, the selected velocity of the atoms used in connection with the collapse and revival measurements in the micromaser leads to fixed interaction times in the cavity; this also helps to reduce the photon number fluctuations since the photon exchange between the atom and cavity field can be exactly controlled. The smallest fluctuations are achieved when the atoms leave the cavity again in the upper state. Of course, it is necessary that energy be deposited in the cavity in order to maintain maser oscillation, but the losses are very small with a high-Q-cavity therefore P_e for atoms leaving the cavity can be adjusted very close to unity.

There are other interesting aspects of Rydberg masers which can only be briefly discussed here. Recently, a two-photon maser was realized by Brune et al.[17]. The two-photon transition was chosen such that there is an intermediate level nearly halfway between the upper and lower maser levels, thus enhancing the transition amplitude. In such a device new features not present in one-photon masers can be observed, e.g. delayed start-up time at threshold or multistable behaviour. Unlike the one-photon maser, which behaves at threshold similarly to a 2nd-order phase transition, the two-photon maser is analogous to a 1st-order phase transition.

Finally we should also mention that it was pointed out by Meystre and co-workers that the micromaser can be used to investigate aspects of chaos and problems of measurement theory [18]. Furthermore the maser can be used for a new test of complementarity [19].

References

1. B.R. Mollow, Phys. Rev. 188, 1969 (1969).

2. J.D. Cresser, J. Häger, G. Leuchs, M. Rateike, H. Walther in Dissipative Systems in Quantum Optics, ed. by R. Bonifacio, pp 21-59, Topics in Current Physics Vol. 27, Springer Verlag (1982).

3. H.J. Kimble, M. Dagenais, L. Mandel, Phys. Rev. Lett. 39:691 (1977).

4. F. Diedrich, H. Walther, Phys. Rev. Lett. 58:203 (1987).

5. R. Short, L. Mandel, Phys. Rev. Lett. 51:384 (1983), and in Coherence in Quantum Optics V, edited by L. Mandel and E. Wolf, Plenum, New York (1984), p. 671.

6. S. Haroche and J. M. Raimond, in: "Advances in Atomic and Molecular Physics", Vol 20, D. Bates and B. Bederson, eds., Academic Press (1985), p. 350.

7. J. A. C. Gallas, G. Leuchs, H. Walther, and H. Figger, in: "Advances in Atomic and Molecular Physics", Vol 20, D. Bates and B. Bederson, eds., Academic Press (1985), p. 414.

8. W. Jhe, A. Anderson, E. A. Hinds, D. Meschede, L. Moi, and S. Haroche, Phys. Rev. Lett. 58:666 (1987).

9. G. Barton, Proc. Roy. Soc., London A 410 (1987) 147 and 175

10. E. T. Jaynes and F. W. Cummings, Proc. IEEE 51:89 (1963).

11. D. Meschede, H. Walther, and G. Müller, Phys. Rev. Lett. 54:551 (1985).

12. H. I. Yoo and J. H. Eberly, Phys. Rep. 118:239 (1985) and P. L. Knight and P. M. Radmore, Phys. Lett. 90A:342 (1982).

13. G. Rempe, H. Walther, and N. Klein, Phys. Rev. Lett. 58:353 (1987).

14. P. Filipowicz, J. Javanainen, and P. Meystre, Phys. Rev. A 34:3077 (1986).

15. L. A. Lugiato, M. O. Scully, and H. Walther, Phys. Rev. A 36:740 (1987).

16. J. Krause, M. O. Scully, and H. Walther, Phys. Rev. A 36:4547 (1987).

17. M. Brune, J. M. Raimond, P. Goy, L. Davidovich, and S. Haroche, Phys. Rev. Lett. 59:1899 (1987).

18. P. Meystre, Opt. Lett. 12:669 (1987).

19. M.O. Scully, H. Walther, Phys. Rev. A 39:5229 (1989).

COHERENCE AND CORRELATION IN QUANTUM OPTICS:

I. FOUNDATIONS

Marlan O. Scully, Heidi Fearn and Briggs W. Atherton

Center for Advanced Studies and
Dept. of Physics and Astronomy
University of New Mexico, Albuquerque, New Mexico 87131
and
Max-Planck Institute für Quantenoptik
D-8046 Garching bei München, West Germany

INTRODUCTION

Coherence and correlation in atomic and radiation physics has many interesting and unexpected consequences. We begin by giving a general discussion of atomic coherence and quantum correlation.

The paper is divided into five sections. Section two is simply a definition of atomic coherence, where we consider both single mode and multimode radiation from one atom and single mode radiation from an ensemble of identical independent atoms. In section three we discuss the Hanle effect, which is a classic example of coherence in quantum mechanics.[1,2] Section four and five are related in that we consider quantum beats in different three level atomic systems. Both systems are treated first semiclassically and then by using quantum electrodynamics methods in order to compare the results of both approaches.

COHERENCE IN ATOMIC PHYSICS

An atom is said to be in a coherent superposition of states when it is described by a wave function corresponding to a linear combination of two or more states. For maximum coherence we expect these states to be equally populated. Generally, the existence of atomic coherence appears in the atomic density matrix off-diagonal elements $\rho_{\mu\mu'}$, between the levels μ and μ'. For example, an atom which is in a coherent superposition of its ground $|b\rangle$ and first excited state $|a\rangle$, possesses an oscillatory charge cloud (and thus an atomic dipole moment) which goes like

$$\rho_{ba} \exp[i(\epsilon_a - \epsilon_b)t/\hbar].$$

In order to physically establish association of coherence with the off-diagonal elements of the atomic density matrix, we consider a two level atom which is in a linear superposition of the ground state $\psi_b(r)$ and the upper state $\psi_a(r)$.[3]

$$\psi(rt) = a(t)\exp(-i\omega_{ab}t)\psi_a(r) + b(t)\psi_b(r) \tag{1}$$

where $\omega_{ab} = (\epsilon_a - \epsilon_b)/\hbar$. The charge cloud associated with this wave function oscillates in a periodic fashion, see Fig. 1.

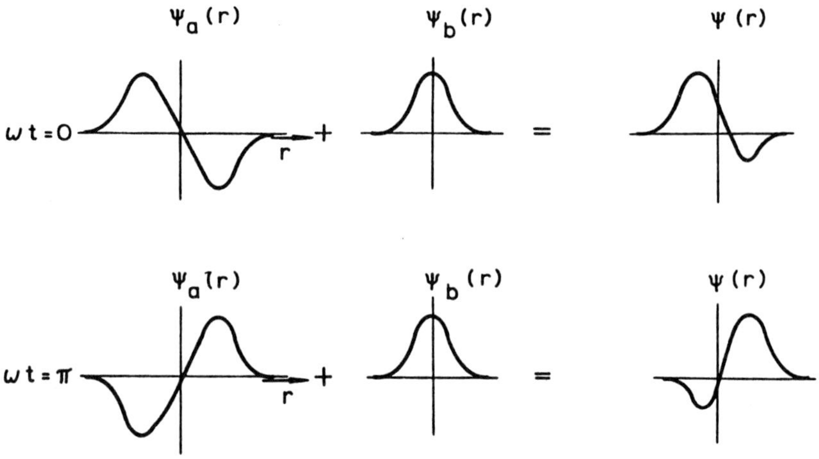

Fig. 1. Coherent oscillation of charge cloud associated with Eq. (1).[5].

The atomic polarization resulting from the linear combination of states is given by

$$\langle \vec{p}(t)\rangle = e\int \psi^*(r,t)r\psi(r,t)\hat{r}dr \tag{2}$$

Inserting Eq. (1) for $\psi(r,t)$ into (2) yields

$$\begin{aligned}\langle \vec{p}(t)\rangle &= e\int \psi_a^*(r)r\psi_b(r)\hat{r}dr[a^*(t)b(t)\exp(i\omega_{ab}t) + c.c.]\\ &= e\vec{r}_{ab}[\rho_{ab}(t) + c.c.]\end{aligned} \tag{3}$$

where $\vec{r}_{ab} = \int \psi_a^*(r)r\psi_b(r)\hat{r}dr$ and ρ_{ba} is an off-diagonal element of the density matrix

$$\rho = \begin{pmatrix} \rho_{aa} & \rho_{ab} \\ \rho_{ba} & \rho_{bb} \end{pmatrix} = \begin{pmatrix} |a|^2 & ab^*\exp(-i\omega_{ab}t) \\ ba^*\exp(i\omega_{ab}t) & |b|^2 \end{pmatrix} \tag{4}$$

Hence the existence of an atomic dipole moment $\langle p\rangle$, or equivalently ρ_{ab}, is associated with atomic coherence. The off-diagonal elements ρ_{ab} would vanish if the atom is completely in the upper or lower states i.e.

$$\rho = \begin{pmatrix} 1 & 0 \\ 0 & 0 \end{pmatrix} \quad \text{or} \quad \rho = \begin{pmatrix} 0 & 0 \\ 0 & 1 \end{pmatrix} \tag{5}$$

Thus, we may equate coherence with the existence of off-diagonal elements of the atomic density matrix.

We have now treated coherence for a single atom. It is also interesting to consider the radiation coming from an ensemble of independent atoms contained in a volume whose dimensions are small compared to the radiation wavelength. If we excite N atoms to some linear superposition of the $|a\rangle$ and $|b\rangle$ states, then the wave function for the ith atom can be written as

$$\begin{aligned}|\psi_i\rangle &= \kappa_i[1 + \beta_i\sigma_i^\dagger]\begin{pmatrix} 0 \\ 1 \end{pmatrix}_i\\ &= \kappa_i\exp(\beta_i\sigma_i^\dagger)\begin{pmatrix} 0 \\ 1 \end{pmatrix}_i\end{aligned} \tag{6}$$

where $|a\rangle = \begin{pmatrix} 1 \\ 0 \end{pmatrix}$, $|b\rangle = \begin{pmatrix} 0 \\ 1 \end{pmatrix}$, β_i is the probability of finding the ith atom in the excited state, κ_i is a constant and $\sigma_i^\dagger$ is the spinor raising operator.

Thus for N atoms we may write,

$$
\begin{aligned}
|\psi\rangle &= \prod_{i=1}^{N} |\psi_i\rangle \\
&= \prod_{i=1}^{N} \kappa_i \exp(\beta_i \sigma_i^\dagger) \begin{pmatrix} 0 \\ 1 \end{pmatrix}_i \\
&= \mathcal{N} \exp\left[\sum_{i=1}^{N} (\beta_i \sigma_i^\dagger)\right] |0\rangle
\end{aligned}
\tag{7}
$$

where

$$
|0\rangle = \begin{pmatrix} 0 \\ 1 \end{pmatrix}_1 \begin{pmatrix} 0 \\ 1 \end{pmatrix}_2 \cdots \begin{pmatrix} 0 \\ 1 \end{pmatrix}_N
$$

and $\mathcal{N}$ is a normalization constant.

A clear example of multiatom coherent excitation is given by superradiance.[4,5] An analysis of superradiance shows that if we have a symmetric wave function and the N atoms are excited to equal populations of the $|a\rangle$ and $|b\rangle$ states then the matrix element for spontaneous emission is proportional to N and the probability of photon emission goes like N^2.

In general a coherent state maybe multimode. For the kth mode of the radiation field the coherent state maybe written as

$$
|\psi_k\rangle = \mathcal{N}_k \exp(\alpha_k a_k^\dagger)|0_k\rangle \tag{8}
$$

thus for for many modes, the general coherent state becomes

$$
\begin{aligned}
|\psi\rangle &= \prod_k |\psi_k\rangle \\
&= \prod_k \mathcal{N}_k \exp(\alpha_k a_k^\dagger)|0_k\rangle \\
&= \mathcal{N} \exp(\sum_k \alpha_k a_k^\dagger)|0\rangle
\end{aligned}
\tag{9}
$$

where

$$
|0\rangle = |0_1\rangle|0_2\rangle \cdots |0_N\rangle
$$

THE HANLE EFFECT

The experiment of Hanle provides one of the most clear and oldest demonstrations of a situation in which atomic coherence plays an important role.[1,2] An ensemble of atoms, situated in a weak magnetic field, was illuminated with a pulse of $\hat{x}$ polarized light. The polarization of the light reradiated in the z-direction was then detected. For a negligibly small magnetic field it is found that the reradiated light is polarized in the x-direction, as depicted in Fig. 2, where $\beta = a_0^{-1}$ and $a_0 =$ Bohr orbital radius.

This is unexpected since we excite only $m = \pm 1$ levels of the atoms. One might expect that only σ^+ and σ^- radiation would be observed in the z-direction, where

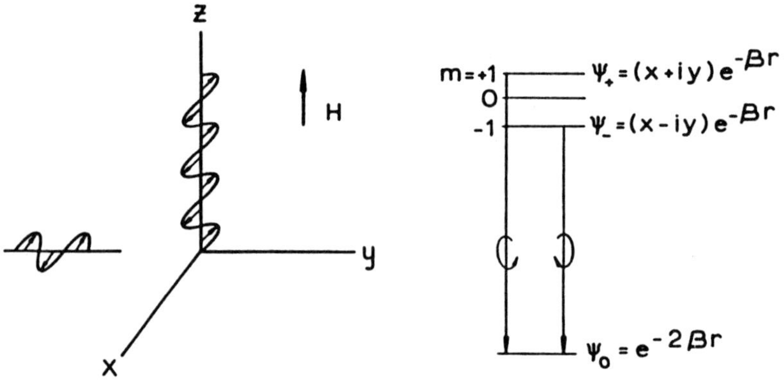

Fig. 2. Schematic illustration of Hanle's experiment and atomic level scheme.[5]

$\sigma^+ = x + iy$ and $\sigma^- = x - iy$ following the notation in Fig. 2. To understand this we must calculate the dipole moment induced by the incident radiation. If we take an atom initially in the ground state

$$\psi(r,0) = \psi_0(r) \tag{10}$$

later in time the electric field

$$\vec{E}(r,t) = \hat{x}E_0 \cos(ky - \omega t) \tag{11}$$

induces transitions to the $m = \pm 1$ levels and the wave function becomes

$$\psi(r,t) = a_+ \exp(i\omega_+ t)\psi_+(r) + a_- \exp(i\omega_- t)\psi_-(r) + a_0\psi_0(r) \tag{12}$$

The atomic frequencies $\omega_\pm$ are given by

$$\omega_\pm = \omega \pm \Delta\omega \tag{13}$$

where $\Delta\omega$ is the splitting of the levels due to the magnetic field. The atomic dipole is then

$$\begin{aligned} \langle \vec{p}(t) \rangle &= e \int \psi^*(r,t)(x\hat{x} + y\hat{y} + z\hat{z})\psi(r,t)dr \\ &= p_+ \{\hat{x}\cos(\omega + \Delta\omega)t + \hat{y}\sin(\omega + \Delta\omega)t\} \\ &+ p_- \{\hat{x}\cos(\omega - \Delta\omega)t - \hat{y}\sin(\omega - \Delta\omega)t\} \end{aligned} \tag{14}$$

where $p_\pm$ is the polarization associated with $\rho_{\pm,0} = a_\pm a_0^*$ and

$$p_\pm = e\langle \pm|x|0\rangle(\rho_{\pm,0} + c.c.) \tag{15}$$

Thus, atomic polarization and hence the radiated light will consist of σ^+ and σ^- components, corresponding to the longitudinal Zeeman effect. In the zero field limit, when $\Delta\omega t \approx 0$ and $p_+ \equiv p_-$, the dipole moment implied by a coherent superposition of + and - states becomes

$$\langle \vec{p}(t) \rangle = (p_+ + p_-)\hat{x}\cos\omega t \tag{16}$$

where $t \simeq 0$ (few nanosec) and $\Delta\omega \simeq 10^{-6}$ Hz.

We observe that the scattered (detected) light is $\hat{x}$ polarized. In the same limit and $\Delta\omega t \simeq \pi/2$ we find

$$\langle \vec{p}(t) \rangle = (p_+ + p_-)\hat{y} \cos \omega t \tag{17}$$

and the scattered light is now $\hat{y}$ polarized. The scattered light therefore changes its direction of polarization from $\hat{x}$ to $\hat{y}$ as time passes. This is the Hanle effect.

Another example of a coherence phenomenon is given by quantum beats.

"V" QUANTUM BEATS

Semiclassical Approach

The phenomenon of quantum beats was first observed by Dr. G. Series and coworkers in the 1960's, but there has recently been a revival of interest resulting in several papers.[6-8] We consider a gas of three level atoms (see Fig. 3) initially in the ground state ψ_c. A short pulse then excites the atoms into a coherent superposition of the three states. The wave function for the atom may then be written as

$$\psi(t) = a \exp(-i\epsilon_a t)\psi_a + b \exp(-i\epsilon_b t)\psi_b + c\psi_c \tag{18}$$

where $\hbar = 1$. We then look at the atomic fluorescence, assuming the system to be on resonance.

The total electric field emitted by this system is given by

$$E = \kappa(p_1 + p_2) \tag{19}$$

where p_1 and p_2 are the off-diagonal atomic density matrix elements, κ is a constant and we have simplified the notation by dropping the vector symbols.

$$p_1 = a^* c \exp[-i(\epsilon_a - \epsilon_c)t] e r_{ac} \tag{20}$$

$$p_2 = b^* c \exp[-i(\epsilon_b - \epsilon_c)t] e r_{bc} \tag{21}$$

the detected photocurrent is proportional to

$$E^* E = \kappa^2 [|p_1|^2 + |p_2|^2 + p_1^* p_2 + c.c.] \tag{22}$$

The modulus squared terms constitute a D.C. current and the remaining terms give the A.C. contributions, which form the beats on the detected signal (see Fig. 4).

The quantum beats arise due to the nonvanishing off-diagonal matrix elements ρ_{ab} resulting from the correlation between states $a \rightarrow b$.

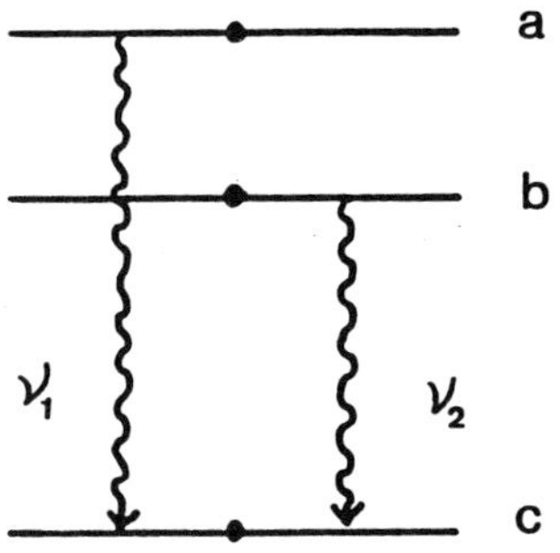

Fig. 3. Three-level atomic structure for "V" type quantum beats.

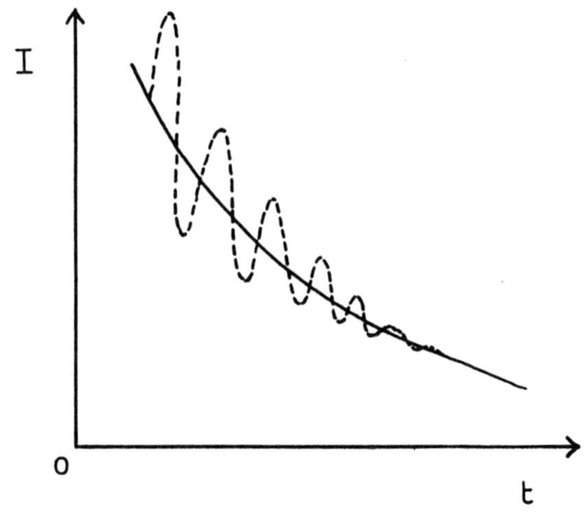

Fig. 4. An illustrative graph of detected light intensity (I) verses time (t). The diagram shows the usual exponential decay (solid line) and the oscillation (dashed line) which one would observe due to quantum beats.

Quantum Mechanical Approach to Quantum Beats

We assume the same ensemble of three-level atoms as above, so that after excitation at time $t = 0$ (few nanosec.) the state of the system is given by

$$\psi(0) = [a\psi_a + b\psi_b + c\psi_c]|0\rangle \tag{23}$$

Using first order perturbation theory we may write the state at time t as

$$\psi(t) = \psi(0) - igt[a\psi_c|1_1\rangle + b\psi_c|1_2\rangle] \tag{24}$$

where the state $|n_k\rangle$ represents n photons of frequency ν_k and g is an interaction coefficient, assumed equal for both $a \to b$ and $a \to c$ transitions. The electric field emitted by the system is now written in operator form as

$$\hat{E}(t) = \epsilon_1 \hat{a}_1 \exp(-i\nu_1 t) + \epsilon_2 \hat{a}_2 \exp(-i\nu_2 t) \tag{25}$$

Using the photodetection theory developed by R. Glauber[9,10] the intensity of the emitted light is given by the expectation value

$$I = \langle\psi(t)|\hat{E}^+(t)\hat{E}(t)|\psi(t)\rangle \tag{26}$$

Substituting Eqs. (24) and (25) into (26) we find

$$I = |gt|^2[|\epsilon_1|^2 a^* a\langle 1_1|\hat{a}_1^+ \hat{a}_1|1_1\rangle + |\epsilon_2|^2 b^* b\langle 1_2|\hat{a}_2^+ \hat{a}_2|1_2\rangle$$
$$+ \epsilon_1^* \epsilon_2 a^* b\langle 1_1|\hat{a}_1^+ \hat{a}_2|1_2\rangle \exp[i(\nu_1 - \nu_2)t] + c.c.]\langle\psi_c|\psi_c\rangle \tag{27}$$

As before we see that the off-diagonal density matrix element is responsible for the oscillatory behavior of the detected light intensity. Looking back at the last section we find that both the semiclasical and the quantum theory give identical results. The D.C. portion of the intensity leads to an exponential decay of the light intensity and the A.C. terms, corresponding to the overlap of states a and b, cause the coherent ringing phenomenon.

In the next section we will apply the above theory to a lower level (Λ) system, wherein correlation is established in the two lower levels. Examining both semiclassical and quantum mechanical approaches, we determine whether or not we should expect to see quantum beats.

LOWER LEVEL "Λ" QUANTUM BEATS

Semiclassical Approach

Consider a three-level atom (as shown in Fig. 5) excited by a short pulse to a coherent superposition of states α, β and γ. The states of the system immediately after excitation can be represented by the wave function

$$\psi(t) = \alpha \exp(-i\epsilon_\alpha t)|\psi_\alpha\rangle + \beta \exp(-i\epsilon_\beta t)|\psi_\beta\rangle + \gamma|\psi_\gamma\rangle \tag{28}$$

where $\hbar = 1$.

We would like to examine the fluorescence, as in the V-type system , in order to determine whether or not we see quantum beats. The output electric field is again given by

$$E = \kappa(p_1 + p_2) \tag{29}$$

where

$$p_1 = \alpha^* \gamma \exp[-i(\epsilon_\alpha - \epsilon_\gamma)t] e r_{\alpha\gamma} \tag{30}$$

$$p_2 = \alpha^* \beta \exp[-i(\epsilon_\alpha - \epsilon_\beta)t] e r_{\alpha\beta} \tag{31}$$

and the intensity is proportional to

$$E^* E = \kappa^2 [|p_1|^2 + |p_2|^2 + p_1^* p_2 + c.c.] \tag{32}$$

Here we see that quantum beats are present due to the off-diagonal matrix element $\rho_{\beta\gamma}$ with exponent $i(\epsilon_\beta - \epsilon_\gamma)t$.

Q. E. D. Approach to Λ Quantum Beats

Using the same level diagram as before the system after excitation is now represented quantum mechanically by the wave function

$$\psi(0) = [\alpha|\psi_\alpha\rangle + \beta|\psi_\beta\rangle + \gamma|\psi_\gamma\rangle]|0\rangle \tag{33}$$

Using first order perturbation theory the state at time t becomes

$$\psi(t) = \psi(0) - igt[\alpha|\psi_\gamma\rangle|1_1\rangle + \alpha|\psi_\beta\rangle|1_2\rangle] \tag{34}$$

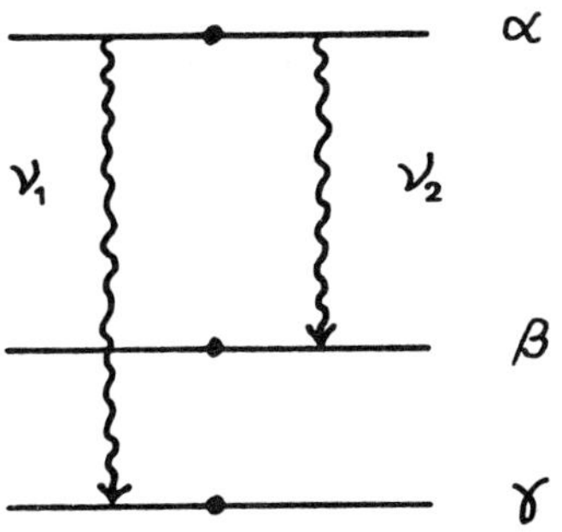

Fig. 5. Three-level atomic structure for "Λ" type quantum beats.

where the notation is similar to the V level Q. E. D. approach. The electric field emitted by the atom is represented by Eq. (25) and so detected intensity is given by

$$\begin{aligned} I &= \langle \psi(t)|\hat{E}^+(t)\hat{E}(t)|\psi(t)\rangle \\ &= |gt|^2 \alpha^* \alpha \Big[(|\epsilon_1|^2 \langle 1_1|\hat{a}_1^+ \hat{a}_1|1_1\rangle + |\epsilon_2|^2 \langle 1_2|\hat{a}_2^+ \hat{a}_2|1_2\rangle)\langle\psi_\gamma|\psi_\gamma\rangle \\ &\quad + \epsilon_1^* \epsilon_2 \langle 1_1|\hat{a}_1^+ \hat{a}_2|1_2\rangle \exp[i(\nu_1 - \nu_2)t]\langle\psi_\gamma|\psi_\beta\rangle + c.c. \Big] \end{aligned} \tag{35}$$

However, $\langle\psi_\gamma|\psi_\beta\rangle = 0$ so that the off-diagonal terms (with oscillatory factors) disappear. [This contrasts with the V system where the off-diagonal terms were multiplied by $\langle\psi_c|\psi_c\rangle \neq 0$.]

Hence, we have shown that the Λ quantum beats do not exist, and the semiclassical treatment above is incorrect. The reason for the discrepancy between the semiclassical and Q.E.D. approaches is that normally one may think of the quantized field as adding (vacuum) fluctuations to the classical Maxwell equations. This is the usual semiclassical hypothesis. For quantum beat theory, it is essential to have a quantized radiation field to take account of the spontaneous emission.[11] Furthermore, no semiclassical theory can explain the absence of quantum beats. In order to explain why quantum beats do not occur we must consider the quantum theory of measurement.

When the upper two levels of an atom were coherently excited, the atom decayed to the ground state $|\psi_c\rangle$. By observing the atom after de-excitation, using the ionization technique we could not determine which frequency photon had been emitted (ν_1 or ν_2). However, for the Λ - system, the atom decays to either level γ or β and if we examine the atom after de-excitation we can tell that a photon of frequency ν_1 was emitted if the atom is in the state $|\psi_\gamma\rangle$, or that a photon of frequency ν_2 was emitted if we measure the atom to be in state $|\psi_\beta\rangle$. Hence we have information about the system which destroys the quantum correlation between the lower levels and also destroys the quantum beats. This is analogous to the destruction of the interference fringes in a single photon Youngs double slit experiment, when we detect which path the photon took. (A better analogy can be made with the "welcher Weg detector."[12,13])

In summary, when quantum correlations are important it is necessary to quantize the radiation field as well as the atoms. Also, one must determine whether or not information has been gained, in the sense of "which path" information, in order to decide if a quantum interference effect will be observed or not.

We reserve discussion on the applications of the above theory for the second paper, II Applications.

REFERENCES

1. V. Weisskopf, Ann. d. Planck **9**, 23 (1931).
2. G. Breit, Rev. Mod. Phys. **5**, 91 (1933).
3. W. E. Lamb, *Quantum Optics and Electronics*, lectures delivered at Les Houches during the 1964 session of the summer school of Theoretical Physics. University of Grenoble, edited by C. Dedwitt, A. Blandin and C. Cohen-Tannoudji.
4. R. H. Dicke, Phys. Rev. **93**, 99 (1954).
5. M. O. Scully, "Coherence," in *Atomic Physics*, Plenum Press, pp. 81-102 (1969).
6. M. O. Scully and M. S. Zubairy, Phys. Rev. **A35**, 52 (1987).
7. M. O. Scully and Shi-Yao Zhu, Phys. Rev. Letts. **62**, 2813 (1989).
8. S. Y. Zhu, M. O. Scully and E. E. Fill, "Lasing without inversion due to initial coherence between lower atomic states." Presented at the Rochester conference (1989).
9. R. J. Glauber, Phys. Rev. **131**, 2766 (1963).
10. R. J. Glauber, *Quantum Optics and Electronics* (1964), see Ref. 3.
11. W. W. Chow, M. O. Scully and J. O. Stoner, Phys. Rev. **A11**, 1389 (1975).
12. B-G. Englert., J. Schwinger and M. O. Scully, Found. Phys. **18**, 1045 (1988).
13. M. O. Scully and H. Walther, Phys. Rev. **A39**, 5229 (1989).

COHERENCE AND CORRELATION IN QUANTUM OPTICS: II. APPLICATIONS

Marlan O. Scully, Heidi Fearn and Briggs W. Atherton

Center for Advanced Studies and
Dept. of Physics and Astronomy
University of New Mexico, Albuquerque, New Mexico 87131
and
Max-Planck Institut für Quantenoptik
D-8046 Garching bei München, West Germany

INTRODUCTION

The effects of coherence and correlation in atomic and radiation physics show up in various different physical systems. In this second paper on the subject of coherence and correlation in quantum optics we wish to analyze several such systems which have been of recent theoretical interest.

Our first example will involve the spin coherence of a polarized beam of spin-1/2 particles. We consider a beam of such particles entering a Stern-Gerlach interferometer. When the two particle beams, produced inside the apparatus, are recombined, it has been shown that, in principle, the spin coherence of the incident beam may be recovered.[1,2] However, if a *welcher Weg* detector[3] is placed in one arm of the interferometer, so that the passage of particles can be detected in one arm but not in the other, it has been found that the coherence in the output beam is totally destroyed. This loss of coherence could be traced to the interaction between the measuring apparatus and the system being observed. Furthermore, it has recently been found that if the "which path" information is erased, the spin coherence of the incident beam can be recovered.[4,5,6] Although the physics of the quantum eraser and the retrieval of spin coherence is explained well by the spin 1/2 particle model, it is better to examine practical experiments in the realm of quantum optics.[4] We shall therefore conclude this section of the paper by a brief outline of a quantum beat analog to the spin quantum eraser problem.

A second and very different example of coherence and correlation in quantum optics is given by quantum beats in atomic systems. In this section we will discuss the effects of quantum beats and how they interfere to cancel absorption and emission. The cancellation of absorption will be discussed in detail, and how it is related to lasing without inversion.

SPIN COHERENCE

Spin coherence and the Stern-Gerlach interferometer (SGI) are treated in more detail by B.-G. Englert elsewhere in this volume so we shall restrict ourselves here to a simple discussion.

Imagine an x-polarized beam of spin 1/2 particles incident to a SGI. The initial polarization of the beam can be represented by

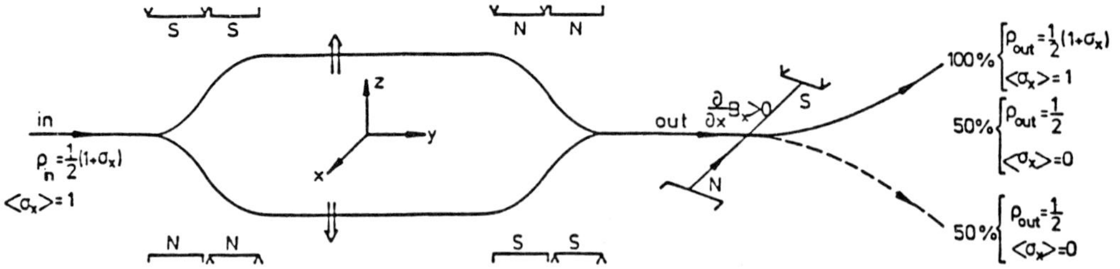

Fig. 1. Side view of the Stern-Gerlach interferometer, showing the final σ_x measurement.

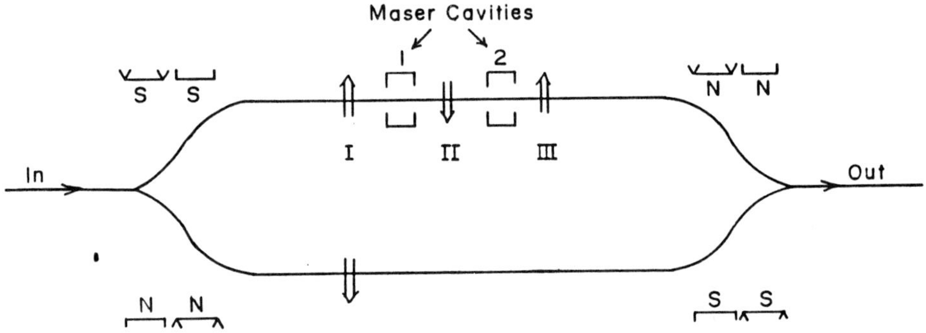

Fig. 2. SGI with micromaser cavities to form a welcher Weg detector.

$$\langle \sigma_x \rangle = 1 \tag{1}$$

or

$$\rho_{spin} = \frac{1}{2}(1 + \sigma_x). \tag{2}$$

The beam is split, by magnetic fields, into two z-polarized beams (spin ↑ and spin ↓) which travel different paths through the interferometer, see Fig. 1. The two beams are then reunited by reversing the magnetic fields. The emerging beam is then analyzed by an external magnetic field. This beam is x-polarized if all the particles can be deflected in the $+x$-direction. The output beam is unpolarized if only 50% of the particles can be deflected in the $+x$-direction. An unpolarized beam corresponds to

$$\langle \sigma_x \rangle = 0 \tag{3}$$

or

$$\rho_{spin} = 1/2 \tag{4}$$

In a series of papers entitled "Is Spin Coherence like Humpty Dumpty ?" (HD) it was found that if we allow ourselves an ideal system, where the magnetic fields can be controlled to great accuracy then spin coherence is preserved.[1] It has been shown in neutron interferometric experiments[7] that two partial beams (spin ↑ and ↓) can be recombined to produce an x-polarized beam. Thus, theory and experiment show that, two partial beams can in principle be combined to produce a coherent polarized spin state.

We now assume that spin coherence may be recovered and move on to consider the effect of a detector placed in the upper arm of the SGI. This detector gives which path (German welcher Weg) information about the particles (see Fig. 2).

The single mode maser cavities are prepared so that it is almost certain that a spin up particle entering cavity 1 is spin down between the cavities and spin up again after leaving cavity 2. We shall take N_j photons to be present in cavity j, where $j = 1$ or 2. We then consider two cases; for the first case we have coherent states of the maser fields and for the second case we have number states. The particle beams are characterized by spin so that the interaction between the cavity photons and the particles correlates the spin degree of freedom to the photon degrees of freedom. The final outcome of the spin measurement depends on the prepared cavity states and the number of photons present in each cavity depends on the initial spin state of the particle. The spin $\uparrow$ and spin $\downarrow$ components are distinguishable in that there is a one-to-one correspondence between spin $\downarrow$ and final photon counts N_1, N_2 as well as spin $\uparrow$ and $N_1 + 1$, $N_2 - 1$. This signifies the correlations set up in the interaction. As a result, which-path information is available provided the photon numbers in the maser fields are changed in a discernible way.

For the coherent maser fields, the standard deviation of photon number in the cavities is given by

$$\sqrt{N_j}. \tag{5}$$

Since N_j is large for a coherent *classical* field we would not be able to detect a change of ± 1 photon, and so no which-path information is available. The spin coherence in the particle beams would therefore be regained.

For number state preparation of the maser fields we have no deviation in photon number and it is, in principle, possible to detect a unit change in photon number. In this case we have which-path information and spin coherence of the output particle beam is lost.[3]

For the final example on spin coherence we shall briefly examine the quantum eraser.[4] We consider an apparatus similar to the above welcher Weg detector and examine the possibility of retrieving spin coherence by erasing the which-path information obtained by the detector. We take two detectors placed symmetrically in the upper and lower paths of the SGI as indicated in Fig. 3.

The first two cavities have a common focus at which point is placed a photodetector. The fields in the first two cavities (1,1′) are initially prepared in the vacuum electromagnetic state. The second cavities (2,2′) hold coherent electromagnetic fields. The magnetic moment of the particle is assumed large so that the vacuum may stimulate spin flip resulting in spontaneous emission of a photon by the particle. Two shutters are arranged to hold the spontaneous radiation in either the upper or lower cavity when the shutters are closed. In this case we would have which-path information. However, when the shutters are opened, light may interact with the photodetector thus eliminating the which-path information. Therefore, when the shutters are opened we can regain the spin coherence in the output particle beam.

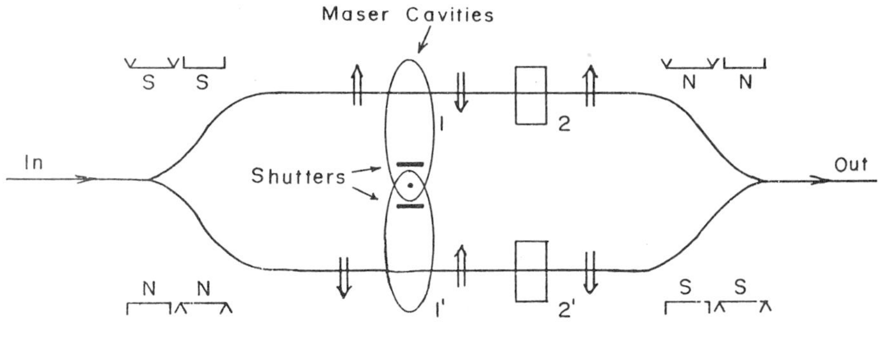

Fig. 3. Side view of the Quantum Eraser.

As mentioned earlier, the vacuum Rabi flopping of a spin 1/2 magnet is not a realistic model. We may however describe an arrangement which is experimentally viable,[4] that of the quantum beat eraser.

Consider a three-level atom (the V-type as discussed in paper I). The upper levels a and b decay to the ground state level c with emission of photons ν_1 and ν_2. If we detect the light emitted by this atom field system we observe quantum beats as discussed previously (paper I).

If we consider a Λ–type three-level atom, whose upper level α decays to level β and γ with emission of photons ν_2 and ν_1, respectively, we observe no quantum beats. It is only necessary to examine the final state of the atom (either $|\beta\rangle$ or $|\gamma\rangle$) to determine which photon (ν_2 or ν_1) was emitted. That is states $|\gamma\rangle$ and $|\beta\rangle$ are welcher Weg detectors for the photons ν_1 and ν_2 with which they are correlated. This is a form of which path information analogous to the Stern-Gerlach spin experiments where ν_1, ν_2 correspond to the spin states $\uparrow$ and $\downarrow$, and the atomic states $|\gamma\rangle$ and $|\beta\rangle$ correspond to the cavity states

$$|N_1 = 1, N_{1'} = 0\rangle \quad \text{or} \quad |N_1 = 0, N_{1'} = 1\rangle \tag{6}$$

where N_1 is the number of photons in the upper cavity and $N_{1'}$ is the corresponding number in the lower cavity. It is possible to outline an experimental procedure, similar in nature to the SGI, using the Λ-type three-level atoms. This quantum eraser has been described by Scully[4] and a calculation together with further details may be found in that reference.

We now move on to consider a further example of coherence in quantum optics.

QUANTUM BEATS AND LASING WITHOUT INVERSION

We will look at two examples, where there are two paths for dipole transitions, E_1 and E_2, to take place. In this paper we will not consider detunings between the fields and the atomic transitions, see Fig. 4.

In the first of these examples the two upper levels are split and they decay into one lower level, this is the V-type. The second and more interesting one, is where the two lower levels are split, and one upper level decays into the two lower levels (Λ-type). The initial state of the system is

$$|\Psi(0)\rangle = A|\Psi_a\rangle + B|\Psi_b\rangle + C|\Psi_c\rangle \tag{7}$$

and the interaction between the levels

$$V = \frac{\hbar}{2}\sum_j N a^\dagger[(g_1 - g_2 e^{i\phi})\sigma_1^j + (g_1 + g_2 e^{-i\phi})\sigma_2^j] + h.a. \tag{8}$$

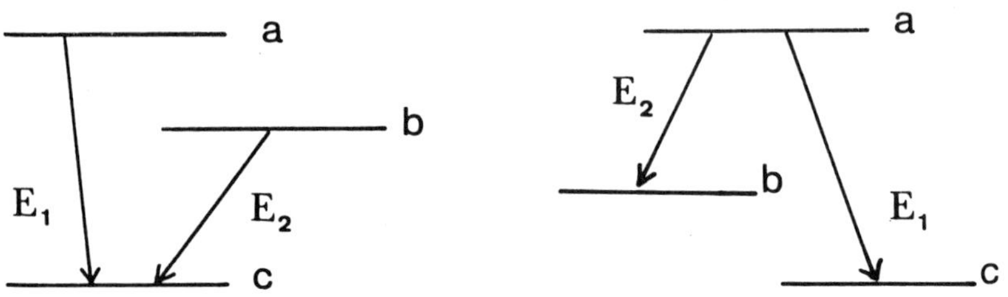

Fig. 4. Three level atomic diagrams for a) the V-type, and b) the Λ-type system.

where the two split states are prepared in a coherent manner, $a^\dagger$ is the photon creation operator, $\sigma_i^j\{i=1,2\}$ are the lowering atomic operators for the jth atom, A, B and C are the probability amplitudes associated with states Ψ_a, Ψ_b and Ψ_c respectively, g_1, g_2 are the dipole coupling constants for the laser fields E_1 or E_2 (see Fig. 4), and ϕ is the relative phase between the split levels. The atoms are injected into the cavity at a rate r and interact with the field in the cavity for a time τ. In the V-type system the equation of motion for the average photon number is[8]

$$\dot{\bar{n}} = \frac{r\tau^2}{4}\{[|G_1A + G_2B|^2](\bar{n}+1) - [|G_1C|^2 + |G_2C|^2]\bar{n}\} \tag{9}$$

where $G_1 = g_1 - g_2e^{i\phi}$ and $G_2 = g_1 + g_2e^{-i\phi}$. We see that the first term is associated with the stimulated and spontaneous emission rates and the second term is associated with the absorption rate. With the aid of a simple example, we will see the effects of the quantum beats in an extreme case. Letting $A = 0.9$, $B = 0$, $C = 0.1$, $G_1 = 0$, $G_2 = 2g_1$ and $\phi = 0$ we get for the average photon number

$$\dot{\bar{n}} = \frac{r\tau^2}{4}\{(0)(\bar{n}+1) - (4g_1^2C^2)\bar{n}\} \tag{10}$$

and we see that in this case we have absorption with no emission. Considering the Λ-type atomic system the rate equation for the average photon number is

$$\dot{\bar{n}} = \frac{r\tau^2}{4}\{[|G_1A|^2 + |G_2A|^2](\bar{n}+1) - [|G_1C + G_2B|^2]\bar{n}\} \tag{11}$$

where G_1 and G_2 are the same as in Eq. 9. We see that the first term is also associated with the stimulated and spontaneous emission rates and the second term is associated with the absorption rate. Illustrating the effects of quantum beats with a further example, letting $A = 0.1$, $B = 0$, $C = 0.9$, $G_1 = 0$, $G_2 = 2g_1$ and $\phi = 0$ the average photon number rate equation becomes

$$\dot{\bar{n}} = \frac{r\tau^2}{4}\{(4g_2^2A^2)(\bar{n}+1) - (0)\bar{n}\} \tag{12}$$

here we have emission without absorption, which results in lasing without inversion. In both cases we see that the quantum beats are responsible for lack of emission or absorption. Quantum beats are a result of the sum of the transition amplitude probabilities from the split levels to the third level. When the relative phases and amplitudes are chosen correctly, we can have enhancement or cancellation of the probability of transition between these levels. In the following we will focus on the latter, cancellation of absorption, in more detail.

Lasing without inversion has a short history. In the early 60's[9] and late 70's[10] separation of the emission and absorption lines through the use of photon recoil were studied. This is only effective when in the X-ray and higher frequency region of the spectrum. Recently Harris has posed the separation of the absorption and emission spectrum through the use of Fano interference.[11,12]

The Λ-quantum beat laser has been studied in great detail.[13] Here we will study the non-degenerate Λ-quantum beat laser with two ways of producing the coherence between the lower two levels; through a microwave field and Raman interaction. Keeping in mind the physical picture described above, we start with the hamiltonian

$$H = \begin{pmatrix} \hbar\omega_a & V_{ab} & V_{ac} \\ V_{ab}^* & \hbar\omega_b & 0 \\ V_{ac}^* & 0 & \hbar\omega_c \end{pmatrix} \tag{13}$$

and the density matrix

$$\rho = \begin{pmatrix} \rho_{aa} & \rho_{ab} & \rho_{ac} \\ \rho_{ba} & \rho_{bb} & \rho_{bc} \\ \rho_{ca} & \rho_{cb} & \rho_{cc} \end{pmatrix} \tag{14}$$

where the field and the macroscopic polarization go as [14]

$$\dot{E}_{a\beta} \propto -Im(\rho_{a\beta}) \tag{15}$$

$$\dot{P} \propto -N\mu_{a,\beta}\rho_{a\beta} \tag{16}$$

where $\beta = \{b, c\}$ is an index for the lower two levels see Fig. 4b. The $b - c$ dipole transition is forbidden. The density matrix follows the usual equation of motion

$$i\hbar\frac{\partial\rho}{\partial t} = [H, \rho] - \text{decay terms and pump terms} \tag{17}$$

Using the master equation and a little algebra, the equations of motion for the off-diagonal matrix elements are

$$\dot{\tilde{\rho}}_{ab} = -\gamma_{ab}\tilde{\rho}_{ab} + iV_{ab}(\rho_{aa} - \rho_{bb}) - iV_{ac}\tilde{\rho}_{ab} \tag{18}$$

$$\dot{\tilde{\rho}}_{ac} = -\gamma_{ac}\tilde{\rho}_{ac} + iV_{ac}(\rho_{aa} - \rho_{cc}) - iV_{ab}\tilde{\rho}_{ac} \tag{19}$$

and for the electric fields

$$\dot{E}_{ab} = \alpha_1[(\rho_{aa} - \rho_{bb})\mu_{ab}E_{ab} - |\tilde{\rho}_{ab}|\mu_{ac}E_{ac}\cos\Phi] \tag{20}$$

$$\dot{E}_{ac} = \alpha_2[(\rho_{aa} - \rho_{cc})\mu_{ac}E_{ac} - |\tilde{\rho}_{ab}|\mu_{ab}E_{ab}\cos\Phi] \tag{21}$$

where $V_{ab} = -\mu_{ab}E_{ab}$, $V_{ac} = -\mu_{ac}E_{ac}$ and

$$\alpha_1 = \frac{V_{ab}N\mu_{ab}}{2\epsilon_0\gamma_{ab}} \tag{22}$$

$$\alpha_2 = \frac{V_{ac}N\mu_{ac}}{2\epsilon_0\gamma_{ac}} \tag{23}$$

$$\Phi = \Phi_{ab} - \Phi_{ac} + \Phi_{bc} \tag{24}$$

the α_1 or α_2 is the linear gain coefficient, $\Phi_{\alpha\beta}$ is the relative phase of the various matrix elements, and $\gamma_{\alpha\beta} = \frac{1}{2}(\gamma_\alpha + \gamma_\beta) + \gamma_{\alpha\beta}^{\text{col}}$ is the decay rate associated with the off-diagonal matrix elements to include collisions. The absorption can be eliminated through interference effects by choosing the appropriate phase Φ. As an instructive example, let us chose $\Phi = \pi$, while having the lower two levels much more populated then the upper level, i.e. let $\rho_{bb} = \rho_{cc} \approx 1/2$; $(\rho_{aa} \ll 1)$, $\mu_{ab} = \mu_{ac}$, and $E_{ab} = E_{ac}$ then we find

$$\begin{aligned} \dot{E}_{ab} &= \alpha_1\mu_{ab}\rho_{aa}E_{ab} \\ \dot{E}_{ac} &= \alpha_2\mu_{ac}\rho_{aa}E_{ac} \end{aligned} \tag{25}$$

here even with ρ_{aa} very small we have gain.

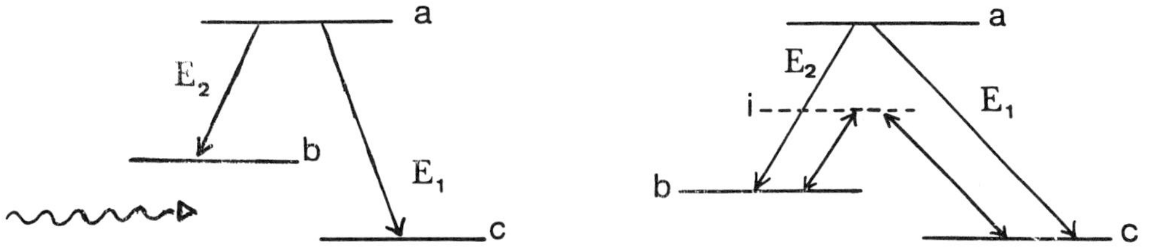

Fig. 5. a) Microwaves to coherently mix the lower two levels, b) Raman to coherently mix the lower two levels.

The off-diagonal matrix element for the microwave field producing the coherence between the lower two levels through quadrapole coupling* is (Fig. 5a),

$$\rho_{bc} = \sqrt{\rho_{bb}\rho_{cc}}\,e^{-i(\nu_\mu t+\Phi_{bc})} \times \frac{1}{2} \tag{26}$$

$$|\rho_{bc}|_{max} = \frac{1}{4} \tag{27}$$

Where ν_μ is the frequency of the microwave field, and Φ_{bc} is the phase factor associated with the microwave field. The factor 1/2 in Eq. (26) is due to collisions. There is still lasing without inversion, with only 1/2 of the atomic absorption being cancelled.

We consider a Raman interaction between levels $b - i$ and $i - c$, where $\rho_{ii} = 0$, which coherently mixes the two lower levels (Fig. 5b). The off-diagonal matrix element for the two lower levels goes as

$$\rho_{bc} = \sqrt{\rho_{bb}\rho_{cc}}\,e^{-i(\nu_j-\nu_k)t+i(\Phi_j-\Phi_k)} \times 1 \tag{28}$$

$$|\rho_{bc}|_{max} = \frac{1}{2} \tag{29}$$

where ν_j or ν_k is the frequency associated with the Raman fields and Φ_j or Φ_k is the phase associated with each of these field respectively. In Eq. 28 a factor of unity appears due to collisions. This cancels all the atomic absorption and lasing without inversion is more collision resilient, then in the case of the coherent mixing of the levels with the microwave field.

The effect of nonabsorption was demonstrated[15,16] using Na-atoms in a magnetic field, see Fig. (6). In the experiment the Raman interaction is created by using two of the longitudinal modes of a mode-locked dye laser. A magnetic field gradient is applied across the sample, ranging from 0 to 150 gauss, to continuously Zeeman split the two lower levels. When the splitting of the two lower level Sodium D-lines coincides to that of the laser lines, we see a lack of the resonance fluorescence "dark lines" perpendicular to that of the laser passing through the medium. In the experiment the positions of the longitudinal modes and the angle between the magnetic field and the incident laser field were changed resulting in the position of the dark lines changed proportionally.

CONCLUSION

The two examples above exemplify the unusual properties of quantum beats. The first example, of coherence in spin-1/2 particles, shows that which-path information destroys coherence and subsequent erasure of this information results in the re-emergence of spin coherence. In the second example, we saw that the effects of quantum beats destroy emission and absorption, in the three-level atomic systems

* Details to be published

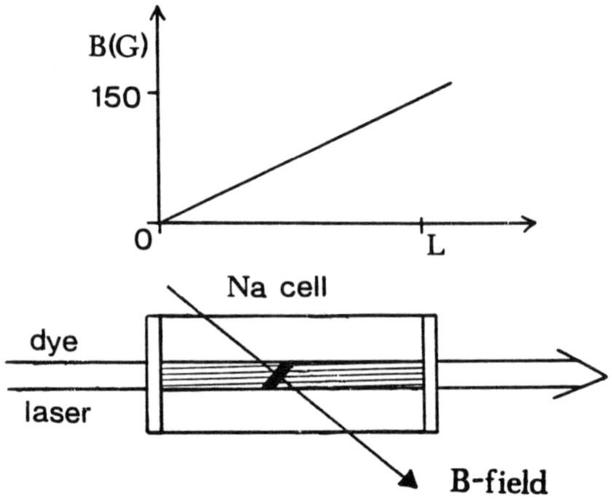

Fig. 6. A magnetic field gradient is applied across the Sodium atoms where at the resonance fluorescence from the dye laser appears as the shaded region.

(see Fig. 4). In the latter case, we have the exciting possibility of producing lasing without inversion. A further application that incorporates both quantum beat effects is the X-ray laser, where nonemission would trap atoms in the upper state, then nonabsorption could be used to produce the lasing.

ACKNOWLEDGMENT

This work was partially supported by the Office of Naval Research.

REFERENCES

1. B-G. Englert, J. Schwinger and M. O. Scully, (HDI) Found. of Phys. **18**, 1045 (1988).
2. J. Schwinger, M. O. Scully and B-G. Englert, (HDII) Z. Phys. D, Atoms, Molecules and Clusters **10**, 135 (1988).
3. M. O. Scully, B-G. Englert and J. Schwinger, (HDIII) Phys. Rev. **A40**, 1775 (1989).
4. (HDIV) Quantum eraser, to be published
5. M. O. Scully and K. Drühl, Phys. Rev. **A25**, 2208 (1982).
6. M. O. Scully and H. Walther, Phys. Rev. **A39**, 5229 (1989).
7. G. Badurck, H. Rauch and D. Tuppinger, Phys. Rev. **A34**, 2600 (1986).
8. M. O. Scully and Shi-Yao Zhu, Phys. Rev. Lett. **62**, 2813 (1989).
9. D. Marcuse, Proc. IEEE **51**, 849 (1963).
10. H. Holt, Phys. Rev. **A16**, 1136 (1976).
11. S. Harris, Phys. Rev. Lett. **62**, 1033 (1989).
12. U. Fano, Phys. Rev. **A124**, 1866 (1961); U. Fano and J. W. Copper, Rev. Mod. Phys. **40**, 441 (1968).
13. M. O. Scully, Phys. Rev. Lett. **55**, 2802 (1985); M. O. Scully and M. S. Zubairy, Phys. Rev. **A35**, 752 (1987); K. Zaheer and M. S. Zubairy, Phys. Rev. **A38**, 5227 (1988); Shi-Yao Zhu and M. O. Scully, Phys. Rev. **A38**, 5433 (1988); J. Bergou, M. Orszag and M. O. Scully, Phys. Rev. **A38**, 768 (1988); M. O. Scully and Shi-Yao Zhu, Phys. Rev. Lett. **62**, 2813 (1989); M. O. Scully, "Lasing Without Inversion Via the Quantum Beat Laser," Proc. NATO Advanced Research Workshop on: *Noise and Chaos in Nonlinear Dynamical Systems*, Torino, March 7-11, (1989).
14. M. Sargent III, M. O. Scully and W. E. Lamb, Jr., *Laser Physics*, Addison-Wesley, Reading, Massachusetts (1974).
15. G. Orriols, Nuovo Cimento **53B**, 1 (1979); G. Alzetta, L. Moi and G. Orriols, Nuovo Cimento **52B**, 209 (1979).
16. G. Alzatta, A. Gozzini, L. Moi and G. Orriols, Nuovo Cimento **36B**, 5 (1976).

ASYMPTOTOLOGY IN QUANTUM OPTICS

W. P. Schleich, J. P. Dowling, R. J. Horowicz, and S. Varro

Max-Planck-Institut für Quantenoptik
D-8046 Garching
Federal Republic of Germany

1. INTRODUCTION AND OVERVIEW

"Start 'er up and see why she don't run!" Shall we follow this advice given to us by the engineer, John Kris? Shall we try to unravel the problems associated with the definition of a Hermitian phase[1] operator in quantum mechanics? But how to "start 'er up," that is, how to gain insight into the obstacles of constructing phase states? Semiclassical quantum mechanics à la Wentzel-Kramers-Brillouin (WKB)[2,3] might be a competent guide in this enterprise. Why? This approach has already given us valuable insight into various problems of quantum optics, and in particular has led to a deeper understanding of the photon count probabilities[4,5] of a coherent,[6—8] and of a highly squeezed state.[9] Motivated by these successes we, in the present lectures, investigate the problem of the phase distribution of a coherent state, and of a quantum mechanical superposition of two coherent states, using WKB techniques. To set the stage, we review briefly the essential ingredients of this approach and illustrate it by rederiving the photon distribution of a coherent and a squeezed state.

The paper is organized as follows: How to describe the periodic motion of a bounded particle of given energy but unknown initial position and momentum; that is one of the questions asked in Sec. 2. Shall we use the notions of classical mechanics[10] and calculate the classical *probability* to find the particle at position x? This probability is proportional to the reciprocal of the particle's momentum. Or shall we follow the 'gospel' of quantum mechanics[11] and evaluate the two *interfering probability amplitudes* corresponding to the right- and the left-running waves? The identical strength of each wave, determined essentially by the square root of the classical probability of locating the particle, creates an oscillatory probability curve. Semiclassical quantum mechanics, which is identified as classical mechanics plus interference effects, is the central result of Sec. 2.

The photon statistics of a coherent state are governed by the overlap in position space between the Gaussian wave function of the coherent state ψ_{coh}, and the oscillatory wave function of the mth energy eigenstate u_m; this is a well-known result. However, to identify this result as a consequence of the Airy function[12] maximum of u_m appearing in the neighborhood of the classical turning point $x \equiv \xi_m = [2(m+1/2)]$ is only possible when we put semiclassical methods to use in Sec. 3. This Airy 'bump' is narrow compared to the width of the Gaussian coherent state wave function and

New Frontiers in Quantum Electrodynamics and Quantum Optics
Edited by A. O. Barut, Plenum Press, New York, 1990

acts as a delta function located at $x = \xi_m$. Hence the overlap $w_m[|\psi_{\text{coh}}\rangle]$ and thus the photon statistics follow the wave function of the coherent state ψ_{coh} at $x = \xi_m$; in other words, as we change m we essentially change x and 'read out' the wave function ψ_{coh}. In the case of a state highly squeezed in the x variable we face just the contrary situation. Here ψ_{sq} is narrow compared to the width of the oscillations and hence the squeezed state wave function reads out the particular value of u_m at $x = \sqrt{2}\alpha$. As we increase m the consecutive wave fronts of u_m move over the point $x = \sqrt{2}\alpha$, and the photon statistics follow u_m in their dependence on x giving rise to the oscillatory photon distribution.

To introduce an additional dimension and consider oscillator phase space, built out of coordinate and momentum rather than position space alone, as a strategy to make the WKB wave look simple, sounds like a crazy idea. However, in Sec. 4 we show that the WKB energy wave function $u_m^{(\text{WKB})} = u_m^{(\text{WKB})}(x)$ can be interpreted as a result of *interfering areas of overlap in phase space*[13]: The mth energy eigenstate, a band in phase space; and the position eigenstate, a narrow phase space 'highway'; are the essential ingredients of this recipe. The square root of their area of cross-over when normalized appropriately represents the magnitude of the two contributing probability amplitudes. The area circumnavigated by the central lines of the two states determines the phase difference between the two amplitudes.

The wedge-shaped 'slice' of phase space—cut out from the Gaussian bell of a coherent state[4,14] by two, angled 'panes of glass' which represent a state of well-defined phase[15]—can be used as a measure of the phase probability of a coherent state. That is the idea presented in Sec. 5, and it is another intriguing application of this concept of *area of overlap and interference in phase space.* We derive the probability amplitude for the phase φ in a coherent state of large displacement and use this amplitude in Sec. 6 to evaluate the phase probability of the quantum mechanical superposition of two coherent states of identical phase, but of different average photon number. Under appropriate conditions the phase uncertainty in this state can fall below the value of a single coherent state. The origin of this surprising result is the superposition principle of ordinary quantum mechanics. We conclude these lecture notes by presenting in Sec. 7 a summary and conclusion.

To avoid any confusion in terminology we prefer to use the expression *asymptotology* over *semiclassical quantum mechanics* to denote the limit of either $\hbar \to 0$ or the limit of large quantum numbers. In the field of quantum optics, *semiclassical theory* stands for a theory[7,8] which treats the atoms quantum mechanically, but the electromagnetic field classically. We adhere to this position in these notes and emphasize that the approach of asymptotology even allows us to treat quantized modes of the radiation field.

2. SEMICLASSICAL QUANTUM MECHANICS IS CLASSICAL MECHANICS PLUS INTERFERENCE

How to describe the motion of a nonrelativistic particle in a binding potential, such as the one shown in Fig. 1a? You may choose classical mechanics as your framework and talk about phase space trajectories; that is one possibility. To find the energy wave functions of this oscillator represents the quantum mechanical counterpart of this classical approach. Unfortunately, an analytic treatment of the corresponding Schrödinger equation is limited to a very few special potentials such as the harmonic oscillator, the Morse potential, and a few more. For the most part we have to resort to numerical solutions. However, the full analytical as well as the numerical solutions often hide striking and remarkable properties of the problems at hand. Such hidden features only come to light in the semiclassical limit of quantum mechanics—the topic of the present section.

2.1. Classical Motion Described by a Probability

Consider a particle with mass of unity, energy of η, and also dimensionless variables x for the coordinate, and p for the momentum, moving in the one dimensional potential $V = V(x)$ shown in Fig. 1a. Conservation of energy can be expressed here as

$$\eta = p^2/2 + V(x) \ , \tag{2.1}$$

and provides immediately the classical, x-p oscillator phase space trajectory

$$p^{(\mathrm{cl})}(x) = \{2\,[\eta - V(x)]\}^{1/2} \tag{2.2}$$

illustrated in Fig. 1b. When we have no knowledge of the initial position of the particle $x_o = x|_{t=0}$, nor of its momentum $p_o = p|_{t=0}$, but only of its energy, we associate with this motion in phase space a classical probability $W_x^{(\mathrm{cl})}$ to find the particle at the position x. This probability is governed by the reciprocal of the classical momentum $p^{(\mathrm{cl})}$, Eq. (2.2). In other words, $W_x^{(\mathrm{cl})}$ is large where the particle's velocity is small, but it is small when $p^{(\mathrm{cl})}$ assumes large values. More precisely we define

$$W_x^{(\mathrm{cl})} = \frac{1}{2}\mathcal{N}^2 \frac{1}{p^{(\mathrm{cl})}(x)} \ . \tag{2.3}$$

At the turning points of the motion, denoted by ϑ and ξ, the momentum $p^{(\mathrm{cl})}$, Eq. (2.2), vanishes and hence $W_x^{(\mathrm{cl})}$, Eq. (2.3), is infinite as shown in Fig. 1c. Nevertheless, the distribution $W_x^{(\mathrm{cl})}$ is still normalizable, that is

$$\int_\vartheta^\xi dx\, W_x^{(\mathrm{cl})} = 1 \ . \tag{2.4}$$

When we substitute Eq. (2.3) into Eq. (2.4) and note that $p^{(\mathrm{cl})}(x) = \frac{dx}{dt}$, we can determine the normalization constant $\mathcal{N}$ by

$$1 = \frac{1}{2}\mathcal{N}^2 \int_\vartheta^\xi \frac{dx}{p^{(\mathrm{cl})}(x)} = \frac{1}{2}\mathcal{N}^2 \int_0^{T/2} dt = \frac{1}{4}\mathcal{N}^2 \oint dt = \frac{1}{4}\mathcal{N}^2 T$$

that is,

$$\mathcal{N} = \frac{2}{\sqrt{T}} \ . \tag{2.5}$$

Hence the symbol

$$T \equiv \oint dt \tag{2.6}$$

denotes the period of the orbit. To identify the reciprocal of the classical momentum as a measure of the position probability—that is the message of this subsection.

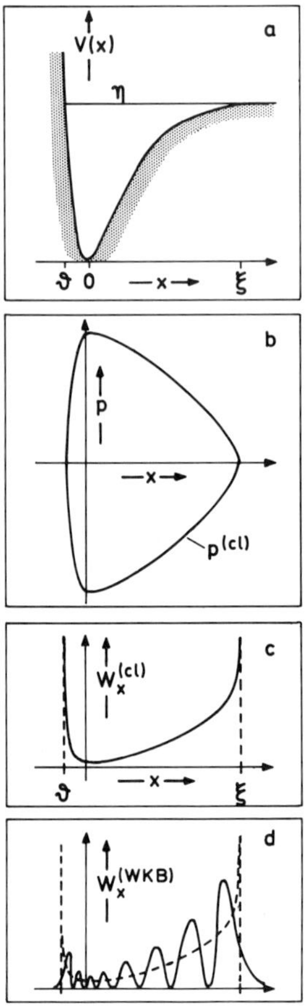

Fig. 1. A particle of unit mass and energy η vibrating in the potential $V = V(x)$ between the turning points ϑ and ξ shown in (a) traverses a closed circuit in oscillator phase space constructed out of dimensionless coordinate x, and momentum p. This phase space trajectory depicted in (b) is given by the classical momentum $p^{(\mathrm{cl})}$. The classical probability of finding the particle at position x, denoted $W_x^{(\mathrm{cl})}$, governed by the inverse of $p^{(\mathrm{cl})}$ and shown in (c) exhibits only a single minimum, namely at the minimum of V. Here the velocity of the particle is at its maximal value. The probability diverges at the turning points ϑ and ξ of the classical motion where the momentum vanishes. The area underneath the curve is normalized to unity. The semiclassical probability $W_x^{(\mathrm{WKB})}$, that is, the solid curve of (d), displays many maxima and minima as a result of interference *between right- and left-going waves, that is, as a consequence of the standing wave corresponding to the energy eigenfunction. The quantity $W_x^{(\mathrm{WKB})}$ does not diverge at the turning points, but exhibits Airy function-type maxima in the neighborhood of ϑ and ξ. Moreover, it reaches into the classically forbidden regime. The envelope of $W_x^{(\mathrm{WKB})}$ is twice the classical probability curve depicted in (d) by the dashed curve, that is, $2W_x^{(\mathrm{cl})}$, to maintain the normalization $\int dx\, W_x^{(\mathrm{WKB})} = 1$, in the presence of the interference-induced oscillations. Classical probability governed by the inverse of the momentum and interference between waves constitute the essential ingredients of semiclassical quantum theory.*

2.2. Quantum Mechanics Versus Classical Mechanics: Interfering Probability Amplitudes Rather Than Probabilities

We achieve the transition from classical mechanics to quantum mechanics by multiplying Eq. (2.1), the conservation of energy expression, by the energy wave function u, and by replacing the momentum p by the differential operator $-i\frac{d}{dx}$. This prescription yields

$$\frac{d^2u}{dx^2} + 2\left[\eta - V(x)\right] u = 0 \; . \tag{2.7a}$$

With the help of Eq. (2.2) the Schrödinger equation reads

$$u''(x) + \left[p^{(\mathrm{cl})}(x)\right]^2 u(x) = 0 \; . \tag{2.7b}$$

Here, and in the remainder of the article, a prime denotes differentiation with respect to the dimensionless position coordinate x.

For the special case of a constant potential $V(x) = V_o$, the classical momentum Eq. (2.2) is constant, that is to say,

$$p^{(\mathrm{cl})}(x) = [2(\eta - V_o)]^{1/2} \equiv p_o^{(\mathrm{cl})} \; .$$

Therefore we can easily integrate the Schrödinger equation Eq. (2.7) to find

$$u(x) = N \cos\left[\int_x^{\xi} d\tilde{x}\, p_o^{(\mathrm{cl})}(\tilde{x}) - \alpha\right]$$

where N and α are constants.

Guided by this example, we approach the case of nonconstant classical momentum $p^{(\mathrm{cl})}$ by proposing the *ansatz*

$$u^{(\mathrm{wave})}(x) \equiv \cos\left[\int_x^{\xi} d\tilde{x}\, p^{(\mathrm{cl})}(\tilde{x}) - \alpha\right] \tag{2.8}$$

which satisfies the differential equation

$$\left[u^{(\mathrm{wave})}\right]'' + \left\{\left[p^{(\mathrm{cl})}\right]^2 - \left[p^{(\mathrm{cl})}\right]' \tan\left[\int_x^{\xi} d\tilde{x}\, p^{(\mathrm{cl})}(\tilde{x}) - \alpha\right]\right\} u^{(\mathrm{wave})} = 0 \; ,$$

an equation similar to the Schrödinger equation Eq. (2.7b). Hence the *ansatz* of Eq. (2.8) is close to the real solution if we can neglect the second contribution in the bracket. However, the *ansatz* of Eq. (2.8) exhibits a more serious deficiency than just not being able to solve Eq. (2.7) in an exact way: It does not satisfy the correspondence principle.[16] In the limit of large quantum numbers the results of quantum mechanics *must* approach those of classical mechanics. In particular, the quantum mechanical probability of finding the particle of energy η at the position x, given by

$$W_x \equiv [u(x)]^2 \; , \tag{2.9}$$

must transmute itself into Eq. (2.3) in this limit.

The correspondence principle and the comparison between the classical and quantum mechanical probabilities, Eqs. (2.3) and (2.9) respectively, suggest an *ansatz* quite different from the wave *ansatz*, Eq. (2.8). This is namely the conjecture that

$$u^{(\mathrm{cl})}(x) = \frac{\mathcal{N}}{\sqrt{p^{(\mathrm{cl})}(x)}} \; . \tag{2.10}$$

Here the constant $\mathcal{N}$ is defined as in Eq. (2.5). Unfortunately $u^{(\mathrm{cl})}$ does not satisfy the Schrödinger equation Eq. (2.7) but instead the rather complicated equation

$$\left[u^{(\mathrm{cl})}\right]'' + \left\{\frac{1}{2}\left(\ln p^{(\mathrm{cl})}\right)'' - \left[\left(\frac{1}{2}\ln p^{(\mathrm{cl})}\right)'\right]^2\right\} u^{(\mathrm{cl})} = 0.$$

Hence we are not allowed to describe the motion of the particle via a wave solution, à la Eq. (2.8) alone, nor are we entitled to apply the concept of "the function $u^{(\mathrm{cl})}$ of Eq. (2.10) being yielded by a classical phase space trajectory." *The quantum mechanical particle is both wave and particle.* Hence the function which describes it best is the *product* of the wave *ansatz*Eq. (2.8) and of the particle *ansatz*Eq. (2.10). In other words

$$\begin{aligned} u^{(\mathrm{WKB})}(x) &\equiv u^{(\mathrm{cl})}(x)\, u^{(\mathrm{wave})}(x) \\ &= \mathcal{N}\left[p^{(\mathrm{cl})}(x)\right]^{-1/2} \cos\left[\int_x^{\xi} d\tilde{x}\, p^{(\mathrm{cl})}(\tilde{x}) - \alpha\right] \end{aligned} \tag{2.11}$$

which satisfies a Schrödinger-type equation

$$\left[u^{(\mathrm{WKB})}\right]'' + \left\{\left[p^{(\mathrm{cl})}\right]^2 - \left[\frac{1}{\sqrt{p^{(\mathrm{cl})}}}\right]'' \sqrt{p^{(\mathrm{cl})}}\right\} u^{(\mathrm{WKB})} = 0\ , \tag{2.12}$$

as well as the correspondence principle. Let us take the probability of finding the particle at the position x, namely

$$W_x^{(\mathrm{WKB})} \equiv \left[u^{(\mathrm{cl})}(x)\right]^2 \left[u^{(\mathrm{wave})}(x)\right]^2\ , \tag{2.13}$$

which follows from Eqs. (2.9) and (2.11). If we now average this probability over the rapid oscillations of $u^{(\mathrm{wave})}$, shown in Fig. 1d, we arrive at the averaged probability function:

$$\begin{aligned} \overline{W}_x^{(\mathrm{WKB})} &\equiv \left[u^{(\mathrm{cl})}(x)\right]^2 \left\langle \left[u^{(\mathrm{wave})}(x)\right]^2 \right\rangle_x \\ &= \frac{\mathcal{N}^2}{p^{(\mathrm{cl})}(x)} \frac{1}{\xi - \vartheta} \int_{\vartheta}^{\xi} dx \frac{1}{2}\left\{1 + \cos\left[2\int_x^{\xi} d\tilde{x}\, p^{(\mathrm{cl})}(\tilde{x}) - 2\alpha\right]\right\}\ , \end{aligned} \tag{2.14}$$

which then reduces to

$$\overline{W}_x^{(\mathrm{WKB})} = \frac{1}{2} \frac{\mathcal{N}^2}{p^{(\mathrm{cl})}(x)} = W_x^{(\mathrm{cl})}\ .$$

In the last step we have made use of the definition of the classical probability density, given in Eq. (2.3). Due to the oscillations of $u^{(\mathrm{WKB})}$ resulting from $u^{(\mathrm{wave})}$, the probability $W_x^{(\mathrm{WKB})}$, Eq. (2.13), never approaches the classical probability Eq. (2.3). Hence it is only the *averaged* probability $\overline{W}_x^{(\mathrm{WKB})}$ which satisfies the correspondence principle.

The $u^{(\mathrm{WKB})}$ *ansatz* is a solution of the Schrödinger equation, provided that

$$\left[p^{(\mathrm{cl})}\right]^2 \gg \left[p^{(\mathrm{cl})}\right]^{1/2} \left|\left\{\left[p^{(\mathrm{cl})}\right]^{-1/2}\right\}''\right|\ . \tag{2.15}$$

When we differentiate the classical momentum, Eq. (2.2), we find

$$\left\{\left[p^{(\mathrm{cl})}\right]^{-1/2}\right\}' = \frac{1}{2}\left[p^{(\mathrm{cl})}\right]^{-5/2} V'$$

and hence

$$\left\{\left[p^{(\mathrm{cl})}\right]^{-1/2}\right\}'' = \frac{5}{4}\left[p^{(\mathrm{cl})}\right]^{-9/2}(V')^2 + \frac{1}{2}\left[p^{(\mathrm{cl})}\right]^{-5/2} V'' ,$$

which when substituted into Eq. (2.15) yields the validity condition of the WKB solution Eq. (2.11). This condition is

$$\left| \frac{5}{32} \frac{V'(x)}{[\eta - V(x)]^3} + \frac{1}{8} \frac{V''(x)}{[\eta - V(x)]^2} \right| \ll 1 . \tag{2.16}$$

Clearly the condition is violated in the neighborhood of the turning points ϑ and ξ, where $V(\vartheta) = V(\xi) = \eta$. Hence the WKB solution (2.11) can only be valid in the domain between the two turning points, that is, in the region where $\vartheta \ll x \ll \xi$. Moreover, according to Eq. (2.16), the potential V—or rather the rate of change of V—puts constraints on the application of the WKB technique. Only slowly varying potentials are allowed. Potentials with cusps, that is, with sharp corners, lie outside of this approach and have to be treated by a different method—a WKB technique which can 'grapple' with a corner.[17]

We now return to the construction of the WKB solution. When we substitute the normalization Eq. (2.5) into the WKB wavefunction Eq. (2.11), we arrive at

$$u^{(\mathrm{WKB})}(x) = 2\,[T\,p(x)]^{-1/2} \cos\left[\int_x^\xi d\tilde{x}\, p(\tilde{x}) - \alpha\right] . \tag{2.17}$$

Here and in the remainder of this article we neglect for simplicity the superscript '(cl)'— the quantity $p(x)$ alone now stands for the classical momentum, Eq. (2.2).

In order for the solution Eq. (2.17) to represent an energy eigenfunction, we have to fix the two remaining free constants: the energy η and the phase α. These quantities we deduce[3] by approximately solving the Schrödinger equation in terms of the Airy function,[12] in the neighborhood of the turning points ϑ and ξ. When we find the asymptotic continuation of this approximate solution into the oscillatory domain of $u_m^{(\mathrm{WKB})}$, i.e. the domain where $\vartheta < x < \xi$, this Airy function solution reduces to the expression of Eq. (2.17)—provided $\alpha = \pi/4$. Since we will perform this calculation for the example of a harmonic oscillator potential in Sec. 3, we refrain now from showing the explicit evaluation but only mention the essential ideas. See Fig. 2.

Accordingly, we know that at the turning points the WKB wave has a phase of $-\pi/4$. Moreover, we recall that the mth energy eigenstate enjoys m zeros. Therefore when we decrease the x variable—starting with ξ and taking it all the way down to ϑ—the total accumulated phase of $u_m^{(\mathrm{WKB})}$ is,

$$\pi/4 + \pi m + \pi/4 = \int_\vartheta^\xi dx\,\sqrt{2\,[\eta_m - V(x)]} \equiv \int_\vartheta^\xi dx\, p_m(x) \tag{2.18}$$

Here η_m denotes the energy which satisfies the resulting Bohr-Sommerfeld-Kramers quantization condition

$$\oint dx\, p_m(x) = 2\pi(m + 1/2) , \tag{2.19}$$

The phase of the WKB wave at x is then given by

$$\phi_m \equiv S_m(x) - \pi/4 = \int_x^\xi dx\, p_m(x) - \pi/4 , \tag{2.20}$$

where $S_m(x) \equiv \int_x^\xi dx\, p_m(x)$.

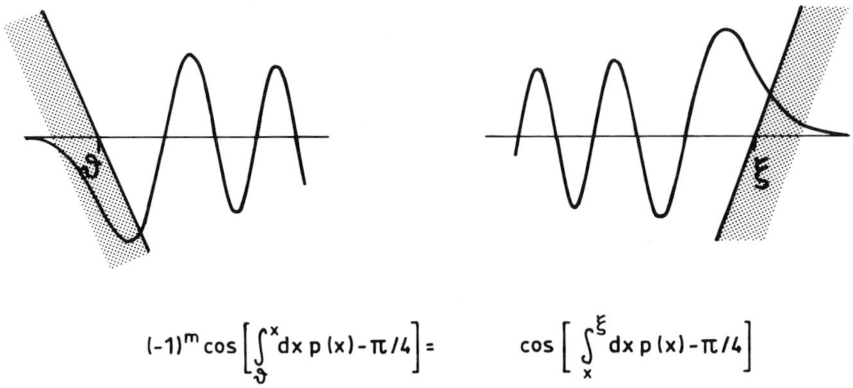

Fig. 2. The phase of a WKB wave function at turning points of the classical motion, ϑ, and ξ, is $(-\pi/4)$, that is, $u_m^{(\mathrm{WKB})}(x = \vartheta, \xi) \sim \cos(-\pi/4)$. Hence the oscillatory part of $u_m^{(\mathrm{WKB})}$ reads $\cos(\int_\vartheta^x dx\, p(x) - \pi/4)$ or $\cos(\int_x^\xi dx\, p(x) - \pi/4)$. To guarantee the uniqueness of the WKB wave function $u_m^{(\mathrm{WKB})}$ at any point x within the classically allowed region, the value of $u_m^{(\mathrm{WKB})}(x)$ obtained by expanding from the right turning point ξ must be identical to the one found from an expansion starting from the left turning point ϑ. Odd quantum numbers imply an odd number of zeros. Hence the wave function in the neighborhood of ϑ has a different sign than at ξ. We have to incorporate an additional phase change of $(-1)^m = \cos(m\pi)$ in the uniqueness condition which leads to the energy quantization $\oint dx\, p(x) = 2\pi(m + 1/2)$.

Thus we have seen in this section a clear exposition of the assertion made earlier. Namely, that semiclassical quantum mechanics is really nothing more than the product of classical mechanics with quantum-like interference terms.

3. PHOTON STATISTICS OF A COHERENT AND A SQUEEZED STATE ILLUMINATED BY WKB WAVE FUNCTIONS

The finding of approximate but known wave functions rather than exact and, for most cases, unknown solutions of the Schrödinger equation—that is a one line summary of the preceding section. But does this WKB procedure also yield deeper insight into the photon statistics for a coherent[6—8] or for a squeezed state[9] of a single mode of the electromagnetic field? Yes! Speaking now in the language of the field mode's mechanical analogue, the harmonic oscillator, in this section we derive the energy spread for these two types of states—a spread which in the field concept corresponds to the photon count probability W_m. This calculation is carried out using the WKB wavefunctions of the harmonic oscillator.

The harmonic oscillator provides a description of the amplitude of one mode of the electromagnetic field, and is a powerful source of insight on what is measurable and what is not. A cylinder of unit mass rolling under the influence of gravity on a metal ruler bent into the shape of a parabola can serve as a model—as shown in Fig. 3. To begin with, the oscillator is in its lowest quantum state

$$\psi_o(x) = \pi^{-1/4} \exp(-x^2/2) \ ,$$

where x denotes the position, and also we have chosen a frequency of unity for the oscillator.

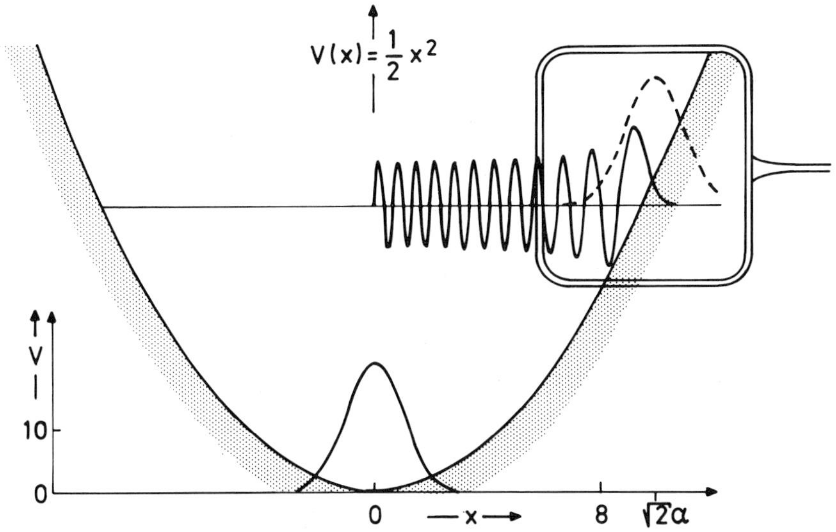

Fig. 3. The sudden displacement of a harmonic oscillator potential by an amount $x_o = \sqrt{2}\alpha$, and its simultaneous lowering corresponding to an energy $x_o^2/2 = \alpha^2$, creates out of its Gaussian ground state $\psi_o = \psi_o(x)$ (shown by the solid curve at the potential minimum) a coherent state $\psi_{\rm coh} = \psi_{\rm coh}(x)$ located at $x = \sqrt{2}\alpha$. This state, whose wavefunction is depicted here by the dashed curve, is not an eigenstate of the oscillator potential. The energy distribution of $\psi_{\rm coh}$, that is, the probability W_m of finding the $m^{\rm th}$ energy eigenstate $u_m = u_m(x)$ in $\psi_{\rm coh}$ is governed by the overlap $w_m[|\psi_{\rm coh}\rangle]$ between u_m and $\psi_{\rm coh}$, as depicted in the 'magnifying glass'. For simplicity we have chosen the specific quantum number $m = 55$ and have show $u_{m=55}$ for only positive x values. The displacement parameter has the value $\alpha = 7$.

3.1. Coherent State

A coherent state is obtained from the ground state given above by a displacement. In the mechanical model this is achieved by suddenly displacing the origin of the harmonic oscillator by an amount $x_o = \sqrt{2}\alpha$, and by simultaneously lowering the potential energy by $\frac{1}{2}x_o = \alpha^2$. The wave function of the state so prepared, reads

$$\psi_{\rm coh}(x) = \pi^{-1/4} \exp\left[-\frac{1}{2}(x - \sqrt{2}\alpha)^2\right] . \tag{3.1}$$

The wave function $\psi_{\rm coh}$ is not an eigenfunction of this oscillator, that is, the coherent state is not a stationary state and thus undergoes a time development.[6—8] The Gaussian wave packet (3.1) —having the potential energy $\frac{1}{2}x_o^2 = \alpha^2$—bounces back and forth between the classical turning points of the vibratory motion corresponding to this energy. It has just the 'right' width to keep its shape while oscillating.

Consequently the coherent state shows a spread in energy. But how large a spread? The energy distribution is not the classically expected delta function located at the classical value

$$\overline{m} = (1/2)(\text{spring constant})(\text{displacement})^2 = (1/2) \cdot 1 \cdot (\sqrt{2}\alpha)^2 = \alpha^2 ,$$

but rather follows from standard quantum mechanics[11] as

$$W_m = |w_m [|\psi_{\rm coh}\rangle]|^2 \ , \tag{3.2a}$$

where

$$w_m [|\psi_{\rm coh}\rangle] \equiv \int_{-\infty}^{\infty} dx\, u_m(x)\, \psi_{\rm coh}(x) \ . \tag{3.2b}$$

Here,

$$u_m(x) = \pi^{-1/4}(2^m m!)^{-1/2} H_m(x) \exp\left[-x^2/2\right] \tag{3.3}$$

denotes the wave function of the $m^{\rm th}$ energy eigenstate,[11] with H_m being the $m^{\rm th}$ Hermite polynomial,[18] and $\psi_{\rm coh}$ is given by Eq. (3.1).

With the help of the relations

$$\frac{1}{2}x^2 + \frac{1}{2}(x - \sqrt{2}\alpha)^2 = \left(x - \frac{1}{\sqrt{2}}\alpha\right)^2 + \frac{1}{2}\alpha^2$$

and[19]

$$\int_{-\infty}^{\infty} dy\, H_m(y) \exp\left[-(y - y_o)^2\right] = \pi^{1/2}(2y_o)^m \ ,$$

we can perform the integration in Eq. (3.2*b*). When this is done, we arrive at

$$w_m [|\psi_{\rm coh}\rangle] = \frac{\alpha^m}{\sqrt{m!}} \exp\left(-\frac{1}{2}\alpha^2\right) \ , \tag{3.4a}$$

or, with the help of the energy spread expression of Eq. (3.2*a*), at

$$W_m [|\psi_{\rm coh}\rangle] = \frac{(\alpha^2)^m}{m!} e^{-\alpha^2} \ . \tag{3.4b}$$

Deeper insight into this result springs forth from a treatment of the overlap integral (3.2*b*), using WKB wave functions $u_m^{\rm (WKB)}$ of the harmonic oscillator. From Fig. 4 we note that the wave function of the coherent state is broad compared to the wavelength of the energy wave function u_m, Eq. (3.3). Hence the oscillatory part of u_m averages out when we evaluate the integral (3.2*b*). The main contribution arises from the last bump of u_m in the neighborhood of the turning point $\xi_m = [2(m + 1/2)]^{1/2}$, a bump which acts essentially as a delta function. Hence, for the present discussion we can approximate u_m by

$$u_m(x) \cong \delta(x - \xi_m) \ ,$$

and the overlap integral (3.2*b*) reduces to

$$w_m [|\psi_{\rm coh}\rangle] \cong \psi_{\rm coh}(\xi_m) = \pi^{-1/4} \exp\left[-\frac{1}{2}(\xi_m - \sqrt{2}\alpha)^2\right] \ .$$

The photon statistics $W_m[|\psi_{\rm coh}\rangle]$ in its dependence on m hence follows the wave function of the coherent state $\psi_{\rm coh}$ in the x variable. Therefore these statistics show a single maximum at $m \cong \alpha^2$ as indicated on the right hand side of Fig. 4.

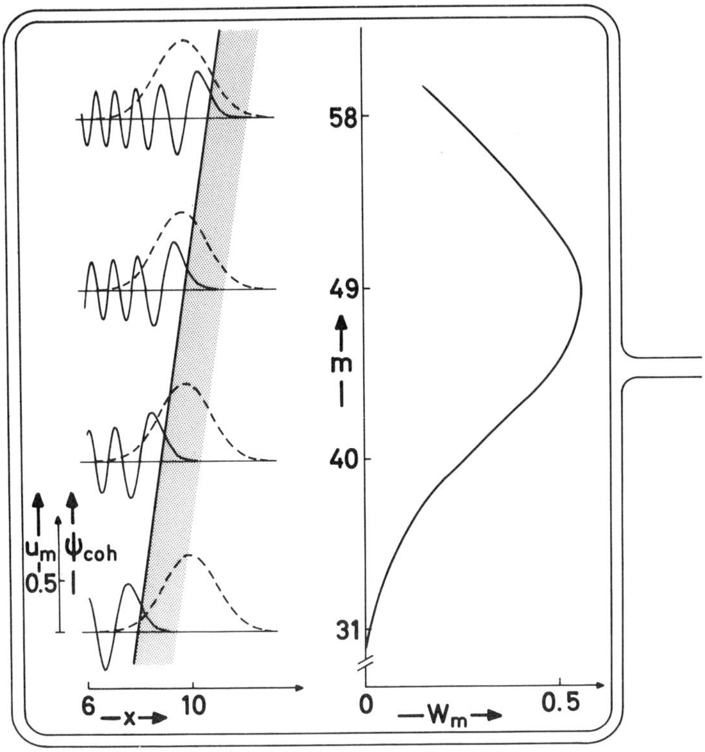

Fig. 4. The probability $W_m = W_m[|\psi_{\rm coh}\rangle]$ of finding the $m^{\rm th}$ energy eigenstate u_m in $\psi_{\rm coh}$, shown on the right of the figure, is governed by the overlap $w_m[|\psi_{\rm coh}\rangle]$ in position space between the two wave functions u_m and $\psi_{\rm coh}$, displayed for particular choices of m on the left hand side of the picture. For quantum numbers $m \cong x_o^2/2 = \alpha^2$, the right wall of the harmonic oscillator $V(x) = x^2/2$ of Fig. 3 is essentially a straight line. Hence the energy wave function in the neighborhood of the classical turning point ξ_m is an Airy function which has its dominant maximum shortly before the potential wall. The width of the coherent state, shown here by the dashed curves, is large compared to the wavelength of the energy eigenwave u_m. Therefore, the oscillatory part of u_m averages out when integrated together with $\psi_{\rm coh}$ over x. Only the last maximum of positive value contributes to the integral $w_m[|\psi_{\rm coh}\rangle]$. This maximum is narrow compared to $\psi_{\rm coh}$ and hence acts as a delta function located at $x = \xi_m$. The overlap integral $w_m[|\psi_{\rm coh}\rangle]$ reduces to the 'reading out' of the value of $\psi_{\rm coh}(x) = \pi^{-1/4}\exp[-(x-\sqrt{2}\alpha)^2/2]$ at the turning point of the $m^{\rm th}$ energy eigenstate ξ_m, that is, $w_m[|\psi_{\rm coh}\rangle] \cong \exp[-(\xi_m - \sqrt{2}\alpha)^2/2]$. This procedure provides in a direct way the Gaussian approximation of the exact Poissonian photon distribution of a coherent state, shown on the right hand side of the figure. Poissonian photon statistics of a coherent state, recognized as the position representation of this state evaluated at the classical turning point of the $m^{\rm th}$ energy eigenstate, that is asymptotology at work.

Tab. 1. Properties of the Harmonic Oscillator.

Potential	$\frac{1}{2}x^2$
Quantized Energy	$m+1/2=\frac{1}{2}p^2+\frac{1}{2}x^2$
Kramers Phase Space Trajectory	$2(m+1/2)=\left[p^{(\mathrm{cl})}(x)\right]^2+x^2$
Turning Points	$\pm\xi_m\equiv\pm\sqrt{2(m+1/2)}$
Energy Wave Function *exact*	$u_m(x)=\frac{1}{\sqrt[4]{\pi}\sqrt{2^m m!}}H_m(x)\,e^{-x^2/2}$
WKB approximation $x\cong\xi_m=\sqrt{2(m+1/2)}$	$u_m^{(\mathrm{WKB})}(x)\cong\frac{\sqrt[3]{2}}{\sqrt[6]{\xi_m}}\mathrm{Ai}\left[-\sqrt[3]{2\xi_m}\,(\xi_m-x)\right]$
$\lvert x\rvert<\xi_m$	$u_m^{(\mathrm{WKB})}(x)\cong\frac{\sqrt{2/\pi}}{\sqrt[4]{\xi_m^2-x^2}}\cos\left[\int_x^{\xi_m}d\tilde{x}\,\sqrt{\xi_m^2-\tilde{x}^2}-\frac{\pi}{4}\right]$
Planck-Bohr-Sommerfeld-Band *inner edge* *outer edge*	$2m=p^2+x^2$ $2(m+1)=p^2+x^2$

To find an asymptotic expansion of the function u_m of Eq. (3.3) in the neighborhood of the classical turning point, to obtain a quantitative evaluation of Eq. (3.2*b*), that is the next task. From Table 1 we recall that the quantized energy of the harmonic oscillator for the potential $V(x)\equiv(1/2)x^2$, reads

$$m+1/2=\frac{1}{2}p^2+\frac{1}{2}x^2\;. \tag{3.5}$$

Hence the two turning points $\pm\xi_m$ corresponding to the mth energy state are given by

$$\xi_m=\left[2\left(m+\frac{1}{2}\right)\right]^{1/2}\;. \tag{3.6}$$

The Schrödinger equation for $u_m=u_m(x)$ follows then from Eq. (3.5) by multiplying through by u_m, and then by replacing the number p by the operator $p=-i\frac{d}{dx}$. This equation is

$$u_m''(x)+\left(\xi_m^2-x^2\right)u_m(x)=0\;. \tag{3.7}$$

In the neighborhood of $x=\xi_m$, that is, for $\xi_m^2-x^2=(\xi_m+x)(\xi_m-x)\cong 2\xi_m(\xi_m-x)$, Eq. (3.7) reduces to

$$u_m''(x)+2\xi_m(\xi_m-x)\,u_m(x)\cong 0\;.$$

When we recall that the Airy function,[12]

$$\mathrm{Ai}(x)=(2\pi)^{-1}\int_{-\infty}^{\infty}dt\exp\left[i\left(\frac{1}{3}t^3+xt\right)\right]\;, \tag{3.8}$$

satisfies the differential equation

$$\mathrm{Ai}''(x)-x\,\mathrm{Ai}(x)=0\;,$$

then we can say that the approximate harmonic oscillator wave function $u_m^{(\mathrm{WKB})}$ in the neighborhood of $x \cong \xi_m$ reads

$$u_m^{(\mathrm{WKB})}(x) \cong \mathcal{N}_m \,\mathrm{Ai}\left[-(2\xi_m)^{1/3}(\xi_m - x)\right] \tag{3.9}$$

where $\mathcal{N}_m$ is a constant. Note that for $x > \xi_m$, corresponding to the classically forbidden regime, the wave function decays appropriately exponentially since in this region the argument of the Airy function assumes positive values. We now substitute the WKB wavefunction Eq. (3.9), together with the coherent wavefunction Eq. (3.1), into the overlap integral given in Eq. (3.2*b*), to find that

$$w_m \cong \pi^{-1/4}\,\mathcal{N}_m \int_{-\infty}^{\infty} dx\,\mathrm{Ai}\left[-(2\xi_m)^{1/3}(\xi_m - x)\right] \exp\left[-\frac{1}{2}(x - \sqrt{2}\alpha)^2\right] .$$

The turning point bump of u_m, described by the first maximum of the Airy function Eq. (3.9), is narrow when compared to the width of ψ_{coh}, as shown in Fig. 4. Hence we evaluate ψ_{coh} at $x = \xi_m$ and factor it out of the integral, which yields

$$\begin{aligned} w_m &= \pi^{-1/4}\mathcal{N}_m \exp\left[-\frac{1}{2}(\xi_m - \sqrt{2}\alpha)^2\right] \int_{-\infty}^{\infty} dx\,\mathrm{Ai}\left[-(2\xi_m)^{1/3}(\xi_m - x)\right] \\ &= \pi^{-1/4}\mathcal{N}_m(2\xi_m)^{-1/3} \exp\left[-\frac{1}{2}(\xi_m - \sqrt{2}\alpha)^2\right] \int_{-\infty}^{\infty} dy\,\mathrm{Ai}(y) . \end{aligned}$$

With the help of Eq. (3.8) for the Airy function we evaluate the integral

$$\begin{aligned} \int_{-\infty}^{\infty} dy\,\mathrm{Ai}(y) &= \int_{-\infty}^{\infty} dt\,\exp\left[(i/3)t^3\right] (2\pi)^{-1} \int_{-\infty}^{\infty} dy\,\exp(ity) \\ &= \int_{-\infty}^{\infty} dt\,\exp\left[(i/3)t^3\right] \delta(t) \\ &= 1 \end{aligned}$$

which reduces the above result for the probability amplitude to

$$w_m[|\psi_{\mathrm{coh}}\rangle] = \pi^{-1/4}\mathcal{N}_m\,(2\xi_m)^{-1/3}\exp\left[-\frac{1}{2}(\xi_m - \sqrt{2}\alpha)^2\right] .$$

The use of the relation

$$\xi_m - \sqrt{2}\alpha = \frac{\xi_m^2 - 2\alpha^2}{\xi_m + \sqrt{2}\alpha} \cong \frac{m + 1/2 - \alpha^2}{\sqrt{2}\alpha}$$

in this expression then yields

$$w_m \cong \pi^{-1/4}\mathcal{N}_m(2\xi_m)^{-1/3} \exp\left[-\frac{1}{2}\left(\frac{m + 1/2 - \alpha^2}{\sqrt{2}\alpha}\right)^2\right] , \tag{3.10a}$$

or finally

$$w_m \cong \pi^{-1/4}\widetilde{\mathcal{N}}(2\sqrt{2}\alpha)^{-1/3} \exp\left[-\frac{1}{2}\left(\frac{m + 1/2 - \alpha^2}{\sqrt{2}\alpha}\right)^2\right] , \tag{3.10b}$$

where in Eq. (3.10*b*) we have evaluated the slowly varying functions $\mathcal{N}_m$ and ξ_m at $m + 1/2 = \alpha^2$. The constant $\widetilde{\mathcal{N}}$, defined as

$$\widetilde{\mathcal{N}} \equiv \mathcal{N}_m\Big|_{m=\alpha^2-1/2} ,$$

follows from the normalization condition

$$1 = \sum_{m=0}^{\infty} W_m \cong \int_0^{\infty} dm\, w_m^2 = \widetilde{\mathcal{N}}^2(2\sqrt{2}\alpha)^{-2/3}\sqrt{2}\alpha$$

that is,

$$\widetilde{\mathcal{N}} = 2^{1/3}(\sqrt{2}\alpha)^{-1/6} ,$$

and consequently the semiclassical limit of the probability amplitude Eq. (3.4a) at last reads

$$w_m[|\psi_{\rm coh}\rangle] = (2\pi)^{-1/4}\alpha^{-1/2}\exp\left[-\frac{1}{2}\left(\frac{m+1/2-\alpha^2}{\sqrt{2}\alpha}\right)^2\right] . \tag{3.11}$$

Why not obtain this result in the more direct way of applying the Stirling expansion[20] to Eq. (3.4a)? Why go through this complex procedure? Two arguments present themselves:

1. This approach brings out most clearly the fact that the photon number probability amplitude Eq. (3.11), and hence the photon distribution of a coherent state, $W_m[|\psi_{\rm coh}\rangle]$, follows essentially the wave function $\psi_{\rm coh}$ given by Eq. (3.1). The last maximum of the energy wave function $u_m^{\rm (WKB)}$, Eq. (3.9), serves as a narrow 'slit' and projects out the value of $\psi_{\rm coh}$ at $x = \xi_m$. This slit effect of the turning point maximum—hidden behind the opaque curtain of the rather complex expression for u_m, Eq. (3.3)— is only transparent in its WKB expression

$$u_m^{\rm (WKB)}(x) \cong 2^{1/3}\xi_m^{-1/6}\,{\rm Ai}\left[-(2\xi_m)^{1/3}(\xi_m - x)\right] , \tag{3.12}$$

valid in the neighborhood of the right turning point x, as summarized in Table 1.

2. When we compare Eq. (3.3) and Eq. (3.12) we find as a by-product of this approach the asymptotic expansion

$$H_m(x)\exp(-x^2/2) \sim 2^{1/3}\pi^{1/4}(2^m m!)^{1/2}\xi_m^{-1/6}{\rm Ai}\left[-(2\xi_m)^{1/3}(\xi_m - x)\right] \tag{3.13}$$

of a Hermite polynomial H_m for $x \cong \xi_m = [2(m+1/2)]^{1/2}$.

3.2. Squeezed State

We now turn to the case of a state highly squeezed in the x variable, that is, a state whose width is smaller than that of the coherent state. In our mechanical model we obtain this state when, in addition to those changes necessary to generate the coherent state, we also suddenly change the frequency of the oscillator from unity to s^{-1}, where $s > 0$. This wave function expressed in the new frequency reads

$$\psi_{\rm sq}(x) = (s/\pi)^{1/4}\exp\left[-(s/2)(x-\sqrt{2}\alpha)^2\right] . \tag{3.14}$$

The coherent state of Eq. (3.1) is thus a special case of Eq. (3.14) for $s = 1$. This squeezed state is not an energy eigenstate of the new potential and hence shows a spread in energy. Its energy distribution—in the language of quantized fields its photon number distribution—is given by

$$W_m = \left|w_m\left[|\psi_{\rm sq}\rangle\right]\right|^2 \tag{3.15a}$$

where the overlap integral reads

$$w_m\left[|\psi_{\rm sq}\rangle\right] \equiv \int_{-\infty}^{\infty} dx\, u_m(x)\,\psi_{\rm sq}(x) . \tag{3.15b}$$

When we substitute Eqs. (3.3) and Eq. (3.14) into Eq. (3.15b) and perform the integral with the help of the relations

$$\frac{1}{2}\left[x^2 + s(x-\sqrt{2}\alpha)^2\right] = \left\{[(s+1)/2]^{1/2}x - s(s+1)^{-1/2}\alpha\right\}^2 + \frac{s}{s+1}\alpha^2$$

and[19]

$$\int_{-\infty}^{\infty} dy \exp\left[-(y-y_o)^2\right] H_m(\lambda y) = \pi^{1/2}(1-\lambda^2)^{m/2} H_m\left[\frac{\lambda}{(1-\lambda^2)^{1/2}} y_o\right]$$

we arrive at[5,21]

$$w_m[|\psi_{\rm sq}\rangle] = \left(\frac{2}{s+1}\right)^{1/2} s^{1/4} \left(\frac{s-1}{s+1}\right)^{m/2} (2^m m!)^{-1/2}$$
$$\times H_m\left[\frac{s}{(s^2-1)^{1/2}}\sqrt{2}\alpha\right] \exp\left(-\frac{s}{s+1}\alpha^2\right) . \qquad (3.16)$$

What a formidable result! Does it allow any insight into how the photon statistics of a squeezed state $W_m[|\psi_{\rm sq}\rangle]$, Eq. (3.15*a*), depends on the quantum number m, the displacement parameter α, or the squeeze s? No! Again it is the evaluation of Eq. (3.15*b*) in terms of WKB wave functions which illuminates the functional dependence of $w_m[|\psi_{\rm sq}\rangle]$.

In Fig. 5 we show the overlap $w_m[|\psi_{\rm sq}\rangle]$ between u_m and the squeezed state wave function Eq. (3.14) for specific quantum numbers m. Here we consider the situation of a state highly squeezed in the x variable, that is, for $s = 2/\epsilon$ where $0 < \epsilon \ll 1$. When we compare Fig. 5 to the corresponding picture for a coherent state, Fig. 4, the wave function $\psi_{\rm sq}$ is now *narrow* compared to the wavelength of the energy wave function u_m. Hence $\psi_{\rm sq}$ acts essentially as a delta function located at $x = \sqrt{2}\alpha$, that is

$$\psi_{\rm sq}(x) \cong \delta(x - \sqrt{2}\alpha) \qquad (3.17)$$

and distills out essentially the value of u_m at $x = \sqrt{2}\alpha$, that is,

$$w_m[|\psi_{\rm sq}\rangle] \cong u_m(x = \sqrt{2}\alpha) . \qquad (3.18)$$

From Fig. 5 and Eq. (3.18) we note that for increasing values of m the photon probability amplitude of a highly squeezed state, $w_m[|\psi_{\rm sq}\rangle]$, and hence its photon statistics follows the energy wave function, u_m, at $x = \sqrt{2}\alpha$ in its dependence on m. For m values smaller than α^2, it is the decaying side of the Airy function approximation of u_m, Eq. (3.12), which creates a small probability of finding m photons as shown on the right side of Fig. 5. For $m \cong \alpha^2$ the turning point hump rests on top of the squeezed state function giving rise to a maximum in the overlap $w_m[|\psi_{\rm sq}\rangle]$ and thus also in $W_m[|\psi_{\rm sq}\rangle]$. For quantum numbers m appropriately larger than α^2 the squeezed state wave function probes *the oscillatory regime of* u_m leading to an *oscillatory behavior* of $W_m[|\psi_{\rm sq}\rangle]$.[5,22] This domain in $W_m[|\psi_{\rm sq}\rangle]$ we now investigate in more detail.

A more complete treatment of the integral Eq. (3.15*b*) does not invoke the delta function approximation of $\psi_{\rm sq}$, Eq. (3.17), but expands[5] $u_m(x)$ into a Taylor series around $x = \sqrt{2}\alpha$, which yields

$$u_m(x) = \sum_{k=0}^{\infty} \frac{1}{k!} \left.\frac{d^k\, u_m(x)}{dx^k}\right|_{x=\sqrt{2}\alpha} (x - \sqrt{2}\alpha)^k .$$

When we substitute this expansion together with Eq. (3.14) into Eq. (3.15*b*) we arrive at

$$w_m[|\psi_{\rm sq}\rangle] = \sum_{k=0}^{\infty} \frac{1}{k!} \left.\frac{d^k\, u_m(x)}{dx^k}\right|_{x=\sqrt{2}\alpha} \left(\frac{2\epsilon}{\pi}\right)^{1/4} \epsilon^{k/2} \int_{-\infty}^{\infty} dy\, y^k\, e^{-y^2} .$$

Using the integral relations[19]

$$\int_{-\infty}^{\infty} dy\, y^{2k+1}\, e^{-y^2} = 0$$

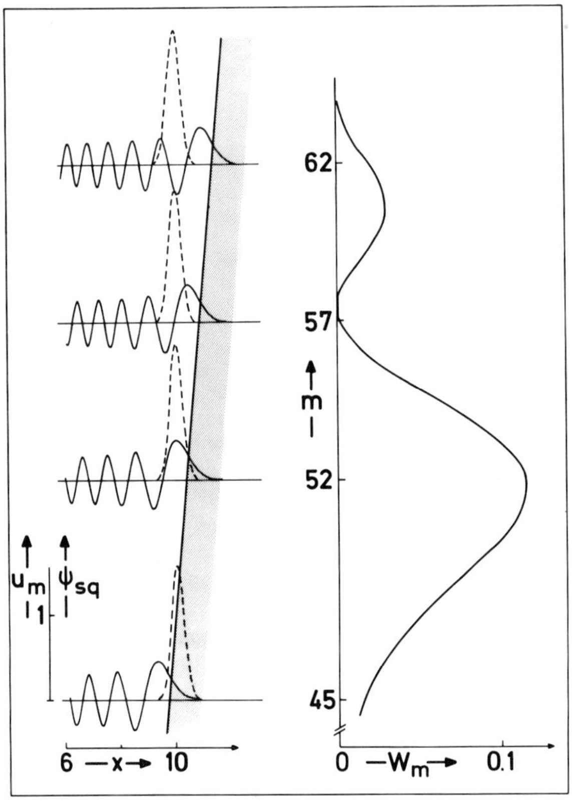

Fig. 5. The energy distribution of a highly squeezed state $W_m[|\psi_{sq}\rangle]$, shown in its oscillatory behavior on the right hand side of the figure, results from the overlap $w_m[|\psi_{sq}\rangle]$ between the mth energy wave function u_m and the squeezed state wave ψ_{sq}, depicted on the left hand side in the neighborhood of $x = \sqrt{2}\alpha$ by solid and by dashed curves respectively. In contrast to the coherent state discussion of Figs. 3 and 4 the Gaussian ψ_{sq} is narrow compared to the wavelength of u_m. This Gaussian acts essentially as a delta function located at $x = \sqrt{2}\alpha$ and maps out the oscillations of u_m in the coordinate x, namely $u_m(x) \sim \cos(S_m(x) - \pi/4)$, onto oscillations in the energy distribution W_m, that is, $W_m[|\psi_{sq}\rangle] \sim \cos[S_m(x = \sqrt{2}\alpha) - \pi/4]$. The oscillatory photon distribution of a highly squeezed state as a result of consecutive wave fronts of u_m moving through the 'narrow slit' provided by the squeezed state, that is the message of semiclassical quantum mechanics à la WKB. For definiteness we have chosen for the values of the displacement and squeezing parameters $\alpha = 7$ and $s = 21$.

and

$$\int_{-\infty}^{\infty} dy\, y^{2k}\, e^{-y^2} = \pi^{1/2} 2^{-2k} (2k)!/k!$$

finally yields

$$w_m[|\psi_{\rm sq}\rangle] = 2^{3/4}\pi^{1/2}\left(\frac{\epsilon}{4\pi}\right)^{1/4} \sum_{k=0}^{\infty} \frac{(\epsilon/4)^k}{k!} \frac{d^{2k}\, u_m(x)}{dx^{2k}}\bigg|_{x=\sqrt{2}\alpha} . \tag{3.19}$$

So far the calculation is exact. However, we now focus on the oscillatory domain of $u_m(x)$, that is, for $|x| < \xi_m$. Since $x = \sqrt{2}\alpha$, this also corresponds to $m > \alpha^2$, the oscillatory domain of $W_m[|\psi_{\rm sq}\rangle]$. According to Sec. 2 the wave function u_m can be approximated in this region by

$$u_m^{(\rm WKB)}(x) \cong \mathcal{N}_m[p_m(x)]^{-1/2} \cos\left[\int_x^{\xi_m} d\tilde{x}\, p_m(\tilde{x}) - \pi/4\right] \tag{3.20a}$$

where we have denoted

$$p_m \equiv \{2\,[\eta_m - V(x)]\}^{1/2} = \left[2(m+1/2) - x^2\right]^{1/2} . \tag{3.20b}$$

The normalization constant $\mathcal{N}_m$ follows from its relation to the oscillation period in Eq. (2.5), as

$$\mathcal{N} = \left(\frac{2}{\pi}\right)^{1/2} .$$

Here we have made use of $T = 2\pi/(\text{frequency} = 1) = 2\pi$. We can now perform the differentiation of u_m using Eq. (3.20). We now neglect the slow variation of $[p_m(x)]^{-1/2}$ compared to the variation of the cosine, that is,

$$\frac{d^{2k}\, u_m^{(\rm WKB)}}{dx^{2k}} \cong (-1)^k\, [p_m(x)]^{2k}\, u_m^{(\rm WKB)}(x)$$

and then Eq. (3.19) for the squeezed state probability amplitude reduces to

$$w_m[|\psi_{\rm sq}\rangle] = 2\left(\frac{\epsilon}{4\pi}\right)^{1/4} \sum_{k=0}^{\infty} \frac{\left[-(\epsilon/2)\,(m+1/2-\alpha^2)\right]^k}{k!} (m+1/2-\alpha^2)^{-1/4}$$
$$\times \cos\left[\int_{\sqrt{2}\alpha}^{\xi_m} d\tilde{x}\, p_m(\tilde{x}) - \pi/4\right]$$

or equivalently,

$$w_m[|\psi_{\rm sq}\rangle] = 2\,\mathcal{A}_m^{1/2} \cos\phi_m \tag{3.21a}$$

where we have defined

$$\mathcal{A}_m \equiv \left(\frac{\epsilon}{4\pi}\right)^{1/2} \frac{\exp\left[-\epsilon(m+1/2-\alpha^2)\right]}{(m+1/2-\alpha^2)^{1/2}} \tag{3.21b}$$

and

$$\phi_m \equiv \int_{\sqrt{2}\alpha}^{\xi_m} dx\, p_m(x) - \pi/4 . \tag{3.21c}$$

We conclude this section by noting that again we could have obtained the result Eq. (3.21) directly. How? Just apply the asymptotic expansion[13]

$$H_m(x)\exp(-x^2/2) \sim \left(\frac{4}{\pi}\right)^{1/4} \frac{(2^m m!)^{1/2}}{(\xi_m^2 - x^2)^{1/4}} \cos\left[\int_x^{\xi_m} d\tilde{x}\, (\xi_m^2 - \tilde{x}^2)^{1/2} - \pi/4\right]$$

for a Hermite polynomial, H_m. This expansion is valid for $|x| < [2(m+1/2)]^{1/2}$ and is obtained by comparing the exact and the WKB solutions to the harmonic oscillator, Eqs. (3.3) and Eq. (3.20) respectively, to Eq. (3.16) for the probability amplitude of a squeezed state. But again such mathematics hides the crucial role of the squeezed state being the narrow slit which reads out the oscillations of the energy wave functions.

4. SEMICLASSICAL WAVE FUNCTIONS AS A RESULT OF INTERFERING AREAS IN PHASE SPACE

"Photon statistics of a coherent or of a squeezed state made simple via a WKB wave function," that is the theme of the preceding section. But how to make the WKB wave itself simple, that is the problem addressed, elaborated on, and solved in the present section. We introduce a pictorial phase space representation of the primitive WKB energy wave function

$$u_m(x) = \langle x|m\rangle \equiv \begin{cases} 0 & \text{for } x < \vartheta_m \text{ or } x > \xi_m \quad (4.1a) \\ \dfrac{\exp[i\phi_m(x)]}{\sqrt{T_m\, p_m(x)}} + \dfrac{\exp[-i\phi_m(x)]}{\sqrt{T_m\, p_m(x)}} & \text{for } \vartheta_m < x < \xi_m \quad (4.1b) \end{cases}$$

where $u_m(x)$ represents a particle of unit mass vibrating in a binding potential V between the two turning points ϑ_m and ξ_m, as shown in Fig. 1.

The WKB approximation associates with an energy eigenstate a Kramers phase space trajectory

$$\eta\Big|_{\mathcal{J}=m+1/2} = \frac{1}{2}p^2(x)\Big|_{\mathcal{J}=m+1/2} + V(x)\,, \tag{4.2}$$

determined by the reduced action

$$\mathcal{J} = (2\pi)^{-1}\oint dx\, p(x) = (2\pi)^{-1}\oint dx\,[2(\eta - V(x))]^{1/2}\,, \tag{4.3}$$

which has the quantized value of $m + 1/2$. However, to depict geometrically the position representation of this state, that is, the quantum mechanical scalar product between this state and the position eigenstate $|x\rangle$, it is more suitable to visualize this state as the Planck-Bohr-Sommerfeld band of Fig. 6a. The edges of this band are defined by the phase space trajectories

$$\eta\Big|_{\mathcal{J}=m} = \frac{1}{2}p^2(x)\Big|_{\mathcal{J}=m} + V(x) \tag{4.4a}$$

and

$$\eta\Big|_{\mathcal{J}=m+1} = \frac{1}{2}p^2(x)\Big|_{\mathcal{J}=m+1} + V(x)\,, \tag{4.4b}$$

corresponding to the reduced actions m and $m+1$, respectively. We represent the position eigenstate $|x\rangle$ by a thin, phase space strip located at x and directed parallel to the momentum axis, as shown in Fig. 6a. Can the area of overlap in phase space between the Planck-Bohr-Sommerfeld band, defined by Eq. (4.4), and this phase space highway be used as a measure for the quantum mechanical scalar product? Yes! No simpler algorithm offers itself. But what are the predictions of such a hypothesis?

- For x values beyond the classical turning points, that is, for $x < \vartheta_m$ and $x > \xi_m$, there is no such overlap and hence

  $$u_m^{(\mathrm{WKB})}(x) \cong 0\,,$$

  in agreement with Eq. (4.1*a*).

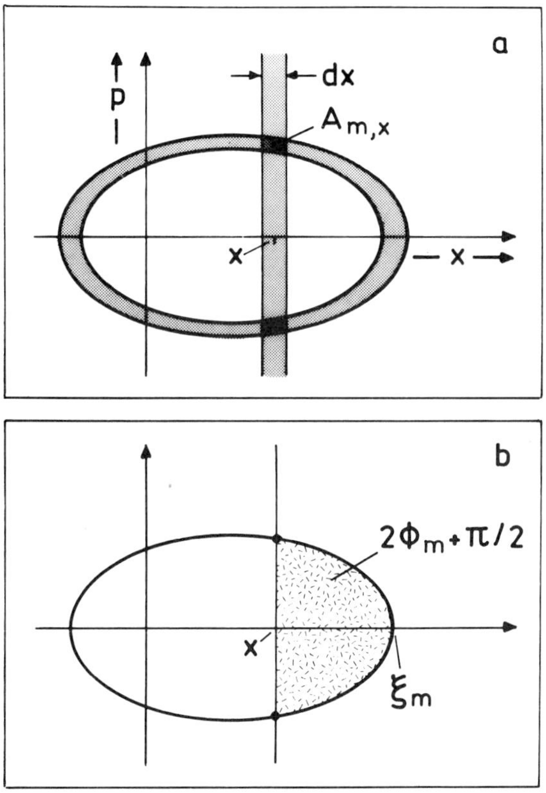

Fig. 6. The area-of-overlap and interference in phase space principle associates with the mth *energy wave function in the position representation,* $u_m(x) \equiv \langle x|m\rangle$*, the area of overlap between the* m^{th} *Planck-Bohr-Sommerfeld band created by the two phase space trajectories corresponding to the two reduced actions* m *and* $m+1$ *and the phase space highway of width* dx *located at* x *and directed parallel to the momentum axis. For* x *values between the two turning points there exist two symmetrically located zones of cross-over* $A_{m,x}$ *corresponding to positive and negative momenta, as shown in (a). Each domain represents a complex-valued probability amplitude: the square root of* $A_{m,x}$ *normalized to the total area of the band, that is, normalized to* 2π*, is its magnitude. The phase difference* $2\phi_m$ *between the two interfering probability amplitudes, in other words, between the interfering areas of (a), is determined up to the constant* $\pi/2$ *by the phase space area circumnavigated by the Kramers trajectory of reduced action* $m+1/2$ *and the center line of the 'highway' at* x *as illustrated in (b).*

- For x values located between the turning points, in other words, in the classically allowed vibratory regime, the highway cuts two symmetrically located diamond shaped zones out of the band. The area of each diamond reads

$$A_{m,x} \equiv \int_{-\infty}^{\infty} d\tilde{x} \int_{p(x)|_{\mathcal{J}=m}}^{p(x)|_{\mathcal{J}=m+1}} dp\, \delta(x-\tilde{x}) = p(x)\Big|_{\mathcal{J}=m+1} - \quad p(x)\Big|_{\mathcal{J}=m} .$$

In accordance with Bohr's correspondence principle[16] we can replace, in the large m limit, a finite difference by a differential. In this case such a procedure allows us to write

$$A_{m,x} \cong \frac{\partial p}{\partial \mathcal{J}}\bigg|_{\mathcal{J}=m+1/2} = \frac{\partial p}{\partial \eta}\bigg|_{\mathcal{J}=m+1/2} \times \frac{\partial \eta}{\partial \mathcal{J}}\bigg|_{\mathcal{J}=m+1/2} . \tag{4.5}$$

From Eq. (4.2) we find

$$\frac{\partial p}{\partial \eta}\bigg|_{\mathcal{J}=m+1/2} = p^{-1}(x)\Big|_{\mathcal{J}=m+1/2} \equiv p_m^{-1}(x)$$

Hence, by differentiating the action integral Eq. (4.3) and comparing it to Eq. (2.6) for the period, we arrive at

$$\frac{d\mathcal{J}}{d\eta}\bigg|_{\mathcal{J}=m+1/2} = (2\pi)^{-1} \oint \frac{dx}{p(x)}\bigg|_{\mathcal{J}=m+1/2} = (2\pi)^{-1} \oint dt\bigg|_{\mathcal{J}=m+1/2} \equiv \frac{T_m}{2\pi} .$$

Here T_m denotes the period of the orbit corresponding to the reduced action $m+1/2$. We may therefore simplify the area of overlap formula Eq. (4.5) to

$$A_{m,x} \equiv 2\pi\, [T_m\, p_m(x)]^{-1} .$$

Equation (4.1*b*) for the WKB wavefunction identifies the square root of the area of cross-over, normalized to the area of the total band, that is,

$$\left[\frac{A_{m,x}}{2\pi}\right]^{1/2} ,$$

as the magnitude of each of the two contributing probability amplitudes. The factor 2π insures the normalization

$$\int dx\, u_m^2(x) = 2\int dx \frac{1}{2\pi} A_{m,x} = \frac{1}{2\pi}\begin{pmatrix}\text{area of} \\ \text{band} = 2\pi\end{pmatrix} = 1 .$$

A complex-valued probability amplitude represented by an area of overlap in phase space? The square root of an appropriately normalized overlap domain as this amplitude's magnitude? Yes! But what is this probability amplitude's phase? In one zone of cross-over the momentum is positive—the particle moves to the right—in the other the momentum is negative indicating motion to the left. The total phase difference between the two *interfering areas* according to Eq. (2.20), reads

$$2\phi_m = 2\left[\int_x^{\xi_m} dx\, p_m(x) - \pi/4\right]$$

which, apart from the constant $\pi/4$, is the area in phase space enclosed by the center lines of the two states given by Eq. (4.2), and the straight line at x propagating parallel to the momentum axis.

Is this formalism of *area-of-overlap and interference in phase space*[13] limited to the quantum mechanical scalar product between an energy eigenstate and a position eigenstate? No! Should we demonstrate the power of this approach by explaining[4] the Poisson-Gauss photon statistics of a coherent state Eq. (3.1) as the area of overlap in phase space between the mth circular Planck-Bohr-Sommerfeld band of a harmonic oscillator (defined by Eq. (4.4) and in Table 1) and a displaced Gaussian bell representing the coherent state? Or should we relate the striking oscillations[5] which appear in the photon distribution of a highly squeezed state Eq. (3.21) to the *interfering* areas of overlap between the mth Planck-Bohr-Sommerfeld band and the Gaussian 'cigar' of the squeezed state—very much in the spirit of Fig. 6? How about a nice interpretation of the rapid modulations in the jump probability between two vibratory states of different electronic levels in a diatomic molecule—as a result of *interference in phase space?*[13,23] Yes! In all three examples—and there are many more—semiclassical quantum mechanics interpreted as *interfering areas of overlap in phase space* provides immediate insight into whatever problem happens to be at hand.

5. PHASE DISTRIBUTION OF A COHERENT STATE FROM AREA OF OVERLAP AND INTERFERENCE IN PHASE SPACE

Consider a given quantum state; take for example a coherent state. What is its distribution in position, $\psi_{\text{coh}} = \psi_{\text{coh}}(x)$, what in momentum, $\Phi_{\text{coh}} = \Phi_{\text{coh}}(p)$, or what in energy, $w_m[|\psi_{\text{coh}}\rangle]$? Each of these questions we can answer right away. Why? There exists a unique operator for these quantities. But what is the phase distribution of a state?[1] How can we answer this question without knowing the phase operator? It is the area of overlap principle of Sec. 4 which allows us to calculate, in the semiclassical limit, a phase distribution of a state without even touching upon the touchy subject of the notion of a phase operator. We illustrate this approach in the present section using the example of a coherent state. For an application of this approach to a squeezed state or a general state we refer to Ref. 15

The question of a phase variable, that is, of a phase operator in quantum mechanics, has been a long standing problem since the decisive year 1926—the dawn of the Bohr-Sommerfeld-*Atommechanik*[16] and the rise of the Schrödinger-Born-Heisenberg-Jordan-*Quantenmechanik.*[24] In contrast to *Atommechanik* which made heavy use of action-angle variables m and φ, this new *Undulationsmechanik* or *Matrizenmechanik* was formulated in terms of the conjugate variables x and p. However, when Dirac[25] in the same year quantized the radiation field by relying on m and φ to be just such conjugate variables, London[26] had already recognized the impossibility of constructing a Hermitian matrix representing the phase operator in the number state bases. In the late fifties and sixties masers and lasers producing electromagnetic waves with relatively well defined phases stimulated the search for such operators. Louisell[27], however, pointed out once again the pathological character of the Dirac phase operator: its matrix elements in the number state bases are indefinite. Since this time many new phase operators have been suggested, as summarized in the review of Ref. 1.

In the present approach we do not start from a *mathematical* definition of a state of well-defined phase in terms of, for example, number states, but from the intuitive semiclassical picture of such a state $|\varphi\rangle$ being represented in x-p oscillator phase space by a diverging beam emerging from the origin.[28] In the most elementary version we associate with the state $|\varphi\rangle$ a phase space distribution

$$P_\varphi(\tilde{\varphi})\, d\tilde{\varphi} = r\, \delta(\varphi - \tilde{\varphi})\, d\tilde{\varphi}\ . \tag{5.1}$$

In this way we avoid the subtleties of the definition of a phase state and the question of an Hermitian phase operator[29] in quantum mechanics.

According to the area of overlap principle, the phase distribution of a coherent state $W_\varphi[|\psi_{\text{coh}}\rangle]$ is identical to the weighted phase space slice cut out of the Gaussian

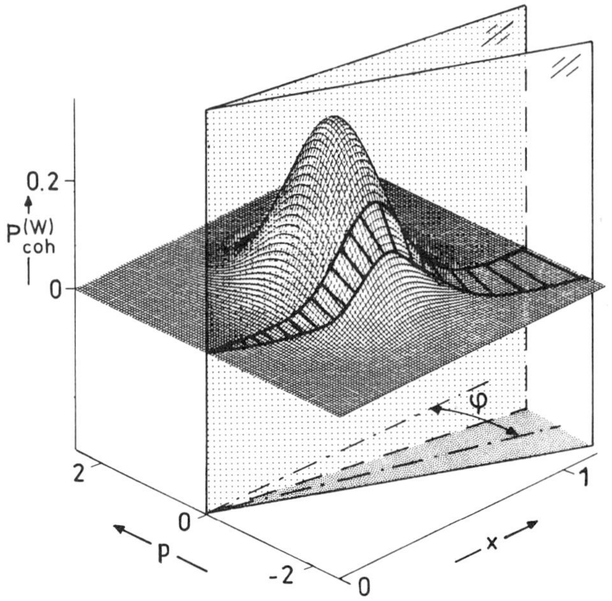

Fig. 7. In its most elementary version we associate with a state, $|\varphi\rangle$, of well-defined phase, φ, a ray propagating in x-p oscillator phase space, emerging from the origin and directed by an angle φ relative to the x axis, shown by dashed-dotted lines. A more appropriate visualization of $|\varphi\rangle$ starts from a divergent beam of solid angle $d\varphi$ propagating along this center line as illustrated by the dark, wedge-shaped, phase space 'slice'. The Gaussian bell $P^{(W)}_{coh}$ of Eq. (5.2) serves as a representation of a coherent state $|\psi_{coh}\rangle$, Eq. (3.1), with the displacement parameter $\alpha = 0.7$. The area-of-overlap in phase space principle associates with the phase probability $W_\varphi[|\psi_{coh}\rangle] = |\langle\varphi|\psi_{coh}\rangle|^2$ of this coherent state, the weighted area, that is, the volume of the shaded phase space slice cut out of the bell by the pie-shaped wedge, formed by the vertical 'panes of glass', which act as the 'knife edges' of the phase state. The volume of this slice acts as a measure of phase probability—no simpler pictorial representation offers itself.

bell

$$P_{\text{coh}}(x,p) = \left|\psi_{\text{coh}}(x)\right|^2 \left|\Phi_{\text{coh}}(p)\right|^2 = \pi^{-1}\exp\left[-(x-\sqrt{2}\alpha)^2 - p^2\right] \tag{5.2}$$

by the diverging beam of the phase state Eq. (5.1), shown in Figs. 7 and 8a.

When we introduce the polar coordinates $x = r\cos\tilde{\varphi}$ and $p = r\sin\tilde{\varphi}$, this area is then given by[30]

$$\begin{aligned} W_\varphi[|\psi_{\text{coh}}\rangle] \cong A_\varphi &= \int_0^\infty dr \int_{-\pi}^{\pi} d\tilde{\varphi}\, r\, \delta(\varphi - \tilde{\varphi})\, P_{\text{coh}}(x = r\cos\tilde{\varphi}; p = r\sin\tilde{\varphi}) \\ &= \pi^{-1}\int_0^\infty dr\, r \exp\left[-(r - \sqrt{2}\alpha\cos\varphi)^2\right] \exp(-2\alpha^2\sin^2\varphi)\ . \end{aligned} \tag{5.3}$$

For φ values such that $\pi/2 < \varphi < 3\pi/2$ we find that $\sqrt{2}\alpha\cos\varphi < 0$, and hence, since $r > 0$, the integral is exponentially small, that is,

$$W_\varphi \cong 0 \qquad \text{for } \pi/2 < \varphi < 3\pi/2\ . \tag{5.4a}$$

For φ values such that $-\pi/2 < \varphi < \pi/2$ the main contribution arises from values of $r \cong \sqrt{2}\alpha\cos\varphi > 0$. When we extend the lower limit of the integral in Eq. (5.3) to $-\infty$ we arrive at

$$W_\varphi[|\psi_{\text{coh}}\rangle] \cong \sqrt{\frac{2\alpha^2}{\pi}}\cos\varphi\exp(-2\alpha^2\sin^2\varphi) \qquad \text{for } -\pi/2 < \varphi < \pi/2\ . \tag{5.4b}$$

Moreover, from Eq. (5.3) we deduce the periodicity property $W_{\varphi+2\pi} = W_\varphi$. The phase distribution of a coherent state given by Eq. (5.4) shows a *single* maximum at $\varphi = 0$ and its width $\delta\varphi_{\text{coh}}$, governed by the phase angle of exponential fall-off, is given by

$$\delta\varphi_{\text{coh}} \cong \arcsin\left[(2\alpha^2)^{-1/2}\right]\ . \tag{5.5}$$

In the limit of large displacements, that is, $2\alpha^2 \gg 1$, we find from Eq. (5.5)

$$\delta\varphi_{\text{coh}} \cong 2^{-1/2}\,\alpha^{-1} \ll 1\ . \tag{5.6}$$

Linearization of Eq. (5.4*b*) yields the Gaussian distribution

$$W_\varphi[|\psi_{\text{coh}}\rangle] \cong \left(\frac{2\alpha^2}{\pi}\right)^{1/2} \exp(-2\alpha^2\varphi^2)\ , \tag{5.7}$$

in agreement with the result of Ref. 31 and the prediction of Fig. 8a.

We are now in a position to calculate the phase probability *amplitude* of a coherent state $w_\varphi[|\psi_{\text{coh}}\rangle] = \langle\varphi|\psi_{\text{coh}}\rangle$. According to the area of overlap approach, this probability amplitude's magnitude is given by the square root of the *probability*, that is, $|\langle\varphi|\psi_{\text{coh}}\rangle| = W_\varphi^{1/2}$, and its phase is determined by the area in phase space enclosed by the central lines of the two states of interest, that is, by the value of the shaded phase space segment shown in Fig. 8b. This area reads $\frac{1}{2}(\sqrt{2}\alpha)^2\varphi = \alpha^2\varphi$, and hence the total amplitude is simply given by[31]

$$\langle\varphi|\psi_{\text{coh}}\rangle = \sqrt[4]{\frac{2\alpha^2}{\pi}}\exp(-\alpha^2\varphi^2)\exp(-i\alpha^2\varphi)\ . \tag{5.8}$$

A single pregnant sentence that summarizes this section would have to be: "One can use the twin concepts of *area of overlap in phase space* and *interference in phase space* as extremely handy tools to calculate the phase probability amplitude of a coherent state."

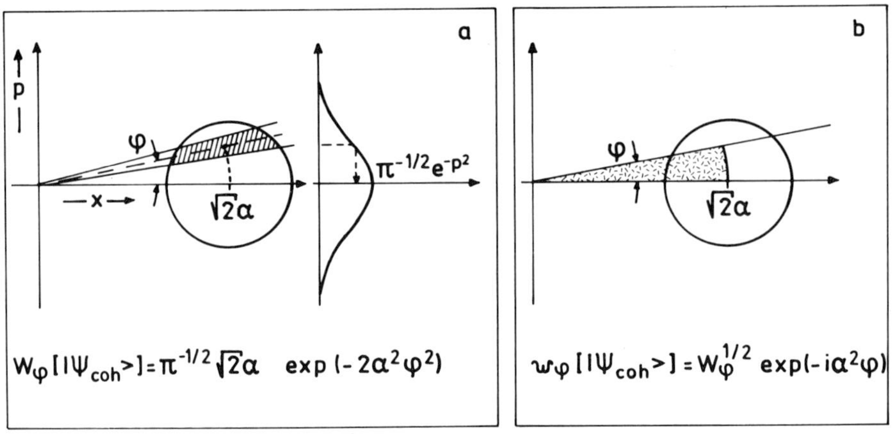

Fig. 8. The probability of finding the phase φ in a coherent state of large displacement parameter α is given by the shaded overlap between the diverging beam representing the phase state and the circle. This circle is the contour line of exponential decay of the coherent state Gaussian bell, Eq. (5.2) *at $x = \sqrt{2}\alpha$. This rectangularly shaped area—approximately determined by its length of unity times its width of $\sqrt{2}\alpha\, d\varphi$ —is weighted by the Gaussian distribution in momentum, evaluated at the center of the rectangle, that is, at $p \cong \sqrt{2}\alpha\varphi$. The magnitude of the corresponding phase probability* amplitude $w_\varphi[|\psi_{\rm coh}\rangle] = \langle\varphi|\psi_{\rm coh}\rangle$ *is then the square root of this slice cut out of the bell. Its phase is governed by the domain enclosed by the central lines of the states of interest. According to (b), the area of this shaded phase space segment is given by $\frac{1}{2}(\sqrt{2}\alpha)^2\varphi = \alpha^2\varphi$.*

6. SQUEEZING THE PHASE VIA QUANTUM INTERFERENCE

At the heart of quantum mechanics lies the superposition principle—to quote from the first chapter of Dirac's[32] classic treatise on quantum mechanics, "... any two or more states may be superposed to give a new state..." Insight into the far-reaching consequences of this principle is offered by the most elementary example of superposing two coherent states of identical phase, but of different average number of photons.[33] Such an example is depicted in Fig. 9. This new superposition state shows quite a remarkable property: Under appropriate conditions its phase uncertainty falls below the value of that of a single coherent state. This striking phenomenon constitutes the topic of the present section.

With the help of the results of Sec. 5, we investigate the phase distribution

$$W_\varphi[|\psi\rangle] = |w_\varphi[|\psi\rangle]|^2 = |\langle\varphi|\psi\rangle|^2 \tag{6.1}$$

of the quantum superposition state

$$|\psi\rangle = \mathcal{N}_\delta\,[|\alpha - \delta\alpha/2\rangle + |\alpha + \delta\alpha/2\rangle]\ . \tag{6.2}$$

For small nonzero displacements $\delta\alpha$, the two states $|\alpha \pm \delta\alpha/2\rangle$ are not orthogonal, i.e.

$$\langle\alpha - \delta\alpha/2|\alpha + \delta\alpha/2\rangle = \exp(-\delta\alpha^2/2) \neq 1\ .$$

The normalization constant $\mathcal{N}_\delta$ was obtained from

$$1 = \langle\psi|\psi\rangle = \mathcal{N}_\delta^2\,[1 + 1 + 2\,\Re(\langle\alpha - \delta\alpha/2|\alpha + \delta\alpha/2\rangle)]\ ,$$

which gave rise to the slightly complicated expression,

$$\mathcal{N}_\delta = \frac{1}{\sqrt{2}}\left[1 + \exp(-\delta\alpha^2/2)\right]^{-\frac{1}{2}}\ . \tag{6.3}$$

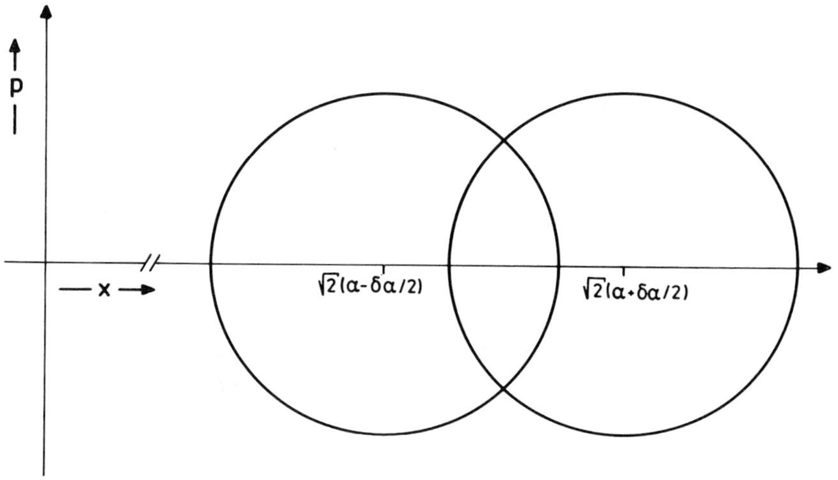

Fig. 9. In a first attempt, we represent the quantum mechanical superposition of two coherent states of identical phase, here for simplicity chosen to be zero, but of different average number of photons $\langle m\rangle = (\alpha \pm \delta\alpha/2)^2$, by two circles of radius unity located at the positive x axis at $x_\pm = \sqrt{2}(\alpha \pm \delta\alpha/2)$. A more suitable approach associates with each coherent state a Gaussian bell: the circle is hence the contour line of exponential decay. This simple minded approach lacks, however, the interference phenomena inherent in the quantum mechanical superposition of two states and expressed in the domain between x_+ and x_- by the Wigner distribution 'hump'.

With the help of Eq. (5.8) we arrive at the phase probability amplitude of the superposition state $|\psi\rangle$, Eq. (6.2). In particular,

$$\begin{aligned} w_\varphi[|\psi\rangle] &= \mathcal{N}_\delta \left[\langle\varphi|\alpha - \delta\alpha/2\rangle + \langle\varphi|\alpha + \delta\alpha/2\rangle\right] \\ &= \mathcal{N}_\delta \sqrt[4]{\frac{2}{\pi}} \Big\{ \sqrt{\alpha - \delta\alpha/2}\exp\left[-(\alpha - \delta\alpha/2)^2\varphi^2\right] \exp\left[-i(\alpha - \delta\alpha/2)^2\varphi\right] \\ &\quad + \sqrt{\alpha + \delta\alpha/2}\exp\left[-(\alpha + \delta\alpha/2)^2\varphi^2\right] \exp\left[-i(\alpha + \delta\alpha/2)^2\varphi\right] \Big\} . \end{aligned}$$

The resulting phase probability of Eq. (6.1) then reads

$$\begin{aligned} W_\varphi[|\psi\rangle] = \mathcal{N}_\delta^2 \sqrt{2/\pi} \Big\{ &(\alpha - \delta\alpha/2)\exp\left[-2(\alpha - \delta\alpha/2)^2\varphi^2\right] \\ &+ (\alpha + \delta\alpha/2)\exp\left[-2(\alpha + \delta\alpha/2)^2\varphi^2\right] \\ &+ 2\sqrt{\alpha^2 - (\delta\alpha/2)^2}\exp\left\{-2\left[\alpha^2 + (\delta\alpha/2)^2\right]\varphi^2\right\}\cos(2\alpha\,\delta\alpha\,\varphi) \Big\} . \end{aligned}$$

When we assume that $\alpha \gg \delta\alpha/2$, namely, that the separation of the two states is small compared to their displacement from the origin—this expression simplifies to

$$W_\varphi[|\psi\rangle] = \frac{\sqrt{2/\pi}}{2\left[1+\exp(-\delta\alpha^2/2)\right]}\Big\{(\alpha-\delta\alpha/2)\exp\left[-2(\alpha-\delta\alpha/2)^2\varphi^2\right]$$
$$+(\alpha+\delta\alpha/2)\exp\left[-2(\alpha+\delta\alpha/2)^2\varphi^2\right]$$
$$+2\alpha\exp(-2\alpha^2\varphi^2)\cos(2\alpha\,\delta\alpha\,\varphi)\Big\} \tag{6.4}$$

where we have made use of Eq. (6.3) for the normalization constant $\mathcal{N}_\delta$. Hence the phase probability of the state $|\psi\rangle$ of Eq. (6.2), which is the superposition of two coherent states of average photon number $(\alpha \pm \delta\alpha/2)^2$, is not only the sum of two Gaussian distributions in the variable φ (one Gaussian for each of the coherent states) but also involves an interference term. This interference contribution is a consequence of the quantum mechanical superposition of the two states.

We now investigate to what extent this superposition-induced term modifies the phase uncertainty of the state $|\psi\rangle$, expressed by the variance

$$\Delta\varphi^2 \equiv \langle\varphi^2\rangle - \langle\varphi\rangle^2 . \tag{6.5}$$

Here we have defined the moments $\langle\varphi^j\rangle$ via

$$\langle\varphi^j\rangle \equiv \int_{-\infty}^{\infty} d\varphi\,\varphi^j\,W_\varphi[|\psi\rangle] . \tag{6.6}$$

We may now substitute the phase probability Eq. (6.4) into the moment equation Eq. (6.6), with the help of the integral relation[19]

$$\int_0^\infty dy\,y^2\exp(-a^2y^2)\cos(by) = \sqrt{\pi}\frac{2a^2-b^2}{8a^5}\exp\left(\frac{-b^2}{4a^2}\right) , \tag{6.7}$$

then Eq. (6.5) for the variance becomes

$$\Delta\varphi^2 = \langle\varphi^2\rangle$$
$$= \frac{1}{2}\frac{1}{1+\exp(-\delta\alpha^2/2)}\left\{\frac{1}{4(\alpha-\delta\alpha/2)^2}+\frac{1}{4(\alpha+\delta\alpha/2)^2}+\frac{1-\delta\alpha^2}{2\alpha^2}e^{-\delta\alpha^2/2}\right\} .$$

Here we have made use of the symmetry $W_{-\varphi}[|\varphi\rangle] = W_\varphi[|\psi\rangle]$ from Eq. (6.4), which implies that $\langle\varphi\rangle = 0$. When we neglect terms of the order $\delta\alpha/2 \ll 1$, this result takes the simple form

$$\Delta\varphi^2 \cong \frac{1}{4\alpha^2}\left[1-\frac{\delta\alpha^2}{1+\exp(\delta\alpha^2/2)}\right] . \tag{6.8}$$

For $\delta\alpha = 0$, when the two coherent states are on top of each other and only a single coherent state located at $x = \sqrt{2}\alpha$ is present, we find that

$$\Delta\varphi^2\Big|_{\delta\alpha=0} = \frac{1}{4\alpha^2} , \tag{6.9}$$

a result which is in agreement with Eq. (5.7) and the integral relation Eq. (6.7). When the two coherent states start to separate, and $\delta\alpha$ increases, the phase uncertianty $\Delta\varphi^2$ falls below the coherent state limit of Eq. (6.9), as shown in Fig. 10. However, for larger values of $\delta\alpha$, the initial decrease in $\Delta\varphi^2$ caused by the increase of $\delta\alpha^2$ is compensated by the exponential growth of the denominator of Eq. (6.8), and thus the phase uncertianty $\Delta\varphi^2$ reapproaches the single coherent state uncertianty, given in Eq. (6.9). In this large $\delta\alpha$ regime the two coherent states are well separated and hence distinguishable. Superpose quantum mechanically two coherent states to find a resultant state which is more well-defined in phase than either of the two original states. Phase squeezing via state superposition—this is the central lesson that we have learned from this example.

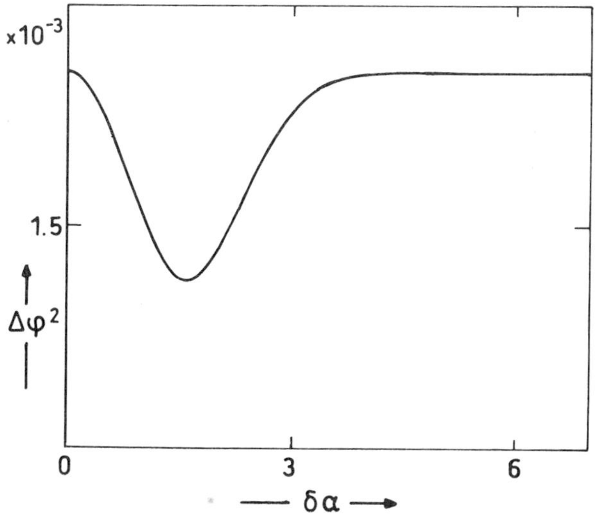

Fig. 10. The phase uncertainty of the quantum state shown in Fig. 9, and consisting of the superposition of coherent states of identical phase but different average number of photons, is depicted here. This uncertainty falls below the value corresponding to the standard phase uncertainty of a single coherent state (indicated here by the horizontal, dashed curve) when we increase the separation of the two states by the amount $\sqrt{2}\delta\alpha$. For $\alpha^2 \gg 1$ the average number of photons stays almost constant. This narrowing of the phase uncertainty results from the interference of the two states. When they are well separated from each other, i.e., when they are distinguishable, the phase uncertainty $\Delta\varphi$ assumes the value of a single coherent state. For definiteness we have chosen for the value of the displacement parameter $\alpha = 7\sqrt{2}$.

A similar phenomenon makes its appearance when we superpose two coherent states of identical mean photon number, but different phases.[33] In this particular case the superposition principle transmutes the Poisson distribution of a single coherent state to sub- and super-Poissonian photon distributions of the new state. These new distributions even exhibit squeezing in the x variable.

We do not know of a better example, which can illustrate in a more striking fashion, the decisive power of this principle which sounds so innocent: Superposition.

7. SUMMARY AND CONCLUSIONS

To illustrate the tools of semiclassical quantum mechanics à la WKB—which is what we mean by *asymptotology*— and to apply them to problems of quantum optics, these are the main themes of the present article. Our long journey started from the known territory of classical mechanics. We studied the map of the unfamiliar country of WKB ("Beware, there be dragons in these lands!"), and reviewed the essential ingredients of this approach. Before we entered foreign land we looked back at our own countryside. Very often, familiar landscapes look quite different when viewed from an unusual place. Such a change in the point of view rewards us with new and beautiful perspectives: The Poissonian photon count probability of a coherent state considered as the 'read out' of its position representation; and the corresponding oscillatory behavior of the photon distribution of a highly squeezed state seen as a result of standing waves marching through the narrow slit of the squeezed state; these are two of the novel insights we have gained from this approach. Our trip finally led us to the exotic world of phase space: The notion of interfering areas of overlap in phase space between two different quantum states, as a measure of their quantum mechanical scalar product, this is one of the attractions of the local folklore of this countryside. The energy wave function of a bounded particle illustrates this principle. Angled panes of glass cut a wedge-shaped slice out of the coherent state Gaussian bell and serve as a measure of phase probability—representing another example of this concept. The strangest creature of this land of phase space that we have met on our travels is that beast, the quantum state consisting of the superposition of two coherent states of identical phase, but of different average photon number. Lots more examples of the successful application of semiclassical techniques to the field of quantum optics could be added. The ones we have presented here are meant only to be an appetizer—an invitation to the reader for many more delightful visits to the Enchanted Land of Asymptotology.

ACKNOWLEDGEMENTS

During the course of this work we have immensely profited from discussions with S. M. Barnett, R. Chiao, C. Cohen-Tannoudji, M. Collett, T. Hänsch, R. F. O'Connell, D. Pegg, A. Schenzle, M. O. Scully, D. F. Walls, and H. Walther. One of us (W. S.) is enormously grateful to J. A. Wheeler for the privilege of being introduced by him to semiclassical phase space physics; and in addition for the many fruitful discussions and collaborations he has had with Professor Wheeler, while being at the Center for Theoretical Physics at the University of Texas at Austin, and also over the past few years. Moreover, we would like to thank A. O. Barut for organizing a most splendid conference. Many thanks go to R. Schilling ('The Wiz') for being infinitely patient with us, while teaching us the arcane secrets of TeX. Without his help, advice, and encouragement, this manuscript would probably never have left the stage of a handwritten draft. Two of us (R. H. and S. V.) thank the Humboldt Foundation for a stipend. One of us (W. S.) is grateful to the Office of Navel Research for its support while being at the Center for Advanced Studies at the University of New Mexico.

REFERENCES

1. For a review on the question of a Hermitian phase operator see for example, P. Car-

ruthers and M. M. Nieto, "Phase and Angle Variables in Quantum Mechanics," Rev. Mod. Phys **40**, 441 (1968); and S. M. Barnett and D. T. Pegg, "Phase in Quantum Optics," J. Phys **A 19**, 3849 (1986); A. Schenzle, "Gibt es doch einen Phasenoperator in der Quantenmechanik?" Phys. Bl. **45**, 84 (1989).

2. The original papers are stimulating to read; G. Wentzel, "Eine Verallgemeinerung der Quantenbedingungen für die Zwecke der Wellenmechanik," Z. Phys. **38**, 518 (1926); H. A. Kramers, "Wellenmechanik und halbzahlige Quantisierung," Z. Phys. **39**, 828 (1926); L. Brillouin, C. R. Acad. Sci. Paris **183**, 24 (1926).

3. Introductory articles and books on WKB include, P. Debye, "Wellenmechanik and Korrespondenzprinzip," Phys. Z. **28**, 170 (1927); H. A. Kramers, "Quantentheorie des Elektrons und der Strahlung," Vol. 2, *Hand-und Jahrbuch der Chemischen Physik,* (Eucken-Wolf, Leipzig, 1938); W. Pauli, "Die allgemeinen Prinzipien der Wellenmechanik," *Handbuch der Physik,* Vol. 24, edited by H. Geiger and K. Scheel (Springer Verlag, Berlin, 1933); for more advanced literature we refer to, J. Heading, *An Introduction to Phase Integral Methods* (Methuen, London, 1962); N. Fröman and P. O. Fröman, *JWKB Approximation, Contributions to the Theory* (North-Holland, Amsterdam, 1965); N. G. de Bruijn, *Asymptotic Methods in Analysis* (North-Holland, Amsterdam, 1958); W. Wasow, *Asymptotic Expansions for Ordinary Differential Equations* (Wiley, New York, 1965); and in particular, M. V. Berry and K. E. Mount, "Semiclassical Approximations in Wave Mechanics," Rep. Prog. Phys. **35**, 315 (1972).

4. W. Schleich, H. Walther, and J. A. Wheeler, "Area in phase space as determiner of transition probability: Bohr-Sommerfeld bands, Wigner ripples and Fresnel zones," Found. Phys. **18**, 953 (1988).

5. W. Schleich and J. A. Wheeler, "Oscillations in photon distribution of squeezed states and interference in phase space," Nature (London) **326**, 574 (1987); W. Schleich and J. A. Wheeler, "Oscillations in photon distribution of squeezed states," J. Opt. Soc. Am. B **4**, 1715 (1987); W. Schleich, D. F. Walls, and J. A. Wheeler, "Area of overlap and interference in phase space versus Wigner pseudo-probabilities," Phys. Rev. **A 38**, 1177 (1988).

6. R. Glauber, "Coherent and incoherent states of the radiation field," Phys. Rev. **131**, 2766 (1963).

7. M. Sargent, M. O. Scully, and W. E. Lamb, *Laser Physics* (Addison-Wesley, Reading, 1974), Appendix H.

8. W. H. Louisell, *Quantum Statistical Properties of Radiation* (Wiley, New York, 1973).

9. For a review on squeezed states see for example, D. F. Walls, "Squeezed states of light," Nature (London) **306**, 141 (1983); see also special issues on squeezed states in, J. Opt. Soc. Am. **B4**(10) (1987); and J. Mod. Opt. **34**, 6–7 (1987); see also, G. Leuchs, "Photon statistics, anti-bunching and squeezed states," *Frontiers of Nonequilibrium, Statistical Physics,* edited by G. T. Moore and M. O. Scully (Plenum, New York, 1986).

10. Throughout these notes we use the dimensionless coordinate $x \equiv (\mu\omega^2/\hbar)^{1/2}q$ and momentum $p \equiv (\mu\hbar\omega)^{-1/2}\tilde{p}$, where q and $\tilde{p}$ denote the position and momentum of a particle of mass μ and ω is a typical frequency of the potential.

11. See for example, D. Bohm, *Quantum Theory* (Prentice Hall, Englewood Cliffs, 1951).

12. M. Abramowitz and I. E. Stegun, *Handbook of Mathematical Functions* (National Bureau of Standards, Washington, D. C., 1964).

13. The concept of *interference in phase space* is spelled-out in, J. A. Wheeler, "Franck-Condon effect and squeezed state physics as double-source interference phenomena," Lett. Math. Phys. **10**, 201 (1985); and W. Schleich, *Interference in Phase Space,* Habilitationsschrift, (Ludwig-Maximilians-Universität, München, 1988); see also, W. Schleich, "Phase Space, Correspondence Principle and Dynamical Phases: Photon Count Probabilities of Coherent and Squeezed States via Interfering Areas in Phase Space," in *Squeezed and Nonclassical Light,* edited by P. Tombesi and E. R. Pike (Plenum Press, New York, 1988).

14. M . Hillery, R. F. O'Connell, M. O. Scully, and E. P. Wigner, " Distribution functions in physics: fundamentals," Phys. Rep. **106**, 121 (1984); V. I. Tatarskii, "The Wigner representation of quantum mechanics," Usp. Fiz. Nauk. **139**, 587 (1983); L. Cohen, "Positive and negative joint quantum distributions," *Frontiers of Nonequilibrium Statistical Physics,* edited by G. T. Moore and M. O. Scully (Plenum Press, New York, 1986).

15. W. Schleich, R. J. Horowicz, and S. Varro, "A Bifurcation in Squeezed State Physics: But How? or Area-of-Overlap in Phase Space as a Guide to the Phase Distribution and the Action-Angle Wigner Distribution of a Squeezed State," *Quantum Optics V,* edited by D. F. Walls and J. Harvey (Springer Verlag, Heidelberg, 1989); W. Schleich, R. J. Horowicz, and S. Varro, " A Bifurcation in the Phase Probability of a Highly Squeezed State," Phys. Rev. **A 40** (15 December 1989).

16. See for example, M. Born, "Vorlesungen über Atommechanik," in *Struktur der Materie in Einzeldarstellungen,* edited by M. Born and J. Franck (Springer Verlag, Berlin, 1925).

17. J. A. C. Gallas, W. Schleich, and J. A. Wheeler, "WKB grapples with a corner" (to be published).

18. G. Szegő, *Orthogonal Polynomials* (American Mathematical Society, New York, 1939).

19. I. S. Gradsteyn and I. M. Ryzhik, *Table of Integrals, Series and Products* (Academic Press, New York, 1965).

20. B. Friedman, *Lectures On Application-Oriented Mathematics* (University of Chicago Press, Chicago, 1957).

21. H. P. Yuen, "Two-photon coherent states of the radiation field, " Phys. Rev. **A 13**, 2226 (1976); D. Stoler, "Equivalence classes of minimum-uncertainty packets," Phys. Rev. **D 1**, 3217 (1970); "Equivalence classes of minimum-uncertainty packets II," **4**, 1925 (1971); M. Nieto, " What are squeezed states really like?" *Frontiers in Nonequilibrium Statistical Physics,* edited by G. T. Moore and M. O. Scully (Plenum Press, New York, 1986).

22. R. S. Bondurant, B. S. thesis, Massachusetts Institute of Technology, 1978.

23. W. Schleich and J. A. Wheeler, "Interference in phase space," Ann. Phys. (Leipzig) (to be published).

24. A. Landé, " Neue Wege der Quantentheorie," Naturwissenschaften **20**, 455 (126).

25. P. A. M. Dirac, "The Quantum Theory of the Emission and Absorption of Radiation," Proc. R. Soc. London, Ser. **114 A**, 243 (1927).

26. F. London, "Über die Jacobischen Transformationen der Quantenmechanik," Z. Phys. **37**, 915 (1927); "Winkelvariable und kanonische Transformationen in der Undulations mechanik," **40**, 193 (1926).

27. W. H. Louisell, "Amplitude and Phase Uncertainty Relations," Phys. Lett. **7**, 60 (1963).

28. For the representation of the phase state as a straight *line* emerging from the origin, see, R. Fanelli and R. E. Struzynski, "The Impossibility of a Quantum Phase Operator," Am. J. Phys. **37**, 928 (1969).

29. For the most recent developments in this field, see, D. T. Pegg and S. M. Barnett, "Unitary Phase Operator in Quantum Mechanics," Europhys. Lett. **6**, 483 (1988); "Phase Properties of the Quantized Single-mode Electromagnetic Field," Phys. Rev. **A 39**, 1665 (1989); and D. T. Pegg, S. M. Barnett, and J. A. Vaccarro "Phase in Quantum Electrodynamics," *Quantum Optics,* edited by D. F. Walls and J. Harvey (Springer Verlag, Heidelberg, 1989); the Hermitian phase operator obtained in this work is based on the Loudon states defined in, R. Loudon, *The Quantum Theory of Light,* 1st ed. (Oxford University Press, Oxford, 1973).

30. This corresponds to expressing the Wigner function of the state in polar coordinates rather than Cartesian coordinates and then integrating over the radial variable. A similar strategy based, however, on the Husimi phase space distribution has been adopted in J. H. Shapiro and S. S. Wagner, "Phase and Amplitude Uncertainties in Heterodyne Detection," IEEE J. Quantum Electron. **20**, 803 (1984).

31. S. M. Barnett and D. T. Pegg, "On the Hermitian Optical Phase Operator," J. Mod. Opt. **36**, 7 (1989).

32. P. A. M. Dirac, *The Principles of Quantum Mechanics* (Clarendon Press, Oxford, 1984).

33. For the related problem of the superposition of two coherent states of different *phases* but identical average number of photons, see, W. Schleich and M. Pernigo, "The principle of superposition in quantum mechanics as the prime source of nonclassical features in the superposition of two coherent states," (to be published).

AN INTRODUCTION TO QUANTUM NOISE

Axel Schenzle

Ludwig-Maximilians-Universität München, Sektion Physik
Theresienstraße 37, 8000 München, West Germany
and
Max-Planck-Institut für Quantenoptik
Ludwig-Prandtl-Straße 10, 8046 Garching, West Germany

Abstract

All real physical systems display noise. Either, because they consist of a multitude of uncontrollable degrees of freedom, as in thermodynamic processes, or because the statistical concepts of quantum mechanics must be applied. Microscopic systems will show both, classical and quantum noise. In this brief overview we demonstrate that quantum fluctuations can have striking similarities to classical noise, but also can show peculiar and characteristic features that are unknown in statistical mechanics.

1. Introduction

It is still a widespread belief that noise is only an annoying aspect of real physical processes, and the presence of fluctuations only indicates a lack of control over the experimental set up. Therefore, usually noise is not considered in any respect as a fundamental or at least interesting phenomenon. Certainly, when experiments become sensitive enough to touch the microscopic regime, noise is picked up from all possible sources and therefore cannot be neglected in theory. But it is still a rather pessimistic view that we reluctantly consider noise in theory only, since it cannot be eliminated in practice. This, however, is just one aspect of the problem. In reality the ubiquity of noise cannot be overlooked and certainly causes tremendous practical problems, but it can also reveal interesting features, features that are impossible for deterministic processes. Obviously noise can reveal an intrinsic instability and can distinguish it from meta-stable stationary states, which are typical for nonlinear dynamic processes. But besides this destabilizing character of noise, it has been demonstrated that under certain conditions, noise can also stabilize a state that is deterministically instable[1]. Such an observation is quite unexpected and almost counter- intuitive to our understanding of the role of fluctuations.

Noise is entirely inevitable when finally the microscopic level is reached, and quantum theory must be applied. In the quantum mechanical regime noise becomes an intrinsic and essential ingredient, required for consistency of the theory. This is the case even in systems over which one has - at least in an idealized sense - total experimental control. It is, however, quite remarkable that quantum noise does bear striking similarities to classical fluctuations, while the physical origin of both is entirely different.

In order to remind ourselves on what we mean by fluctuations and what the mathematical concepts are, for treating classical as well as quantum noise, we want to begin with an extremely simple but probably instructive example. A device wellknown to practically all physics students is the mirror galvanometer, which is an oldfashioned but nevertheless sensitive meter for measuring tiny electrical currents. It demonstrates in an obvious way the

New Frontiers in Quantum Electrodynamics and Quantum Optics
Edited by A. O. Barut, Plenum Press, New York, 1990

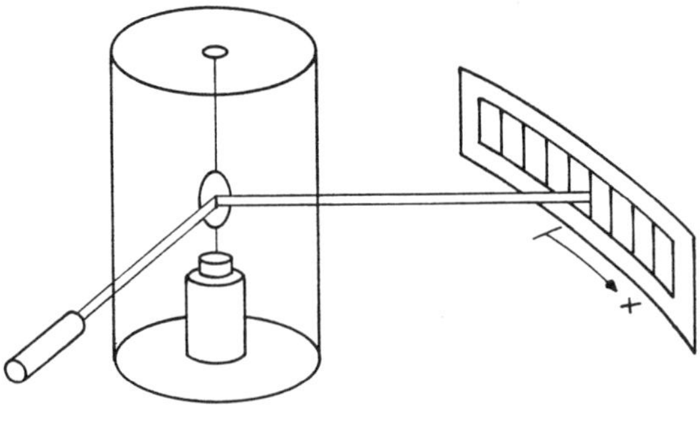

Fig.1 The shaking image of the light spot on a distant screen, reflected from the mirror, illustrates noise in a sensitive measuring device.

gradual loss of control and, as a consequence, the random behaviour of a measuring device, when its sensitivity is pushed towards the limits. In this device the indicator is a light beam that is reflected from a small mirror, which is suspended almost freely on a thin thread. This is illustrated in figure 1. The sensitivity of this device makes it susceptible for all sorts of perturbations, to sonic vibrations, vibrations of the support and fluctuations of the ambient air. On a distant screen, the indicator beam creats a light spot that undergoes random excursions about an average position. There are two basically different pictures that both can be used for describing such a statistical phenomenon. On the one hand, one can specify the position of the spot on the screen directly as a function of time by a random trajectory $x=x(t)$. On the other hand one may define a probability measure $P(x)dx$ on the accessible interval. $P(x)$ measures the relative frequency by which the trajectory returns to the neighbourhood of a chosen position x, while $x(t)$ is a continuous function of time that is only defined in a statistical sense. These two pictures are indicated in figure 2. The trajectories describe in an intuitive way what we actually 'see', but from a mathematical point of view the probabilistic description is by far more useful, especially when it comes to solving nonlinear problems.

The process that has marked the beginning of stochastic modelling was the problem of Brownian motion. In the Langevin picture the dynamic evolution is described by trajectories which satisfy ordinary differential equations. The statistical behaviour is generated by the addition of a randomly fluctuating force. This concept of a freely diffusing particle, as observed under the microscope, has been generalized in the past to a random walk in phase spaces with arbitrary dimensions, under nonlinear forces, for systems far from thermal equilibrium and finally to processes with state dependent noise:

$$d/dt\, x_j(t) = F_j(\{x_l\}) + G_{j,k}(\{x_l\})\, \xi_k(t) \tag{1}$$

$j=1,2,3...,D$ runs over the number of independent degrees of freedom. The random forces $\xi_j(t)$ are not given in explicit form, but are only defined through their statistical properties. The most commonly used model assumes white noise:

$$\begin{aligned} < \xi_j(t) > &= 0 \\ < \xi_j(t)\, \xi_k(t') > &= \delta_{j,k}\, \delta(t-t') \end{aligned} \tag{2}$$

Noise with a finite correlation time like an Ornstein-Uhlenbeck process can be incorporated by including auxiliary variables. $F(\{x_j\})$ summarizes the deterministic forces that act on the sytem. They can be reversible and originate from a Hamiltonian, or can be dissipative and result from the interaction with reservoirs. $G_{j,k}(\{x_l\})$ is a measure of the strength of noise, which can be different for different degrees of freedom, and may even depend on the actual position in phase space.

For the general case of a nonlinear diffusion process, eq.(1) cannot be solved in analytical terms and one has to resort to numerical simulation. The physical observables are the ensemble averages over the canonical variables, like the moments $\langle x_j \rangle$, $\langle x_j{}^2 \rangle$ etc. or the

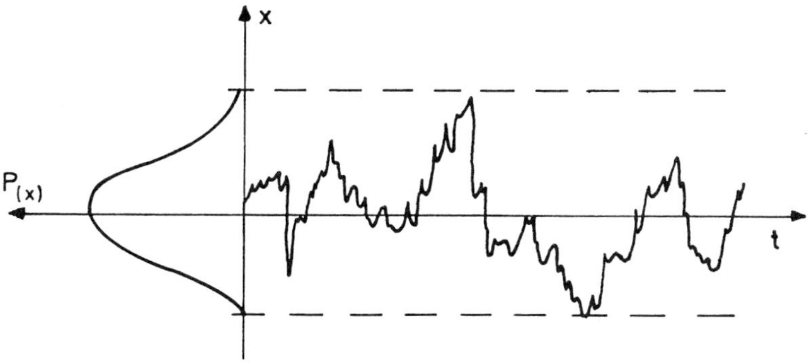

Fig.2 There are two ways for characterizing a stochastic process, either through an ensemble of random trajectories, or through a probability density. One approach is the basis for the Langevin -, the other for the Fokker-Planck description.

correlation coefficients $\langle x_j x_k \rangle - \langle x_j \rangle \langle x_k \rangle$. In a numerical treatment they are obtained by averaging over a large ensemble of equivalent trajectories:

$$\langle x_j^n x_k^m \rangle = \frac{1}{N} \sum_{r=1}^{N} (x_j^r)^n (x_k^r)^m \tag{3}$$

where x_j^r denotes an individual realization. Thereby we mean a trajectory with an individual history of random kicks chosen from a Gaussian ensemble of random numbers. N is the number of trials, which should be chosen as large as practically possible. Similarly, time dependent correlation functions can be simulated by averaging over an ensemble of trajectories that are displaced in time relative to each other.

Nonlinear stochastic differential equations - with a few possible exceptions - can only be simulated numerically. Instead of following the fate of an individual trajectory in course of time, as indicated in figure 2, one may also characterize a stochastic process by the probability that trajectories will visit a certain neighbourhood in phase space. The time evolution of the probability density is determined by the Fokker-Planck equation. While the trajectories satisfy nonlinear ordinary differential equations, the evolution of the probability density is governed by a linear partial differential equation. The two descriptions in terms of trajectories or probabilities are entirely equivalent:

$$\{x_j(t)\} \rightarrow P(\{x_j\},t)$$

The Fokker-Planck equation related to the stochastic differential equation eq.(1) can be written in the form[2]:

$$\begin{aligned} \frac{\partial P(\{x_j\},t)}{\partial t} &= -\frac{\partial}{\partial x_j} F_j P + \frac{1}{2} \frac{\partial}{\partial x_j} G_{jl}(\{x_s\}) \frac{\partial}{\partial x_k} G_{kl} P \\ &\equiv L\, P(\{x_j\},t) \\ &= -\partial/\partial x_j\, J_j(\{x_j\},t) \end{aligned} \tag{4}$$

With the definition of the probability current J, the Fokker-Planck equation assumes the form of a continuity equation, which guarantees that in course of time total probability is conserved. The Fokker-Planck equation must be solved subject to appropriate boundary conditions. A random process naturally occurs in a certain domain of phase space, which may be finite or infinite. At the boundaries of that domain, the normal component of the probability current must vanish, otherwise probability would be exchanged between the physically allowed and the forbidden regions. Depending on the details of the problem and the chosen representation, this condition must be imposed explicitly on the solution or may be satisfied automatically by the Fokker-Planck equation[3]. This latter case is rather troublesome when it comes to numerical integration, where the boundary conditions are needed in explicit mathematical form[4].

If there is no external time dependent influence on the system, we have an autonomous process and the functions G and F do not dependent on time explicitly. For such a stationary process the Fokker-Planck equation can be cast into the form of an eigenvalue problem:

$$L\, P_{\{n\}} = -\lambda_{\{n\}}\, P_{\{n\}} \tag{5}$$

where {n} stands for a set of s separation parameters or "quantum numbers", which are specified by the boundary conditions. The general time dependent solution is then written in the form:

$$P(\{x_j\},t) = \sum_{\{n\}} c_{\{n\}}\, P_{\{n\}}(\{x_j\}) \exp(-\lambda_{\{n\}} t) \tag{6}$$

The expansion coefficients $c_{\{n\}}$ are determined uniquely by the choice of the initial distribution $P(\{x_j\},t{=}0)$. Since L is not a selfadjoint operator we also must consider the adjoint problem:

$$L^+\, W_{\{n\}} = -\lambda^*_{\{n\}}\, W_{\{n\}} \tag{7}$$

The orthonormality condition in terms of the adjoint problems:

$$\int W_{\{n\}}\, P_{\{m\}}\, dx^s = \delta_{\{n\},\{m\}} \tag{8}$$

allows one to determine the expansion coefficients c for any given positive initial condition. There must always exist a stationary distribution that will be assumed in the long time limit $t \to \infty$:

$$\lambda_{\{0\}} = 0 \quad \text{with} \quad P_{\{0\}}(\{x_j\}) \quad \text{and} \quad W_{\{0\}}(\{x_j\}) = 1 \tag{9}$$

For the special choice of {n} = 0 the orthogonality relation eq.(7) leads to:

$$\int P_{\{m\}}\, d^D x = \delta_{\{0\},\{m\}} \tag{10}$$

As a consequence:

- the steady state distribution is normalized:

$$\int P_{\{0\}}(\{x_j\})\, d^D x = 1 \tag{11}$$

- the eigenfunctions corresponding to non-trivial eigenvalues $\lambda \neq 0$ must assume negative values, since:

$$\int P_{\{m\}}(\{x_j\})\, d^D x = 0 \quad \text{for } \{m\} \neq 0 \tag{12}$$

and therefore cannot be identified individually with a probability density.

- the general time dependent solution is normalized as well. By integrating over eq.(6) it follows that :

$$c_{\{0\}} = 1 \tag{13}$$

and the steady state is always included in the expansion of an arbitrary probability density. These conditions together guarantee that the solution of the Fokker-Planck equation remains positive for all times. Since the Fokker-Planck equation posesses formal similarities to the Schrödinger equation, it may be useful to notice an essential difference. While a quantum system can always be prepared, at least in principle, in an arbitrary eigenstate, this is impossible in case of a diffusion problem. The individual "eigenstates" have no probability interpretation and any normalized state is a superposition that contains at least the steady state. The eigenvalues in general are complex λ . In order to guarantee relaxation into steady state it is necessary that Re $\lambda_{\{n\}} > 0$ except for {n=0}.

Returning to our pedagogical example of the suspended mirror, one may argue that the fluctuations could be reduced by evacuating the glass container which surrounds the device. Thereby the fluctations of the air are greatly reduced, but the slight vibrations of the building are still transmitted through the suspension points. By more and more sophisticated insulation techniques one may reduce and finally get rid of the classical fluctuations, only to find out that there are still fluctuations due to the quantum mechanical nature of the process. Certainly, for a macroscopic mechanical system this randomness would be extremely hard to observe, but not impossible in the field of quantum optics. The vibrating mirror is practically a harmonic oscillator and so is a quantized field mode. For a pair of canonically conjugate variables like x,p even in the ground or the vacuum state uncertainties remain, that are expressed by the variances:

$$
\begin{aligned}
&< \Delta^2 x > = < x^2 > - < x >^2 = \frac{1}{2}\left(\frac{\hbar}{m\omega}\right) \\
&< \Delta^2 p > = < p^2 > - < p >^2 = \frac{1}{2}(\hbar m\omega)
\end{aligned}
\tag{14}
$$

On top of these fluctuations that are entirely due to the quantum nature of the problem, classical noise can be superimposed and the variances may exceed the vacuum level. Classical noise is either of thermal origin and an intrinsic property of the system or imposed from the outside. In both case it can be manipulated to a certain extent - sometimes even be eliminated. This is not so in the quantum case, where the randomness is an intrinsic property that reflects the operator character of physical observables. They are practically uncontrollable from the outside. Nevertheless it is surprizing that inspite of the entirely different origin of the two forms of noise, for a large class of processes quantum noise can be associated with the action of a random force. This is the case when the considered system interacts with a thermal reservoir, even at T=0, where noise is purely quantum mechanical - but qualitatively indistinguishable from thermal noise. On the other hand, the non-commutability of operators in case of a nonlinear dynamic process introduces noise that has no classical analog and cannot be simulated by any kind of sophisticated random force. In general, both sorts of randomness are present simultaneously and superimposed. The most interesting and puzzling situation occurs, however, when the non-classical properties dominate over the classical ones and new features emerge. As we will see, these similarities and dissimilarities are most intuitively understood on the basis of "quasi-probabilities".

A realistic model for a quantum optical process must include non-linear forces and dissipation. The energy flux through the system that drives the process out of equilibrium can either be the result of reversible interactions or due to the coupling with non-thermal reservoirs. The state of the quantum system is then characterized by the statistical operator ρ and the dynamic evolution follows from the master equation:

$$
\frac{d}{dt}\rho(t) = \frac{i}{\hbar}[H,\rho] + \left(\frac{\partial}{\partial t}\rho\right)_{irr} \tag{15}
$$

It can be a forbiddingly complex task to solve this equation in a certain representation even for simple physical systems when realistic parameters are used. Instead of treating such a complex high dimensional matrix problem, it is in many cases more convenient to reformulate the dynamic equations in terms of quasi-probabilities. Thereby one replaces the bulky matrix equation by a more convenient partial differential equation. In many cases this equation has a rather intuitive appearance - it is of the form of a Fokker-Planck equation, but now for a pure quantum mechanical problem. However, the essential property that qualifies an equation to be identified with a Fokker-Planck equation is the positive semi-definiteness of the diffusion matrix, and this is by no means a priori guaranteed for any quantum mechanical process. It is a necessary condition, however, for establishing a random stochastic process. A quantum process with positive diffusion is mathematically indistinguishable from a classical stochastic process. A master equation with "negative diffusion" has no relation to stocastic process, it causes non-classical behaviour and cannot be simulated by random trajectories like Brownian motion.

The paper is organized as follows: In chapter 2 we briefly introduce the statistical concepts, which are used to discuss noise in quantum optical systems. For the understanding of more complex and realistic fields, it may be helpful to include a list where the properties of certain basic reference states of the light field are summarized. This will be done for Fock-, coherent-, squeezed- and the thermal equilibrium state. In chapter 3 the concept of quasi-probabilities is introduced and discussed as much as it will be needed later. In the last chapter we present three examples in greater detail. We want to show the continuous transition from processes with quantum noise of pure classical appearance, systems where classical and quantum noise compete and finally nonlinear processes, where the non-classical features dominate. The first example is a nonlinear oscillator with damping, which is used to illustrate the peculiarities of different quasi-probability concepts. We will see that notion of noise contains an ambiguity and depends in form and strength on the used representation. The second example is the laser, but with a non-classical dissipative reservoir - a reservoir that is prepared in a broadband squeezed state. When no squeezing is applied we recover the traditional laser with pure spontaneous emission noise. The classical stochastic appearance in this example is maintained even for finite squeezing, but the noise is distributed unequally over the quadratures. The last example is a nonlinear parametric process, where the gain originates from the reversible dynamics and 'quasi-classical' gain noise is absent. Here the non-classical features dominate, the "diffusion matrix" becomes negative and there exists no equivalent stochastic model.

2. Statistical Properties of the Quantized Field

Generally speaking, quantum optics is the science of the interaction of light and matter. In a more limited sense, this term is restricted to the interaction of matter with the quantized electromagnetic field. In both, the semiclassical and the quantum case, physical observables are calculated as ensemble averages taken either over the externally imposed classical or the intrinsic quantum fluctuations. In the following, the averages either represent integrals over the classical probability density or the trace over the statistical operator - or both, when the two noise sources are present simultaneously. Experimentally we observe either the properties of light, such as the averages over the field, over the intensity, their correlations, the photon counting distribution, or one is interested in measuring atomic properties, such as level populations or the atomic coherence. For the purpose of quantization it is convenient to expand the field $E(x,t)$ in eigenmodes of a suitable cavity:

$$E(x,t) = \sum_{k,j} \left(\frac{2\pi\hbar\omega}{V}\right)^{1/2} e_{kj} \,[\, b_{kj}^{\dagger}\, e^{i\omega_k t} + b_{kj}\, e^{-i\omega_k t} \,] \tag{16}$$

where e_{kj} is the polarization vector, ω_k the resonance frequency of the mode with wave number k, V is the quantization volume. For the understanding of elementary processes it is often sufficient to consider a few modes only - sometimes just a single one. In quantized form $E(x,t)$ represents an operator valued field, while the amplitudes b_{kj} and $b_{kj}^{\dagger}$ turn into the corresponding boson operators. In the present chapter we consider only a single quantized mode and therefore abbreviate the field in the following form:

$$E(x,t) = E_0 \,(\, b^{\dagger} + b \,) \tag{17}$$

where E_0^2 is the field intensity corresponding to a single photon. In homo- or heterodyne experiments it is possible to measure the amplitude of the field directly, by beating it with a local oscillator. Thereby it is possible to obtain phase sensitive information about the field e.g. the average of the quadrature components is defined as:

$$X^+(t) = \frac{1}{2} E_0 < \{b(t) + b^\dagger(t)\} > \quad \text{and} \quad X^-(t) = \frac{1}{2i} E_0 < \{b(t) - b^\dagger(t)\} > \tag{18}$$

A photo detector measures the average intensity of the field :

$$I(t) = E_0^2 < b^\dagger(t)\, b(t) > \tag{19}$$

For a stationary process these averages are time independent. A measure of the randomness of the fields is provided by appropriately defined variances or correlation coefficients. The fluctuations of the quadrature components of the field are characterized by:

$$\begin{aligned} < \Delta^2 X^+ > &= \frac{1}{4} E_0^2 [\langle (b^\dagger + b)^2 \rangle - < b^\dagger + b >^2] \\ < \Delta^2 X^- > &= -\frac{1}{4} E_0^2 [\langle (b^\dagger - b)^2 \rangle - < b^\dagger + b >^2] \end{aligned} \tag{20}$$

while the noisyness of the field intensity follows from:

$$< \Delta^2 I > = E_0^2 \{ \langle b^\dagger b\, b^\dagger b \rangle - \langle b^\dagger\, b \rangle^2 \} \tag{21}$$

For a classical process the ordering of $b^\dagger$ and b has no significance, but it is essential in the quantum case. Two photon properties are naturally described by:

$$g_2 = \frac{< b^\dagger\, b^\dagger\, b\, b >}{< b^\dagger\, b >^2} \tag{22}$$

a function that vanishes in the vacuum and the one-photon state. A photon multiplier turns the intensity of the incoming field into a sequence of counting events. The number of events, registered during certain counting interval, is proportional to the observation time and the intensity : $\langle n \rangle = \langle b^\dagger b \rangle \propto \langle I \rangle$. If the intensity is completely stable, then the random counts are distributed according to the Poisson law: $< \Delta^2 n > = < n >$. Deviations from that law are a measure for the fluctuations of the field and are conveniently characterized by Mandel's Q parameter[5]:

$$Q = \frac{< \Delta^2 n > - < n >}{< n >} \tag{23}$$

In a classical description, the intensity fluctuations can only broaden the counting distribution and Q necessarily is positive. It is also positive for quantum noise that has a classical appearance like spontaneous emission noise. The quantum nature of the field is unmistakably indicated by a negative value of Q. The two parameters g_2 and Q are naturally related since they mainly differ by the chosen ordering:

$$g_2 = 1 + \frac{Q}{< n >} \quad \text{or} \quad Q = < n > (g_2 - 1) \tag{24}$$

One of the fundamental questions of quantum optics is to determine the statistical properties of the field from those of the emitting atomic source. In general the properties of these fields are rather complex and only in limiting case will they be approximated by one of the simple elementary states. For comparison, we list here the basic statistical properties of a number of reference states i.e. Fock state $|n >$, coherent state $|\alpha >$, squeezed vacuum state $|r >$ and the thermal equilibrium ρ_{th} :

	$\|n\rangle$	$\|\alpha\rangle$	$\|r\rangle$	ρ_{th}
$\langle b^\dagger + b\rangle$	0	$\mathrm{Re}(\alpha)$	0	0
$\pm\langle\Delta^2(b^\dagger \pm b)\rangle$	2n+1	1	$e^{\pm r}$	$2n_{th}+1$
$\langle b^\dagger b\rangle$	n	$\|\alpha\|^2$	$\sinh^2(r)$	n_{th}
$\langle\Delta^2 b^\dagger b\rangle$	0	$\|\alpha\|^2$	$\frac{1}{2}\sinh^2(r)$	$n_{th}(n_{th}+1)$
Q	-1	0	$\cosh(2r)$	n_{th}
g_2	1-1/n	1	$3+\frac{1}{\sinh^2(r)}$	2
W_m	$\delta_{n,m}$	$\frac{\|\alpha\|^{2m}}{m!}\, e^{\|\alpha\|^2}$		$\frac{n_{th}^{\ m}}{(1+n_{th})^{m+1}}$

Here $n_{th} = (\exp(\hbar\omega/kT - 1)^{-1}$ and $W_m = |\langle n|\psi\rangle|^2$ is the photon distribution. The photon statistics of the squeezed vacuum was not included in the list since the expression didn't fit into the box:

$$W_{2m+1} = 0 \quad \text{and} \quad W_{2m} = \frac{(-1)^m}{2^m m!}\,[\tanh(r)]^m \sqrt{\frac{2m!}{\cosh(r)}} \tag{25}$$

When inspecting this table more carefully it is immediately evident that it contains a number of apparent inconsistencies hidden in the rows for Q and g_2. One can approach the vacuum in a number of ways, but not in each way do we get the same value for the Q parameter: The vacuum can be defined as a

Fock state	$\|n\rangle$	for	$n\to 0$	then	$Q = -1$
coherent state	$\|\alpha\rangle$	for	$\alpha\to 0$	then	$Q = 0$
squeezed state	$\|r\rangle$	for	$r\to 0$	then	$Q = 1$
thermal state	ρ_{th}	for	$T\to 0$	then	$Q = 0$

The reason for these discrepancies is easily understood. First of all, the Fock states have discrete labels and there exists no continuous vacuum limit. The remaining discrepancy is traced back to the limiting behaviour $Q \to 0/0$ which obviously is not welldefined.

That the Q value for the squeezed state differs from that of the coherent state for α=0 or the thermal state for T=0 is understood as follows: The statistical operator ρ for an arbitrary state is characterized by its matrix elements $\rho_{n,m}$ in Fock representation. Then Q is written in the form:

$$Q = \frac{\Sigma n^2\rho_{nn} - \{\Sigma\, n\, \rho_{nn}\}^2}{\Sigma\, n\, \rho_{nn}} - 1 \tag{26}$$

We now assume that the matrix elements depend continuously on a certain parameter λ such that in the limit $\lambda\to 0$ the state approaches the vacuum. In case that close to λ=0 the two leading contributions are:

$$\rho_{00} = \frac{1}{1+\lambda}\ , \quad \rho_{mm} = \frac{\lambda}{1+\lambda} \quad \text{and} \quad \rho_{nn} = O(\lambda^2) \text{ for } n\neq m$$

where m could be any integer $m \geq 1$ then it is easy to derive following general expression:

$$Q = (m - 1) \tag{27}$$

For the coherent and the thermal state : m=1, while the squeezed vacuum only contains even photon numbers, therefore: m=2. One certainly may construct states with any specific λ dependence and one will find every desired value for the vacuum Q parameter.

Since Q and g_2 are closely related we expect a similar ambiguity for g_2. Inspection of the previous list reveales that the vacuum levels of g_2 for the states $|\alpha\rangle$, $|r\rangle$ but also for ρ_{th} are all different i.e. Q = 1, ∞, 2 respectively. This fact can be understood in the same way as above:

$$g_2 = \frac{\Sigma\, n^2\rho_{nn} - \Sigma\, n\rho_{nn}}{\Sigma\, n\rho_{nn}} \tag{28}$$

In case that for $\lambda \to 0$ $\rho_{22} >> \rho_{nn}$ for all n > 2 the result only depends on the ratio of the two leading elements:

$$g_2 = 2 \left[\frac{\rho_{22} + \ldots}{\rho_{(11)^2} + \ldots} \right] \tag{29}$$

- For the coherent state, the required ratio is $1/2\, e^{|\alpha|^2}$, and the vacuum limit is: $g_2 = 1$.

- For the squeezed state g_2 doesn't exist, since $\rho_{11} \equiv 0$ for any r, therefore g_2 grows like ρ_{22}^{-1}.

- Characteristic for the thermal state is the relation : $\rho_{nn} \propto (\rho_{11})^n$ and the vacuum limit therefore is $g_2 = 2$.

That the discrepancies are even more severe for g_2 is not at all surprizing, since the two photon correlation function g_2 has no real physical meaning for the vacuum - one might have expected, however, that g_2 would vanish.

Intrinsic time dependent properties, independent of the arbitraryness of the initial preparation, are described by the stationary correlation functions. Phase information like the spectral properties and the linewidth are contained in the amplitude correlation function:

$$G_1(t) = \langle b^\dagger(t)\, b(0) \rangle \tag{30}$$

while the two photon correlation function describes temporal excursions of the field intensity or, in a different language, the photon correlations:

$$G_2(t) = \langle b^\dagger(0)\, b^\dagger(t)\, b(t)\, b(0) \rangle \tag{31}$$

In analogy one defines multi-photon correlations of arbitrary order that contain n creation and n destruction operators in normal order. They provide a systematic way of describing the statistical properties of a given light field in increasing detail. The higher order functions, however, are experimentally less important while increasingly harder to measure. Nevertheless, there exists one statistical property that combines all correlations up to infinite order, the photon counting statistics:

$$W(n,T) = \mathrm{tr}\, \rho\, T\, \frac{1}{n!} \left[\eta \int_0^T b^\dagger b\, dt \right]^n \exp\left[-\eta \int_0^T b^\dagger b\, dt \right] \tag{32}$$

W(n,T) is the probability that a photon detector registers n counting events when exposed T seconds to the light. ρ is the statistical operator of the investigated field, η is proportional to the quantum efficiency and T is an ordering prescription[6,7]. While rapid fluctuations of the field are most conveniently measured in frequency space through spectral functions, slow fluctuations are better described by W(n,T). For example, the problem of Quantum Jumps [8] with its macroscopically long time scale can be characterized by the counting statistics, while in frequency space the features of the intermittent signal would appear only at extremely low frequencies.

3. Quasi-Probabilities

We have emphasized over and over that quantum noise and classical noise have some features in common, and some not. In the present theoretical description, however, this connection is not at all obvious, since totally different mathematical tools are used in both fields. In quantum theory the statistical operator ρ contains all the physical information, while classical stochastic processes are described by a probability measure $P(x,p,t)dxdp$. The formal difference between the classical and the quantum description, which is most evident for deterministic systems, is much less obvious, when classical stochastic processes are compared with quantum mechanical ones. In both cases the physical observables are ensemble averages:

$$< b^\dagger b > \; = \; \mathrm{tr}\, \rho(t)\, b^\dagger b$$

and (33)

$$< x\, p > \; = \; \int x\, p\, P(x,p,t)\, dxdp$$

but the mathematical concepts for calculating those averages are widely different. It has first been demonstrated by E.Wigner[9] that it is possible to cast quantum mechanics into a probabilistic formalism using c-numbers by defining a suitable probability distribution. In quantum optics a similar concept has been proposed R.Glauber[10], based on the coherent states of the field, and which is taylor made for quantum optical problems. Since these "probabilities" are not necessarily positive, it is safer to use the term quasi-probabilities.

In statistical physics, a tool quite useful in many applications is the characteristic function that allows one to calculate arbitrary ensemble averages by mere differentiation:

$$\chi_c(q,k) \; = \; \int e^{i(qx+kp)}\, P(x,p)\, dxdp \tag{34}$$

With the knowledge of the characteristic function of the process, the ensemble averages are conveniently calculated as:

$$< x\, p > \; = \; \frac{\partial}{\partial iq}\frac{\partial}{\partial ik}\, \chi_c(q,k) \Big|_{q=0,\, k=0} \tag{35}$$

The usefulness of the classical characteristic function raises the question if it is possible to construct a similar c-number concept for the quantum case. It seems natural to try the following replacement: $P(x.p,t) \rightarrow \rho(t)$, $x,p \rightarrow b^\dagger, b$, and $\int dxdp \rightarrow \mathrm{tr}$ which actually leads us to the definition:

$$\chi_q(\beta,\beta^*) \; = \; \mathrm{tr}\, \rho(t)\, e^{i\beta^* b^\dagger}\, e^{i\beta\, b} \tag{36}$$

The quantum averages are now obtained in complete analogy to the classical ones by differentiation:

$$< b^\dagger\, b > \; = \; \frac{\partial}{\partial i\beta^*}\frac{\partial}{\partial i\beta}\, \chi_q(\beta,\beta^*) \Big|_{\beta=0,\beta^*=0} \tag{37}$$

and now on the basis of a pure c-number formalism. So far we have succeeded in constructing a probabilistic formalism for quantum processes. Since the charcteristic function χ_c is the Fourier transform of the probability density P, we can always reconstruct the probability by the inverse transformation. Just for curiosity, one could be tempted to use the same idea also to the quantum case, by defining:

$$P(\alpha,\alpha^*,t) = \int e^{-i\beta^*\alpha^* - i\beta\alpha} \chi_q(\beta^*,\beta,t)\, d\beta d\beta^* \tag{38}$$

While the inverse transformation of χ_c merely cancels the previous Fourier-transform, it certainly couldn't return the density operator in the quantum case - since the Fourier transform of a c-number remains a c-number. Therefore the question arises, what physical meaning is behind this artificially constructed function $P(\alpha,\alpha^*)$, and what can it be used for? In a bold step, one is tempted to stretch the previous analogies even further and formulate integrals over the variables α,α^* similar to the ensemble averages eq.(33), e.g. :

$$< \alpha^*\alpha > = \int \alpha^*\alpha\, P(\alpha,\alpha^*,t)\, d\alpha d\alpha^* \tag{39}$$

It is a quite easy to show that this quantity is identical to the quantum mechanical average:

$$< \alpha\alpha^* > = < b^\dagger b > \tag{40}$$

and more generally $< \alpha^n \alpha^{*m} > = < b^{\dagger m} b^n >$. At this point we have finally found a way to describe classical and quantum processes by using very similar languages.

As a matter of principle, a c-number formalism is unable to distinguish between different operator orderings, and a specific order can only be associated with a specifically defined quasi-probability. Without explicitly mentioning it, we have already taken the liberty to choose a special ordering, when we defined the quantum characteristic function: We used normal order of the b's and $b^\dagger$'s in eq.(36). As a consequence, the so defined quasi-probability - the Glauber-Sudarshan P function - allows one to calculate normally ordered moments of B and $b^\dagger$. Similarly, we could have added the b's and $b^\dagger$'s in the exponent, which amounts to symmetric ordering. The corresponding quasi-probability is the Wigner-function, while the Q-function is obtained, when the operator order in eq.(36) is inverted. The Q-function can then be used for calculating anti-normally ordered moments[11,12]:

$P(\alpha,\alpha^*)$:	$\int \alpha\alpha^*\, P(\alpha,\alpha^*)\, d\alpha d\alpha^*$	=	$< b^\dagger b >$
$W(\alpha,\alpha^*)$:	$\int \alpha\alpha^*\, W(\alpha,\alpha^*)\, d\alpha d\alpha^*$	=	$\frac{1}{2} \langle b^\dagger b + b b^\dagger \rangle$
$Q(\alpha,\alpha^*)$:	$\int \alpha\alpha^*\, Q(\alpha,\alpha^*)\, d\alpha d\alpha^*$	=	$< b\, b^\dagger >$

Before we discuss special physical problems in terms of quasi-probabilities in the next chapter, it may be interesting to mention some general features. A classical harmonic oscillator with damping comes to rest, when no external forces are applied. As a result its coordinate and its momentum vanish. If noise is included, phase sensitive momets still vanish, but phase-invariant ones remain finite. Similarly, for a damped quantum oscillator, all normally ordered moments vanish in the long time limit, since the oscillator relaxes into the vacuum. Therefore, the corresponding Fokker-Planck equation for $P(\alpha,\alpha^*)$ has no diffusion term, and conceptually there is no noise. However, some of the moments calculated with the Wigner-or the Q-representation relax towards finite values, since $< b\, b^\dagger >$ remains finite even in vacuum. With a dissipative equation this can only be achieved if noise is added, and the Fokker-Planck equations for W and Q must always contain a diffusion term.

Deterministic processes are represented by δ-distributions, stochastic ones by regular probability functions. Since the strength of noise increases from P through W to Q, one expects that Q displays the smoothest dependence, while P may be singular and can contain δ-functions. It has been shown in general[13] that W and Q always exist as regular functions, while $P(\alpha,\alpha^*)$ may contain δ-functions and derivatives of arbitrary order. If one attempts to solve the 'Fokker-Planck' equation by analytical methods, the singular behaviour is rather unpleasant and numerical integration is impossible. Unfortunately, however, it are just those physical problems that are most interesting, since it is exactly this singular behaviour and the non-positivity that distinguishes quantum process from classical ones.

4. Physical Examples

It is now finally time to illustrate the general ideas discussed in the previous chapters on the basis of a number of physical examples. This will be done in three separate paragraphs. Each model will be characterized by a Hamiltonian for the modes of the light field only, and will contain dissipative interactions that either cause a random energy flow into the system, or take energy out by means of dissipation. These last contributions are merely added to the reversible evolution equation and turn it into an irreversible dynamic process:

$$\left[\frac{\partial \rho}{\partial t} \right]_{irr} = \begin{cases} \text{dissipative loss}: & \Gamma_1 [\, b, \rho\, b^\dagger \,] + \Gamma_1 [\, b\, \rho, b^\dagger \,] \\ \text{irreversible gain}: & \Gamma_2 [\, b^\dagger, \rho\, b \,] + \Gamma_2 [\, b^\dagger \rho, b \,] \end{cases} \tag{41}$$

All linear resevoir interactions can be cast into this general form[14], and the various possible physical mechanisms behind these interactions determine only the parameters Γ_1, Γ_2.

We begin with the formal example of a nonlinear oscillator with damping, which allows us to introduce the concept, and to emphasize the typical differences between the most commonly used quasi-probabilities. It will become obvious that the notion of noise in quantum mechanics varies from representation to representation, and it is interesting to notice that the same physical problem can be described by quite different equations. The next paragraph contains the basic quantum optical problem: the laser, but with a generalized reservoir. This example is chosen to demonstrate the competition of the "quasi-classical" spontaneous emission noise and the non-classical noise of a squeezed reservoir. In the final paragraph, a model is used which is taken from the field of nonlinear optics, where gain results from the reversible, nonlinear interactions among the field modes, and the quantum features are no longer hidden behind classical noise.

4.1 The Nonlinear Dissipative Oszillator

In terms of the creation and destruction operators for the harmonic oscillator, the simplest nonlinear model that one can think of is described by the following Hamiltonian:

$$H = \hbar\omega b^\dagger b + \hbar g\, b^\dagger b^\dagger b\, b \tag{42}$$

which certainly has a rather unusual appearance in tems of the canonical variables x,p. This shouldn't bother us at the moment, since we don't want to predict any specific experiment, but use the model only for demonstrating some properties of quasiprobabilities. The master equation we start from is of the general form of eq.(15) and turns into a partial differential equation, when we use one of the representations i.e. P, W, or Q . Since we don't want to present separate equations for the three cases individually, we write the equation in condensed form, by introducing the following notation[(13)]:

$$Z(\alpha,\alpha^*,t,\epsilon) \;=\; \frac{1}{\pi\epsilon} \int \exp(-|\alpha-\beta|^2/\epsilon\,)\; P(\beta,\beta^*,t)\; d^2\beta \tag{43}$$

$Z(\alpha,\alpha^*,\epsilon,t)$ is a generalized quasi-probability that contains the three familiar ones as limiting cases :

for $\epsilon = 0$	:	$Z(\alpha,\alpha^*)$	=	$P(\alpha,\alpha^*)$
for $\epsilon = 1/2$	:	$Z(\alpha,\alpha^*)$	=	$W(\alpha,\alpha^*)$
for $\epsilon = 1$	:	$Z(\alpha,\alpha^*)$	=	$Q(\alpha,\alpha^*)$

The master equation is normally written as an operator relation for ρ. Using the quasi-probability representation $Z(\alpha,\alpha^*,t)$ one obtains for the dissipative, nonlinear oscillator the following partial differential equation:

$$
\begin{aligned}
\frac{\partial Z(\alpha,\alpha^*,t,\epsilon)}{\partial t} = & \frac{i\partial}{\partial\alpha}(\omega -2g\epsilon-i\gamma +2g|\alpha|^2)\alpha\ Z \\
& - \frac{i\partial}{\partial\alpha^*}(\omega -2g\epsilon+i\gamma +2g|\alpha|^2)\alpha^*\ Z \\
& + 2\epsilon\gamma \frac{\partial^2}{\partial\alpha\partial\alpha^*} Z \\
& +ig(2\epsilon-1)\left[\frac{\partial^2}{\partial\alpha^2}\alpha^2 - \frac{\partial^2}{\partial\alpha^{*2}}\alpha^{*2}\right] Z \\
& +2ig\epsilon(\epsilon-1)\left[\frac{\partial^3}{\partial\alpha^2\partial\alpha^*}\alpha^* - \frac{\partial^3}{\partial\alpha^{*2}\partial\alpha}\alpha\right]Z
\end{aligned} \tag{44}
$$

In complex notation the structure and therefore the interpretation of the second derivatives is not immediately obvious. Therefore we rewrite them in terms of real and imaginary parts: $\alpha = x + iy$, $\alpha^* = x - iy$. While :

$$\partial^2/\partial\alpha\partial\alpha^* = 1/4\ (\partial^2/\partial x^2 + \partial^2/\partial y^2) = 1/4\ \Delta$$

has the form of a classical diffusion term (45)

$$\partial^2/\partial\alpha^2 + \partial^2/\partial\alpha^{*2} = 1/2\ (\partial^2/\partial x^2 - \partial^2/\partial y^2)$$

by itself has no stochastic interpretation.

For the case of Glauber's P representation i.e. $\epsilon=0$, the process involves no classical noise, since the corresponding diffusion term vanishes as well as the third order derivatives. If it were not for the nonlinearity $\propto g$, which is responsible for the second derivatives with a non-positive diffusion matrix, the equation would have the form of a deterministic process artificially written in stochastic notation. In that case, i.e. $g=0$ an initial δ-distribution would remain sharp and would evolve along the deterministic trajectory. For the nonlinear problem, however, in course of time the non-positive diffusion terms will turn the P function into a highly singular object that is difficult to treat analytically, and which could never be handled numerically.

For $\epsilon = 1$ we obtain the evolution equation for the Q-function. Here we do observe classical noise that is generated by the interaction with the dissipative reservoir, and which guarantees e.g. that in the long time limit $\langle \alpha^*\alpha \rangle = \langle b\, b^\dagger \rangle \neq 0$. The third order derivatives also vanish in this case. As a consequence, the only non-classical features come from the same terms as in the previous example. The fact that reservoir noise as well as quantum noise from the nonlinear interaction contribute both to the "diffusion terms" is typical for the P- and Q-representations. This isn't the case for the Wigner-distribution.

The equation for the Wigner-distribution is obtained for $\epsilon = 1/2$. Here the second order derivatives represent pure diffusive behaviour and contain only the interaction with the reservoir. Quantum noise in this case is clearly separated from reservoir fluctuations and is represented by the cubic derivatives. From a mathematical point of view this is a rather unpleasant observation, since it makes the analytical as well as the numerical solution of such an equation extremely difficult. Many authors have taken the liberty to simply drop these troublesome higher order derivatives, for no reason other than to get rid of terms that are difficult to interpret. But at the same time one looses the quantum mechanical consistency of the equation and typical quantum features are lost. In many cases this unsystematic manipulation can lead to serious problems like negative intensities - in any case it leads to entirely uncontrolable results.

The question may arise, what the typical features of quantum noise really are, or why after all do we associate noise with cubic derivatives or the "non-positive diffusion terms". Obviously, they cannot be simulated by random forces. But nevertheless, they have something in common with classical noise. For example, let us imagine that the point $x=0$ is a stationary, instable point of a deterministic problem. Then in the purely deterministic case $x^n(t) \equiv 0$ for all times if $x(t=0) = 0$ initially. In the presence of classical noise, however, $< x^2(t) >$ will not vanish, and the coordinate will start to evolve even if we start from the same initial state $x(t=0)=0$. This, in the picture of the Fokker-Planck equation is caused by the presence of second order derivatives. However, the same could be said also if the diffusion matrix would be non-positive, as it is the case for the quantum processes. Therefore, the non-classical terms in the master-equation nevertheless can "kick" a particle off a potential hill in the same way as it is done by classical random forces. The third derivatives in principle have the same effect only in higher orders. What makes these processes differ from the classical Fokker-Planck dynamics is the fact that the P- as well as the W-distributions must not remain positive, even if they started that way initially. A negative probability, however, is impossible in classical statistics and therefore the quantum system can exhibit features that can never be obtained by classical noise.

4.2 Laser with a Squeezed Vacuum

Formally speaking, the laser is an oscillator, excited by the energy flow between two reservoirs. Since the energy is supplied by an ensemble of inverted two-level atoms, the gain must saturate and the interaction becomes nonlinear. The loss of energy is usually described by a reservoir that is held in the vacuum state in case of optical processes and in thermal equilibrium at a finite temperature for the maser. Here we will generalize the state of the reservoir, by assuming that it is prepared in a multi-mode squeezed vacuum state[15,16]. This squeezed vacuum contains usable energy and therefore the gain of the system is increased. At the same time, also the fluctuations are increased and thereby the losses. As a result, the deterministic part of the laser Fokker-Planck equation remains unchanged, and only the diffusion terms are modified. Gain noise, which results from the atomic ensemble, competes with the noise of the squeezed reservoir which is entirely non-classical. Gain noise is phase invariant, while noise from the squeezed reservoir adds in one quadrature but subtracts from gain noise in the other one. As long as the laser is not operated too far below threshold, a region which isn't of particular interest anyway, the diffusion matrix remains positive. Therefore we still obtain a Fokker-Planck equation in this generalized case[17,18]:

$$\begin{aligned}\frac{\partial P(\alpha,\alpha^*,t)}{\partial t} &= -\frac{\partial}{\partial\alpha}\{A - C\cosh^2(r) + C\sinh^2(r) - B|\alpha|^2\}\,\alpha P \\ &-\frac{\partial}{\partial\alpha^*}\{A - C\cosh^2(r) + C\sinh(r) - B|\alpha|^2\}\,\alpha^* P \\ &- C\cosh(r)\sinh(r)\left\{\frac{\partial^2}{\partial\alpha^2} + \frac{\partial^2}{\partial\alpha^{*2}}\right\}P \\ &+ \{2A + 2C\sinh^2(r)\}\frac{\partial^2}{\partial\alpha\partial\alpha^*}P \end{aligned} \tag{46}$$

We have not combined all the terms yet in order to show the different contributions separately. Similar to the equation in the previous chapter we recognize classical noise and non-classical contributions. Decomposed into real and imaginary parts $\alpha = x + iy$ we find :

$$\begin{aligned}\frac{\partial P(x,y,t)}{\partial t} &= -\frac{\partial}{\partial x}(a - x^2 - y^2)x P - \frac{\partial}{\partial y}(a - x^2 - y^2)y P \\ &+ \left\{1 + \frac{b}{2}(e^{-2r} - 1)\right\}\frac{\partial^2}{\partial x^2}P + \left\{1 + \frac{b}{2}(e^{2r} - 1)\right\}\frac{\partial^2}{\partial y^2}P\end{aligned} \tag{47}$$

In proceeding from eq.(46) to eq.(47) we have also renormalized the variables, such that in the limit $r \to 0$ one obtains the traditional laser equation. For arbitrarily large squeezing $r \to \infty$ the noise terms become $1-b/2$ for the direction of the real part and $1+b/2$ for the imaginary direction. Since the physical meaning of the parameter b is: b = loss/gain, the condition $b<2$ encloses the entire physically interesting parameter region.

As a consequence of the loss of symmetry of the Fokker-Planck equation the condition of detailed balance is violated, and even the steady state solution can no longer be obtained in analytical form. Therefore we can only present here the results of a numerical integration,

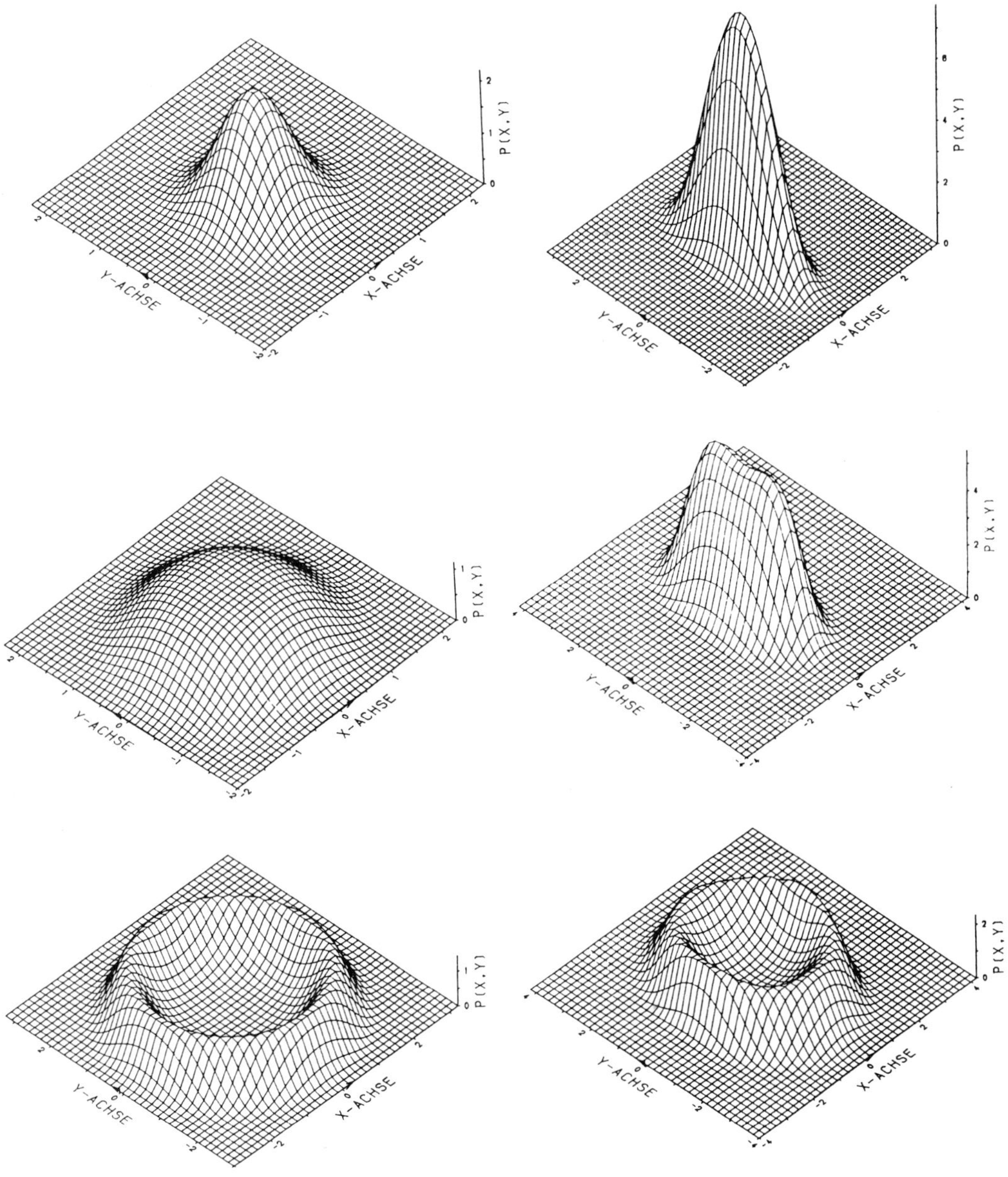

Fig.3 Steady state probability for the laser with a squeezed reservoir r=1, (B) compared with the regular laser r=0, (A). The parameters were chosen below: a=-4, at: a=0, and above threshold: a=4.

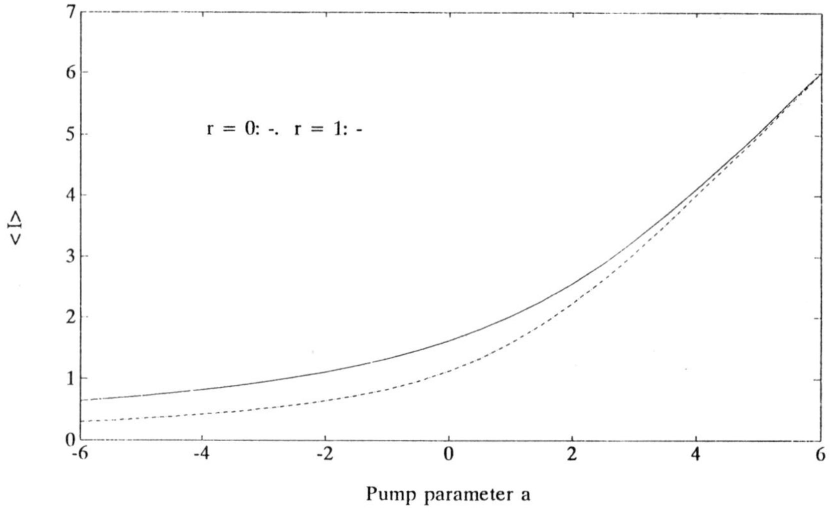

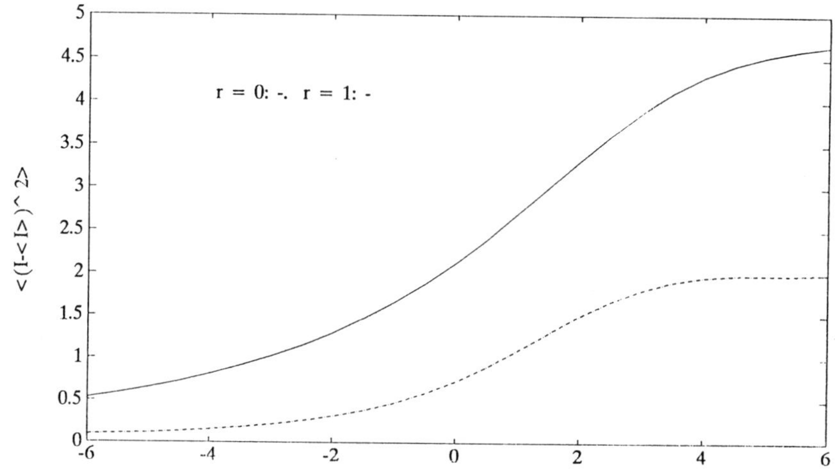

Fig.4 Average of the intensity and its variance as a function of the pump parameter a. Solid line r=1, dashed line r=0.

based on the method of continued fractions[2]. It is illustrating to compare the case of the traditional laser, i.e. r=0 with the generalized model r≠0. We have plotted in fig. (3) the steady state distribution P as a function of x,y while the gain parameter a is varied through the threshold regime. The sequence (A) displays the normal laser with r=0 for differnt values of the gain parameter, while the sequence (B) shows the same for the squeezed case for r=1. The steady state distribution is obviously largely modified and contains considerable fluctuations even above threshold. While the 3D plots give us an intuitive picture of the process, the actual physical observables are the ensemble averages like the average intensity or its variance. This information is contained in P(x, y) e.g.:

$$\langle I \rangle = E_0^2 \langle b^\dagger b \rangle = E_0^2 \int (x^2+y^2)\, P(x,y)\, dxdy$$

and displayed in fig.(4) . The solid line is the squeezed laser compared with the usual laser represented by the dashed line. While the intensities very soon approach each other closely above threshold, the variances, as expected, don't meet. Since here we describe a quantum process entirely in classical stochastic terms, P will remain positive. Therefore the inequalities of classical statistics remain valid, and the variances of the quadrature components exceed those of the vacuum or the coherent state. In this model, squeezing below the vacuum level is impossible above threshold.

Without the theoretical results, it wouldn't be so obvious how the modification of the Fokker-Planck equation would change the properties of the laser, since it is only the noise part that was changed, while the more easily interpreted deterministic forces remained unaltered. Our intuitive picture is even more vague, when it comes to predicting the dynamic properties of the system like the time scale of intensity fluctuations or the laser linewidth. As the eigenvalues of a linear stochastic process are independent of the strength of noise, the effects of a modified diffusion can only become visible through the nonlinearity of the process. A detailed discussion of the time dependent properties will be presented in a seperate paper[19].

4.3 Sub-Harmonic Generation

A last example is chosen where the non-classical features dominate the statistical properties entirely, and the evolution equation is no longer of Fokker-Planck typ. Here the diffision matrix is not only asymetric but has a negative eigenvalue. The parametric interaction of two modes of the light field in a nonlinear crystal with a frequency ration of 1:2 can also display gain. As already indicated in chapter 4.1, the P representation in this case is expected not to contain any classical noise and the quantum fluctuations due to the nonlinear interaction must become clearly visible. With a negative diffusion term, the positive definiteness of the P-distribution isn't guaranteed, it isn't even guaranteed that P exists as a regular function or a tempered distribution. The basic Hamiltonian for this process is interpreted easily:

$$H = \sum_{j=1}^{2} \hbar\omega_j b_j{}^\dagger b_j + \hbar g \left(b_2 b_1{}^{\dagger 2} + b_2{}^\dagger b_1{}^2 \right) \tag{48}$$

$b_1, b_1{}^\dagger$ represent the sub-harmonic mode with frequency $\omega_1=\omega$, $b_2, b_2{}^\dagger$ the mode with frequency ω_2 where $\omega_2 = 2\omega_1 = 2\omega$ and g is proportional to the nonlinear susceptibility of the medium. The interaction term either creates two photons of the lower frequency ω_1 on the expense of one of the harmonic frequency ω_2 or it creates the second harmonic field by destroying two sub-harmonic photons. This process takes place in a nonlinear optical crystal, which is usually placed inside a resonator for increasing the field intensity. In order to drive the generation of the sub-harmonic field, a field of the harmonic frequency ω_2 must be applied from the outside. In the theoretical model this can be done by adding the following term to the Hamiltonian:

$$-i\hbar \left\{ F^* e^{i\omega_2 t}\, b_2 - F\, e^{-i\omega_2 t}\, b^{\dagger 2} \right\} \tag{49}$$

The loss of the two cavity modes again is described by the coupling to two individual vacuum reservoirs as indicated in eq.(41). When combining all these effects, the master equation assumes the following form [20,21,22]:

$$\begin{aligned} \frac{\partial P(\alpha_1,\alpha_1{}^*,\alpha_2,\alpha_2{}^*,t)}{\partial t} &= \frac{\partial}{\partial\alpha_1}\{\gamma_1\alpha_1 - 2g\,\alpha_2\alpha_1{}^*\}\,P + \text{c.c.} \\ &+ \frac{\partial}{\partial\alpha_2}\{\gamma_2\alpha_2 + g\alpha_1{}^2 - F\}\,P + \text{c.c.} \\ &+ g\left\{\frac{\partial^2}{\partial\alpha_1{}^2}\,\alpha_2 + \frac{\partial^2}{\partial\alpha_1{}^{*}}\,\alpha_2{}^*\right\}P \end{aligned}$$

The evolution equation in the previous example was always of the form of a Fokker-Planck equation, and the corresponding problem could be interpreted in terms of a classical stochastic processes. In the entire interesting parameter range there always existed an equivalent formulation in terms of a complex Langevin equation. For the present problem this is no longer the case and stochastic simulation of this process is impossible. We also expect that the P-function will be highly singular and not suitable for a numerical treatment of the problem. We are not forced to use the P-representation, but could make also use of one of the other quasi-probabilities, which are known to exist as regular analytic functions. However, since the phase space of the parametric process is now 4-dimensional, no reliably numerical method

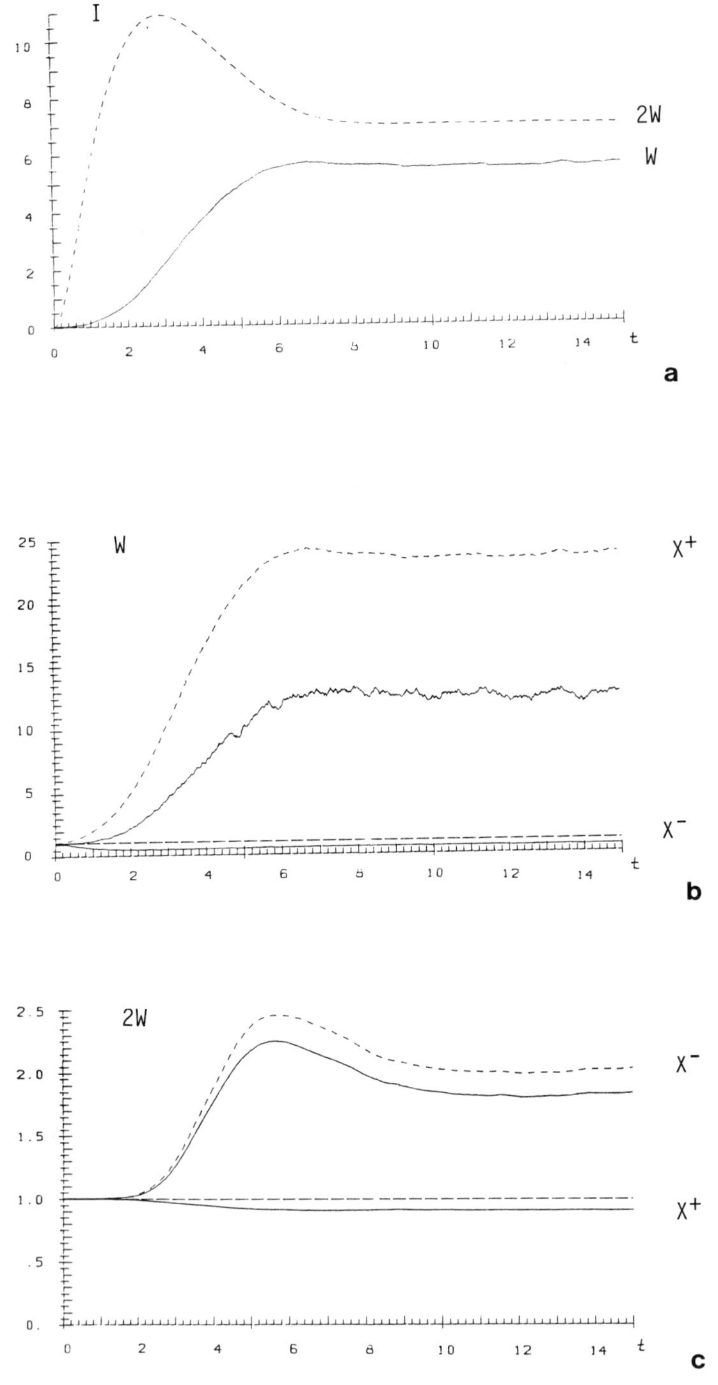

Fig.5 Time dependence of the average field for the sub-harmonic process above threshold. The average intensities of the sub-harmonic field (ω) and the second harmonic one (2ω) are plotted in the top figure (a), and the variances of the two quadrature components are shown in (b) for the sub-harmonic, in (c) for the second harmonic field.

exists to solve a partial differntial equation with 5 variables. The method of continued fractions, which has been so successful in many stochastic problems, becomes forbiddingly complex already when applied to a problem with three independent variables.

Here we touch the main technical problem of quantum optics. Presently there exists no systematic method for solving nonlinear, dissipative quantum processes. Only in certain regimes, when linearization is possible, does one obtain absolutely reliable results. In the neighbourhood of instabilities, however, linearization is impossible, and in each individual case one has to search for a suitable perturbation method. The instability region is of special interest in the present context, since it seems that quantum properties become most visible there. The amount of squeezing e.g. that can be obtained theoretically is optimized at threshold. Finally, there is one promising approach that had been suggested to work quite generally, which is based on simulation of a stochastic process[23] . But how is this possible, when the evolution equation for $P(\alpha,\alpha^*,t)$ is not of Fokker-Planck form as in our present example ?

In cases where the higher order derivatives can be neglected by a perturbative argument, Drummond and Gardiner have found a way to turn this equation into the form of a Fokker-Planck equation, by doubling the dimensions of phase space. At a first glance, this step seems to make things even worse, since any extra dimension for a non-trivial partial differential equation is disastrous as the numerical difficulties increase "exponentially". The statistically equivalent set of stochastic differential or Langevin equations, however, can still be simulated straight forewardly. Here the doubling of the number of equations only doubles the amount of numerical work. For a while it seemed that this approach could provide a universal tool for attacking most nonlinear processes, where the quantum nature is essential. Unfortunately, this didn't turn out to be true in practice. At the moment it is not generally clear when to expect correct, and when erroneous results. The only thing that is certain is that both can happen.

The problem of sub-harmonic generation, successfully resisted all attempts for solving it by any other method. Therefore we applied the socalled positive P-representation method of Drummond and Gardiner to this problem, and simulated the corresponding dissipative nonlinear process in eight dimensions, i.e. two for α_1, two for α_2 and after doubling this makes eight. Initially, the system is assumed to be in the vacuum state, which in semi-classical approximation is an instable equilibrium state. Quantum noise, even the peculiar kind with negative diffusion, does trigger the time evolution and the intensity of the fields grow from zero. After averaging over a large sum of individual trajectories, we obtain the smooth transient behaviour of the intensities, or the variances of the two quadrature components. The averages over the stochastic trajectories in the extended space are equivalent to the averages over the quantum mechanical ensemble. The presence of non-classical noise manifests itself in sub-Poissonian photon statistics, or in squeezing.

In fig.5 we have plotted the transient evolution of the typical statistical properties of the two coupled fields. The intensities as they emerge from the vacuum by the action of noise are plotted in (a). The sub-harmonic field is delayed with respect to the second harmonic one. This is a result of the hierarchy of the dynamic interactions. Initially, there is no light in the resonator. Then the external field of the second harmonic frequency enters the cavity and builds up an intensity at that frequency, before the nonlinear process, triggered by the quantum fluctuations, sets in to generate the sub-harmonic field. The variances of the two quadrature components X^+,X^- of the two fields $\omega,2\omega$ are plotted in the two following figures, for the second harmonic in (b) and for the sub-harmonic in (c). The non-classical behaviour is obvious, since in each field there is always one component, for which the fluctuations fall below the vacuum level. For curiosity, we have multiplied the variances of the two conjugate components, in order to check if they exceed the boundary of Heisenberg's uncertainty relation, and fortunately, they do !

After all we have mentioned about the approach used here, the natural question arises: How reliable are these results after all? Another approach for treating the problem approximately consists of the quantum analog of the classical cumulant expansion[24] . We have found that below, at and in a reasonable range above the threshold, both methods are in very good agreement, which suggests that the results cannot be too far from the truth. Since both methods are based on entirely different concepts, it is hard to imagine that the agreement could be in any way accidental. Therefore we are quite confident that the obtained results are in principle correct solution - solutions of a time dependent, nonlinear, dissipative process, where quantum noise pays the essential role.

References

1. R.Graham and A.Schenzle, Phys.Rev.A26, 1676 (1982)
2. H.Risken, "The Fokker-Planck Equation", (Springer Verlag Berlin 1984)
3. W.Feller, "An Introduction to Probability Theory and its Applications", (John Wiley, N.Y. 1968)
4. Ch.Ginzel and A.Schenzle, to be published.
5. L.Mandel, Proc.Phys.Soc. London 72, 1037 (1958) and 74, 233 (1959)
6. R.J.Glauber in "Quantum Optics and Electronics", ed. C.DeWitt, A.Balandin and C.Cohen-Tannoudji (Gordon and Breach, N.Y., 1965)
7. P.L.Kelly and W.H.Kleiner, Phys.Rev.136, 316 (1964)
8. A.Schenzle and R.G.Brewer, Phys.Rev.A34, 3127 (1986)
9. E.P.Wigner, Phys.Rev.40, 749 (1932)
10. R.J.Glauber, Phys.Rev.Lett. 10, 84 (1963), Phys.Rev.131, 2766 (1963)
11. H.Haken in "Encyclopedia of Physics", XXV/2c, ed. by L.Genzel (Springer Verlag, Berlin 1969)
12. M.Hillery, R.F.O'Connel, M.O.Scully, and E.P.Wigner, Phys. Rep. 106, 121 (1984)
13. K.E.Cahill and R.J.Glauber, Phys.Rev.177, 1882 (1969)
14. P.N.Argyres and P.L.Kelley, Phys.Rev.A134,98 (1964)
15. M.J.Collet and C.W.Gardiner, Phys.Rev.A31, 3761 (1985)
16. M.A.Dupertuis, S.M.Barnett and S.Stenholm, J.Opt.Soc.Am.B,4, 1102 (1987) B,4 1124 (1987)
17. J.Gea-Banacloche, Phys.Rev.Lett. 59, 543 (1987)
18. Ch.Ginzel,J.Gea-Banacloche and A.Schenzle,to appear in Acta Phys. Polonica.
19. Ch.Ginzel, R.Schack and A.Schenzle, to be published
20. P.D.Drummond, K.J.McNeil and D.F.Walls, Optica Acta, 27, 321 (1980)
21. P.D.Drummond, K.J.McNeil and D.F.Walls, Optica Acta, 28, 211 (1981)
22. M.Dörfle and A.Schenzle, Z.Phys.B 65, 113 (1986)
23. P.Drummond and C.W.Gardiner, J.Phys.A13, 2353 (1980)
24. R.Schack and A.Schenzle, to appear in Phys.Rev.A.

TWO PHOTON INTERFERENCE

H. Fearn

Max-Planck Institut für Quantenoptik
D-8046 Garching, Federal Republic of Germany
and
Center for Advanced Studies and
Department of Physics and Astronomy
University of New Mexico, Albuquerque, New Mexico 87131

INTRODUCTION

Two photon interference can be studied by detecting the output light from a beamsplitter (bs) using incident radiation in the form of a photon pair excitation. When the photon pair is excited in a single input at the bs, the output state resembles that expected for independent classical particles, whereas quantum interference effects occur when the photon pair is divided between the two input arms.

The theory of two-photon interference is presented here in a tutorial fashion with the emphasis on simplicity. Interested readers will be directed to more detailed accounts in the literature. Including the introduction, the paper is divided into five sections. The second section is a brief history of the photon and photon localization which is included in order to establish the concept of a photon wavepacket and to give credit where credit is due. The third section contains Fock (number) state calculations for both single photon and two photon interference at a beamsplitter (bs). These results represent a special case of the time-dependent wavepacket calculations, simultaneous arrival at the bs and perfect overlap. Diracs famous statement [Dirac 1958] that a single photon interferes only with itself, is found to be in disagreement with the two photon results.

Section four gives the time dependent analysis of two-photon interference and is split into three subsections in order to accommodate three different light sources. The first is an ideal two atom light source, which is not very realistic but exemplifies the form of the results. The second case is a non-degenerate parametric oscillator, which has been used experimentally, and the third is cascade emission light source.

The last section contains a summary of the results and conclusions.

THE PHOTON

In 1900 Planck explained the blackbody radiation curve in terms of discrete quanta. Using a blackbox to represent the blackbody, he assumed that the box walls consisted of harmonic oscillators which could absorb or emit energy only in discrete units $nh\nu$, where n is an integer, ν is the frequency of light and h is Plancks constant. This led him to write the famous blackbody energy distribution formula

$$E(\nu)d\nu = \frac{8\pi^2\nu^2}{c^3}\frac{h\nu}{\exp(h\nu/K_B T) - 1}d\nu \tag{1}$$

Later, in 1905 Einstein wrote the paper "On a heuristic viewpoint concerning the production and transformation of light." In this paper Einstein re-examines Plancks

New Frontiers in Quantum Electrodynamics and Quantum Optics
Edited by A. O. Barut, Plenum Press, New York, 1990

work and makes the bold suggestion that it is not only the harmonic oscillators in the box wall that are associated with discrete packets of energy, but the radiation field as well.

Four years later Taylor performed the first single-photon interference experiment using a flame light source, diffraction grating and a photographic plate. In order to think in terms of a photon traveling in one direction or another [Frisch 1965] and not split at the diffraction grating, (or bs) photon interference can be explained in terms of the uncertainty in detection of each photon. [Pfleegor and Mandel 1967] The detection process itself then causes the interference since localizing the photon at the detector makes it intrinsically uncertain from which direction the photon came. This blurring of the detector resolution can be illustrated by a simple example, summarized from the last reference.

Let the output from an interferometer converge at an angle θ onto a detector, so that the photon emerges in slightly different directions depending on which path was taken through the interferometer. Take the fringe spacing to be L and the wavelength of the light to be λ. The experimental arrangement is shown in Fig. 1.

From Fig. 1(b) it is clear that

$$\lambda = L\theta \tag{2}$$

where order of magnitude values of L and θ are approximately 10^{-3}m and 10^{-7}m respectively. The uncertainty in the x direction is given by Heisenberg's relation

$$\Delta x \Delta p_x \geq h \tag{3}$$

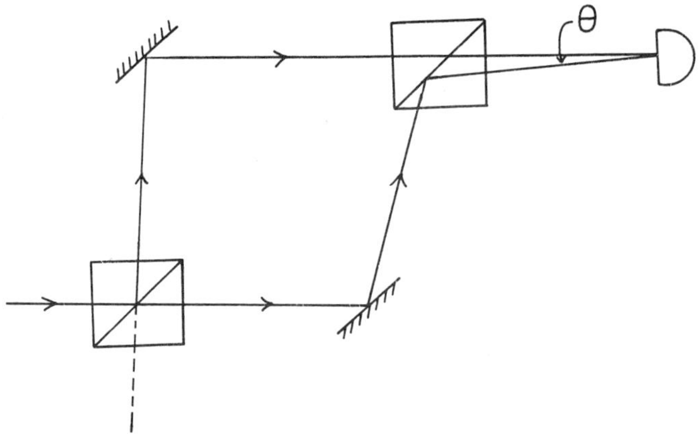

Fig. 1a. Interference with angled output, viewed from above.

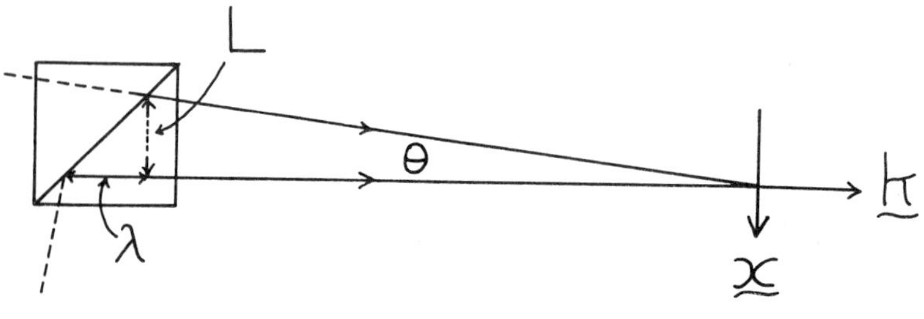

Fig. 1b. Enlargement of output beamsplitter.

In order to see fringes we require

$$\Delta x < L = \frac{\lambda}{\theta} \tag{4}$$

where Δx is normal to the fringes and the wavevector. Hence the uncertainty in the photon momentum is given by

$$\Delta p_x \geq \frac{h\theta}{\lambda} = \hbar k\theta \tag{5}$$

which rules out knowledge of from which direction the photon came. If one tries to detect the photon in either of the interferometer arms it is well known that the interference pattern disappears.

In order to think of the photon in an arm of an interferometer, or localized at some point, it is desirable to define a wavepacket for the photon. The idea of zero mass particles originated from Einstein (1905), who described light as

> "consisting of a finite number of energy quanta which are localized at points in space, which move without dividing, and which can only be produced and absorbed as complete units."

This description of photons illustrates the need for localization of position and energy density. Classically, the concept of localizability is so fundamental in the description of particles that little thought is given to a detailed analysis of the problem. Quantum mechanically, position is defined as a linear operator in Hilbert space, conjugate to the momentum operator. Extending this concept to the relativistic case is not obvious and the first attempts appeared in the late 1960's. An early relativistic treatment of elementary particles was given by Newton and Wigner (1949). They established that only particles which can exist in superpositions of states with opposite helicities ± 1 can be localized in a finite region. But their results appeared invalid for mass $= 0$ particles. Later, in 1967 Jauch and Piron introduced the idea of weak-localizability in order that mass=0 particles could be treated. These ideas were subsequently collected together by Amrein (1969). In his extensive paper he introduces an operator $\hat{\phi}$ which transforms a one-photon state $|\phi\rangle$ into the vacuum state multiplied by the wavefunction corresponding to $|\phi\rangle$.

$$\langle 0|\hat{\phi}(x,t)|\phi\rangle = \phi(x,t) \tag{6}$$

It should be noted that the positional dependence of ϕ refers to the field associated with the photon rather than the photon itself. The wavefunction $\phi(x,t)$ corresponds to the Newton and Wigner wavepacket in the limit mass $\rightarrow 0$.

A similar relativistic treatment has been given by Pike and Sarkar (1986) and a non-relativistic treatment was written by Mandel in 1966. It is this kind of photon wavepacket which we shall exploit later.

In section four we employ a "continuous number state" representation for the photon. This is similar to the approach of Mandel and to the non-relativistic treatment of Amrein. It involves a wavepacket which can be created or destroyed using photon creation/annihilation operators. We allow the wavepacket to spread over all frequencies with a lorentzian lineshape. All photons are taken to have the same polarization so we neglect polarization vectors throughout.

PHOTON INTERFERENCE: TIME INDEPENDENT APPROACH

Consider the symmetric lossless bs in Fig. 2. Let us define the complex reflection and transmission coefficients to be r and t respectively where for example

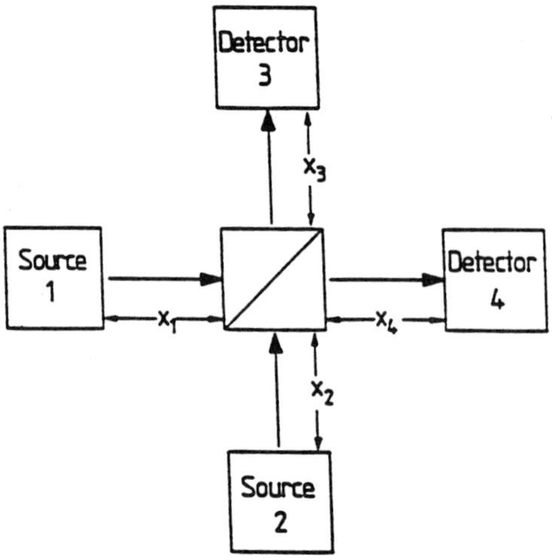

Fig. 2. Beamsplitter configuration showing notation for input and output arms.

$$r = |r| \exp(i \arg r) \tag{7}$$

Due to energy conservation considerations we find that

$$|r|^2 + |t|^2 = 1 \tag{8}$$

and

$$r^* t + r t^* = 0 \tag{9}$$

(which is equivalent to $\arg r - \arg t = \pi/2$).

The output photon annihilation operators are related to the input operators via the following matrix equation

$$\begin{pmatrix} a_3 \\ a_4 \end{pmatrix} = \begin{pmatrix} r & t \\ t & r \end{pmatrix} \begin{pmatrix} a_1 \\ a_2 \end{pmatrix} \tag{10}$$

where a_i is the photon annihilation operator for arm i of the bs.

Let us send one photon into arm 1, so that the input state can be defined as

$$|n_1, n_2\rangle = |1_1, 0_2\rangle \tag{11}$$

where n_i are the number of photons in arm i and $n_1 + n_2 = 1$. The probability that the photon will be reflected into arm 3 is given by

$$\begin{aligned} P_{n_3=1,n_4=0} &= |\langle 1_1, 0_2 | a_3^\dagger | 0_3, 0_4 \rangle|^2 \\ &= |\langle 1_1, 0_2 | r^* a_1^\dagger + t^* a_2^\dagger | 0_1, 0_2 \rangle|^2 \\ &= |r|^2 \end{aligned} \tag{12}$$

The probability that the photon will be transmitted into arm 4 is given by

$$\begin{aligned} P_{n_3=0,n_4=1} &= |\langle 1_1, 0_2 | a_4^\dagger | 0_3, 0_4 \rangle|^2 \\ &= |\langle 1_1, 0_2 | t^* a_1^\dagger + r^* a_2^\dagger | 0_1, 0_2 \rangle|^2 \\ &= |t|^2 \end{aligned} \tag{13}$$

Notice that the sum of the probabilities add up to unity. If we wish to know the likelihood of forming a superposition state, the probability is then given by

$$\begin{aligned} P_{r,t} &= |\langle 1_1, 0_2|\{r|1_3, 0_4\rangle + t|0_3, 1_4\rangle\}|^2 \\ &= |\langle 1_1, 0_2|ra_3^\dagger + ta_4^\dagger|0_3, 0_4\rangle|^2 \\ &= |\langle 1_1, 0_2|(|r|^2 + |t|^2)a_1^\dagger + (r^*t + rt^*)a_2^\dagger|0_1, 0_2\rangle|^2 \end{aligned} \tag{14}$$

This implies that inputting a single photon will always give rise to a superposition state which agrees with Diracs dictum that a single photon interferes only with itself. However, we may now consider a 2-photon problem assuming that the photons always hit the bs simultaneously. [Fearn and Loudon 1987] The photon pair may enter the bs in the same input, arm 1. This case has been treated experimentally by Brendel et al. (1988) and Lange et al. (1988). Alternatively the photons can enter separate inputs, which has been experimentally observed by Hong, Ou and Mandel (1987) and Rarity and Tapster (1988). It is found that the photon distribution resembles that of independent particles when they arrive in the same arm, but that quantum mechanical interference effects, characteristic of Bose-Einstein statistics, are observed when the two photons arrive in different arms.

Suppose that the combined input state $|n_1, n_2\rangle$ has n_1 photons in arm 1 and n_2 photons in arm 2, such that $n_1 + n_2 = 2$. There are clearly three distinct input states of this kind and three distinct photocount readings for the two output arms. The probabilities of the photocount readings denoted P_{n_3,n_4} are easily obtained and the results are shown in the table below. To illustrate the calculations, we show how to obtain possible output states for the input $|1_1, 1_2\rangle$, which corresponds to the middle column of the table.

(a) For the output $|n_3 = 2, n_4 = 0\rangle = 2^{-\frac{1}{2}} a_3^\dagger a_3^\dagger |0_3, 0_4\rangle$ the probability of obtaining 2 photons in output 3 is given by

$$\begin{aligned} P_{2,0} &= |\langle 1_1, 1_2|2_3, 0_4\rangle|^2 \\ &= |\langle 1_1, 1_2|2^{\frac{-1}{2}}(r^*a_1^\dagger + t^*a_2^\dagger)(r^*a_1^\dagger + t^*a_2^\dagger)|0_1, 0_2\rangle|^2 \\ &= \frac{1}{2}|2r^*t^*|^2 \\ &= 2|r|^2|t|^2 \end{aligned} \tag{15}$$

Table 1. Output Photon Probabilities for Two Input Photons. [Taken from Fearn and Loudon 1987.]

Probability	$\|2_1, 0_2\rangle$	$\|1_1, 1_2\rangle$	$\|0_1, 2_2\rangle$
$P_{2,0}$	$\|r\|^4$	$2\|r\|^2\|t\|^2$	$\|t\|^4$
$P_{1,1}$	$2\|r\|^2\|t\|^2$	$(\|r\|^2 - \|t\|^2)^2$	$2\|r\|^2\|t\|^2$
$P_{0,2}$	$\|t\|^4$	$2\|r\|^2\|t\|^2$	$\|r\|^4$

(b) The probability of obtaining 2 photons in output 4 is similar to the above calculation and we obtain

$$P_{0,2} = 2|r|^2|t|^2 \tag{16}$$

(c) The probability of obtaining a single photon in each output ($|n_3 = 1, n_4 = 1\rangle = a_4^\dagger a_3^\dagger |0_3, 0_4\rangle$) is given by

$$\begin{aligned} P_{1,1} &= |\langle 1_1, 1_2 | 1_3, 1_4 \rangle|^2 \\ &= |\langle 1_1, 1_2 | (t^* a_1^\dagger + r^* a_2^\dagger)(r^* a_1^\dagger + t^* a_2^\dagger) | 0_1, 0_2 \rangle|^2 \\ &= |t^{*2} + r^{*2}|^2 \\ &= (|t|^2 - |r|^2)^2 \end{aligned} \tag{17}$$

where we have used Eq. (9) in the last step.

We may also prove that the 2-photon superposition state

$$2^{-\frac{1}{2}}(|2_3, 0_4\rangle + |0_3, 2_4\rangle) \tag{18}$$

is the only output for the input state $|1_1, 1_2\rangle$ when we have a 50:50 bs $|r|^2 = |t|^2 = 1/2$. In terms of creation operators, the input is written as

$$|1_1, 1_2\rangle = a_1^\dagger a_2^\dagger |0_1, 0_2\rangle \tag{19}$$

Using the time reversal property of a symmetric lossless bs we have

$$\begin{pmatrix} a_1 \\ a_2 \end{pmatrix} = \begin{pmatrix} r^* & t^* \\ t^* & r^* \end{pmatrix} \begin{pmatrix} a_3 \\ a_4 \end{pmatrix} \tag{20}$$

Thus

$$\begin{aligned} |1_1, 1_2\rangle &= (ra_3^\dagger + ta_4^\dagger)(ta_3^\dagger + ra_4^\dagger)|0_3, 0_4\rangle \\ &= 2^{\frac{1}{2}} rt(|2_3, 0_4\rangle + |0_3, 2_4\rangle) + (r^2 + t^2)|1_3, 1_4\rangle \\ &= 2rt2^{-\frac{1}{2}}(|2_3, 0_4\rangle + |0_3, 2_4\rangle) + \exp(2i \arg r)(|r|^2 - |t|^2)|1_3, 1_4\rangle \end{aligned} \tag{21}$$

So, for a 50:50 bs we see that the only possible outcome is the normalized superposition state of Eq. (18). This is clearly a 2-photon effect since the superposition state cannot be generated by a single mode (single input) excitation. One may also assume that too great a significance has been placed upon Diracs statement.

Referring to the tabulated results, we note that the distribution of photocounts is the same as would be predicted for the splitting of a beam of classical distinguishable particles, when both photons are excited in the same input arm. This agrees with the theory of Richter, Brunner and Paul (1964,1966) and Prasad, Scully and Martienssen (1987) who also predict this "classical particle" behavior. However, when the photons are excited one in each input, the probability of both counts occurring in the same output is twice the value for classical distinguishable particles, in agreement with Feynman (1965). It is interesting to note that the probability for one photon in each output vanishes in the case of a 50:50 bs with $|r| = |t|$. These results agree with the experimental evidence for which references are cited above.

TWO-PHOTON INTERFERENCE: TIME DEPENDENT

The discrete mode theory outlined in the previous section is incapable of describing time-dependent correlations between the input photons. The aim of this section is a more complete description of 2-photon interference effects by means of a continuous mode theory for various kinds of light source. This section summarizes some of the results of a more detailed paper on two-photon interference by Fearn and Loudon (1989).

The most elementary source has only two atoms, whose excitation at appropriate times can provide input photons with arbitrary time separation. This rather unrealistic case clarifies the role of the Bose-Einstein statistics, particularly in regard to the distinction between the results obtained in the same arm and each photon in a different arm.

Practical two-photon light sources are based on the parametric down-converter, or on atomic cascade emission. With suitably low levels of excitation, the emissions from both of these sources consist of highly correlated photon pair events, as was first shown experimentally by Burnham and Weinberg (1970) and by Clauser (1974) respectively. Experiments performed to date include Brendle et al. (1988), Lange et al. (1988), Hong, Ou and Mandel (1988) and Rarity and Tapster (1988) who have used parametric down conversion in free space as a light source. Atomic cascade experiments have been carried out by Friberg, Hong and Mandel (1985). The detailed theoretical descriptions of the interference effects in these experiments must include the filters that are used to limit the optical bandwidths. The light sources treated below have bandwidths that are limited by their intrinsic properties and there is no need to include the effects of external filters. We shall treat light sources whose linewidth γ and mean frequency ω_0 are related to the detector bandwidth T^{-1} by

$$T^{-1} \ll \gamma \ll \omega_0 \tag{22}$$

Futhermore, we shall assume that both photons will be detected during a counting period T, so that the photon wavepackets maybe assumed negligibly small outside this time interval. As before, all photons have the same polarization and we neglect polarization vectors.

Two-Atom Light Source

This is an idealized source consisting of two-atoms which are excited at arbitrary times. The bs theory employed here is essentially one-dimensional. [Fearn and Loudon 1987]

One-dimensional optical excitations can be described by creation/annihilation operators which obey the commutator

$$[a_i(\omega), a_j^{\dagger}(\omega')] = \delta_{ij}\delta(\omega - \omega') \tag{23}$$

For a single photon we may introduce the creation operator

$$A^{\dagger}(\alpha) = \int_0^{\infty} d\omega \alpha(\omega) a^{\dagger}(\omega) \tag{24}$$

where $A^{\dagger}(\alpha)|0\rangle = |1\rangle$ and the integral maybe extended down to $-\infty$ in view of the narrow bandwidth excitations we shall use [see Eq. (22)] This continuous photon operator behaves in an identical manner to the usual discrete number operator. The amplitude $\alpha(\omega)$ is normalized as follows

$$\int d\omega |\alpha(\omega)|^2 = 1 \tag{25}$$

and represents the photon wavepacket. The matrix element of the single photon operator

$$\langle 0|a(\omega)|1\rangle = \alpha(\omega) \tag{26}$$

is reminiscent of the photon annihilation operator of Amrein (1969).[See Eq. (6)]

The nature of the single photon state depends on $\alpha(\omega)$. We shall derive this amplitude by consideration of the far field produced by spontaneous emission from an atom of transition frequency ω_i at time t_i. The single photon electric field matrix element at a distance x_i from the atom is then given by [Loudon 1973],

$$\langle 0|E(x,t)|1\rangle = \frac{-e\omega_i^2 D \sin\alpha}{4\pi\epsilon_0 c^2 x_i} \exp[-(i\omega_i+\gamma_i)(t-t_i-x_i/c)]\theta(t-t_i-x_i/c) \tag{27}$$

where α is the angle between the direction of the transition dipole moment D and the observation direction x_i, in the plane of polarization of the electric field, θ is the unit step function, and the decay rate has been assumed much smaller that the transition frequency $\gamma_i \ll \omega_i$ in agreement with the assumption of Eq. (22).

The far field matrix element can be Fourier transformed and normalized to obtain the corresponding frequency function given by

$$\alpha(\omega) = \frac{(\gamma_i/\pi)^{\frac{1}{2}}}{\omega - \omega_i + i\gamma_i} \exp[i\omega_i(t_i + x_i/c)] \tag{28}$$

which has the form of a Lorentzian frequency distribution. If an associated time dependent function is defined by

$$\alpha(t) = (2\pi)^{\frac{-1}{2}} \int d\omega \alpha(\omega) \exp(-i\omega t) \tag{29}$$

then insertion of (28) into (29) produces

$$\alpha(t) = -i(2\gamma_i)^{-\frac{1}{2}} \exp[-(i\omega_i+\gamma_i)(t-t_i-x_i/c)]\theta(t-t_i-x_i/c) \tag{30}$$

This function is a normalized replica of the electric field matrix element (27). The single photon creation operator (24) can be written in the form

$$A^\dagger(\alpha) = \int_0^\infty dt \alpha(t) a^\dagger(t) \tag{31}$$

where $\alpha(t)$ is defined by (30). The analogue of the matrix element (26) is clearly

$$\langle 0|a(t)|1\rangle = \alpha(t) \tag{32}$$

so that for the number state $|n\rangle$ we have

$$a(t)|n\rangle = \alpha(t) n^{\frac{1}{2}} |n-1\rangle \tag{33}$$

The time dependent function $\alpha(t)$ satisfies the normalization condition

$$\int_0^T dt |\alpha(t)|^2 = 1 \tag{34}$$

for $\gamma T \gg 1$

We now consider the use of pairs of atoms for the light source in the experiment outlined in the previous section. The notation here is similar to our previously defined notation. A state with a single excitation in arm 1 and none in arm 2 is defined by

$$|1_1, 0_2\rangle = A_1^\dagger(\alpha)|0_1, 0_2\rangle \tag{35}$$

For two atoms in the light source of arm 1, producing in general excitations with different frequency functions $\alpha_1(\omega)$ and $\alpha_2(\omega)$, the field state is represented by

$$|2_1, 0_2\rangle = N A_1^\dagger(\alpha_1) A_1^\dagger(\alpha_2)|0_1, 0_2\rangle \tag{36}$$

where N is a normalization constant. The normalization condition gives

$$\langle 2_1, 0_2 | 2_1, 0_2 \rangle = 1 = N^2 \{1 + | \int dt |\alpha_1^*(t) \alpha_2(t)|^2\} \tag{37}$$

where the normalization property (25) and the time development equivalent of the commutation relation (23) has been used. Thus the usual boson normalization constant $2^{-1/2}$ is recovered for identical excitation functions $\alpha_1(t) = \alpha_2(t)$ (as assumed in the last section) but N otherwise takes on a larger value, and it becomes equal to unity for sufficiently different excitation times t_1 and t_2 of the atoms in the same source. The overlap integral in Eq. (37), using different atomic parameters ω_1, γ_1, x_1 and $\omega_2, \gamma_2 x_2$ for α_1 and α_2 respectively becomes

$$\int_0^T dt \alpha_1^*(t) \alpha_2(t) = \frac{2i(\gamma_1\gamma_2)^{\frac{1}{2}} \exp[(i\omega_1 - \gamma_1)(t_2 - t_1 + \delta\tau)]}{\omega_1 - \omega_2 + i(\gamma_1 + \gamma_2)} \tag{38}$$

where $\delta\tau = (x_2 - x_1)/c$

This expression applies when the atomic excitation times and distances from the bs are such that $t_1 < t_2 + \delta\tau$. When the reverse inequality is satisfied subscripts 1 and 2 should be interchanged. For photons incident in the same arm $x_1 = x_2$ so that $\delta\tau = 0$. Thus for the special case of two identical atoms in the source, the field normalization constant is

$$N = \{1 + \exp(-2\gamma|t_1 - t_2|)\}^{-\frac{1}{2}} \tag{39}$$

where 2γ is the common decay rate.

A two-photon excitation with one photon in each of the two input arms is correspondingly represented by the state

$$|1_1, 1_2\rangle = A_1^\dagger(\alpha_1) A_2^\dagger(\alpha_2) |0_1, 0_2\rangle \tag{40}$$

The state is already normalized on account of the independence of the operators for the two input arms and the normalization property of the single-photon states (24).

In order to work out the photocount expectations analogous to the previous section we use the photodetection theory originally developed by Glauber (1963). The mean photon number and the mean square photon number expectation values are given by

$$\langle m_i(T) \rangle = \eta_i \int_0^T dt \langle a_i^\dagger(t) a_i(t) \rangle \tag{41}$$

and

$$\langle [m_i(T)]^2 \rangle = \langle m_i(T) \rangle + \eta_i^2 \int_0^T dt \int_0^T dt' \langle a_i^\dagger(t) a_i^\dagger(t') a_i(t) a_i(t') \rangle \tag{42}$$

where the angled brackets denote averages over counting periods T, η_i are the detector efficiencies and i=3 or 4 for the output arms.

The correlation between the two outputs is given by

$$\langle m_3(T) m_4(T) \rangle = \eta_3 \eta_4 \int_0^T dt \int_0^T dt' \langle a_3^\dagger(t) a_4^\dagger(t') a_3(t) a_4(t') \rangle \tag{43}$$

The output annihilation operators are related to the input operators by matrix Eq. (10). We may now consider two specific cases (1) the photon pair enters the same input arm of the bs and (2) the photons enter separate arms.

(1) Both incident photons enter arm 1.

The annihilation operator in output 3 can be written as

$$a_3(t) = Nr(a_1(\alpha_1, t) + a_1(\alpha_2, t)) + ta_2(t) \tag{44}$$

where $a_2(t)$ represents the vacuum mode.

Using Eq. (32) and restricting the arrival of both photons to the photocount period the mean number of photons detected in output 3 becomes

$$\langle m_3(T)\rangle = \eta_3|r|^2N^2\int_0^T dt|\alpha_1(t) + \alpha_2(t)|^2 \tag{45}$$

From Eqs. (34) and (37) we see that the integral is equal to $2N^2$ hence we may write,

$$\langle m_3(T)\rangle = 2\eta_3|r|^2 \tag{46}$$

similarly

$$\langle m_4(T)\rangle = 2\eta_4|t|^2 \tag{47}$$

The second order moments and correlation between the outputs are evaluated in a similar fashion. The results are

$$\langle m_3(T)(m_3(T)-1)\rangle = 2\eta_3^2|r|^4 = 2\eta_3^2P_{2,0} \tag{48}$$

$$\langle m_4(T)(m_4(T)-1)\rangle = 2\eta_4^2|t|^4 = 2\eta_4^2P_{0,2} \tag{49}$$

$$\langle m_3(T)m_4(T)\rangle = \eta_3\eta_4 2|r|^2|t|^2 = \eta_3\eta_4P_{1,1} \tag{50}$$

The above correlation functions have been written in terms of the output probabilities P_{n_3,n_4} defined previously. It should be noted that the temporal overlap between α_1 and α_2 does not effect the results due to a similar overlap dependence of the normalization constant N. Lange et al. (1988) have shown both theoretically and experimentally that a similar cancellation occurs for spatial overlap. The time dependence of the wavefunctions therefore does not contribute to the results in this case and the probabilities above correspond to the time independent values of the last section.

(2) One incident photon in each arm.

The mean number of photons in output 3 in this case becomes

$$\langle m_3(T)\rangle = \eta_3\int_0^T dt\{|r|^2|\alpha_1(t)|^2 + |t|^2|\alpha_2(t)|^2\} \tag{51}$$

Using Eqs. (8) and (34) we find

$$\langle m_3(T)\rangle = \eta_3 \tag{52}$$

Similarly, the mean number of the photons in output 4 becomes

$$\langle m_4(T)\rangle = \eta_4 \tag{53}$$

The second factorial moments and correlation between detectors are given by

$$\langle m_i(T)(m_i(T)-1)\rangle = 2\eta_i^2|r|^2|t|^2(1 + |\int_0^T dt\alpha_1^*(t)\alpha_2(t)|^2) \tag{54}$$

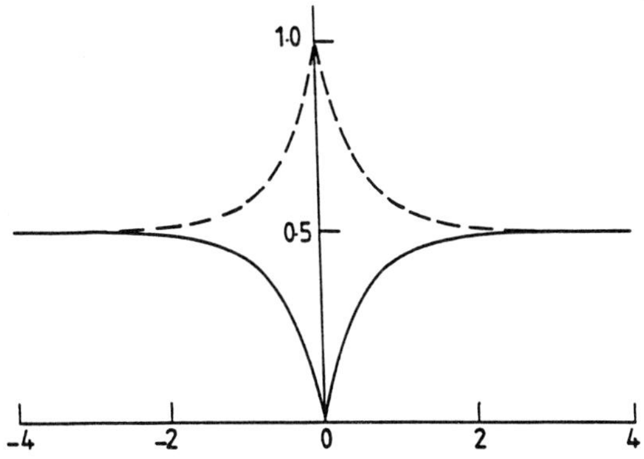

$\gamma(t_2 - t_1 + \delta\tau)$

Fig. 3. The photocount second factorial moment $\langle m_i(m_i - 1)\rangle/\eta_i^2$ (broken curve) and the correlation $\langle m_3 m_4\rangle/(\eta_3\eta_4)$(continuous curve) for excitation of each input arm by identical atoms.

where i=3,4

$$\langle m_3(T)m_4(T)\rangle = \eta_3\eta_4\left(|r|^4 + |t|^4 - 2|r|^2|t|^2\left|\int_0^T dt\alpha_1^*(t)\alpha_2(t)\right|^2\right) \tag{55}$$

where the overlap integral is given by Eq. (38).

For identical simultaneous excitations, $\alpha_1(t) = \alpha_2(t)$ and the overlap integral is unity which gives

$$P_{1,1} = (|r|^2 - |t|^2)^2 \tag{56}$$

$$P_{2,0} = 2|r|^2|t|^2 = P_{0,2} \tag{57}$$

in agreement with table I. [Feynman 1965].

If there is no overlap between the photon states, the overlap integral is zero and we obtain

$$P_{1,1} = |r|^4 + |t|^4 \tag{58}$$

$$P_{2,0} = |r|^2|t|^2 = P_{0,2} \tag{59}$$

which is the classical distinguishable particle result [Hong,Ou and Mandel 1987].

For the special case of $\omega_1 = \omega_2$, $\gamma_1 = \gamma_2\gamma$ and a 50:50 bs ($|r|^2 = |t|^2 = 1/2$) the correlation and second factorial moments can easily be plotted against the time delay $\gamma(t_2 - t_1 + \delta\tau)$ see Fig. 3. It is seen that the correlation (55) vanishes in this case, when the atoms are excited at times t_1 and t_2 such that $t_1 = t_2 + \delta\tau$. The phenomenon is a form of quantum mechanical interference effect that arises from the Bose-Einstein commutation properties of the photon creation and annihilation operators. The width of the normalized correlation line graph plotted against time delay is related to the linewidth of the individual photons. The lineshape if turned upside down, is approximately Lorentzian which is the expected frequency distribution for the photons emitted from atoms in spontaneous decay events [see Eq. (28)].

In practice the experiments are performed using continuous sources in which the light is emitted as a stream of correlated photon pairs. For the next two light sources, we shall treat only the input state in which photons enter separate arms.

Two-Photon Interference with light from a parametric down-conversion process

The output properties of this kind of parametric oscillator have been summarized by Collett and Loudon (1987) and we shall use their notation and a generalization of their results in order to calculate the second order moments and the correlation between the detectors.

We shall consider a non-degenerate parametric oscillator, driven at a rate denoted by $\xi/2$ located within a single ended cavity with transmission rate $\gamma/2$. The oscillator emits two output beams, whose frequency spectra are centered on ω_1 and ω_2 directed respectively into inputs 1 and 2 of the bs. The output beams have equal steady fluxes, where the flux is defined as

$$f_i = \langle a_i^\dagger(t) a_i(t) \rangle$$

and

$$f_1 = f_2 = f = \frac{1}{2} \frac{\xi^2 \gamma}{(\gamma^2/4 - \xi^2)} \tag{60}$$

[Collett and Loudon 1987].

The mean photocounts at each detector are evaluated using Eq. (41) and are given by

$$\langle m_3(T) \rangle = \eta_3 f T \tag{61}$$

and

$$\langle m_4(T) \rangle = \eta_4 f T \tag{62}$$

These results can be compared with (52) and (53) where the single photon is replaced by the photon flux in time T.

At this point the reader interested in a detailed analysis of the calculation is referred to the paper by Fearn and Loudon (1989). For the purposes of this more tutorial write up, we shall skip over the details and examine the results. The second order moments and the correlation term are found to be

$$\begin{aligned} \langle m_i(T)(m_i(T) - 1) \rangle = \eta_i^2 f T \Bigg\{ & f T + 2|r|^2 |t|^2 \\ & + \frac{2|r|^2 |t|^2 \gamma^2 \exp(-\gamma|\delta\tau|)}{\gamma^2 + (\omega_1 - \omega_2)^2} \\ & \times \left(\cos[(\omega_1 - \omega_2)|\delta\tau|] + \frac{\gamma \sin[(\omega_1 - \omega_2)|\delta\tau|]}{(\omega_1 - \omega_2)} \right) \Bigg\} \end{aligned} \tag{63}$$

for i=3 or 4 and

$$\begin{aligned} \langle m_3(T) m_4(T) \rangle = \eta_3 \eta_4 f T \Bigg\{ & f T + |r|^4 + |t|^4 \\ & - \frac{2|r|^2 |t|^2 \gamma^2 \exp(-\gamma|\delta\tau|)}{\gamma^2 + (\omega_1 - \omega_2)^2} \\ & \times \left(\cos[(\omega_1 - \omega_2)|\delta\tau|] + \frac{\gamma \sin[(\omega_1 - \omega_2)|\delta\tau|]}{\omega_1 - \omega_2} \right) \Bigg\} \end{aligned} \tag{64}$$

These results apply to the limit of long integration time consistent with Eq. (22). They show similar two-photon interference effects to those inherent in the corresponding single atom results (54) and (55), for a single photon in each input. The dependence of the relative timing of the two photons has now been replaced by

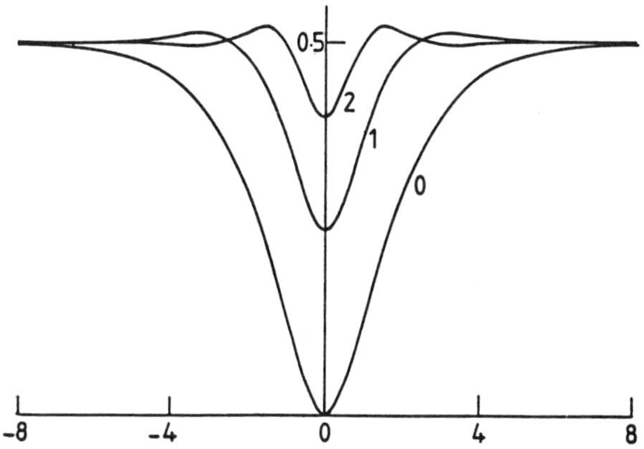

Fig. 4. The contribution to the scaled photocount correlation $\langle m_3 m_4 \rangle / (\eta_3 \eta_4 fT)$ given by Eq. (64) that is independent of T for values of $(\omega_1 - \omega_2)/\gamma$ shown against the curves.

a more complicated function. This has the effect of rounding the sharp feature in Fig. 3 for simultaneous photon arrivals at the bs. This rounding can be attributed to the line broadening within the source. Figure 4 shows some representative curves of the photocount correlation for a 50:50 bs and for different values of $\omega_1 - \omega_2$, where it has been assumed that $fT \ll 1$ and only the term linear in fT in Eq. (64) has been retained. For $(\omega_1 - \omega_2)/\gamma = 2$, we notice the onset of oscillation or ringing which is a manifestation of quantum beats.

It is seen in Fig. 4 that maximum interference should be observed for $\omega_1 = \omega_2$. One should note that the one-dimensional theory used here is not strictly valid for the degenerate case because of the practical difficulties in separating the two photons of a pair to route them into separate arms of the bs. Futhermore, the quadratic terms f^2T^2, which represent the contributions of photons from different parametric events, tend to obscure the two-photon interference effect. The interference effect is most clearly observable for conditions such that $fT \ll 1$.

Two-Photon Interference with light from Atomic Cascade Emission

A different source of light with strong two-photon correlations is provided by atomic cascade emission. The correlations were measured a long time ago by Clauser (1974). Again, the reader interested in all the details is referred to Fearn and Loudon (1989), only a brief outline of the calculation is given below.

We consider a light source consisting of identical atoms with a three level structure. Each atom is assumed to be incoherently excited from its ground state $|1\rangle$ to an upper level $|3\rangle$ at a steady rate R. The atom returns to its ground state via an intermediate level $|2\rangle$ emitting photons of frequency ω_1 and ω_2 in cascade sequence, see Fig. 5.

The field correlation functions for the emitted light, which are needed to evaluate the outcome of a two-photon interference experiment, are determined by the corresponding functions of the atomic transition operators. It is a simple matter to solve the equations of motion of the atomic density matrix, using a rate equation approach, and hence determine the time dependence of the low order expectation values of the transition operators. More complicated multi-time expectation values can then be calculated by using the quantum regression theorem [Loudon 1973] It is assumed that the atomic excitation mechanism has been turned on long enough to reach steady state conditions. It is important to note that the transition operators do not commute and the required fourth order correlation functions depend upon their

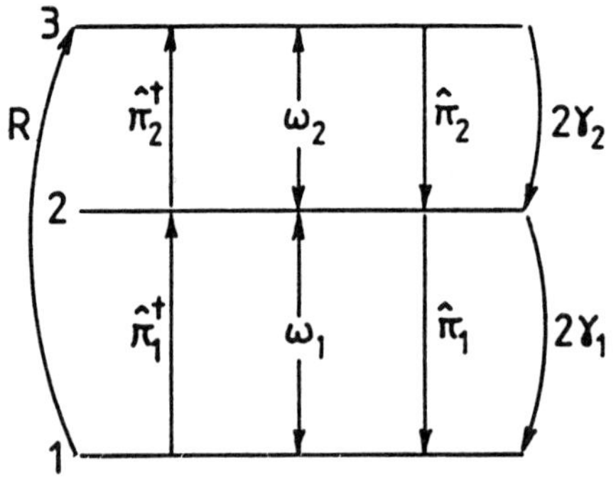

Fig. 5. Notation for atomic energy levels and transition operators used in cascade emission theory.

ordering. It is a fairly lengthy procedure to calculate these time ordered correlations and the reader is spared the details, [Fearn 1989]

We now consider two-photon interference with a cascade emission light source in which the light of frequency ω_1 enters the bs via arm 1 and the light of frequency ω_2 enters via arm 2. The electric field operators are related to atomic transition operators by the source field expression [Loudon 1973]

$$E_1^\dagger(x,t) = -\frac{e\omega_1^2 D_{12}\sin\alpha_1}{4\pi\epsilon_0 c^2 x_1}\pi_1(t_1) \tag{65}$$

where α_1 is the angle between the transition dipole moment and the observation direction. The field matrix element (27) is obtained by a suitable application of this operator relation. A similar expression relates the field operator $E_2^\dagger$ for the light of frequency ω_2 to the transition operator π_2.

The photon flux through a small section of the spherical wavefront of area A_1 is

$$f_1 = \frac{RA_1}{4\pi x_1^2} \tag{66}$$

where the spherical wavefront has the surface area $4\pi x_1^2$ at a distance x_1 from the atom and the flux is proportional to the rate of atomic excitation R as expected. A similar expression applies for the light of frequency ω_2. If the two beams are collected from equal solid angles we can set

$$f_1 = f_2 = f \tag{67}$$

The mean photon number at detectors 3 or 4 respectively become

$$\langle m_3(T)\rangle = \eta_3 fT \tag{68}$$

$$\langle m_4(T)\rangle = \eta_4 fT \tag{69}$$

The correlation and second factorial moments at the two detectors can now be obtained with further use of the photodetection theory outlined earlier and conversion of expectation values of field operators to expectation values of atomic operators according to the prescriptions

$$a_i^\dagger a_i \rightarrow \frac{2\gamma_i f}{R}\pi_i^\dagger \pi_i \tag{70}$$

where i=1,2.

Analogous to the parametric oscillator we assume a small driving rate with $R \ll \gamma_1$ and γ_2 so that the effects of the photon pairs emitted in the same cascade emission are enhanced. The second moments and correlation are given by

$$\langle m_i(T)(m_i(T)-1)\rangle = \eta_i^2 fT\left\{ fT + 2|r|^2|t|^2 \frac{fT}{R}[1+2\gamma_1 \exp(-2\gamma_1 \delta\tau)\Lambda\theta(\delta\tau)]\right\} \tag{71}$$

for i=3,4 and

$$\langle m_3(T)m_4(T)\rangle = \eta_3\eta_4 fT\left\{ fT + \frac{f}{R}\left[|r|^4+|t|^4 - 8|r|^2|t|^2\gamma_1 \exp(-2\gamma_1\delta\tau)\Lambda\theta(\delta\tau)\right]\right\} \tag{72}$$

where

$$\Lambda = \frac{2\gamma_1+\gamma_2}{(\omega_1-\omega_2)^2+(2\gamma_1+\gamma_2)^2} - \frac{\exp[-(2\gamma_1+\gamma_2)\delta\tau](2\gamma_1+\gamma_2)\cos[(\omega_1-\omega_2)\delta\tau]}{(\omega_1-\omega_2)^2+(2\gamma_1+\gamma_2)^2} + \frac{\exp[-(2\gamma_1+\gamma_2)\delta\tau](\omega_1-\omega_2)\sin[(\omega_1-\omega_2)\delta\tau]}{(\omega_1-\omega_2)^2+(2\gamma_1+\gamma_2)^2} \tag{73}$$

These results are similar in form to the parametric light source. However, these terms are *not* symmetric in $\delta\tau$. The results show independent particle behavior when $\delta\tau$ is negative. A negative $\delta\tau$ corresponds to an input arm 1 that has a longer path length to the bs than the input in arm 2 ($x_1 > x_2$). Thus the photon of frequency ω_1, emitted second in the cascade sequence, is delayed further by its additional path length and there is no likelihood of both photons arriving simultaneously at the bs.

Two-photon interference effects do however occur for positive $\delta\tau$, where the photon of frequency ω_1 has a shorter distance to travel so that both photons may arrive simultaneously at the bs. The negative contribution in Eq. (72) is maximized for $\omega_1 \simeq \omega_2$ and we illustrate the interference by considering the case $\omega_1 = \omega_2$ where (72) reduces to

$$\langle m_3(T)m_4(T)\rangle = \eta_3\eta_4 fT\left\{ fT + \frac{f}{R}\left[|r|^4+|t|^4 - \frac{8|r|^2|t|^2}{(2+\gamma_1/\gamma_2)}\exp(-2\gamma_1\delta\tau) \times \{1-\exp[-(2+\gamma_2/\gamma_1)\gamma_1\delta\tau]\}\theta(\delta\tau)\right]\right\} \tag{74}$$

Figure 6 shows the part of this correlation function which is linear in T for a 50:50 bs and for two relative values of the spontaneous emission rates.

It is seen that, in contrast to the results found for the two-atom light source and the parametric oscillator, as represented by Figs. 3 and 4 respectively, there is

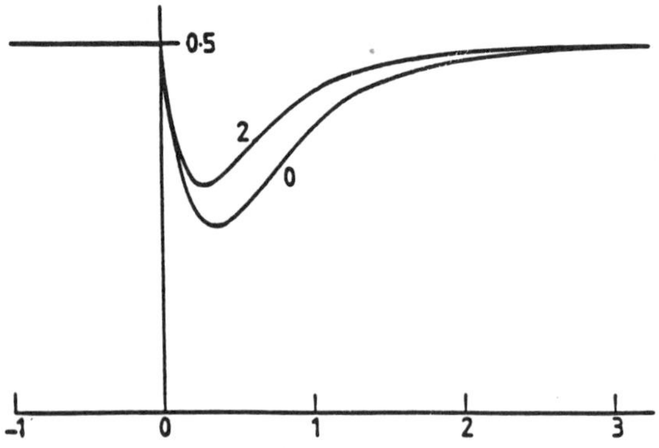

Fig. 6. The contribution to the scaled photocount correlation $R\langle m_3 m_4\rangle/(\eta_3\eta_4 f^2 T)$ given by Eq. (74) that is independent of T for values of γ_2/γ_1 shown against the curves.

no value of the relative time delay $\delta\tau$ for which the photocount correlation vanishes. The minimum value of the normalized correlation in Fig. 6 is 0.25 and it occurs at

$$\gamma_1 \delta\tau = \frac{1}{2} \ln 2 \tag{75}$$

which is found by minimizing Eq. (74) with respect to $\gamma_1\delta\tau$ in the limit $\gamma_1 \gg \gamma_2$. This limit corresponds physically to a cascade sequence in which the emission of photon ω_1 follows as rapidly as possible after the emission of photon ω_2. There are no values of the parameters for a cascade emission source such that identical and simultaneous photon states can be produced in the two input arms. This limits the amount of output interference that can be observed with a cascade emission source.

CONCLUSIONS

We have seen that for arrangements in which the photons in a pair enter the bs via two different inputs the output states for which one photon is detected in each arm, are strongly inhibited for simultaneous arrival of the input photons. This is clearly displayed by the photocount correlation which may vanish for certain values of the parameters for input light from two atom and parametric oscillator sources, and can be reduced to one half of its independent-particle value for light from a cascade emission source.

Calculations have assumed photodetection integration times much longer than any correlation times that characterize the input light. This analysis is therefore incapable of resolving any intrinsic localization properties of the photons.

Two-photon interference provides an example of non-classical effects in the statistical properties of light. It illustrates a text book property [Feynman et al. 1965] of particles that obey Bose-Einstein statistics, and it clearly shows the continuous transition from boson to independent particle behavior as the temporal[oral overlap of the excitations is removed by steady increase in their relative time delay.

ACKNOWLEDGMENTS

HF would like to thank Prof. M. O. Scully for the opportunity to work at the Max-Planck institute and at U.N.M. The author also wishes to thank Prof. A. Barut for the invitation to present this work at the N.A.T.O Summer School, and Prof. R. Loudon for collaboration in the original calculations. Finally, HF thanks B. Atherton for proof reading the final draft of the manuscript. This work was partially supported by the Office of Naval Research.

REFERENCES

Amrein, W. O., Helv. Phys. Acta **42**, 149 (1969).

Brendle, J., Schütrumpf, S., Lange, R., Marteinssen, W. and Scully, M. O., Europhys. Letts. **5**, 223 (1988).

Burnham, D. C. and Weinberg, D. L., Phys. Rev. Letts. **25**, 84 (1970).

Clauser, J. F., Phys. Rev. **D9**, 853 (1974).

Collett, M. J. and Loudon, R., J. Opt. Soc. Am **B4**, 1525 (1987).

Dirac, P. A. M., *The Principles of Quantum Mechanics*, Clarendon Press, Oxford, p. 9, (1958).

Einstein, A., Annl. Phys. **17**, 132 (1905); English translation by A. B. Arons and M. B. Peppard, Am. J. Phys. **33**, 367 (1965).

Fearn, H., *Ph.D Thesis Essex University (1989)* Wivenhoe Park, Colchester, Essex, England.

Fearn, H. and Loudon, R., Optics Commun. **64**, 485 (1987); and J. Opt. Soc. Am. **B6**, 917 (1989).

Feynman, R. P., Leighton, R. B. and Sands, M., *The Feynman Lecture on Physics*, Addison-Wesley, Reading (1965), Vol. 3, Section 4-2.

Friberg, S, Hong, C. K and Mandel, L., Phys. Rev. Letts. **54**, 2011 (1985).

Frisch, O. R., Contemp. Phys. **45**, (1965).

Glauber, R. J., Phys. Rev. **131**, 2766 (1963).

Hong, C. K, Ou, Z. Y and Mandel, L., Phys. Rev. Letts. **59**, 2044 (1987).

Jauch, J. M and Piron, C., Helv. Phys. Acta **40**, 559 (1967).

Lange, R, Brendel, J, Mohler, E and Marteinssen, W., Europhys. Letts **5**, 619 (1988).

Loudon, R., *The quantum theory of light*, Clarendon Press, Oxford, (1973).

Newton, T. D. and Wigner, E. P., Rev. Mod. Phys **21**, 400 (1949).

Mandel, L., Phys. Rev. **144**, 1071 (1966).

Pfleegor, R. L. and Mandel, L., Phys. Rev **159**, 1084 (1967).

Pike, E. R. and Sarkar, S., " Photons and Interference" in *Frontiers in Quantum Optics*, Adam Hilger, Bristol, (1986).

Planck, M., Verh. dt. Phys. Ges. **2**, 202 + 237 (1900).

Prasad, S., Scully, M. O. and Marteinssen, W., Optics Commun. **62**, 139 (1987).

Rarity, J. G. and Tapster, P. R., "Nonclassical effects in parametric downconversion" in *Photon localisation, detection and amplification and antibunching*, editors E. R. Pike and H. Walther, Adam Hilger pp122-150 (1988).

Richter, G., Brunner, W. and Paul, H., Ann. Phys. (N. Y.) **14**, 239 (1964) and Ann. Phys. (N. Y.) **17**, 262 (1966).

Taylor, G. I., Proc. Cam. Phil. Soc **15**, 114 (1909).

SQUEEZED LIGHT

Elisabeth Giacobino

Laboratoire de Spectroscopie Hertzienne de l'ENS, Université P. et M. Curie,75252 PARIS Cedex 05, FRANCE

1. INTRODUCTION

The accuracy in any optical measurement is limited by the noise, e.g. the random fluctuations of the light beams which are monitored. Various techniques have been successfully implemented to reduce the "technical noise", or "instrumental noise", but they are ineffective in decreasing the basic noise, inherent to the nature of light, the quantum noise. As far as intensity measurements are concerned, this noise can be viewed as the result of the random fluctuations of the photons in the beam (shot noise).

The suppression of instrumental causes of noise has gone so far in fundamental as well as in technical applications that the shot noise floor is actually encountered in many situations. Moreover, investigations in the field of quantum optics, initiated a decade ago, have shown that, if quantum noise cannot be suppressed, its presence can be circumvented.

The quantum fluctuations of light are determined by the properties of the electric field operator. A monochromatic light wave with frequency ω is characterized by the electric field:

$$E(t) = E_1 \cos \omega t + E_2 \sin \omega t \tag{1}$$

where E_1 and E_2 are operators which do not commute, since they are proportional to linear combinations of the creation and annihilation operators a^+ and a

$$E_1 \simeq a + a^+$$

$$E_2 \simeq i\,(a - a^+)$$

As a result, the product of the fluctuations of E_1 and E_2 obey a Heisenberg inequality:

$$\Delta E_1 . \Delta E_2 \geqslant E_0^{\,2} \tag{2}$$

where E_0 is a constant. The ground state of the field, or vacuum state is associated with a field with zero mean value and fluctuations $\Delta E_1 = \Delta E_2 = E_0$. Other fields also have minimum fluctuations on both quadratures and non zero mean values for E_1 and

New Frontiers in Quantum Electrodynamics and Quantum Optics
Edited by A. O. Barut, Plenum Press, New York, 1990

E_2. They are the coherent states introduced by Glauber[1](Fig.1). They are the best approximation to a classical field.

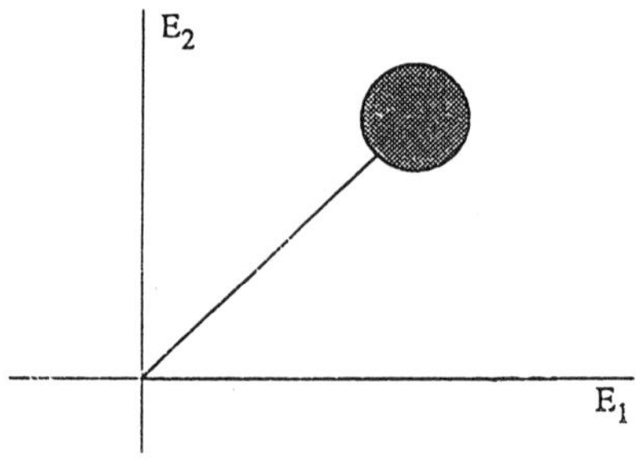

Figure 1. Coherent field.

In most light sources, there is no preferred phase for the quantum fluctuations and $\Delta E_1 = \Delta E_2$. But it has been proposed to generate "squeezed" states of the light in which the fluctuations in one quadrature are reduced with respect to the standard quantum limit E_0, at the expense of increased fluctuations in the other one[2]. Alternatively, one can consider reducing the amplitude fluctuations; the phase fluctuations are then increased (Fig 2).

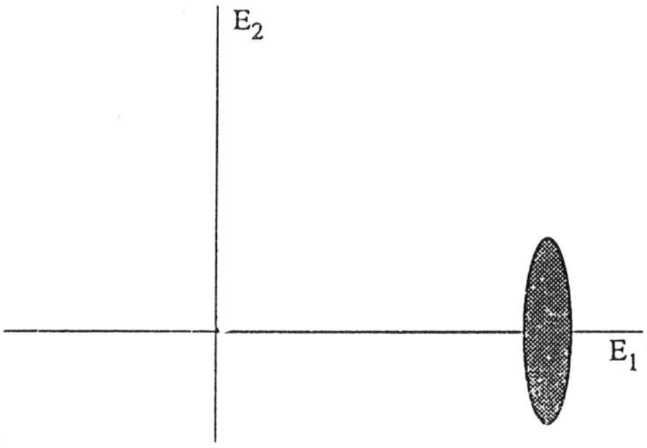

Figure 2. Squeezed field (squeezing is in the amplitude quadrature)

The concept of squeezing can be extended to other quantities involving several light beams. Among those, twin beams are of particular interest[3]. Twin beams ideally have perfectly correlated intensities: the noise in the difference between their intensities is zero. The conjugate quantity, which is the phase difference is then expected to have large quantum fluctuations.

2. PROCESSES INVOLVED IN SQUEEZING

For a few years, several schemes have been used to generate squeezed light. They can be broken up into two main categories. On the one hand there are all-optical processes, involving nonlinear optical interaction, such as four-wave mixing[4-5-6], three-wave mixing (parametric down-conversion[7-8-9-10] or second harmonic generation[11]), or optical bistability[12] (Fig.3).

On the other hand, one can "regulate" the photon flow of a laser by reducing the fluctuations of the pumping process[13]. This produces amplitude squeezed states. In the following we will concentrate on the all-optical systems.

Figure 3. Three-wave mixing ($\chi^{(2)}$ process): one photon is annihilated, two photons are created (a); four-wave mixing ($\chi^{(3)}$ process): two photons are annihilated, two photons are created (b).

Several processes rely on the creation of two signal photons while one or two pump photons are annihilated. When the process is "degenerate", the two photons are created in the same mode of the field, and squeezing of the emitted field occurs in some quadrature component. When the process is non degenerate, the signal photons are created in different modes, and the squeezed quantity is the difference of the intensities of the two fields: the systems generates twin beams.

Parametric conversion using a nonlinear $\chi^{(2)}$ crystal has been the most successful process to date to generate squeezed light. In this paper, we will give some more details about the expriments involving parametric generation in a resonant cavity, which have produced the largest quantum noise reductions.

3. SQUEEZED LIGHT DETECTION WITH A BEAMSPLITTER

To show evidence for the squeezing, the (squeezed) signal field is combined with a strong laser field at the same frequency (which acts as a local oscillator) in a balanced homodyne detector (Fig. 4). The balanced homodyne detector is made of a 50-50 beamsplitter and of a symmetrical detection.

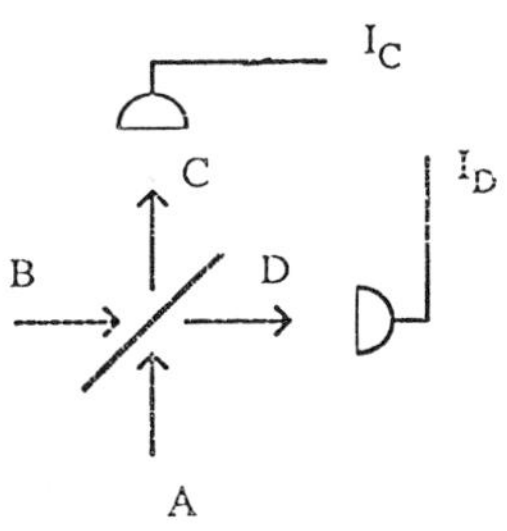

Figure 4. Balanced homodyne detector.

The signal E_B field and the local oscillator E_A, which is assumed to be much stronger than the signal, are combined on the beamsplitter. The outgoing fields are proportional to E_C and E_D

$$E_C = E_A + E_B, \qquad E_D = E_A - E_B \tag{3}$$

The minus sign in E_D arises from a π phase shift between the fields reflected off the two sides of the beamsplitter; one is reflected off the air-glass interface, the other is reflected off the glass-air interface.

If we now consider that E_B is the component of the signal in phase with the local oscillator, we have:

$$E_C^2 - E_D^2 = 4E_A.E_B$$

By subtracting the intensities in both arms, only the interference terms survive, and the fluctuation of the difference intensity is:

$$\Delta(I_C - I_D) = \sqrt{I_A} <\Delta E_B> \tag{4}$$

where $<\Delta E_B>$ are the mean fluctuations of the field operator E_B. If E_B is in a coherent state or in the vacuum state, $<\Delta E_B> = E_0$, and the balanced detector sees a noise which is the standard quantum noise, or shot noise of a beam with total intensity I_A, proportional to $\sqrt{I_A}$. This property is independent of the actual noise of the local oscillator beam, which is eliminated in the difference process.

If the signal beam is in a squeezed state, the fluctuations in the difference signal can become smaller or larger than the shot noise , depending on which component E_1 or E_2 is squeezed.

4. SQUEEZED LIGHT GENERATION IN A DEGENERATE OPTICAL PARAMETRIC OSCILLATOR

In Ref[7], a degenerate parametric oscillator (OPO) made of a $LiNbO_3$ nonlinear $\chi^{(2)}$ crystal in an optical cavity and pumped with a single mode doubled Nd:YAG laser is operated below threshold (Fig 5).

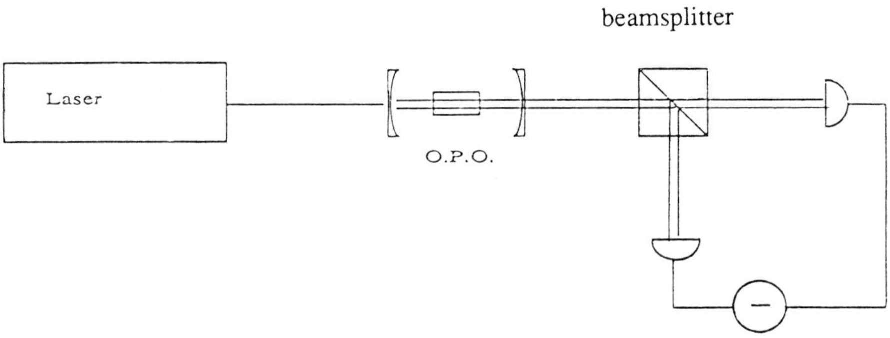

Figure 5. Typical experimental set-up.

The occurence of squeezing in this case can be understood in terms of a semiclassical transformation by the parametric oscillator of the vacuum fluctuations[14] entering the cavity. The equation for the signal field α inside the optical cavity is:

$$\tau\dot{\alpha} = -\gamma\alpha + g\alpha^* + t\alpha^{in} \qquad (5)$$

where τ is the cavity roundtrip time, γ is the damping coefficient of the field in the cavity; γ is related to the amplitude transmision coefficient t of the output mirror by $\gamma = t^2/2$ (t is assumed to be small); g is the parametric gain, and α^{in} is the vacuum field entering the cavity through the output mirror.

Writing Eq. 5 and its complex conjugate in the frequency domain and using the phase quadratures $\alpha_1 = \alpha + \alpha^*$ and $\alpha_2 = \alpha - \alpha^*$ (with similar notations for α_1^{in} and α_2^{in}), one gets:

$$\alpha_1 = \frac{t\,\alpha_1^{in}}{\gamma - g + i\Omega} \qquad \alpha_2 = \frac{t\,\alpha_2^{in}}{\gamma + g + i\Omega} \qquad (6)$$

The outgoing field α^{out} is related to the field inside the cavity and to the incoming vacuum field by:

$$\alpha^{out} = t\alpha - \alpha^{in} \qquad (7)$$

where the reflection coefficient r has been approximated by one. Using Eqs. 6 and 7, one gets:

$$\alpha_1^{out} = \frac{\gamma + g - i\Omega}{\gamma - g + i\Omega}\,\alpha_1^{in} \qquad (8)$$

$$\alpha_2^{out} = \frac{\gamma - g - i\Omega}{\gamma + g + i\Omega}\,\alpha_2^{in} \qquad (9)$$

Equations 8 and 9 clearly show that the vacuum field is amplified in quadrature α_1, while quadrature α_2 is squeezed and even goes to zero for $\Omega = 0$ and $\gamma = g$ that is for zero noise frequency and when threshold is approached. The vacuum field is squeezed. The noise reduction obtained in Ref.[7] is 63% .

5. TWO-FIELD SQUEEZING IN A NON DEGENERATE PARAMETRIC OSCILLATOR

Squeezing also occurs in a parametric oscillator operated above threshold in the non degenerate regime[9]. The nonlinear medium is a KTP crystal pumped by an Ar^+ laser. When the pumping is strong enough, the system operates above threshold and emits two laser-like beams, called signal and idler beams , which have orthogonal polarizations They are separated at the output of the optical cavity with a polarizing beamsplitter (Fig.6).

It is expected that the two beams exiting the cavity are strongly correlated. The photons are produced by pairs (see Fig. 3) one in each beam, so even though each beam may have significant flutuations, these flutuations are identical. Actually, this is only true in the absence of an optical cavity, for the optical cavity introduces some decorrelation between the twin beams. This can be understood with a simple qualitative argument.

The twin photons are emitted at the same time, but they may make different numbers of roundtrips inside the cavity before going out. To detect the two photons of a pair, one has to wait for a time which is of the order of the cavity storage time. As a result, the fluctuations of the two beams are well correlated on a time scale long compared to the cavity storage time, badly correlated on shorter times. In the frequency domain, the fluctuations are well correlated if their frequencies are smaller than the inverse of the cavity storage time, badly correlated for larger frequencies.

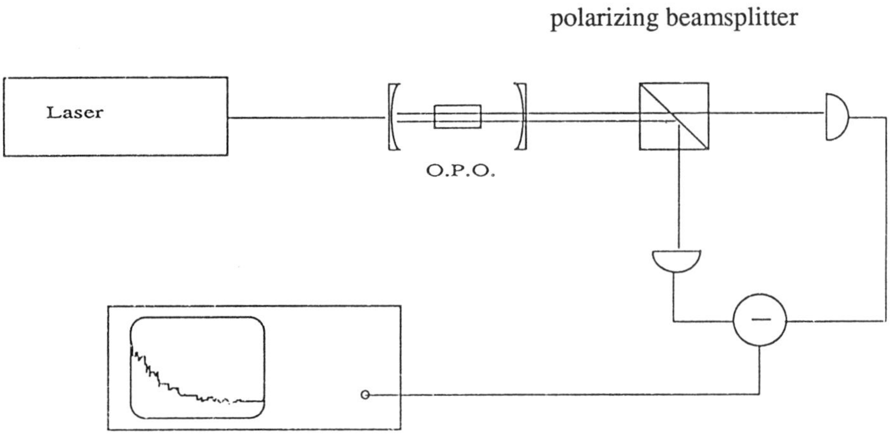

Figure 6. Twin beam generation in a non degenerate parametric oscillator.

The intensities of the two beams are measured and subtracted. The resulting current reflects the fluctuations on the intensity difference. If the beams were perfectly correlated, there would be no noise on the intensity difference. If they are not correlated at all, and in the absence of classical noise, the noise of the intensity difference is expected to be the same as the noise of a single beam separated in two parts by a beamsplitter (see Eq. 4); it is the shot noise of a beam which would have an intensity equal to the sum of the intensities of the two beams.

The presence of correlation between the twin beams is thus linked to a noise level in the intensity difference below the shot noise, that is to squeezing in the intensity difference. To check for this property, the signal corresponding to the difference between the currrents of the two photodiodes is fed into a spectrum analyser. The noise in the intensity difference is found to be squeezed on a frequency range of the order of the inverse of the cavity storage time (Fig.7).

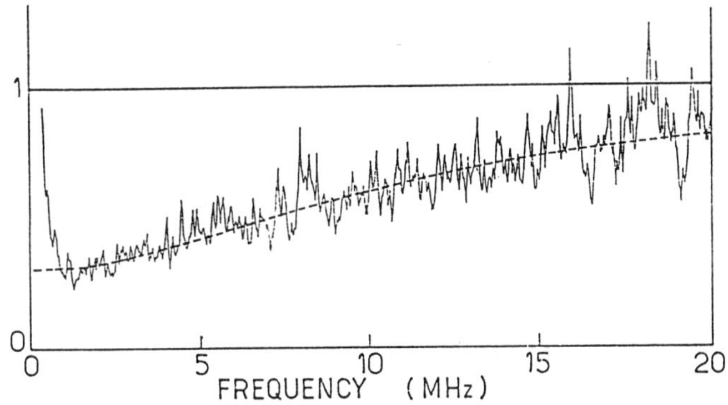

Figure 7. Squeezing spectrum in the intensity difference of the twin beams emitted by a non degenerate parametric oscillator

The level of the sandard quantum noise has been determined using the properties of the 50-50 beamsplitter. To do so, a half-wave plate is placed before the polarizing beamsplitter which separates the twin beams, with its axes in such a position that the polarization of each beam is rotated by 45°. Then the polarizing beamsplitter acts like a 50-50 beamsplitter for each beam. If each beam was alone, the noise detected in the difference of the intensities of the two channels would be its shot noise. Since the beams have different wavelengths, they don't interfere, and everything is as if each of them was alone. The corresponding noises simply add their powers, thus very directly yielding the shot noise of a beam which would have the same total intensity.

The semiclassical method used for the OPO below threshold can be extended to the derivation of the above threshold case. The equations of motion for the classical amplitudes α_1 and α_2 of the signal and idler fiels, and α_0 of the pump field inside the cavity obey the following equations

$$\tau \dot{\alpha}_1 = -\gamma \alpha_1 + 2\chi \alpha_2^* \alpha_0 + t\, \alpha_1^{in} \tag{10}$$

$$\tau \dot{\alpha}_2 = -\gamma \alpha_2 + 2\chi \alpha_1^* \alpha_0 + t\, \alpha_2^{in} \tag{11}$$

$$\tau \dot{\alpha}_0 = -\gamma_0 \alpha_2 - 2\chi \alpha_1 \alpha_2 + t_0\, \alpha_0^{in} \tag{12}$$

In contrast to Eq. 5, these equations take into account the saturation of the pump field (Eq. 12), which is necessary if we want to consider above threshold operation.

In these equations, γ is the cavity damping coefficient for the signal and idler fields, γ_0 is the cavity damping coefficient for the pump field; χ is the nonlinear coefficient of the crystal. α_1^{in} and α_2^{in} are the vacuum fluctuations entering the cavity through the coupling mirror, and α_0^{in} is the incoming pump field, assumed to be a coherent field, with a non zero mean value, and fluctuations equal to the vacuum fluctuations. t and t_0 are the transmission coefficients of the coupling mirror for the signal and pump fields.

The steady state for the mean values $\overline{\alpha}_1$ and $\overline{\alpha}_2$ is determined by solving the equations with $\overline{\alpha_1^{in}}$ and $\overline{\alpha_2^{in}}$ equal to zero. If the pump intensity $(\overline{\alpha_0^{in}})^2$ is large enough, there is a non zero solution:

$$\bar{\alpha}_1^{\,2} = \bar{\alpha}_2^{\,2} = (\gamma_0\gamma/4\chi^2)\,(\sigma-1)$$

$$\bar{\alpha}_0 = \gamma/2\chi$$

where σ, the pump parameter, is given by

$$\sigma = 8(\chi^2/\gamma_0\gamma^2/4\chi^2)\,(\overline{\alpha_0^{in}})^2$$

Eqs. 10-12 are then linearized around this steady state solution, to find the fluctuations. The steady state solutions are taken to be real. Then the real part of the field fluctuations gives the amplitude fluctuations (which are proportional to the intensity fluctuations), and the imaginary part of the field fluctuations give the phase fluctuations.

As before, one takes the Fourier transform of these equations and uses the input-output relationship (7). In additon the normalization factor for the input fluctuations is chosen so that

$$< |\delta\alpha_i(\Omega) + \delta\alpha_i^*(\Omega)|^2 > = 1 \qquad \text{for i= 0,1,2.}$$

The noise spectrum on the intensity difference is found to be[3][15]:

$$S_I(\Omega) = \frac{\Omega^2\tau^2}{\Omega^2\tau^2 + 4\gamma^2} \tag{10}$$

where $S_I(\Omega) = < \Delta I(\Omega)\,.\,\Delta I(\Omega)^* >$ and the shot noise has been normalized to 1. At zero noise frequency, perfect squeezing is expected in the intensity difference. The exprimental spectrum is in good agreement with theory if losses are taken into account. Those are shown to degrade the maximum squeezing, explaining the measured 69% noise reduction[9].

The fluctuations in the phase difference between the two fields can also be calculated. One finds

$$S_s(\Omega) = 1 + 4\gamma^2/\Omega^2\tau^2$$

This expression diverges when Ω goes to zero, which is characteristic of a phase diffusion process.

On the other hand, the noise on the sum of the phases can be shown to be squeezed (Fig. 8). But this property only holds close to threshold, contrary to the squeezing of the intensity difference.

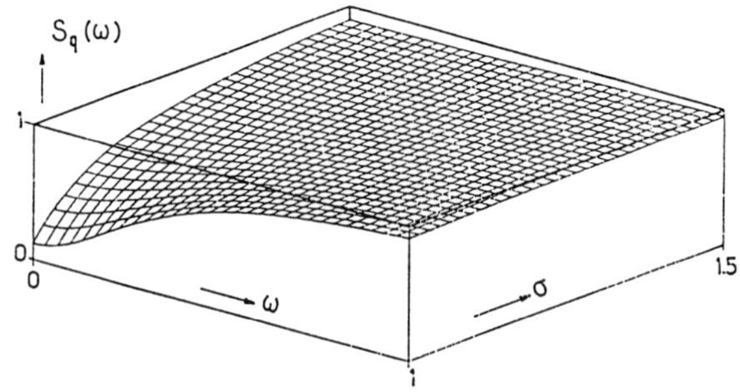

Figure 8. Squeezing of the sum of the phases of the twin fields

6. SPECTROSCOPY WITH SQUEEZED LIGHT

Squeezed light having lower than shot noise fluctuations is expected to improve the signal to noise ratio in high resolution spectroscopy experiments. However, squeezed light is fragile and squeezing is easily destroyed by transmision losses. Losses can be modeled in a general way with a beamsplitter having amplitude transmission and reflexion coefficients t and r. In the same way as above, the squeezed field E_S (assumed here to have a non zero mean value) combines on the beamsplitter with the vacuum field E_0, yielding an output field E given by:

$$E = tE_S + rE_0 \tag{11}$$

If the input field has an amplitude squeezing spectrum $S_A(\Omega)$ with:

$$S_A(\Omega) = < \Delta E_A(\Omega) \,.\, \Delta E_B(\Omega)^* > \tag{12}$$

the squeezing spectrum $S_I(\Omega)$ of the intensity I measured after the beamsplitter, e.g. the lossy element, is:

$$S_I(\Omega) = t^2 S_A(\Omega) + r^2 \tag{13}$$

or if η is the loss coefficient ($\eta = t^2$)

$$S_I(\Omega) = \eta S_A(\Omega) + (1-\eta) \tag{14}$$

It is clearly seen in Eq. 14 that the losses couple back some of the vacuum fluctuations into the squeezed light and tend to bring back the noise spectrum to shot noise (normalized to 1 in the above expression).

Quadrature squeezed light as well as twin photons can be used to improve the signal-to-noise ratio in spectroscopy. Quadrature squeezed light is particularly well suited to interferometric measurements of phase shifts[16]. A Mach-Zender is illuminated with coherent light on one port of the input beamsplitter and with squeezed light on the other input port. The phase difference $\Phi(t)$ between the two arms is modulated at frequency Ω : $\Phi(t) = \Phi_0 + 2\delta \cos\Omega t$, with Φ_0 close to $\pi/2$

The interference signal is detected as the difference between the intensities in the two output ports. The rms electric current at frequency Ω is:

$$i = 2 e \eta I_1 \delta \tag{14}$$

where I_1 is the intensity of the incoming coherent light and η is the loss coefficient. If only the vacuum field is incident at the second input port, the noise current i_N is given by:

$$i_N{}^2 = 2 e^2 \eta I_1 B \tag{15}$$

where B is the detection bandwidth. If a squeezed vacuum field with a noise spectrum $S(\Omega)$ ($S(\Omega) < 1$) is used instead at the second input port, Eq. 14 shows that the noise spectrum becomes:

$$i_S{}^2 = i_N{}^2 \left(\eta S(\Omega) + (1-\eta) \right) = R\, i_N{}^2 \tag{16}$$

The signal-to -noise ratio is then improved by a factor R . R factors on the order of 2 have been obtained experimentally[16].

Twin photons are very promising in difference measurements. The simplest one involves ultrasensitive absorption measurement[17]. The sample with an absorption coefficient $\gamma(t) = \gamma \cos\Omega t$ is placed on one of the twin beams, while the other one propagates freely to the detector. The signal at frequency Ω, measured on the intensity difference is:

$$i = (1/2\,2)\, e\, \eta\, I\, \gamma \tag{17}$$

where I is the sum of the intensities of the two beams. If uncorrelated beams are used, the signal will be detected on a noise background equal to:

$$i_N^2 = 2\, e^2\, \eta\, I\, B \tag{18}$$

whereas, with twin beams, it is:

$$i_S^2 = i_N^2\, (\, \eta\, S_I(\Omega) + (1-\eta)\,) = R\, i_N^2 \tag{19}$$

where $S_I(\Omega)$ is the squeezing spectrum of the intensity difference between the twin beams. Here also the signal-to-noise ratio is improved by a factor R .

1 R.J. Glauber, Phys. Rev. 130 2529 (1963); 131 2766 (1963)

2 D.F. Walls, Nature 306 141 (1983)

3 S. Reynaud, C.Fabre and E.Giacobino, J. Opt. Soc. Am. B 4 1520 (1987)

4 R.E. Slusher, L.W.Hollberg, B. Yurke, J.C.Mertz and J.F.Valley, Phys. Rev. Lett. 55 2409 (1985).

5 R.M.Shelby, M.D.Levenson, S.H.Perlmutter, R.G.DeVoe and D.F.Walls, Phys. Rev. Lett. 57 691 (1986); B.L.Schumaker, S.H.Perlmutter, R.M.Shelby and M.D.Levenson, Phys. Rev. Lett. 58 357 (1987).

6 S.T.Ho, N.C.Wong, J.H.Shapiro and P.Kumar, OSA Meeting Technical Digest (1988) TUL1.

7 L.A. Wu, H.J. Kimble, J.L. Hall and Huifa Wu, Phys. Rev. Lett. 57 2520 (1986)

8 P. Grangier, R.E. Slusher, B. Yurke and A. La Porta, Phys. Rev. Lett. 59 2153 (1987); R.E. Slusher, P. Grangier, A. La Porta, B. Yurke and M.J. Potasek, Phys. Rev. Lett. 59 2566 (1987)

9 A.Heidmann, R.J.Horowicz, S.Reynaud, E.Giacobino and C.Fabre, Phys. Rev. Lett. 59 2555 (1987); T.Debuisschert, S.Reynaud, A.Heidmann, E.Giacobino and C.Fabre, Quantum Opt. 1 (1989).

10 J.C.Rarity and P.R.Tapster, ONR seminar on "Photons and Quantum Fluctuations" eds E.R.Pike and H.Walther p122 (1988).

11 S.F.Pereira, M.Xiao, H.J.Kimble and J.L.Hall, Phys. Rev. A 38 4931 (1988).

12 M.G.Raizen, L.A.Orozco, M.Xiao, T.L.Boyd and H.J.Kimble, Phys. Rev. Lett. 59 198 (1987)

13 S.Machida, Y.Yamamoto and Y.Itaya, Phys. Rev. Lett. 58 1000 (1987) and 60 792 (1988).

14 S.Reynaud and A.Heidmann, Optics Comm. 71 209 (1989)

15 C.Fabre, E.Giacobino, A.Heidmann and S.Reynaud, J.Phys. France 50 1209 (1989)

16 M.Xiao, L.A.Wu and H.J.Kimble, Phys. Rev. Lett. 59 278 (1987); P. Grangier, R.E. Slusher, B. Yurke and A. La Porta, Phys. Rev. Lett. 59 2153 (1987).

17 E.Giacobino, C.Fabre, S.Reynaud, A.Heidmann and R.Horowicz, ONR seminar on "Photons and Quantum Fluctuations" eds E.R.Pike and H.Walther p81 (1988).

MULTIPHOTON EFFECTS IN LASER-ATOM INTERACTIONS

Wilhelm Becker

Center for Advanced Studies and
Department of Physics and Astronomy
University of New Mexico, Albuquerque, New Mexico 87131, USA

INTRODUCTION

Multiphoton ionization and the closely related process of higher harmonic production provide a unique opportunity to study the interaction between light and matter in a situation where the effective coupling is highly nonlinear with respect to the electromagnetic field. If the atom has to absorb N photons in order to be ionized then a theoretical description will have to involve perturbation theory at least of the Nth order. For typical experiments, viz. ionization of rare gases with lasers in the near infrared, N is quite large, between 10 and 20. Up to the early eighties, perturbation theory of the lowest possible order had been sufficient to describe all available data. Total ion yields, for example, always exhibited a proportionality to the Nth power of the intensity (below saturation). Similarly, rates for harmonic production showed the same behavior until very recently. The situation changed dramatically when the attention turned to quantities other than total ion yields. For intensities of the order of 10^{13} W/cm^2 the energy spectra of the ejected electrons start to exhibit features which can no longer be explained by lowest order perturbation theory (LOPT). Very recently, the rates for production of very high harmonics (up to the 33rd harmonic of the incident laser frequency) have been measured with unprecedented precision and have been observed not to follow the expectations from LOPT either.

Remarkably, although the interpretation of multiphoton ionization constantly makes reference to "photons," the theoretical description does not require a quantized electromagnetic field. Rather, we have a typical external field or background field problem. The periodic time dependence of the external field has the consequence that the energy of a charged particle is only conserved up to multiples of $\hbar\omega$ (with ω the laser frequency). This, along with the fact that each occurrence of the interaction Hamiltonian ($e\vec{r}\cdot\vec{E}$ or $e\vec{p}\cdot\vec{A}/m$) changes the energy by $\pm\hbar\omega$, allows for the usage of "photons" in the interpretation of the results although they are not, strictly speaking, part of the theoretical framework. The external field approximation is excellent as long as the laser field is not significantly depleted during the interaction with the gas. For the intensities employed in current experiments (up to 10^{15} W/cm^2) a nonrelativistic description is still safe. In a first attempt at further simplifying the problem one considers an electron subject to a fixed binding potential and coupled to the classical external field. The problem then is to solve the Schrödinger equation in a situation where neither of the two interactions can be treated as weak compared with the other.

In this paper the nonperturbative phenomena observed thus far will be reviewed and simple models for their explanation will be discussed. First, we will consider kinematical effects of a free electron in a laser field which are related to the concept of the ponderomotive potential. We will then turn to simple Keldysh-type models of above-threshold ionization with the goal of understanding the envelope of the peaks of the electron spectrum. In particular, we consider a model atom consisting of a three-dimensional attractive delta-function well. This model then provides the wave functions for the calculation of higher harmonic emission by a single atom. Throughout this paper we use units such that $\hbar = c = 1$.

ABOVE-THRESHOLD IONIZATION

When an atom is ionized by a comparatively weak field, in most cases the electron will just absorb the minimum number N of photons which are required for ionization. With an increasingly smaller rate, the electron may occasionally absorb $N+1$ or even $N+2$ photons which leads to well separated peaks in the electron energy spectrum with a spacing of $\hbar\omega$. This was first observed by Agostini et al.[1] Figure 1a shows such a spectrum which agrees with what one expects from perturbation theory. With increasing intensity, however, this picture begins to change. Figure 1b displays a spectrum recorded at an intensity five times that of Fig. 1a. The spectrum now extends to much higher intensities, the lowest peak is no longer the dominant one, and, most remarkably, the lowest peak has actually disappeared or become in-

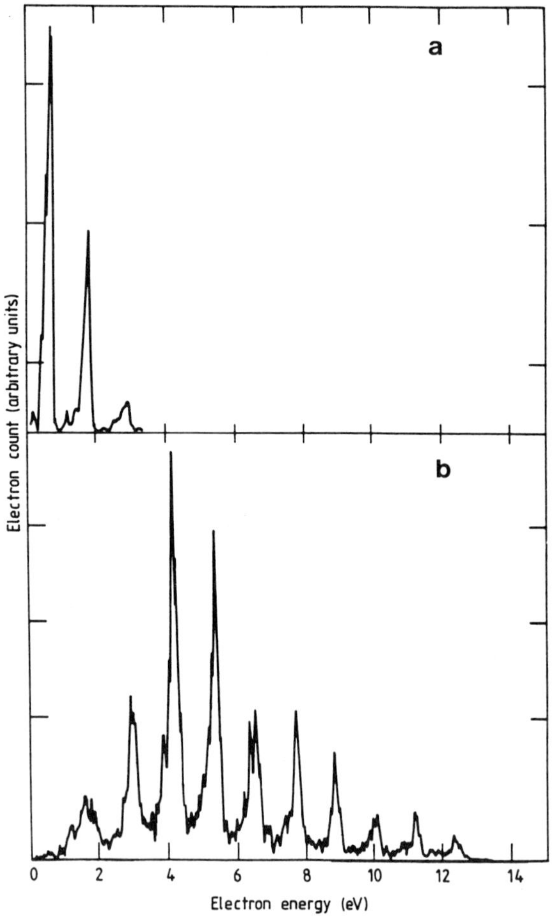

Fig. 1. Electron energy spectra for multiphoton ionization of xenon at 1064 nm ($N = 11$) and a pulse length of 120 psec.; (a) intensity $I = 2.2 \times 10^{12}$ W/cm^2; (b) $I = 1.1 \times 10^{13}$ W/cm^2 [from Ref. 3].

discernible from the background. Clearly, the features exhibited in Fig. 1b defy an explanation based on perturbation theory. Notice that the position of the peaks is unchanged from Fig. 1a to Fig. 1b. These phenomena which were first observed by Kruit et al.[2] have become known under the heading "above-threshold ionization" (ATI). Particularly the suppression of the low-energy peaks (for higher intensities more and more of them disappear) generated a great deal of interest in the effect. We will see below that the two main features observed in Fig. 1b – peak suppression on one hand vs. peak switching, i.e. the fact that some high-order peak becomes the dominant, one on the other – have different origins. Peak suppression turns out to be a kinematical effect completely independent of the atom.

The Ponderomotive Potential

A number of features observed in ATI (including peak suppression) can be understood just in terms of the behavior of an otherwise free electron in a laser field. Consider, for the time being, a monochromatic laser field $\vec{A} = \vec{A}(t)$ where the space dependence is neglected. The velocity of an electron in this field is the sum of a constant and an oscillating part:

$$\vec{v} = \vec{v}_0 + \vec{v}_q, \qquad \vec{v}_q = -\frac{e\vec{A}(t)}{m}. \tag{1}$$

The time average of the quiver velocity $\vec{v}_q$ over a period of the laser field vanishes, viz.

$$\langle \vec{v}_q \rangle = 0. \tag{2}$$

Hence, the time average of the kinetic energy

$$\langle \frac{m}{2}\vec{v}^2 \rangle = \frac{m}{2}\vec{v}_0^2 + \frac{m}{2}\vec{v}_q^2 = \frac{m}{2}\langle \vec{v} \rangle^2 + \frac{e^2}{2m}\langle \vec{A}^2 \rangle \geq \frac{e^2}{2m}\langle \vec{A}^2 \rangle \tag{3}$$

consists again of two terms and we can identify the second with the quiver energy. The latter is proportional to the laser intensity and provides a lower limit of the kinetic energy: the kinetic energy of an electron in a laser field must always exceed this quiver energy.

Let us now consider the more general situation where the laser field has a spatial variation, i.e. $\vec{A} = \vec{A}(\vec{r}, t)$. It can be shown[4,5] that to lowest order in $e|\vec{A}|/m$ the average velocity satisfies the equation

$$m\frac{d}{dt}\langle \vec{v} \rangle = -\vec{\nabla}\left(\frac{e^2}{2m}\langle \vec{A}^2 \rangle\right), \tag{4}$$

where the average is over the period associated with the center of the frequency distribution. If $\langle \vec{A}^2 \rangle$ is independent of time, then the r.h.s. specifies a conservative force (with respect to the average motion) and we have

$$\frac{d}{dt}\left(\frac{m}{2}\langle \vec{v} \rangle^2 + \frac{e^2}{2m}\langle \vec{A}^2 \rangle\right) = 0. \tag{5}$$

This force and the associated potential are referred to as the ponderomotive force and potential, respectively. Comparing the conservation law (5) with the time-averaged kinetic energy (3) we notice that the latter is conserved whenever the ponderomotive force can be considered as conservative.

We are now in a position to understand the origin of peak suppression.[6–9] In the presence of the laser field, the quiver energy adds to the field-free ionization energy leading to an effective field-dependent ionization energy which increases linearly with

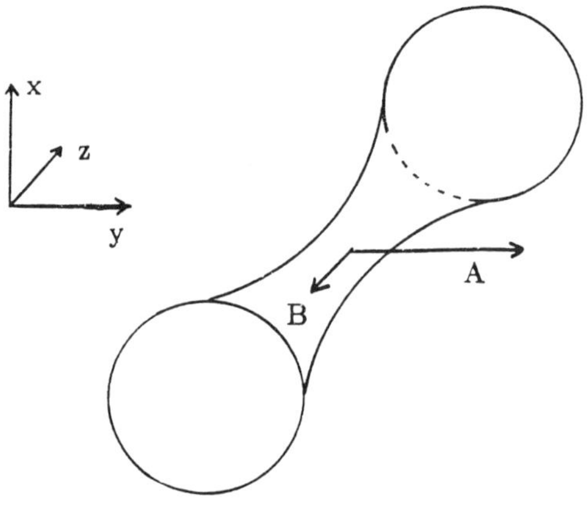

Fig. 2. The laser pulse propagates in the z-direction through the double-ended trumpet. Ionization takes place predominantly in the region where the intensity is highest, i.e. near the waist. Case A: the electron leaves the pulse on one side. Case B: the electron is swept over by the pulse and never gets into the region where there is a strong spatial gradient.

the laser intensity.* Hence, the electron has to absorb more photons for ionization than in a weak field. It is now of vital importance in which way the electron exits the laser pulse. Figure 2 indicates two extreme possibilities. In case A the electron leaves the pulse on one side. During this process the temporal variation of the laser pulse envelope is negligible and the electron mainly experiences the strong spatial gradient of the pulse envelope. The ponderomotive force is conservative, and the conservation law (5) makes sure that the kinetic energy is conserved while the electron leaves the pulse. Consequently, no electron reaches the detector with an energy smaller than $e^2\langle\vec{A}^2\rangle/2m$ (cp. Eq. (3)), where $\langle\vec{A}^2\rangle$ refers to the position where the electron was set free. In most cases this has happened at the peak intensity of the laser.

In contrast, in case B the electron never experiences the spatial gradient of the pulse envelope. Instead, in this case the field is well described by $\vec{A} = \mathrm{Re}(A(t)\hat{e}e^{-i\omega t})$ with $A(t)$ going to zero as the pulse moves on. The ponderomotive force is not conservative and the conservation law (5) does not hold. Rather, the average velocity $\langle\vec{v}\rangle$ is conserved. When the electron is left behind by the pulse it loses its quiver energy whose magnitude in electron volts is

$$\frac{e^2}{2m}\langle\vec{A}^2\rangle = 9.3\times10^{-14}\lambda^2(\mu m)I(W/cm^2)eV, \tag{6}$$

* The question arises of why not the energies of all states including the initial bound state of the electron are raised by the same amount $e^2\langle\vec{A}^2\rangle/2m$. In fact, the term $e^2\langle\vec{A}^2\rangle/2m$ can be eliminated from the Schrödinger equation by a gauge transformation.[10] However, this obscures the eminent physical meaning of this term which describes the quivering motion of the unbound electron whose amplitude is $(e|\vec{A}|/m)(\lambda/2\pi)$ with λ the laser wavelength. For the fields of interest this is large compared with the radius of the orbit of the bound electron. Hence it is intuitively plausible that the bound electron being unable to exert the quiver motion will not suffer the ponderomotive energy shift. Formally, this is brought about by the $\vec{p}\cdot\vec{A}$-term almost cancelling in second order the first order contribution from the A^2-term.[11] What is left is the ac Stark shift of the ground state which results form the static polarizability of the atom. This shift is small compared with the ponderomotive shift and is here neglected. If the aforementioned gauge transformation is carried out, the cancellation is offset and the ground state is shifted down by the ponderomotive energy.[10] Hence, in either gauge, the energy difference between the ground state and a continuum state contains the ponderomotive energy.

where $\vec{A}^2$ and I refer to the time and position at which ionization took place. The positions of the peaks in the electron energy spectrum, which is recorded by the detector, are now intensity dependent in such a way that they are red-shifted with increasing intensity. The magnitude of this redshift provides an accurate diagnostic means to measure the intensity of an ultrashort laser pulse.[13] Figure 3 depicts the effect of the two different cases A and B. In experiments, the shorter the pulse and the slower the electrons are, the more case B will be dominant. The early experiments such as Ref. 2 employed rather long pulses leading to dominance of case A and intensity-independent positions of the peaks. Figure 4 confronts two spectra taken at the same intensity, but for different pulse lengths. Some of the peaks in Fig. 1b and Fig. 4 show the effect of boundary conditions intermediate between the extreme cases A and B: these peaks drop precipitously to zero on the high-energy side while the drop on the low-energy side is more gradual. The latter comes from electrons which lost only a part of their ponderomotive energy.

The electron in a laser pulse has been compared with a surfer[14] who may be lifted up or let down by a wave (case B) or surf down from a crest into a valley converting potential into kinetic energy (case A). These concepts have been put to test in a beautiful experiment by Bucksbaum et al.,[14] where two lasers were used. A first pulse employed ATI to generate a beam of electrons with the ATI-series of discrete energies. These were then given the chance to surf up or down a second pulse which arrived later by some variable relative delay time.

The ponderomotive potential also plays the dominant role in the observed angular distributions of the low-energy electrons.[14] Notice that all of these effects are completely independent of the atom whose only function is to inject electrons into the field environment.

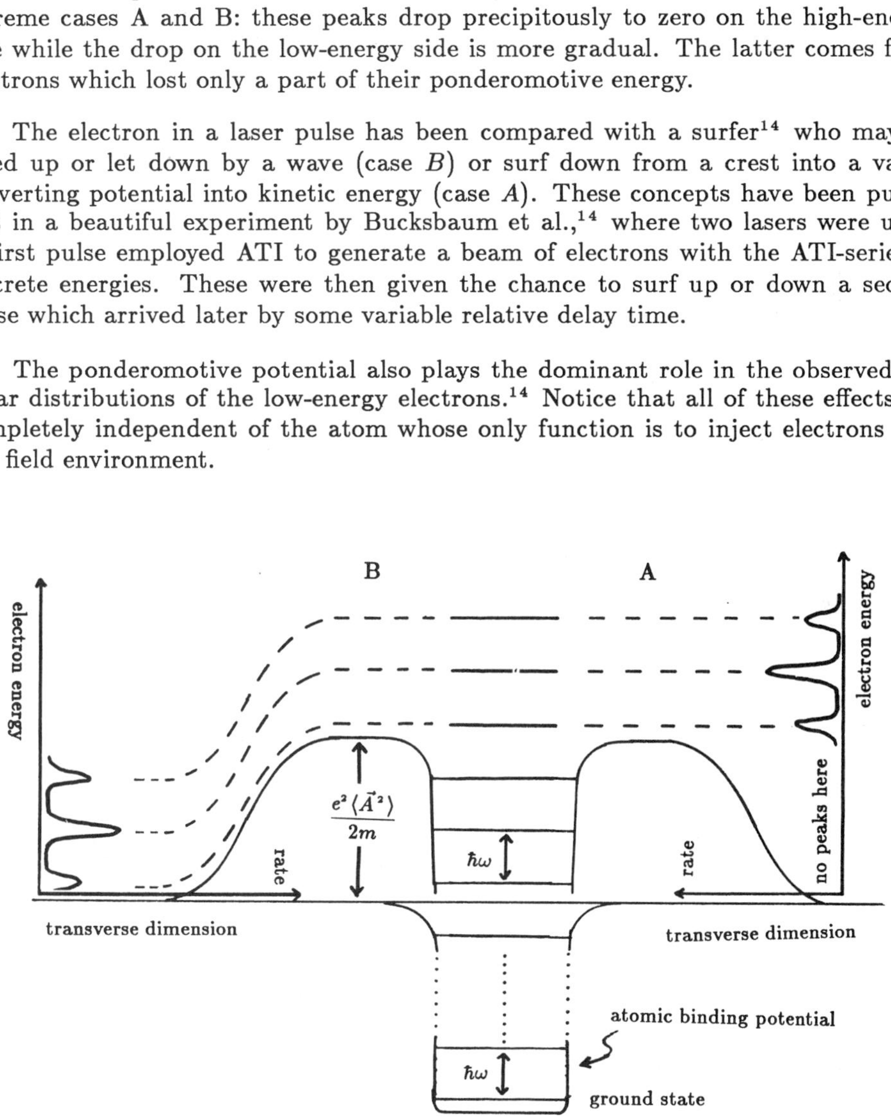

Fig. 3. Effect of the ponderomotive barrier on the ATI-spectrum. Case A: kinetic energy is conserved while the electron leaves the pulse; no peaks are observed with an energy smaller than the ponderomotive energy. Case B: the electron loses the ponderomotive energy when it leaves the pulse; no peak suppression. The small ac-Stark shift of the ground state is not indicated in the figure.

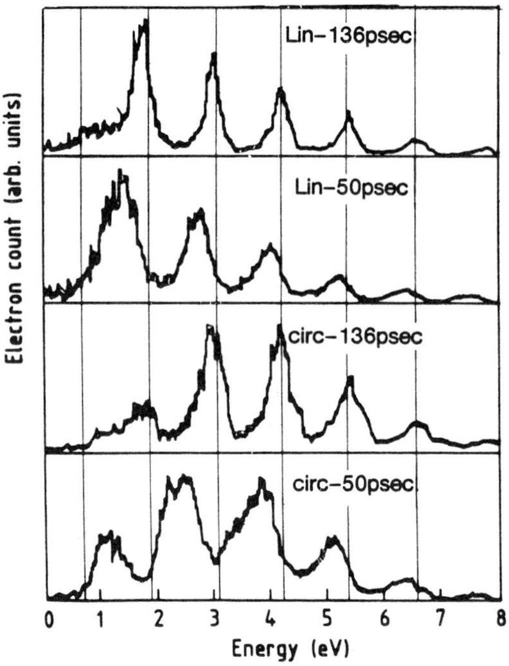

Fig. 4. Multiphoton ionization of xenon at 1054 nm with $I = 7.5 \times 10^{12}$ W/cm^2 and different polarizations and pulse lengths [from Ref. 12]. The low-energy peaks of the spectra taken with the 50 psec pulse are redshifted by the ponderomotive energy $(e^2/2m)\langle \vec{A}^2 \rangle = 0.79eV$ from Eq. (6).

Recent experiments, however, did show effects which provide evidence of the atomic level structure.[15,16] Figure 5 displays spectra recorded in multiphoton ionization of xenon at 615 nm by a very short pulse ($\sim$ 100 fs). The spectrum of Fig. 5a which is due to a linearly polarized laser exhibits a pronounced substructure which repeats itself from peak to peak. Such structure is absent from the spectrum of Fig. 5b which was obtained with circular polarization. The substructure is caused by intermediate resonances with Rydberg levels[15,16] as explained in what follows. In many cases the Rydberg states near the continuum threshold exhibit about the same ponderomotive shifts as the continuum states. Figure 6 depicts the field-dependent ionization threshold and a particular Rydberg state (dashed line) as a function of the laser intensity along with the "photon ladder" built on the ground state whose ac Start shift is neglected. For a certain intensity I_R there is a $(N-1)$-photon resonance with the shifted Rydberg state. Let us compare several pulses with increasingly higher intensities each having an intensity distribution between $I = 0$ and some I_{max} as indicated below the intensity axis in Fig. 6. For a short laser pulse, most electrons created by N-photon ionization experience type B-boundary conditions. Hence when they leave the pulse, they are downshifted in energy by the ponderomotive shift corresponding to the intensity at which they were set free. The electron spectra generated by each of the pulses are sketched in Fig. 7. Electrons set free at the intensity I_R are resonantly enhanced which shows as a peak at the intensity independent energy $N\omega - |E_0| - (ea_R)^2/2m$. Figure 8 shows spectra from seven-photon ionization (five or six-photon resonant) of xenon at 615 nm with 120 fsec pulses of various intensities. The ponderomotive potential sweeps over many Rydberg states (two of which are indicated by vertical dashed lines and several more are visible). In contrast, a long but otherwise identical pulse would favor type A-boundary conditions and thereby lump the entire structure into one essentially structureless peak. Figure 8 shows the spectrum generated by absorption of the minimum number of photons (seven in the present case). As shown in Fig. 5a, the next peak corresponding to eight-photon ionization repeats the same structure. The preceding interpretation of

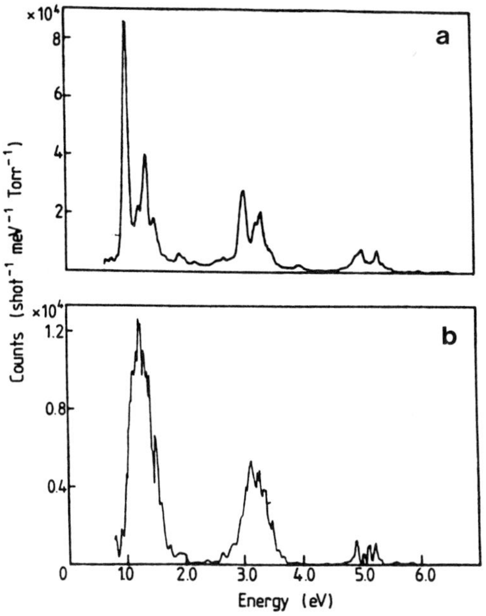

Fig. 5. Seven-, eight-, and nine-photon ionization of xenon with 120 fs pulses a 615 nm. The spectra in (a) and (b) were obtained at the same intensity, but with linear (a) versus circular (b) polarization [from Ref. 16].

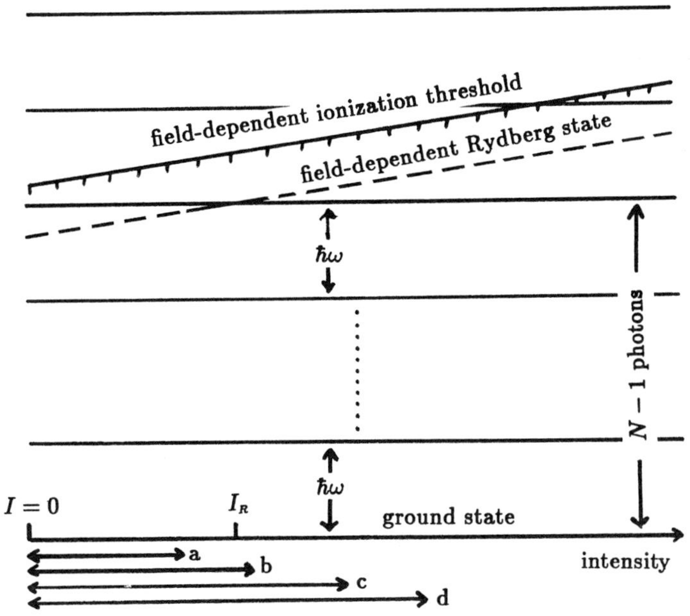

Fig. 6. $(N-1)$-photon-resonant N-photon ionization (schematic): ionization at intensity I_R is $(N-1)$-photon resonant with the Rydberg state indicated by the dashed line. It is assumed that the intensity-dependent energy shift of the Rydberg state agrees with the ponderomotive shift of a continuum state. The intensity range of several pulses to be discussed in Fig. 7 is marked below the abscissa. The ac-Stark shift of the ground state is neglected.

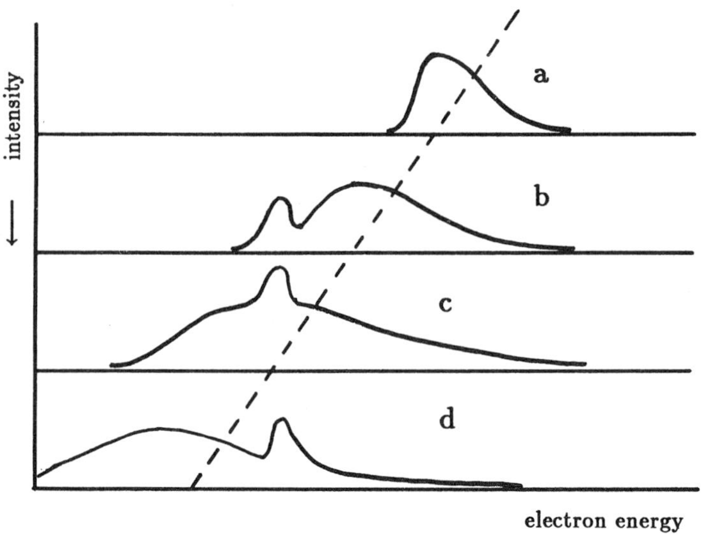

Fig. 7. Electron energy spectra corresponding to the pulses marked by (a), (b), (c) and (d) in Fig. 6. The peak energy of pulse (a) is below the resonant intensity I_R, and no resonance enhancement occurs. In cases (b), (c), and (d) the peak energy exceeds I_R, and ionization at this intensity is enhanced resulting in a resonance peak at the intensity-independent energy $N\omega - |E_0| - e^2\langle \vec{A}_R^2\rangle/2m$. Here $|E_0|$ is the field-independent ionization energy and $\langle \vec{A}_R^2\rangle$ is related to the resonance intensity I_R by Eq. (6).

the peak structure is supported by the fact that a laser with circular polarization generates hardly any structure (Fig. 5b). The reason is that the electron having absorbed a specific number of circularly polarized photons is in a state with large angular momentum. Angular momentum conservation then largely restricts the number of possible resonances. In contrast, absorption of linearly polarized photons does not generate an eigenstate of angular momentum.

Peak Switching

While peak suppression can be understood solely by the action of the ponderomotive potential, the explanation of peak switching, viz. the fact that the entire energy spectrum along with its maximum moves to higher and higher energies, requires some more effort. We will approximate the atom by a static potential $V(\vec{r})$, which binds one electron. In this context we will derive an exact expression for the matrix element that governs multiphoton ionization.

Prior to the arrival of the laser pulse the electron is in the ground state with the wave function $\psi_0^{(0)}(\vec{r},t) = \psi_0^{(0)}(\vec{r})\exp(i|E_0|t)$ where $|E_0|$ denotes the binding energy. The probability that the electron is ionized into a final state with momentum $\vec{p}$ and wave function $\psi_{\vec{p}}^{(0)}(\vec{r},t)$ is determined by the square of the matrix element

$$M_p = \lim_{t\to\infty} \int d^3r\,\psi_{\vec{p}}^{(0)}(\vec{r}t)^*\,\psi^{(0)}(\vec{r}t), \tag{7}$$

where $\psi^{(0)}(\vec{r},t)$ is given by the time evolution of the initial state $\psi_0^{(0)}(\vec{r},t)$, viz.

$$\psi^{(0)}(\vec{r},t) = \lim_{t'\to-\infty} \int d^3r'\,iG_+(\vec{r}t,\vec{r}'t')\psi_0^{(0)}(\vec{r}'t'). \tag{8}$$

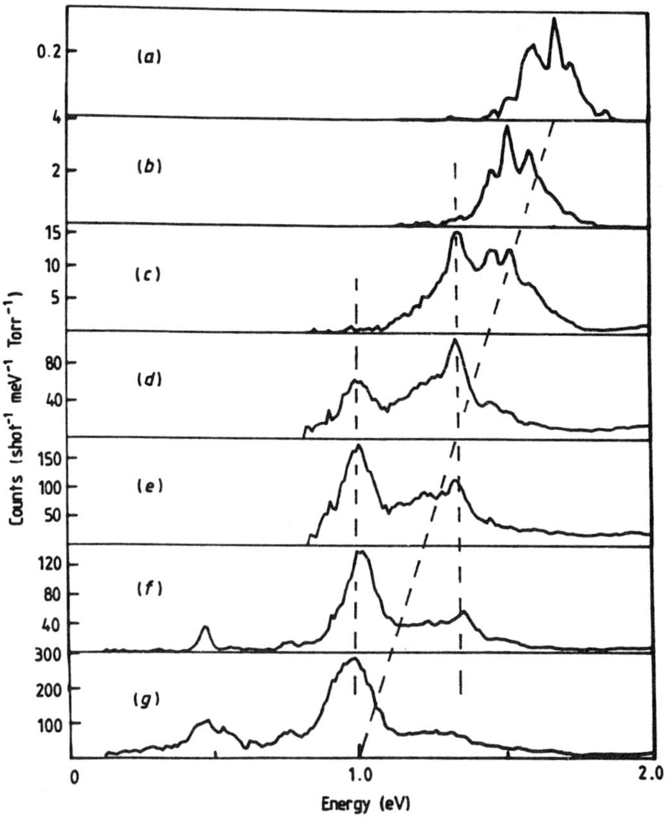

Fig. 8. Seven-photon ionization of xenon with 120 fs pulses at 615 nm for linear polarization. The intensity increases from 1×10^{13} W/cm^2 (a) to 3×10^{13} W/cm^2 (g). Several resonances with upshifted Rydberg states are visible (two are indicated by vertical dashed lines). The ponderomotive red-shift is indicated by the inclined dashed line [from Agostini et al.[16])].

Here $G_+(\vec{r}t, \vec{r}'t')$ denotes the retarded propagator in the presence of both the ionizing field $\vec{E}(t)$ described in the dipole approximation and the binding potential $V(\vec{r})$:

$$\left(i\frac{\partial}{\partial t} - \frac{\vec{p}^2}{2m} + e\vec{r}\cdot\vec{E}(t) - V(\vec{r})\right) G_+(\vec{r}t, \vec{r}'t') = \delta(t-t')\delta(\vec{r}-\vec{r}'). \tag{9}$$

The unperturbed propagator $G_+^{(0)}(\vec{r}t, \vec{r}'t')$ satisfies the same equation with the term $e\vec{r}\cdot\vec{E}(t)$ missing. The full propagator then obeys the integral equation

$$G_+ = G_+^{(0)} + G_+^{(0)} H_I G_+ = G_+^{(0)} + G_+ H_I G_+^{(0)} \tag{10}$$

written in shorthand operator notation with $H_I \to -e\vec{r}\cdot\vec{E}(t)$. Using Eq. (8) and the second version of Eq. (10) in the matrix element (7) and exploiting the orthogonality of the eigenfunctions of $H_0 = \vec{p}^2/2m + V(\vec{r})$ we obtain

$$M_p = \lim_{t\to\infty} \int d^3r d^3\vec{r}' dt' \psi_{\vec{p}}^{(0)}(\vec{r}t)^* G_+(\vec{r}t, \vec{r}'t')(-e\vec{r}'\cdot\vec{E}(t'))\psi_0^{(0)}(\vec{r}'t'). \tag{11}$$

Notice that Eq. (11) contains the exact propagator in the presence of everything while both wave functions are in the presence of the binding potential but in the absence of the field. Equation (11) is still exact. In what follows we will discuss two approximations based on it.

The Keldysh-Faisal-Reiss(KFR) Approximation[17–20]

In this approximation, the binding potential is ignored everywhere except in the initial bound state wave function $\psi_0^{(0)}(\vec{r}'t')$. This is accomplished by replacing $\lim_{t\to\infty}\int d^3r\psi_{\vec{p}}^{(0)}(\vec{r}t)^* iG_+(\vec{r}t,\vec{r}'t')$ by the Volkov wave function $\psi_{\vec{p}}^{(E)}(\vec{r}'t')^*$ in the electric field gauge:

$$\psi_{\vec{p}}^{(E)}(\vec{r},t) = (2\pi)^{-3/2}\left[\exp i\left(\vec{\pi}(t)\cdot\vec{r} - \frac{1}{2m}\int^t d\tau\vec{\pi}^2(\tau)\right)\right] \tag{12}$$

with

$$\vec{\pi}(t) = \vec{p} - e\vec{A}(t) \tag{13}$$

the gauge-invariant mechanical momentum. Thus the approximation is

$$M_p^{(K)} = -i\int d^3r dt\psi_{\vec{p}}^{(E)}(\vec{r}t)^*(-e\vec{r}\cdot\vec{E}(t))\psi_0^{(0)}(\vec{r}t). \tag{14}$$

With the help of several integrations by part this can also be written in the form

$$M_p^{(K)} = -i\int d^3r dt\psi_p^{(E)}(\vec{r}t)^* V(\vec{r})\psi_0^{(0)}(\vec{r}t). \tag{15}$$

The matrix element $M_p^{(K)}$, both in the form (14) or (15), is in view of Eqs. (12) and (13) manifestly gauge invariant as was Keldysh's original expression.[17] What is usually referred to as the KFR-approximation differs from the above in that the $\vec{p}\cdot\vec{A}$ gauge is used. As a consequence, in Eq. (14) $-e\vec{r}\cdot\vec{E}(t)$ is replaced by $-(e/m)\vec{p}\cdot\vec{A}(t) + (e^2/2m)\vec{A}^2(t)$ and the Volkov solution is used in the $\vec{p}\cdot\vec{A}$ gauge. The latter differs from Eq. (12) by the absence of the factor $\exp(-ie\vec{r}\cdot\vec{A}(t))$. This allows for an explicit evaluation of the $\vec{p}\cdot\vec{A}$ gauge analog of the matrix element (14) for hydrogen wave functions in the form of a sum over Bessel functions.[19,20] A correspondingly simple evaluation is not possible for the forms (14) and (15) in the $\vec{r}\cdot\vec{E}$ gauge that we derived here, due to the presence of the gauge factor $\exp(-ie\vec{r}\cdot\vec{A}(t))$. One must, however, be aware of the fact that this commonly used form is not gauge invariant. The gauge invariant form derived here and various nongauge invariant analogs have been compared quantitatively[21] and found to behave markedly different already for moderately high fields.

In the KFR-approximation (14) the initial bound-state wave function is projected on the Volkov solution (12) after just one interaction with the external field. Clearly, this can only be justified if intermediate resonances (e.g., of the kind discussed at the end of the preceding section) never play any role. In fact, it has been shown[22] that the Keldysh approximation is (almost) exact for the case of a circularly polarized field and a delta-function potential which has only one bound state.

A Final-State-Interaction Approximation [9,23]

Here one concentrates on the interaction of the ejected electron with the laser field in the continuum trying to bypass as much as possible the details of how it got there. To that end, we replace in Eq. (11) (as before in the KFR-approximation)

$\psi_{\vec{p}}^{(0)}(\vec{r},t)$ by a plane wave and notice that using the same operator notation as in Eq. (10) we can write

$$G_{+} = G_{+}^{(E)}(1 + VG_{+}) \tag{16}$$

where the Volkov propagator $G_{+}^{(E)}$ satisfies Eq. (9) with the term $-V(\vec{r})$ missing. We then define an effective potential that lifts the electron into the lowest continuum state, by the N-photon absorption part of $(1 + VG_{+})H_I\psi_0^{(0)}$. The model is then summarized by the matrix element

$$M = \int d^3r dt \psi_{p}^{(E)}(\vec{r},t)^* V_{\text{eff}}(\vec{r}) e^{i(|E_0| - N\omega)t}. \tag{17}$$

The time-dependence $\exp i(|E_0| - N\omega)t$ takes into account the initial bound state and the absorption of N photons. The effective potential is not calculated *ab initio* from its above definition, but rather we assume that

$$\int d^3r V_{\text{eff}}(\vec{r}) e^{-i\vec{\pi}(t)\cdot\vec{r}} = \text{const} \times I^{N/2}. \tag{18}$$

The chosen intensity dependence is such that it yields the factor of I^N of lowest order perturbation theory. We then have modeled ATI as a two-step process: by absorbing N photons the electron reaches in an otherwise unspecified way the continuum. From there on it is described by the Volkov wave function (12), i.e. the binding potential is neglected. Since the effective potential is only specified up to a constant this model does not allow for the calculation of a total ionization rate, it only yields the relative heights of the ATI-peaks.

If the laser field is taken as a plane wave, with sinusoidal time dependence $\vec{A} = \text{Re}(A_0 \hat{e} \exp(-i\omega t))$, both models produce very simple expressions for the ionization rates into the individual ATI-peaks, cp. Refs. 19, 20 and 9, 23. Essentially, they are given by squares of Bessel functions, ordinary Bessel functions for a circularly polarized field and generalized Bessel functions for linear polarization. In both models, the finite extent of the laser pulse and the question of whether the electron exits the pulse with boundary conditions of case A or case B as discussed above, are taken care of by hand. In case B, the canonical momentum $\vec{p}$, which enters the Volkov solution (12) as a parameter, is conserved so that $\vec{p}^2/2m$ is the energy of the electron at the detector. In contrast, in case A, it follows from the conservation law (5), since $\vec{p} = m\vec{v}_0 = m\langle\vec{v}\rangle$ inside the field, that the quantity $(\vec{p}^2 + e^2\langle\vec{A}^2\rangle)/2m$ has to be identified with the energy at the detector. In place of the Volkov function (12) which only holds for a purely time-dependent field $\vec{A}(t)$, an approximate wave function can be used that allows for a laser field $\vec{A}(\vec{r},t)$ of finite spatial extent.[24] This leads to much more complicated expressions, but avoids the just mentioned necessity to impose the appropriate boundary conditions by hand.

Both of the two simple models discussed above give qualitatively similar results which reproduce the general features of Fig. 2. In general, it appears that the KFR-model works better for circular polarization where intermediate near-resonant states play a lesser role, while the final-state-interaction two-step model gives better results for linear polarization where near-resonant states are certain to be around. A careful comparison of these two models (as well as others) with experiment for the case of linear polarization is reported in Ref. 25. Both models do not agree very well with one-dimensional computer simulations of model atoms.[26,27] Further progress can be made by a systematic determination of the wave function in terms of quasi-energy (Floquet) theory.[28]

The Three-Dimensional Delta-Function Potential

None of the preceding two models required the knowledge of the wave function $\psi_0(\vec{r},t)$ (introduced in Eq. (7)) that develops out of the initial bound state in the presence of the laser field. We will need such a wave function for the calculation of higher harmonic production. In general, it cannot be obtained in analytical form. However, for a delta-function potential a largely analytical solution can be found.[29]

We will consider the potential[30]

$$V(\vec{r}) = \frac{2\pi}{\kappa m}\delta(\vec{r})\frac{\partial}{\partial r}r, \tag{19}$$

which has exactly one bound state* with binding energy $|E_0| = \kappa^2/2m$ and wave function $\exp(-i\kappa r)/r$. The potential is subject to a laser field described by a vector potential $\vec{A}(t)$ which depends only on time. The Schrödinger equation to be solved is equivalent to the integral equation

$$\psi_0(\vec{r},t) = \psi_0^{(0)}(\vec{r},t) + \int d^3r'dt'G^{(E)}(\vec{r}t,\vec{r}'t')V(\vec{r}')\psi_0(\vec{r}'t') \tag{20}$$

where

$$G^{(E)}(\vec{r}t,\vec{r}'t') = \theta(t-t')\exp(-i\mathcal{R}(\vec{r}t,\vec{r}'t'))\exp(-i\mathcal{M}(t,t'))G^{(0)}(\vec{r}-\vec{r}',t-t') \tag{21}$$

denotes the Volkov propagator in the $\vec{r}\cdot\vec{E}$ gauge. Here

$$\mathcal{R}(\vec{r}t,\vec{r}'t') = e\left(\vec{A}(t)\cdot\vec{r} - \vec{A}(t')\cdot\vec{r}' - \frac{\vec{r}-\vec{r}'}{t-t'}\cdot\int_{t'}^{t}d\xi\vec{A}(\xi)\right), \tag{22}$$

$$\mathcal{M}(t,t') = \frac{e^2}{2m}\left(\int_{t'}^{t}d\xi\vec{A}^2(\xi) - \frac{1}{t-t'}\left(\int_{t'}^{t}d\xi\vec{A}(\xi)\right)^2\right), \tag{23}$$

and

$$G^{(0)}(\vec{r}-\vec{r}',t-t') = -\left(\frac{im}{2\pi(t-t'-i\epsilon)}\right)^{3/2}\exp\left(i\frac{m}{2}\frac{(\vec{r}-\vec{r}')^2}{t-t'}\right) \tag{24}$$

is the free Schrödinger propagator. We will actually look for a quasi-energy solution, viz. a solution of the homogeneous analog of Eq. (20) with the first term on the r.h.s. omitted.

Owing to the delta-function potential (19), Eq. (20) (without the inhomogeneous term) allows for the computation of the wave function $\psi_0(\vec{r},t)$ provided it is known at the origin. One can easily convince oneself that near the origin this wave function differs from the wave function in the absence of the field only by a time-dependent factor $u(t)$, viz.

$$\psi_0(\vec{r},t)\xrightarrow{r\to 0}\left(\frac{1}{r}-\kappa\right)u(t). \tag{25}$$

Using Eqs. (25) and (21) to (24) in Eq. (20) one can carry out the integration over $\vec{r}'$. In the limit $\vec{r}\to 0$, one then obtains an integral equation for $u(t)$:

$$\kappa u(t) = -\left(\frac{im}{2\pi}\right)^{1/2}\int_0^\infty\frac{d\tau}{\tau^{3/2}}(\exp i\mathcal{M}(t+\tau,t)u(t+\tau)-u(t)). \tag{26}$$

The ansatz

$$u(t) = e^{-iEt}w(t), \qquad w(t) = \sum_{n=-\infty}^{\infty}a_n\exp(2in\omega t) \tag{27}$$

* Without the regularizing factor $(\partial/\partial r)r$ which is understood to be applied to the ensuing wave function this potential does not have any normalizable bound state; for an interesting discussion in n dimensions, see Refs. 31,32.

with periodic $w(t)$, viz. $w(t) = w(t + \pi/\omega)$, determines both the quasi-energy E and the function $w(t)$, i.e. the Fourier coefficients a_n. The quasi-energy E is complex. This is, in a sense, the reaction of the system to the omission of the inhomogeneous term in Eq. (20). Also, *a priori*, the quasi-energy is only specified modulo multiples of $\hbar\omega$. We have to select that particular branch

$$E = -|E_0| + \Delta - \frac{1}{2}\, i\Gamma \tag{28}$$

where Δ and Γ tend to zero when the field is turned off. The real part Δ is the small Stark shift of the ground state and Γ is the total ionization rate per time. The quasi-energy approach only makes sense when both quantities are very small compared with $|E_0|$.

For a circularly polarized field

$$\vec{A}(t) = a(\hat{x}\cos(\omega t + \phi) + \hat{y}\sin(\omega t + \phi)) \tag{29}$$

the function $\mathcal{M}(t, t')$ depends only on the difference of the arguments, viz.

$$\mathcal{M}(t + \tau, t) = \frac{(ea)^2}{2m}\tau\left(1 - \left(\frac{\sin(\omega\tau/2)}{\omega\tau/2}\right)^2\right) \equiv \mathcal{M}(\tau). \tag{30}$$

In this case, $w(t) =$ const. solves Eq. (26),[33] and the quasi-energy E is a solution of the equation

$$\kappa = -\left(\frac{im}{2\pi}\right)^{1/2}\int_0^\infty \frac{d\tau}{\tau^{3/2}}(\exp i(\mathcal{M}(\tau) - E\tau) - 1). \tag{31}$$

For circular polarization, the determination of the wave function $\psi_0(\vec{r}, t)$ is thereby reduced to finding the quasi-energy. For some applications, $E \approx -|E_0|$ is a sufficient approximation. For a field which is not circularly polarized, $w(t)$ is no longer strictly constant, but for not too intense fields a constant is still a good approximation.[29]

Let us then evaluate the matrix element (7) for circular polarization. If we use the integral equation (20) and identify $\lim_{t\to\infty} \int d^3r\psi_{\vec{p}}^{(0)}(\vec{r}, t)^* iG^{(E)}(\vec{r}t, \vec{r}'t')$ with the Volkov wave function (12) (this is an approximation, since $\psi_{\vec{p}}^{(0)}(\vec{r}, t)$ is a wave function in the presence of the binding potential) we obtain

$$M_{\vec{p}} = -i\int d^3r dt \psi_{\vec{p}}^{(E)}(\vec{r}, t)^* V(\vec{r})\psi_0(\vec{r}, t). \tag{32}$$

Since for the delta-function potential

$$V(\vec{r})\psi_0(\vec{r}, t) = V(\vec{r})\psi_0^{(0)}(\vec{r}, t)\exp\left[-i\left(\Delta - \frac{i}{2}\Gamma\right)t\right], \tag{33}$$

it turns out that M_p is identical with its Keldysh-approximation (15) inasmuch as the ac-Stark shift Δ and the width Γ can be neglected. For the fields currently employed in ATI, this is an excellent approximation. In this sense, for a delta-function potential the Keldysh-approximation is almost exact.[22]

A short calculation for the field (29) with $\phi = 0$ then yields the explicit result

$$M_p \approx M_p^{(K)} = i\frac{\sqrt{\kappa}}{m}\sum_n e^{in\delta} J_n\left(\frac{eap_T}{m\omega}\right)\delta\left(|E_0| + \frac{1}{2m}(\vec{p}^2 + (ea)^2) - n\omega\right) \tag{34}$$

with $\tan\delta = p_y/p_x$ and $p_T = (p_x^2 + p_y^2)^{1/2}$ the momentum transverse to the propagation direction of the laser. Notice that in Eq. (34) only terms with $n \geq N$ contribute. The final-state interaction approximation discussed above yields essentially the same result[9,23] except that the order of the Bessel functions is $n - N$ instead of n.

PRODUCTION OF VERY HIGH HARMONICS

Besides above-threshold ionization, another effect that occurs when a gas sample is irradiated by a high-intensity laser is the emission of harmonics of the incident laser light. Standard nonlinear optics based on LOPT would lead one to expect a sequence of odd harmonics with intensities decreasing as $(I/I_0)^N$ where I is the intensity of the incident laser and I_0 a constant. The first experiment[34] that indicated a deviation from this expectation dealt with higher harmonic emission of neon irradiated by a laser with a wavelength of 248 nm and an intensity of about 10^{15} W/cm^2: The intensities of the harmonics drop as expected from the third to the eleventh harmonic by several orders of magnitude. This is followed, however, by a plateau extending up to the seventeenth harmonic along which the harmonic intensities are more or less equal. A series of recent experiments[35,36] on xenon, krypton, and argon with a laser wavelength of 1064 nm and intensities in the range of 10^{13} W/cm^2 at a gas pressure of 15 Torr has established this behavior as a general pattern of harmonic emission at high intensities. The following features are revealed: (1) odd harmonics of the incident frequency (up to the 33rd in Ar at 3×10^{13} W/cm^2), (2) the fifth harmonic is considerably less intense that the third(by approximately two orders of magnitude at 3×10^{13} W/cm^2), (3) starting with about the seventh harmonic the intensities of the higher harmonics remain comparable establishing a "plateau"-region, (4) the plateau ends at a fairly well defined harmonic number (having a pronounced dependence on the atom and the laser intensity) beyond which the intensities of the harmonics quickly drop below detectability decreasing by about one order of magnitude from one harmonic to the next higher one, (5) the absolute intensity of the harmonics decreases sharply with increasing ionization potential of the atom. Feature (2) is in agreement with what one expects from LOPT, features (3) and (4) apparently are not. (The existence of a plateau at a fixed intensity I is not necessarily incompatible with LOPT: the reference intensity I_0 introduced above may depend on N.[37,38] However, all existing evidence at this time suggests that the experiments of Refs. 34-36 cannot be explained on the basis of LOPT). Figure 9 reproduces a harmonic intensity distribution typical of those reported in Ref. 36.

A theoretical description of the observed production of higher harmonics is much more complicated than a theory of above-threshold ionization. For ATI, collective effects could be suppressed completely by performing the experiments at sufficiently low pressures. In contrast, production of higher harmonics requires comparatively high pressures. As a consequence, the atoms no longer emit (or absorb) independently. Maximum net emission of the entire gas sample in a given direction requires "phase matching"[39] which in a possibly complicated way depends on details of the atom, the laser pulse, and the gas jet. Phase matching becomes less effective for increasing harmonic number. It has been estimated that (in the case of a gas jet with Lorentzian shape) the phase-matching factor decreases by almost five orders of magnitude from the fifth to the 31st harmonic.[36] This would imply that the response of the individual atom has to *increase* by a corresponding factor in order to yield the observed plateau in the collective response. However, the collective aspects of higher harmonic emission at high intensities are at present only poorly understood.

Because of the complexity of the collective behavior, theoretical attempts have thus far only dealt with harmonic emission by a single atom using LOPT,[38,40] essential state models,[41] or computer simulations of one-dimensional model atoms.[42] A fairly realistic xenon atom has been treated[43] as well as, recently, the hydrogen atom in terms of the quasi-energy method.[38] There is also an analytically solvable model where the external field couples to the Lenz vector rather than the position vector.[44] References 38, 42 and 43 show unequivocal evidence of a plateau region.

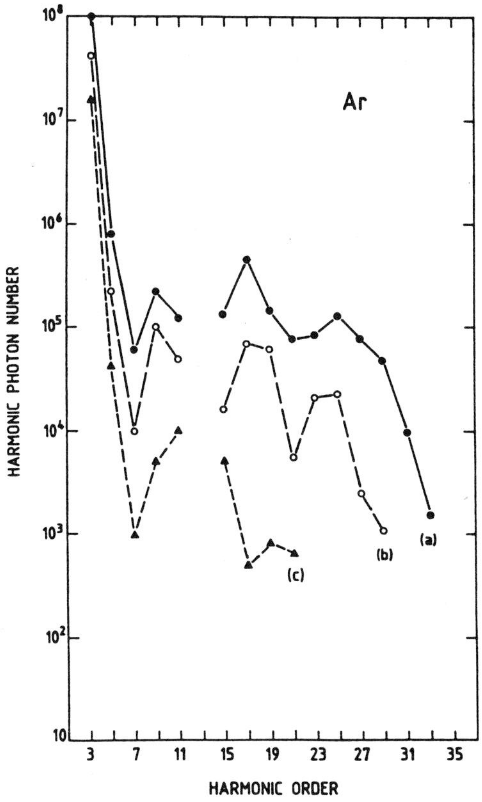

Fig. 9. Harmonic intensity distribution in argon at 15 Torr irradiated by a 1064-nm laser field at 3×10^{13} W/cm^2 (a), 2.2×10^{13} W/cm^2 (b) and 1.6×10^{13} W/cm^2 (c). No emission of the 13th harmonic was observed [from Li et al.[36]].

In the following we report results on harmonic production obtained for the three-dimensional delta-function well (19). We can obtain the quasi-energy wave function $\psi_0(\vec{r}, t)$ as described above. We can then evaluate the matrix element

$$M = \sqrt{\frac{2\pi\Omega}{V}} \int d^3 r dt e^{i\Omega t} \psi_0(\vec{r}, t)^* e\vec{r} \cdot \vec{\epsilon} \psi_0(\vec{r}, t) \tag{35}$$

describing emission of a photon with frequency Ω and polarization $\vec{\epsilon}$ by the atom in the field-dressed ground state. The quantity V denotes a normalization volume. Since we are dealing with quasi-energy states which do not have a fixed energy the atom can be in the same state before and after emission of a photon as implied by Eq. (35). We notice that the matrix element (35) is essentially the Fourier transform of the ground-state expectation value of the dipole moment of the ground state. As we already did for the description of the laser field, we here adopted the dipole approximation for the emitted photon, too. In general, we would have to replace $\exp(i\Omega t)$ by $\exp i(\Omega t - \vec{K} \cdot \vec{r})$ in order to include radiation from higher atomic multipole moments. These terms would also generate even harmonics. Along the axis of the incident laser radiation, however, they do not contribute. In the context of the dipole approximation, no higher harmonics are radiated in a circularly polarized laser field.

The actual calculation of the matrix element (35) for the delta-function well and a linearly polarized laser field is described elsewhere.[45] The calculation is exact except for letting $\Delta = \Gamma = 0$ and approximating the function $w(t)$ (Eq. (27)) by a constant. It can be performed analytically up to one remaining quadrature which has to be done numerically. A typical harmonic spectrum is shown in Fig. 10 for

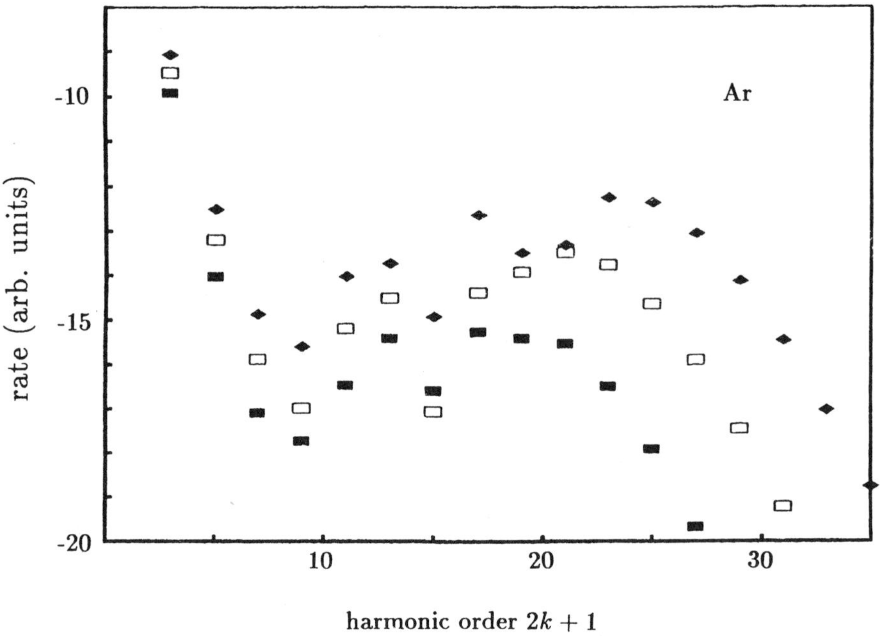

Fig. 10. Single-atom harmonic emission calculated for the three-dimensional delta-function potential (19) with the binding energy corresponding to argon. The parameters are the same as in Fig. 9. Black diamond: 3×10^{13} W/cm^2; open rectangle: 2.2×10^{13} W/cm^2; black rectangle: 1.55×10^{13} W/cm^2.

the case of argon. An initial rapid decrease of the harmonic intensities is followed by a somewhat ragged plateau which is on the average sloping upwards. From some order on, the intensities start to drop quickly, by about one order of magnitude per harmonic order. The calculated spectrum is for a single atom. For this reason it cannot directly be compared with the experimental spectrum of Fig. 9. Taking phase matching into account will reduce the intensities of the high harmonics and possibly lead to a flat or even downward sloping plateau in the collective response.

In any event, the outcome of this calculation shows that the observed plateau is not as such a collective effect since it is exhibited already in the emission of the single atom. Moreover, since the plateau is manifestly visible for a delta-function potential, it should be a very general effect not dependent on specific features of the atomic species.

REFERENCES

1. P. Agostini, F. Fabre, G. Mainfray, G. Petite, and N. K. Rahman, Phys. Rev. Lett. **42**, 1127 (1979).
2. P. Kruit, J. Kimman, H. G. Muller, and M. J. Van der Wiel, Phys. Rev. A**28**, 248 (1983).
3. G. Petite, P. Agostini, and H. G. Muller, J. Phys. B **21**, 4097 (1988).
4. T. W. B. Kibble, Phys. Rev. **150**, 1060 (1966).
5. P. Mulser, J. Opt. Soc. Am. B **2**, 1814 (1985).
6. H. G. Muller, A. Tip, and M. J. Van der Wiel, J. Phys. B **16**, L679 (1983).
7. E. Fiordilino and M. H. Mittleman, J. Phys. B **18**, 4425 (1985).
8. A. Szöke, J. Phys. B **18**, L427 (1985).
9. W. Becker, R. R. Schlicher, and M. O. Scully, J. Phys. B **19**, L785 (1986).
10. P. W. Milonni, and J. R. Ackerhalt, Phys. Rev. A **39**, 1139 (1989).
11. T. W. B. Kibble, A. Salam, and J. Strathdee, Nucl. Phys. B **96** (1975).
12. G. Petite, P. Agostini, and I. Yergeau, J. Opt. Soc. Am. B **5**, 765 (1987).

13. H. G. Muller, H. B. van Linden van den Heuvell, P. Agostini, G. Petite, A. Antonetti, A. Franco, and A. Migus, Phys. Rev. Lett. **60**, 565 (1988).
14. P. H. Bucksbaum, R. R. Freeman, M. Bashkansky, and T. J. McIlrath, J. Opt. Soc. Am. B **4**, 760 (1987).
15. R. R. Freeman, P. H. Bucksbaum, H. Milchberg, S. Darack, D. Schumacher, and M. E. Geusic, Phys. Rev. Lett. **59**, 1092 (1987).
16. P. Agostini, A. Antonetti, P. Breger, M. Crance, A. Migus, H. G. Muller, and G. Petite, J. Phys. B **22**, 1971 (1989).
17. L. V. Keldysh, Zh. Eksp. Teor. Fiz. **47**, 1945 (1964) [Sov. Phys. JETP **20**, 1307 (1965)].
18. F. H. M. Faisal, J. Phys. B **6**, L312 (1973).
19. H. R. Reiss, Phys. Rev. A **22**, 1786 (1980).
20. H. R. Reiss, J. Phys. B **20**, L79 (1987); and H. R. Reiss, J. Opt. Soc. Am. B **4**, 726 (1987).
21. R. Burlon, C. Leone, F. Trombetta, and G. Ferrante, Nuovo Cimento 9D, **1033** (1987).
22. W. Becker, J. K. McIver, and M. Confer, Phys. Rev. A (1989).
23. W. Becker, R. R. Schlicher, M. O. Scully, and K. Wódkiewicz, J. Opt. Soc. Am. B **4**, 743 (1987).
24. S. P. Goreslavsky, N. B. Narozhny, and Y. P. Yakovlev, J. Opt. Soc. Am. B **6**, 1752 (1989).
25. G. Petite, P. Agostini, and H. G. Muller, J. Phys. B **21**, 4097 (1988).
26. J. Javanainen and J. H. Eberly, Phys. Rev. A **39**, 458 (1989).
27. L. A. Collins and A. L. Merts, Phys. Rev. A **37**, 2415 (1988); J. N. Bardsley, A. Szöke, and M. J. Comella, J. Phys. B **21**, 3899 (1988).
28. A. Szöke, *Atomic and Molecular Processes with Short Intense Laser Pulses*, 1988, ed. A. D. Bandrauk, NATO ASI Series, Vol. 171 (Plenum, New York), p. 207; R. M. Potvliege and R. Shakeshaft, Phys. Rev. A **38**, 4597 (1988).
29. N. L. Manakov and A. G. Fainshtein, Zh. Eksp. Teor. Fiz. **79**, 751 (1980) [Sov. Phys. JETP **52**, 382 (1981)].
30. E. Fermi, Ric. Sci. **7**, 13 (1936).
31. S. Grossmann and T. T. Wu, J. Math. Phys. **25**, 1742 (1984).
32. K. Wódkiewicz, to be published.
33. I. J. Berson, J. Phys. B **8**, 3078 (1975).
34. A. McPherson, G. Gibson, H. Hara, U. Johann, T. S. Luk, I. A. McIntyre, K. Boyer, and C. K. Rhodes, J. Opt. Soc. Amer. B **4**, 595 (1987).
35. M. Ferray, A. L'Huillier, X. F. Li, L. A. Lompré, G. Mainfray, and C. Manus, J. Phys. B **21**, L31 (1988).
36. X. F. Li, A. L'Huillier, M. Ferray, L. A. Lompré, and G. Mainfray, Phys. Rev. A **39**, 5751 (1989).
37. B. Gao and A. F. Starace, Phys. Rev. A **39**, 4550 (1989).
38. M. Potvliege and R. Shakeshaft, Phys. Rev. A **40**, 3061 (1989).
39. J. F. Reintjes, *Nonlinear Optical Parametric Processes in Liquids and Gases*, (Academic, New York, 1984); G. C. Bjorklund, IEEE J. Quantum Electron. **QE-11**, 287 (1975).
40. Y. Gontier and M. Trahin, IEEE J. Quantum Electron. **QE-18**, 1137 (1982); R. M. Potvliege and R. Shakeshaft, Z. Phys. D **11**, 93 (1989); L. Pan, K. T. Taylor, and C. W. Clark, Phys. Rev. A **39**, 4894 (1989).
41. B. W. Shore and P. L. Knight, J. Phys. B **20**, 413 (1987).
42. J. H. Eberly, Q. Su, and J. Javanainen, Phys. Rev. Lett. **62**, 881 (1989); J. H. Eberly, Q. Su, and J. Javanainen, J. Opt. Soc. Amer. B **6**, 1289 (1989).
43. K. C. Kulander, and B. W. Shore, Phys. Rev. Lett. **62**, 524 (1989).
44. L. C. Biedenharn, G. A. Rinker, and J. C. Solem, J. Opt. Soc. Amer. B **6**, 221 (1989).
45. W. Becker, S. Long, and J. K. McIver (to be published).

VIRTUAL CLOUDS IN QUANTUM OPTICS

G. Compagno, G.M. Palma, R. Passante, F. Persico

Istituto di Fisica dell' Universita' and IAIF-CNR
Palermo, Italy

I. DRESSED SOURCES IN THE GROUND STATE

I.1. Qualitative Description.

The sources of a field, even at absolute zero temperature, are generally surrounded by a cloud of virtual particles constituted by quanta of the field [1]. From a qualitative point of view, the origin of the cloud can be understood as follows. The source is subjected to quantum fluctuations because of its interaction with the field. In the course of a fluctuation one or more "virtual" quanta of the field may appear, depending on the form of the source-field coupling. The unperturbed energy E of the source-field system is not necessarily conserved during such a fluctuation, and states of unperturbed energy $E' \neq E$ may be attained. The magnitude of the energy unbalance $\delta E = |E'-E|$, however, is constrained by the Heisenberg uncertainty principle

$$\delta E \approx \hbar / \tau \tag{1.1}$$

where τ is the duration of the fluctuation, at the end of which the energy must balance again and the extra quanta of the field are to be reabsorbed. Because of their quantum nature these processes take place also at 0°K.

Since fluctuations take place continuously, the source can be described as surrounded by a steady-state cloud of virtual quanta continuously emitted into, and reabsorbed from, the field. We shall call "dressed source" the bare source together with its virtual cloud. In contrast with the behaviour of a real quantum, which is emitted by an energy-conserving process ($E'=E$) and which can abandon the source in view of its infinite lifetime, a virtual quantum can only reach out at finite distance from the source

$$r \approx c\tau \approx \hbar c/\delta E \tag{1.2}$$

where c is a typical velocity of the field. Consequently one should expect the linear dimensions of the virtual cloud surrounding the source to coincide roughly with (1.2) for virtual transitions characterized by an energy unbalance δE. Well-known examples of applicability of (1.2) are the following.

New Frontiers in Quantum Electrodynamics and Quantum Optics
Edited by A. O. Barut, Plenum Press, New York, 1990

- A meson field of mass m coupled to a static source fixed at the origin, described by Hamiltonian

$$H = H_0 + H' \; ; \quad H_0 = \sum_{\underline{k}} \hbar\, \omega_k \, a^+_{\underline{k}} a_{\underline{k}}$$

$$H' = -g \sum_{\underline{k}} \sqrt{\frac{\hbar}{2V\omega_k}} \left(\rho_{\underline{k}}\, a_{\underline{k}} + \rho^*_{\underline{k}}\, a^+_{\underline{k}} \right) \; ; \; \omega_k = c\sqrt{k^2 + \mu^2} \tag{1.3}$$

where V is the field quantization volume, g is the source-field coupling constant, a_k and a^+_k are the Bose operators for the meson field, $\rho_{\underline{k}}$ is the Fourier transform of the source density $\rho(\underline{r})$ and $\mu^{-1} = \hbar/mc$ is the Compton wavelength of the meson. Clearly the source is dressed by a cloud of mesons. Since the minimum energy required to create a meson is $\delta E = mc^2$, use of (1.2) yields for the meson cloud surrounding the source $r \approx \hbar/mc$, which is the meson Compton wavelength. For π-mesons ($m \approx 140$ MeV), $r \approx 1.4 \cdot 10^{-13}$ cm, which is of the right order of magnitude for the experimental mean-square radius of the proton and of the neutron [2].

- An electron in the conduction band of a polar lattice interacts with the longitudinal optical phonons according to the Hamiltonian

$$H = H_0 + H' \; ; \quad H_0 = \hbar\, \omega_0 \sum_{\underline{k}} a^+_{\underline{k}} a_{\underline{k}} + \frac{1}{2m} P^2$$

$$H' = \sum_{\underline{k}} \left(V_{\underline{k}}\, a^+_{\underline{k}}\, e^{-i\underline{k}\cdot\underline{r}_e} + V^*_{\underline{k}}\, a_{\underline{k}}\, e^{i\underline{k}\cdot\underline{r}_e} \right) \tag{1.4}$$

where a_k and a^+_k are Bose operators for the longitudinal phonon field modes of frequency ω_0, $\underline{r}_e$ and $\underline{p}$ are the position and momentum operators of the electron of bare mass m, and V_k is the electron-phonon coupling constant. Here the electron is dressed by a cloud of phonons which may be visualized as a lattice distortion accompanying the source in its motion. The electron being a light particle, and in practice the only mobile one due to the low group velocity of the optical phonons, its recoil cannot be neglected like in the previous example. Thus the energy fluctation δE in (1.2) is $\hbar\omega_0 + p^2/2m$. The strongest interaction is with phonons of wavelength λ such that in a time $2\pi/\omega_0$ the electron travels over a distance corresponding to λ; this yields $2\pi p/m\,\omega_0 = \lambda$ or $p = m\,\omega_0/k$. Moreover, from momentum conservation during the fluctuation $p = \hbar k$ one gets $p^2/2m = \hbar\omega_0/2$, and $\delta E \approx 3\hbar\omega_0/2$. Using for c in (1.2) the velocity of the electron $c = p/m = [\hbar\omega_0/m]^{1/2}$, one obtains $r \approx (2/3)[\hbar/\omega_0 m]^{1/2}$. Using for m the free electron mass, one gets $r \approx 10^{-7}$ cm, in good agreement with the so-called "polaron radius" evaluated from experimental data on various semiconductors and insulators [3].

- A two-level atom interacts with the vacuum photon field in the Coulomb gauge according to the Hamiltonian [4]

$$H = H_0 + H' \; ; \quad H_0 = \hbar\omega_0 S_z + \sum_{\underline{k}j} \hbar\, \omega_{kj}\, a^+_{\underline{k}j} a_{\underline{k}j}$$

$$H' = \sum_{\underline{k}j} \left(\varepsilon_{\underline{k}j}\, a_{\underline{k}j}\, S_+ + \varepsilon^*_{\underline{k}j}\, a^+_{\underline{k}j}\, S_- \right) - \sum_{\underline{k}j} \left(\varepsilon_{\underline{k}j}\, a^+_{\underline{k}j}\, S_+ + \varepsilon^*_{\underline{k}j}\, a_{\underline{k}j}\, S_- \right) \tag{1.5}$$

where ω_0 is the energy difference of the two atomic levels, $S_{z,\pm}$ are pseudospin atomic S=1/2 operators, a_{kj} and a^+_{kj} are Bose operators for the transverse photons of momentum $\underline{k}$ and polarization index j (frequency =ck) and ε_{kj} are the atom-photon coupling constants. Thus the ground state atom is surrounded by a cloud of photons. The energy unbalance in one of the processes leading the atom from the ground to the excited state is $\delta E = (\hbar\omega_0 + \hbar\omega_k)$. This can be substituted in (1.2), which in the limit of long-wavelength photons ($\omega_k \approx 0$) yields $r \approx c/\omega_0 = \lambda_0/2\pi$, where λ_0 is the wavelength of the atomic frequency. Thus we obtain typical linear dimensions for the dressed atom which are of the order of $10^{-4} \div 10^{-5}$ cm.

I.2. The Virtual Cloud

For all the examples discussed above, the eigenstates $|\phi_n\rangle$ of H_o will be called the "bare" states of the system. The "dressed" states $|\tilde{\phi}_n\rangle$ are obtained as appropriate linear combinations of $|\phi_n\rangle$, coinciding with the eigenstates of H if these exist. The dressed ground state $|\tilde{0}\rangle$ of H in particular, apart from suitable normalization constants, can be expressed as

$$|\tilde{0}\rangle = |0\rangle + |0'\rangle \tag{1.6}$$

where $|0\rangle$ is the ground state of H_o and $|0'\rangle$ is a correction due to H' which is a mixture of eigenstates of H_o with various numbers of field quanta.

In order to obtain a quantitative description of the virtual cloud, one introduces an appropriate operator $\Psi(\underline{r})$, whose quantum average on the dressed ground state $\langle\tilde{0}|\Psi(\underline{r})|\tilde{0}\rangle$ is deemed to yield a reliable idea of the virtual field as a function of the distance $\underline{r}$ from the source. Thus the criterion for $\Psi(\underline{r})$ is slightly ambiguous in principle, but by considering the previous three examples we shall realize that this ambiguity does not seem to have undesirable consequences in practise.

- For a static point source-meson field described by Hamiltonian (1.3) with $\rho_{\underline{k}} = 1$, we may choose

$$\Psi(\underline{r}) = \Phi(\underline{r}) = \sum_{\underline{k}} \sqrt{\frac{\hbar}{2V\omega_k}} \left(a_{\underline{k}}^{+} e^{-i\underline{k}\cdot\underline{r}_e} + a_{\underline{k}} e^{i\underline{k}\cdot\underline{r}_e} \right) \tag{1.7a}$$

where $\Phi(\underline{r})$ is the meson field amplitude and V is the quantization volume. This choice yields [2]

$$\langle \tilde{0} | \Phi(\underline{r}) | \tilde{0} \rangle = g \frac{e^{-\mu r}}{4\pi c^2 r} \tag{1.7b}$$

Alternatively, we may choose for $\Psi(\underline{r})$ the energy density as given [5] by the time-component of the energy-momentum tensor T^{kl}

$$\Psi(\underline{r}) = T^{oo}(\underline{r}) = \frac{1}{2}\{\Pi^2(\underline{r}) + c^2[\nabla\Phi(\underline{r})]^2 + c^2\mu^2\Phi^2(\underline{r})\} \tag{1.8a}$$

where $\pi(\underline{r}) = \dot{\Phi}(\underline{r})$ is the canonical momentum conjugate to $\Phi(\underline{r})$. Choice (1.8a) yields [6]

$$\langle \tilde{0} | T^{oo}(\underline{r}) | \tilde{0} \rangle = g^2 \frac{e^{-2\mu r}}{32\pi^2 c^2 r^4} (2\mu^2 r^2 + 2\mu r + 1) \tag{1.8b}$$

Both (1.7b) and (1.8b) describe a dress which falls exponentially with a characteristic distance of the order of μ^{-1}, which is the same result as that obtained on the basis of the qualitative argument leading to (1.2).

- For the polaron case described by Hamiltonian (1.4) one can choose [7]

$$\Psi(\underline{r}) = V(\underline{r}) = -\frac{1}{e} \sum_{\underline{k}} \left(V_{\underline{k}}^{*} a_{\underline{k}}^{+} e^{-i\underline{k}\cdot\underline{r}_e} + V_{\underline{k}} a_{\underline{k}} e^{i\underline{k}\cdot\underline{r}_e} \right) \tag{1.9}$$

where $V(\underline{r})$ is the potential induced by the longitudinal optical phonons at point $\underline{r}$ and e is the electronic charge. Alternatively, one can choose

$$\Psi(\underline{r}) = \varrho(\underline{r}) = -(1/4\pi)\nabla^2 V(\underline{r})$$

where $\varrho(\underline{r})$ is the charge density induced in the lattice [8] by the phonon distortion. The two operators yield equivalent results. For intermediate coupling and assuming the kinetic energy to be negligible with respect to $\hbar\omega_0$, one has [8]

$$< \tilde{0} | \rho(\underline{r}) | \tilde{0} > = -\frac{em\omega_0}{2\pi h}\left(\frac{1}{\varepsilon_\infty} - \frac{1}{\varepsilon_0}\right)\exp\{-\frac{2m\omega_0}{h}|\underline{r} - \underline{r}_e|^2\} \tag{1.10}$$

where ε_∞ and ε_0 are the values of the dielectric constant $\varepsilon(\omega)$ of the crystal at infinite and at zero frequency respectively. Expression (1.10) exhibits good agreement with the qualitative conclusions obtained in I.1.

- For a multilevel atom, and in the minimal coupling scheme, several forms of $\Psi(\underline{r})$ have been used; in particular the so-called "Coarse Grained Energy Density" (CGED) [9]

$$W(\underline{r}) = \frac{1}{4\pi}[\underline{E}_\perp^{(-)}(\underline{r})\cdot\underline{E}_\perp^{(+)}(\underline{r}) + \underline{B}^{(-)}(\underline{r})\cdot\underline{B}^{(+)}(\underline{r})] \tag{1.11a}$$

where superscript (±) indicates positive and negative energy components of the transverse fields, as well as the total energy density [10]

$$H(\underline{r}) = \frac{1}{4\pi}[\underline{E}^2(\underline{r}) + \underline{B}^2(\underline{r})] \tag{1.11b}$$

For the multilevel atom in the multipolar scheme [11], matrix elements of the Poynting vector [12]

$$S_i(\underline{r}) = \frac{c}{8\pi}\varepsilon_{ijk}[E_{\perp j}(\underline{r})B_k(\underline{r}) + B_k(\underline{r})E_{\perp j}(\underline{r})] \tag{1.11c}$$

as well as of the magnetic energy density [13]

$$H_{mag}(\underline{r}) = \frac{1}{8\pi}\underline{B}^2(\underline{r})] \tag{1.11d}$$

were obtained. In all cases considered $<\tilde{0}|\Psi(\underline{r})|\tilde{0}>$, with the atom at r=0, is characterized by a r^{-n} dependence in the far region, which turns into a more complicated r-dependence in the near region, where n is a postive integer. For the two-level atom model of sec. I.1, taken in the multipolar representation and with dipole approximation, the choice $\Psi(\underline{r}) \equiv H_{el}(\underline{r})$, with

$$H_{el}(\underline{r}) = \frac{1}{8\pi}\underline{E}^2(\underline{r}) \tag{1.11e}$$

and with

$$\varepsilon_{\underline{k}j} = i\sqrt{\frac{2\pi\hbar\omega_k}{V}}\,\hat{e}_{\underline{k}j}\cdot\underline{d} \tag{1.12}$$

yields

$$<0|H_{el}(\underline{r})|0> = \begin{cases} \dfrac{\hbar c}{32\pi^2}\dfrac{2d_l d_n}{h\omega_0}\dfrac{1}{r^7}(13\delta_{ln} + 7\hat{r}_l\hat{r}_n) & (r >> \omega_0/c) \\ \dfrac{1}{2\pi}d^2\dfrac{1}{r^6} & (r << \omega_0/c) \end{cases} \tag{1.13}$$

where $\underline{d}$ is the matrix element of the electron dipole moment operator $e\underline{r}_e$ between the bare ground and excited atomic states. Also this result is in qualitative agreement with the simple considerations leading to (1.2).

Thus we conclude that quantitative calculations yield results in agreement with a simple qualitative model for the main features of the cloud of virtual particles surrounding ground-state field sources.

I.3. Number of Virtual Quanta in the Cloud

Information about the dressing of the source is contained in $|0'\rangle$ appearing in (1.6). A relevant question is if $|0'\rangle$ can be evaluated by perturbation expansion in powers of the coupling constant appearing in H'. In view of the structure of H', which connects the bare ground state (the "bare vacuum") to excited states of H_0 with different numbers of quanta in steps of one, it would appear reasonable to require, for the velidity of the perturbation expansion, that the average number $\langle N\rangle_0$ of quanta in the dressed ground state $|\tilde{0}\rangle$ be much smaller than unity. Thus

$$\langle N\rangle_0 \equiv \langle\tilde{0}|\sum_{\underline{k}} a_{\underline{k}}^{+} a_{\underline{k}}|\tilde{0}\rangle \quad \ll 1 \tag{1.14}$$

is a condition which we expect the number of quanta in the virtual cloud should satisfy for the validity of perturbation treatment of H'.

- For the meson model described by Hamiltonian (1.3), the total number of mesons in the cloud can be evaluated exactly, and it is given by

$$\langle N\rangle_0 = \frac{1}{2V} g^2 \frac{1}{\hbar} \sum_{\underline{k}} |g_{\underline{k}}|^2/\omega_k^3 = \frac{1}{4\pi^2\hbar c^3} g^2 \int_0^\infty \frac{|\rho_{\underline{k}}|^2}{(k^2+\mu^2)^{3/2}} k^2 dk$$

Assuming $|\rho_{\underline{k}}| = 1$ ($k<k_M$) and negligible for $k>k_M$ is equivalent to assigning a radius $a \approx k_M^{-1}$ to the source. Furthermore, if $k_M \gg \mu$, then a is much smaller than the Compton wavelength of the mesons. In these conditions we may approximate

$$\langle N\rangle_0 \sim \frac{1}{4\pi^2\hbar c^3} g^2 \int_\mu^{k_M} \frac{dk}{k} = \frac{1}{4\pi^2\hbar c^3} g^2 \ln\frac{k_M}{\mu} \tag{1.15}$$

Assuming [2] $k_M/\mu \approx 3$ and $g \approx 10^3$ cgs for a nucleon case, we obtain $\langle N\rangle_0 \approx 1$, which clearly shows that convergence of the perturbation expansion is not guaranteed.

- For the optical polaron model described by Hamiltonian (1.3) with

$$V_{\underline{k}} = \frac{\hbar\omega_0}{k}\left(\frac{\hbar}{2m\omega_0}\right)^{1/4}\left(\frac{4\pi\alpha}{V}\right)^{1/2} \quad ; \quad \alpha = \frac{e^2}{\hbar}\left(\frac{1}{\varepsilon_\infty} - \frac{1}{\varepsilon_0}\right)\left(\frac{m}{2\hbar\omega_0}\right)^{1/2} \tag{1.16}$$

a straightforward second-order perturbation theory yields

$$\langle N\rangle_0 = \sum_{\underline{k}} \frac{V_{\underline{k}}^2}{\Delta_{\underline{k}}^2} \sim \frac{2\alpha}{\pi} A^{1/2} \int_0^\infty \frac{dk}{(1+Ak^2)^2} = \frac{\alpha}{2} \tag{1.17}$$

$$A = \frac{\hbar}{2m\omega_0}$$

where $\Delta_{\underline{k}}$ is the energy difference between the phonon vacuum and the state where a scattering process has taken place with the creation of a photon of wavevector $\underline{k}$. α in (1.16) is only dependent on the crystal. For a typical covalent semiconductor its value ranges from 0.02 to 0.6; in these cases perturbation theory is appropriate since $\langle N\rangle_0 \ll 1$. On the other hand for alkali halides such as NaCl $\alpha \simeq 5$: perturbation theory breaks down and other methods are necessary.

- For the two-level atom model described by (1.5), and using (1.12), perturbation theory yields

$$\langle N \rangle_0 = \frac{2\pi}{\hbar V} d_l d_m \sum_{\underline{k}} (\delta_{lm} - \hat{k}_l \hat{k}_m) \frac{\omega_k}{(\omega_0 + \omega_K)^2}$$

where the sum rule

$$\sum_j (\hat{e}_{\underline{k}j} \cdot \underline{d})^2 = (\delta_{lm} - \hat{k}_l \hat{k}_m)\, d_l d_m$$

has been used. Transforming sums into integrals yields

$$\langle N \rangle_0 \sim \frac{2}{3} \frac{d^2}{\pi\hbar c} \int_0^{k_M} \frac{k^3}{(k_0 + k)^2} dk \sim \frac{1}{3} \frac{d^2 k_M^2}{\pi\hbar c} = \frac{4\pi}{3} \frac{e^2}{\hbar c} \sim 3 \cdot 10^{-2} \tag{1.18}$$

where $k_0 = \omega_0 / c$ and k_M is a cut-off frequency that one introduces to account for the finite atomic dimensions. In fact, we take $k_M \simeq a_0^{-1}$, which entitles us to neglect k_0-dependent terms in the integral in (1.18). We see that in the QED case use of perturbation theory is permitted in view of the smallness of the fine structure constant $e^2/\hbar c$.

I.4. Energy Shifts and Virtual Clouds

We now evaluate in (1.3)

$$\langle \tilde{0} | H_0 | \tilde{0} \rangle = g^2 \frac{1}{2V} \sum_{\underline{k}} |\rho_{\underline{k}}|^2 / \omega_k^2 \quad ; \quad \langle \tilde{0} | H' | \tilde{0} \rangle = -g^2 \frac{1}{V} \sum_{\underline{k}} |\rho_{\underline{k}}|^2 / \omega_k^2 \tag{1.19}$$

Thus the total energy shift of the dressed ground-state with respect to the bare one amounts to

$$\Delta = -g^2 \frac{1}{2V} \sum_{\underline{k}} |\rho_{\underline{k}}|^2 / \omega_k^2 \tag{1.20}$$

This shift is generally ascribed to a change δ_M in the mass of the source [2], in the sense that one puts $\Delta = -\delta_M c^2$ and considers the Hamiltonian

$$H_0 + H' + \delta_M c^2$$

whose ground-state is unshifted with respect to the bare ground state of H_0. The first of (1.19) shows that an important contribution to the mass renormalization of the meson source comes from the energy stored in the virtual cloud. Thus the meson virtual cloud contributes to the physical mass of the nucleon.

In the optical polaron case, and assuming α small enough in order to apply perturbation theory, the energy of an eigenstate of H_0 in (1.4) with the electron having an energy $p^2/2m$, is changed by H' into [8]

$$\frac{1}{2m} P^2 + \sum_{\underline{k}} \frac{|V_{\underline{k}}|^2}{\Delta_{\underline{k}}} \sim -\alpha\hbar\omega_0 + \left(1 - \frac{\alpha}{6}\right) \frac{1}{2m} P^2 \tag{1.21}$$

Thus the energy shift in (1.21) can be interpreted in terms of a renormalized (or dressed) mass $m^* = m/(1-\alpha/6)$; part of this shift can be ascribed to the elastic energy contained in the virtual cloud.

The two-level atom Hamiltonian (1.5) can be written as

$$H = H_A + H_F + H' \ ; \quad H_A = \hbar\omega_O S_z \ ; \quad H_F = \sum_{\underline{k}j} \hbar\, \omega_{\underline{k}j}\, a^+_{\underline{k}j} a_{\underline{k}j} \tag{1.22}$$

As it is well known, independently of the atom-field coupling scheme [14], the energy shift of the ground-state of the system due to this coupling is

$$-\frac{\hbar\omega_0}{2} + <\tilde{0}|H_A|\tilde{0}> + <\tilde{0}|H_F|\tilde{0}> + <\tilde{0}|H'|\tilde{0}> =$$

$$= -\frac{2}{3\pi c^3} d^2 \omega_0^2 \left(\omega_M - \omega_0 \ln\frac{\omega_M}{\omega_0} \right) + \frac{e^2\hbar}{2\pi m c^3}\omega_M^2 \tag{1.23}$$

The terms in ω_M are electron mass renormalization terms, whereas the term in $\ln(\omega_M/\omega_0)$ is the Lamb shift of the ground state [15]. It is evident that the virtual cloud contributes directly to the total shift. In (1.23)

$$\omega_M = c k_M = c a_0^{-1} \tag{1.24}$$

is a cut-off frequency which entails neglect of wavelengths shorter than the Bohr radius a_0 and which accounts for finite atomic dimensions.

We conclude that the presence of the virtual cloud contributes to shift the bare levels. In the atom-photon case part of the total shift is called the Lamb shift, whereas in the meson and in the polaron cases the shifts are interpreted only in terms of mass renormalization.

I.5. Forces and Virtual Clouds

Suppose we have two static meson sources at points 0 and $\underline{r}_0$. Then it is convenient to put $\rho(\underline{r}) = \rho_1(\underline{r}) + \rho_2(\underline{r})$ with $\rho_1(\underline{r})$ and $\rho_2(\underline{r})$ localized around 0 and $\underline{r}_0$ respectively. Consequently the Hamiltonian is

$$H = \sum_{\underline{k}} \hbar\,\omega_{\underline{k}}\, a^+_{\underline{k}} a_{\underline{k}} - g\sum_{\underline{k}} \sqrt{\frac{\hbar}{2V\omega_k}}\, [\rho^*_{1\underline{k}}(1 + e^{i\underline{k}\cdot\underline{r}_0}) a_{\underline{k}} + \text{h.c.}] \tag{1.25}$$

and

$$|\rho_{\underline{k}}|^2 = 2|\rho_{1\underline{k}}|^2 (1 + \cos \underline{k}\cdot\underline{r}_0)$$

If the spread of ρ_1 and ρ_2 is small, then $|\rho_{1\underline{k}}|^2 \approx 1$. Substituting this in (1.20) yields

$$\Delta \sim 2\Delta_0 - g^2 \frac{1}{V}\sum_{\underline{k}} \frac{\cos \underline{k}\cdot\underline{r}_0}{\omega_k^2} = 2\Delta_0 - \frac{1}{4\pi c^2} g^2 \frac{1}{r_0} e^{-\mu r_0} \tag{1.26}$$

where Δ_0 is the energy shift of each source, evaluated as if the other did not exist. The r_0-dependence of Δ in (1.26) produces an attractive force $\partial\Delta/\partial r_0$ between the two sources. This is described in terms of the Yukawa potential $\Delta - 2\Delta_0$ between the two nucleons [2]. At the same time we see from the first of (1.19) that the appearance of the Yukawa forces is related to the presence of interference terms in the energy density of the virtual cloud. Pictorially these interference terms become more evident if one considers the Feynman graphs contributing to Δ

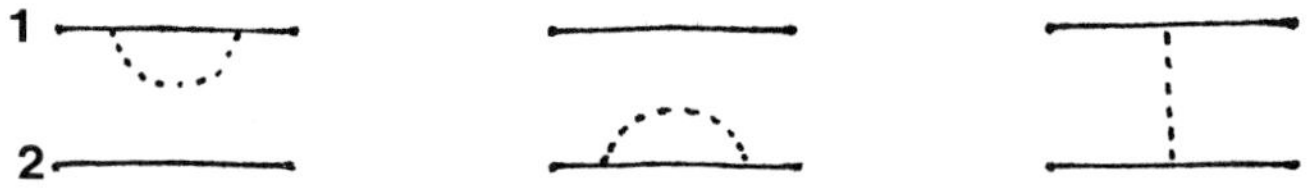

It is the third of these graphs which originates the interference terms and yields the Yukawa forces.

The case of two polarons is much more complicated because of the electron recoil. It is interesting, however, to consider the result that is obtained if the electron mass is considered infinitely large, even if this model is not very realistic. If we have two electrons at 0 and $\underline{r}_0$, from (1.4) we have

$$H = \hbar\omega_0 \sum a^+_{\underline{k}} a_{\underline{k}} + \sum [V^*_{\underline{k}} a_{\underline{k}} (1 + e^{i\underline{k}\cdot\underline{r}_0}) + h.c.] \tag{1.27}$$

Thus the structure of the Hamiltonian is formally the same as in (1.25) and we can use result (1.26) for the total ground-state shift. One obtains [16]

$$\Delta = -\frac{2}{\hbar\omega_0}\sum_{\underline{k}}|V_{\underline{k}}|^2 - \frac{2}{\hbar\omega_0}\sum_{\underline{k}}|V_{\underline{k}}|^2 \cos\underline{k}\cdot\underline{r}_0$$
$$= -\frac{2}{\hbar\omega_0}\sum_{\underline{k}}|V_{\underline{k}}|^2 - \left(\frac{1}{\varepsilon_\infty} - \frac{1}{\varepsilon_0}\right)\frac{e^2}{r_0} \tag{1.28}$$

where (1.16) has been used and the integrations performed. The first of the two terms on the RHS of (1.28) is twice the ground-state shift of the isolated polaron, and it is independent of r_0. The second term is an attractive interaction of the Coulomb form which opposes the repulsion between the two bare electrons. Because of the formal identity of (1.25) and (1.26), the same considerations about the interference terms in the energy density of the virtual cloud apply. The main difference between (1.26) and (1.28) is the difference in the potential range. This is obviously related to the finite mass of the virtual meson, in contrast with the zero mass of the phonons. Reintroducing electron recoil, the Feynman graphs contributing to Δ in (1.28) are

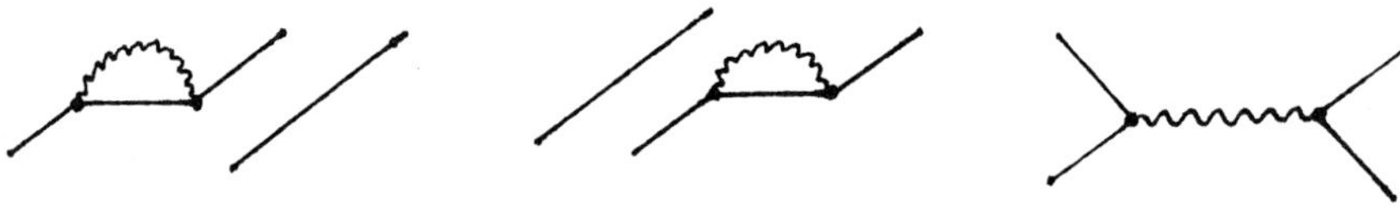

For two identical two-level atoms at points $\underline{r}_1$ and $\underline{r}_2$, the generalization of Hamiltonian (1.5) can be shown to be approximately transformed unitarily into

$$H = -\hbar\omega_0 + \sum_{\underline{k}j}\hbar\,\omega_{\underline{k}j}\, a^+_{\underline{k}j} a_{\underline{k}j} + H'$$
$$H' = -\frac{1}{2}\alpha^{(1)}_{lm}\underline{E}_{\perp l}(\underline{r}_1)\underline{E}_{\perp m}(\underline{r}_1) - \frac{1}{2}\alpha^{(2)}_{lm}\underline{E}_{\perp l}(\underline{r}_2)\underline{E}_{\perp m}(\underline{r}_2) \tag{1.29}$$

by a transformation due to Craig and Power [17]. In (1.29)

$$\underline{E}_\perp(\underline{r}) = i\sum_{\underline{k}}\sqrt{\frac{\hbar 2\pi\omega_k}{V}}\,(\hat{e}_{\underline{k}j} a_{\underline{k}j} e^{i\underline{k}\cdot\underline{r}} - \hat{e}^*_{\underline{k}j} a^+_{\underline{k}j} e^{-i\underline{k}\cdot\underline{r}})$$
$$\alpha^{(i)}_{lm} = 2d^{(i)}_l d^{(i)}_m/(\hbar\omega_0) \tag{1.30}$$

This Hamiltonian cannot be expected to treat correctly the modes of frequency $\omega_k > \omega_0$. Consequently the atom-atom interaction is treated correctly only for distances $|\underline{r}_1-\underline{r}_2| \gg \lambda_0$ (far zone). A second-order perturbation treatment yields the ground-state shift as

$$\Delta = -\sum_{\underline{k}_2 j_2, \underline{k}_1 j_1} \frac{\langle 0 | H' | \underline{k}_2 j_2, \underline{k}_1 j_1 \rangle \langle \underline{k}_2 j_2, \underline{k}_1 j_1 | H' | 0 \rangle}{\hbar(\omega_{k_1} + \omega_{k_2})} =$$

$$= -\frac{\hbar c}{8\pi} \frac{1}{R^7} C_{lm} \alpha_{lm}^{(1)} \alpha_{lm}^{(2)} \tag{1.31}$$

where C_{lm} are appropriate numerical coefficients $C_{11}=C_{22}=13$; $C_{33}=20$; $C_{12}=26$; $C_{23}=C_{13}=-30$ and where $R = |\underline{r}_1 - \underline{r}_2|$. This is the celebrated Van der Waals attractive potential (in the far zone). Feynman graphs contributing to (1.31) are of the form

Interference terms in the energy density of the photon cloud have been related to the Van der Waals force [18]. For electrically isotropic bodies $\alpha_{lm} = \alpha\, \delta_{lm}$, and the interaction term in (1.29) becomes

$$H' = -\frac{1}{2}\alpha^{(1)} \underline{E}_\perp^2(\underline{r}_1) - \frac{1}{2}\alpha^{(2)} \underline{E}_\perp^2(\underline{r}_2) \tag{1.32}$$

which displays rather clearly the role of the energy density in Van der Waals forces.

II. HALF-DRESSED SOURCES

II.1. Introduction

We have seen that normally a source at 0°K is surrounded by a cloud of quanta of its own virtual field. This composite particle is called a "dressed source". We have also discussed the importance of this virtual cloud for the interaction between two sources. In fact, we have seen that this interaction is mediated by the exchange of virtual quanta between the two sources. It follows that bare sources, insofar as they are completely deprived of their virtual clouds, are noninteracting objects and as such difficult to observe.

In between these two extreme conditions, namely a completely bare and a completely dressed source, it is possible however to imagine intermediate situations in which the virtual cloud of a source at a given time t=0 is not the same as in equilibrium because it has been perturbed in some way. Then we may legitimately ask in which way the dress of the source at t=0 will regain its equilibrium configuration and what will the physical properties of a source be during the process of regeneration of the virtual cloud. A simple example is that of an atom in a local static field which can be switched on and off. This changes the atomic energy levels via Stark shift [19] and hence the virtual cloud, which obviously can become time-dependent.

A source whose virtual dress is incomplete, or more generally out of equilibrium, has been called a "half-dressed source" by Feinberg [20], who has also provided the first preliminary discussion of their physics by examples from QED, QMD and QCD. Noticeably, he interprets the Landau-Pomeranchuk effect [21], whereby a relativistic electron experiencing multiple scattering has a reduced bremsstrahlung cross-section, in terms of an electron which in the first scattering event loses part of its virtual cloud, in such a way that in the successive scattering acts it is not able to yield the expected bremmstrahlung radiation. Similarly, Feinberg interprets the very weak dependence on the atomic number A of the fraction K of energy lost by a high-energy proton in a proton-nucleus collision ($K \propto A^{0.06}$) in terms of the nucleon losing part of its external cloud of virtual pions in the first scattering event within a nucleus, and thereby losing its ability to interact strongly with the rest of the nuclear constituents.

In both the examples mentioned above, reduction of the cross section takes place during the regeneration time, which is defined as the time in the laboratory frame of reference which is necessary to reconstruct the virtual cloud. In the particle centre of mass frame, this time is in general given by the time taken by a signal to traverse the fully dressed particle.

II.2. Models of Bare and Half-dressed Sources

In what follows we shall consider various models of half-dressed sources. From what has been said it is evident that a great variety of half-dressed states exist for the same source-field system, corresponding to the infinitely many ways in which the virtual cloud can be distorted from its equilibrium configuration. In order to simplify the task and to make the model mathematically tractable, we shall find it useful to define an abstract entity which we call the "bare source". This entity, generally speaking, should be included in the same class of abstract concepts such as point mass or point charge, which are obtained by a conceptual limiting process of masses or charges of smaller and smaller dimensions, but which do not really exist in the sense that they cannot be attained in practice; nonetheless they are extremely useful, for example, in the realm of classical physics. Masses and charges may be treated as point-like when their physical dimensions are negligible with respect to the typical lengths which characterize the experiment or the calculation in which they are thought to be involved. For example, in a calculation involving the electrostatic potential, a set of classical charges can be considered as point-like if in a multipolar expansion one can neglect all the multipolar terms and retain the monopole only; as it is well known, this happens when the distance r at which the potential is being evaluated is much larger than all the mutual intercharge distances R. The point-charge limit is attained either for $R \to 0$ or for $r \to \infty$. Consider now a hydrogen atom at the centre of a perfect cavity [22]. Moreover, suppose that the cavity is spherical of radius L. This is obviously also the linear dimension of the cloud of virtual photons in the ground state of the system. Note that for $L \gg a_0$ the changes induced in the vacuum physical electron mass and in the Lamb shift by the cavity are negligible for our purposes [23], since mostly high-frequency virtual photons, which are not influenced much by the cavity, contribute to these effects. In fact the system, with respect to a normal dressed atom, is deprived of the low-frequency part of the spectrum of virtual photons with wavelengths larger than L, which can reach out at distances $r>L$. We now imagine that the cavity walls disappear suddenly at $t=0$. In this way at $t=0$ we have a fairly well defined atomic structure, a virtual energy density which vanishes at $r>L$ and a spectrum of virtual photons which is missing of the low-frequency part (wavelengths $>L$): this is strictly speaking a half-dressed hydrogen atom, but it may be considered an effectively bare hydrogen atom (in analogy with an effective classical point-charge) if viewed from a distance $r \gg L$ which can be attained only by the missing long-wavelength photons, or if one is concerned with the low-frequency part of the spectrum only.

In the case of a polaron in a semiconductor one is confronted with a simpler problem. The phonon spectrum of the system is in fact bound by the

Debye frequency $\omega_D \approx 10^{13}$ sec^{-1}, corresponding to a wavelength λ_D of the order of interatomic distances. Thus if one penetrates the phonon cloud to a distance of the order of λ_D from the electron one should be able to observe the source as phononwise bare. Furthermore the mechanism of creation of the polaron, which can be obtained by shining light of frequency $\approx 10^{15}$ sec^{-1} corresponding to the gap between valence and conduction band and thereby creating an unbound electron-hole pair at t=0, provides a model of an initially bare electron, since the lattice around the electron cannot adapt to the new situation faster than ω_D^{-1}, that is long after the "creation" of the electron.

II.3. Dressing and Undressing Processes

In this section we consider the time-dependence of the virtual cloud of some of the models discussed above.

- Polaron. This case has been discussed in terms of the Hamiltonian

$$H = \sum_m E_{mn}\,\sigma_{mn} + \sum_k \hbar\omega_k\, a_k^+ a_k + \sum_{k,m} V_k\,(a_k^+ + a_{-k})\,\sigma_{mn} \qquad (2.1)$$

where the operators satisfy the commutation rules

$$[\sigma_{ij},\sigma_{mn}]=\delta_{jm}\sigma_{in}-\delta_{in}\sigma_{mj}; \quad [a_k,a_{k'}]=\delta_{kk'}.$$

Hamiltonian (2.1) is different from (1.4) in that it represents a polaron bound to a one-dimensional lattice site. E_m are the bare eigenvalues of the localized electron and σ_{mn} is the operator which takes the electron from state |m> to state |n>. Moreover, ω_k in (2.1) belongs to an acoustic branch with $\omega_k = c|k|$. The electron is assumed to be bare at t=0, and the time evolution of the quantum average of the lattice displacement operator

$$U_n = \sqrt{\frac{\hbar}{2mN}}\sum_k \frac{1}{\sqrt{\omega_k}} e^{-ikn}\,(a_k^+ + a_k) \qquad (2.2)$$

on state |0> is evaluated [24]. The results can be illustrated as follows: two counterpropagating symmetrical pulses appear at t=0 at the electron site from the unperturbed lattice. These two pulses diverge at the velocity of sound c, leaving a static distortion behind which is typical of the normal polaron and which is bound to the localized electron.

- Nucleon. The model Hamiltonian is the same as in (1.3). The point-source is assumed to be bare at r=0 and at t=0. The time-dependent quantum average of the virtual energy density operator (1.8a) at point r is evaluated [25] on state |0>, with the following result

$$<0|\Pi^2(r)|0> = \frac{1}{16\pi^2c^2}\, g^2\, \frac{1}{r^2}\left[\frac{\partial F(r,t)}{\partial r}\right]^2$$

$$<0|\Phi^2(r)|0> = \frac{1}{16\pi^2c^2}\, g^2\, \frac{1}{r^2}\left[\frac{\partial}{\partial r}\int_0^t F(r,t')\,dt'\right]^2$$

$$<0|[\nabla\Phi(r)]^2|0> = \frac{1}{16\pi^2c^2}\, g^2\, \frac{1}{r^2}\left[\frac{\partial}{\partial r}\frac{1}{r}\frac{\partial}{\partial r}\int_0^t F(r,t')\,dt'\right]^2 \qquad (2.3)$$

Since

$$F(t,r) = J_0\left(\mu c\sqrt{t^2 - r^2/c^2}\right)\Theta(t - r/c) \qquad (2.4)$$

the energy density in (2.3) behaves causally, and the effect of creation of the bare source at t=0 is felt only at points within a sphere of radius ct centered on the source. This we shall call the "causality sphere" for obvious reasons. We note that the source of the sphere is the site of a singularity of the energy density which we will not discuss here [25]. Asymptotically

$$F(r,\infty) = 0 \qquad ; \int_0^t F(r,t)\,dt = \frac{1}{\mu c} e^{-\mu r} \tag{2.5}$$

and the total energy density for $t \rightarrow \infty$ readjusts to the dressed ground-state value (1.8b). At each point $\underline{r}$ in space this readjustment is practically completed after a time $\tau \approx (1/c)[r^2+1/\mu^2]^{1/2}$. The disappearance of a fully dressed source has also been investigated. In this case at t=0 the energy density within a sphere of radius ct decreases to zero, whereas the energy density outside this causality sphere remains unchanged.

- Two-level atom. Here we describe this by a Craig-Power (CP) Hamiltonian of the form (1.29)

$$H = -\frac{1}{2}\hbar\omega_o + \sum_{\underline{k}j}\hbar\omega_k a^+_{\underline{k}j}a_{\underline{k}j} - \frac{1}{2}\alpha_{mn}E_{\perp m}(0)E_{\perp n}(0) \tag{2.6}$$

where the atom is placed at the origin and α_{mn} is given by (1.30). The atom is assumed to be bare at t=0, and the time-development of the energy density is investigated. The result is the following [25]

$$< H_{el}(r) > = \frac{\hbar c\alpha_{mn}}{32\pi^2}\frac{1}{r^7}(13\delta_{mn} + 7\hat{r}_m\hat{r}_n)[1 - \Theta(r - ct)]$$

$$< H_{mag}(r) > = -\frac{7\hbar c\alpha_{mn}}{32\pi^2}\frac{1}{r^7}(\delta_{mn} - \hat{r}_m\hat{r}_n)[1 - \Theta(r - ct)] \tag{2.7}$$

In the limit $t \rightarrow \infty$ the first of (2.7) yields the far-zone static expression obtained in (1.13). In view of the limitations of the CP approximation, the near-zone expression cannot be obtained from Hamiltonian (2.6). At finite times, the energy density at $\underline{r}$ vanishes as long as t<r/c and it attains the ground-state value after time r/c. The surface of the causality sphere is the site of a singularity of the energy density. The energy deposited into the virtual field is contained in this singularity, whose energy varies with time (in the isotropic case $\alpha_{mn}=\alpha\delta_{mn}$) as

$$\frac{\alpha hc}{\pi} \cdot \frac{1}{(ct)^4} \tag{2.8}$$

In the complementary case where a fully dressed source is suddenly decoupled from the field at t=0, a sphere of radius ct expands outward from the source. The energy density within this causality sphere vanishes, whwreas outside it has the normal ground-state value. The singularity of the energy density on the surface of the sphere has the same form as in the previous case, but opposite sign. Thus the singularity sweeps out, during its motion, all the ground-state virtual field, leaving the bare vacuum behind.

II.4. Detection of the Dressing Process

We consider a gedanken experiment based on the following model. The coupling of a source S to the field can be suddendly switched on or off. S is fixed at the origin of the reference system. A second source T placed at $\underline{r}_o$ and constantly coupled to the field, serves as detector. In fact T is bound by harmonic forces to an oscillation centre situated at point $\underline{R}_o$, and its motion is used to monitor the changes in the time-dependent virtual field of S.

- Virtual mesons. On the basis of (1.3), the Hamiltonian of the S-T system is

$$H = \sum_{\underline{k}} \hbar\,\omega_k\, a^+_{\underline{k}} a_{\underline{k}} - g\sum_{\underline{k}} \sqrt{\frac{\hbar}{2V\omega_k}}(a^+_{\underline{k}} + a_{\underline{k}}) -$$

$$- g\sum_{\underline{k}} \sqrt{\frac{\hbar}{2V\omega_k}}(a_{\underline{k}}\, e^{-i\underline{k}.\underline{R}_o} + a^+_{\underline{k}}\, e^{i\underline{k}.\underline{R}_o}) + \frac{1}{2m_o}P_o^2 + \frac{1}{2}Kx_o^2 +$$

$$+ ig\underline{x}_0 \cdot \sum_{\underline{k}} \sqrt{\frac{\hbar}{2V\omega_k}}\, \underline{k}\,(a_{\underline{k}}\, e^{-i\underline{k}.\underline{R}_o} - a^+_{\underline{k}}\, e^{i\underline{k}.\underline{R}_o}) \tag{2.9}$$

where $\underline{x}_o=\underline{R}_o-\underline{r}_o$, $\underline{p}_o$ and m_o are position, momentum and mass of the test oscillator T, K being its elastic constant. If S is completely bare at t=0, the following radial force [26] can be shown to act on oscillator T

$$< F_{OR} >_t \;=\; \frac{1}{4\pi c^2}\, g^2 \frac{\partial}{\partial R_o}\frac{1}{R_o}\frac{\partial}{\partial R_o}\int_0^t F(R_0, t')\, dt'$$

$$F(t, R_o) = cJ_o\,(\mu c \sqrt{t^2 - R_o^2/c^2}\,)\Theta(t - R_o/c) \tag{2.10}$$

As $t \to \infty$ this tends to

$$< F_{OR} >_\infty = \frac{\partial}{\partial R_o}\left[\frac{1}{4\pi c^2}\, g^2 \frac{1}{R_o}\, e^{-\mu R_o}\right] \tag{2.11}$$

From (2.10) we see that when the bare source appears at t=0, the force acting on T vanishes until $t=R_o/c$. After this time the force oscillates until it settles to (2.11), which is the normal Yukawa force between two dressed nucleons. The motion of T along the radial direction under the action of $\langle F_{OR}\rangle_t$ is described by

$$X_O(t) = 0 \qquad (t < R_o/c)$$

$$X_O(t) = \frac{1}{4\pi c^2}\frac{1}{\omega_T^2 m_O}g^2\frac{1}{R_o}\{(\mu + \frac{1}{R_o})e^{-\mu R_o} +$$

$$+ \mathrm{Im}[\,i\,e^{i\omega_o t}(\sqrt{\mu^2 - \omega_T^2/c^2} + 1/R_o)e^{-R_o\sqrt{\mu^2 - \omega_T^2/c^2}}]\} \quad (t \to \infty)$$

$$\tag{2.12}$$

Here $\omega_T = [K/m_o]^{1/2}$. The first term within {} in (2.12) represents a shift of the oscillation centre due to the static asymptotic part of the force (2.11). The second term represents oscillations which survive for large t. Thus the simple detector model discussed is capable of detecting regeneration of the virtual meson cloud which develops around an initially bare source. A similar analysis can be developed for the case of the source S disappearing at t=0.

- Virtual photons. From (1.29), we take the Hamiltonian to be

$$H = -\hbar\omega_0 + \hbar\sum_{\underline{k},j}\omega_k a^+_{\underline{k}j}a_{\underline{k}j} + \frac{1}{2m}P_0^2 + \frac{1}{2}Kx_0^2$$

$$-\frac{1}{2}\alpha^s_{mn}E_{\perp m}(0)E_{\perp n}(0) - \frac{1}{2}\alpha^T_{mn}E_{\perp m}(r_o)E_{\perp n}(r_o) \qquad (2.13)$$

Here the α_{mn} are the anisotropic static ground-state polarizabilities of the two two-level atoms S and T. In order to simplify the mathematics we shall average out this anisotropy and use $\alpha_{mn}=\alpha\delta_{mn}$. After the averaging process the force acting on T due to the sudden appearance of the bare S at t=0 is

$$\langle F_{OR}\rangle_t = \frac{1}{4\pi}\alpha^S\alpha^T ch\{-\frac{161}{R_o^8}[1-\Theta(R_o-ct)] - \frac{161}{R_o^7}\delta(R_o-ct)$$

$$-\frac{83}{R_o^7}\delta'(R_o-ct) - \frac{58}{3R_o^5}\delta''(R_o-ct) - \frac{10}{3R_o^4}\delta'''(R_o-ct)$$

$$-\frac{7}{15R_o^3}\delta^{iv}(R_o-ct) - \frac{1}{15R_o^2}\delta^{v}(R_o-ct)\} \qquad (2.14)$$

As $t\to\infty$ this tends to the usual Van der Waals force, but for times smaller than that needed by a signal to connect the two sources the force vanishes. The motion of T along the radial direction under the action of $\langle F_{OR}\rangle_t$ is described by [26]

$$x_o(t) = -(4\pi)^{-1}\alpha^S\alpha^T\hbar/m_0\omega_T\{(161c/R_o^8\omega_T)[1-\cos\omega_T(t-R_o/c)] +$$

$$+[161/R_o^7 - 58\omega_T^2/3c^2R_0^5 + 7\omega_T^4/15c^4R_o^3]\sin\omega_T(t-R_o/c)$$

$$-[83\omega_T/cR_o^6 - 10\omega_T^3/3c^3R_o^4 + \omega_T^5/15c^5R_o^2]\cos\omega_T(t-R_o/c)\}\Theta(ct-R_o) \qquad (2.15)$$

As in the meson case, there is no motion of the detector for $t<R_o/c$. For $R_o<c/\omega_T$ the oscillation amplitude is proportional to the static R_o^{-8} Van der Waals force. This is what one should expect as a consequence of the virtual nature of the expanding photon cloud. For $R_o>c/\omega_T$, however, the oscillation amplitude becomes proportional to R_o^{-2}, which is remindful of the behaviour of a detector under the action of the electric field emitted by an oscillating dipole [27].

In conclusion, we have shown that it is possible in principle to detect the growth of virtual clouds which develop around initially bare meson and photon sources.

II.5. Orders of Magnitude

In order to get a physical feeling of the orders of magnitude involved in less idealized circumstances than those examined hitherto, we now consider a positronium atom which annihilates at t=0. This p-atom plays the role of the source S. We neglect all the subtleties of positronium physics, and we characterize the ground-state positronium by a static electric polarizability $\alpha_p\equiv\alpha_H$. For our test atom T we take a Na atom trapped in an optical molasse. Thus $m_o\simeq 3.82\cdot 10^{-23}$ gr, and the atom within the molasse oscillates [28] at very low frequency $\omega_T\simeq 6\cdot 10^3$ rad sec^{-1}. These oscillation can be revealed by observing the resonance fluorescence of the Na atom, which is modulated by the oscillations in the atom rest frame [29]. If a positronium atom annihilates at a distance $R_o\simeq 10^{-4}$ cm from the Na oscillator, it should release γ-photons plus the virtual cloud. The γ-photon impact on Na is likely to cause ionization, whose effects can be eliminated by sweeping away the ions by appropriate electric fields. The effect of the virtual cloud after annihilation can be shown to contribute Na oscillations at frequency of amplitude

$$x_o = -\frac{1}{4\pi}\alpha_P\alpha_{Na}\frac{\hbar c}{m_O\omega_T^2}\frac{161}{R_o^8} \tag{2.16}$$

because $R_o \ll c/\omega_T$. Assuming [30] $\alpha_p=\alpha_H=0.67$ A^3, $\alpha_{Na}\approx 23.58$ A^3, one gets $x_o \approx 4.62\cdot 10^{-16}$ cm. This is an unrealistically small amplitude, much smaller than the thermal oscillation amplitude of the Na atom inside the molasse trap, which is given by $[\hbar n/m_o\omega_T]^{1/2}\approx 5\cdot 10^{-3}$ cm $(n=KT/\hbar\omega_T)$. Since ω_T is very low, however, one can conceivably sum the effects of many different p-atoms annihilating within one half oscillation period of 10^{-3} sec. It is immediate to see that 10^{14} p-atoms annihilating within 10^{-3} sec should give an amplitude of $\approx 4\cdot 10^{-2}$ cm. This effect could be measured, since this amplitude is about one order of magnitude larger than the Na zero-point amplitude. Obviously the question arises about the feasibility of getting the necessary number of p-atoms. Production of p-atoms is a technically complicated matter, but it would seem that the available flux is limited [31] by an initial flux of positrons p^+ of the order of $4\cdot 10^5$ p^+/sec. This yields an upper limit of $4\cdot 10^2$ p ready for annihilation in 10^{-3} sec. Thus it would seem that with the available positronium technique it is not possible to observe undressing. One cannot exclude, however, that the necessary flux of virtual photons can be obtained using other dressing/undressing sources.

III. DRESSING IN PROCESSES INVOLVING REAL PHOTONS

III.1. Introduction

We shall now discuss some new results concerning the behaviour of the dressing cloud of virtual photons during energy-conserving processes, in which real photons are exchanged between an atom and the e.m. field. In particular we discuss the absorption of a single delocalized real photon by an isolated atom. This may also be considered as a prototype quantum mechanical study of the absorption process of a photon by a photodetector. It is interesting to remark that while the behaviour of the field in the phenomenon of spontaneous emission has been discussed in great detail both for a scalar [32] and for the e.m. field in vacuum [33], and also in the case of "ping-pong" emission of two atoms [34] and in a cavity [35], the absorption does not seem to have attracted much attention, although the role of the form of the atom-radiation coupling in photodetectors has been investigated [36] and although the scattering of a localized wavepacket by an atom has been discussed in detail [37]. The reason is probably that absorption has been perceived as complementary to emission and that because of this little new information could be obtained by such a study. A little thought, however, will reveal a feature that should make absorption of a delocalized photon by an atom worth studying per se. This is the possibility of long-range atom-field correlations playing a role, which in emission does not generally exist. In particular, during emission an atom generates a photon in a region localized within the coherence length, and the wavepacket is in general a superposition of many different field modes. In absorption, the initial one-photon state of the field can be taken as being in a single-mode state, or at least delocalized over a region large compared with the emission coherence length. Thus while in emission the effective atom-photon interaction lasts only the emission time, after which the group velocity of light takes the photon wavepacket out of atomic reach, in absorption from a very delocalized photon an interaction time between the atom and the photon which is being absorbed may not be so obviously defined. A recent paper [38] reports numerical results bearing on the field absorbed by a Rydberg atom in a cavity; the expression used for the intensity of the field, however, does not ensure strict causality [32].

In order to keep the mathematics as simple as possible we describe the atom by a two-level formalism. For those who, having in mind a typical photodetector in which the final state lays in the continuum, feel that the two-level approximation is too rash, we point out that the mathematical structure of the two-level atom can be complicated at a later stage by putting the upper atomic level in contact with the levels in a band, e.g. by tunneling processes which introduces the desired continuum in the final atomic state.

A final remark is that in most of the work available on absorption or emission of real photons it is traditional to assume the atom to be initially in a bare state and totally uncorrelated with the vacuum, with the noticeable exception of the work by Davidovich and Nussenzweig [37] which, however, is concerned with scattering of a localized photon. We shall take this usual point of view too, and we will see where it leads when the dressing processes are taken into account. It should nonetheless be mentioned that the possibility of attaining such an initial bare state is by no means obvious from a conceptual point of view.

III.2. Model and Perturbation Approach

We slightly modify Hamiltonian (1.5) and use the form

$$H = \hbar\omega_0 S_z + \sum_{\underline{k},j} \hbar\, \omega_k\, a^{+}_{\underline{k}j} a_{\underline{k}j} +$$

$$+\sum_{\underline{k},j} (\varepsilon_{\underline{k}j}\, a_{\underline{k}j}\, S_+ + \varepsilon^{*}_{\underline{k}j}\, a^{+}_{\underline{k}j}\, S_-) - \lambda \sum_{\underline{k},j} (\varepsilon_{\underline{k}j}\, a^{+}_{\underline{k}j}\, S_+ + \varepsilon^{*}_{\underline{k}j}\, a_{\underline{k}j}\, S_-) \tag{3.1}$$

where λ can be 0 or 1. For $\lambda=0$ we get the so-called Rotating Wave Approximation (RWA), whereas for $\lambda=1$ all terms which yield dressing in the atomic ground state are retained. Thus λ helps us in keeping track of the terms which generate the ground-state virtual cloud. Moreover, using the multipolar form for the atom-field coupling, in the dipole approximation we obtain

$$\varepsilon_{\underline{k}j} = i\sqrt{\frac{2\pi\hbar\omega_k}{V}}\, \hat{e}_{\underline{k}j} \cdot \underline{d} \tag{3.2}$$

where $\underline{d}$ is the matrix element of the electron dipole moment operator $e\underline{r}_e$ between ground and excited atomic state and $\underline{e}_{\underline{k}j} = -\underline{e}_{-\underline{k}j}$ is the photon polarization vector. The photon equations of motion obtained from (3.1) and their solutions are

$$\dot{a}_{\underline{k}j} = -\frac{i}{\hbar}[a_{\underline{k}j}, H] = -i\omega_k a_{\underline{k}j} - \frac{i}{\hbar}\varepsilon^{*}_{\underline{k}j} S_- + \lambda \frac{i}{\hbar}\varepsilon_{\underline{k}j} S_+$$

$$a_{\underline{k}j}(t) = a_{\underline{k}j}(0)\, e^{-i\omega_k t} -$$

$$-\frac{i}{\hbar}\varepsilon^{*}_{\underline{k}j}\int_0^t S_-(t')\, e^{-i\omega_k (t-t')}dt' + \frac{i}{\hbar}\lambda\varepsilon_{\underline{k}j}\int_0^t S_+(t')\, e^{-i\omega_k(t-t')}dt' \tag{3.3}$$

The solution of the analogous equation for atomic operators is

$$S_+(t) = S_+(0)\, e^{i\omega_0 t} -$$

$$-\frac{i2}{\hbar}\sum_{\underline{k}j}\varepsilon^{*}_{\underline{k}j}\, \{\int_0^t a^{+}_{\underline{k}j}(t')S_z(t')\, e^{i\omega_0(t-t')}dt' - \lambda\int_0^t a_{\underline{k}j}(t')S_z(t')\, e^{i\omega_0(t-t')}dt'\,\} \sim$$

$$\sim S_+(0)\, e^{i\omega_0 t} - \frac{2}{h}\sum_{\underline{k}j} e^{i\omega_0 t}\, \varepsilon^*_{\underline{k}j}\, S_Z(0)\{a^+_{\underline{k}'j'}(0)F^*(k_0 - k') - \lambda a_{\underline{k}'j'}(0)F^*(k_0 + k')\} \qquad (3.4)$$

where

$$F(k_a + k_b) = -i\int_0^t e^{i(\omega_{ka}+\omega_{kb})t'}dt' = \frac{1 - e^{i(\omega_{ka}+\omega_{kb})t'}}{(\omega_{ka}+\omega_{kb})} \qquad (3.5)$$

We remark that the approximation made on the RHS of (3.4) consists in assuming small changes of S_z and a_{kj}. Consequently the validity of any result obtained using (3.4), and involving large changes of these operators, is limited to those parts of space-time where the influence of small changes only is dominant, and in particular to a zone not too far from the wavefront.

Substitution of (3.4) into (3.3) yields expressions for $a_{kj}(t)$, and consequently for the electric energy density operator at point $\underline{r}$

$$E^2_\perp(\underline{r},t) = -\frac{2\pi\hbar}{V}\sum_{\underline{k}\underline{k}'jj'}\sqrt{\omega_k\omega_{k'}}\,\hat{e}_{\underline{k}j}\hat{e}_{\underline{k}'j'}\{a_{\underline{k}j}a_{\underline{k}'j'}e^{i(\underline{k}+\underline{k}')\cdot\underline{r}} + a^+_{\underline{k}j}a^+_{\underline{k}'j'}e^{-i(\underline{k}+\underline{k}')\cdot\underline{r}} - a_{\underline{k}j}a^+_{\underline{k}'j'}e^{i(\underline{k}-\underline{k}')\cdot\underline{r}} - a^+_{\underline{k}j}a_{\underline{k}'j'}e^{-i(\underline{k}-\underline{k}')\cdot\underline{r}}\} \qquad (3.6)$$

This expressions can be used to evaluate matrix elements of the energy density up to $O(e^2)$ on various bare states of the form $|S_z\ \{n_{kj}\}\rangle$, where S_z corresponds to a state of the bare atom and $\{n_{kj}\}$ yields the photon distribution.

III.3. Dressing of the Bare Ground State

In particular the dressing of the bare ground state, which we have discussed previously, is described by

$$\langle\downarrow\{0_{\underline{k}j}\}|E^2_\perp(\underline{r},t)|\downarrow\{0_{\underline{k}j}\}\rangle = \frac{4\pi}{V}\sum_{\underline{k}j}\frac{1}{2}\hbar\omega_k - \frac{8\pi}{V}\lambda^2\sum_{\underline{k}j}(\hat{e}_{\underline{k}j}\cdot\bar{d})(\hat{e}_{\underline{k}j})_m\bar{d}_n\frac{k}{k+k_0}\times$$
$$\times\mathrm{Re}\,(e^{i\underline{k}\cdot\underline{r}}D^r_{mn}\{[e^{-ikr} - e^{-i(k_0+k)ct}e^{ik_0r}]\Theta(ct - r)\}) \qquad (3.7)$$

where

$$D^r_{mn} = \frac{1}{r}[(\delta_{mn} - \hat{r}_m\hat{r}_n)\frac{\partial^2}{\partial r^2} + (\delta_{mn} - 3\hat{r}_m\hat{r}_n)(\frac{1}{r^2} - \frac{1}{r}\frac{\partial}{\partial r})] \qquad (3.8)$$

The first term on the RHS of (3.7) is space abd time-indpendent and it coincides with the zero-point energy density of the vacuum. The second term is obviously causal, and it shoul describe the development of the virtual cloud at $\underline{r}$ after creation of the bare ground state source at $\underline{r}=0$, $t=0$. As expected, this term vanishes for $\lambda=0$. In the limit of large t, and for $r<ct$, (3.7) yields

$$\lim_{t\to\infty} \langle\downarrow \{0_{kj}\}| E_\perp^2(r,t) |\downarrow \{0_{kj}\}\rangle = \frac{4\pi}{V}\sum_{kj}\frac{1}{2}\hbar\omega_k -$$

$$-\lambda\frac{2}{\pi}d_l d_n\frac{1}{r}\{\delta_{ln}[-\frac{9}{4}\frac{k_0}{r^4}+\frac{1}{2}\frac{k_0^3}{r^2}-(\frac{k_0^4}{r}-3\frac{k_0^2}{r^3}+\frac{1}{r^5})f(2k_0r)+(2\frac{k_0^3}{r^2}-2\frac{k_0}{r^4})g(2k_0r)]$$

$$-\hat{r}_l\hat{r}_n[-\frac{1}{4}\frac{k_0}{r^4}+\frac{k_0^3}{r^2}-(\frac{k_0^4}{r}+\frac{k_0^2}{r^3}-\frac{3}{r^5})f(2k_0r)+(2\frac{k_0^3}{r^2}+6\frac{k_0}{r^4})g(2k_0r)]\} \quad (r<ct) \tag{3.9}$$

where

$$f(z) = Ci(z)\sin z - si(z)\cos z; \qquad g(z) = -Ci(z)\cos z - si(z)\sin z$$

Expanding (3.9) in powers of $(k_0r)^{-1}$ and retaining only the lowest order terms yields

$$\lim_{t\to\infty} \langle\downarrow \{0_{kj}\}| E_\perp^2(r,t) |\downarrow \{0_{kj}\}\rangle \sim$$

$$\sim \frac{4\pi}{V}\sum_{kj}\frac{1}{2}\hbar\omega_k + \lambda^2\frac{\hbar c}{4\pi}\frac{2d_l d_n}{\hbar\omega_0}\frac{1}{r^7}(13\delta_{ln}+7\hat{r}_l\hat{r}_n) \quad (r<ct; k_0r \gg 1) \tag{3.10}$$

This result coincides with the first of (27), which was obtained from the CP Hamiltonian (2.6). In our case (3.10) has been obtained from the original Hamiltonian (3.1) as a far-region approximation. In fact, one can obtain also the near-zone approximation for $k_0r \ll 1$ from (3.9) as

$$\lim \langle\downarrow \{0_{kj}\}| E_\perp^2(r,t) |\downarrow \{0_{kj}\}\rangle \sim \frac{4\pi}{V}\sum_{k,j}\frac{1}{2}\hbar\omega_k + \lambda^2 4d^2\frac{1}{r^6} \quad ;(r<ct; k_0r \ll 1) \tag{3.11}$$

which behaves according to the static expression (1.13).We rermark that validity of expressions (3.7) to (3.10) is not restricted to the $r \approx ct$ region, because during the dressing process of the ground state large changes in S_z and in a_{kj} are not expected.

The atomic dynamics can be obtained using Heisenberg equations and (3.1). In particular, the time-dependent atomic energy during the dressing process is given by

$$\langle\downarrow \{0_{kj}\}| \hbar\omega_0 S_z(t) |\downarrow \{0_{kj}\}\rangle = -\frac{\hbar\omega_0}{2} + \frac{2\lambda^2\omega_0}{\hbar}\sum_{kj}|\varepsilon_{kj}|^2\frac{1-\cos(\omega_k+\omega_0)t}{(\omega_k+\omega_0)^2} \tag{3.12}$$

It is convenient to evaluate (3.12) in the minimal coupling scheme with

$$\varepsilon_{kj} = -i\sqrt{\frac{2\pi\hbar\omega_0^2}{V\omega_k}}\,\hat{e}_{kj}\cdot\bar{d} \tag{3.13}$$

because in this scheme (3.12) is representative of what might be called the "time-dependent Lamb shift" due to the reconstruction of the virtual cloud outside the atom [15]. One obtains, for $\omega_M \gg \omega_0$,

$$\langle\downarrow \{0_{kj}\}| h\omega_0 S_z(t) |\downarrow \{0_{kj}\}\rangle = -\frac{h\omega_0}{2} + \lambda^2\frac{\gamma\hbar}{\pi}\{\ln\frac{\omega_M}{\omega_0} - 1 - Ci(\omega_M t) +$$

$$Ci(\omega_0 t) - \omega_0 t[\frac{\cos\omega_M t}{\omega_M t} + Si(\omega_M t) - \frac{\cos\omega_0 t}{\omega_0 t} - Si(\omega_0 t)]\} \tag{3.14}$$

where $\omega_M = ck_M$ is the cut-off frequency and

$$\gamma = 4|\bar{d}|^2\omega_0^3/3\hbar c^3$$

is the spontaneous relaxation rate. For $\omega_0 t>1$ expression (3.14) reduces to

$$<\downarrow \{0_{kj}\} \mid \hbar\omega_0 S_z(t) \mid \downarrow \{0_{kj}\}> \sim -\frac{\hbar\omega_0}{2} + \lambda^2\frac{\gamma\hbar}{\pi}\{\ln\frac{\omega_M}{\omega_0} - 1 - \frac{\sin\omega_M t}{\omega_M t} + \frac{\cos\omega_0 t}{(\omega_0 t)^2}\} \tag{3.15}$$

which displays oscillations of decreasing amplitude at ω_0-frequency similar to those obtained in the theory of spontaneous decay [39], as well as an asymptotic atomic energy proportional to the Lamb shift [40] similar to that previously reported in the literature [41]. The high-frequency oscillations in (3.15) are likely to be an artifact of the model, since they are strongly cut-off dependent.

III.4. Absorption of a Real Photon

We assume one real delocalized photon to exist in the $\underline{k}_1 j_1$ mode, all other modes being empty. From (3.6) we obtain [42]

$$\begin{aligned}<\downarrow, \underline{k}_1 j_1 \mid E_\perp^2(\underline{r},t) \mid \downarrow \underline{k}_1 j_1> &= <\downarrow \{0_{kj}\} \mid E_\perp^2(\underline{r},t) \mid \downarrow \{0_{kj}> + \frac{4\pi}{V}\hbar\omega_{k_1} - \\ &- \frac{8}{V}(\hat{e}_{k_1 j_1}\cdot\bar{d})(\hat{e}_{k_1 j_1})_m \bar{d}_n \times \\ &\times \mathrm{Re}\,(\frac{k_1}{k_1 - k_0}\{e^{i\underline{k}_1\cdot\underline{r}}[I_{mn}(k_1,\underline{r},t) - e^{i(k_0 - k_1)ct} I_{mn}(k_0,\underline{r},t)] + \\ &+ \lambda e^{-i\underline{k}_1\cdot\underline{r}}[I_{mn}(-k_1,\underline{r},t) - e^{-i(k_0 - k_1)ct} I_{mn}(-k_0,\underline{r},t)]\} - \\ &- (\frac{k_1}{k_1 + k_0}\{\lambda e^{i\underline{k}_1\cdot\underline{r}}[I_{mn}(k_1,\underline{r},t) - e^{-i(k_0 + k_1)ct} I_{mn}(-k_0, r, t)] + \\ &+ \lambda e^{-ik_1 r}[I_{mn}(-k_1,r,t) - e^{i(k_0 + k_1)ct} I_{mn}(k_0,\underline{r},t)]\})\end{aligned} \tag{3.16}$$

where

$$I_{mn}(k_i,\underline{r},t) = -D^r_{mn}\int_0^\infty \frac{1 - e^{i(k - k_i)ct}}{k - k_i}\sin kr\,dk \tag{3.17}$$

$$I_{mn}(k_i,\underline{r},t) + I^*_{mn}(-k_i,\underline{r},t) = -D^r_{mn}\int_{-\infty}^\infty \frac{1 - e^{i(k - k_i)ct}}{k - k_i}\sin kr\,dk = -\pi D^r_{mn}\,e^{-ik_i r}\Theta(ct - r)$$

The first term on the RHS of (3.16) is the energy density of the bare vacuum plus that of the half-dressed ground-state atom, which is causal as from (3.7). The second term is the energy density of the only real photon initially present, and it is inversely proportional to the quantization volume, as it should be. The third rather complicated term must be interpreted as due to the absorption process. Due to (3.17), this term is causal only for $\lambda=1$. in other words, a RWA Hamiltonian gives rise to noncausal behaviour in the energy density, and only taking into account the counterrotating terms restores causality. For $\lambda=1$ in fact (3.16) and (3.17) yield

$$<\downarrow, \underline{k}_1 j_1 \mid E_\perp^2(\underline{r},t) \mid \downarrow \underline{k}_1> = <\downarrow \{0_{kj}\} \mid E_\perp^2(\underline{r},t) \mid \downarrow \{0_{kj}\}> + \frac{4\pi}{V}\hbar\omega_{k_1}$$

$$+ \frac{8\pi}{V}(\hat{e}_{k_1 j_1}\cdot\bar{d})(\hat{e}_{k_1 j_1})_m \bar{d}_n \times \mathrm{Re}\,(\frac{k_1}{k_1 - k_0}\; e^{i\underline{k}_1\cdot\underline{r}} D^r_{mn}\{[e^{-ik_1 r} - e^{-i(k_0 - k_1)ct}e^{-ik_0 r}]\,\Theta(ct - r)$$

$$- \frac{k_1}{k_1 + k_0} e^{-i\underline{k}_1 \cdot \underline{r}} D^r_{mn} \{ [e^{-i k_1 r} - e^{-i(k_0 + k_1)ct} e^{i k_0 r}] \Theta(ct - r) \}) \quad (3.18)$$

Thus the action of the counterrotating terms in (3.1) is not limited to dressing the ground state, as in the case with no real photon present; in addition these terms interfere with the rotating ones to modify the real absorption process and to make it compatible with the requirement of causality. In fact (3.18) shows that for r>ct the energy density is that due to the real photon only and that it is not changed by the absorption process. For $k_1 \neq k_0$ enough to ensure only minor changes in S_z and in a_{kj}, the last term in (3.18) for $t \to \infty$ tends to a time-independent limit at any point in space, given by

$$\frac{16\pi}{V} (\hat{e}_{k_1 j_1} \cdot \bar{d})(\hat{e}_{k_1 j_1})_m \bar{d}_n \frac{k_1 k_0}{k_1^2 - k_0^2} \mathrm{Re}(e^{i\underline{k}_1 \cdot \underline{r}} D^r_{mn} e^{-i k_1 r}) \quad (3.19)$$

which may be regarded as the "shadow" of the two-level atom under illumination of the single photon.

Also in this case the atomic dynamics can be obtained from the Heisenberg equations of motion. Here it is convenient to discuss the rate of change of atomic population. Within the same approximations as those leading to (3.12), one finds from (3.1)

$$<\downarrow, \underline{k}_1 j_1 | \dot{S}_z(t) | \downarrow, \underline{k}_1 j_1 > =$$
$$= \frac{2}{\hbar^2} |\varepsilon_{\underline{k}_1 j_1}|^2 [\frac{\sin(\omega_{k_1} - \omega_0)t}{(\omega_{k_1} - \omega_0)} + \lambda^2 \frac{\sin(\omega_{k_1} + \omega_0)t}{(\omega_{k_1} + \omega_0)}]$$
$$\frac{2\lambda^2}{\hbar^2} \sum_{\underline{k} j} |\varepsilon_{\underline{k} j}|^2 \frac{\sin(\omega_k + \omega_0)t}{(\omega_k + \omega_0)} \quad (3.20)$$

Comparison with (3.12) immediately shows that the last term on the RHS of (3.20) is the contribution of vacuum fluctuations to the rate of change of S_z. The first two terms in (3.20) are due to the interaction of the atom with the single real delocalized photon. For $\omega_{k_1} \approx \omega_0$, the second term on the RHS of (3.20) can be neglected with respect to the first. Then, using the minimal coupling scheme and assuming $\omega_M >> \omega_0$, $\omega_0 t>1$, (3.20) yields

$$<\downarrow, \underline{k}_1 j_1 | \dot{S}_z(t) | \downarrow, \underline{k}_1 j_1 > \sim \frac{2}{\hbar^2} |\varepsilon_{\underline{k}_1 j_1}|^2 \frac{\sin(\omega_{k_1} - \omega_0)t}{(\omega_{k_1} - \omega_0)} - \lambda^2 \frac{\gamma}{\pi} [\frac{\sin \omega_0 t}{(\omega_0 t)^2} + \frac{\cos \omega_M t}{\omega_M t}] \quad (3.21)$$

The last term in (3.21) is strongly model-dependent and of such a high frequency to be undetectable in practice. Thus on resonance ($\omega_{k_1} = \omega_0$) and using (3.13), it is easy to see that what is left of the λ^2 term prevails on the term due to the interaction with the real photon if $(ct)^3 << KV$, where K is a numerical factor of O(1) and V is the quantization volume. Thus the main contribution to the rate of change of S_z comes from the interaction with the vacuum fluctuations, at least until the causality sphere expands into a sizeable fraction of the quantization volume.

III.5. Remarks

The result that the virtual photons originating from energy nonconserving terms in (3.1) are essential in order to preserve causal behaviour of the electric energy density around an absorbing atom, seems to have implications which are worth pointing out.

Suppose in fact that a local measurement is performed on the atom at time t and that the atom is found to have $S_z=\frac{1}{2}$. Since in order to preserve causality one must include in H the counterrotating terms, one cannot be sure if the atom is ↑ because it has absorbed the real photon $\underline{k}_1 j_1$ or because it has emitted a virtual photon. On the other hand, one should require that a detector of photons be such that, by performing local observations on it alone, one should be able to obtain univocal information on the number of photons interacting with the detector itself. It is clear that our two-level atom fails to satisfy this requirement in the case of a single delocalized photon. Thus causality for a one-photon field does not seem to be compatible with the basic requirement of photodetection operation.

Another noticeable consequence of our results is that finding the atom excited at time t does not necessarily involve changes in the quantum average of the energy density at points r>ct. This is to be ascribed to the fact that the atom can attain the excited state by following two independent paths, one involving absorption of the real photon and the other involving emission of a virtual photon. Interference between these two paths leads to the causal behaviour of the energy density, and at the same time introduces many-photon intermediate states thereby avoiding the paradox of the real photon being absorbed by the atom and still contributing a finite energy density at points r>ct.

Finally, we remark that in connection with experiments on the foundations of quantum mechanics and with the theory of measurement it is quite often implicitly assumed that excitation of a detector corresponds to absorption of the real photon and to collapse of the photon wavefunction. As we have seen, however, such a collapse does not take place when the virtual photons absorbed and emitted by the detector are taken into account. In fact, the collapse is smoothed out into a gradual change over all space of the photon energy density. Naturally in these conditions it seems difficult to define an absorption time analogous to the spontaneous emission time.

IV. CONCLUSIONS

The sources of a radiation field in the true or physical ground state are generally dressed by a cloud of virtual quanta of the field. The structure of this cloud yields information on the dynamical structure of the source. Moreover, this virtual cloud contributes to physically observable quantities such as the renormalized mass of the source or the Lamb shift of the ground state, and it is the vehicle through which different sources may interact via exchange of virtual quanta.

Apart from these ground-state, fully dressed sources, it is possible to obtain sources whose cloud is out of equilibrium. The virtual cloud of these sources tends to attain the equilibrium configuration, and in these cases the virtual cloud is time-dependent. Half-dressed sources are defined as sources initially deprived of the external portion of their virtual cloud. This leads to a plausible definition of an effectively bare source as a source which is dressed only by high-frequency, effectively unobservable quanta. Regeneration of the virtual cloud of these effectively bare sources is shown to take place within a sphere of radius ct centered on the source, i.e. the causality sphere. Two initially bare sources of the same field do not interact until their regenerating virtual clouds overlap with the sources. Using this effect it is possible, at least in principle, to detect the regenerating virtual cloud. Moreover, the field of a source which is annihilated at t=0 disappears gradually within the causality sphere, while the field outside this sphere is the same as in the presence of the source.

The conclusions above are valid for dressed and half-dressed electrons in a polar crystal, for nucleons interacting with a meson field and for atoms interacting with photons. Specializing to the latter case, one can investigate the role of virtual photons in the absorption and emission of real photons by atoms. The virtual photons are found to play a fundamental role, in that they ensure causality in the behaviour of the field during the real process. They also show up in the dynamics of the photodetector, where they provide alternative paths to the excited atomic state, different from the direct absorption of the real photon. This fact seems to have consequences relevant to the theory of photodetection and possibly to the theory of quantum measurement.

The authors are grateful to Istituto Nazionale di Fisica della Materia (CNR) and to Comitato Regionale Ricerche Nucleari e Struttura della Materia for partial financial support.

REFERENCES

1. L. Van Hove, Physica 18,145 (1952); 21,901 (1955); 22,343 (1956)
2. E.M. Henley, W. Thirring, Elementary Quantum Field Theory (McGraw-Hill 1962)
3. G.C. Kuper, G.D. Whitfield ed.s, Polarons and Excitons, (Oliver and Boyd 1963)
4. C. Leonardi, F. Persico, G. Vetri, Riv. Nuovo Cim. 9,1 (1986)
5. N.N. Bogoljubov, D.V. Shirkov, Introduction to the Theory of Quantized Fields, (Interscience 1959)
6. F. Persico, E.A. Power, Phys. Rev. A36,475 (1987)
7. F.M. Peeters, J.T. Devreese, Phys. Rev. B31,4890 (1985)
8. T.D. Lee, F.E. Low, D. Pines, Phys. Rev. 90,297 (1953)
9. R. Passante, G. Compagno, F. Persico, Phys. Rev. A31,2827 (1985)
10. R. Passante, E.A. Power, Phys. Rev. A35,188 (1987)
G. Compagno, R. Passante, F. Persico, Phys. Lett. A121,19 (1987)
11. E.A. Power, S. Zienau, Phil, Trans. Roy. Soc. A251,427 (1959)
12. E.A. Power, T. Thirunamachandran, Phys. Rev. A28,2663 (1983)
13. F. Persico, E.A. Power, Phys. Lett. A114,309 (1986)
14. E.A. Power, T. Thirunamachandran, JOSA B2,1100 (1985)
J.R. Ackerhalt, P.W. Milonni, JOSA B1,116 (1984)
15. G. Compagno, R. Passante, F. Persico, Phys. Lett. A98,253 (1983)
J. Dalibard, J. Dupont-Roc, C. Cohen-Tannoudji, J. de Phys. 43,1617 (1982)
16. C. Kittel, Quantum Theory of Solids (J. Wiley & Sons, 1963)
17. D.P. Craig, E.A. Power, Int. J. Quantum Chem. 3,903 (1969)
18. G. Compagno, F. Persico, R. Passante, Phys. Lett. A112,215 (1985)
19. G. Compagno, G.M. Palma, Phys. Rev. A37,2979 (1988)
20. E.L. Feinberg, Sov. Phys. Usp. 23,629 (1981)
21. L.D. Landau, I.Ya. Pomeranchuk, Dokl. Akad. Nauk SSSR 92,535,735 (1953)
22. E.A. Power, T. Thirunamachandran, Phys. Rev. A25,2473 (1982)
23. P. Dobiasch, H. Walther, Ann. de Phys. 10,825 (1985)
24. D.W. Brown, K. Linderberg, B.J. West, J. Chem. Phys. 84,1574 (1986)
25. F. Persico, E.A. Power, Phys. Rev. A36,475 (1987)
26. G. Compagno, R. Passante, F. Persico, Phys. Rev. A38,600 (1988)
27. M. Born, E. Wolf Principles of Optics (Pergamon, 1980)
28. E.L. Raab, M. Prentiss, A. Cable, S. Chu, D.E. Pritchard, Phys. Rev. Lett. 59,2631 (1987)

29. F. Diedrich, H. Walther, Phys. Rev. Lett. 58,203 (1987)
30. J.M. Standard, P.R. Certain, J. Chem. Phys. 83,3002 (1985)
31. S. Chu, A.P. Miller, Phys. Rev. Lett. 48,1333 (1982)
32. M. De Haan, Physica A132,375 (1985)
33. V.P. Bykov, A.A. Zadernovskii, Opt. Spectr. 48,130 (1980)
V.P. Bykov, Sov. Phys. Usp. 27,631 (1984)
34. P.W. Milonni, P.L. Knight, Phys. Rev. A10,1096 (1974)
35. J. Parker, C.R. Stroud, Phys. Rev. A35,4226 (1987)
R.J. Cook, P.W. Milonni, Phys. Rev. A35,5081 (1987)
36. P.D. Drummond, Phys. Rev. A35,4253 (1987)
37. L. Davidovich, Ph.D. thesis, University of Rochester (1975)
38. M. Mallalieu, J. Parker, C.R. Stroud, Phys. Rev. A37,4765 (1988)
39. K. Wodkiewicz, J.H. Eberly, Ann. Phys. 101,574 (1976)
P.L. Knight, P.W. Milonni, Phys. Lett. A56,275 (1976)
40. H. Bethe, Phys. Rev. 72,339 (1947)
41. J. Seke, Coherence and Quantum OpticsV, L. Mandel, E. Wolf eds. (Plenum, 1984), p.557
42. G. Compagno, G.M. Palma, R. Passante, F. Persico, Eur. Lett. 9,215 (1989)

DEVELOPMENTS IN THE THEORY OF MULTIPHOTON ABSORPTION BY MOLECULES (BOUND-BOUND): APPLICATIONS OF A CHIROPTIC CHARACTER

E.A. Power

Department of Mathematics
University College London
Gower Street
London WC1E 6BT
England

ABSTRACT

A survey is given of the recent work by Meath and Power[1] on multiphoton absorption in the intensity range below saturation. Generalizations of the Einstein B-coefficient are introduced and the results for many absorption processes expressed in terms of these. Several special generalizations are considered, for example the dependence on the polarization character of the light beam, the dependence on multipolar transitions of higher moment than the electric dipole and the effect of permanent moments in addition to transition moments. Finally the known connection between absorption and scattering in dispersion relations allow a brief discussion of non-linear optical rotation by asymmetric molecules.

1 INTRODUCTION

In many graduate courses on radiation physics or quantum theory the Einstein B-coefficient

$$B = \frac{2\pi}{3\hbar^2} |\underline{\mu}^{no}|^2 \qquad (1)$$

is obtained, following Dirac's dynamical theory of the quantized electromagnetic field, from Fermi's golden rule applied to the coupled system, molecules and radiation. This contrasts with Einstein's original derivation from equilibrium statistical mechanics for radiation. The B-coefficient is deduced from the rate for single-photon absorption by a molecule which, initially in it's ground state $|0\rangle$ with energy E_0 , jumps to the electronic excited state $|n\rangle$ with energy E_n on absorption of the photon. The molecular transition is assumed to be electric-dipole allowed and $\underset{\sim}{\mu}^{no}$ in equation (1) is the transition moment

New Frontiers in Quantum Electrodynamics and Quantum Optics
Edited by A. O. Barut, Plenum Press, New York, 1990

$$\underset{\sim}{\mu}^{no} = -e \int \bar{\psi}_n(\underline{q})(\underline{q} - \underline{R})\psi_o(\underline{q})dV \ . \tag{2}$$

It is simplest to use the multipolar Hamiltonian to discuss to the dynamics. In this Hamiltonian the interaction term is

$$H_{Int} = -\underset{\sim}{\mu}.\underset{\sim}{D}^{\perp}(\underset{\sim}{R}) \tag{3}$$

where $\underset{\sim}{D}^{\perp}$ is the transverse Maxwell displacement field and $\underset{\sim}{R}$ the position vector locating the molecule. [If the molecule is large and complex the absorption may well be confined to a localized atomic cluster in which case $\underset{\sim}{R}$ would locate this chromophore.]

Elementary quantum electrodynamics[2] requires the vector potential and its conjugate momentum field to be expanded into terms linear in the creation and annihilation operators for photons. If these operators are denoted by $a^{+}(\underline{k},\lambda)$ and $a(\underline{k},\lambda)$ for the quanta corresponding to mode $(\underline{k},\lambda)$ we have

$$\underset{\sim}{D}^{\perp}(\underline{r}) = i \sum \left(\frac{2\pi\hbar\omega}{V}\right)^{1/2} \left(a(\underline{k},\lambda)\underline{e}(\underline{k},\lambda)e^{\underline{ik}.\underline{r}} - a^{+}(\underline{k},\lambda)\bar{\underline{e}}(\underline{k},\lambda)e^{-\underline{ik}.\underline{r}}\right) , \tag{4}$$

since $\underline{D}^{\perp}$ is proportional to the conjugate momentum field in the multipolar theory. In equation (4) $\underline{e}(\underline{k},\lambda)$ is the polarization vector, in general complex, for the mode $(\underline{k},\lambda)$ and ω the circular frequency of the photon. The matrix element for the single photon absorption is

$$\langle H_{Int} \rangle - -\underset{\sim}{\mu}^{no}.\underline{e}(\underline{k},\lambda) \left(\frac{2\pi\hbar\omega\mathcal{N}}{V}\right)^{1/2} \tag{5}$$

and the absorption rate is

$$\tilde{\Gamma} = \frac{2\pi}{\hbar} |\underline{\mu}^{no}.\underline{e}|^2 \ \frac{2\pi\hbar\omega\mathcal{N}}{V} \rho \tag{6}$$

where ρ is the density of states and $\mathcal{N}$ the occupancy number for photons of state $(\underline{k},\lambda)$ in the incident beam. It is usual to express $\tilde{\Gamma}$ in terms of the radiation energy density per unit frequency $\mathcal{I}$

$$\mathcal{I} = \frac{\hbar\omega\mathcal{N}}{V}(2\pi\hbar\rho) \ . \tag{7}$$

Using the notation

$$\tilde{\Gamma} = \tilde{B}\mathcal{I} \tag{8}$$

where

$$\tilde{B} = \frac{2\pi}{\hbar^2} |\underline{\mu}^{no}.\underline{e}|^2 \tag{9}$$

is the oriented or locked-in B-coefficient. If the molecule is freely rotating relative to the laboratory system, in which the beam of radiation is fixed, as for a gas or dilute solution, we find for the average rate Γ ,

$$\Gamma = B\mathcal{I} \tag{10}$$

where B is given by equation (1). In this way of factoring Γ the dependence of the rate on the molecular structure is separated from its dependence on the radiation beam. $\mathcal{I}$ depends only on the characteristics of the radiation and B depends only on the molecular electric-dipole moment. In the unaveraged, locked, case $\tilde{B}$ does have a weak dependence on the radiation in so far as it is the component of the dipole moment in the direction of polarization that appears in equation (9).

In a series of papers (W.J. Meath and E.A. Power)[1] we have extended this procedure to consider multiphoton-absorption and, in addition, we have computed the absorption rates where the effective molecular moments go beyond the electric-dipole approximation. The results have been presented through a set of *generalized* B-coefficients. This is possible as any particular absorption rate factors into terms dependent only on the incident beam and one, a generalized B-coefficient, which contains all the molecular information. To be consistent with the notation used previously for the single-photon case we use a tilde i.e. $\tilde{\Gamma}^{(N)}$ and $\tilde{B}^{(N)}$ for N-photon absorption for fixed orientation of the molecule with respect to the light beam and $\Gamma^{(N)}$, $B^{(N)}$ after rotational averaging for random orientation of the molecule.

II MULTIPHOTON ABSORPTION RATES

For Fermi's golden rule to be a valid approximation the intensity must be below that for which saturation effects are relevant. However, even in this regime, the problem is highly non-linear. Unlike the case of single-photon absorption, where the absorption rate is linear in the intensity and where the Lambert-Beer law holds so that the intensity falls exponentially in a homogeneous sample, the rates depend on the electromagnetic field intensities to a high power. In all cases there is the factor $\mathcal{I}$ as in equations (8) and (10) but each additional photon absorbed implies an additional factor of the amplitude of the Maxwell field within the matrix element and so it's square in the rate. Hence for N-photon absorption the rate varies as I^{N-1} where I is the irradiance of the beam

$$I = \langle \text{beam} | a^{+}(\underline{k},\lambda) a(\underline{k},\lambda) | \text{beam} \rangle \, \frac{\hbar \omega c}{V} \,. \tag{11}$$

Furthermore, since we are investigating multiphoton absorption from a single beam, the annihilation operators $a(\underline{k},\lambda)$ occur N-times in the matrix element and Γ is proportional to

$$g^{(N)} = \frac{\langle \text{beam} | \, [a^{+}(\underline{k},\lambda)]^{N} [a(\underline{k},\lambda)]^{N} \, | \text{beam} \rangle}{\langle \text{beam} | a^{+}(\underline{k},\lambda) a(\underline{k},\lambda) | \text{beam} \rangle^{N}} \tag{12}$$

the N'th order coherence factor.

In analogy with the single-photon case we define $B^{(N)}$ by explicitly extracting the factors that depend on the light beam from $\Gamma^{(N)}$; i.e.

$$\Gamma^{(N)} = \mathcal{I} I^{N-1} g^{(N)} B^{(N)} \tag{13}$$

with a similar relationship for the orientated rates and coefficients $\tilde{\Gamma}^{(N)}$, $\tilde{B}^{(N)}$. If electric-dipole couplings are the only ones operative

$$\tilde{B}^{(N)} = \frac{(2\pi)^2}{\hbar^2 c^{N-1}} \left| \sum_{\alpha_1 \beta \ldots \zeta} \frac{\mu_i^{n\alpha} \mu_j^{\alpha\beta} \mu_k^{\beta\gamma} \ldots \mu_p^{\zeta o} e_i e_j \ldots e_p}{(E_{n\alpha} - \hbar\omega)(E_{n\beta} - 2\hbar\omega)\ldots(E_{n\zeta} - (N-1)\hbar\omega)} \right|^2 \tag{14}$$

as discussed in a brief paper in these proceedings. Averaging for random orientations of the absorbing molecules is carried out by expressing $\tilde{B}$ in terms of the Euler angles (θ,ϕ,χ) which determine the orientation of the body axes of the molecules with respect to laboratory axes and then carrying out the integration

$$B = \frac{1}{8\pi^2} \int \tilde{B}(\theta,\phi,\chi) \sin\theta \, d\theta \, d\phi \, d\chi \, . \tag{15}$$

This averaging procedure is often the most technical problem involved in determining the generalized B-coefficients. As $\tilde{B}$ depends on many vectors or tensors fixed in the body there are many Euler matrices $R_{i\lambda}$ involved in $\tilde{B}$ and a systematic method for carrying out (15) is to determine

$$I^{(n)}_{i_1 i_2 \ldots i_n ; \lambda_1 \lambda_2 \ldots \lambda_n} = \frac{1}{8\pi^2} \int R_{i_1\lambda_1} R_{i_2\lambda_2} \ldots R_{i_n\lambda_n} \sin\theta \, d\theta \, d\phi \, d\chi \, . \tag{16}$$

These are tabulated for up to $n = 6$ in Craig and Thirunamachandran's[2] book and see Andrews[3] for higher values of n .

III DESCRIPTION OF POLARIZATION

The most general choice for $\underline{e}(\underline{k},\lambda)$ corresponds to elliptically polarized light. It can be constructed by a loaded, in both magnitude and phase, linear superposition of right- and left-handed circularly polarized modes. If α is the phase difference and β a radian measure of the amplitude difference we write, for $0 \le \beta \le \pi$ and $-\pi \le \alpha \le \pi$,

$$\underline{e} = \exp\left(\frac{1}{2} i\alpha\right) \cos\left(\frac{1}{2}\beta\right) \underline{e}^L + \exp\left(-\frac{1}{2} i\alpha\right) \sin\left(\frac{1}{2}\beta\right) \underline{e}^R \, . \tag{17}$$

Special cases are circularly polarized light for $\beta = 0$ and π and plane polarized light for $\beta = \pi/2$. An alternative description in terms of cartesian coordinates with the z-axis as the direction of propagation is

$$\underline{e} = \frac{\varepsilon_x e^{-i\delta_x} \underline{e}_x + \varepsilon_y e^{-i\delta_y} \underline{e}_y}{\left(\varepsilon_x^2 + \varepsilon_y^2\right)^{1/2}} . \tag{18}$$

This form gives direct connections to the classical field amplitudes and phases. Normally it is only the phase difference $\delta = \delta_y - \delta_x$ which is relevant. If we use the symmetrical relationships connecting $\underline{e}^L$ and $\underline{e}^R$ with $\underline{e}_x$ and $\underline{e}_y$, namely

$$\underline{e}^L = \bar{\underline{e}}^R = \frac{1}{\sqrt{2}} \left(\exp(-i\pi/4)\ \underline{e}_x + \exp(i\pi/4)\ \underline{e}_y\right) \tag{19}$$

we have

$$\frac{\varepsilon_x^2}{\varepsilon_x^2 + \varepsilon_y^2} = \frac{1 + \sin\alpha\ \sin\beta}{2} , \quad \frac{\varepsilon_x^2}{\varepsilon_x^2 + \varepsilon_y^2} = \frac{1 - \sin\alpha\ \sin\beta}{2} \tag{20}$$

and

$$\tan\delta = \frac{\cot\beta}{\cos\alpha} . \tag{21}$$

The inverse relations are

$$\cot\alpha = \frac{2\varepsilon_x \varepsilon_y \cos\delta}{\varepsilon_x^2 + \varepsilon_y^2} , \quad \cos\beta = \frac{-2\varepsilon_x \varepsilon_y \sin\delta}{\varepsilon_x^2 + \varepsilon_y^2} \tag{22}$$

IV ELECTRIC-DIPOLE CONTRIBUTIONS FOR N = 2

Before discussing the $N = 2$ case it is worthwhile quickly working through the single-photon result. From equation (14) we have

$$\tilde{B} = \frac{2\pi}{\hbar^2} \mu_i^{no} \bar{\mu}_j^{no} e_i \bar{e}_j \tag{23}$$

so that, in terms of the components of the electric-dipole moments in the body frame μ_λ ,

$$\tilde{B} = \frac{2\pi}{\hbar^2} R_{i\lambda} R_{j\mu} \mu_\lambda^{no} \bar{\mu}_\mu^{no} e_i \bar{e}_j . \tag{24}$$

Since

$$I^{(2)}_{ij;\lambda\mu} = \frac{1}{3} \delta_{ij} \delta_{\lambda\mu} \tag{25}$$

The rotational average of $\tilde{B}$ is

$$B = \frac{2\pi}{3\hbar^2} |\underline{\mu}^{no}|^2 e_i \bar{e}_j \delta_{ij}$$

$$= \frac{2\pi}{3\hbar^2} |\underline{\mu}^{no}|^2 , \qquad (26)$$

the result quoted in equation (1). As is to be expected this result is independent of the polarization of the radiation as $\underline{e}.\bar{\underline{e}} = 1$ for any polarization.

For $N = 2$, we have from equation (4)

$$\tilde{B}^{(2)} = \frac{(2\pi)^2}{\hbar^2 c} \sum_{\alpha,\alpha^1} \frac{\mu_i^{n\alpha} \mu_j^{\alpha o} \bar{\mu}_k^{n\alpha'} \bar{\mu}_l^{\alpha' o} e_i e_j \bar{e}_k \bar{e}_l}{(E_{n\alpha} - \hbar\omega)(E_{n\alpha^1} - \hbar\omega)} . \qquad (27)$$

We note several possibilities that arise which are not present for $N = 1$.
(i) There are resonances in the generalized B-coefficients if the molecular spectra includes levels $|\alpha\rangle$ between $|0\rangle$ and $|n\rangle$ such that $E_{n\alpha} \sim \hbar\omega$.
(ii) There are contributions to the B-coefficients from *permanent* electric moments in polar molecules. These usually depend on the vector difference between the permanent moments in the excited and ground states

$$\underline{d} = \underline{\mu}^{nn} - \underline{\mu}^{oo} . \qquad (27)$$

For example the terms in the matrix element which have such permanent moment factors are proportional to

$$\left(\frac{\mu_i^{nn} \mu_j^{no}}{-\hbar\omega} + \frac{\mu_i^{no} \mu_j^{oo}}{\hbar\omega} \right) e_i e_j = - \frac{d_i \mu_j^{no}}{\hbar\omega} e_i e_j \qquad (28)$$

(iii) Following from (ii) we can hope to make a valid 2-level approximation if all the other energy level E_α are greatly in excess of E_n and E_o. In this case

$$\tilde{B}^{(2)} \simeq \frac{(2\pi)^2}{\hbar^2 c} \frac{d_i \mu_j d_k \bar{\mu}_l}{(\hbar\omega)^2} e_i e_j \bar{e}_k \bar{e}_l . \qquad (29)$$

We note $\underline{d}$ is always real.
(iv) The value of $B^{(2)}$ depends on the polarization character of the light. This dependence survives random averaging of the molecular orientation. The reason for this is that isotropic four-tensors include $\delta_{ij}\delta_{kl}$ in addition to $\delta_{ik}\delta_{jl}$ and $\delta_{il}\delta_{jk}$. The latter pair both give unity in the context of equation (29) but the former gives $(\underline{e}.\underline{e})(\bar{\underline{e}}.\bar{\underline{e}})$ and we have from equation (17)

$$\underline{e}.\underline{e} = \sin\beta \tag{30}$$

where β is the colatitude angle for the representative point on the Poincare sphere. This implies $\underline{e}.\underline{e} = 0$ for circularly polarized light but $\underline{e}.\underline{e} = 1$ for plane polarized light.

To illustrate the methodology the calculations for the two-level approximation are now given. For the locked in case with notation of (18) equation (29) becomes

$$\tilde{B}^{(2)} = \frac{4\pi^2}{\hbar^4\omega^2 c}\,\frac{\left|\underline{d}.\left(\varepsilon_x e^{-i\delta_x}\underline{e}_x + \varepsilon_y e^{-i\delta_y}\underline{e}_y\right)\right|^2 \left|\underline{\mu}.\left(\varepsilon_x e^{-i\delta_x}\underline{e}_x + \varepsilon_y e^{-i\delta_y}\underline{e}_y\right)\right|^2}{\left(\varepsilon_x^2 + \varepsilon_y^2\right)^2} \tag{31}$$

$$= \frac{4\pi^2}{\hbar^2\omega^2 c}\,\frac{1}{\left(\varepsilon_x^2 + \varepsilon_y^2\right)^2}\Big[d_x^2\mu_x^2\varepsilon_x^4 + d_y^2\mu_y^2\varepsilon_y^4 + d_x^2\mu_y^2\varepsilon_x^2\varepsilon_y^2 + d_y^2\mu_x^2\varepsilon_x^2\varepsilon_y^2$$

$$+ 2d_xd_y\mu_x\mu_y(1 + \cos2\delta)\,\varepsilon_x^2\varepsilon_y^2 + 2d_xd_y\cos\delta\;\varepsilon_x\varepsilon_y\left(\varepsilon_x^2\mu_x^2 + \varepsilon_y^2\mu_y^2\right)$$

$$+ 2\mu_x\mu_y\cos\delta\;\varepsilon_x\varepsilon_y\left(\varepsilon_x^2d_x^2 + \varepsilon_y^2d_y^2\right)\Big]\;. \tag{32}$$

On the other hand for the freely rotating case

$$B^{(2)} = \frac{4\pi^2}{\hbar^4\omega^2 c}\, I^{(4)}_{ijkl;\lambda\mu\upsilon o}\; d_\lambda\mu_\mu d_\upsilon\bar{\mu}_o\; e_ie_j\bar{e}_k\bar{e}_l$$

and, since

$$I^{(4)} = \frac{1}{30}\Big[\delta_{ij}\delta_{kl}\left(4\delta_{\lambda\mu}\delta_{\upsilon o} - \delta_{\lambda\upsilon}\delta_{\mu o} - \delta_{\lambda o}\delta_{\mu\upsilon}\right)$$

$$+ \delta_{ik}\delta_{jl}\left(-\delta_{\lambda\mu}\delta_{\upsilon o} + 4\delta_{\lambda\upsilon}\delta_{\mu o} - \delta_{\lambda o}\delta_{\mu\upsilon}\right)$$

$$+ \delta_{il}\delta_{jk}\left(-\delta_{\lambda\mu}\delta_{\upsilon o} - \delta_{\lambda\upsilon}\delta_{\mu o} + 4\delta_{\lambda o}\delta_{\mu\upsilon}\right)\Big]\;, \tag{33}$$

$$B^{(2)} = \frac{2\pi^2}{15\hbar^4\omega^2 c}\Big[\sin^2\beta\left(4\delta_{\lambda\mu}\delta_{\upsilon o} - \delta_{\lambda\upsilon}\delta_{\mu o} - \delta_{\lambda o}\delta_{\mu\upsilon}\right)$$

$$+ \left(-\delta_{\lambda\mu}\delta_{\upsilon o} + 3\delta_{\lambda\upsilon}\delta_{\mu o} + 3\delta_{\lambda o}\delta_{\mu\upsilon}\right)\Big]d_\lambda\mu_\mu d_\upsilon\bar{\mu}_o$$

$$= \frac{2\pi^2}{15\hbar^4\omega^2 c}\Big[\sin^2\beta\left(3|\underline{d}.\underline{\mu}|^2 - d^2|\underline{\mu}|^2\right) + |\underline{d}.\underline{\mu}|^2 + 3d^2|\mu|^2\Big]\;. \tag{34}$$

Without loss of generality we can take $\underline{\mu}$ to be real so that if ε is the angle between $\underline{d}$ and $\underline{\mu}$

$$B^{(2)} = \frac{2\pi^2}{15\hbar^4\omega^2 c} d^2\mu^2\left[\left(3 - \sin^2\beta\right) + \cos^2\varepsilon\left(3\sin^2\beta + 1\right)\right] . \qquad (35)$$

The special cases follow immediately

circular $$B^{(2)} = \frac{2\pi^2 d^2\mu^2}{15\hbar^4\omega^2 c}\left(3 + \cos^2\varepsilon\right) \qquad (36)$$

plane $$B^{(2)} = \frac{4\pi^2 d^2\mu^2}{15\hbar^4\omega^2 c}\left(1 + 2\cos^2\varepsilon\right) . \qquad (37)$$

In a similar way we can find $B^{(2)}$ for the realistic many level situation. The detailed results are given in reference (1d). Here we quote the result after averaging.

$$B^{(2)} = \frac{2\pi^2}{15\hbar^2 c}\sum_{\alpha,\alpha'} \frac{1}{(E_{n\alpha} - \hbar\omega)(E_{n\alpha'} - \hbar\omega)} \times \left[2\underline{\mu}^{n\alpha}.\underline{\mu}^{\alpha o}\,\underline{\mu}^{n\alpha'}.\underline{\mu}^{\alpha' o}\left(1 - 2\cos^2\beta\right)\right.$$

$$\left. + \left\{\underline{\mu}^{n\alpha}.\underline{\mu}^{n\alpha'}\,\underline{\mu}^{\alpha o}.\underline{\mu}^{\alpha' o} + \underline{\mu}^{n\alpha}.\underline{\mu}^{\alpha' o}\,\underline{\mu}^{\alpha o}.\underline{\mu}^{n\alpha'}\right\}\left(1 + \cos^2\beta\right)\right] . \qquad (38)$$

If a single energy E_α resonance predominates

$$B^{(2)} = \frac{2\pi^2}{15\hbar^2 c}\frac{|\underline{\mu}^{n\alpha}|^2|\underline{\mu}^{\alpha o}|^2}{(E_{n\alpha} - \hbar\omega)^2}\left[\cos^2\gamma\left(4 - 3\cos^2\beta\right) + \left(2 + \cos^2\beta\right)\right] \qquad (39)$$

where γ is the body-fixed angle between $\underline{\mu}^{n\alpha}$ and $\underline{\mu}^{\alpha o}$. Again we have the special cases:-

circular $$B^{(2)} = \frac{2\pi^2}{15\hbar^2 c}|\underline{\mu}^{n\alpha}|^2|\underline{\mu}^{\alpha o}|^2\frac{3 + \cos^2\gamma}{(E_{n\alpha} - \hbar\omega)^2} \qquad (40)$$

plane $$B^{(2)} = \frac{4\pi^2}{15\hbar^2 c}|\underline{\mu}^{n\alpha}|^2|\underline{\mu}^{\alpha o}|^2\frac{1 + 2\cos^2\gamma}{(E_{n\alpha} - \hbar\omega)^2} . \qquad (41)$$

V CHIROPTICAL PROPERTIES

In section IV it has been emphasized that the generalised B-coefficients depend on the ellipticity of the light but that for $N = 1$ the Einstein B-coefficient is independent of the polarization. In fact $B^{(1)}$ does depend on the ellipticity if the calculation is taken beyond the electric-dipole approximation. We can include higher multipoles by extending the interaction Hamiltonian (3) and the matrix element shown in equation (5). To take the calculation up to the order of the electric quadripole and the magnetic dipole we use

$$H_{Int} = - \underline{\mu}.\underline{D}^{\perp} - \underline{m}.\underline{B} - \frac{1}{2}\underline{\underline{Q}}.\underline{\nabla}\underline{D}^{\perp} - \ldots \tag{42}$$

and the matrix element can be obtained from equation (5) by replacing $\underline{\mu}$

$$\underline{\mu} \longrightarrow \underline{\mu} + \underline{m} \times \hat{\underline{k}} + \frac{1}{2} e\underline{\underline{Q}}.\underline{k} . \tag{43}$$

Thus

$$\tilde{B}^{(1)} = \frac{2\pi}{\hbar^2}\left(\underline{\mu} + \underline{m} \times \hat{\underline{k}} + \frac{1}{2} i\underline{\underline{Q}}.\underline{k}\right)_i \left(\bar{\underline{\mu}} + \bar{\underline{m}} \times \hat{\underline{k}} - \frac{1}{2} i\underline{\underline{Q}}.\underline{k}\right)_j e_i \bar{e}_j \tag{44}$$

and the first order correction to equation (1) will arise from terms in (44) linear in $\underline{\mu}$ and $\underline{m}$ or linear in $\underline{\mu}$ and $\underline{\underline{Q}}$. Rotational averaging is straightforward; we have

$$\left.\begin{aligned}
\mu_i \mu_j &\longrightarrow \frac{1}{3}\delta_{ij} |\underline{\mu}|^2 \\
\mu_i \bar{m}_l \epsilon_{jlk} &\longrightarrow \frac{1}{3}\underline{\mu}.\bar{\underline{m}} \epsilon_{jik} \\
m_l \bar{\mu}_j \epsilon_{ilk} &\longrightarrow \frac{1}{3}\underline{\mu}.\bar{\underline{m}} \epsilon_{ijk} \\
\mu_i Q_{jk} &\longrightarrow 0
\end{aligned}\right\} \tag{45}$$

The electric quadripole does not contribute to the single-photon absorption rate on average but the magnetic dipole does. The use of

$$e_i \bar{e}_j \epsilon_{ijk} = - i \cos\beta \hat{k}_k \tag{46}$$

yields

$$B^{(1)} = \frac{2\pi}{3\hbar^2}|\underline{\mu}|^2 + \frac{4\pi}{3\hbar^2}(\underline{\mu}.\mathrm{Im}\underline{m}) \cos\beta . \tag{47}$$

A simple consequence of equation (47) is the differential absorption rate between left- and right-handed circularly polarized light, namely,

$$B_L^{(1)} - B_R^{(1)} = \frac{8\pi}{3\hbar^2}(\underline{\mu}.\text{Im}\underline{m}) \ . \tag{48}$$

Equation (48) is the Condon-Rosenfeld result[4] for circular dichroism. The optical rotary strength for the transition $0 \longrightarrow n$ is defined to be

$$R^{no} = \underline{\mu}^{no}.\text{Im}\underline{m}^{no} \ . \tag{49}$$

It is clear, since the operator $\underline{m}$ is perpendicular to $-e\underline{r} = \underline{\mu}$, that the optical rotary strengths obey the sum-rule

$$\sum_n R^{no} = 0 \ . \tag{50}$$

The chirality effects can be considered in the context of multiphoton spectroscopy. For 2-photon absorption and taking into account only the lead term in the magnetic dipole transition moments

$$\tilde{B}_L^{(2)} - \tilde{B}_R^{(2)} = \frac{8\pi^2}{\hbar^2 c}\sum_{\alpha,\beta}\left[\left(E_{n\alpha} - \hbar\omega\right)\left(E_{n\beta} - \hbar\omega\right)\right]^{-1} \times \left\{\left[(\text{Im}\underline{m})_i^{n\alpha}\,\mu_j^{\alpha o} + \mu_j^{n\alpha}(\text{Im}\underline{m})_i^{\alpha o}\right]\mu_k^{n\beta}\,\mu_l^{\beta o}\right.$$

$$\left. + \mu_i^{n\alpha}\,\mu_j^{\alpha o}\left[(\text{Im}\underline{m})_k^{n\beta}\,\mu_l^{\beta o} + \mu_l^{\beta\alpha}(\text{Im}\underline{m})_k^{\beta o}\right]\right\} e_i e_j \bar{e}_k \bar{e}_l \ . \tag{51}$$

Considerable simplification occurs for a two-level system and we have, for propagation in the z-direction,

$$\tilde{B}_L^{(2)} - \tilde{B}_R^{(2)} = \frac{4\pi^2}{\hbar^4\omega^2 c}\left(\underline{\mu}.\text{Im}\underline{m} - \mu_z(\text{Im}\underline{m})_z\right)\left(\underline{d}^2 - d_z^2\right) \ . \tag{52}$$

The resulting differential B-coefficient for a randomly oriented molecule is

$$B_L^{(2)} - B_R^{(2)} = \frac{8\pi^2}{15\hbar^4\omega^2 c}\left[3(\underline{\mu}.\text{Im}\underline{m})\underline{d}^2 + (\underline{\mu}.\underline{d})(\underline{d}.\text{Im}\underline{m})\right] \ . \tag{53}$$

It should be noted that a non-zero difference of the B's for the two helicities depends on the existence of permanent electric moments and also the difference is not a function of R^{no} alone but also depends on the component of $\text{Im}.\underline{m}$ in the direction of $\underline{d}$. A molecule with $\text{Im}\underline{m}$ perpendicular to $\underline{\mu}$ can still show 2-photon circular dichroism in contrast to the situation in the 1-photon case.

For a realistic many-level system equation (51) has terms independent of $\underline{d}$ these being, of course, the only terms for atoms or molecules with no permanent moments. The average of these terms is[5]

$$B_L^{(2)} - B_R^{(2)} = \frac{4\pi^2}{15\hbar^2 c} \sum_{\alpha,\beta} \left[\left(E_{n\alpha} - \hbar\omega\right)\left(E_{n\beta} - \hbar\omega\right)\right]^{-1} \times$$

$$\left\{-2\left[(\mathrm{Im}\underline{m})^{n\alpha}.\underline{\mu}^{\alpha o} + (\mathrm{Im}\underline{m})^{\alpha o}.\underline{\mu}^{n\alpha}\right] \underline{\mu}^{n\beta}.\underline{\mu}^{\beta o} + 3\left[(\mathrm{Im}\underline{m})^{n\alpha}.\underline{\mu}^{\beta o} + (\mathrm{Im}\underline{m})^{\beta o}.\underline{\mu}^{n\alpha}\right] \underline{\mu}^{n\beta}.\underline{\mu}^{\alpha o}\right.$$

$$\left. + 3(\mathrm{Im}\underline{m})^{n\alpha}.\underline{\mu}^{\alpha\beta}\ \underline{\mu}^{\alpha o}.\underline{\mu}^{\beta o} + 3(\mathrm{Im}\underline{m})^{\alpha o}.\underline{\mu}^{\beta o}.\underline{\mu}^{n\alpha}.\underline{\mu}^{n\beta}\right\} \qquad (54)$$

Similar calculations have been carried out for elliptically polarized light. As for $N = 1$ the elliptical B-coefficient can be obtained from the circular by multiplication by $\cos\beta$. In the appendix of our paper(1d) we prove that (for $N = 1$ and 2) the differential rates for elliptically polarized light

$$\Delta B^{(N)}_{\text{elliptical}} = \Delta B^{(N)}_{\text{circular}} \cos\beta = -\Delta B^{(N)}_{\text{circular}} \frac{2\varepsilon_x \varepsilon_y \sin\delta}{\varepsilon_x^2 + \varepsilon_y^2} . \qquad (55)$$

This is true for other sources of optical rotation such as an effective electrical quadrupole moment. For $N = 2$ the electric quadrupole terms survive random averaging. As an example the two-level electric quadrupole term in the differential rate prediction is

$$\Delta B^{(2)}_{\text{elliptical}} = \frac{4\pi^2}{15\hbar^4\omega^2 c^2} \underline{d} \times \underline{\mu}.\left(\underline{\underline{Q}}.\underline{d} - \underline{\underline{\Delta}}.\underline{\mu}\right) \cos\beta \qquad (56)$$

where

$$\underline{\underline{\Delta}} = \underline{\underline{Q}}^{nn} - \underline{\underline{Q}}^{oo} \qquad (57)$$

is any permanent quadrupole moment difference between the electronic excited state $|n\rangle$ and the ground state $|o\rangle$.

For $N \geq 3$ equation (55) does not apply. We have calculated(1d) the generalized B-coefficient for $N = 3$. The algebraic manipulations are tedious and the resulting expressions for the absorption rates complicated. One simple case is the averaged 3-photon differential rate for a two-level system with no permanent moments and no quadrupole effects, then

$$B_L^{(3)} - B_R^{(3)} = \frac{4}{35} \frac{\pi^3}{\hbar^6 \omega^4 c^2} R^{no} |\mu^{no}|^4 . \qquad (58)$$

Another facet of optical activity is optical rotation. Plane polarized light has its direction of polarization rotated as it passes through an active medium. The degree of specific rotation ϕ depends on the difference between the refractive indices for left- and right-circularly polarized light

$$\phi(\omega) = \frac{\omega}{2c} \mathrm{Re}\left[n_L(\omega) - n_R(\omega)\right] \tag{59}$$

These refractive indices are related to the absorption rates by dispersion relations. The Kramers-Krönig relations

$$\mathrm{Re}\; n(\omega) = 1 + \frac{c}{\pi} P\int_0^\infty \frac{\varepsilon(\omega')}{\omega'^2 - \omega^2} d\omega' \tag{60}$$

have to be generalized[6] to take into account non-trivial crossing relations involving opposing helicities. The correct dispersion relations are

$$\mathrm{Re}\; n_L(\omega) = 1 + \frac{c}{\pi} P\int_0^\infty \frac{\varepsilon_L(\omega')}{\omega'(\omega' - \omega)} d\omega' + \frac{c}{2\pi} P\int_0^\infty \frac{\varepsilon_R(\omega')}{\omega'(\omega' + \omega)} d\omega' \tag{61}$$

and the corresponding equation with L and R interchanged. Inserting these into equation (59) yields (ϕ is in radians/metre)

$$\phi(\omega) = \frac{\omega^2}{2\pi} P\int_0^\infty \frac{\varepsilon_L(\omega') - \varepsilon_R(\omega')}{\omega'(\omega'^2 - \omega^2)} d\omega' . \tag{61}$$

Now circular dichroism for $N = 1$ gives as the differences of the extinction coefficients

$$\varepsilon_L(\omega) - \varepsilon_R(\omega) = 2\pi \sum_n \hbar\omega_{no}\left(B_L^{no} - B_R^{no}\right)\delta\left(\omega - \omega_{no}\right) \tag{62}$$

assuming a sharp zero-width line. Thus

$$\phi(\omega) = \sum_n \frac{\hbar\omega^2\left(B_L^{no} - B_R^{no}\right)}{\omega_{no}^2 - \omega^2} = \frac{8\pi}{3}\hbar \sum_n \frac{\omega^2 R^{no}}{\omega_{no}^2 - \omega^2} = \frac{8\pi\hbar}{3} \sum_n \frac{\omega_{no}^2 R^{no}}{\omega_{no}^2 - \omega^2} \tag{63}$$

The last equality in (63) follows from the sum rule equation (50), and it is of interest to note in view of the companion paper in these proceedings, the factor $\left(\frac{\omega_{no}}{\omega}\right)^2$ is irrelevant in this case.

In the multiphoton regime a general development of dispersion relationships appears not to be possible according to the work of Smet and van Groenendael[7]. However a limited application where all photons are from a single beam (as is discussed here) and where no harmonic generation is involved is available. For non-linear absorption from a single mode of circular frequency ω, so that $N\hbar\omega \sim E_{no}$, the intensity falls away in transmission through a homogeneous low density solute according to the law

$$\frac{dI}{dz} = -\sum_{n} (N\hbar\omega)\Gamma_{no}^{(N)}\Delta\omega\delta\left(N\omega - \omega_{no}\right) . \tag{64}$$

The Lambert-Beer law holds only for $N = 1$ and a more complicated absorption law follows in general[8]. However the imaginary part of the complex refraction index can be approximated

$$\text{Im } n(\omega) = \frac{c}{2\omega}\varepsilon^{N}(\omega) = \frac{\pi}{\omega} g^{(N)} I^{N-1} \sum_{n} (N\hbar\omega)B_{no}^{(N)}\Delta\omega\delta\left(N\omega - \omega_{no}\right) \tag{65}$$

whence

$$\phi(\omega) = \frac{\omega^2}{2\pi}\sum_{n} P\int_0^{\infty} \frac{2\pi N\hbar\omega' g^{(N)} I^{N-1}\Delta B^{N}\delta\left(N\omega' - \omega_{no}\right)}{\omega'\left(\omega'^2 - \omega^2\right)} d\omega'$$

$$= \sum_{n} \frac{(N\hbar\omega)^2 g^{N} I^{N-1}\Delta B^{N}}{E_{no}^2 - (N\hbar\omega)^2} . \tag{66}$$

Unlike the $N = 1$ case (equation (63)) we cannot replace $(N\hbar\omega)^2$ by E_{no}^2 as the sum rule does not work and the slope of optical rotary dispersion is distorted by the $\left(\frac{N\hbar\omega}{E_{no}}\right)^2$ factor.

The experimental signature of non-linear optical rotation is the intensity dependence of the rotation angle. There is preliminary experimental evidence of this in the work of Gedanken and Tamir[9]. A beam from an excimer laser at 456nm passes through a solution of camphorsulfonic acid in water. By examining the rotation of the plane of polarization for +,- and a racemic mixture for the acid and by focusing the beam to obtain a high intensity they show a dependence of ϕ on I^N (over an intensity range of one order of magnitude) which is close to $N = 4$ in equation (66).

References

1. a. W.J. Meath and E.A. Power, *Mol. Phys.* 51:585 (1984).
   b. W.J. Meath and E.A. Power, *J. Phys. B* 17:763 (1984).
   c. W.J. Meath and E.A. Power, *J. Phys. B* 20:1945 (1987).
   d. W.J. Meath and E.A. Power, *J. Mod. Optics* 36:977 (1989).
2. D.P. Craig and T. Thirunamachandran, "Molecular Quantum Electrodynamics," Academic Press, London (1984).
3. D. Andrews and T. Thirunamachandran, *J. Chem. Phys.* 67:5026 (1977).
4. L. Rosenfeld, *Z. Phys.* 52:161 (1928).
   E.U. Condor, *Rev. Mod. Phys.* 9:432 (1937).
5. J. Tinoco, *J. Chem. Phys.* 62:1006 (1975).
   E.A. Power, *J. Chem. Phys.* 63:1348 (1975).
6. C.A. Emeis, L.J. Oosterhoff and G. de Vries, *Proc. Roy. Soc.* A297:54 (1967).
   W.P. Healy and E.A. Power, *Am. J. Phys.* 42:1070 (1974).
7. F. Smet and A. van Groenendael, *Nuovo Cimento* B20:273 (1974), and *Phys. Rev.* A19:334 (1979).
8. R. Loudon, "The Quantum Theory of Light," Clarendon Press, Oxford (1973).
9. A. Gedanken and M. Tamir, *Rev. Sci. Inst.* 58:950 (1987).

CORRELATED EMISSION LASERS

B.J. Dalton

Physics Department
University of Queensland
Brisbane, Australia

Blackett Laboratory
Imperial College of Science, Technology and Medicine
London, England

Clarendon Laboratory
Oxford University
Oxford, England

INTRODUCTION

The basic idea of a Correlated Emission Laser (CEL) was introduced by Scully (1985). Consider a laser containing an active medium of three level atoms (as shown in Fig. 1 with vee, lambda or cascade configurations) contained in a cavity for which either one or two modes are important. In this system two atomic transitions 1–3, 2–3 share a common state 3. Each transition is coupled to a cavity mode, which may be the same as in a one mode CEL, or which may be different as in a two mode CEL. The operation of the CEL depends on coherence between states 1 and 2. Such coherence could be produced via certain optical pumping processes or in the vee or lambda cases via a strong RF or microwave field. Fundamentally, the 1–2 coherence results in a correlation in the 1–3, 2–3 atomic transitions, thereby producing a correlation in the contributions to the electric fields produced in the cavity mode or modes due to these transitions.

The effect on the cavity fields in pictured in Fig. 2 for both the one and two mode cases (Schleich and Scully, 1988; Orszag et al., 1988). A typical two mode case is the Hanle Effect CEL (Scully, 1985) involving a three level vee system in which coherent pumping produces the 1–2 (Zeeman) coherence, and in which the 1–3 and 2–3 transitions are coupled to cavity modes a and b respectively. a and b may have the same frequency, but orthogonal polarizations. A typical one mode case is the Two Photon CEL (Scully and Zubairy, 1988; Scully et al., 1988) involving a three level cascade system in which atoms are prepared in a coherent superposition of the two outer states 1,2 (optical coherence) and in which the 2–3 and 3–1 transitions are coupled to the same cavity mode a. In the two mode case transitions associated with stimulated emission events result in colinear additions associated with laser action to the electric fields in the a and b cavities. Transitions associated with spontaneous emission events result in random contributions δE_a, δE_b to the a and b cavities, and which are associated with laser noise. However, due to the correlation between the 1–3, 2–3 atomic transitions, the angle between δE_a, δE_b is not random thereby reducing the fluctuation in the <u>phase difference</u> Φ between the a, b

New Frontiers in Quantum Electrodynamics and Quantum Optics
Edited by A. O. Barut, Plenum Press, New York, 1990

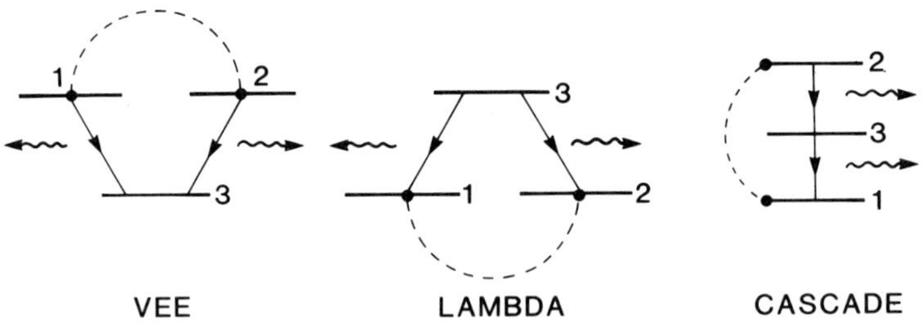

Fig. 1 Three level atomic systems as active media for correlated emission lasers. Atomic coherence between states 1,2 .______. results in correlations between 1–3, 2–3 transitions ⟶, thereby producing correlations in the photon ∿∿∿▸ emission.

cavity fields. The two mode CEL thus has reduced noise in the phase difference and this manifests itself in terms of narrow spectral features in the light field of the two moded cavity. Reductions in line width below the Townes–Schawlow limit for a one mode laser may occur. In the one mode case, again transitions associated with stimulated emission events result in the colinear additions to the electric field of the cavity that are associated with laser action. Contributions δE_{23}, δE_{31} to the electric field in the cavity associated with successive spontaneous emission processes that would contribute to laser noise are again correlated, so that the overall fluctuations in the laser cavity <u>phase</u> Φ is reduced. This reduced noise in the laser phase may again result in narrow spectral features, or even in a squeezed light field in which the fluctuations in one quadrature component of the cavity field are reduced below that for a coherent state. (For a recent review of squeezing, see Loudon and Knight, 1987).

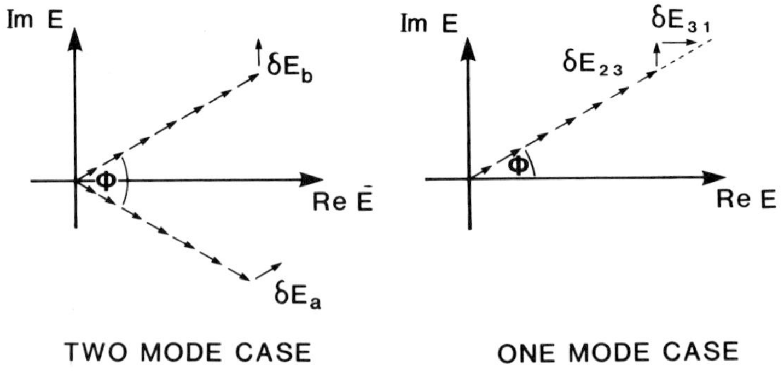

Fig.2 Pictorial representation of correlated emission laser action for two mode and one mode cases. Contributions to cavity electric fields due to stimulated emission processes (colinear arrows) and spontaneous emission processes (δE_a, δE_b, δE_{23}, δE_{31}) are shown. Reductions in fluctuations in the phase difference (two mode case) or in the phase (one mode case) results.

Apart from its role in terms of improving the fundamental limits of laser performance, the CEL is significant to the basic theory of quantum noise and its modification. Applications of CEL to Ultra High Resolution Interferometry, such as in Gravity Wave Detectors, Laser Gyroscopes and measurements of Ultra Small Displacements, have been proposed (Scully, 1985; Orszag et al., 1988; Scully and Gea-Banacloche, 1986). Applications to Optical Communications via the production of bright squeezed light in the Two Photon CEL may also be possible. The production of two moded light fields with correlations in the phase fluctuations for each mode, as in the Hanle Effect or Quantum Beat CEL could be of interest in spectroscopic studies of atomic coherence (Swain, 1988a; Dalton and Knight, 1982a, 1982b; Dalton et al., 1985). The study of CEL has been reviewed previously by Orszag et al., (1988), and the general behaviour of CEL studied by Schleich and Scully (1988), Schleich et al. (1988), Bergou et al., (1989a) and Lu et al., (1989a).

The various types of CEL that have been examined are listed in Table 1. Of the two mode CEL, the Quantum Beat CEL has been studied by Scully and Zubairy (1987), Swain (1988a), Bergou and Orszag (1988a) and Bergou et al., (1988a). The Quantum Beat CEL Micromaser has been discussed by Bergou and Orszag (1988b), the Free Electron CEL by Orszag et al., (1987) and the Holographic CEL by Krause and Scully (1987). Previous studies of the Hanle Effect CEL have been made by Bergou et al., (1988b) and Zaheer and Zubairy (1988). In the case of one mode CEL, the Two Photon CEL has been treated by Scully and Zubairy (1988), Scully et al., (1988), Lu et al., (1989b), Bergou et al., (1989b). Related work on two photon lasers has been carried out by Swain (1988b), Boone and Swain (1989a, 1989b, 1990). The Degenerate Quantum Beat Laser is discussed by Scully et al., (1989) and the Polarization CEL by Benkert and Scully (1990).

Experimental studies of CEL have been reported by Toschek and Hall (1987) in He-Ne lasers, by Ohtsu and Liou (1987) in semi conductor lasers and by Winters and Hall (1990), again in He-Ne lasers.

Most theoretical work on CEL has been developed on the basis of the Scully-Lamb theory of the laser (see Sargent, Scully and Lamb, 1974 for an overview). In this approach a master equation for the density operator describing the cavity mode(s) is obtained by tracing out the atomic variables. In linear versions of the theory the atom-cavity coupling is treated correct to the second order, whilst in the non-linear versions the atomic motion is first solved for in terms of a no-back reaction approximation. In the Quantum Beat CEL case, coherence is included via a classical RF field coupling the 1,2 atomic states, whilst in the Hanle Effect CEL coherence is included via treating the initial atomic state as a coherent superposition of states 1,2. Cavity damping and spontaneous emission decay are included and the atoms are injected at (usually) random times into the cavity, similarly to the situation in masers. In many respects the atoms are treated like a reservoir. The master equation for the cavity density operator can then be converted into a Fokker-Planck equation for the Glauber-Sudarshan P representation distribution function (Glauber 1963a, 1963b; Sudarshan, 1963) and using amplitude and phase variables for the cavity mode(s). Quenching of the noise in the phase difference for the two mode case or the phase for the one mode case is then discussed in terms of the behaviour of the diffusion matrix elements associated with these variables.

Another approach to laser theory is that of Haken and co-workers (see Haken, 1970 for an overview). Beginning with a master elevation for the combined atom plus cavity systems and including atomic relaxation, atomic pumping and cavity damping contributions (see Louisell, 1973) a Fokker-Planck equation is derived for a Glauber-Sudarshan P representation distribution function extended to describe atomic variables and assuming the number of atoms is large. The atomic motion is then eliminated in the high Q cavity regime (where atomic variables relax much faster than cavity variables) via an adiabatic approximation either in terms of the distribution function itself, or more conveniently after converting the Fokker-Planck equation into Langevin equations for stochastic variables

Table 1. Types of Correlated Emission Lasers

Type	Features	Behaviour
(1) Two Mode CEL		
Quantum Beat CEL	.Medium of N(≫1) three level vee systems .Coherence produced by RF field .Cavity frequencies may be different	Quenching of noise in phase difference between cavity fields
Hanle Effect CEL	.Medium of N(≫1) three level vee systems .Coherence produced by optical pumping .Cavity frequencies may be same .Cavity polarizations may be orthogonal	As for Quantum Beat CEL
Quantum Beat Micromaser CEL	.As for Quantum Beat CEL but N~1	Photon statistics of coupled mode same as for the two level micromaser
Free Electron CEL	.Electron beam medium .Coherence between two momentum states	As for Quantum Beat CEL
Holographic CEL	.Gain medium with spatial modulation	As for Quantum Beat CEL
(2) One mode CEL		
Two Photon CEL	.Medium of N(≫1) three level cascade systems .Atoms prepared in coherent superposition of outer states .System may be Rydberg atoms	.Quenching of noise in cavity phase .Squeezing of cavity field
Degenerate Quantum Beat CEL	.Medium of N(≫1) three level lambda systems .Coherence produced by RF field	Lasing without inversion
Degenerate Quantum Beat CEL	.Medium of N(≫1) three level vee systems .Coherence produced by RF field	Inversion without lasing
Polarization CEL	.Medium of N(≫1) two levels systems .Atoms prepared in coherent superposition of states .Atoms injected regularly into cavity	

describing both atoms and cavities. Finally, the discussion proceeds as in the Scully–Lamb theory in terms of cavity variables.

A variant of the Haken approach that has proved particularly useful in describing non classical light fields (where the replacement of the Fokker–Planck equation with c–number Langevin equations is not possible for the ordinary P representation) is to make use of the positive P representation. Originally introduced by Drummond and Gardiner (1980) for bosonic systems, it has now been extended and widely used for atomic systems also (see for example, Reid, 1988; Carmichael et al., 1986; Reid and Walls, 1985; Reid and Walls, 1983; Drummond and Walls, 1981).

In the CEL situation we fundamentally have a situation where phase dependent atomic noise affects the overall fluctuations in cavity phase variables. It may, therefore, be of some interest to treat the CEL problem via the Haken approach, modified via the use of a positive P representation to guarantee the use of c–number Langevin equations, and to then compare the predictions with that of the Scully–Lamb approach where the atomic motion is eliminated much earlier. A fully non–linear theory can be developed or converted into a linear approximation if required. In the present paper the case to be treated is that of the Hanle Effect CEL.

Before developing this theory in Section Three it is however appropriate to first discuss the measured quantities of interest for the CEL and this is done in the next section.

MEASUREMENTS IN THE CORRELATED EMISSION LASER

At position $\underset{\sim}{R} = 0$ the electric field operator in the Heisenburg picture can be expressed in terms of positive, negative frequency components $\underset{\sim}{E}^+$, $\underset{\sim}{E}^-$ as

$$\underset{\sim}{E}(t) = \underset{\sim}{E}^+(t) + \underset{\sim}{E}^-(t) \tag{2.1}$$

$$\underset{\sim}{E}^- = (\underset{\sim}{E}^+)^\dagger \tag{2.2}$$

For a two moded cavity of volume V associated with annihilation operators a, b, polarisation vectors $\underset{\sim}{\epsilon}_a$, $\underset{\sim}{\epsilon}_b$ and frequencies ω_a, ω_b

$$\underset{\sim}{E}^+ = i\lambda_a\underset{\sim}{\epsilon}_a a + i\lambda_b\underset{\sim}{\epsilon}_b b \tag{2.3}$$

where

$$\lambda_c = \sqrt{\frac{\hbar\omega_c}{2\epsilon_0 V}} \qquad c = a,b \tag{2.4}$$

gives the electric field per photon. For the one moded cavity the second term in (2.3) is dropped

If $\rho(0)$ is the density operator for the CEL at t=0 then the mean value of a quantity associated with the operator $\hat{\Omega}_H(t)$ in the Heisenberg picture or $\hat{\Omega}_S(0)$ in the Schrodinger picture can be written in terms of a trace as

$$\langle\hat{\Omega}\rangle = \mathrm{Tr}\ \hat{\Omega}_H(t)\rho(0) \tag{2.5}$$

$$= \mathrm{Tr}\ \hat{\Omega}_S(0)\rho(t) \tag{2.6}$$

Measurements of the intensity and spectrum for the CEL can be discussed in terms of the Glauber (see Glauber, 1964; Sargent et al., 1974) theory of a model two level atom photon detector, whose transition dipole matrix element is $\underset{\sim}{\mu}$.

The intensity is given in terms of the components of $\underset{\sim}{E}^+$ along $\underset{\sim}{\mu}$ as a one time average

$$I = \langle E_{\mu}^{-}(t)E_{\mu}^{+}(t)\rangle \tag{2.7}$$

which in the case of the two moded cavity becomes

$$I = \bar{\lambda}^2 \left\{\cos^2\theta_a\langle a^\dagger a\rangle + \cos^2\theta_b\langle b^\dagger b\rangle + \cos\theta_a.\cos\theta_b \; (\langle a^\dagger b\rangle + \langle b^\dagger a\rangle)\right\} \tag{2.8}$$

where $\lambda_a, \lambda_b \approx \bar{\lambda}$ and $\cos\theta_c = \underset{\sim}{\epsilon}_c\cdot\hat{\underset{\sim}{\mu}}$ (c=a,b). For the one moded cavity only the first term remains.

The unnormalized spectrum at spectral frequency $\bar{\omega}$, measured over a time interval 0–T involves a two time average

$$S(\bar{\omega},T) = \int_0^T\!\!\int dt_1 dt_2 e^{-i\bar{\omega}(t_1-t_2)}\langle E_{\mu}^{-}(t_1)E_{\mu}^{+}(t_2)\rangle. \tag{2.9}$$

In the case of the two moded cavity we have

$$\begin{aligned}\langle E_{\mu}^{-}(t_1)E_{\mu}^{+}(t_2)\rangle \\ &= \bar{\lambda}^2\left\{\cos^2\theta_a\langle a^\dagger(t_1)a(t_2)\rangle + \cos^2\theta_b\langle b^\dagger(t_1)b(t_2)\rangle\right. \\ &\quad \left. + \cos\theta_a.\cos\theta_b(\langle a^\dagger(t_1)b(t_2)\rangle + \langle b^\dagger(t_1)a(t_2)\rangle)\right\}\end{aligned} \tag{2.10}$$

For the one moded cavity only the first term remains.

Thus we see that for the two moded cavity the intensity and spectrum contain terms involving both modes such as $\langle b^\dagger a\rangle$, $\langle b^\dagger(t_1)a(t_2)\rangle$as well as terms such as $\langle a^\dagger a\rangle$, $\langle a^\dagger(t_1)a(t_2)\rangle$ involving one mode alone. In terms of cavity intensities variables I_a, I_b, phase difference Φ and phase average μ (see later) we will have

$$\langle a^\dagger a\rangle = \langle I_a^{\frac{1}{2}}(t)I_a^{\frac{1}{2}}(t)\rangle \tag{2.11a}$$

$$\langle a^\dagger(t_1)a(t_2)\rangle = \langle I_a^{\frac{1}{2}}(t_1)e^{\frac{1}{2}i\Phi(t_1)}e^{i\mu(t_1)}I_a^{\frac{1}{2}}(t_2)e^{-\frac{1}{2}i\Phi(t_2)}e^{-i\mu(t_2)}\rangle \; e^{i\omega_0(t_1-t_2)} \tag{2.11b}$$

$$\langle b^\dagger a\rangle = \langle I_b^{\frac{1}{2}}(t)I_a^{\frac{1}{2}}(t)e^{-i\Phi(t)}\rangle \tag{2.11c}$$

$$\langle b^\dagger(t_1)a(t_2)\rangle = \langle I_b^{\frac{1}{2}}(t_1)e^{-\frac{1}{2}i\Phi(t_1)}e^{i\mu(t_1)}I_a^{\frac{1}{2}}(t_2)e^{-\frac{1}{2}i\Phi(t_2)}e^{-i\mu(t_2)}\rangle \; e^{i\omega_0(t_1-t_2)} \tag{2.11d}$$

Thus we see the different role the phase difference plays a different role in the one mode and two mode terms.

For the one mode cavity case, only terms like $\langle a^\dagger a \rangle$ and $\langle a^\dagger(t_1)a(t_2)\rangle$ are involved. A single cavity intensity I_a and phase variable Φ_a is involved, and we can utilize (2.11a), (2.11b) via the replacement $\mu \to \Phi_a$ and ignoring Φ. Here the behaviour of the cavity phase Φ_a determines the spectral features.

The role of atomic coherence may be illustrated for a two mode Hanle effect or Quantum Beat CEL (see Fig 1, vee case). If atomic coherence is present we may take the initial state to be of the form

$$|\psi(0)\rangle = (A_1|1\rangle + A_2|2\rangle)|0_a 0_b\rangle \tag{2.12}$$

where $|n_a n_b\rangle$ are states with n_a n_b photons in cavity modes a, b respectively. Ignoring all interactions except the atom-cavity couplings described earlier, this state evolves to

$$\begin{aligned}|\psi(t)\rangle &= A_1(t)e^{-i\omega_1 t}|10_a0_b\rangle + A_2(t)e^{-i\omega_2 t}|20_a0_b\rangle \\ &+ A_{3a}(t)e^{-i(\omega_3+\omega_a)t}|31_a0_b\rangle \\ &+ A_{3b}(t)e^{-i(\omega_3+\omega_b)t}|30_a1_b\rangle\end{aligned} \tag{2.13}$$

A straight forward calculation gives

$$\begin{aligned}I(t) &= \lambda^2\Big\{\cos^2\theta_a|A_{3a}|^2 + \cos^2\theta_b|A_{3b}|^2 \\ &+ \cos\theta_a\cos\theta_b(A_{3a}^*A_{3b}e^{i(\omega_a-\omega_b)t} + c.c)\Big\}\end{aligned} \tag{2.14}$$

showing the presence of oscillating contributions due to $\langle b^\dagger a\rangle$, $\langle a^\dagger b\rangle$ terms. These are only present if both A_{3a} and A_{3b} are non-zero. This in turn requires both A_1 and A_2 in the initial state vector to be non-zero also, since for example $|20_a0_b\rangle$ only evolves into a linear combination of itself and $|30_a1_b\rangle$.

THEORY OF HANLE EFFECT CEL

In the laser model (see Fig 3) we have:

. $N(\gg 1)$ three level atoms (Vee system) coupled near resonance to two optical cavity modes a, b (modelled as quantum harmonic oscillators).

. the atoms undering coherent and incoherent pumping due to interaction with the laser energy source inducing populations and coherences in the upper states of the atomic systems.

. the cavity modes being subject to cavity damping and thermal noise due to interaction with external heat baths (mirrors).

. the atoms undergoing spontaneous emission relaxation, due to interaction with the near empty spontaneous emission modes.

The model involves various parameters, atomic and cavity operators:

. a, $a^\dagger$, b, $b^\dagger$ are annihilation, destruction operators for cavity modes a, b of frequencies ω_a, ω_b, polarizations $\underset{\sim}{\epsilon}_a$, $\underset{\sim}{\epsilon}_b$ and wave vectors $\underset{\sim}{k}_a$, $\underset{\sim}{k}_b$.

. Atomic operators for the N atom system are defined in terms of the transition operators $(\sigma_{ij})_\mu = (|i\rangle\langle j|)_\mu$ for the μth atom located at position $\underset{\sim}{R}_\mu$ in the cavity of volume V as

$$S_a^+ = \sum_\mu e^{-i\underset{\sim}{k}_a\cdot\underset{\sim}{R}_\mu}(\sigma_{31})_\mu \tag{3.1a}$$

$$S_a = \sum_\mu e^{i\underset{\sim}{k}_a\cdot\underset{\sim}{R}_\mu}(\sigma_{13})_\mu \tag{3.1b}$$

$$S_{az} = \sum_\mu \tfrac{1}{2}\left[(\sigma_{33})_\mu - (\sigma_{11})_\mu\right] \tag{3.1c}$$

$$S_b^+ = \sum_\mu e^{-i\underset{\sim}{k}_b\cdot\underset{\sim}{R}_\mu}(\sigma_{32})_\mu \tag{3.1d}$$

$$S_b = \sum_\mu e^{i\underset{\sim}{k}_b\cdot\underset{\sim}{R}_\mu}(\sigma_{23})_\mu \tag{3.1e}$$

$$S_{bz} = \sum_\mu \tfrac{1}{2}\left[(\sigma_{33})_\mu - (\sigma_{22})_\mu\right] \tag{3.1f}$$

$$S^+ = \sum_\mu e^{-i(\underset{\sim}{k}_a-\underset{\sim}{k}_b)\cdot\underset{\sim}{R}_\mu}(\sigma_{21})_\mu \tag{3.1g}$$

$$S = \sum_\mu e^{i(\underset{\sim}{k}_a-\underset{\sim}{k}_b)\cdot\underset{\sim}{R}_\mu}(\sigma_{12})_\mu \tag{3.1h}$$

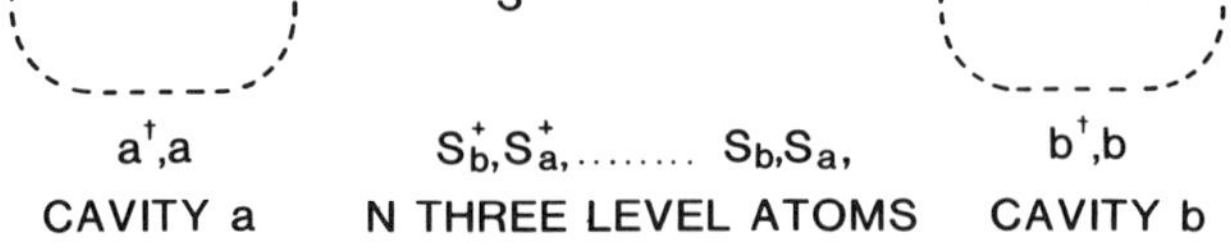

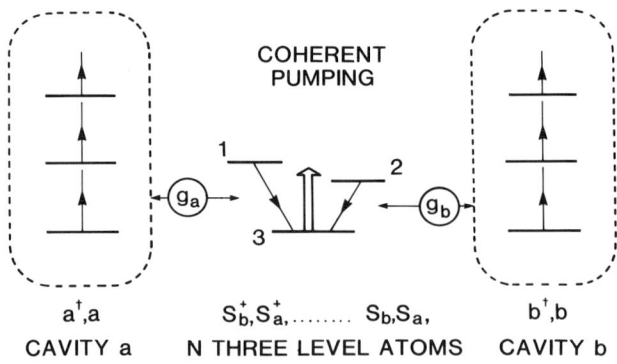

Fig. 3 Hanle effect Correlated Emission Laser model. N three level vee systems subject to a coherent excitation process are coupled to two cavity modes. The 1–3 atom transition is coupled to the a mode, the 2–3 transition to the b mode. Relaxation processes are not shown.

. $\omega_{ij} = \omega_i - \omega_j$ are atomic transition frequencies

. g_a, g_b are one photon Rabi frequencies given in terms of the matrix elements of the dipole operator $\underset{\sim}{d}$ as

$$g_a = \sqrt{\frac{\omega_a}{2\epsilon_0 \hbar V}} < 1|\underset{\sim}{d}.\underset{\sim}{\epsilon}_a|3 > \qquad \text{real} \tag{3.2a}$$

$$g_b = \sqrt{\frac{\omega_b}{2\epsilon_0 \hbar V}} < 2|\underset{\sim}{d}.\underset{\sim}{\epsilon}_b|3 > \qquad \text{real} \tag{3.2b}$$

.w_c is the complex coherent pumping rate for the process 3 → 1,2

.K is the cavity damping rate

.$\bar{n}$ is the mean thermal photon number

.γ_{13}, γ_{23} are the spontaneous emission rates for the transitions 1→3, 2→3

.w_{31}, w_{32} are the incoherent pumping rates for the transitions 3→1, 3→2

The master equation for the system of two cavity modes, N three level atoms is given as

$$\frac{\partial \rho}{\partial t} = \frac{1}{i\hbar} \left[H_0 + H_1, \rho \right] + L_C\rho + L_F\rho + L_A^{\downarrow}\rho + L_A^{\uparrow}\rho \tag{3.3}$$

where

$$H_0 = - S_{az} \frac{2}{3}\hbar (\omega_{13} + \omega_{12}) - S_{bz} \frac{2}{3}\hbar (\omega_{23} - \omega_{12}) + a^{\dagger}a\, \hbar\omega_a + b^{\dagger}b\, \hbar\omega_b \tag{3.4}$$

is the free atom–free field Hamiltonian

$$H_1 = i\hbar \left[g_a^* S_a^+ a^{\dagger} + g_b^* S_b^+ b^{\dagger} - g_a S_a a - g_b S_b b \right] \tag{3.5}$$

is the atom–cavity interaction in the rotating wave approximation

$$\begin{aligned} L_C\rho = \tfrac{1}{2}w_c &\left\{ \sum_{\mu} \left[[(\sigma_{13})_\mu \rho, (\sigma_{32})_\mu] + [(\sigma_{13})_\mu, \rho(\sigma_{32})_\mu] \right] \right\} e^{i(\underset{\sim}{k}_a - \underset{\sim}{k}_b).\underset{\sim}{R}_\mu} \\ + \tfrac{1}{2}w_c^* &\left\{ \sum_{\mu} \left[[(\sigma_{23})_\mu \rho, (\sigma_{31})_\mu] + [(\sigma_{23})_\mu, \rho(\sigma_{31})_\mu] \right] \right\} e^{-i(\underset{\sim}{k}_a - \underset{\sim}{k}_b).\underset{\sim}{R}_\mu} \end{aligned} \tag{3.6}$$

describes coherent atomic pumping

$$\begin{aligned} L_F\rho = K\Big\{ &(\bar{n} + 1)\left[[a\rho, a^{\dagger}] + [a, \rho a^{\dagger}] \right] + \bar{n} \left[[a^{\dagger}\rho, a] + [a^{\dagger}, \rho a] \right] \\ &+ (\bar{n} + 1) \left[[b\rho, b^{\dagger}] + [b, \rho b^{\dagger}] \right] + \bar{n} \left[[b^{\dagger}\rho, b] + [b^{\dagger}, \rho b] \right] \Big\} \end{aligned} \tag{3.7}$$

is the cavity damping-thermal noise term

$$. \ L_A^{\downarrow}\rho = \tfrac{1}{2}\gamma_{13}\,(\bar{n}+1)\left\{\sum_{\mu}\left[\left[(\sigma_{31})_\mu\,\rho,(\sigma_{13})_\mu\right] + \left[(\sigma_{31})_\mu,\,\rho(\sigma_{13})_\mu\right]\right]\right\}$$

$$+ \tfrac{1}{2}\gamma_{23}\,(\bar{n}+1)\left\{\sum_{\mu}\left[\left[(\sigma_{32})_\mu\,\rho,(\sigma_{23})_\mu\right] + \left[(\sigma_{32})_\mu,\,\rho(\sigma_{23})_\mu\right]\right]\right\} \quad (3.8)$$

describes atomic relaxation

$$. \ L_A^{\uparrow}\rho = (\tfrac{1}{2}\gamma_{13}\,\bar{n} + \tfrac{1}{2}w_{31})\left\{\sum_{\mu}\left[\left[(\sigma_{13})_\mu\,\rho,(\sigma_{31})_\mu\right] + \left[(\sigma_{13})_\mu,\,\rho(\sigma_{31})_\mu\right]\right]\right\}$$

$$+ (\tfrac{1}{2}\gamma_{23}\,\bar{n} + \tfrac{1}{2}w_{32})\left\{\sum_{\mu}\left[\left[(\sigma_{23})_\mu\,\rho,(\sigma_{32})_\mu\right] + \left[(\sigma_{23})_\mu,\,\rho(\sigma_{32})_\mu\right]\right]\right\} \quad (3.9)$$

is the incoherent atomic pumping term.

In the master equation the coherent atomic pumping term for the case of a single atom at the origin is

$$\dot{\rho}_{12} = w_c\,\rho_{33} \quad (3.10a)$$

$$\dot{\rho}_{21} = w_c^{*}\,\rho_{33} \quad (3.10b)$$

showing state 3 population being turned into 1-2 atomic coherence at a rate w_c. The coherent atomic pumping term is based on a theory of optical pumping (Barrat and Cohen-Tannoudji, 1961; Dalton, 1987) for a system with two upper states and one lower state. The usual broad band excitation assumptions are involved. As an example, a pumping process from a J=0 level with light polarized in the x direction could be used to excite the Zeeman coherence between the M=-1 and M=+1 states of a J=1 level.

Analysis of the laser model involves the following steps:

. The master equation is converted into a Fokker-Planck equation for a generalised positive P distribution function describing appropriate c-number atomic and cavity variables and using an expansion in powers of 1/N.

. The Fokker-Planck equation is replaced by c-number Langevin equations in the Ito interpretation (Gardiner, 1985) for stochastic atomic and cavity variables.

. The fast relaxing atomic variables are adiabatically eliminated and expressed as functions of cavity variables and noise terms in the high Q cavity regime. This results in non-linear Langevin equations for the cavity variables which depend only on atomic parameters.

. A steady state for certain cavity variables is then obtained from the deterministic part of the Langevin equations.

. Langevin equations for small fluctuations about the steady state are then derived using the usual linear noise theory approach.

. A formal solution of the stochastic differential equations of linear noise theory is then obtained.

. Expressions for measured quantities are then evaluated using stochastic averages and introducing appropriate dimensionless parameters.

. Numerical analysis of the results for various parameter choices can then be performed.

In order to set up the Fokker–Planck equation we first define a characteristic function χ via

$$\chi(\underline{\lambda}) = \mathrm{Tr}\,(\Omega\rho) \tag{3.11}$$

where

$$\underline{\lambda} = \left[\, \lambda_b^+ \; \lambda_a^+ \; \lambda^+ \; \eta_a \; \eta_b \; \lambda \; \lambda_a \; \lambda_b \; \beta_b^+ \; \beta_b \; \beta_a^+ \; \beta_a \,\right]$$

is a set of complex variables, and Ω is defined by

$$\Omega = \Omega_A \, \Omega_F \tag{3.12a}$$

$$\Omega_A = e^{i\lambda_b^+ S_b^+} \, e^{i\lambda_a^+ S_a^+} \, e^{i\lambda^+ S^+} \, e^{i\eta_a S_{az}} \, e^{i\eta_b S_{bz}} \, e^{i\lambda S} \, e^{i\lambda_a S_a} \, e^{i\lambda_b S_b} \tag{3.12b}$$

$$\Omega_F = e^{i\beta_b^+ b^\dagger} \, e^{i\beta_b b} \, e^{i\beta_a^+ a^\dagger} \, e^{i\beta_a a} \tag{3.12c}$$

where Ω_A, Ω_F relate to atomic and cavity operators respectively.

The characteristic function is then related to the distribution function f in terms of the generalised positive P representation as

$$\chi(\underline{\lambda}) = \int d^2\underline{\alpha} \; e^{+i\underline{\lambda}\cdot\underline{\alpha}} \, f(\underline{\alpha}) \tag{3.13}$$

where

$$\underline{\alpha} = \left[\, v_b^+ \; v_a^+ \; v^+ \; D_a \; D_b \; v \; v_a \; v_b \; \alpha_b^+ \; \alpha_b \; \alpha_a^+ \; \alpha_a \,\right]$$

is a second set of complex variables, each representing an atomic or cavity operator.

Expectation values of normally ordered products of these operators are given in terms of integrals of products of the related complex variables with the distribution function. The integrals are taken over the entire complex planes. Thus

$$\left\langle [S_b^+]^p [S_a^+]^q [S^+]^r [S_{az}]^s [S_{bz}]^t [S]^u [S_a]^v [S_b]^w [b^\dagger]^a [b]^b [a_a^\dagger]^c [a]^d \right\rangle$$

$$= \int d^2\underline{\alpha} \; [v_b^+]^p [v_a^+]^q [v^+]^r [D_a]^s [D_b]^t [v]^u [v_a]^v [v_b]^w [\alpha_b^+]^a [\alpha_b]^b [\alpha_a^+]^c [\alpha_a]^d f(\underline{\alpha}) \tag{3.14}$$

with

$$1 = \int d^2\underline{\alpha}\ f(\underline{\alpha}) \tag{3.15}$$

After deriving the Fokker–Planck equation for $f(\underline{\alpha})$ we than transform to new variables

$$\underline{K} = \left[\ x_b^+\ x_a^+\ u^+\ W_a\ W_b\ u\ x_a\ x_b\ I_b\ I_a\ \Phi\ \mu\ \right]$$

which allow for:

. An overall rotation at the average atomic optical frequency ω_0

. The rotation of atomic optical coherences to follow the instantaneous average phase μ of the fields a, b

. The introduction of the physically important phase difference Φ of the fields a, b

. The introduction of the cavity intensities.

The new variables are as follows:

. I_a, I_b	Intensities of cavities a, b
. Φ	Phase difference $\Phi_a - \Phi_b$
. μ	Average phase $\frac{1}{2}(\Phi_a + \Phi_b)$
. x_b, x_b^+	Atomic optical coherences 23, 32
. x_a, x_a^+	Atomic optical coherences 13, 31
. u, u^+	Atomic Zeeman coherences 12, 21
. W_a, W_b	Atomic population differences 31, 32

The transformation from old $\underline{\alpha}$ to new $\underline{K}$ variables is given by:

$$v_b^+ = e^{-i\omega_0 t}\ e^{-i\mu}\ x_b^+ \tag{3.16a}$$

$$v_a^+ = e^{-i\omega_0 t}\ e^{-i\mu}\ x_a^+ \tag{3.16b}$$

$$v^+ = u^+ \tag{3.16c}$$

$$D_a = \tfrac{1}{2}\ W_a \tag{3.16d}$$

$$D_b = \tfrac{1}{2}\ W_b \tag{3.16e}$$

$$v = u \tag{3.16f}$$

$$v_a = e^{i\omega_0 t}\ e^{i\mu}\ x_a \tag{3.16g}$$

$$v_b = e^{i\omega_0 t}\ e^{i\mu}\ x_b \tag{3.16h}$$

$$\alpha_b^+ = e^{i\omega_0 t}\, e^{i\mu}\, e^{-\frac{1}{2}i\Phi}\, I_b^{\frac{1}{2}} \tag{3.16i}$$

$$\alpha_b = e^{-i\omega_0 t}\, e^{-i\mu}\, e^{\frac{1}{2}i\Phi}\, I_b^{\frac{1}{2}} \tag{3.16j}$$

$$\alpha_a^+ = e^{i\omega_0 t}\, e^{i\mu}\, e^{\frac{1}{2}i\Phi}\, I_a^{\frac{1}{2}} \tag{3.16k}$$

$$\alpha_a = e^{-i\omega_0 t}\, e^{-i\mu}\, e^{-\frac{1}{2}i\Phi}\, I_a^{\frac{1}{2}} \tag{3.16l}$$

where

$$\omega_0 = \tfrac{1}{2}(\omega_{13} + \omega_{23}) \tag{3.17}$$

The new distribution function ψ then satifies a Fokker-Planck equation of the form

$$\frac{\partial\psi}{\partial t} = \left\{ \frac{\partial}{\partial K_i}\,[A_i] + \tfrac{1}{2}\,\frac{\partial}{\partial K_i}\,\frac{\partial}{\partial K_j}\,[D_{ij}] \right\} \psi \tag{3.18}$$

where the drift (A_i) and diffusion (D_{ij}) matrices are all <u>independent</u> of the average phase μ. This feature is the important consequence of introducing new atomic coherences that follow the average cavity phase. Examples of one and two time averages have been given earlier (in equation (2.11)). As the Fokker-Planck equation is very lengthy we will not give it here.

The Fokker-Planck equation is equivalent to c number Langevin equations of the form

$$\dot{K}_i = - A_i + F_i \tag{3.19}$$

where $-A_i$ is the deterministic term related to the drift matrix and F_i the random force term. This may in turn be written in terms of fundamental Kronecker delta correlated Gaussian Markoff white noise terms Γ_j as

$$F_i = d_{ij}\,\Gamma_j \tag{3.20}$$

where

$$\langle\Gamma_j(t)\,\Gamma_i(t')\rangle = \delta_{ij}\,\delta(t-t') \tag{3.21}$$

and the expansion coefficients d_{ij} are related to the diffusion matrix via

$$D_{ij} = (dd^T)_{ij} \tag{3.22}$$

The expansion coefficients are also independent of the average phase μ.

The Langevin equations for the Hanle effect CEL are:

$$\dot{I}_b = 2K\left[\bar{n}-I_b\right] + g_b I_b^{\frac{1}{2}}\left[x_b^+\, e^{-\frac{1}{2}i\Phi} + x_b\, e^{\frac{1}{2}i\Phi}\right] \quad +F_{I_b} \tag{3.23a}$$

$$\dot{I}_a = 2K\left[\bar{n}-I_a\right] + g_a I_a^{\frac{1}{2}}\left[x_a^{\frac{1}{2}}\, e^{\frac{1}{2}i\Phi} + x_a\, \bar{e}^{\frac{1}{2}i\Phi}\right] \quad +F_{I_a} \tag{3.23b}$$

$$\dot{\Phi} = \Delta-\delta_a+\delta_b-\frac{ig_b}{2} I_b^{-\frac{1}{2}}\left[x_b^+ e^{-\frac{1}{2}i\Phi} - x_b e^{\frac{1}{2}i\Phi}\right] + \frac{ig_a}{2} I_a^{-\frac{1}{2}}\left[x_a^+ e^{\frac{1}{2}i\Phi} - x_a e^{-\frac{1}{2}i\Phi}\right] + F_\Phi \tag{3.23c}$$

$$\dot{\mu} = -\omega_0 + \tfrac{1}{2}(\omega_a + \omega_b) + \frac{ig_b}{4} I_b^{-\frac{1}{2}}\left[x_b^+ e^{-\frac{1}{2}i\Phi} - x_b e^{\frac{1}{2}i\Phi}\right] + \frac{ig_a}{4} I_a^{-\frac{1}{2}}\left[x_a^+ e^{\frac{1}{2}i\Phi} - x_a e^{-\frac{1}{2}i\Phi}\right] + F_\mu \tag{3.23d}$$

$$\begin{aligned}\dot{x}_b = {} & -\gamma_b x_b + g_a I_a^{\frac{1}{2}} u^+ e^{\frac{1}{2}i\Phi}\left[1 + \frac{1}{4I_a}\right] - g_b I_b^{\frac{1}{2}} W_b e^{-\frac{1}{2}i\Phi}\left[1 + \frac{1}{4I_b}\right] \\ & + \tfrac{1}{4} x_b\left[g_b I_b^{-\frac{1}{2}}\left[x_b^+ e^{-\frac{1}{2}i\Phi} - x_b e^{\frac{1}{2}i\Phi}\right] + g_a I_a^{-\frac{1}{2}}\left[x_a^+ e^{\frac{1}{2}i\Phi} - x_a e^{-\frac{1}{2}i\Phi}\right]\right] \\ & - \tfrac{1}{8} K \bar{n} x_b \left[\frac{1}{I_a} + \frac{1}{I_b}\right] + F_{x_b}\end{aligned} \tag{3.23e}$$

$$\begin{aligned}\dot{x}_b^+ = {} & \gamma_b^* x_b^+ + g_a I_a^{\frac{1}{2}} u e^{-\frac{1}{2}i\Phi}\left[1 + \frac{1}{4I_a}\right] - g_b I_b^{\frac{1}{2}} W_b e^{\frac{1}{2}i\Phi}\left[1 + \frac{1}{4I_b}\right] \\ & - \tfrac{1}{4} x_b^+\left[g_b I_b^{-\frac{1}{2}}\left[x_b^+ e^{-\frac{1}{2}i\Phi} - x_b e^{\frac{1}{2}i\Phi}\right] + g_a I_a^{-\frac{1}{2}}\left[x_a^+ e^{\frac{1}{2}i\Phi} - x_a e^{-\frac{1}{2}i\Phi}\right]\right] \\ & - \tfrac{1}{8} K \bar{n} x_b^+ \left[\frac{1}{I_a} + \frac{1}{I_b}\right] + F_{x_b^+}\end{aligned} \tag{3.23f}$$

$$\begin{aligned}\dot{x}_a = {} & -\gamma_a x_a + g_b I_b^{\frac{1}{2}} u e^{-\frac{1}{2}i\Phi}\left[1 + \frac{1}{4I_b}\right] - g_a I_a^{\frac{1}{2}} W_a e^{\frac{1}{2}i\Phi}\left[1 + \frac{1}{4I_a}\right] \\ & + \tfrac{1}{4} x_a \left[g_b I_b^{-\frac{1}{2}}\left[x_b^+ e^{-\frac{1}{2}i\Phi} - x_b e^{\frac{1}{2}i\Phi}\right] + g_a I_a^{-\frac{1}{2}}\left[x_a^+ e^{\frac{1}{2}i\Phi} - x_a e^{-\frac{1}{2}i\Phi}\right]\right] \\ & - \tfrac{1}{8} K \bar{n} x_a \left[\frac{1}{I_a} + \frac{1}{I_b}\right] + F_{x_a}\end{aligned} \tag{3.23g}$$

$$\begin{aligned}\dot{x}_a^+ = {} & -\gamma_a^* x_a^+ + g_b I_b^{\frac{1}{2}} u^+ e^{\frac{1}{2}i\Phi}\left[1 + \frac{1}{4I_b}\right] - g_a I_a^{\frac{1}{2}} W_a e^{-\frac{1}{2}i\Phi}\left[1 + \frac{1}{4I_a}\right] \\ & - \tfrac{1}{4} x_a^+\left[g_b I_b^{-\frac{1}{2}}\left[x_b^+ e^{-\frac{1}{2}i\Phi} - x_b e^{\frac{1}{2}i\Phi}\right] + g_a I_a^{-\frac{1}{2}}\left[x_a^+ e^{\frac{1}{2}i\Phi} - x_a e^{-\frac{1}{2}i\Phi}\right] \\ & - \tfrac{1}{8} K \bar{n} x_a^+ \left[\frac{1}{I_a} + \frac{1}{I_b}\right] + F_{x_a^+}\end{aligned} \tag{3.23h}$$

$$\begin{aligned}\dot{u} = {} & -\gamma\left[u - u^e\right] - g_a I_a^{\frac{1}{2}} x_b^+ e^{\frac{1}{2}i\Phi} - g_b I_b^{\frac{1}{2}} x_a e^{\frac{1}{2}i\Phi} \\ & + r_c\left[W_a - W_a^e\right] + r_c\left[W_b - W_b^e\right] + F_u\end{aligned} \tag{3.23i}$$

$$\dot{u}^+ = -\gamma^*\left[u^+ - u^{+e}\right] - g_a\, I_a^{\frac{1}{2}}\, x_b\, e^{-\frac{1}{2}i\Phi} - g_b\, I_b^{\frac{1}{2}}\, x_a^+\, e^{-\frac{1}{2}i\Phi}$$

$$+ r_c^*\left[w_a - w_a^e\right] + r_c^*\left[w_b - w_b^e\right] \qquad + F_{u^+} \qquad (3.23j)$$

$$\dot{w}_a = -\gamma_a^{\parallel}\left[w_a - w_a^e\right] - \gamma_{ba}\left[w_b - w_b^e\right]$$

$$+ 2g_a\, I_a^{\frac{1}{2}}\left[x_a^+\, e^{\frac{1}{2}i\Phi} + x_a\, e^{-\frac{1}{2}i\Phi}\right] + g_b\, I_b^{\frac{1}{2}}\left[x_b^+\, e^{-\frac{1}{2}i\Phi} + x_b\, e^{\frac{1}{2}i\Phi}\right] + F_{w_a} \qquad (3.23k)$$

$$\dot{w}_b = -\gamma_{ab}\left[w_a - w_a^e\right] - \gamma_b^{\parallel}\left[w_b - w_b^e\right]$$

$$+ g_a\, I_a^{\frac{1}{2}}\left[x_a^{\frac{1}{2}}\, e^{\frac{1}{2}i\Phi} + x_a\, e^{-\frac{1}{2}i\Phi}\right] + 2g_b\, I_b^{\frac{1}{2}}\left[x_b^+\, e^{-\frac{1}{2}i\Phi} + x_b\, e^{\frac{1}{2}i\Phi}\right] + F_{w_b} \qquad (3.23l)$$

The parameters involved in the Langevin equations are as follows:

. $\Delta = \omega_{12}$ — Upper states transition frequency (3.24a)

. $\delta_a = \omega_{13} - \omega_a$ — Cavity a detuning (3.24b)

$\delta_b = \omega_{23} - \omega_b$ — Cavity b detuning (3.24c)

. $$w_a^e = N\,\frac{(\bar{\gamma}_{23}\bar{\gamma}_{13} - \bar{\gamma}_{31}\bar{\gamma}_{23})}{(\bar{\gamma}_{23}\bar{\gamma}_{13} + \bar{\gamma}_{32}\bar{\gamma}_{13} + \bar{\gamma}_{31}\bar{\gamma}_{23})}$$ Equilibrium population differences allowing for relaxation and pumping only (3.24d)

$$w_b^e = N\,\frac{(\bar{\gamma}_{23}\bar{\gamma}_{13} - \bar{\gamma}_{32}\bar{\gamma}_{13})}{(\bar{\gamma}_{23}\bar{\gamma}_{13} + \bar{\gamma}_{32}\bar{\gamma}_{13} + \bar{\gamma}_{31}\bar{\gamma}_{23})} \qquad (3.24e)$$

. $$u^e = \frac{N\, w_c^*\bar{\gamma}_{23}\bar{\gamma}_{13}}{\left[\frac{1}{2}(\bar{\gamma}_{13} + \bar{\gamma}_{23}) - i\Delta\right]\left[\bar{\gamma}_{23}\bar{\gamma}_{13} + \bar{\gamma}_{32}\bar{\gamma}_{13} + \bar{\gamma}_{31}\bar{\gamma}_{23}\right]}$$ Equilibrium upper states 1-2 coherences (3.24f)

$$u^{+e} = \frac{N\, w_c\bar{\gamma}_{23}\bar{\gamma}_{13}}{\left[\frac{1}{2}(\bar{\gamma}_{13} + \bar{\gamma}_{23}) + i\Delta\right]\left[\bar{\gamma}_{23}\bar{\gamma}_{13} + \bar{\gamma}_{32}\bar{\gamma}_{13} + \bar{\gamma}_{31}\bar{\gamma}_{23}\right]}$$ allowing for relaxation and pumping only (3.24g)

. $\gamma_\perp^b = \frac{1}{2}\,(\bar{\gamma}_{23} + \bar{\gamma}_{31} + \bar{\gamma}_{32})$ — Transverse relaxation rate for 2-3 system (3.24h)

$\gamma_\perp^a = \frac{1}{2}\,(\bar{\gamma}_{13} + \bar{\gamma}_{32} + \bar{\gamma}_{31})$ — Transverse relaxation rate for 1-3 system (3.24i)

$$\gamma_\perp = \tfrac{1}{2}(\bar{\gamma}_{13} + \bar{\gamma}_{23})$$ Transverse relaxation rate for 1-2 system (3.24j)

$$\cdot\ \gamma_b^{\parallel} = \frac{1}{3}(\bar{\gamma}_{31} - \bar{\gamma}_{13} + 2\bar{\gamma}_{32} + 4\bar{\gamma}_{23})$$ Longitudinal relaxation rate for 2-3 system (3.24k)

$$\gamma_a^{\parallel} = \frac{1}{3}(\bar{\gamma}_{32} - \bar{\gamma}_{23} + 2\bar{\gamma}_{31} + 4\bar{\gamma}_{13})$$ Longitudinal relaxation rate for 1-3 system (3.24l)

$$\cdot\ \gamma_{ab} = \frac{1}{3}(\bar{\gamma}_{31} + 2\bar{\gamma}_{13} + 2\bar{\gamma}_{32} - 2\bar{\gamma}_{23})$$ Cross coupling relaxation rate for 1-3 → 2-3 (3.24m)

$$\gamma_{ba} = \frac{1}{3}(\bar{\gamma}_{32} + 2\bar{\gamma}_{23} + 2\bar{\gamma}_{31} - 2\bar{\gamma}_{13})$$ Cross coupling relaxation rate for 2-3 → 1-3 (3.24n)

$$\cdot\ \gamma = -i\Delta + \gamma_\perp \tag{3.24o}$$

$$\gamma_b = \tfrac{1}{2}i\Delta - \tfrac{1}{2}i(\delta_a + \delta_b) + \gamma_\perp^b \tag{3.24p}$$

$$\gamma_a = -\tfrac{1}{2}i\Delta - \tfrac{1}{2}i(\delta_a + \delta_b) + \gamma_\perp^a \tag{3.24q}$$

$$\cdot\ K_b = -i(\delta_b + \tfrac{1}{2}\Delta) + K \tag{3.24r}$$

$$K_a = -i(\delta_a - \tfrac{1}{2}\Delta) + K \tag{3.24s}$$

$$\cdot\ r_c = \frac{1}{3} w_c^*$$ Coherence pumping rate parameter (3.24t)

$$\cdot\ \bar{\gamma}_{13} = \gamma_{13}(\bar{n} + 1)$$ Total transition rate 1 → 3 (3.24u)

$$\bar{\gamma}_{23} = \gamma_{23}(\bar{n} + 1)$$ Total transition rate 2 → 3 (3.24v)

$$\bar{\gamma}_{31} = \gamma_{13}\,\bar{n} + w_{31}$$ Total transition rate 3 → 1 (3.24w)

$$\bar{\gamma}_{32} = \gamma_{23}\,\bar{n} + w_{32}$$ Total transition rate 3 → 2 (3.24x)

After adiabatically eliminating the atomic variables by ignoring $\dot{x}_b, \ldots, \dot{W}_b$ in (3.23) we derive Langevin equations for the cavity variables of the form

$$\dot{I}_a = 2K(\bar{n} - I_a) + \alpha + F_{I_a} \tag{3.25a}$$

$$\dot{I}_b = 2K(\bar{n} - I_b) + \beta + F_{I_b} \tag{3.25b}$$

$$\dot{\Phi} = \Delta - \delta_a + \delta_b + p + F_{\Phi} \tag{3.25c}$$

$$\dot{\mu} = -\omega_0 + \tfrac{1}{2}(\omega_a + \omega_b) + m + F_{\mu} \tag{3.25d}$$

Here the quantities α, β, p and m are functions of I_a, I_b and Φ and depend linearly on the atomic noise terms

$$F_{x_b^+}, \ldots\ldots\ldots, F_{x_b}.$$

The equations for α, β, p, m are of the form:

$$\begin{pmatrix} A & B & E & F \\ C & D & G & H \\ A^+ & B^+ & E^+ & F^+ \\ C^+ & D^+ & G^+ & H^+ \end{pmatrix} \begin{pmatrix} \alpha \\ \beta \\ p \\ m \end{pmatrix} = \begin{pmatrix} s_e \\ t_e \\ s_e^+ \\ t_e^+ \end{pmatrix} + \begin{pmatrix} s_n \\ t_n \\ s_n^+ \\ t_n^+ \end{pmatrix} \tag{3.26a}$$

or in matrix notation

$$M\,X = b_e + b_n \tag{3.26b}$$

In these equations:

. A, B,...., H^+ are functions of intensities I_a, I_b and phase difference Φ, but are not dependent on average phase μ.

. B, C, E, F, G, H, B^+, C^+, E^+, F^+, G^+, H^+ are linearly dependent on m. Thus the equations (3.26) are non-linear.

. s_e, t_e, s_e^+, t_e^+ are functions of I_a, I_b, Φ and the equilibrium population differences and Zeeman coherences W_a^e, W_b^e, u^e, u^{+e}.

. s_n, t_n, s_n^+, t_n^+ are functions of I_a, I_b, Φ and depend linearly on atomic noise terms $F_{x_b^+}$, $F_{x_a^+}, \ldots\ldots\ldots, F_{x_b}$

. Explicit expressions for the quantities in (3.26) are given in the appendix.

Overall we thus have

$$M = M(I_a, I_b, \Phi, m) \tag{3.27a}$$

$$b_e = b_e(I_a, I_b, \Phi) \tag{3.27b}$$

$$b_n = b_n(I_a, I_b, \Phi, F_{x^+}, F_{x^+}, \ldots\ldots\ldots, F_x, F_x) \tag{3.27c}$$

The deterministic equations for the field variables I_a, I_b, Φ, μ are obtained by ignoring the noise terms

$$F_{x_b^+}, \ldots, F_{x_b}, \ldots, F_\mu$$

in (3.26), and are of the form:

$$\dot{I}_a = 2K(\bar{n} - I_a) + \alpha(I_a, I_b, \Phi) \tag{3.28a}$$

$$\dot{I}_b = 2K(\bar{n} - I_b) + \beta(I_a, I_b, \Phi) \tag{3.28b}$$

$$\dot{\Phi} = \Delta - \delta_a + \delta_b + p(I_a, I_b, \Phi) \tag{3.28c}$$

$$\dot{\mu} = -\omega_o + \tfrac{1}{2}(\omega_a + \omega_b) + m(I_a, I_b, \Phi) \tag{3.28d}$$

where now

$$\begin{bmatrix} & & \\ & M & \\ & & \end{bmatrix} \begin{bmatrix} \alpha \\ \beta \\ p \\ m \end{bmatrix} = \begin{bmatrix} s_e \\ t_e \\ s_e^+ \\ t_e^+ \end{bmatrix} \tag{3.29}$$

The steady state is found by putting $\dot{I}_a = \dot{I}_b = \dot{\Phi} = 0$ in (3.28) and solving for I_{as}, I_{bs}, Φ_s giving the steady state for the intensities and phase difference. Since $\dot{\mu}$ cannot also be zero there is no steady state for the average phase.

Writing $$X_s^T = \left[\alpha_s, \beta_s, p_s, m_s\right] \tag{3.30a}$$

$$M_s = M\left[I_{as}, I_{bs}, \Phi_s, m_s\right] \tag{3.30b}$$

$$b_e^s = b_e\left[I_{as}, I_{bs}, \Phi_s\right] \tag{3.30c}$$

as the expressions for X, M, b_e associated with the steady state values, we then have

$$M_s X_s = b_e^s \tag{3.31}$$

and

$$0 = 2K\left[\bar{n} - I_{as}\right] + \alpha_s \tag{3.32a}$$

$$0 = 2K\left[\bar{n} - I_{bs}\right] + \beta_s \tag{3.32b}$$

$$0 = \Delta - \delta_a + \delta_b + p_s \tag{3.32c}$$

as the two sets of equations that define the steady state.

The linear noise theory expressions for small fluctuations ΔI_a, ΔI_b, $\Delta\Phi$ about the steady state values are obtained by considering small variations $\Delta\alpha$, $\Delta\beta$, Δp, Δm associated with ΔI_a, ΔI_b, $\Delta\Phi$ and with the noise terms, which are included with the substitution of I_{as}, I_{bs}, Φ_s in the expressions for s_n, t_n, s_n^+, t_n^+ and in the random forces

$$F_{x_b^+}, \ldots, F_{x_b}, F_{I_b}, \ldots, F_\mu.$$

The basic result for linear noise theory for the Hanle effect CEL is

$$\begin{bmatrix} \dot{\Delta I}_a \\ \dot{\Delta I}_b \\ \dot{\Delta\Phi} \\ \dot{\mu} \end{bmatrix} = \begin{bmatrix} 0 \\ 0 \\ 0 \\ -\omega_o+\frac{1}{2}(\omega_a+\omega_b) \\ + m_s \end{bmatrix} + \begin{bmatrix} H_s^{-1}K_s - 2K \begin{bmatrix} 100 \\ 010 \\ 000 \\ 000 \end{bmatrix} \end{bmatrix} \begin{bmatrix} \Delta I_a \\ \Delta I_b \\ \Delta\Phi \end{bmatrix}$$

$$+ \left[H_s^{-1} C_s d_A^s + d_F^s \right] \begin{bmatrix} \Gamma_{x_b^+} \\ \vdots \\ \Gamma_{x_b} \\ \Gamma_{I_b} \\ \vdots \\ \Gamma_\mu \end{bmatrix} \qquad (3.33)$$

where the right side contains a constant term, a linear deterministic term and a noise term.

In the basic noise theory result:

$$.H_s = M_s + \left[\frac{\partial M}{\partial m} \right]_s X_s \; [0001] \qquad (3.34a)$$

$$.K_s = \left[\left[\frac{\partial b_e}{\partial I_a} \right]_s - \left[\frac{\partial M}{\partial I_a} \right]_s X_s \right] \; [100]$$

$$+ \left[\left[\frac{\partial b_e}{\partial I_b} \right]_s - \left[\frac{\partial M}{\partial I_b} \right]_s X_s \right] \; [010]$$

$$+ \left[\left[\frac{\partial b_e}{\partial \Phi} \right]_s - \left[\frac{\partial M}{\partial \Phi} \right]_s X_s \right] \; [001] \qquad (3.34b)$$

$$.b_n^s = b_n \left[I_{as},\; I_{bs},\; \Phi_s,\; F_{x_b^+},\; \ldots\ldots,\; F_{x_b} \right] \qquad (3.34c)$$

$$= C_s \begin{bmatrix} F_{x_b^+} \\ \vdots \\ F_{x_b} \end{bmatrix} \qquad (3.34d)$$

$$\begin{bmatrix} F_{x_b^+} \\ \vdots \\ F_{x_b} \\ \hline F_{I_a} \\ \vdots \\ F_{\mu} \end{bmatrix} = \begin{bmatrix} d_A^s \\ \hline d_F^s \end{bmatrix} \begin{bmatrix} \Gamma_{x_b^+} \\ \vdots \\ \vdots \\ \Gamma_{\mu} \end{bmatrix} \tag{3.34e}$$

Thus the quantities appearing in the linear noise theory equation (3.33) involve the matrix m, its derivatives and that of b_e with respect to I_a, I_b, Φ and evaluated at the steady state values, the steady state value of the matrix C that couples the noise matrix b_n to the random atomic noise terms

$F_{x_b^+}, \ldots, F_{x_b}$

and also the steady state values of the transformation matrix d.

For the fluctuations in the cavity intensities I_a, I_b and the phase difference Φ we have

$$\begin{bmatrix} \Delta\dot{I}_a \\ \Delta\dot{I}_b \\ \Delta\dot{\Phi} \end{bmatrix} = -B \begin{bmatrix} \Delta I_a \\ \Delta I_b \\ \Delta\Phi \end{bmatrix} + G \begin{bmatrix} \Gamma_{x_b^+} \\ \vdots \\ \Gamma_{\mu} \end{bmatrix} \tag{3.35a}$$

or in matrix notation

$$\Delta\dot{I} = -B\,\Delta I + G\,\Gamma \tag{3.35b}$$

where (1.2.3 means first three rows of)

$$B = 2K \begin{bmatrix} 100 \\ 010 \\ 000 \end{bmatrix} - \left[H_s^{-1} K_s \right]_{1.2.3} \tag{3.36a}$$

$$G = \left[H_s^{-1} C_s d_A^s + d_F^s \right]_{1.2.3} \tag{3.36b}$$

In the above result:

.B is the stability matrix. Its eigenvalues must have positive real parts for the steady state to be stable.

.G is the coupling matrix.

.Equation (3.36) shows how the intensity and phase difference fluctuations are coupled to each other and linked to the fundamental noise terms.

For the average phase μ we have

$$\dot{\mu} = \Omega_0 - b \begin{pmatrix} \Delta I_a \\ \Delta I_b \\ \Delta\Phi \end{pmatrix} + g \begin{pmatrix} \Gamma_{x_b^+} \\ \vdots \\ \Gamma_\mu \end{pmatrix} \tag{3.37a}$$

or in the matrix notation

$$\dot{\mu} = \Omega_0 - b\,\Delta I + g\,\Gamma \tag{3.37b}$$

where (4 means the fourth row of)

$$\Omega_0 = -\omega_0 + \tfrac{1}{2}(\omega_a + \omega_b) + m_s \tag{3.38a}$$

$$b = -(H_S^{-1}K_S)_4 \tag{3.38b}$$

$$g = (H_S^{-1}C_S d_A^S + d_F^S)_4 \tag{3.38c}$$

In this result:

. The average phase is linked to the intensity and phase difference fluctuations and to the noise.

. As the intensities and phase difference also depend on the noise terms there will be a second dependence on the noise in the equation for $\dot{\mu}$.

The equation (3.35) for the intensities and phase difference may be solved as

$$\Delta I(t) = e^{-Bt}\Delta I(0) + \int_0^t dt_1\, e^{-B(t-t_1)} G\,\Gamma(t_1) \tag{3.39}$$

Substituting this solution into (3.37) we have

$$\dot{\mu} = \Omega_0 - b e^{-Bt}\Delta I(0) - b\int_0^t dt_1\, e^{-B(t-t_1)} G\,\Gamma(t_1) + g\,\Gamma \tag{3.40}$$

This shows that the average phase will in general be subject to a coloured noise term (the integral over t_1).

The equation (3.40) for the average phase can also be solved as:

$$\mu(t) = \mu(0) + \Omega_0 t + bB^{-1}(e^{-Bt}-1)\,\Delta I(0) + g\int_0^t dt_1\,\Gamma(t_1)$$
$$-b\int_0^t dt_2 \int_0^{t_2} dt_1\, e^{-B(t_2-t_1)} G.\Gamma(t_1) \tag{3.41}$$

Both solutions (3.39), (3.41) have transient terms and noise terms. The average phase also has a deterministic linear variation with time $\mu(0) + \Omega_0 t$. From (3.38a) the quantity m_s is a frequency shift. The properties of Γ can be used to calculate the stochastic averages.

For the mean values:

$$\langle \Delta I(t) \rangle = e^{-Bt} \, \Delta I(0) \tag{3.42}$$

$$\rightarrow 0 \qquad \text{as } t \rightarrow \infty$$

$$\langle \mu(t) \rangle = \mu(0) + \Omega_0 t + bB^{-1} \, (e^{-Bt} - 1) \, \Delta I(0) \tag{3.43}$$

$$\rightarrow \mu(0) + \Omega_0 t - bB^{-1} \, \Delta I(0) \qquad \text{as } t \rightarrow \infty$$

Defining the variations $\delta(\Delta I)$ and $\delta\mu$ via

$$\delta(\Delta I) = \Delta I - \langle \Delta I \rangle \tag{3.44a}$$

$$\delta\mu = \mu - \langle \mu \rangle \tag{3.44b}$$

We can then easily show (Gardiner, 1985) that at long times the correlation matrix for the intensities and phase difference,

$$\langle \delta(\Delta I)_\infty \delta(\Delta I)_\infty^T \rangle,$$

satisfies the equation:

$$B \langle \delta(\Delta I)_\infty \delta(\Delta I)_\infty^T \rangle + \langle \delta(\Delta I)_\infty \delta(\Delta I)_\infty^T \rangle \, B^T = GG^T \tag{3.45}$$

Writing

$$G = \begin{bmatrix} G \\ g \end{bmatrix} \tag{3.46a}$$

$$= H_S^{-1} \, C_S d_A^S + d_F^S \tag{3.46b}$$

we see using (3.22), (3.36b), (3.37c) that:

$$G \, G^T = \begin{bmatrix} GG^T & Gg^T \\ gG^T & gg^T \end{bmatrix} \tag{3.47a}$$

$$= H_S^{-1} \, C_S D_{AA}^S C_S^T \, (H_S^{-1})^T + H_S^{-1} C_S D_{AF}^S$$

$$+ D_{FA}^S C_S^T \, (H_S^{-1})^T + D_{FF}^S \tag{3.47b}$$

which enables GG^T to obtained from the coupling matrix C_S, the matrix H_S and the diffusion matrix for the steady state D_S.

Other quantities of physical interest and numerical evaluations will be considered later.

ACKNOWLEDGMENTS

The author is grateful for helpful discussions with P.L. Knight, M.O. Scully, D.F. Walls, S. Barnett, P. Drummond, A. Ekert, B. Kennedy, M. Orszag, M. Reid, W. Schleich, S. Swain, M. Winters, S. Zubairy. Financial support from Australian Research Council, and Wolfson College, Oxford is also acknowledged.

APPENDIX

(A) 4 x 4 matrix M:

$$\mathcal{A} = \underline{A} + \frac{IC_a}{(\gamma^{\parallel}_a\gamma^{\parallel}_b - \gamma_{ab}\gamma_{ba})}(2\gamma^{\parallel}_b - \gamma_{ba}) + \frac{JC_b}{(\gamma^{\parallel}_a\gamma^{\parallel}_b - \gamma_{ab}\gamma_{ba})}(-2\gamma_{ab} + \gamma^{\parallel}_a)$$

$$\mathcal{B} = \underline{B} + \frac{IC_a}{(\gamma^{\parallel}_a\gamma^{\parallel}_b - \gamma_{ab}\gamma_{ba})}(\gamma^{\parallel}_b - 2\gamma_{ba}) + \frac{JC_b}{(\gamma^{\parallel}_a\gamma^{\parallel}_b - \gamma_{ab}\gamma_{ba})}(-\gamma_{ab} + 2\gamma^{\parallel}_a)$$

$$\mathcal{C} = \underline{C} + \frac{KC_a}{(\gamma^{\parallel}_a\gamma^{\parallel}_b - \gamma_{ab}\gamma_{ba})}(2\gamma^{\parallel}_b - \gamma_{ba}) + \frac{LC_b}{(\gamma^{\parallel}_a\gamma^{\parallel}_b - \gamma_{ab}\gamma_{ba})}(-2\gamma_{ab} + \gamma^{\parallel}_a)$$

$$\mathcal{D} = \underline{D} + \frac{KC_a}{(\gamma^{\parallel}_a\gamma^{\parallel}_B - \gamma_{ab}\gamma_{ba})}(\gamma^{\parallel}_b - 2\gamma_{ba}) + \frac{LC_b}{(\gamma^{\parallel}_a\gamma^{\parallel}_b - \gamma_{ab}\gamma_{ba})}(-\gamma_{ab} + 2\gamma^{\parallel}_a) \qquad \text{(A1)}$$

where

$$\underline{A} = \frac{f_a C_b^2}{2\gamma^*} \qquad \underline{B} = \tfrac{1}{2}\Gamma_b + \frac{f_a C_a^2}{2\gamma^*}$$

$$\underline{C} = \tfrac{1}{2}\Gamma^+_a + \frac{f_b C_b^2}{2\gamma^*} \qquad \underline{D} = \frac{f_b C_a^2}{2\gamma^*}$$

$$E = -\tfrac{1}{2}\, iI_b\Gamma'_b - \frac{if_a}{2\gamma^*}(I_bC_a^2 + I_aC_b^2) \qquad F = +\, iI_b\Gamma_b + \frac{if_a}{\gamma^*}(I_bC_a^2 - I_aC_b^2)$$

$$G = -\tfrac{1}{2}\, iI_a\Gamma^+_a - \frac{if_b}{2\gamma^*}(I_aC_b^2 + I_bC_a^2) \qquad H = -\, iI_a\Gamma^+_a - \frac{if_b}{\gamma^*}(I_aC_b^2 - I_bC_a^2)$$

$$I = -\frac{f_a r_c^* C_b e^{i\Phi}}{\gamma^*} \qquad J = f_bC_b - \frac{f_a r_c^* C_a e^{i\Phi}}{\gamma^*}$$

$$K = f_aC_a - \frac{f_b r_c^* C_b e^{i\Phi}}{\gamma^*} \qquad L = -\frac{f_b r_c^* C_a e^{i\Phi}}{\gamma^*} \qquad \text{(A2)}$$

$$C_a = g_a I_a^{\frac{1}{2}} \qquad C_b = g_b I_b^{\frac{1}{2}} \qquad \text{(A3)}$$

$$f_a = 1 + \frac{1}{4I_a} \qquad f_b = 1 + \frac{1}{4I_b} \qquad \text{(A4)}$$

and

$$\Gamma_a = \gamma_a + im + \tfrac{1}{8} K\bar{n} \left(\frac{1}{I_a} + \frac{1}{I_b}\right)$$

$$\Gamma_b = \gamma_b + im + \tfrac{1}{8} K\bar{n} \left(\frac{1}{I_a} + \frac{1}{I_b}\right)$$

$$\Gamma_a^+ = \gamma_a^* - im + \tfrac{1}{8} K\bar{n} \left(\frac{1}{I_a} + \frac{1}{I_b}\right)$$

$$\Gamma_b^+ = \gamma_b^* - im + \tfrac{1}{8} K\bar{n} \left(\frac{1}{I_a} + \frac{1}{I_b}\right) \qquad \text{(A5)}$$

The $\Gamma_a, \ldots \Gamma_b^+$ are m dependent generalized atomic decay, detuning parameters

(B) 1 x 4 deterministic matrix b_e:

$$s_e = -f_b C_b^2 W_b^e + f_a C_a C_b u^{+e} e^{i\Phi}$$

$$t_e = -f_a C_a^2 W_a^e + f_b C_a C_b u^{+e} e^{i\Phi} \qquad \text{(A6)}$$

(C) 1 x 4 noise matrix b_n:

$$s_n = \underline{s}_n - \frac{I(\gamma_b^{||} G_{W_a} - \gamma_{ba} G_{W_b})}{C_b(\gamma_a^{||}\gamma_b^{||} - \gamma_{ab}\gamma_{ba})} - \frac{J(-\gamma_{ab} G_{W_a} + \gamma_a^{||} G_{W_b})}{C_a(\gamma_a^{||}\gamma_b^{||} - \gamma_{ab}\gamma_{ba})}$$

$$t_n = \underline{t}_n - \frac{K(\gamma_b^{||} G_{W_a} - \gamma_{ba} G_{W_b})}{C_b(\gamma_a^{||}\gamma_b^{||} - \gamma_{ab}\gamma_{ba})} - \frac{L(-\gamma_{ab} G_{W_a} + \gamma_a^{||} G_{W_b})}{C_a(\gamma_a^{||}\gamma_b^{||} - \gamma_{ab}\gamma_{ba})} \qquad \text{(A7)}$$

where

$$\underline{s}_n = G_{x_b} + \frac{f_a}{\gamma^*} G_{u^+} \qquad \underline{t}_n = G_{x_a^+} + \frac{f_b}{\gamma^*} G_{u^+} \qquad \text{(A8)}$$

and

$$G_{x_b} = g_b I_b^{\frac{1}{2}} e^{\frac{1}{2}i\Phi} F_{x_b} \qquad G_{x_b^+} = g_b I_b^{\frac{1}{2}} e^{-\frac{1}{2}i\Phi} F_{x_b^+}$$

$$G_{x_a} = g_a I_a^{\frac{1}{2}} e^{-\frac{1}{2}i\Phi} F_{x_a} \qquad G_{x_a^+} = g_a I_a^{\frac{1}{2}} e^{\frac{1}{2}i\Phi} F_{x_a^+}$$

$$G_u = g_a I_a^{\frac{1}{2}} g_b I_b^{\frac{1}{2}} e^{-i\Phi} F_u \qquad G_{u^+} = g_a I_a^{\frac{1}{2}} g_b I_b^{\frac{1}{2}} e^{i\Phi} F_{u^+}$$

$$G_{W_a} = g_a I_a^{\frac{1}{2}} g_b I_b^{\frac{1}{2}} F_{W_a} \qquad G_{W_b} = g_a I_a^{\frac{1}{2}} g_b I_b^{\frac{1}{2}} F_{W_b} \qquad \text{(A9)}$$

(D) The elements $A^+, \ldots . t_n^+$

$$\underline{A}^+ = \frac{f_a C_b^2}{2\gamma} \qquad \underline{B}^+ = \tfrac{1}{2}\Gamma_b^+ + \frac{f_a C_a^2}{2\gamma}$$

$$\underline{C}^+ = \tfrac{1}{2}\Gamma_a + \frac{f_b C_b^2}{2\gamma} \qquad \underline{D}^+ = \frac{f_b C_a^2}{2\gamma}$$

$$E^+ = \tfrac{1}{2} i I_b \Gamma_b^+ + \frac{i f_a}{2\gamma}(I_b C_a^2 + I_a C_b^2) \qquad F^+ = -\, i I_b \Gamma_b^+ - \frac{i f_a}{\gamma}(I_b C_a^2 - I_a C_b^2)$$

$$G^+ = \tfrac{1}{2} i I_a \Gamma_a + \frac{i f_b}{2\gamma}(I_a C_b^2 + I_b C_a^2) \qquad H^+ = i I_a \Gamma_a + \frac{i f_b}{\gamma}(I_a C_b^2 - I_b C_a^2)$$

$$I^+ = -\frac{f_a r_c C_b e^{-i\Phi}}{\gamma} \qquad J^+ = f_b C_b - \frac{f_a r_c C_a e^{-i\Phi}}{\gamma}$$

$$K^+ = f_a C_a - \frac{f_b r_c C_b e^{-i\Phi}}{\gamma} \qquad L^+ = -\frac{f_b r_c C_a e^{-i\Phi}}{\gamma} \tag{A10}$$

$$s_e^+ = -f_b C_b^2 W_b^e + f_a C_a C_b u^e e^{-i\Phi} \qquad t_e^+ = -f_a C_a^2 W_a^e + f_b C_a C_b u^e e^{-i\Phi} \tag{A11}$$

$$\underline{s}_n^+ = G_{x_b^+} + \frac{f_a}{\gamma} G_u \qquad \underline{t}_n^+ = G_{x_a} + \frac{f_b}{\gamma} G_u \tag{A12}$$

To obtain A^+, B^+, C^+, D^+ replace $\underline{A}$, $\underline{B}$, $\underline{C}$, $\underline{D}$, I, J, K, L, by $\underline{A}^+$, $\underline{B}^+$, $\underline{C}^+$, $\underline{D}^+$, I^+, J^+, K^+, L^+ respectively in (A1).

To obtain s_n^+, t_n^+ replace $\underline{s}_n$, $\underline{t}_n$, I, J, K, L by $\underline{s}_n^+$, $\underline{t}_n^+$, I^+, J^+, K^+, L^+ respectively in (A7).

REFERENCES

Barrat J.P. and Cohen-Tannoudji C., 1961, J. Phys. Rad., 22: 329.
Benkert C. and Scully M.O., 1990, (to be published).
Bergou J. and Orszag M., 1988a, J. Opt. Soc. Amer., B5: 249.
Bergou J. and Orszag M., 1988b, Phys. Rev., A38: 763.
Bergou J., Orszag M. and Scully M.O., 1988a, Phys. Rev., A38: 754.
Bergou J., Orszag M. and Scully M.O., 1988b, Phys. Rev., A38: 768.
Bergou J., Orszag M., Scully M.O. and Wodkiewicz K., 1989a, Phys. Rev., A39: 5136.
Bergou J., Lu N. and Scully M.O., 1989b, Opt. Comm., 73: 57.
Boone A.W. and Swain S., 1989a, Opt. Comm., 73: 47.
Boone A.W. and Swain S., 1989b, Quant. Opt., 1: 27.

Boone A.W. and Swain S., 1990, Phys. Rev., A41:
Carmichael H.J., Satchell J.S. and Sarkar S., 1986, Phys. Rev., A34: 3166.
Cohen-Tannoudji C., 1977, in: "Frontiers in Laser Spectroscopy I (Les Houches 1975)", edited by R.Balian, S.Haroche and S.Liberman (North Holland, Amsterdam) p7.
Dalton B.J., 1987, J. Phys., B20: 251.
Dalton B.J. and Knight P.L., 1982a, J. Phys., B15: 3997.
Dalton B.J. and Knight P.L., 1982b, Opt. Comm., 42: 411.
Dalton B.J. McDuff R. and Knight P.L., 1985, Optica Acta., 32: 61.
Drummond P.D. and Gardiner C.W., 1980, J. Phys., A13: 2353.
Drummond P.D. and Walls D.F., 1981, Phys. Rev., A23: 2563.
Gardiner C.W., 1985, "Handbook of Stochastic Methods"., (Springer-Verlag, Berlin).
Glauber R.J., 1963a, Phys. Rev., 130: 2529.
Glauber R.J., 1963b, Phys. Rev., 131: 2766.
Glauber R.J., 1964, in: "Quantum Optics and Electronics, (Les Houches)", edited by C. DeWitt, A. Blandin and C. Cohen-Tannoudji, (Gordon and Breach, New York).
Haken H., 1970, in: "Handbuch der Physik 25 (2c) Licht und Materie" (Springer-Verlag, Berlin).
Krause J. and Scully M.O., 1987, Phys. Rev., A36: 1771.
Loudon R. and Knight P.L., 1987, J. Mod. Opt., 34: 709.
Louisell W.H., 1973, "Quantum Statistical Properties of Radiation", (Wiley, New York).
Lu N., Zhu S-Y. and Agarwal G.S., 1989a, Phys. Rev., A40: 258.
Lu N., Zhao F-X. and Bergou J., 1989b, Phys. Rev., A39: 5189.
Ohtsu M. and Liou K-Y., 1987, App. Phys. Lett., 52: 10.
Orszag M., Becker W. and Scully M.O., 1987, Phys. Rev., A36: 1310.
Orszag M., Bergou J., Schleich W. and Scully M.O., 1988, The Correlated Emission Laser: Theory and Recent Developments in: "Squeezed and Non-classical Light", Vol 190, NATO ASI Series, edited by P. Tombesi and E.R. Pike (Plenum, New York) p287.
Reid M.D., 1988, Phys. Rev., A37: 4792.
Reid M.D. and Walls D.F., 1983, Phys. Rev., A28: 332.
Reid M.D. and Walls D.F., 1985, Phys. Rev., A31: 1622.
Sargent M., Scully M.O. and Lamb W.E. Jr., 1974, "Laser Physics" (Addison-Wesley, Reading, Ma.).
Schleich W. and Scully M.O., 1988, Phys. Rev., A37: 1261.
Schleich W., Scully M.O. and Von Garssen H.G., 1988, Phys. Rev., A37: 3010.
Scully M.O., 1985, Phys. Rev. Lett., 55: 2802.
Scully M.O. and Gea-Banacloche J., 1986, Phys. Rev., A34: 4043.
Scully M.O. and Zubairy M.S., 1987, Phys. Rev., A35: 752.
Scully M.O. and Zubairy M.S., 1988, Opt. Comm., 66: 303.
Scully M.O., Wodkiewicz K., Zubairy M.S., Bergou J., Lu N and Ter Vehn J.M., 1988, Phys. Rev. Lett., 60: 1832.
Scully M.O., Zhu S-Y. and Gavrielides A., 1989, Phys. Rev. Lett., 62: 2813.
Sudarshan E.C.G., 1963, Phys. Rev. Lett., 10: 277.
Swain S., 1988a, J. Mod. Opt., 35: 1.
Swain S., 1988b. J. Mod. Opt., 35: 103.
Toschek P.E. and Hall J., 1987, Abstract in 15th International Conference on Quantum Electronics, J. Opt. Soc. Amer., B4: 124.
Winters M. and Hall J., 1990, in: "New Frontiers in QED and Quantum Optics", Vol , NATO ASI Series, edited by A.O. Barut, (Plenum, New York).
Zaheer K. and Zubairy M.S., 1988, Phys. Rev., A38: 5227.
Toschek P.E. and Hall J., 1987, Abstract in 15th International Conference on Quantum Electronics, J. Opt. Soc. Amer., B4: 124.
Winters M. and Hall J., 1990, in: "New Frontiers in QED and Quantum Optics", Vol , NATO ASI Series, edited by A.O. Barut, (Plenum, New York).
Zaheer K. and Zubairy M.S., 1988, Phys. Rev., A38: 5227.

CORRELATED SPONTANEOUS EMISSION IN A ZEEMAN LASER

M. P. Winters and J. L. Hall

Joint Institute for Laboratory Astrophysics
University of Colorado, and
National Institute of Standards and Technology
Boulder, Colorado 80309

INTRODUCTION

In recent years there has been great interest in the development of high sensitivity laser interferometers for use in the detection of gravitational waves. Ordinarily, the sensitivity of these devices is limited by quantum fluctuations that take the form of shot noise in passive interferometers and spontaneous emission noise in active interferometers. Proposals for increasing the sensitivity of these interferometers include squeezed states for passive devices[1] and the correlated emission laser (CEL) for active devices.[2,3] Generation of squeezed states has now been demonstrated by several groups[4,5] and it is the purpose of this article to describe an experimental investigation of the CEL.

THE CORRELATED EMISSION LASER AND LASER PHASE NOISE

The "quantum beat" version of the CEL utilizes a gain medium with two laser transitions sharing a common lower level (Fig. 1). A coherent superposition of the upper states $|a>$ and $|b>$ is created by driving an rf transition between these levels. This superposition results in the correlation of the spontaneous emission into the two laser modes and under appropriate conditions this correlation leads to the reduction (and possible elimination) of phase diffusion noise in the relative phase of the two modes.[2] This noise reduction in the relative phase makes the CEL ideally suited for use in an active interferometer.

Present theory does not allow us to predict the extent of noise reduction obtainable in our particular laser system, but we can say quite generally what its signature will be. For this discussion, a useful background is to look at the phase noise in free-running and phase locked laser (PLL) systems. The reason for examining the PLL is twofold. First, phase locking does occur in the CEL and so it will form the foundation of our simple CEL picture and, second, we need to know how to distinguish between correlated emission and phase locking in case only the latter exists in our laser.

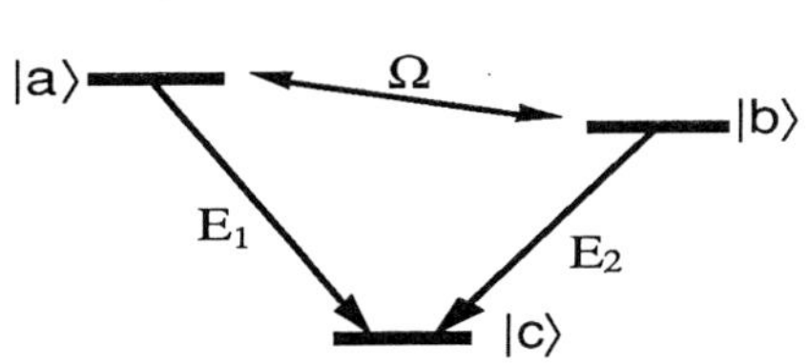

Fig.1. CEL level diagram. Two laser transitions share a common lower level and the upper levels are coupled by an rf magnetic transition with an effective Rabi rate Ω.

Phase Noise in a Free Running Laser

It is well known that in a free-running laser the phase will undergo a random walk[6] with mean square phase given by $<\Delta\phi^2>= Dt$, where

$$D=\frac{2\pi^2 h\nu\left(\Delta\nu_{cav}\right)^2}{P_{laser}}\frac{N_2}{N_2\text{-}N_1}=2\pi\Delta\nu_{laser} \tag{1}$$

is the diffusion rate corresponding to the standard Schawlow-Townes linewidth. In a typical HeNe laser this linewidth is of the order of some milliHertz and is completely masked by the abundance of technical noise. Only by measuring noise between two modes in the same laser is it feasible to see this phase diffusion.

Phase Noise in a Phase Locked Laser

In the case of a two-mode PLL the relative phase behavior is described by Adler's equation[7]:

$$\frac{\partial\phi}{\partial t}=a - b\ sin\ \phi + F(t) \tag{2}$$

where a is a detuning parameter, b is the locking strength (or locking range), and $F(t)$ is a delta correlated noise term due to spontaneous emission into the two modes. This is a very general equation - it can be used to describe injection locking, laser gyro lockup, and even some electronic phase locked loops. For instance, in the case of external injection locking, a would be the frequency difference between the injection laser and the free-running laser, and b would depend on the relative intensities of the injection and locked laser fields. Eq. 2 suggests a steady state lockup phase, $\phi_0 = sin^{-1}(a/b)$ with fluctuations around it given by:

$$\left\langle \Delta\phi^2 \right\rangle = \frac{D}{2b cos\ \phi_0}\left(1 - e^{-2bt\ cos\ \phi_0}\right) \quad . \tag{3a}$$

It is interesting to look at the limits of Eq. 3:

$$\left\langle \Delta\phi^2 \right\rangle = Dt \qquad t << \frac{1}{b} \tag{3b}$$

$$\left\langle \Delta\phi^2 \right\rangle = \frac{D}{2b cos\ \phi_0} \qquad t >> \frac{1}{b} \quad . \tag{3c}$$

At times short compared to the inverse of the injection locking strength, the laser behaves exactly like a free-running laser, but at long times the fluctuations saturate to a constant value. It is apparent then that a PLL will show noise reduction in the relative phase only if observed in the long time limit. In this limit the noise will appear to be "sub Schawlow-Townes," and will show a corresponding decrease in slope in an Allan Variance versus time plot.

Phase Noise in a Correlated Emission Laser

For the purpose of discussing the experiment, a useful model for describing the behavior of the CEL is the "geometric picture."[8] We start with a phasor diagram (Fig. 2) showing two laser fields E_1 and E_2 with relative phase ϕ. Each field fluctuates under the influence of its own driving term F_i, but we allow for a nonzero cross correlation between the terms. Using simple geometry the relative phase can be written as:

$$\delta\phi = \left[cos\frac{\phi}{2}\ Im\left(F_1\text{-}F_2\right) + sin\frac{\phi}{2}\ Re\left(F_1+F_2\right)\right]\delta t \quad . \tag{4}$$

This can be simplified further by rewriting it in terms of two new driving terms, $F_- \equiv Im(F_1\text{-}F_2)$ and $F_+ \equiv Re(F_1+F_2)$ that have the following properties:

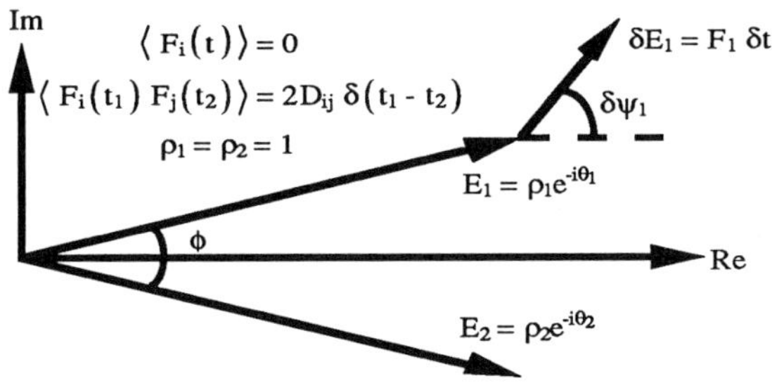

Fig. 2. Phasor diagram for the "geometric picture" of the CEL (Ref. 8). The laser fields E_i have a relative phase ϕ and fluctuate under the influence of F_i.

$$\left\langle F_{\pm}(t_1) F_{\pm}(t_2) \right\rangle = (D_{11} + D_{22})(1 \pm \varepsilon)\, \delta(t_1 - t_2) \tag{5a}$$

$$\left\langle F_{+}(t_1) F_{-}(t_2) \right\rangle = 0 \qquad \varepsilon \equiv \frac{2\, Re\, D_{12}}{D_{11} + D_{22}} \quad . \tag{5b,c}$$

This is then plugged into our locking equation (Eq. 2) to yield an appropriate Langevin equation for the CEL:

$$\frac{\partial \phi}{\partial t} = a - b \sin \phi + \cos \frac{\phi}{2} F_{-} + \sin \frac{\phi}{2} F_{+} \quad . \tag{6}$$

Insight can already be gained from this equation. Under the condition of maximum correlation ($\varepsilon = 1$) it can be shown that $F_{-} \equiv 0$. The remaining F_{+} term is then multiplied by zero if $\phi = 0$ (i.e. when E_1 and E_2 line up in the phasor diagram) and the quantum noise in the relative phase is completely suppressed. The steady state solution is identical to the PLL case but the fluctuations now contain an additional term:

$$\left\langle \Delta\phi^2 \right\rangle = \frac{D}{2b \cos \phi_0}\left(1 - e^{-2bt \cos \phi_0}\right)\left(1 - \varepsilon \cos \phi_0\right) \tag{7a}$$

$$\left\langle \Delta\phi^2 \right\rangle = Dt\left(1 - \varepsilon \cos \phi_0\right) \qquad t << \frac{1}{b} \tag{7b}$$

$$\left\langle \Delta\phi^2 \right\rangle = \frac{D}{2b \cos \phi_0}\left(1 - \varepsilon \cos \phi_0\right) \qquad t >> \frac{1}{b} \tag{7c}$$

where we have written $D_{11} = D_{22} \equiv D$. It is this additional $(1-\varepsilon \cos \phi_0)$ term which reduces the phase locked laser noise and thus provides the signature of the CEL effect. In principle, an experiment could operate in either time limit of Eq. 7, but there are problems interpreting the data at long times. Effects that tend to increase ε also tend to increase b, making it difficult to tell whether any observed noise reduction is a result of CEL or PLL. We could look for the $\cos \phi_0$ dependence of the CEL term but limited signal to noise makes it difficult to distinguish it from the $\cos \phi_0$ in the denominator of the PLL term. As we will discuss later, technical noise will also become a problem at long times. For these reasons we have decided to work in the short time limit where any noise reduction and ϕ_0 dependence can be safely attributed to the CEL effect. (Some measurements have been reported in the long time limit.[9,10])

EXPERIMENT

Atomic System

It is useful to first consider the "ideal" atomic system for our experiment. The lower level has angular momentum J=0 and the upper level has J=1. An axial dc magnetic field Zeeman splits the upper level into three m sublevels and causes the laser to run in two oppositely circularly polarized modes at different frequencies. The absence of a Δm=0 transition makes the level scheme look like that in Fig. 1. Coherence between the upper states is achieved by driving stepwise resonant, two photon magnetic dipole transitions between the m = -1,+1 states through the m=0 state. An effective single photon matrix element for the two photon $|\Delta m|$=2 transition can then be written[11]:

$$M_{1,-1} = \frac{\langle -1 | \mu \cdot \mathbf{B}_1 | 0 \rangle \ \langle 0 | \mu \cdot \mathbf{B}_1 | 1 \rangle}{h\left(\omega_{10} - \omega_{rf} + i\Gamma\right)} 2\pi \tag{8}$$

where $\boldsymbol{B}=\boldsymbol{B_1} \cos \omega_{rf} t$ is the driving field, ω_{10} is the Δm=1 frequency difference, and Γ is the decay rate of the m=0 level.

The actual transition is the 633 nm line in a commercial HeNe laser. The natural linewidth is 17.5 MHz, Doppler broadening is 1500 MHz, and the collisional broadening is ~150 MHz. The upper level ($3s_2$) has J=1, g=1.295 and the lower level ($2p_4$) has J=2, g=1.301. The nearly identical g factors have the fortuitous effect that the laser still runs in two circularly polarized modes, but the level diagram is no longer as simple as that in Fig. 1. Instead of the previous V system, we now have three Λ's and a V with several degenerate laser transitions. The transverse field drives magnetic dipole transitions between the lower m levels as well as the upper m levels. The existence of these extra lower levels is the main reason why present theory does not apply directly to our system. We must simply assume that correlated emission can exist in our system and then rely on the geometric picture to tell us what to look for.

Experimental Apparatus

Fig. 3 shows the layout of the experiment. The Zeeman laser consists of a commercial HeNe plasma tube enclosed in a hermetically sealed can which, in turn, is mounted on a floating optical table to isolate it from the environment. The plasma discharge is run from a high voltage power supply in conjunction with a low noise active current regulator. The output power is 50 μW per mode maximum and the free spectral range is 550 MHz. The walls of the can provide mounts for the required electromagnets as well as water cooling to extract heat generated by the magnets and the plasma tube. The axial dc magnetic field ($\hat{z} B_0$) is generated by a precisely controlled solenoidal electromagnet winding and can be varied to give a Δm=1 splitting of 75-125 MHz. The two laser modes share a common cavity resonance and gain-splitting/cavity-pulling effects give a resulting laser splitting of ~100 kHz. The transverse ac magnetic field, $\hat{x} B_1 \cos \omega_{rf} t$, is generated with a low phase noise, direct digital synthesizer and a pair of coils inside the laser can. A high power audio amplifier boosts the output of the synthesizer and drives the coils through a resonant circuit. The large homogeneous linewidth makes the exact value of ω_{rf} unimportant, so it is set mainly by technical considerations. At $\omega_{rf}/2\pi$ = 50 kHz it is possible to generate a field strength of B_1= 15 G, a value much higher than we could achieve at a frequency corresponding to the 75-125 MHz atomic splitting. A windfall of this

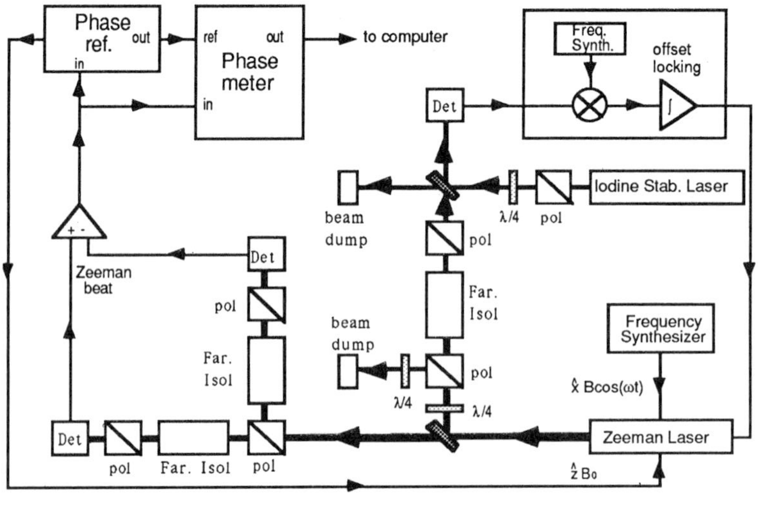

Fig. 3. Layout of the CEL experiment

lower drive frequency is that we get additional (sideband injection) phase locking and this allows us to vary the lockup phase of the laser by detuning ω_{rf} with respect to the laser frequency splitting. This additional locking also sets a reasonable short time limit for the system ($t < 10$ ms) before phase locking effects begin to dominate.

Laser output from the sealed can passes through an anti-reflection coated window and feedback from external optical components is reduced by extensive optical isolation. Faraday isolators are used before all of the photodiodes and low-scatter gyro mirrors are used immediately after the laser. The gyro mirrors are mounted on piezoelectric transducers driven by random noise to scramble the phase of any light that does scatter back into the laser, thus reducing instability from "self injection locking." The unused output from the back of the laser is absorbed by a black glass filter contacted to the rear mirror using index matching fluid.

The laser frequency splitting in a Zeeman laser shows a quadratic dependance on the detuning of the cavity resonance[12] from the center of the laser gain curve. Noise in the relative phase due to cavity instabilities can therefore be minimized by centering the cavity resonance on the gain curve. This is accomplished by offset locking one of the laser modes to an I_2 stabilized laser with the corrections to the cavity length of the Zeeman laser done with a piezoelectric transducer and a small heater mounted on the plasma tube.

Data Acquisition

The two sigma-polarized laser modes are combined with a (linear) polarizing beamsplitter and are detected with a pair of photodiodes. The 100 kHz beatnote in the difference of the photodiode currents contains all the information about the relative phase of the two modes. It is a non-trivial matter to extract this information under the time and resolution constraints mentioned earlier ($t < 10$ ms, $\Delta\nu \sim 1$ milliHertz). This is done by measuring the phase of the beatnote relative to a stable external phase reference. We use a specially-built analog phase meter which measures the time interval between a zero crossing of the beatnote and that of a low noise, slightly-tunable crystal oscillator. In essence, a capacitor is charged by a stable current source during the measured interval and the voltage on the capacitor is sampled by a fast A/D convertor. Subtracting a stable half-scale dc current doubles the analog range. The capacitor charge state is reset after the voltage is measured. Phase information is obtained from this voltage via premeasured conversion factors. The phase reference is a quartz crystal-based phase-locked loop, locked onto the beatnote with a ~10 ms attack time constant. This long time constant allows the phase locked loop to "flywheel" over the (fast) phase fluctuations of interest, but causes it to track out any long term phase drift or frequency error. Due to the high-Q nature of the quartz crystal voltage controlled oscillator (VCO), the phase-locked loop has a tracking range of only about 20 Hz when operating at 100 kHz and this range can be exceeded in the course of a few minutes. To prevent loss of lock, the low frequency part of the VCO error signal is sent as a slow servo control to the dc magnetic field to keep the beatnote at the center of the VCO range. The time constant of this loop is on the order of a second to insure that there are no noticeable field changes during the course of a data run. This phase measurement system can measure the relative phase once every cycle of the 100 kHz beatnote for up to 10^4 consecutive cycles with 10^{-4} radian single point resolution. This makes it possible to measure phase noise corresponding to a 100 μHz linewidth after only 10 μs, easily satisfying the above-noted constraints.

The analog phase meter/phase reference measures the relative phase as a function of time, but it is relative phase *fluctuations* as a function of time that are needed. This can be computed from the phase vs. time record using the Allan 2-point variance[13]:

$$\left\langle \Delta\phi^2(\tau) \right\rangle = \frac{1}{2}\left\langle \left[\phi(t+\tau) - \phi(t)\right]^2 \right\rangle \quad . \tag{9}$$

DATA ANALYSIS

Constant Lockup Phase

Fig. 4 shows the measured diffusion constant as a function of the applied field (B_1) with ϕ_0=0. In this part, each data run consisted of measuring the phase for 10^4 cycles with the ac field on (CEL) and then with the field off (free run). The diffusion constant for each was calculated from the Allan

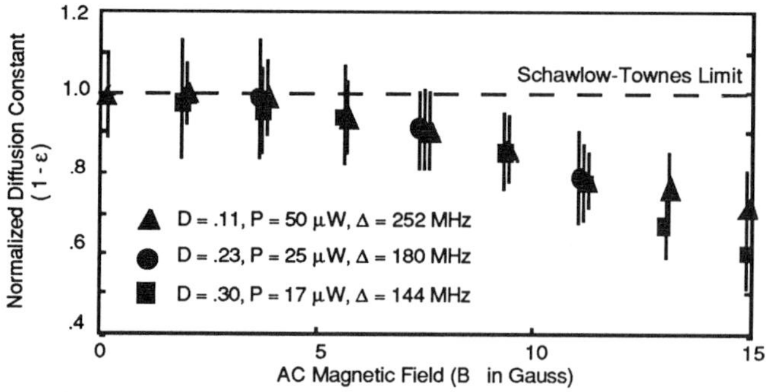

Fig. 4. Noise quenching in the CEL vs. ac field strength. Data are normalized to the measured free-running diffusion constant (D). P is the laser output power per mode and Δ is the Δm=2 Zeeman splitting.

variance plot and a normalized diffusion constant was computed by dividing the CEL value by the free run value. Between 50 and 100 of these normalized values were then averaged for each data point shown in the figure. Data were taken under several conditions with different values of free-running diffusion constant, laser output power, and Zeeman splitting. These values were changed in a very indirect way due to the limited range of the phase reference. Normally, a reduction in B_0 would cause the beatnote to fall below 100 kHz, out of the range of VCO. A way to bring the beat frequency back up to 100 kHz is to spoil the Q of the laser cavity via thermally induced stress (and hence misalignment) which reduces the amount of cavity pulling. Consequences of this approach are a reduction in output power (due to higher losses) and an increase in the free running diffusion constant (due to reduced power and lower cavity Q). Maximum noise reduction was observed for the case of the smallest Zeeman splitting and was about 40%.

Constant Coupling Field

Fig. 5 shows phase diffusion as a function of lockup phase for B_1=15 G (constant ε). Small detunings of the injected frequency ω_{rf} with respect to (half) the free-running beatnote (measured with the ac field on but strongly detuned) change the parameter a in Eq. 2 and lead to corresponding changes in ϕ_0. For a given ϕ_0, the relative phase was sampled for 10^3 cycles and the Allan variance calculated and plotted. This was repeated for 200 values of ϕ_0 to build up a 3-dimensional phase diffusion picture. Twelve of these data sets were then averaged resulting in the curve shown in Fig. 5.

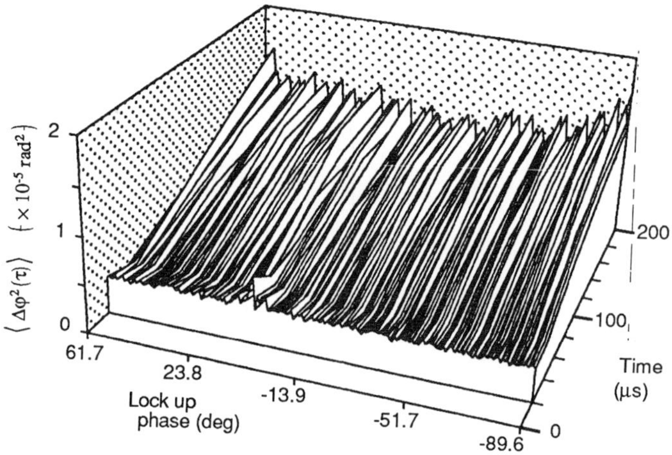

Fig. 5. Phase fluctuations as a function of lockup phase and time for an ac field strength of B_1=15 G. Computer fit yields $\varepsilon = .319 \pm .121$.

The phase meter as described above is insensitive to the actual value of ϕ_0 because the phase reference tracks out any phase offsets in the system. In order to measure ϕ_0 directly, the output from the synthesizer generating the coupling field at ω_{rf} was used as the phase reference. This was possible since the beatnote was always locked to the coupling field during this part of the experiment. The beatnote, at twice ω_{rf}, must first be divided by 2 before comparison with the reference, thus reducing the sampling rate by a factor of 2.

Both the t and ϕ_0 dependence are clearly visible in the figure. Fitting an equation of the form of (7b) yields a value $\varepsilon = .319 \pm .121$.

Limitations Due to Technical Noise

A question that arises is whether we are really measuring spontaneous emission noise or simply technical noise. Technical noise is reduced many orders of magnitude by using a common cavity mode for the two transitions and this is what makes it feasible even to look for quantum noise. Furthermore, the experiments were performed at night in a new vibration- and sound-isolated laboratory. Even so, at long times technical noise must finally dominate due to its t^2 dependance (the signature of frequency flicker, a $1/f$ type of noise in the laser frequency[14]). Fig. 6 shows the phase noise of a free running laser from 10 μs to 20 s. The transition from phase diffusion to frequency flicker occurs at about 10 ms, conveniently similar to the time at which phase locking begins to control the phase evolution behavior. Fig 6 also shows the identical behavior of the free-running laser and the PLL in the short time limit, and the noise reduction of the PLL at long times. The signature of the CEL effect would be a downward shift of the entire curve, but for a variety of reasons the effect is too small to be seen in this particular case.

From Fig. 3, we see that the noise we measure has the correct time dependance, but we can go one step further and estimate a diffusion constant from Eq. 1 using measurable parameters. Some care must be taken when measuring the cavity linewidth because it should be done without any gain medium. Simply turning off the laser and scanning over the cavity resonance with a second laser is unsatisfactory due to drifts and possible thermally-induced alignment (and hence loss) differences. Instead, we used a second HeNe laser, filled with a different isotope of Ne (^{22}Ne) to scan over a cavity resonance two orders (1100 MHz) from the one used by the Zeeman laser, thus probing just outside the gain curve. We measured $\Delta\nu_{cav} = (570 \pm 26)$ kHz and $P_{laser} = 45\mu W$ per mode, resulting in $D = (.045 \pm .004)\ s^{-1}$ for the calculated value per mode, and $D = (.089 \pm .008)\ s^{-1}$ for the beat. This agrees quite well with the measured diffusion constant of the beat, $D = (.083 \pm .008)\ s^{-1}$ taken under the same conditions.

GRAVITATIONAL WAVE DETECTION WITH THE CEL

One of the clear results of Einstein's General Relativity is the prediction of gravitational-wave

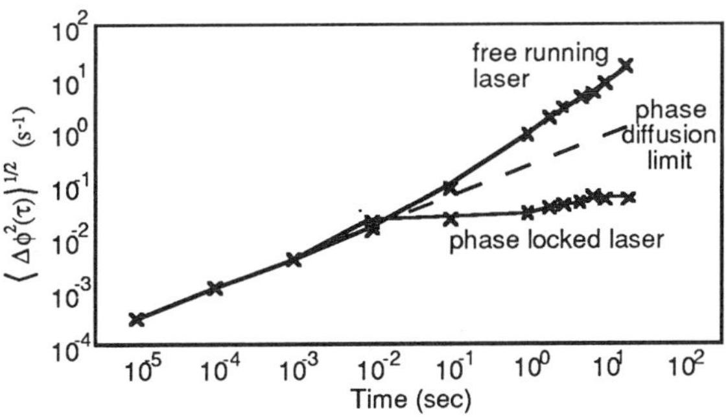

Fig. 6. Phase fluctuations vs. time for a free running laser and PLL. Pure quantum phase diffusion is evident in both up to 10 ms where the free running laser starts to show the effects of technical noise (variance increases with t^2) and phase locking becomes effective in the PLL. Technical noise increases PLL variance slightly beyond 1 s. Correlated emission would shift the curves downward, but the effect is too small to be seen in this particular case.

radiation. The predicted measureable effects are exceedingly weak and up to the present have not been detected directly. However one cannot seriously doubt the existence of such radiation, particularly after the discovery by Taylor and Weisberg[15] that the measured orbital period decrement of the binary pulsar PSR 1913+16 is in agreement -- at the 1% level -- with the predicted rate due to power lost to gravitational radiation. Still the gravitational radiation, due to known sources, produces such weak signals at the earth that only with the highest degree of invention and diligence will detection be possible. This challenge has been taken up seriously by several groups worldwide, and is serving as a stimulation to "fixes" and sensitivity-enhancing "tricks" from the quantum optics field. Indeed, the CEL was first proposed by Scully[2] in the G-W and inertial-sensor context. In the following we briefly summarize one view of the G-W / CEL connection for an active detector.

As shown by Schleich and Scully[8] and noted above, the phase-locked relationship between the two phasors representing the optical dipole moments causes the spontaneously-emitted photons dumped into the two modes to carry a strong correlation. To the extent that the two phasors are parallel, $(1-\cos\phi_0) \rightarrow 0$, and the simple theory predicts the *relative* phase diffusion goes to zero. In the language in which the spontaneous emission is attributed to "stimulation emission" due to the fluctuating vacuum field, the noise suppression is clearly expected. When the two phasors are parallel, the vacuum fields cannot distinguish between them to induce relative phase-changing emission. The diffusion of the relative phase goes to zero, and the diffusion of the (uninteresting) average phase is increased. The phase noise is less -- 40% less in our present experiments -- than the "minimum possible" phase diffusion associated with the fundamental Schawlow-Townes linewidth process.

Unfortunately, there are some problems in the application of the CEL to inertial sensing or gravitational wave detection. In the steady state, any phase difference (signal) impressed onto the laser beat is reduced by the phase-locking action. Although spontaneous emission noise is also reduced by this factor, it is not attractive to reduce the detected signal size in the presence of other technical noise sources that will also set sensitivity limits. It is clear, however, that CEL and phase-locking are working on different time scales. By constraining the phasors representing the optical dipole moments to be parallel in an appropriate rotating frame, the vacuum field can cause no differential phase evolution *on any time scale.* In contrast, the classical process of phase-locking requires some time to take effect. Inspection of Eq. 2 shows that, for small deviations from equilibrium, recovery is an exponential process with time constant $1/b$. It is in this transient time domain, $t<1/b$, where we have both low noise and maximum signal -- a situation just right for enhancing the sensitivity for detection of millisecond pulses of gravitational radiation! Our detailed analysis of this sensitivity will be presented elsewhere.[16]

SUMMARY

In conclusion, we have measured noise reduction 40% below the Schawlow-Townes limit in a HeNe laser due to the CEL effect. This noise reduction is in the relative phase of a two mode laser and appears to have potential implications for the design of optimally sensitive interferometers for gravity wave detection: The choice may possibly be between a passive interferometer with a squeezed vacuum reference or an active interferometer with a CEL.

We would like to thank G. Agarwal, C. Benkert, J. Bergou, M. Orszag, W. Schleich, M. O. Scully, S. Swartz, and P. Zoller for many useful discussions. This research began as a collaboration with Professor P. E. Toschek and we would like to thank him for his enthusiasm and numerous contributions to earlier stages of the work. Dr. B. Petley also must be thanked for his useful refinements during the initial stages of the project. This research has been supported in part by the National Institute of Standards and Technology under its program of research into possible techniques for precise measurement, and in part by the Office of Naval Research and the National Science Foundation.

REFERENCES

1. C. M. Caves, Phys. Rev. D **23**, 1693 (1981)

2. M. O. Scully, Phys. Rev. Lett. **55,** 2802 (1985)

3. M. O. Scully and J. Gea-Banacloche, Phys. Rev. A **34**, 4043 (1986)

4. R. E. Slusher, L. W. Hollberg, B. Yurke, J. C. Mertz, and J. F. Valley, Phys. Rev. Lett. **55**, 2409 (1985)

5. See JOSA B ,volume 4, October 1987, special issue on squeezed light

6. A. E. Siegman, *Lasers* (Mill Valley: University Science Books, 1986) chap 11

7. R. Adler, Proc. IRE **34**, 351 (1946)

8. W. Schleich and M. O. Scully, Phys. Rev. A **37**, 1261 (1988)

9. P. E. Toschek and J. L. Hall, Abstract in XV International Conference on Quantum Electronics, JOSA B **4**, 124 (1987)

10. M. Ohtsu and K-Y. Liou, Appl. Phys. Lett. **52**, 10 (1987)

11. M. Weissbluth, *Photon-Atom Interactions* (New York:Academic Press, 1988) chap 7

12. T. Baer, F. V. Kowalski, and J. L. Hall, Appl. Opt. **19**, 3173 (1980)

13. D. W. Allan, Proc. IEEE **54**, 221 (1966)

14. D. W. Allan, J. H. Shoaf, and D. Halford, *Time and Frequency:Theory and Fundamentals*, NBS Monograph 140

15. J. H. Taylor and J. M. Weisberg, Astrophys. J. **354**, 434 (1989).

16. M. P. Winters and J. L. Hall, to be published.

QUANTUM THEORY OF LINEAR AMPLIFIERS

K.Zaheer and M.S.Zubairy

Department of Electronics
Quaid-i-Azam University
Islamabad,Pakistan

I. INTRODUCTION

A linear amplifier is an amplifier whose output is linearly related to its input. This broad definition requiring only the linearity of operation enables one to give a unified account of the quantum limits for such devices without going into the details of their internal working.

The principal use of amplifiers is in the measurement process; that they can bridge the gap between the quantum theory and classical theory by strengthening a quantum signal sufficient enough to bring it into the classical domain, thus allowing its measurement without any significant disturbance. Apart from this, there are numerous other applications. In communication systems, e.g., it may be advantageous to replace a repeater unit in an optical fibre link by a direct amplification unit or an amplifier unit could be used in a predetection capacity at the receiver.

A question of interest is whether photon cloning or accurate reproduction of an incident photon by an amplifier

New Frontiers in Quantum Electrodynamics and Quantum Optics
Edited by A. O. Barut, Plenum Press, New York, 1990

is possible. Such precise duplication of photons would present opportunities for overcoming the limitations imposed by uncertainty principle in interference and photon correlation experiments[1-3]. Wooters and Zurek[4] have however pointed out that it is not possible to "clone" a photon of arbitrary polarisation. Since amplification is achieved through an emissive process in optical amplifiers, it is invariably influenced by spontaneous emission. Milonni and Hardies[5] have discussed the role of spontaneous emission in cloning. The amplifier does indeed amplify the photons of arbitrary input polarisation but the output also includes a large proportion of unpolarised "noise photons" due to spontaneous emission. The amplification process, therefore, has the same fundamental quantum limitations.

Until recently due to a dearth of applications and the difficulty of designing of such quantum-limited amplifiers, the interest in quantum limits in such devices was only academic in nature. However, the envisaged application of amplifiers circumventing the quantum limits for gravity-wave detection[6], the development of amplifiers based on the dc SQUID[7] and generation of squeezed states[8] have coincided well to stimulate a great deal of interest in linear amplifiers. Because of their inherent low noise properties[9], squeezed states are naturally advantageous to be used as optical information carriers. In this connection, two questions arise. First, can the output of a linear amplifier still display nonclassical features if these are imposed at the input?[10-13] Second, can an amplifier generate a squeezed output? Answers to these questions divide the amplifiers into two general classes, namely, the phase-insensitive and phase-sensitive amplifiers[14] which are discussed in the subsequent sections.

II. LINEAR AMPLIFICATION PROCESS: GENERAL DESCRIPTION

The prperties of linear amplifiers have been studied from the early days of quantum optics; the method of analysis ranging from a phenomenological rate equation approach by Shimoda *et al.*[15] to density matrix formalism by Friberg and Mandel[11] and Heisenberg operator picture by Loudon and Shephered[13]. We first consider the linear input-output equation for an amplifier. For simplicity, throughout this paper, we assume that the input and output signals are carried by single bosonic modes. The evolution equation for the signal can be written as

$$a_{out} = M\, a_{in} + L\, a_{in}^{\dagger} + \mathcal{F} \, . \tag{1}$$

where a_{in} ($a_{in}^{\dagger}$) and a_{out} ($a_{out}^{\dagger}$) represent the annihilation (creation) operators for the input and output field mode respectively and M and L are the amplitudes of the amplifier gain which depend on the internal degrees of freedom. For simplicity, we assume M and L to be real. The operator $\mathcal{F}$ is introduced in Eq.(1) so as to satisfy the boson commutation relation for the output mode. As a consequence, $\mathcal{F}$ itself is a boson operator satisfying

$$\left[\mathcal{F}, \mathcal{F}^{\dagger}\right] = 1 - M^2 + L^2 \, . \tag{2}$$

Introducing the Hermitian operators X_i (i=1,2) for the signals as

$$X_1 = \frac{1}{2}\left(a^{\dagger} + a\right) \, , \tag{3a}$$

$$X_2 = \frac{i}{2}\left(a^{\dagger} - a\right) \, , \tag{3b}$$

the evolution equation (1) leads to the following equation:

$$\left(X_i\right)_{out} = \sqrt{G_i}\left(X_i\right)_{in} + \mathcal{F}_i \quad ; \quad i=1,2 \quad . \tag{4}$$

Here G_1 and G_2 are the gains for the two quadrature phases respectively, given by

$$G_1 = (M+L)^2 \quad ; \quad G_2 = (M-L)^2 , \tag{5}$$

and

$$\mathcal{F}_1 = \frac{1}{2}\left(\mathcal{F}^{\dagger}+\mathcal{F}\right) \quad , \tag{6a}$$

$$\mathcal{F}_2 = \frac{i}{2}\left(\mathcal{F}^{\dagger}-\mathcal{F}\right) \quad . \tag{6b}$$

Since we are interested in the fluctuations in $\mathcal{F}$ and not in its mean value, we can assume $\langle\mathcal{F}\rangle=\langle\mathcal{F}^{\dagger}\rangle=0$ without any loss of generality. Then from Eq.(4), we have

$$\langle X_i\rangle_{out} = \sqrt{G_i}\,\langle X_i\rangle_{in} \quad ; \quad i=1,2 , \tag{7}$$

$$\left(\Delta X_i^2\right)_{out} = G_i\left(\Delta X_i^2\right)_{in} + \Delta\mathcal{F}_i^2 \quad ; \quad i=1,2 \quad . \tag{8}$$

The signal to noise ratio (SNR) in the output quadrature components is

$$\mathrm{SNR}_{out}\left(X_i\right) = \frac{\langle X_i\rangle_{in}}{\left[\left(\Delta X_i^2\right)_{in} + \Delta\mathcal{F}_i^2/G_i\right]} \quad ; \quad i=1,2 \quad . \tag{9}$$

Clearly, SNR at the output of a linear amplifier, for any phase quadrature, is less than the SNR in the input due to the uncorrelated additive noise $\Delta\mathcal{F}_i^2$.

Note that, in the foregoing discussion, we have not considered the internal working of the amplifier. The complexities of the amplifier and in particular, the role of the internal degrees of freedom is buried in the gains G_1, G_2 and the operator $\mathcal{F}$. In fact, the only considerations have been, (a) linearity of operation and (b) consistency of quantum mechanics, i.e., preservation of the commutation relations at the output. Such an approach was introduced by Haus and Mullen[16] to study the fundamental quantum limits in linear amplification process. Equation (8) shows that in a linear amplification process, not only the input noise is amplified, also the noise at the output has a component of "added noise" due to amplification. This is a manifestation of the fluctuation dissipation theorem.

It is interesting to consider the quantum limits on the added noise in the two quadratures. From Eqs.(2) and (5) together with the definitions (6a) and (6b), we have,

$$\left[\mathcal{F}_1, \mathcal{F}_2\right] = \frac{i}{2}\left(1 \pm \sqrt{G_1 G_2}\right) , \tag{10}$$

where the lower sign holds from $M \geq L$ and the upper sign for $M \leq L$. The corresponding uncertainty principle is

$$\Delta\mathcal{F}_1 \Delta\mathcal{F}_2 \geq \frac{1}{4} \left|1 \pm \sqrt{G_1 G_2}\right| . \tag{11}$$

Defining the total noise for the two quadratures as

$$\Delta\mathcal{F} = \left[\Delta\mathcal{F}_1^2 + \Delta\mathcal{F}_2^2\right]^{1/2} , \tag{12}$$

the lower limit on $\Delta\mathcal{F}$ obtained from the generalized uncertainty principle is

$$\Delta\mathcal{F} > \left[1/2\left|1 \pm \sqrt{G_1 G_2}\right|\right]^{1/2} . \tag{13}$$

This inequality involving the total added noise and the gains for the two quadratures is usually referred to as the Caves theorem[14]. The inequalities (11) and (13) show that, in principle, it is possible to reduce the added noise in one quadrature at the expense of a larger noise in the other quadratue; an idea in essence similar to the concept of squeezed states.

Based on the amplification properties of the two quadratures, linear amplifiers can be subdivided into two general classes. An amplifier that treats the two quadratures equally, i.e., it amplifies both the quadratures by the same factor and adds equal (phase-insensitive) noise to the two quadratures. In such an amplifier, there is no preferred phase and a phase shift at the input results in an equal or opposite phase shift at the output. Correspondingly, an amplifier obeying such an invariance is either phase preserving or phase conjugating respectively. For a phase preserving amplifier, L=0 and for a phase conjugating amplifier, M=0. In contrast, a phase. sensitive amplifier treats the two quadratures differently in the form of unequal gains or unequal added noise (or both).

III. PHASE-INSENSITIVE AMPLIFICATION IN A TWO-LEVEL SYSTEM

In order to construct an amplifier, one needs a reservoir, full of energy that it can supply to the signal. But an ordinary thermal reservoir or a collection of harmonic oscillators would just feed thermal noise into the signal. The requirement, therefore is apparently contradictory that the reservoir be extremely hot and yet it should not be highly excited, i.e., the expectation values $\langle b_k^\dagger b_k \rangle$ and $\langle b_k b_k^\dagger \rangle$ be minimal where b_k are the reservoir operators.

A simple model proposed by Glauber[17] is obtained by making the reservoir out of a collection of inverted harmonic oscillators. Such a reservoir has no ground state but the states of the inverted harmonic oscillators have an upper bound for which the above mentioned expectation values are minimal.

A practical implementation of this model consists of a group of partly excited two level atoms. In this case, the linearity of operation means that only the one-photon processes are taking place. This situation could correspond to a laser with the end mirrors removed, being perturbed by a weak external field. As would be seen in the following, this still adds phase-insensitive (thermal) noise to the signal since the state of the reservoir resembles a canonical distribution with negative temperature.

The interaction Hamiltonian for any given atom in the interaction picture is

$$V=\hbar g\left[a^{\dagger}\sigma_{-} + \sigma_{+}a\right] \quad , \tag{14}$$

where $\sigma_{-}=|b\rangle\langle a|$ and $\sigma_{+}=|a\rangle\langle b|$ are the atomic flipping operators in terms of the upper $(|a\rangle)$ and lower $(|b\rangle)$ states of the atom. The field is assumed to be in resonance with the atom and the rotating-wave approximation has been made. The system under consideration can be simulated by assuming that the atoms are being injected inside the cavity at a rate r where they interact with the field for a time $\tau=1/\gamma$ during their lifetime and then are removed from the cavity[18]. The change in the reduced density matrix for the field ρ_F during the time τ due to one atom is

$$\delta\rho_F(t)=\rho_F(t+\tau)-\rho_F(t) \quad , \tag{15}$$

where t is the time the interaction starts. A "coarse

grained" change in the radiation field due to N atoms over a time Δt is

$$\Delta\rho_F(t)=N\delta\rho_F(t) \quad , \tag{16}$$

where $N=r\Delta t$. Assuming a distribution $P(\tau)=\gamma e^{-\gamma\tau}$ for the interaction time of the atoms, the coarse grain derivative is

$$\dot{\rho}_F=r\int_{\infty}^{0} \gamma e^{-\gamma\tau}\, \delta\rho_F(t)d\tau \quad , \tag{17}$$

Here it is assumed that the upper limit "∞" is much longer than the interaction time τ, but short compared to the interval Δt.

In the interaction picture, the equation of motion for the density matrix is

$$\dot{\rho}_{AF}=-i/\hbar \left[V(t),\rho_{AF}(t)\right] \quad . \tag{18}$$

Equation (18) can be formally integrated and iterated once to give, in the Born approximation

$$\delta\rho_F(t)=-i/\hbar\ Tr_A\int_t^{t+\tau} dt_1\left[V(t_1),\rho_{AF}(t)\right]$$

$$-1/\hbar^2\ Tr_A\int_t^{t+\tau} dt_1\int_t^{t_1} dt_2\left[V(t_1),\left[V(t_2),\rho_{AF}(t)\right]\right] \quad , \tag{19}$$

where a trace over the atomic variables has been taken and Eq.(15) has been used. In order to evaluate the commutators in Eq.(19), we need to specify, the atom-field density matrix $\rho_{AF}(t)$ at the injection time. We consider a situation in which the atoms are pumped incoherently to states $|a\rangle$ and $|b\rangle$ with probabilities ρ_{aa} and ρ_{bb} respectively. Assuming that the atomic and the field part of the density matrix are uncorrelated at the initial time t,

$$\rho_{AF}(t)=\left(\rho_{aa}|a\rangle\langle a| + \rho_{bb}|b\rangle\langle b|\right)\otimes\rho_F(t) \quad . \tag{20}$$

Using Eqs.(14), (19) and (20) in Eq.(17), we obtain the equation of motion for the reduced density operator as[19-21]

$$\dot{\rho}_F=-\frac{\alpha}{2}\rho_{aa}\left(aa^\dagger\rho_F-2a^\dagger\rho_F a+\rho_F aa^\dagger\right)-\frac{\alpha}{2}\rho_{bb}\left(a^\dagger a\rho_F-2a\rho_F a^\dagger+\rho_F a^\dagger a\right), \tag{21}$$

where $\alpha=rg^2/\gamma^2$ is a rate constant. In the above equation, the terms proportional to ρ_{aa} and ρ_{bb} correspond to gain (due to emission) and loss (due to absorption) respectively. As would be seen in the following, net amplification is achieved when $\rho_{aa}>\rho_{bb}$.

Equation (21) has been known for a long time and has even been solved for arbitrary times[10,20-22]. For the present purpose however, we can use it to obtain equations of motion for various moments of the field operators a and $a^\dagger$,

$$\frac{d}{dt}\langle a\rangle=-\frac{\alpha}{2}\left(\rho_{aa}-\rho_{bb}\right)\langle a\rangle \ , \tag{22a}$$

$$\frac{d}{dt}\langle a^\dagger a\rangle=\alpha\left(\rho_{aa}-\rho_{bb}\right)\langle a^\dagger a\rangle + \alpha\rho_{aa} \ , \tag{22b}$$

$$\frac{d}{dt}\langle a^2\rangle=\alpha\left(\rho_{aa}-\rho_{bb}\right)\langle a^2\rangle \ , \tag{22c}$$

The set of linear equations (22a)-(22c) can be solved exactly to obtain

$$\langle a\rangle_t=\sqrt{G}\,\langle a\rangle_0 \ , \tag{23a}$$

$$\langle a^\dagger a\rangle_t=G\,\langle a^\dagger a\rangle_0+(G-1)\,\frac{\rho_{aa}}{\rho_{aa}-\rho_{bb}} \ , \tag{23b}$$

$$\langle a^2\rangle_t=G\,\langle a^2\rangle_0 \ , \tag{23c}$$

where

$$G=\exp[\alpha(\rho_{aa}-\rho_{bb})t] \quad , \tag{24}$$

is the gain factor. Equations (23a)-(23c) can be viewed as input-output equations for the amplifier; the expectation value at time t being the output from the amplifier in terms of the input (expectation value at time t=0). It is clear that amplification takes place (G>1) when $\rho_{aa}>\rho_{bb}$ and the system acts as an attenuator (G<1) if $\rho_{aa}<\rho_{bb}$.

Similar to the definitions (3a) and (3b), we now define the two quadratures X_θ and $X_{\theta+\pi/2}$ through the relation,

$$X_\theta=\frac{1}{2}\left(a^\dagger e^{i\theta}+ae^{-i\theta}\right) \; . \tag{25}$$

Here the multiplication by the exponential factors merely means an arbitrary rotation in the complex X-plane and it leaves the commutation relations unaltered. It follows from Eqs.(23a)-(23c) that

$$\langle X_\theta\rangle_t=\sqrt{G}\;\langle X_\theta\rangle_o \quad , \tag{26}$$

$$\left(\Delta X_\theta^2\right)_t=G\;\left(\Delta X_\theta^2\right)_o\;+\;(G-1)\mathcal{N} \quad , \tag{27}$$

where

$$\mathcal{N}=\frac{1}{\rho_{aa}-\rho_{bb}} \; . \tag{28}$$

On comparing Eqs.(26) and (27) with Eqs.(7) and (8), we see that the two quadrature components X_θ and $X_{\theta+\pi/2}$ experience equal gains ($G_1=G_2=G$) and that the added noise in the two quadratures is equal and independent of the phase angle θ, i.e., $\Delta\mathcal{F}_1^2=\Delta\mathcal{F}_2^2=(G-1)\mathcal{N}$. The amplifier under consideration is therefore, a phase-preserving, phase-insensitive amplifier.

From Eq.(27), we also see that the maximum gain preserving any squeezing at the output is

$$G_{max} = \frac{1/4 + \mathcal{N}}{(\Delta X_{\theta}^{2})_{o} + \mathcal{N}} , \tag{29}$$

which for a highly squeezed input, $(\Delta X_{\theta}^{2})_{o}=0$, gives $G_{max}=2$. This is usually referred to as the "cloning limit". Any squeezing imposed at the input therefore, disappears during the process of amplification due to phase-insensitive "added noise". A direct correspondence between the quantum limits obtained in Sec. II, based on consistency requirements and the results obtained in this section for a realistic amplifier is obvious. But it is interesting to ask, "what is the origin of added noise in a practical amplifier?". The answer to this question lies hidden in Eq.(20) which reasonably decorrelates the signal and the amplifier's internal degres of freedom, represented by the atomic part of the density matrix, at the time the interaction starts. It has been shown by Loudon and Shephered[13] that the added noise is uncorrelated to the signal and obeys thermal statistics.

IV. DEGENERATE PARAMETRIC AMPLIFIER

In a parameteric amplifier, two modes, usually called the signal and the idler at frequencies ω_s and ω_i are coupled in a nonlinear crystal having a $\chi^{(2)}$ coefficient[23], by a pump mode at frequency ω_p such that

$$\omega_p = \omega_s + \omega_i . \tag{30}$$

When the signal and idler frequencies are equal, the amplifier is said to operate in the degenerate mode. The Hamiltonian for the system in the interaction picture is

$$H = \hbar\varkappa\left(a^{\dagger 2} b + a^{2} b^{\dagger}\right) \quad . \tag{31}$$

Here b and a are the annihilation operators for the pump and signal (idler) modes respectively and $\varkappa$ is a coupling constant which depends upon the second order susceptibility tensor. In the parametric approximation, the pump is assumed to be excited to a large amplitude coherent state and the pump depletion is neglected. The interaction Hamiltonian in Eq.(31) can then be written as

$$H_p = \hbar\varkappa\beta_p\left(a^{\dagger 2} e^{-i\varphi} + a^{2} e^{i\varphi}\right) \quad . \tag{32}$$

where β_p and φ are the real amplitude and phase of the coherent pump field.

The Heisenberg equations of motion for the signal mode operators are

$$\dot{a} = -iE_p\ a^{\dagger} e^{-i\varphi} \quad , \tag{33a}$$

$$\dot{a}^{\dagger} = iE_p\ a\ e^{i\varphi} \quad . \tag{33b}$$

where $E_p = 2\varkappa\beta_p$ is the dimensionless nonlinearity. Solution of Eqs.(33a) and (33b) gives the following evolution equations[24,25]:

$$a_t = a_0 \cosh(E_p t) - ia_0^{\dagger} \sinh(E_p t) e^{-i\varphi} \quad , \tag{34a}$$

$$a_t^{\dagger} = a_0^{\dagger} \cosh(E_p t) + ia_0 \sinh(E_p t) e^{i\varphi} \quad . \tag{34b}$$

For the phase-angle choice $\varphi=\pi/2$, Eqs. (34a) and (34b) can be compared with Eq.(1) and its hermitian conjugate to note that $\mathcal{F}=0$, $M=\cosh(E_p t)$ and $L=-\sinh(E_p t)$.

The two quadratures experience unequal gains which can be found from Eq.(5) or directly from Eqs.(34a) and (34b), given by

$$G_1=\exp(-2E_p t) \quad ; \quad G_2=\exp(2E_p t) \quad . \tag{35}$$

A degenerate parametric amplifier is therefore, a phase-sensitive amplifier which amplifies one quadrature at the expense of an attenuation in the other quadrature. Since $\mathcal{F}=0$, it does not add any noise during amplification.

V. PHASE-SENSITIVE AMPLIFICATION VIA COHERENT SUPERPOSITION

As discussed in Sec.III, a reservoir of inverted harmonic oscillators adds phase-insensitive noise to the signal since it corresponds to a canonical distribution with negative temperature. Clearly, the operating state of the amplifier or the state of the reservoir determines the properties of the output signal. One therefore needs to modify or "rig" the reservoir in some way. Dupertuis *et al.*[26,27] first proposed the "rigging" of the reservoir by preparing it in a multimode squeezed vacuum state centered around the frequency of the boson mode to be amplified. They also considered a reservoir consisting of multiatom squeezed states[28] or simply correlated pains of atoms. Both these models represent phase sensitive amplifies in which the added noise in the two quadratures is unequal. As a practical counterpart, in the following, we discuss a two-photon linear amplifier considered by Scully and Zubairy[29] in which phase-sensitivity is introduced by preparing the gain medium, i.e., the atoms, in a coherent superposition of states.

The system consists of three level atoms in cascade configuration with the levels $|a\rangle$, $|b\rangle$ and $|c\rangle$. The boson mode of frequency ω is assumed to be in resonance with the two atomic transitions. The interaction Hamiltonian in the interaction representation and in the rotating-wave approximation is

$$V=\hbar g\left[a^{\dagger}\left(|b\rangle\langle a|+|c\rangle\langle b|\right) + a\left(|a\rangle\langle b|+|b\rangle\langle c|\right)\right] . \tag{36}$$

We consider a situation in which the atoms are injected at a rate r, in a coherent superposition of the states $|a\rangle$ and $|c\rangle$. The atomic wave function at time t is therefore,

$$|\psi(t)\rangle_A=C_a|a\rangle + C_c|c\rangle , \tag{37}$$

and the atom-field density operator at the initial time is

$$\rho_{AF}(t)=\left[\rho_{aa}|a\rangle\langle a|+\rho_{ac}|a\rangle\langle c|+\rho_{ca}|c\rangle\langle a|+\rho_{cc}|c\rangle\langle c|\right]\otimes\rho(t). \tag{38}$$

Here $\rho_{aa}=\rho^{*}_{ca}$ is the initial coherence between the levels $|a\rangle$ and $|c\rangle$ and ρ_{aa} and ρ_{cc} are the corresponding populations for the two levels.

Following the same procedure as in Sec.III, the equation of motion for the reduced density operator is obtained as

$$\begin{aligned}\dot{\rho}_F= &-\frac{\alpha}{2}\rho_{aa}\left(aa^{\dagger}\rho_F - 2a^{\dagger}\rho_F a + \rho_F aa^{\dagger}\right)\\ &-\frac{\alpha}{2}\rho_{ca}\left(aa\rho_F - 2a\rho_F a + \rho_F aa\right)\\ &-\frac{\alpha}{2}\rho_{ac}\left(a^{\dagger}a^{\dagger}\rho_F - 2a^{\dagger}\rho_F a^{\dagger} + \rho_F a^{\dagger}a^{\dagger}\right)\\ &-\frac{\alpha}{2}\rho_{cc}\left(a^{\dagger}a\rho_F - 2a\rho_F a^{\dagger} + \rho_F a^{\dagger}a\right) .\end{aligned} \tag{39}$$

Similar to Eq.(21), the terms proportional to ρ_{aa} and ρ_{cc} correspond to the usual gain and absorption. But now, the anomalous terms proportional to ρ_{ac} and ρ_{ca} are also present and are responsible for the phase-sensitive operation of the amplifier.

From Eq.(39), the equations of motion for the operator expection values same as Eqs.(22a)-(22c) can be obtained. The solution of these equations gives,

$$\langle a\rangle_t=\sqrt{G}\ \langle a\rangle_o \quad , \tag{40a}$$

$$\langle a^\dagger a\rangle_t=G\ \langle a^\dagger a\rangle_r + (G-1)\frac{\rho_{aa}}{\rho_{aa}-\rho_{cc}} \quad , \tag{40b}$$

$$\langle a^2\rangle_t=G\ \langle a^2\rangle_o - (G-1)\frac{\rho_{ac}}{\rho_{aa}-\rho_{cc}} \quad , \tag{40c}$$

with $G=\exp[\alpha(\rho_{aa}-\rho_{cc})t]$ as the gain factor.

The evolution of the quadrature components defined in Eq.(25) can now be obtained from Eq.(40a) and its complex conjugate as

$$\langle X_\theta\rangle_t=\sqrt{G}\ \langle X_\theta\rangle_o \quad . \tag{41}$$

The noise in these quadratures is

$$\left(\Delta X_\theta^2\right)_t=G\ \left(\Delta X_\theta^2\right)_o + (G-1)\mathcal{N}_\theta \quad , \tag{42}$$

where

$$\mathcal{N}_\theta=\frac{\rho_{aa}+\rho_{cc}-2|\rho_{ac}|\cos(2\theta+\phi)}{4(\rho_{aa}-\rho_{cc})} \quad , \tag{43}$$

with

$$\rho_{ac}=|\rho_{ac}|\exp(i\phi) \quad . \tag{44}$$

Equation (43) is similar in form, to Eq.(27), but now the added noise, i.e., the second term depends upon the phase angle θ. Defining a parameter ϵ such that

$$\rho_{aa}=(1+\epsilon)/2 \ ; \qquad \rho_{cc}=(1-\epsilon)/2 \ ; \qquad \rho_{ac}=\frac{1}{2}(1-\epsilon^2)^{1/2} \ , \tag{45}$$

$$\mathcal{N}_\theta=\frac{1-(1-\epsilon^2)^{1/2}\cos(2\theta+\phi)}{4\epsilon} \ . \tag{46}$$

It then follows that the signal $\langle X_\theta \rangle$ can be amplified $(G>>1)$ without introducing the added noise $(\mathcal{N}_\theta<<1)$ under the following limits:

$$(2\theta+\phi)=0 \ ; \ \epsilon \longrightarrow 0 \ ; \ \alpha t \longrightarrow \infty \ ; \ \alpha t\epsilon=\text{finite} \ . \tag{47}$$

Under these conditions, $N_{\theta+\pi/2} \longrightarrow \infty$, i.e., the added noise in one quadrature is reduced at the expense of increased added noise in the second quadrature such that $(\mathcal{N}_\theta+\mathcal{N}_{\theta+\pi/2})=1/2\epsilon>1/2$. This agrees with the Caves Theorem (c.f. Eq.(13)).

A rather interesting result is obtained when one considers the fluctuations in the number of photons, $:\Delta n^2:$ at the output. The photon statistics are called Poisson, sub- or super-Poisson depending upon whether the normally ordered number fluctuations are equal to, less than or greater than zero respectively. For this amplifier, it is possible to have sub-Poisson statistics at the output for a particular class of super-poisson states at the input[29,30]. However, for this counter-intuitive process, the initial field state is required to satisfy the condition

$$-1 < \left[2\langle a^\dagger a\rangle_0-\langle a^2\rangle_0-\langle a^{\dagger 2}\rangle_0\right] < 0 \ , \tag{48}$$

and the amplifier is limited to finite ranges of gain. So far no explicit expression for such a field has been found.

Zaheer and Zubairy[31] discussed a two-level

phase-sensitive amplifier in which the atoms are injected in pairs in a coherent superposition of the two atomic states and phase sensitive operation is achieved through the "cooperativity" of the atoms. They also showed that the added noise is in a squeezed state, the squeezing parameter being a function of the initial atomic variables. The amplification properties however depend upon the particular scheme employed to prepare the atoms.

Ansari *et al.*[32] have considered a situation in which the coherence is generated between the upper and lower levels in a three-level cascade system by an external, classical driving field of frequency 2ω. They show that for sufficiently strong driving field, this system becomes an ideal parametric amplifier. However, for weak driving field, it behaves as a phase-insensitive amplifier. In between the two extremes, one obtains phase-sensitive amplification and squeezing for certain values of various parameters such as detuning etc.

REFERENCES

1. A.Garuccio, V.Rapisarda, and J.P.Vigier, Phys. Lett. **90A**:17 (1982).
2. Y.Cantelaube, Phys. Lett. **101A**:7 (1984).
3. A.Garuccio, A.Kypriandis, D.Sardelis, and J.P.Vigier, Nuovo Cim. Lett. **39**:225 (1984).
4. W.K.Wooters and W.H.Zurek, Nature **299**:802 (1982).
5. P.W.Milonni and M.L.Hardies, Optics Commun. **92A**:321 (1982).
6. K.S.Thorne, Rev. Mod. Phys. **52**:285 (1980).
7. For a survey of progress in dc SQUID, see J.Clark, IEEE Trans. Elec. Dev. **ED-27**:1896 (1980).

8. R.E.Slusher, L.W.Hollberg, B.Yurke, J.C.Mertz and J.F.Valley, Phys. Rev. Lett. **55**:2409 (1985); M.D.Levenson, R.M.Shelby, S.Perlmutter, R.DeVoe, and D.F.Walls, Phys. Rev. Lett. **57**:691 (1986); L.A.Wu, H.J.Kimble, J.L.Hall, and H.Wu, Phys. Rev. Lett. **57**:2520 (1986).
9. For a review of squeezed states, see K.Zaheer and M.S.Zubairy, in *Advances in Atomic, Molecular, and Optical Physics*, Vol. 27, eds. D.R.Bates and B.Bederson, Academic Press (in press).
10. S.Stenholm, Optics Commun. **58**:177 (1986).
11. S.Friberg and L.Mandel, Optics Commun. **46**:141 (1983).
12. C.K.Hong, S.Friberg, and L.Mandel, J. Opt. Soc. Am. B **2**:494 (1985).
13. R.Loudon and T.J.Shephered, Optica Acta **31**:1243 (1984).
14. C.M.Caves, Phys. Rev. D **26**:1817 (1982).
15. R.Shimoda, H.Takahasi, and C.H.Townes, J. Phys. Soc. Japan **12**:687 (1957).
16. H.A.Haus and J.A.Mullen, Phys. Rev. **128**:2407 (1962).
17. R.J.Glauber, in *Frontiers in Quantum Optics*, eds. E.R.Pike and S.Sarkar, Hilger, London (1986).
18. M.Sargent III, M.O.Scully, and W.E.Lamb Jr., *Laser Physics*, Addison-Wesely, Reading, MA (1974).
19. M.O.Scully and W.E.Lamb Jr., Phys. Rev. **159**:208 (1967).
20. S.Carusotto, Phys. Rev. A **11**:1629 (1975).
21. N.B.Abraham and S.R.Smith, Phys. Rev. A **15**:421 (1977).
22. E.B.Rockower. N.B.Abraham and S.R.Smith, Phys. Rev. A **17**:1100 (1978).
23. A.Yariv, *Quantum Electronics*, Wiley, New York (1973).
24. W.H.Louisell, A.Yariv, and A.E.Siegmann, Phys. Rev. **124**:1646 (1961).
25. J.P.Gordon, W.H.Louisell, and L.R.Walker, Phys. Rev. **129**:481 (1963).

26. M.A.Dupertuis and S.Stenholm, J. Opt. Soc. Am. B **4**:1094 (1987).

27. M.A.Dupertuis, S.M.Barnett, and S.Stenholm, J. Opt. Soc. Am. B **4**:1102 (1987).

28. M.A.Dupertuis, S.M.Barnett, and S.Stenholm, J. Opt. Soc. Am. B **4**:1124 (1987).

29. M.O.Scully and M.S.Zubairy, Optics Commun. **66**:303 (1988).

30. M.A.Dupertuis and S.Stenholm, Phys. Rev. A **37**:1226 (1988).

31. K.Zaheer and M.S.Zubairy, Optics Commun. **69**:37 (1988).

32. N.A.Ansari, J.Gea-Banacloche, and M.S.Zubairy, Phys. Rev. A (to be published).

INTERFERENCE OF LIFETIME BROADENED RESONANCES: NONRECIPROCAL GAIN AND LOSS PROFILES

A. Imamoğlu, J. J. Macklin, and S. E. Harris

Edward L. Ginzton Laboratory
Stanford University
Stanford, CA 94305

INTRODUCTION

It is commonly believed that laser systems have reciprocal absorption and emission profiles, and that population inversion is a necessary condition for obtaining laser amplification. Recently, it has been shown that interferences in lifetime broadened systems can result in nonreciprocal absorption and emission profiles.[1,2] At frequencies where the absorption rate goes to zero, light amplification is possible even though the lower level population greatly exceeds the upper level population. Figure 1 shows a prototype system: The two upper states $|2\rangle$ and $|3\rangle$ must have the same parity, total angular momentum J, and m_j quantum numbers;[3] and must decay to an identical continuum. The decay could be due to spontaneous emission to the states of a final level $|f\rangle$ (not shown in figure) or due to autoionization. The steady state loss mechanism for atoms initially in state $|1\rangle$ is Raman scattering (or virtual autoionization) into a final continuum. This process takes place through two intermediate states, $|2\rangle$ and $|3\rangle$, and is subject to quantum interference effects. The emission profile of an atom initially excited into state $|2\rangle$ or $|3\rangle$ from a reservoir does not show such interference effects. We have shown that these results also hold for two dressed states produced by using an additional electromagnetic field.[4] Selection rules simplify for such a system, making it possible to make lasers that operate without inversion.

ANALYSIS OF A RADIATIVELY-BROADENED SYSTEM

The basis set that we use in the analysis consists of the eigenstates of the non-interacting atom plus radiation field Hamiltonian. Assuming no photons in any but the laser mode of the radiation field initially, we can write the state vector of the total system in the interaction representation as[3,5]

New Frontiers in Quantum Electrodynamics and Quantum Optics
Edited by A. O. Barut, Plenum Press, New York, 1990

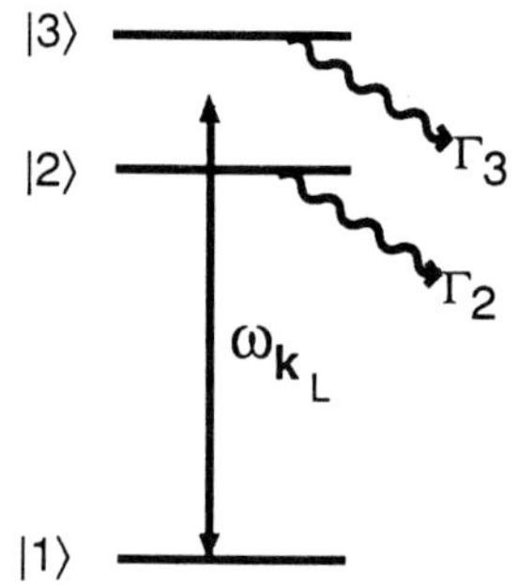

Fig. 1 Schematic of a prototype system. The selection rule for this type of system is: states $|2\rangle$ and $|3\rangle$ must have the same parity, J, and m_j; and must decay to a common continuum.

$$|\Psi_I(t)> = a_1(t)\Big|1, n_{\vec{k}_L}> + a_2(t)\Big|2, n_{\vec{k}_L} - 1> \\ + a_3(t)\Big|3, n_{\vec{k}_L} - 1> + \sum_{\sigma\vec{k}} a_{f,\sigma\vec{k}}(t)\Big|f, n_{\vec{k}_L} - 1, 1_{\sigma\vec{k}}> \tag{1}$$

where a_1, a_2, a_3, $a_{f,\sigma\vec{k}}$ are the probability amplitudes; $n_{\vec{k}_L}$ is the number of photons in the laser mode ; $1_{\sigma\vec{k}}$ represents the fact that the number of photons in the radiation mode $\vec{k}\sigma$ is one; $\vec{k}$ and σ are the wavevector and polarization of the spontaneously emitted (or Raman scattered) photon, respectively. The expansion in Eq. (1) assumes that the radiative decay into atomic state $|1\rangle$ is negligible and that only the eigenstates with approximately equal energy are coupled by the interactions. The latter assumption is equivalent to the rotating wave approximation in the semiclassical approach.

By substituting Eq. (1) in Schrodinger's equation[1,3] and integrating over the continuum states, we arrive at the coupled equations for the probability amplitudes

$$\frac{\partial a_1'(t)}{\partial t} = \kappa_{12}a_2'(t) + \kappa_{13}a_3'(t) \tag{2a}$$

$$\frac{\partial a_2'(t)}{\partial t} + i\Delta\tilde{\omega}_2 a_2'(t) = \kappa_{21}a_1'(t) + \kappa_{23}a_3'(t) \tag{2b}$$

$$\frac{\partial a_3'(t)}{\partial t} + i\Delta\tilde{\omega}_3 a_3'(t) = \kappa_{31}a_1'(t) + \kappa_{32}a_2'(t) \tag{2c}$$

where :

$$a_1(t) = ia_1'(t)\,;\; a_2)(t) = a_2'(t)e^{i(\omega_{21}-\omega_{\vec{k}_L})t}\,;\; a_3(t) = a_3'(t)e^{i(\omega_{31}-\omega_{\vec{k}_L})t} \tag{3a}$$

$$\Delta\tilde{\omega}_2 = \omega_{21} - \omega_{\vec{k}_L} - i\frac{\Gamma_2}{2} \\ \Delta\tilde{\omega}_3 = \omega_{31} - \omega_{\vec{k}_L} - i\frac{\Gamma_3}{2} \tag{3b}$$

$$\kappa_{12} = \kappa_{21} = i g_{12\vec{k}_L} n_{\vec{k}_L}^{1/2} = \frac{i\Omega_{12}}{2}$$

$$\kappa_{13} = \kappa_{31} = i g_{13\vec{k}_L} n_{\vec{k}_L}^{1/2} = \frac{i\Omega_{13}}{2} \tag{3c}$$

$$\kappa_{23} = \kappa_{32} = -\frac{(\Gamma_2\Gamma_3)^{1/2}}{2}$$

The real diagonal terms Γ_2 and Γ_3 are the direct radiative decay rates of the states $|2\rangle$ and $|3\rangle$, respectively. The real cross-coupling terms κ_{23} and κ_{32} give the finite amplitude for the absorption of a virtual photon emitted from state $|3\rangle$ ($|2\rangle$), by state $|2\rangle$ ($|3\rangle$). We note that Eqs. 2 have the same form when the two upper states decay by ionization.

The solution of Eqs. 2, subject to the appropriate initial condition, determine the single atom contribution to the gain or loss of the laser field $\omega_{\vec{k}_L}$. If the atom is in state $|1\rangle$ initially, then the probability for absorption of a laser photon is $1-|a_1(t)|^2$. If the atom is initially excited into state $|2\rangle$, then the probability of emission of a laser photon is $|a_1(t)|^2$. In Fig. 2, we numerically solve Eqs. 2 to show the absorption behavior of a three-level system (Fig. 2b) and compare it to that of the two-level system (Fig. 2a). The parameters of the three-level system are chosen to cause a perfect cancellation of the steady-state absorption. The two-level system exhibits a non-zero slope of $|a_1(t)|^2$ in steady-state which is absent in the three-level system. In both cases the atom undergoes a transient response due to the $t=0$ initial condition; this produces a loss which cannot be canceled by interference.

If an atom at $t=0$ is put into an upper level, it has some probability of being stimulated down to level $|1\rangle$ before if decays out of the system. The magnitude of this stimulated response is on the order but always less than the magnitude of the transient absorptive response. As pumping into lower level $|1\rangle$ invokes a transient absorption, lasing without population inversion can only occur if the excitation rate of the upper level atoms exceed that of the lower level atoms.

DENSITY MATRIX

In order to obtain the behavior of an ensemble of atoms, we use the density matrix equations.[7] We assume that the atomic state $|2\rangle$ is incoherently pumped at a constant rate R_2 from a reservoir (*i.e.* not from any one of the states considered so far in the analysis), and that the other atomic states are not pumped. The steady-state ($\gg \Gamma_2^{-1}, \Gamma_3^{-1}$) solution of the density matrix equations then give:

$$\frac{d < n_{\vec{k}_L} >}{dt} = W_{abs}\rho_{11}^{(o)} + W_{em}(\rho_{22}^{(o)} + \rho_{33}^{(o)}) \tag{4}$$

where :

$$W_{abs} = \frac{4(\kappa_{31}\Delta\omega_2\Gamma_3^{1/2} + \kappa_{21}\Delta\omega_3\Gamma_2^{1/2})^2}{4\Delta\omega_2^2\Delta\omega_3^2 + (\Delta\omega_2\Gamma_3 + \Delta\omega_3\Gamma_2)^2)} \tag{5a}$$

$$W_{em} = \frac{\omega_{32}^2}{(\Gamma_2+\Gamma_3)^2 + 4\omega_{32}^2} \cdot \frac{4\kappa_{21}^2\Delta\omega_3^2\Gamma_2 + \Gamma_2\Gamma_3(\kappa_{31}\Gamma_2^{1/2} - \kappa_{21}\Gamma_3^{1/2})^2}{4\Delta\omega_2^2\Delta\omega_3^2 + (\Delta\omega_2\Gamma_3 + \Delta\omega_3\Gamma_2)^2} \tag{5b}$$

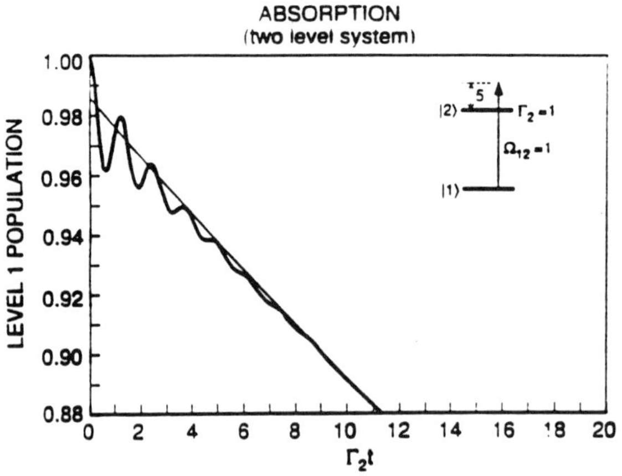

Fig. 2(a) Response of a two level system which at $t = 0$ is excited into level $|1\rangle$. Boldface curve is numerical solution of the coupled equations 2a,b,c describing the evolution of the system. The straight line is the steady state solution from theory. The offset and slope of the straight line are the transient and steady state loss, respectively (Ref. 6)

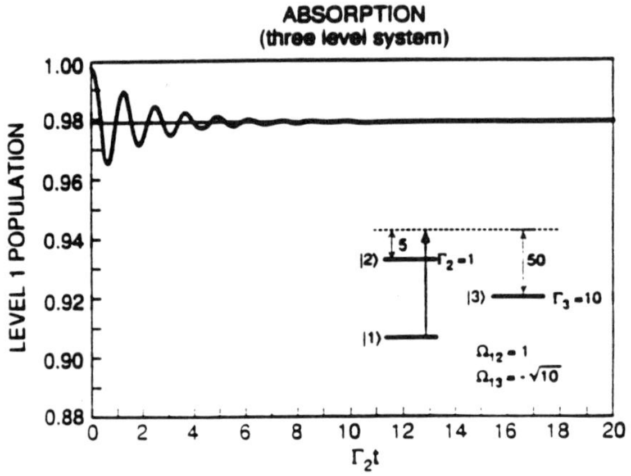

Fig. 2(b) Response of a three level system which at $t = 0$ is excited into level $|1\rangle$. Curves have the same meaning as in Fig. 2(a). Zero slope means zero steady state loss, though a finite transient loss (non-zero intercept) occurs for the values of the parameters chosen.

$$\Delta\omega_j = \omega_{j1} - \omega_{\vec{k}_L} \qquad j = 2,3$$

The terms W_{abs} and W_{em} in Eq. 4 are identified as the stimulated absorption and emission rates, respectively. Note that when $\kappa_{31}\Delta\omega_2\Gamma_3^{1/2} = -\kappa_{21}\Delta\omega_3\Gamma_2^{1/2}$, $W_{abs} = 0$ but $W_{em} \neq 0$. If there is non-zero pumping into atomic state $|3\rangle$, the form of the Eqs. 3 and 5b will be different,[7] but their predictions will remain the same on a qualitative basis.

So far we have only considered the presence of a single decay channel for the upper states. In general there are other non-cancellable channels (for example spontaneous emission on the laser channel) which change the zero in the absorption to a minimum. We have shown that if the decay rate of the cancellable channel is much larger than the decay rate of the non-cancellable channel, then the ratio of the stimulated emission cross section to the absorption cross section is equal to the ratio of the cancellable and non-cancellable decay rates.

The major problem with these type of lasers is that it is not easy to find nearby states which decay strictly to the same continuum. For states which decay by autoionization there are almost always several channels with comparable amplitudes into which the state may decay. States which decay by radiation are generally separated by many inverse lifetimes giving negligible gain cross section at the frequency where the absorption is zero.

In a recent Letter, we have shown that similar interference effects can be induced by using an additional electromagnetic field to create a pair of dressed states which apriori decay to the same continuum.[4] Figure 3a shows the energy level diagram in the bare state picture: We take the probe Rabi frequency Ω_{13} to be very small as compared to the coupling strength Ω_{23} and Γ_3, where Γ_3 is the decay rate (to an arbitrary continuum) of state $|3\rangle$. The frequency detunings of the bare system are defined as $\Delta\omega_c = \omega_3 - \omega -_2 - \omega_c$; and $\Delta\omega_p = \omega_3 - \omega_p - \omega_1$.

By applying the dressed state transformation:

$$\begin{aligned} |2d\rangle &= \cos\theta|2\rangle - \sin\theta|3\rangle \\ |3d\rangle &= \sin\theta|2\rangle + \cos\theta|3\rangle \\ \tan 2\theta &= \frac{-\Omega_{23}}{\Delta\omega_c} \end{aligned} \tag{6}$$

we diagonalize the interactions with the coupling laser. The corresponding dressed state picture is shown in Fig. 3b. The system in this basis set is equivalent to the prototype system shown in Fig. 1. By using the dressed state parameters in the equations (1)-(3) we can obtain the absorption and emission rates corresponding to the dressed system:

$$W_{\text{abs}} = \frac{4\Omega_{13}^2(\Delta\omega_p - \Delta\omega_c)^2\Gamma_3}{(\Delta\omega_c^2 \sec^2 2\theta - (2\Delta\omega_p - \Delta\omega_c)^2)^2 + 4\Gamma_3^2(\Delta\omega_p - \Delta\omega_c)^2} \tag{7a}$$

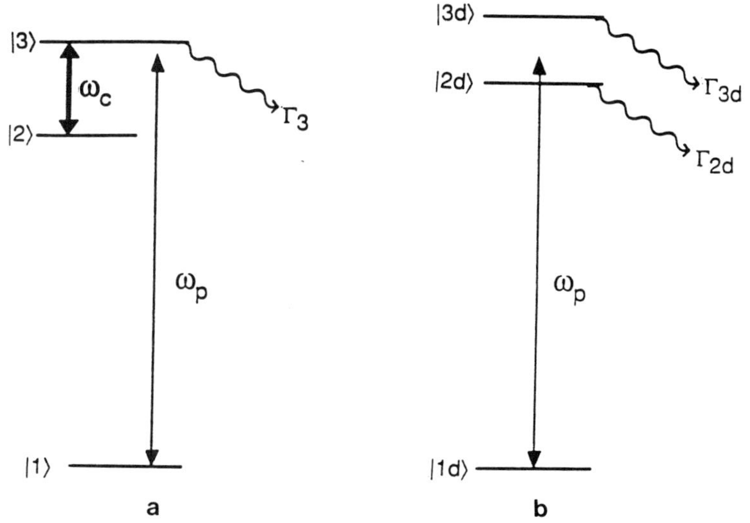

Fig. 3 Energy level diagram of electromagnetic field coupled system. ω_c is the frequency of the coupling field, and ω_p is the frequency of the probe. Γ_3 is the decay rate to an arbitrary continuum. a) bare state system b) equivalent dressed state system.

$$W_e = \Gamma \left[\frac{\Omega_{13}^2 \Delta\omega_c^2 \tan^2 2\theta}{(\Delta\omega_c^2 \sec^2 2\theta - (2\Delta\omega_p - \Delta\omega_c)^2)^2 + 4\Gamma_3^2(\Delta\omega_p - \Delta\omega_c)^2} \right] \tag{7b}$$

where $\Gamma = \dfrac{\Delta\omega_c^2 \Gamma_3 \tan^2 2\theta}{(4 + 2\tan^2 2\theta)\Delta\omega_c^2 + \Gamma_3^2}$

The emission rate given above is obtained for an incoherent pumping into bare state $|2\rangle$. Pumping of state $|3\rangle$ results in an emission profile which is identical to that of the absorption. We explain this effect by noting that the pumping of any one of the bare states result in a superposition excitation of both dressed states. If the initial pumping is into state $|2\rangle$, the interference in stimulated emission that results is constructive in between the dressed states, whereas for pumping into state $|3\rangle$, it is destructive.

Figure 4 shows the absorptive and emissive transition rates of Eqs. 7a and 7b as a function of the detuning (normalized to Γ_3) of the probe frequency. The transition rates are normalized to the peak transition rate of the uncoupled $|3\rangle - |1\rangle$ system. We have taken the coupling field to be on resonance with the bare $|2\rangle - |3\rangle$ transition, and let $\Omega_{23} = 0.5\Gamma_3$. The normalized emission rate for this system is 0.6, whereas the absorption rate is zero. We note that the hole in the absorption profile has been observed experimentally.[8]

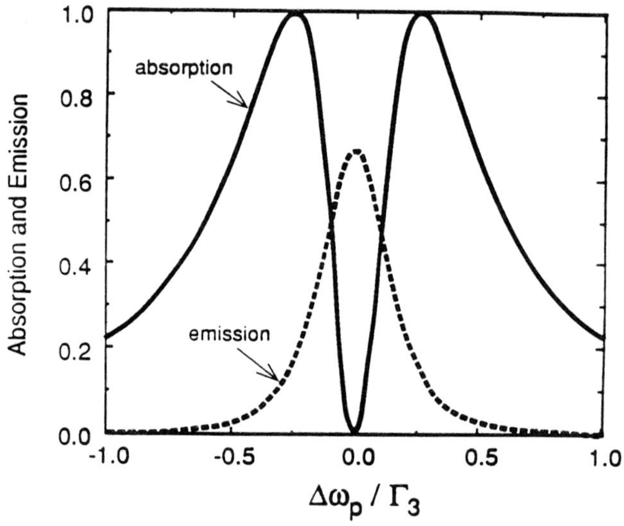

Fig. 4–Emission and absorption transition rates [Eqs. (7a) and (7b)] normalized to the transition rate of the uncoupled $|1\rangle - |3\rangle$ systems. The drive strength $\Omega_{23} = .5\Gamma_3$ and $\omega_c = \omega_3 - \omega_2$.

DISCUSSION AND APPLICATIONS

We have already pointed out the two main advantages of the dressed system: The absence of noncancellable channels and the possibility of getting stimulated emission cross-sections on the order of that of the uncoupled system, at the zero absorption frequency. With these features, we believe that the results of this work will be of importance in regions of the spectrum where lifetime decay rates exceed collisional widths. This is generally the case for autoionization broadened states, and for radiatively broadened transitions which have wavelengths less than about 100 Å. In such systems, using a visible laser might remove the population inversion requirement and result in gain at soft x-ray wavelengths.

Acknowledgements

The authors thank K. H. Hahn and D. A. King for helpful discussions. This work was supported by the Strategic Defense Initiative Organization, the U.S. Army Research Office, the U.S. Air Force Office of Scientific Research and the U.S. Office of Naval Research.

References

1. S. E. Harris, *Phys. Rev. Lett.* **62**, 1033 (1989).
2. V. G. Arkhipkin and Yu. I. Heller, *Phys. Lett.* **98A**, 12 (1983).
3. A. Imamoğlu, *Phys. Rev. A* **40**, 2835 (1989).
4. A. Imamoğlu and S. E. Harris, *Opt. Lett.* (to be published).
5. M. Sargent III, M. O. Scully, and W. E. Lamb, Jr., *Laser Physics* (Addison-Wesley, Reading, MA, 1974), pp. 236-241.
6. S. E. Harris and J. J. Macklin, *Phys. Rev.A* **40**, 4135 (1989).
7. J. J. Macklin and S. E. Harris, (unpublished).
8. H. R. Gray, R. M. Whitley, and C. R. Stroud, Jr., *Opt. Lett.* **3**, 218 (1978).

QUASIPROBABILITIES AND PHOTON NUMBER DISTRIBUTIONS FOR NON–CLASSICAL LIGHT FIELDS

M.S. Kim, F.A.M. de Oliveira* and P.L. Knight

Optics Section, Blackett Laboratory, Imperial College,
London, SW7 2BZ, England.

INTRODUCTION

There has been considerable interest in quasiprobabilities[1–2] and photon number distributions[3] for non–classical fields[4]. In quantum mechanics it is not possible for a particle simultaneously to have a well defined position and momentum, according to the uncertainty principle so that one cannot define a true probability distribution in phase space, which is composed of the position and the momentum axes, and instead we are forced to employ quasiprobabilities. Although the quasiprobabilities are neither measurable nor true probabilities, they are of great use in the study of quantum mechanical systems[2]. The photon number distribution can also exhibit properties determined by the quantum nature of light. The photon number distribution $P(\ell)$ is the probability of there being ℓ photons in the field. The photon number distribution for the coherent state, for example, is Poissonian while that for the photon number state is a delta function. This paper relates the nonclassical properties of quantum quasiprobabilites to the photon number distributions.

Theoretically we can rotate, displace and/or squeeze states in phase space by applying appropriate operators[5]. Applying a Glauber displace operator[1] $\hat{D}(\beta)$ to the vacuum we obtain the coherent state; this is implemented by driving the quantised field by a classical current. The vacuum can be used as an input field for squeezing systems such as a parametric oscillator[6], where the squeezing is generated by a quadratic operator[4] $\hat{S}(r)$. Recent developments in quantum optics have lead to suggestions of how nonclassical states of light, particularly number states of the electromagnetic field may be prepared[4]. At microwave frequencies, the Rydberg atom micromaser[7] is highly sensitive to the quantised nature of the radiation field in a cavity and has "trap" states where the field approaches a number state with a large degree of sub–Poisson photon statistics[8]. These trap states are those in which the quantum field has the required photon number to generate multiples of full Rabi cycles in subsequent atoms entering the cavity, leaving the field in consequence unchanged. A distribution of interaction times, analogous to normal laser pump fluctuations, will wash out such effects. Experimental observation of sub–Poissonian trap states, with the cavity field prepared to good approximation in a number state, has been reported when care was taken to ensure uniform atom–field interaction times in the micromaser[7].

Localised one–photon states have been constructed by optical shutter techniques using photons generated in pairs in parametric down conversion, where the signal photon opens a photoelectric detection gate to the idler photon[9]. Related aspects of such conditioned measurements have been proposed for photon number state preparation in unitary transformation. What we have shown is that this is valid only for the Wigner function. [See also Ref.14]

Using the definitions of the density operators and the quasiprobability functions we find the quasiprobabilities for the squeezed number and the displaced number states and plot them in Fig.2. As we expect from the earlier discussion, we see that the quasiprobabilities for the displaced number state have the same form as for the number state in the displaced coordinates. Fig.2(a) shows the cylindrical shape of the Q–representation for the photon number state is squeezed as the squeeze parameter r increases. Comparing Fig.2(a) with Fig.2(b) representing the Q–representations, we see that a simple contraction cannot transform one into the other. Nevertheless, it is easy to see that the Wigner function of the squeezed number state is exactly the expected contraction of that of the displaced number state.

Glauber[1] and Sudarshan[15], independently, have introduced the diagonal P–representation for the probability density. The P–representation is defined as the Fourier transform of the normal–ordered characteristic function. Using the definition (5) of the squeezed thermal state we obtain the characteristic function $C_p(\eta)$ for the squeezed thermal state as

$$C_p(\eta) = \exp[-(e^r\sinh r + \bar{n}e^{2r})\eta_x^2 - (\bar{n}e^{-2r} - e^{-r}\sinh r)\eta_y^2]. \tag{17}$$

Considering the squeezing in one direction, namely r > 0, the coefficient of η_x^2 is always negative but that of η_y^2 can become positive, which makes the characteristic function diverge. Thus only when

$$\bar{n}e^{-2r} - e^{-r}\sinh r > 0 \tag{18}$$

can we Fourier transform $C_p(\eta)$ and find the P–representation $P_{st}(\alpha)$ for the squeezed thermal state as follows

$$P_{st}(\alpha) = \frac{1/\pi}{[(\sinh r + \bar{n}e^{r})(\bar{n}e^{-r} - \sinh r)]^{\frac{1}{2}}} \times\exp[-\alpha_y^2 e^{-r}/(\sinh r+\bar{n}e^{r}) - \alpha_x^2 e^{r}/(-\sinh r+\bar{n}e^{-r})]. \tag{19}$$

The diagonal P–representation is well–defined for a classical state, but either it is negative or does not exist for states exhibiting non–classical behaviour. The condition (18) can be written as

$$(2\bar{n} + 1)\, e^{-2r} > 1. \tag{20}$$

The left hand side of eq.(20) is the variance of the quadrature operator for the squeezed thermal state (see eqs(8)) and the factor 1 of the right hand side is the quadrature variance of the vacuum. If the quadrature variances are larger than the minimum uncertainty limit, it is possible to describe the squeezed thermal state in terms of a well–behaved P–representation which is positive everywhere. The squeezed thermal state can show either a classical or a quantum behaviour depending on the sizes of $\bar{n}$ and r.

PHOTON NUMBER DISTRIBUTION

The photon number distribution $P(\ell) \equiv \langle \ell \mid \hat{\rho} \mid \ell \rangle$ gives the probability of there being ℓ photons in the field. The photon number distribution $P(\ell) \equiv \langle \ell \mid \hat{\rho} \mid \ell \rangle$ for the squeezed number state is obtained using the factorisation of the squeeze operator

$$P_{sn}(\ell) = \frac{n!\ell!}{(\cosh r)^{2\ell+1}}\left(\tfrac{1}{2}\tanh r\right)^{n-\ell} S(r, \ell, n)\cos^2\frac{(n-\ell)\pi}{2}, \tag{21}$$

where

$$S(r, \ell, n) = \left| \sum_m \frac{(-1)^m(2^{-1}\tanh r)^{2m}\,(\cosh r)^{2m}}{m!(\ell-2m)![m+(n-\ell)/2]!} \right|^2. \tag{22}$$

where n' = n for the squeezed number state and n' = $\bar{n}$ for the squeezed thermal state. The variances are independent of the coherent amplitude of the state for the displaced number state as $< (\Delta X_{1,2})^2 > = 2n + 1$, and for $n \geq 1$ are always greater than the ones for the vacuum.

THE SECOND–ORDER CORRELATION FUNCTION

The Glauber second–order correlation, which describes the intensity fluctuations of the radiation field, is defined as[1]

$$g^{(2)} \equiv 1 + \frac{< (\Delta n)^2 > - < \hat{n} >}{< \hat{n} >^2} \quad (9)$$

For Poissonian statistics, $g^{(2)} = 1$. If $g^{(2)} < 1$, the light is said to be sub–Poissonian, otherwise, it is super–Poissonian. When β is assumed real for convenience, it is a straightforward calculation to find that for the squeezed coherent state

$$g^{(2)} = 1 - \frac{1}{< \hat{n} >} + \frac{1}{< \hat{n} >^2} [\beta^2 e^{-2r} + 2\sinh^2 r \cosh^2 r], \quad (10)$$

where the mean photon number $< \hat{n} > = \beta^2 + \sinh^2 r$.

Similarly we obtain the second–order correlations for the squeezed thermal and squeezed number states and plot the dependence of $g^{(2)}$ on the squeeze parameter r in Fig.1. When $r \gg 1$, for the squeezed number state[5],

$$g^{(2)} \approx 1 + \frac{2(n^2 + n + 1)}{(2n^2 + 1)^2}. \quad (11)$$

When a large–photon number state is squeezed, $g^{(2)}$ approaches to 1.5. Squeezing the thermal field actually increases the fluctuation in the field intensity. For a limiting case, as $r \gg 1$, $g^{(2)} \longrightarrow 3$. The factor 3 of super–Poissonicity was earlier discussed by Ekert et al. and Janzsky et al. in the study of the statistical properties of the squeezed vacuum[13].

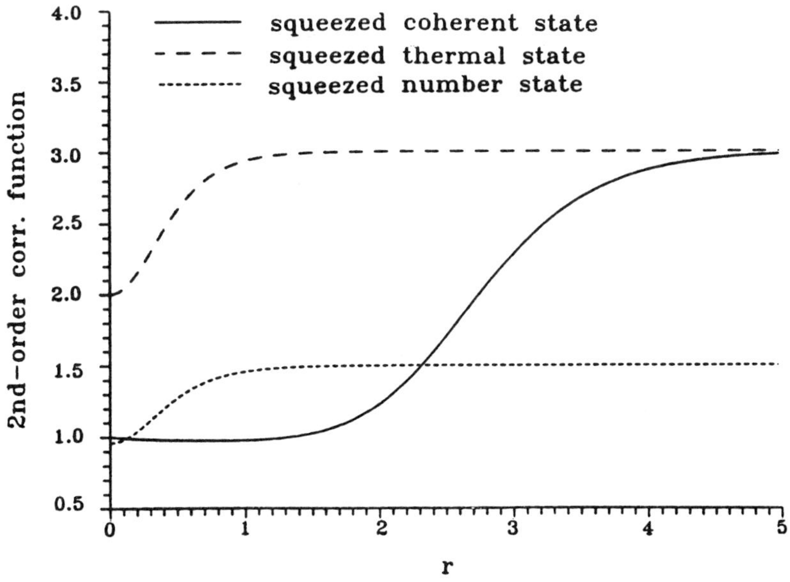

Fig.1 The second–order correlation function for the squeezed coherent, the squeezed thermal and the squeezed number state when the average photon number is 25, i.e. n = 25, $\bar{n}$ = 25 and α = 5. r is the squeeze parameter.

Fig.1 compares the second–order correlation functions of the various squeezed states. The squeezed coherent state shows sub–Poissonicity when r is small but becomes super–Poissonian as the squeezing increases. When the photon number state is squeezed, the photon number distribution is blurred and the photon statistics immediately become super–Poissonian. Squeezing the thermal field results in larger fluctuations in the field intensity. As the photon number state is squeezed, the photon number distribution is broadened and the photon statistics very rapidly become super–Poissonian.

We find the second–order correlation function, measuring the deviation from Poisson statistics for the displaced number state $|\beta, n>$

$$g^{(2)} = 1 + n \left[\frac{2\ |\beta|^2 - 1}{n + |\beta|^2}\right]. \tag{12}$$

The displaced number state has sub–Poissonian photon statistics if $|\beta|^2 < \frac{1}{2}$, in other words if the coherent contribution adds more than half a photon to the average photon number, the state is sub–Poissonian independent of the initial photon number n.

QUASIPROBABILITY FUNCTIONS

A unified discussion of quasiprobabilities and ordering of operators has been given by Cahill and Glauber[1]. A quasiprobability is defined as

$$W(\alpha, p) = \pi^{-2} \int \mathrm{Tr}\ \{\hat{\rho}\ \hat{D}(\lambda)\}\ e^{p|\lambda|^2} \exp(\alpha\ \lambda^* - \alpha^*\ \lambda)\ d^2\lambda \tag{13}$$

where $\hat{D}(\lambda)$ is the Glauber displacement operator, defined in eq.(2). For p = 1, we obtain the Glauber P–representation, p = 0 gives the Wigner function and p = –1 gives the Q–representation. Based on this definition, very general conclusions can be reached about a given new state of light, if such state can be demonstrated to be obtained by a transformation from a previous one. We denote the original functions and operators with a subscript 0 and the displaced ones with a subscript d. Let us consider the effect of the displacement $\hat{D}\ (\beta)$ on the original state described by the density matrix $\hat{\rho}_0$. The displaced quasiprobabilities are given by

$$W_d\ (\alpha, p) = \pi^{-2} \int \mathrm{Tr}\ \{\hat{\rho}_0\ \hat{D}(\lambda)\}\ e^{p|\lambda|^2} \exp\ [(\beta - \alpha)\ \lambda^* - (\beta^* - \alpha^*)\ \lambda]\ d^2\lambda$$

or

$$W_d(\alpha, p) = W_0(\alpha - \beta, p) \tag{14}$$

and we see that for any ordering the new quasiprobability is obtained by a simple displacement of the quasiprobability of the original state. This is a very important result, because we can immediately say that if a state does not have a given quasiprobability then a displacement of that state will produce a new state which does not have that quasiprobability as well. From this, we conclude that the displaced number state does not have a well–behaved Glauber P–function. One could argue that this should be expected, but curiously, we know from the results of the quadrature variances and the second–order correlation function that this state can be simultaneously super–Poissonian and have more quadrature fluctuations than the vacuum.

We will derive the effect of squeezing for comparison purposes. In this case, the new state density matrix $\hat{\rho}_s$ is obtained from the original density matrix $\hat{\rho}_0$ by the transformation

$$\hat{\rho}_s = \hat{S}\ (r)\ \hat{\rho}_0\ \hat{S}^\dagger\ (r). \tag{15}$$

We denote the initial functions and operators with a subscript 0 and the squeezed ones with a subscript s. Using the Bogoliubov transformation (eq.(3)), it can be shown that $\hat{S}\ \hat{D}\ (\lambda)\ \hat{S}^\dagger = \hat{D}\ (\zeta)$, where $\zeta = \lambda\cosh r + \lambda^*\sinh r$. Using eq.(15) in eq.(13), we obtain

$$W_s(\alpha, p) = \pi^{-2}\int \mathrm{Tr}\,\{\hat{\rho}_0\,\hat{D}(\zeta)\}\, e^{p|\lambda|^2} \exp(\bar{\alpha}\,\zeta^* - \bar{\alpha}^*\,\zeta)d^2\zeta, \tag{16}$$

where $\bar{\alpha} = \alpha\cosh r + \alpha^*\sinh r$. Comparison of eq.(16) with eq.(13) shows no simple connection between them, because of the presence of the term $e^{p|\lambda|^2}$ which is not transformed like the other terms. But for p = 0, this factor disappears, and we can see that $W_s(\alpha, 0) = W_0(\bar{\alpha}, 0)$, i.e. the result is a scaling transformation, with lengths in one direction being contracted and the orthogonal direction being expanded[14]. The Wigner function is the only quasiprobability allowing a simple description of such transformation, probably due to the fact that it is associated with hermitean symmetrical and antisymmetrical combinations of the annihilation and creation operators. It might be thought that we could obtain all the quasiprobabilities from the simpler ones by a simple

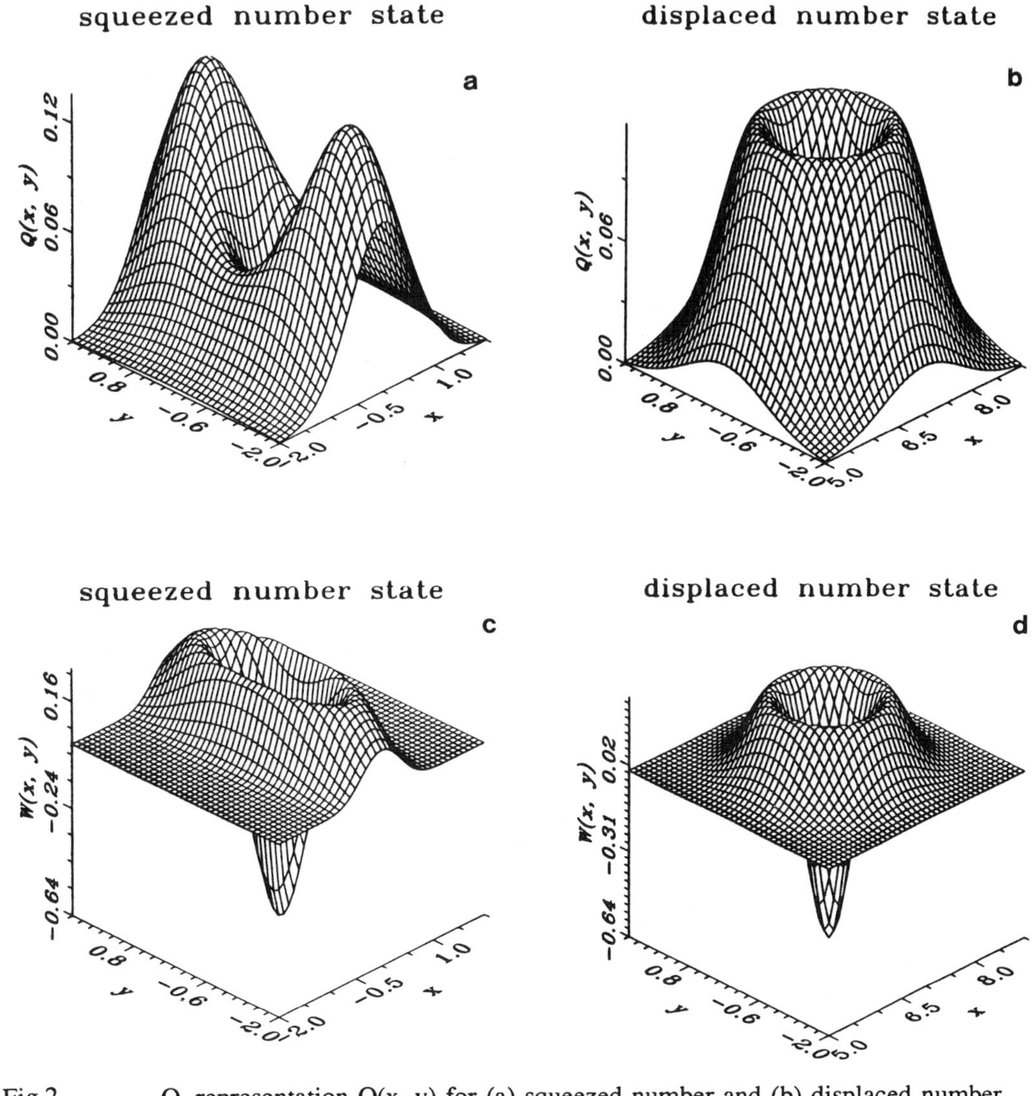

Fig.2 Q–representation Q(x, y) for (a) squeezed number and (b) displaced number state. Wigner function W(x, y) for (c) squeezed number and (d) displaced number state. The initial photon number is 1, r = 1 and β = 7. Here x = Re(α) and y = Im(α).

non–degenerate parametric amplification. Given that photon number states (at least those with modest occupation numbers) can be generated, it is natural to ask whether they can be amplified or squeezed. Rather than address this problem directly, we turn our attention to the simpler problems of displacing and squeezing a number state by the displacement operator $\hat{D}(\beta)$ and the squeeze operator $\hat{S}(r)$.

Finally, instead of a vacuum state or a number state input to a squeezing device, we can also consider a thermal input field[10]. Indeed in many early squeezing experiments, a thermal field with excess fluctuations was the actual input, and phase–sensitive quadrature variances of the thermal–like input measured. This kind of input is of great importance in studies of squeezed microwave radiation using Josephson devices, where at the lower frequencies, thermal noise in the input is inevitable and hard to eradicate[11]. We study the photon statistics of the squeezed thermal state in terms of the Bose–Einstein weighted sum of squeezed number states described above[12].

We consider the properties of the squeezed number, the squeezed coherent, the squeezed thermal and the displaced number states in this paper. We study the statistical properties of the various squeezed states and the displaced number states using the mean variances and the Glauber second–order correlation functions[1]. We then calculate the quasiprobabilities for these nonclassical states, and use them to determine the physical characteristics of nonclassical light. The photon number distribution is extensively discussed, and especially the pairwise oscillations resulting from the two–photon nature of the squeeze operator.

The squeezed coherent state defined as

$$|\beta, r> \equiv \hat{D}(\beta)\hat{S}(r)\ |0> \tag{1}$$

where the squeeze[4] $\hat{S}(r)$ and the displacement[1] $\hat{D}(\beta)$ operators are given by

$$\hat{S}(r) \equiv \exp(\frac{1}{2}r\hat{a}^2 - \frac{1}{2}r\hat{a}^{\dagger 2}),\ \ \hat{D}(\beta) \equiv \exp(\beta\hat{a}^\dagger - \beta^*\hat{a}). \tag{2}$$

Here the squeeze parameter r has been assumed real for convenience. The squeezing operators provide a Bogoliubov transformation of the annihilation operator as

$$\hat{S}^\dagger(r)\hat{a}\hat{S}(r) = \hat{a}\cosh r - \hat{a}^\dagger \sinh r. \tag{3}$$

We define the squeezed number and the displaced number state density matrices by

$$\hat{\rho}_{sn} \equiv \hat{S}(r)\,|\,n><n\,|\,\hat{S}^\dagger(r), \qquad \hat{\rho}_{dn} \equiv \hat{D}(\beta)\,|\,n><n\,|\,\hat{D}^\dagger(\beta). \tag{4}$$

The squeezed thermal state is the Bose–Einstein weighted sum of the squeezed number states, with density matrix[12]

$$\hat{\rho}_{st} \equiv (1+\bar{n})^{-1}\sum_{n=0}^{\infty}\left[\frac{\bar{n}}{1+\bar{n}}\right]^n \hat{\rho}_n \tag{5a}$$

where $\bar{n}$ is the average photon number of the thermal input field. Alternatively, the density matrix $\hat{\rho}_{th}$ can be expressed in a coherent state basis as

$$\hat{\rho}_{st} = \frac{1}{\pi\bar{n}}\int d^2\beta\, \exp(-\bar{n}^{-1}|\beta|^2)\ \hat{S}(r)\,|\,\beta><\beta\,|\,\hat{S}^\dagger(r). \tag{5b}$$

The quadrature operators are defined by

$$\hat{X}_1 = \hat{a} + \hat{a}^\dagger,\ \ \hat{X}_2 = -i(\hat{a} - \hat{a}^\dagger). \tag{6}$$

For the squeezed coherent state, the variances $<(\Delta X_i)^2> \equiv <\hat{X}_i^2> - <\hat{X}_i>^2$ are

$$<(\Delta X_1)^2> = \exp(-2r), \qquad <(\Delta X_2)^2> = \exp(2r). \tag{7}$$

One quadrature variance has been reduced at the expense of the expansion of the other quadrature variance, due to the squeezing. For the squeezed number and the squeezed thermal states,

$$<(\Delta X_1)^2> = (2n'+1)e^{-2r}, \qquad <(\Delta X_2)^2> = (2n'+1)e^{2r}, \tag{8}$$

The factorials in eq.(21) are valid for non–negative integers, so that $\frac{1}{2}(\ell - n) \leq m \leq \frac{1}{2}\ell$. The cosine term is responsible for the vanishing of $P_{sn}(\ell)$ when $|n - \ell|$ is odd. For the squeezed thermal state the photon distribution $P_{st}(\ell)$ using eq.(5) is

$$P_{st}(\ell) = (1+\bar{n})^{-1} \sum_{n} P_{sn}(\ell, n) \left[\frac{\bar{n}}{1 + \bar{n}}\right]^{n}. \tag{23}$$

For the squeezed vacuum

$$P_{sv}(\ell) = \frac{(\frac{1}{2}\tanh r)^{\ell}\ell!}{(\frac{1}{2}\ell)!^{2}\cosh r} \cos^{2}\frac{\ell}{2}\pi. \tag{24}$$

When ℓ is odd $P_{sv}(\ell)$ is zero, otherwise $P_{sv}(\ell)$ may be non–zero. These pairwise oscillations are the result of the quadratic, or two–photon nature of the squeeze operator $\hat{S}(r)$. We have plotted the photon distributions $P_{sn}(\ell)$, $P_{st}(\ell)$ as functions of photon number ℓ in Fig.3. Both distributions show pairwise oscillations. More noticeable is the large–scale macroscopic oscillations of the squeezed number state photon distribution, analogous to those found by Schleich and Wheeler for the squeezed coherent state. Under the condition $r \gg 1$ and $n > \ell \gg 1$ we can approximate eq.(21) [The proof is given in ref.5] as

$$P_{sn}(\ell) \approx \frac{2/\sqrt{\pi}}{(2\ell+1)\cosh r} \cos^{2}\frac{(n-\ell)\pi}{2} \left[\cos\frac{\ell\pi}{2}\cos\alpha_{\ell} - \sin\frac{\ell\pi}{2}\sin\alpha_{\ell}\right]^{2} \tag{25}$$

where $\alpha = \sqrt{2\ell + 1}/\sinh r$. We have plotted $P_{sn}(\ell)$ for $n = 20$ and $r = 2$ in Fig.3(a). When n and ℓ are even,

$$P_{sn}(\ell) \propto \frac{1}{\ell} \cos^{2}\frac{\sqrt{2\ell + 1}}{\sinh r}. \tag{26}$$

As ℓ grows the oscillations slow down (Fig.3(a)). When n and ℓ are odd,

$$P_{sn}(\ell) \propto \frac{1}{\ell} \sin^{2}\frac{\sqrt{2\ell + 1}}{\sinh r}. \tag{27}$$

When the input average photon number is small, i.e. $\bar{n} \leq 1$, the only important contribution to the field state is from the squeezed number state $n \approx 0$ (squeezed

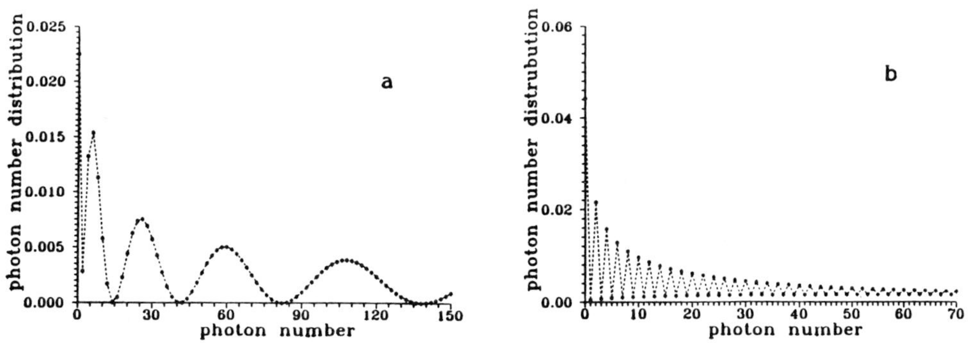

Fig.3 Photon number distribution $P_{sn}(\ell)$ of (a)the squeezed number state when $r = 2$ and $n = 20$, valid for even integers, $\ell = 0, 2, 4, \cdots$ (b)the squeezed thermal state when $r = 3$ and $\bar{n} = 2$.

vacuum). Thus the photon number distribution for the squeezed thermal state of $\bar{n} \le 1$ is similar to the squeezed vacuum and oscillates between zero and non–zeros in a pairwise pattern. However when $\bar{n}$ is large we cannot disregard the contributions from the squeezed number states with n large. Eventually the contributions from appropriately large photon number states are more important. When n is appropriately large and $\ell < n$, the approximation (25) is applicable. The sum of odd and even number photon states will show the pairwise oscillation as in Fig.3 (b).

The photon number distribution of the displaced number state, $P_{dn}(\ell)$ can be obtained from

$$P_{dn}(\ell) = |\langle \ell|\beta, n\rangle|^2 = |\langle \ell|\, \hat{D}(\beta)\, |n\rangle|^2 \tag{28}$$

where

$$\langle \ell|\hat{D}(\beta)|n\rangle = (n!/\ell!)^{\frac{1}{2}}\, \beta^{\ell - n}\, e^{-\frac{1}{2}|\beta|^2} \mathcal{L}_n^{(\ell - n)}(|\beta|^2) \tag{29}$$

where $\mathcal{L}_n^{(\ell - n)}(x)$ is an associated Laguerre polynomial. For $\beta = 0$, we recover the photon number state, where $P_{dn}(\ell) = \delta_{\ell n}$. In the squared modulus, we have a polynomial in ℓ of degree n, so extending this function for real values of ℓ, one would have eventually n zeros. Therefore it can be expected that $P_{dn}(\ell)$ have up to n minima between (n + 1) maxima. In fig.4, we plot $P_{dn}(\ell)$ for $\beta = 7$, and n = 1, 10. We see that there is a strong similarity between these functions and the Hermite polynomials of order n. Let us discuss the case with n = 1. In this case, it is readily seen that eq.(28) can be rewritten as

$$P_{dn}(\ell) = (\ell!)^{-1}\, e^{-|\alpha|^2}\, |\beta|^{2(\ell - 1)}\, (\ell - |\beta|^2)^2 \tag{30}$$

This distribution has a zero at $\ell = |\alpha|^2$, if $|\alpha|^2$ is an integer. It is interesting to note that adding a coherent amplitude to a one photon state, the probability of having zero photons becomes non–zero. We see in fig.4(a) that there are two peaks, around the minimum, as opposed to a coherent state, where there is only one peak value for the number of photons. For n = 2, we have a second minimum appearing, and the number of maxima increases, being four for n = 3, and eleven, for n = 10. We report elsewhere the connection between the photon statistics and interference in phase–space[5].

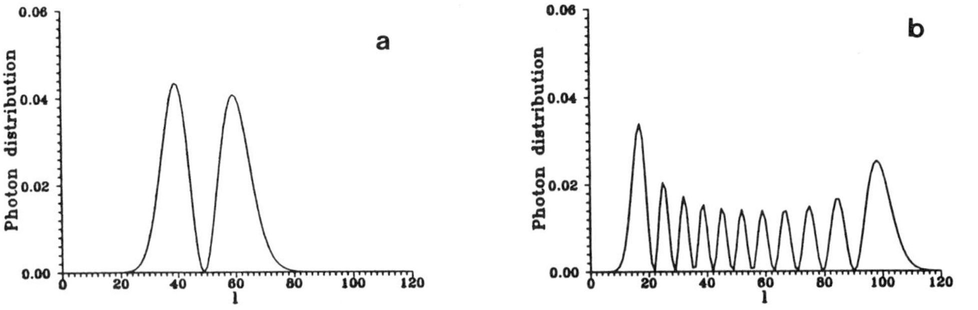

Fig.4 Photon number distribution for the displaced number state $|7, n\rangle$. (a) n = 1, (b) n = 10.

ACKNOWLEDGEMENTS

This work was supported in part by the U.K. Science and Engineering Research Council.

REFERENCES

*Permanent address: Departamento de Física, Universidade Federal do Rio Grande do Norte, 59000 Natal RN, Brazil.

1. R.J. Glauber, in: Quantum Optics, eds. S.M. Kay and A. Maitland (Academic Press, London, 1970) p.53; K.E. Cahill and R.J. Glauber, Phys.Rev. 177(1969), 1857; 1883.
2. M. Hillery, R.F. O'Connell, M.O. Scully and E.P. Wigner, Phys.Rep. 106 (1984), 121.
3. W. Schleich and J.A. Wheeler, Nature 326 (1987), 574.
4. R. Loudon, Rep.Prog.Phys. 43 (1980), 913; R. Loudon and P.L. Knight, J.Mod.Opt. 34 (1987), 709; F.A.M. de Oliveira and P.L. Knight, Phys.Rev.Lett. 61 (1988), 830.
5. M.S. Kim, F.A.M. de Oliveira and P.L. Knight, Opt. Comm. 72 (1989), 99; Phys.Rev.A. in press; F.A.M. de Oliveira, M.S. Kim, P.L. Knight and V. Bužek, submitted to Phys.Rev.A.
6. L.–A. Wu, H.J. Kimble, J.L. Hall and H. Wu, Phys.Rev.Lett. 57 (1986), 2520.
7. D. Meschede, H. Walther and G. Müller, Phys.Rev.Lett. 54 (1985), 551; H. Walther, Private communication.
8. F. Cummings and A.K. Rajagopal, Phys.Rev.A 39 (1989), 3414.
9. C.K. Hong and L. Mandel, Phys.Rev.Lett. 56 (1986), 58.
10. H. Fearn and M. Collett, J.Mod.Opt. 35 (1988), 553
11. B. Yurke et al., Phys.Rev.Lett. 60 (1988), 764.
12. P.L. Knight and L. Allen, Concepts of Quantum Optics (Pergamon Press, Oxford, 1983).
13. A. Ekert and K. Rzążewski, Opt.Comm. 65 (1988), 225; J. Janszky and Y. Yushin, Phys.Rev.A 36 (1987), 1288.
14. A. Ekert and P.L. Knight, to be published.
15. E.C.G. Sudarshan, Phys.Rev.Lett. 10 (1963), 277.

A MODEL FOR LASER DAMAGE:
AN EXAMPLE OF HAMILTONIAN CHAOS

Miguel Orszag
Facultad de Fisica
Pontificia Universidad Catolica de Chile
Casilla 6177 Santiago,Chile

1.Introduction

I present here a laser damage model for a transparent medium, where one has impurities, which are responsible for the damage process, these impurities being modelled by square wells, with electrons trapped in them.[1,2]

One can use Classical Mechanics and find a threshold value for the electromagnetic field for which the KAM surfaces, of this non-integrable problem, break, in a one dimensional model.

In the second section, a tutorial description of the KAM theory is given, for a simple problem with two degrees of freedom (a pair of coupled harmonic oscillators).

We show, that for a weak coupling, in a sense(to be specified more accurately below), the natural frequencies of the system, for most initial conditions, are only slightly altered by the presence of such a weak non-integrable perturbation.

The argument presented here is based on perturbation theory. If we assume a Hamiltonian of the form:

$$H = H_0 + V, \tag{1}$$

where H_0 represents an integrable system and V a weak non-linear and non-integrable perturbation, then historically two types of solutions have been found[3]:

(1) Poincare [4] and others found, via perturbation expansions, that a weak perturbation changes only slightly the unperturbed motion, in the sense that the frequencies of the system suffer only minor changes and they also observed the appearence of harmonics.

(2) Fermi [5], on the other hand, found that even small non-integrable perturbations can produce dramatic effects on the unperturbed motion. He even observed ergodic behavior in a totally deterministic problem.

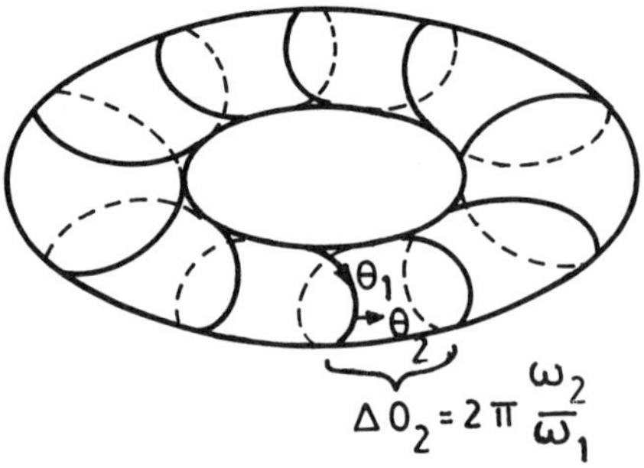

Figure 1.Torus in phase space

These divergent views can be unified by the so called KAM theorem. But before doing so, let's define integrability.

A Hamiltonian system is called integrable if one can find a canonical transformation such that if the original Hamiltonian is expressed in terms of action-angle variables , the new Hamiltonian depends only on the new actions. If one has N degrees of freedom, then we find N actions which are constants of motion, and the corresponding angles that are linear functions of time.

If we take a simple example of an integrable system with two degrees of freedom, the phase space is four dimensional and the trajectories move on a two-torus, where the two radii are J_1 and J_2. (Figure 1)

Closed orbits occur only when the ratio of the two frequencies is a rational number, that is :

$$\frac{\omega_1}{\omega_2} = \frac{m}{n}.$$

In this case, let's assume that θ_1 performs a complete cycle and $\delta\theta_2$ is the corresponding shift in θ_2. Then:

$$\delta\theta_2 = 2\pi\frac{\omega_2}{\omega_1}.$$

On the other hand, if the ratio of the two frequencies is an irrational, the orbit never repeats itself and approaches every point on the torus.

In section 3, a model for laser damage is discussed and some new insight is given as to describe the destruction of the KAM tori in terms of instabilities of periodic orbits.

2.The KAM Theorem and beyond.

It was originally stated by Kolmogorov[6] in 1954 and was proven later, independently, by Arnold[7] and Moser[8]. There are several good review articles in this subject [9,10,11]

.Let's assume a physical system with two degrees of freedom, with a Hamiltonian given by:

$$H = H_0(J_1, j_2) + V(J_1, J_2, \phi_1, \phi_2), \tag{2}$$

expressed in terms of angle-action variables.

If $V = 0$, J_1 and J_2 are constants of motion, and the system has two characteristic frequencies given by $\omega_i = \frac{\partial H_0}{\partial J_i}$ and$\phi_i = \omega_i t + \phi_{i0}$ (i =1,2).

As mentioned earlier, the trajectories move on a two-torus with radii J_1 and J_2 and rotation angles ϕ_1 and ϕ_2.

The KAM theorem addresses the question of what happens to the invariant tori of the integrable Hamiltonian H_0, when V is small, but different from zero. The theorem says that most of the unperturbed tori with incommensurate frequencies ($\frac{\omega_1}{\omega_2}$ = irrational number) continue to exist, being only slightly distorted by the perturbation. On the other hand, the tori bearing periodic motion or very nearly periodic motion, with commensurate frequencies ($\frac{\omega_1}{\omega_2} = \frac{m}{n}$) are greatly deformed by the perturbation, no matter how small is it and the trajectories no longer remain close to the unperturbed tori. The KAM theorem states that the vast majority of the initial conditions (in the sense of measure theory) , for V small, lie on the preserved tori. However, there is a small set of initial conditions, for which the trajectories are not in the preserved tori, but rather pathologically interspersed between the preserved tori.

We say that the tori corresponding to conmmensurate frequencies are "destroyed" and ergodic motion is generated in small regions of phase space.

Now, we will study the conditions under which the tori are destroyed, based on small denominator arguments. Following Walker and Ford[3], we expand V in Fourier series, keeping only one component:

$$H = H_0(J_1, J_2) + f_{mn}(J_1, J_2)\cos(m\phi_1 + n\phi_2) + ... \tag{3}$$

We try now, to eliminate the angle dependence of the particular Fourier term using a generating function:

$$F = j_1\phi_1 + j_2\phi_2 + B_{mn}(j_1, j_2)\sin(m\phi_1 + n\phi_2), \tag{4}$$

which only differs slightly from the identity transformation ($F = j_1\phi_1 + j_2\phi_2$) by the term proportional to B_{mn}.

The canonical transformation is:

$$(J_1, J_2, \phi_1, \phi_2) \longrightarrow (j_1, j_2, \phi_1, \phi_2) \tag{5}$$

With the generating function given in eq(4), we get:

$$J_1 = \frac{\partial F}{\partial \phi_1} \longrightarrow J_1 = j_1 + mB_{mn}\cos(m\phi_1 + n\phi_2),$$

$$J_2 = \frac{\partial F}{\partial \phi_2} \longrightarrow J_2 = j_2 + nB_{mn}\cos(m\phi_1 + n\phi_2),$$

$$\theta_1 = \frac{\partial F}{\partial j_1} \longrightarrow \theta_1 = \phi_1 + \frac{\partial B_{mn}}{\partial j_1}\sin(m\phi_1 + n\phi_2),$$

$$\theta_2 = \frac{\partial F}{\partial j_2} \longrightarrow \theta_2 = \phi_2 + \frac{\partial B_{mn}}{\partial j_2}\sin(m\phi_1 + n\phi_2). \tag{6}$$

Now we can also expand $H_0(J_1, J_2)$ in terms of the new action variables:

$$H_0(J_1, J_2) = H_0(j_1, j_2) + \frac{\partial H_0}{\partial j_1}\delta j_1 + \frac{\partial H_0}{\partial j_2}\delta j_2 + ..., \tag{7}$$

or:

$$H_0(J_1, J_2) = H_0(j_1, j_2) + [-(m\omega_1 + n\omega_2)B_{mn}]\cos(m\phi_1 + n\phi_2) \tag{8}$$

where $\omega_1 = \frac{\partial H_0}{\partial j_1}, \omega_2 = \frac{\partial H_0}{\partial j_2}$ and$\delta j_1, \delta j_2$ were obtained from eq.(6).

The total Hamiltonian now becomes:

$$H = H_0(j_1, j_2) + [-(m\omega_1 + n\omega_2)B_{mn} + f_{mn}]\cos(m\phi_1 + n\phi_2). \tag{9}$$

The idea is to eliminate the m-n Fourier component(resonance) by setting the coefficient of the cosine term in eq.(9) to zero, and H becomes once more, phase independent. Therefore:

$$B_{mn} = \frac{f_{mn}}{m\omega_1 + n\omega_2}. \tag{10}$$

Now we recognize the two opposite cases mentioned in the introduction.

If $\mid m\omega_1(J_1, J_2) + n\omega_2(J_1, J_2) \mid \gg f_{mn}$, B_{mn} is small and the generating function required to eliminate the m-n Fourier component of the perturbation, differs only slightly from the identity transformation. This is the case when the ratio of the two frequencies is an irrational number or a rational with very large m,n.

The opposite case, that is when:

$$\mid m\omega_1(J_1, J_2) + n\omega_2(J_1, J_2) \mid \ll f_{mn}, \tag{11}$$

is called a resonance (m-n resonance).

When f_{mn} (the perturbation) is small, there is a small range of frequencies ω_1 and ω_2 (or a small range of initial conditions) for which the inequality (11) issatisfied. This defines a narrow band or layer in phase space. We will call this band, the *stochastic layer.As it turns out, the presence of stochastic layers seems to be a universal property of non-integrable Hamiltonians.*

To ilustrate this point with an example, we show in Figure 2(a), the phase space trajectories of the well known pendulum.The separatrix, separates oscillations from rotations.[12] It also contains points of unstable equilibrium, which are the intersections of the separatrix with the x-axis (these points are often called hyperbolic points). Now, in Figure 2(b), we perturb the pendulum with a non-linear, non-integrable perturbation. We immediately notice the appearence of a stochastic layer replacing the separatrix. If the pendulum has an initial condition within the stochastic layer, *it will be trapped in the layer, in its subsequent motion.*

In figure 2(b), we also observe curves above and below the stochastic layer, that look very similar to the unperturbed ones. These correspond to irrational ratios of the frequencies and their shapes are only slightly disturbed. We will refer to them as KAM curves or KAM surfaces.

There are, in every non-integrable Hamiltonian problem, KAM surfaces and stochastic layers between them. The KAM surfaces act as "barriers" that prevent the particles to escape from a given stochastic layer.

A popular view of the Hamiltonian chaos that goes beyond the validity of the KAM theory and it is mainly supported by a large body of numerical work, goes as follows: when the perturbation is small, the phase space contains many thin stochastic layers, corresponding to various (m,n) resonances and KAM surfaces which isolate these resonances from each other.

A given particle, trapped by its initial condition, in one of these layers, will stay in that layer and cannot cross a KAM surface to a different layer. This trapping is in the vertical direction of the phase space (p) and these particles can gain or loose a very limited amount of momentum.

To summarize, for systems with two degrees of freedom, KAM surfaces block the flow of trajectories over wide ranges of the action(momentum) variables.

However, as the size of the perturbation increases, the inequality (11) corresponds to a broader zone and, according to Chirikov[13], the destruction of the KAM surfaces and the ulterior onset of global chaos is due to the groth and eventual overlap of these resonances. Later on, Escande and Dorveil[14] pointed out that the resonance zones are self-similar, and developped a renormalization group scheme, in order to predict the critical values of the perturbation, for which the KAM surfaces get destroyed.

What do we mean by the destruction of a KAM surface?

It means, that a particle, originally trapped in a stochastic layer, can now *cross*

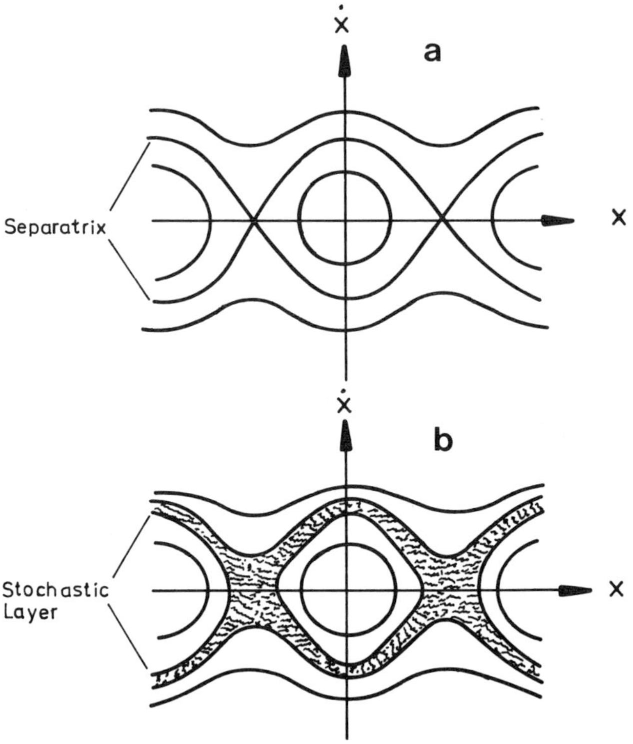

Figure 2. Phase portraits of unperturbed (a) and (b)perturbed pendulums. The separatrices divide the oscillatory from the rotational trjectories. As seen from the curve b, the perturbation destroys the separatrix and a stochastic layer is formed in it's vecinity. The region occupied by the stochastic layers has holes that a stochastic trajectory cannot enter. There is an infinite number of stability islands, within which, an infinite number of still thinner stochastic layers reside. After ref.12

the KAM surface to a different resonance. The KAM's, as seen in a Poincare surface section, appear to have a distribution of holes with a Cantor set structure.

There is a very interesting conjecture put forward by Greene[15]. He addressed the problem of the standard map of a plane onto itself, representing, for example, the Hamiltonian problem of two coupled oscillators. The main hypothesis, here, is that the dissapearence of a KAM surface is associated to the change in stability of nearby(in phase space) periodic orbits.

He explored extensively, by numerical methods, the relation between the KAM surfaces and the periodic orbits.

If we call $\alpha = \frac{\omega_1}{\omega_2}$ the rotation number and since the KAM's correspond to irrational rotation numbers, while the periodic orbits have rational α, the problem here reduces to how to approach an irrational number by a sequence of rationals.

A good way of representing irrationals is by continued fractions.
If α is an irrational, it has a unique continued fraction representation of the form:

$$\alpha = \cfrac{1}{a_1 + \cfrac{1}{a_2 + \cfrac{1}{a_3 + \cfrac{1}{a_4 + \cdots}}}} \tag{12}$$

or in shorthand notation:

$$\alpha = [a_1, a_2, a_3 ...], \tag{13}$$

where the a_n's are positive integers.

The succesive iterates of a continued fraction $\frac{r_n}{s_n}$, where a_n is the last term taken, approximates α well in the sense that no other $\frac{r}{s}$ is closer to α, when $s \leq s_n$.

Now, if we want to explore the destruction of the last KAM, we would expect it to correspond to an irrational furthest away from a rational, or the one that converges the slowest. That irrational has been known for a long time with the name of *golden mean*:

$$\alpha_I = [1, 1, 1, ...] = \frac{\sqrt{5} - 1}{2} \tag{14}$$

For the standard mapping, where the resonances have equal amplitudes, frequencies and phases, there is plenty of numerical evidence that the last KAM to be destroyed corresponds to the golden mean. Unfortunately, in other maps, where the resonances have different amplitudes and frequencies, the last KAM does not correspond to the golden mean. This will be the case of the example discussed in the last section of this article.[16]

3. A Model for Laser Damage.

Here we study the effects of the laser radiation on dielectrics with small metallic inclusions.[2] These could typically be dielectric mirrors with inclusions having diameters of 20 to 200 A.

This is schematically represented in Figure 3.

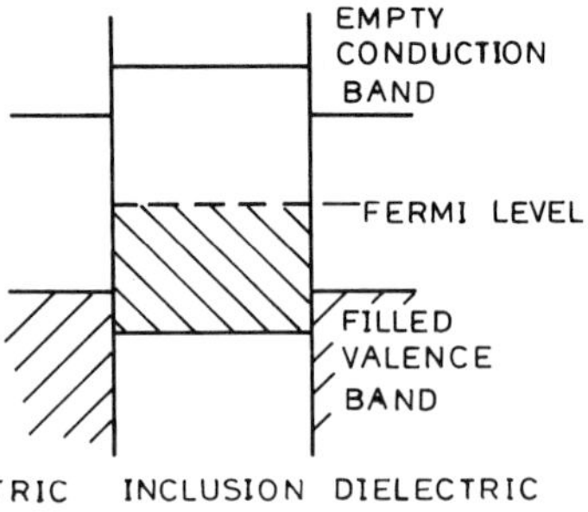

Figure 3. The partially filled conduction band of a metallic inclusion in a dielectric. The access for the electrons into the empty conduction band of the dielectric is from the conduction band of the impurity, provided they can gain the required energy from an externally applied field. After ref.2

We have a partially filled conduction band of the metallic inclusion, that could provide the access to the electrons into the empty conduction band of the dielectric.

In a first step of the damage process, the radiation field should manage to bring the electrons up from the conduction band of the impurity to the empty conduction band of the dielectric.

In a second stage, these electrons, assisted by phonons, should produce an avalanche process. We will study the first stage only, namely how to bridge the gap between the Fermi level of the impurity and the bottom of the conduction band in the dielectric. Typical energy gaps are of the order of 5 to 10 eV and our photons have a small fraction of an eV. (Figure 3).

The metallic inclusion can be, as a first model, represented by a one dimensional square well of diameter 2a, and given the large size of the inclusion, as compared to the laser wavelenght, we can consider, in the dipole approximation, that the laser field is only time varying, neglecting its spatial dependence.

The Hamiltonian of the system can be written as:[17]

$$H = p^2 + b[\eta(x-1) + \eta(-x-1)] + \epsilon x \cos(\omega t) \tag{15}$$

$(b \longrightarrow \infty)$
where:

$$\eta(x) = \begin{cases} 1, & \text{for } x > 0 \\ 0, & \text{for } x < 0. \end{cases} \tag{16}$$

is the step function, the mass was taken to be $\frac{1}{2}$ and $\epsilon x \cos(\omega t)$ is the interaction term in the dipole approximation, and, a was chosen to be 1.(the two walls are located at the positions ± 1).

In the free case, the position and momentum of the particle can be written as:

$$x(t) = -1 + 2t\sqrt{(E_0)}sign[\sin(\pi t\sqrt{(E_0)})],$$

$$p(t) = \sqrt{(E_0)}sign[\sin(\pi t\sqrt{(E_0)})], \tag{17}$$

where E_0 is the energy and sign[ ] is the sign of the argument.

Now, we perform the canonical transformation:

$$x = -1 + \frac{2\theta}{\pi}sign[\sin(\theta)],$$

$$p = \frac{\pi I}{2}sign[\sin(\theta)], \tag{18}$$

with:

$$I = \frac{2\sqrt{(E_0)}}{\pi},$$

$$\theta = \frac{\pi^2 It}{2}. \tag{19}$$

The Hamiltonian, in terms of the new variables, becomes:

$$H = \frac{\pi^2 I^2}{4} + \epsilon x(I,\theta)\cos(\omega t). \tag{20}$$

If we expand x in Fourier series (x is a periodic function of time), we readily get:

$$H = \frac{\pi^2 I^2}{4} - \frac{4\epsilon}{\pi^2} \sum_{\substack{n=-\infty \\ n\,odd}}^{\infty} \frac{1}{n^2}\cos(n\theta - \omega t). \tag{21}$$

It is apparent, from eq(21), that the perturbation produces an infinite number of resonance zones, characterized by:

$$n\frac{\partial\theta}{\partial t} = \omega = \frac{n\pi^2 I}{2} = n\pi\sqrt{(E_0)}, \tag{22}$$

or

$$p_0 = \frac{\omega}{n\pi}. \tag{23}$$

As per eq(23), the resonances for large n converge to $p_0 = 0$. In the Figure 4 we show a stroboscopic plot ($\omega = 5$) of the phase space, where a stochastic layer was formed near the origin for $\epsilon = 0.05$.

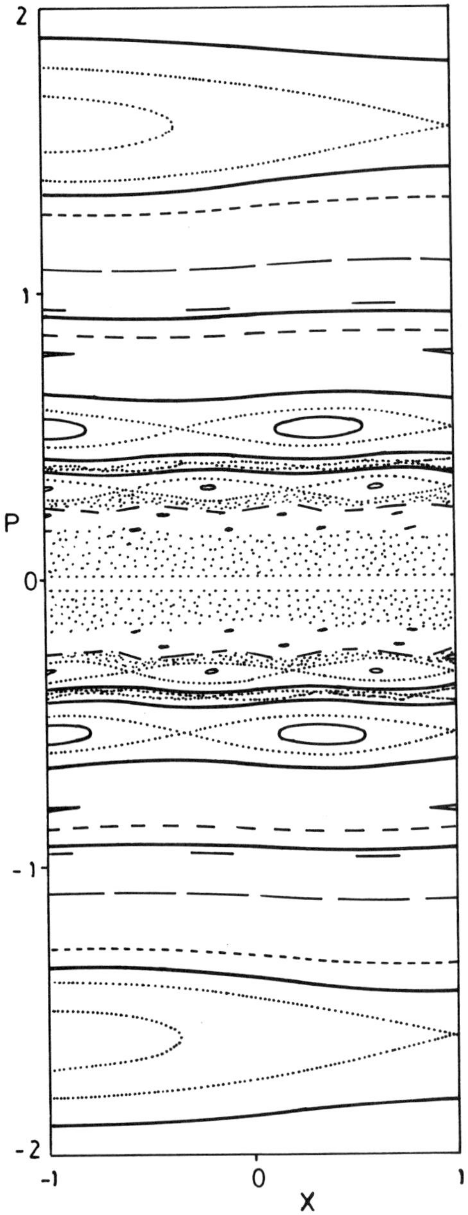

Figure 4. Phase space trajectories for the Hamiltonian H at $\omega = 5, \epsilon = 0.05$. Aft. Ref. 17

As the field is increased, the stochastic layer around the origin grows, due to the breakdown of the KAM's away from the origin, on the p-axis.

In the Figure 5, 2 curves are shown, corresponding to the n=3 resonance. The particle started from the right wall ($x_0 = 1$) with an initial momentum of $p_0 = -0.300$ The curves correspond to field amplitude values of $\epsilon = 0.035$ and $\epsilon = 0.041$. The breakdown of the KAM curve occured at some value in between.

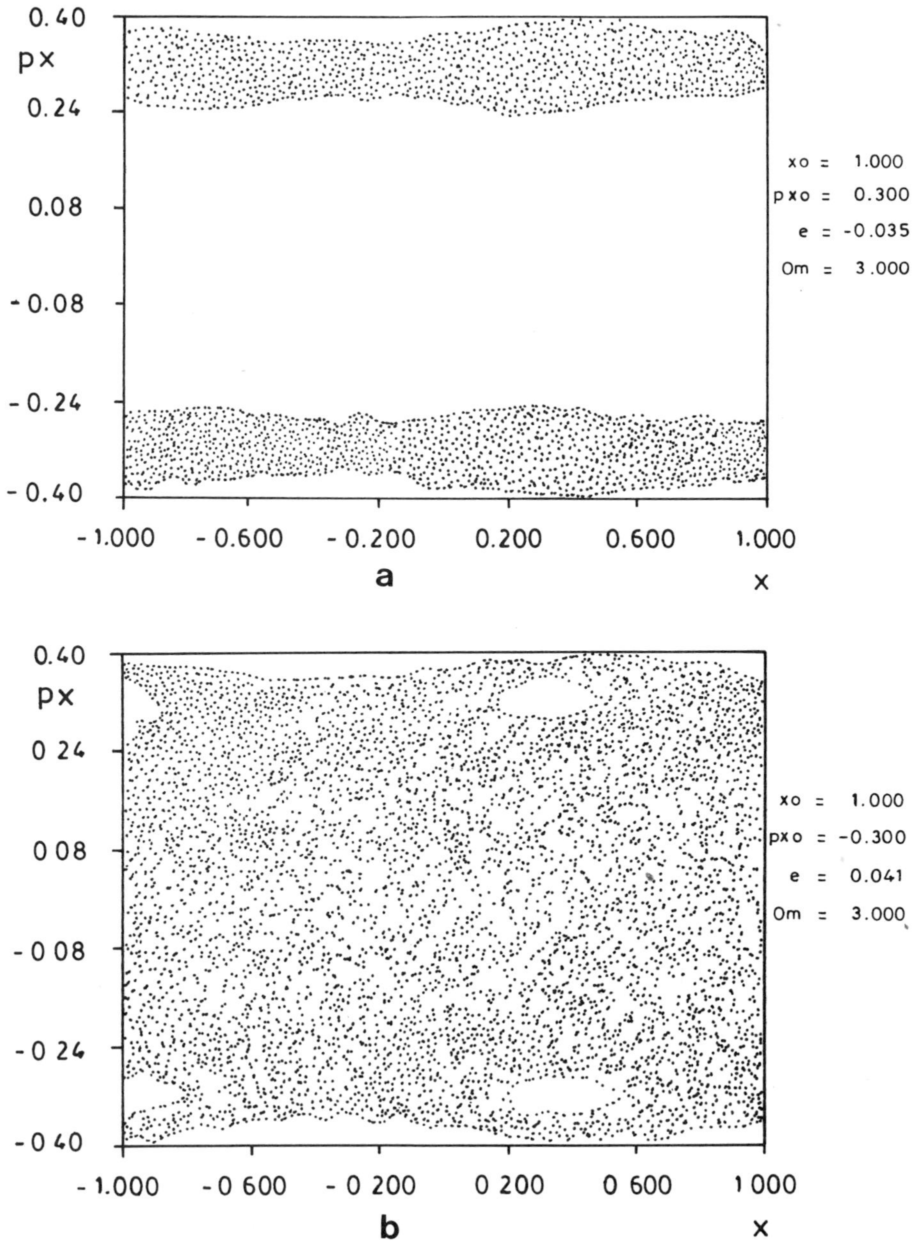

Figure 5. The n= 3 resonance and its stochastic layer. The field $\epsilon = 0.035$ is below the critical value (a) and $\epsilon = 0.041$ is above (b).

Now, we will look at this problem from a different perspective. Instead of trying to find phase space plots for the trajectories, we will construct a map for the hit times at the walls.[1]

The position of the particle at time t is:

$$x(t) = [q\frac{\epsilon}{m\omega^2}\cos(\omega t_0) - (\dot{x}(t_0) - \frac{q\epsilon}{m\omega}\sin(\omega t_0))t_0 + x(t_0)]$$
$$+t[\dot{x}(t_0) - \frac{q\epsilon}{m\omega}\sin(\omega t_0)] - \frac{q\epsilon}{m\omega^2}\cos(\omega t), \qquad (24)$$

We enphasize here, that the initial conditions will have to be imposed, every time the particle hits the wall, with the corresponding momentum reflection.

In a shorthand notation, lets write:

$$x_s(t) = a_s + b_s t + K\cos(\omega t), \qquad t_s \leq t \leq t_{s+1}, \qquad (25)$$

where:

$$K = \frac{q\epsilon}{m\omega^2},$$

t_s being the the hit time number s in one of the walls.
Now if $x_s = x_s(t_s)$, then:

$$x_s = (-1)^s = a_s + b_s + K\cos(\omega t_s). \qquad (26)$$

On the other hand:

$$p_s = \frac{x'_s}{2} = \frac{b_s}{2} - \frac{K\omega}{2}\sin(\omega t_s), \qquad m = 0.5$$

and one can write:

$$p_s = \frac{b_s}{2} - \frac{K\omega}{2}\sin(\omega t_s) = -\frac{b_{s-1}}{2} + \frac{K\omega}{2}\sin(\omega t_s),$$

or

$$b_{s-1} + b_s = 2K\omega\sin(\omega t_s). \qquad (27)$$

In the last equation, we have used the fact that the momentum is reversed when the particle hits the wall.

The solution to eq.(27) is:

$$b_s = (-1)^s b_0 + 2K\omega\sum_{r=1}^{s}(-1)^{s+r}\sin(\omega t_r). \qquad (28)$$

Now we can wrire also the position of the particle slightly before and after hitting a wall:

$$(-1)^s = a_{s-1} + b_{s-1}t_s + K\cos(\omega t_s) = a_s + b_s + K\cos(\omega t_s). \qquad (29)$$

From equations (28) and (29), we readily get:

$$a_s = (-1)^s[a_0 + 2s] - \sum_{r=1}^{s}(-1)^{r+s}(2K\omega t_r\sin(\omega t_r) + 2K\cos(\omega t_r)). \qquad (30)$$

Finally, if we replace eq's (28) and (30) into (29), we get the map for the hit times:

$$\begin{aligned}1 = a_0 + 2s + \sum_{r=1}^{s-1}(-1)^r[-2K\omega t_r \sin(\omega t_r) - 2K\cos(\omega t_r)] \\ +t_s[b_0 + 2K\omega \sum_{r=1}^{s-1}(-1)^r \sin(\omega t_r)] - K(-1)^s \cos(\omega t_s).\end{aligned} \tag{31}$$

The equation (31) is a recursion relation involving all the hit times up to t_s. It is possible to iterate numerically the eq.(31), with an initial condition p_0 in one of the resonance zones.

In the figures 6 (a) and 6 (b) we plot $a = \frac{t_s}{s}$ versus s, with the initial p_0 within the n=3 zone. For values of the field amplitude ϵ less than $\epsilon = 0.03528$, $\frac{t_s}{s}$ behaves as shown in Fig.6(a), that is, assymptotically goes to a constant. In other words, t_s behaves like:

$$t_s = as + \text{oscillating term}, \tag{32}$$

However, for a critical field (in this case $\epsilon = 0.03529$), the ansatz (32) changes abruptly.

The precise value of p_0 in the previous example was obtaind by exploring various initial momentums, as to obtain the maximum ϵ critical. This behavior is in agreement with Greene's idea that the instability of a periodic solution with a very log period, within a given zone corresponds to the breakdown of the relevant KAM surface.

The results presented here (Fig.6), are in excelent agreement with the critical fields obtained from a renormalization group theory [14].

Furthermore, it has the advantage of finding, in an economical way the critical fields. Also, conceptually, it gives a more transparent picture of the problem of Hamiltonian chaos in the sense that the "breakdown of the KAM's ", a concept hard to digest, is replaced the instability of very long periodic orbits.

Further analytical and numerical work is under way, at the present time, to investigate the precise connection between the present results and the dissapearence of the KAM surfaces in phase space.

ACKNOWLEDGEMENTS

The author would like to thank M.O.Scully and A.O.Barut for the invitation and warm reception in Istanbul. This work was partially funded by the project FONDECYT, Project 0363/88.

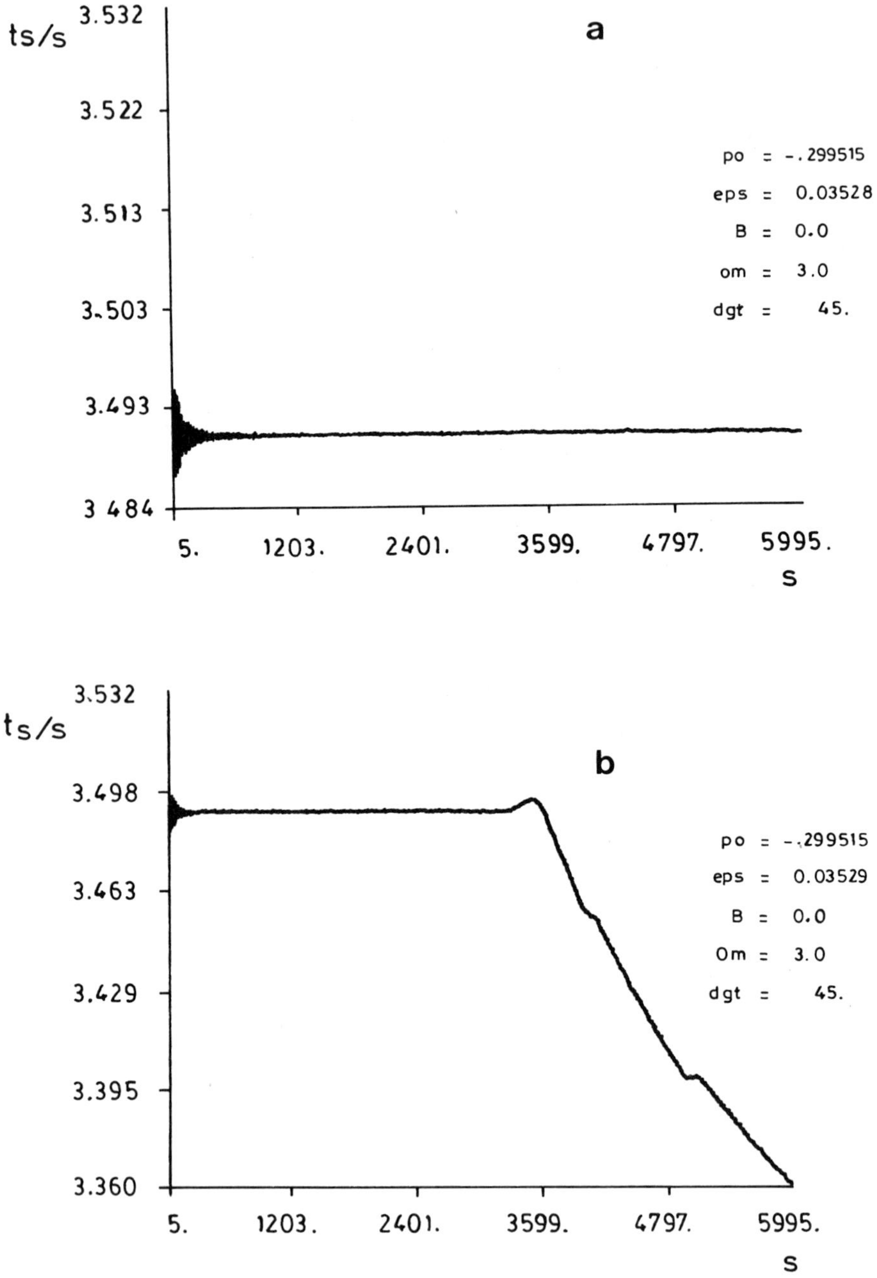

Figure 6. $\frac{t_s}{s}$ versus s with $\epsilon = 0.03528$(a) and $\epsilon = 0.03529$ (b).The breakdown of the ansatz corresponds to the instability of a periodic cycle.

REFERENCES

1.- M.Orszag, M.Fuka, W.Becker, R.Ramirez, J.McIver, J.Alfaro (in preparation).

2.- W. Becker and J.K. McIver, Phys. Lett. **121**, 286(1987).

3.- A detailed discussion and various examples are found in : W.H. Walker and J.Ford, Phys.Rev. **188**, 416 (1969).

4.- H.Poincare, *Methodes Nouvelles de la Mechanique Celeste*, (Dover. Publ. Inc. NY 1957).

5.- E.Fermi Z.Physik,**24**,261 (1923).

6.- A.N.Kolmogorov, Dokl.Akad.Nauk, SSSR,**98**, 527(1954).

7.- V.I.Arnold, Russian.Math.Surveys, **18**, 9(1963).

8.- J.Moser, Nachr.Akad.Wiss, 2, Math.Physik **K1**,1 (1962).

9.- V.I.Arnold, *Mathematical Methods of Classical Mechanics*, (Springer, Heildelberg, NY. 1978).

10.- M.V.Berry in S.Jorna (ed) *Topics on Nonlinear Dynamics*, Am. Inst.Phys.Proc. **46**, (1978).

11.- R.H.G.Helleman in E.G.D.Cohen(ed) *Fundamental Problems in Statistical Mechanics*,**5** (North.Holland, Amsterdam, 1980).

12.- A.A.Chernikov, R.Z.Sagdeev, G.M.Zaslavsky, Physics.Today, **41**, 27 (1988).

13.- B.V.Chirikov, Phys.Rep.,**52**, 263(1979).

14.- F.Escande and F.Dorveil, Journ.Stat.Phys, **26**, 257(1981).

15.- J.M.Greene, J.Math.Phys, **20**,1183(1979) A previous work by the same author is: J.M. Greene, J.Math.Phys, **9**, 760 (1968).

16.- A general discussion on the convergence to the golden mean is found in A.J. Lichtenberg and M.A. Lieberman *Regular and Stochastic Motion*, (Springer,NY, 1083).

17.- W.A.Lin and L.E.Reichl, Physica D, **19**,145 (1986).

Tests of Quantum Electrodynamics and Related Symmetry Principles Using Positronium

A. Rich, R. S. Conti, D. W. Gidley, J.S. Nico
M. Skalsey, J. Van House and P. W. Zitzewitz*

Randall Laboratory of Physics
The University of Michigan, Ann Arbor
MI 48109 USA
*Department of Natural Sciences
The University of Michigan-Dearborn
Dearborn, MI 48128 USA

1. INTRODUCTION

Positronium (Ps) is a particle-anti-particle bound system in which the only interaction present (at currently verifiable levels of calculational and experimental precision) is the electromagnetic interaction. It possesses self-annihilation channels not present in any other atomic system. These features make it ideal for the study of the relativistic 2-body problem in QED as embodied in the Bethe-Salpeter equation and in variants of this equation. In addition, because Ps is a particle-anti-particle system, it is an eigenstate of the charge conjugation operator C as well as CP (Figure 1). Thus, its decay into various numbers of gamma rays which are also C eigenstates, as well as radiative transitions between its different states, can be used to set limits on the validity of conservation of the discrete symmetries C, CP, and CPT. One may also search for proposed axion-like particles (X) in the decay of Ps, since if its mass is less than $2m_ec^2$, X may be observable[1] through the decay $1^3S_1 \rightarrow X + \gamma$.

In this article we will review recent research related to the above general features of Ps, however we will not discuss in detail all the interesting research of a fundamental nature being done using positrons and Ps. We mention here several such topics and the latest references from which the previous literature can be obtained: antihydrogen formation[2,3], ultrafast Ps[4,5], Ps in astrophysics[6], search for the electron helicity density in optically active materials[7], precision

New Frontiers in Quantum Electrodynamics and Quantum Optics
Edited by A. O. Barut, Plenum Press, New York, 1990

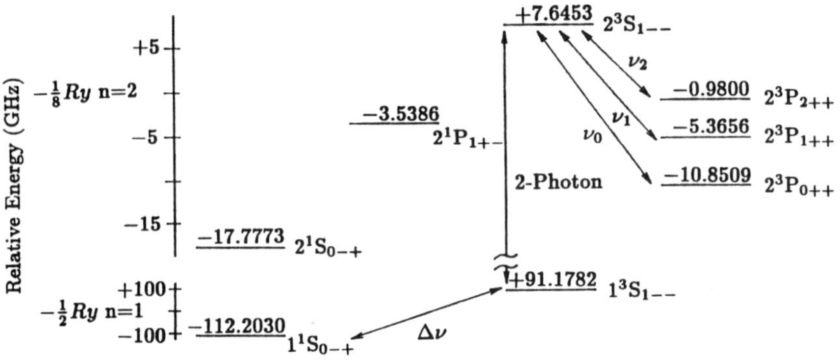

Fig. 1. Energy levels of $n^{2S+1}\mathrm{L}_{\mathrm{J}PC}$ Ps for $n = 1$ and 2.

positron polarimetry[8], Ps diffraction[9], positron imaging[10,11,12], and the search for resonances in Bhabha scattering that are predicted by models which seek to explain the observed emission of electron-positron pairs in heavy-ion collisions[13] .

2. TESTS OF QED USING POSITRONIUM

2.1 Positronium Spectroscopy

2.1.1 Theory At the present time, precision spectroscopic measurements have been made on three different sets of Ps energy levels, the ground-state fine structure transition[14], the $n = 1$ to $n = 2$ transition[15], and the three fine structure transitions[16] in $n = 2$. A goal of all experiments is to test the QED corrections to the Ps system. To help determine the ability of each measurement to test QED, we first explore the QED contributions to each energy level.

Theoretical energy levels of the $n = 1$ and 2 states of Ps are given by

$$E(n^{2S+1}\mathrm{L}_J) \equiv E_q = \frac{Ry}{2}(\frac{-1}{n^2} + A_q\alpha^2 + B_q\alpha^3 + C_q\alpha^4 + \ldots) \qquad (1)$$

and displayed in Figure 1. Here the index q ($q = 1 \ldots 8$) refers to one of the 8 states (not including magnetic substates) in the $n = 1$ and 2 manifold as shown in Table 1 and terms involving $ln\ \alpha$ are absorbed into the A_q, B_q, and C_q coefficients. The first term, the non-relativistic coulombic ("gross") energy, is reduced from that of hydrogen by a factor of two that represents the reduced mass correction. The $A_q\alpha^2$ terms[17] come from relativistic corrections, spin-spin interactions, and virtual annihilation. The $B_q\alpha^3$ contributions[18,19] include recoil and retardation terms, radiative corrections, and Ps-specific effects such as virtual annihilation. Some contributions to the $C_q\alpha^4$ terms have been calculated[20,21,22] but no complete set for any energy level is yet available. Table 1 summarizes the results of existing calculations done using $Ry/2 = 1,644,920,981.0(0.5)$ MHz[23] and $\alpha^{-1} = 137.0359895(61)$[24].

Table 1. Theoretical Energy Levels of the $n = 1$ and 2 Ps states. For each state, the upper line contains the coefficients of the terms, while the lower line is the energy of the corresponding term in MHz.

q (State)	Coulomb term	A_q	B_q	(Total Energy)
1 (1^1S_0)	-1	-21/16	4.324921	
	-1,644,920,981.0	-114,967.5	2,764.5	$(-1,645,033,184.0 + 4.7C_1)$
2 (1^3S_1)	-1	49/48	2.751882	
	-1,644,920,981.0	89,419.2	1,759.0	$(-1,644,829,802.8 + 4.7C_2)$
3 (2^1S_0)	-1/4	-53/256	0.559204	
	-411,230,244.5	-18,134.8	357.4	$(-411,248,021.9 + 4.7C_3)$
4 (2^3S_1)	-1/4	65/768	0.362449	
	-411,230,244.5	7,413.6	231.7	$(-411,222,599.2 + 4.7C_4)$
5 (2^1P_1)	-1/4	-31/768	-0.004549	
	-411,230,244.5	-3,535.7	-2.9	$(-411,233,783.1 + 4.7C_5)$
6 (2^3P_0)	-1/4	-95/768	-0.024444	
	-411,230,244.5	-10,835.2	-15.6	$(-411,241,095.3 + 4.7C_6)$
7 (2^3P_1)	-1/4	-47/768	-0.007865	
	-411,230,244.5	-5,360.6	-5.0	$(-411,235,610.1 + 4.7C_7)$
8 (2^3P_2)	-1/4	-43/3840	0.001420	
	-411,230,244.5	-980.9	0.9	$(-411,231,224.5 + 4.7C_8)$

The theoretical and experimental values of the measured transitions are given in Table 2. Each experimental result listed has a unique sensitivity to the various effects. In Ps, where the constituent masses are equal, the recoil contributions are as large as any other order $\alpha^4 Ry$ correction, so these terms are all equally subject to measurement. Also annihilation terms, which are not present in hydrogen, are important in Ps. Among the Ps tests, the $1^3S_1 \rightarrow 2^3S_1$ measurement provides a determination of the Ps Rydberg and the equality of e^+ and e^- masses, by means of comparisons between Ps and H spectroscopy[15], while the $2^3S_1 \rightarrow 2^3P_J$ measurements are the only ones to probe P state terms. Of all present experiments, the $n = 1$ fine structure measurement places the tightest constraints on the uncalculated terms.

Table 2. Theoretical and Experimental Values of the Measured Transitions in Positronium.

Transition	ν_{theory}(MHz)	ν_{expt}(MHz)
$1^3S_1 \rightarrow 1^1S_0$	$203{,}381.2 + 4.7(C_2 - C_1) + \ldots$	203,389.1(0.7)
$1^3S_1 \rightarrow 2^3S_1$	$1{,}233{,}607{,}203.6 + 4.7(C_4 - C_2) + \ldots$	1,233,607,218.9(10.7)
$2^3S_1 \rightarrow 2^3P_2$	$8{,}625.3 + 4.7(C_4 - C_8) + \ldots$	8,619.6(2.7)
$2^3S_1 \rightarrow 2^3P_1$	$13{,}010.9 + 4.7(C_4 - C_7) + \ldots$	13,001.3(3.9)
$2^3S_1 \rightarrow 2^3P_0$	$18{,}496.1 + 4.7(C_4 - C_6) + \ldots$	18,504.1(10.0)

Assuming that the uncalculated order $\alpha^4 Ry$ terms account for any differences between the experimental and the theoretical values, the following constraints are placed on the C_q coefficients: $C_2 - C_1 = +1.7 \pm 0.16$, $C_4 - C_2 = +3.3 \pm 2.3$, $C_4 - C_8 = -1.2 \pm 0.6$, $C_4 - C_7 = -2.0 \pm 0.8$, and $C_4 - C_6 = +1.7 \pm 2.1$.

Based on the B_q terms, we expect the coefficients for the P states, C_5 through C_8, to be small compared to unity. Using this assumption, we perform a weighted average of the three $n = 2$ fine structure experimental results to obtain $C_4 = -1.34 \pm 0.46$. If the next round of experiments for the 2^3S_1 to 2^3P_J transitions all reach the subMHz level, as we now anticipate (Section 2.1.4), and the differences $C_4 - C_q$ for $q = 6, 7, 8$ are all the same within the experimental uncertainties, then, as above, those differences could be interpreted as a measurement of C_4 to ± 0.1. A new two-photon measurement (see below) would then fix C_2 and, in turn, the existing $n = 1$ fine structure measurements would fix C_1. On the other hand, any differences in the $C_4 - C_q$ measurements could be interpreted as the effect of significant C_6, C_7, and C_8, P state terms.

Although the current frequency interval measurements in Ps are not as precise as those available for hydrogen[25,26,27], Ps tests are uniquely sensitive to recoil and annihilation effects and insensitive to the nuclear size effects which limit the sensitivity of hydrogen experiments to QED effects. Ultimately, given the very intense slow positron beams that we feel can be attained, Ps $n = 2$ fine structure measurements could reach the 10 kHz level of precision obtained in hydrogen[25], thus permitting an unambiguous measurement of order α^5 Ry contributions.

2.1.2 The 1^3S_1 -2^3S_1 separation The most recent[15,28] $1^3S_1 \rightarrow 2^3S_1$ transition measurement used a two-photon excitation technique and is precise to 9 ppb. In this experiment, a 10 ns burst of 50-100 positrons forms ground-state thermal Ps on the surface of an Al(111) target. Two photons from counterpropagating 486-nm laser pulses excite the Ps to the 2^3S_1 state. A third photon from the same laser ionizes the Ps atom, producing a detectable positron. The Ps resonance frequency is measured with respect to a Te_2 absorption line.

The 1 MHz natural linewidth of the transition is broadened to approximately 40 MHz, primarily due to the ≈34 MHz bandwidth resulting from short-term laser pointing instabilities, although statistics are sufficient to determine the line center of a single sweep to ≈3 MHz. Major systematic effects in determining the Ps line in relation to the Te_2 reference are the second-order Doppler shift due to Ps motion (determined to ± 3-4 MHz) and the ac Stark shift (determined to ± 5-7 MHz) resulting from the high-intensity laser. In the original measurement[28] the Te_2 reference line was calibrated in a separate measurement relative to a deuterium (D_β) line with a 10.6 MHz uncertainty. A subsequent recalibration of the Te_2 line[28,15] altered the reference frequency by 39.8 MHz. In addition, it was discovered that the output frequency of the pulsed amplifier was shifted relative to the injected CW fundamental by ~ 30 MHz, resulting in an ~ 60 MHz shift in the 2 photon transition frequency. After these corrections were made, the result obtained for the transition frequency is[16] 1,233,607,218.9(10.7) MHz. This result places a constraint on the combinations of terms of the coefficients of the uncalculated $\alpha^4 Ry$ terms, $C_4 - C_2 = +3.3 \pm 2.3$. Improvements of one order of magnitude in precision as a first step, and two orders of magnitude in a final stage apparatus are anticipated[15].

2.1.3 The 1^3S_1 - 1^1S_0 separation The most recent measurements of the triplet-singlet separation frequency, $\Delta\nu$, completed in 1983, are:

Group	$\Delta\nu$(GHz)	Uncertainty (ppm)
Brandeis[29]	203.3875(16)	8
Yale[14]	203.38910(74)	3.6

Because the radiation needed to drive the direct transition is in the millimeter range where appropriate rf sources cannot be obtained, both experiments used essentially the same technique based on the mixing of the singlet and triplet $m = 0$ Ps substates by a magnetic field. A microwave Zeeman transition at frequency f_0 is driven between the triplet $m = \pm 1$ and $m = 0$ substates, and the triplet-singlet separation is calculated from the Breit-Rabi formula.

The Ps is formed with 30% efficiency when positrons from a ^{22}Na source slow down in gas (N_2 or SF_6), confined in a microwave cavity in the static magnetic field. An rf magnetic field (10 G amplitude) at constant frequency is applied perpendicular to B, and the static field is swept through the resonance. When the Zeeman frequency, f_0, equals the rf frequency, transitions are driven from the $m = \pm 1$ to the $m = 0$ substate. As the result of the magnetic field mixing, the $m = 0$ state decays predominantly by means of 2-γ events, resulting in a 10% resonant increase in the two-gamma coincidence rate as detected by pairs of back-to-back scintillators and photomultiplier tubes.

Precision determination of the line center within the large natural line width (6200 ppm) necessitates exact theoretical knowledge and experimental control of the line shape as well as a large amount of data (in the Yale experiment 300 resonance line sweeps were made, each comprising three million counts). The most recent correction to the line shape theory was to include the effects of annihilation. Inclusion of this correction shifted $\Delta\nu$ by an amount at least as large as the uncertainties in the measurements. A second systematic effect results from the formation of Ps in a gas. Frequency shifts due to Ps-gas collisions must be removed by extrapolation to zero gas density. Finally, the magnetic field inhomogeneities over the region must be reduced to sub part-per-million levels. The uncertainties in the Yale measurement can be grouped into the following categories: statistical (2.8 ppm), magnetic field-related (1.9 ppm), microwave related (0.5 ppm), gas related (0.3 ppm), and line-shape related (0.7 ppm) for a total of 3.6 ppm. To our knowledge no new experiment is in progress at this time.

2.1.4 <u>The $n = 2$ fine structure</u> The first measurement of a frequency interval in $n = 2$ Ps, done at Brandeis[30] for the 2^3S_1 - 2^3P_2 transition, yielded the result $\nu_2 = [8631 \pm 7]$ MHz. Subsequently, a series of experiments at Michigan[16] resulted in an improved measurement of ν_2 and the first measurements of the 2^3S_1 - 2^3P_1 and 2^3S_1 - 2^3P_0 transitions ν_1 and ν_0. Both experiments used the same technique to determine the transition frequencies. Ps in the 2^3S_1 state is irradiated with microwaves whose frequency can be varied and a resonant increase in 2^3P_J states is observed as the result of a stimulated emission. The signature for the 2^3P_J state is its emission of a Lyman-α photon [$\tau_p = 3.2$ ns, $\lambda(\mathrm{L}\alpha) = 243$ nm] leaving the Ps in the 1^3S_1 state which subsequently annihilates to three gammas with a lifetime $\lambda_T^{-1} = 142$ ns.

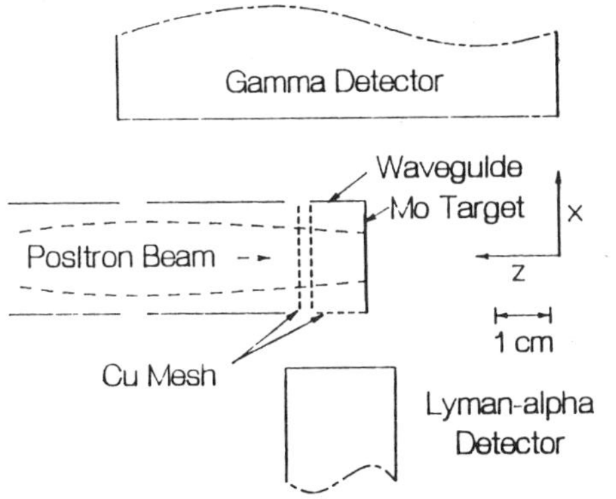

Fig. 2 Experimental Apparatus. (See text for a detailed description.)

A schematic representation of the Michigan apparatus[16] is shown in Fig. 2. An electrostatically focused beam (10^5 e^+/s at 65 eV) enters a section of waveguide (2.3cm × 1.0cm) and strikes a polycrystalline molybdenum target attached to the opposite inner wall of the waveguide. A fraction (3×10^{-4}) of the incident positrons is emitted from the target as $n = 2$ Ps. Assuming equal distribution in all of the $n = 2$ magnetic substates, 3/16 of them will be in the 2^3S_1 state, and 9/16 will be in the 2^3P_J ($J = 0, 1, 2$) states. Lyman-α photons are detected in a solar-blind photomultiplier with an overall 1% detection efficiency. One or more of the gamma rays from the annihilation of the 1^3S_1 states are detected in two Pilot-B plastic scintillators with a combined detection efficiency of 0.15.

In the absence of rf in the waveguide, only those 2^3P_J states originally formed contribute to the signal rate $R(0,\nu)$ of a Lyman-α photon with a delayed γ ray. As the rf frequency is scanned the signal rate $R(I,\nu)$ will increase resonantly at the transition frequencies ν_J. The theoretical expression[16] for the ratio $r \equiv [R(I,\nu) - R(0,\nu)]/R(0,\nu)$ is fitted to data taken in the vicinity of all three transition frequencies and at a variety of rf intensities I. The results are displayed in Table 3. In addition to the results for the ν_J, this fit also provides the first measurement of the radiative decay rate (γ) for 2^3P_J Ps. This result is in agreement with the expected rate, which is, to lowest order, half that of the corresponding decay rate in hydrogen.

Table 3. Results of fit to r. T is the 2^3S_1 transit time across the waveguide and $\gamma/4\pi$ is the natural half-width of the transition.

Parameter	Value	σ_{stat}	σ_{syst}	Theory
ν_o	18504.1 MHz	10.0 MHz	1.7 MHz	18496.1 MHz
ν_1	13001.3 MHz	3.9 MHz	0.9 MHz	13010.9 MHz
ν_2	8619.6 MHz	2.7 MHz	0.9 MHz	8625.3 MHz
T	17.3 ns	1.2 ns	4.0 ns	—
$\gamma/(4\pi)$	22.6 MHz	2.7 MHz	5.0 MHz	24.9 MHz

At Michigan a new experiment to measure ν_J to ten times higher precision is now underway. The transitions will be detected efficiently ($> 50\%$) by photoionization and the resonance will be observed as a decrease in the 2^3S_1 states which exit the waveguide region rather than an increase in 2^3P_J states (detection efficiency $\sim 10^{-3}$ for both Lyman α and delayed γ detection). The theoretical implications of both the recent fine-structure results and of a possible improved experiment are discussed in Section 2.1.1.

2.2 Decay Rates in Positronium

2.2.1 Measurement of the 1^3S_1 decay rate of Ps In recent publications[31,32] a new 200 ppm measurement of the vacuum decay rate, λ_T of triplet Ps formed in a gas was presented. The result (ref. 32), $\lambda_T = 7.0514 \pm 0.0014 \mu s^{-1}$, represents a factor of four improvement over previous measurements and is in substantial agreement with existing experimental results, the most recent of which are (see bibliography in reference 32) $7.056 \pm 0.007\mu s^{-1}$, $7.045 \pm 0.006\mu s^{-1}$, $7.051 \pm 0.005\mu s^{-1}$, and $7.050 \pm 0.013\mu s^{-1}$. These latter values are all 1-2.5 standard deviations above the present theoretical value and the new measurement exceeds theory by 9.4 experimental standard deviations.

The theoretical value of λ_T may be expressed as the sum of decay rates into three gammas (λ_3), five gammas (λ_5), *etc.*: $\lambda_T = \lambda_3 + \lambda_5 + \ldots$. The contribution of λ_5 has been calculated[33,34] to be $\lambda_5 \sim 10^{-6}\lambda_3$, and is thus negligible. The leading term is:

$$\lambda_3 = \frac{\alpha^6 mc^2}{\hbar}\frac{2(\pi^2-9)}{9\pi}\left[1 + A_3(\frac{\alpha}{\pi}) - \frac{1}{3}\alpha^2 \ln \alpha^{-1} + B_3(\frac{\alpha}{\pi})^2 + \ldots\right]. \qquad (2)$$

The two most recent calculations give $A_3 = -10.266 \pm 0.011$[20] and $A_3 = -10.282 \pm 0.003$[35]. The coefficient B_3 is still uncalculated, and one obtains through order $\alpha^2 \ln \alpha$, $\lambda_3 = 7.03830 \pm 0.00005\mu s^{-1}$ ($1 \times (\alpha/\pi)^2$ adds only 0.00005 μs^{-1} to λ_3). If one assumes that the disagreement between the theoretical and experimental values of λ_T is due to the $(\alpha/\pi)^2$ term, then $B_3 = 340 \pm 33$ is required to bring theory and experiment into agreement. We note here that it may be more appropriate[36] to write the second order term as a coefficient times α^2 rather than $(\alpha/\pi)^2$, so that (the still rather large) coefficient of 34 would explain the difference. Exotic, non-QED decay modes of o-Ps have been considered[37,38] to account for the discrepancy but recent axion searches place severe restrictions on such decays[39].

In the experimental technique used in our new measurement[32] Ps is formed in a gas in a magnetic field of 6.8 kG (Fig. 3). The field confines positrons to the axis of the chamber and reduces the $1^3S_1(m = 0)$ lifetime to 13 ns. The $1^3S_1(m = \pm 1)$ states are unperturbed and continue to decay in the field with a rate, λ, which depends on the particular gas and its density. The decay rate is determined by fitting of the annihilation lifetime spectrum and λ_T is then determined by extrapolating λ to zero gas density. The results of this extrapolation are: isobutane 7.0524 ± 0.0013 μs^{-1}, neopentane 7.0551 ± 0.0026 μs^{-1}, nitrogen 7.0487 ± 0.0018 μs^{-1}, and neon 7.0501 ± 0.0023 μs^{-1} with the weighted average value being 7.0514 ± 0.0013 μs^{-1}. The uncertainty is obtained from the isobutane result with the other gases treated as systematic tests.

The measurement of λ_T has included extensive systematic tests[32,40]. We have recently performed an extrapolation of λ *vs* gas density using only high density (pressure greater than one atmosphere) N_2 and Ne gases. Using only

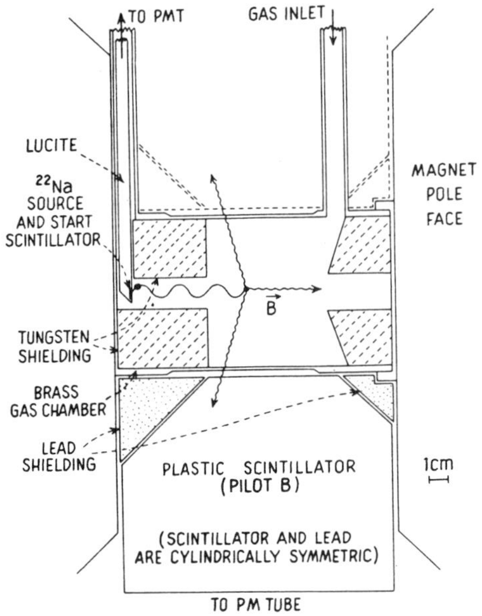

Fig. 3. The gas-filled Ps formation chamber and detector arrangement.

our data with P> 1 atmosphere and published[41] high pressure data for N_2 (7-36 atmospheres) and Ne (7-39 atmospheres) we extrapolate to zero density to find $\lambda_T = 7.0491 \pm 0.0021\mu s^{-1}$ and $7.0483 \pm 0.0029\mu s^{-1}$ respectively. The close agreement of these values with the values measured entirely below two atmospheres is a strong systematic check on any gas related effects.

All of the systematic tests[32,40] to date support the reference 31 results. For example, using the apparatus for n=2 fine structure measurements collisional quenching of the n=2 state was shown[40] to be sufficient to preclude it from appreciably affecting λ_T. The 1900 ppm difference with theory remains unresolved at this time. A new and systematically very different experiment[42] designed to obtain comparable precision is currently in progress and should definitively check the present 9.4 sigma difference between the theoretical calculations and the measurement of λ_t in gases.

This new experiment utilizes a slow positron beam to produce positronium in an MgO-lined evacuated cavity similar to the technique of Gidley et al[43]. Two of the more significant improvements are the remoderator and time tagging[44] of the beam. These produce improved focussing and eliminate the need for a start detector inside the Ps formation cavity. An important systematic effect regarding the possibility of o-Ps decaying in the MgO powder grains, rather than in the vacuum, has been shown to be negligible at the 100 ppm level.

The observed decay rate, λ_0, is assumed to have the form

$$\lambda_0 = \lambda_T + c\frac{\bar{v}}{4}\frac{A}{V} + P_a\frac{\bar{v}}{4}\frac{S}{V} \tag{3}$$

where V and S are the cavity volume and surface area respectively, A is the area of the e^+ entrance aperture, $\bar{v}$ is the average velocity of the o-Ps atoms, c is the probability that escaped o-Ps will not be detected, and P_a is the probability of annilihation during a wall encounter. Hence, the vacuum decay rate, λ_T, is obtained as an extrapolation in A/V, which accounts for the disappearance of o-Ps through the aperture, and S/V, which accounts for any collisional quenching that may occur at the surface of the cavity. Data is taken using several different cavity sizes to perform this extrapolation. At this time, preliminary results are in agreement with current experiments and are not in agreement with the theoretical prediction. A publishable result is anticipated in early 1990.

2.2.2 Measurement of the 1^1S_0 Ps Decay Rate There is only one precision measurement of the singlet ground state (parapositronium) decay rate[45] with sufficient accuracy to test the first order radiative connections to λ_S. The singlet decay rate may be expressed as $\lambda_S = \lambda_2 + \lambda_4 + \lambda_6 \ldots$. Since $\lambda_4 \sim 1.5 \times 10^{-6}\lambda_2$ is small[46,34], we need only concentrate on λ_2 in the present discussion. The expression[20] for λ_2 is:

$$\lambda_2 = \frac{mc^2}{2\hbar}\alpha^5\left[1 + A_2(\frac{\alpha}{\pi}) + \frac{2}{3}\alpha^2 \ln\alpha^{-1} + B_2(\frac{\alpha}{\pi})^2 \ldots\right] \tag{4}$$

where $A_2 = -(5 - \pi^2/4) = -2.532$ and B_2 is, as yet, uncalculated. Through order $\alpha^2 \ln\alpha^{-1}$ the decay rate is $\lambda_2 = 7.9866\ \mathrm{ns}^{-1}$ ($1 \times (\alpha/\pi)^2$ would add $0.0004\ \mathrm{ns}^{-1}$).

In the experiment, Ps is formed in isobutane gas in a uniform magnetic field of about 4 kG (the experimental arrangement is almost identical to that shown for λ_T in Fig. 3). The magnetic field mixes the $m = 0$ triplet and singlet states and thus the annihilation lifetime spectrum has 2 exponential components: the unperturbed decay from the $m = \pm 1$ states and the "quenched" decay from the $m = 0$ state (at 4 kG the lifetime is about 30 ns). Measurement of these decay rates, λ_T and λ_T', allows one, after extrapolation to zero gas density, to solve for

λ_S. The result is $\lambda_S = 7.994 \pm 0.011$ ns^{-1}, in agreement with the λ_2 calculation at the 1400 ppm level.

Considering the 1900 ppm difference between the measured value of λ_T and λ_3, it would be interesting to measure λ_S at comparable precision ($\sim$ 200 ppm). The current $\lambda_S - \lambda_2$ difference is (1100 ± 1400) ppm and cannot distinguish such an effect. A measurement at the 200 ppm level would be of immediate interest since the computationally simpler B_2 will probably be calculated before B_3. We will shortly begin construction of a λ_S experiment that is designed to reach the 200 ppm level of precision.

2.2.3 Measurement of the Ps^- Decay Rate The Ps negative ion, consisting of two electrons and one positron, is a relatively simple system for testing many-electron calculational schemes[47]. There is recently additional interest[48] in measuring the ground state decay rate of Ps^-, Γ, to see if the 1900 ppm discrepancy surrounding λ_T enters into Γ by way of λ_S. Approximately 98% of Γ is given simply by the spin average, $\Gamma \approx 0.25\lambda_S + 0.75\lambda_T$. Thus, if there is a sizable discrepancy in λ_S (see previous discussion) it would show up at the same level in Γ.

The best theoretical value[49,50] of Γ is 2.0861 ns^{-1}. This two photon, Hylleraas-type calculation also includes order-α corrections for 3-photon annihilation and 2-photon radiative corrections[49].

The Ps^- ion was first observed by Mills[51], who has reported the only measurement of Γ to date[52]. In this experiment, Ps^- is formed on a thin carbon film and accelerated by applying a constant potential, V, to two grids separated by a distance, d. Measurements of the number of ions reaching a the second grid as a function of d (and hence the proper time since Ps^- emission) yields $\Gamma = 2.09 \pm 0.09$ ns^{-1}. A remeasurement[48] of Γ using an improved variation of this time-of-flight technique is presently underway with the goal of achieving 1000 ppm accuracy.

3. TESTS OF DISCRETE SYMMETRIES USING POSITRONIUM

3.1 Introduction

The interest in discrete symmetry tests originates from the observed P (parity) violation in weak interactions and CP (charge conjugation-parity) violation in neutral kaon decays. Positronium presents a unique opportunity for the testing of symmetries in a purely leptonic, particle-antiparticle system. Positronium states are eigenstates of C and CP, as well as P. Hence a variety of symmetry tests are possible[53]. The predicted size, using the standard model, of P-violating, CP-conserving effects in Ps is far too small to be detectable[54]. However, there is no way to preclude observable CP-violating or CPT violating effects for Ps in a model-independent way from other experiments[53,55].

3.2 C-Invariance and Forbidden Decay Modes in Positronium

The 1^1S_0 and 1^3S_1 states of Ps are eigenstates of C with eigenvalues $(-1)^{L+S}$ of $+1$ and -1, respectively. Since a single photon has a C eigenvalue of -1, singlet Ps is constrained to decay into an even number of γ rays and triplet Ps to an odd number, if C is a good symmetry. By searching for Ps decays into the "wrong" number of γ rays, one can test the principle of C invariance in the Ps system.

The most recent experiment[56] performed to search for the C-violating decay $1^1S_0 \rightarrow 3\gamma$, measured a branching ratio defined as $b = \lambda_S^{3\gamma}/\lambda_S^{2\gamma}$ where $\lambda_S^{n\gamma}$ is the singlet decay rate into n γ rays. The experiment required the subtraction of a substantial background due to primarily $1^3S_1 \rightarrow 3\gamma$ events. The result for b is $b < 2.8 \times 10^{-6}$ (68% confidence interval).

The search for the C-violating decay, $1^3S_1 \rightarrow 4\gamma$, is potentially almost free of direct backgrounds due to normal decays. One experiment has directly searched for the above decay mode[57]. A tetrahedral detector geometry was used to suppress the $1^1S_0 \rightarrow 4\gamma$ allowed events [a phase space argument gives the suppression[58]] which are in any case predicted to be only about 10^{-6} of the $1^1S_0 \rightarrow 2\gamma$ rate. From this experiment, a limit can be set on the branching ratio $b \equiv \lambda_T^{4\gamma}/\lambda_T^{3\gamma} < 8 \times 10^{-6}$ at the 68% confidence interval.

3.3 Search for CP Non-Conservation in $n = 2$ Ps

A one-photon transition ($C = -1$) between the 2^3S_1 ($C = -1$) and 2^1P_1 ($C = -1$) Ps levels is forbidden by C conservation. A search for such a transition is especially sensitive to a C-violating interaction, $H_{\not{C}P}$, that conserves parity and thus violates CP[53]. This sensitivity is due to the close proximity ($\Delta E = 1.8$ GHz) of the 2^3P_1 and 2^1P_1 states which causes a relatively large 2^3S_1 to 2^1P_1 transition amplitude for a given $H_{\not{C}P}$:

$$M_{\not{C}P} = \frac{\langle 2^1P_1|H_{\not{C}P}|2^3P_1\rangle\langle 2^3P_1|E1|2^3S_1\rangle}{\Delta E - i\Gamma/2} \tag{5}$$

Here $\langle E1 \rangle$ is the electric dipole matrix element and $\Gamma/2 = 25$ MHz.

CP non-conservation (CPNC), thus far observed only in the kaon system, has not been explained in any fundamental sense. Several models[59,60] have been developed to explain the observed CP violations, but evidence in favor of one model to the exclusion of all others is lacking. Ps is the only atomic system that is an eigenstate of CP and, moreover, shares with the kaon system the property of having two near-lying states with the same J, the same P, and opposite C. For Ps these are the 2^3P_1 and 2^1P_1 states with $\Delta E = 1.8$ GHz, while in the kaon system they are K_L and K_S with $\Delta m - i\Gamma/2 = (0.85 - 0.89i)$ GHz. The kaon system illustrates the advantage of a small energy denominator where tiny CPNC effects are amplified as relatively large mixings between K_L and K_S.

At the University of Michigan we are taking data in an experiment to place limits on the existence of the 2^3S_1 to 2^1P_1 transition. Positronium in the 2^3S_1 state is irradiated with microwaves near the predicted 11.184 GHz transition frequency while a search is made for the resonant production of the 2^1P_1 state. The 2^1P_1 state will be observed by its distinctive decay scheme, a Lyman-α optical decay at 243 nm ($\tau = 3.18$ ns) to the 1^1S_0 state which quickly ($\lambda_S^{-1} = 0.12$ ns) decays to a pair of back-to-back 511 keV gamma rays. A Zeeman-induced transition at the same frequency is used to calibrate our sensitivity. Preliminary data have been taken, the Zeeman-induced 2^3S_1 to 2^1P_1 transition has been observed, and we expect to obtain the first limits on $\langle H_{\not{C}P} \rangle$ next.

3.4 CPT and CP Tests in Polarized and Aligned Ps using Angular Correlations

A novel CPT test that measures a T-odd, CP-even angular correlation in the decay of polarized Ps has recently been completed[61]. The angular correlation[54] is $C_n[\hat{S} \cdot \hat{k}_1 \times \hat{k}_2]$, where S is the triplet Ps spin, $|k_1| > |k_2| > |k_3|$ are the momenta of the three decay γ rays, and C_n, the amplitude of the correlation, is expected to be zero if CPT is conserved. Polarized Ps is produced using a polarized low-energy e^+ beam incident on a MgO-covered CEMA plate. The Ps is confined in an 11 cm^3 MgO-lined cavity and three NaI scintillators detect the decay γ rays.

The final result for C_n, including both statistical and systematic errors, was found to be $C_n = +0.020 \pm 0.023$, consistent with CPT invariance. This result, although a far less precise comparison in an absolute sense than other well known CPT tests using leptons (*e. g.* e^+ - e^- or μ^+ - μ^- g-factor comparisons) is not precluded by them in a model-independent way and so constitutes an independent test of CPT. A measurement of C_n to an accuracy thirty times better than the present result can be envisioned. Other discrete symmetry tests using spin aligned Ps, including tests of CP invariance, are also possible[54,55,61].

4. EXOTIC PARTICLE SEARCHES USING POSITRONIUM

It is possible[1] for triplet Ps to decay into a single photon and a neutral, pseudoscalar particle, if the particle mass is less than $2m_e$. The once-popular axion is an example of this type of pseudoscalar, although searches for this decay mode would be sensitive to any neutral pseudoscalar particle in this mass range[37,62]. With the assumption that the pseudoscalar particle escapes from the apparatus undetected, the signal for this decay mode is a single, mono-energetic γ ray from the decay of triplet Ps. The observed energy of the γ ray can then be related, through the kinematics of the decay, to the mass of the neutral pseudoscalar.

Searches for the single γ-ray triplet decay have used Ge detectors[63,64,65] and the Heidelberg "Crystal Ball" NaI array[39]. These experiments have set limits on the pseudoscalar emission branching ratio relative to the allowed 3γ decay of order $10^{-6} - 10^{-7}$ in the mass range of 200 – 1000 keV, and of order 10^{-4} - 10^{-5} in the mass range of 1-200 keV.

Another possible decay mode of triplet Ps is to weakly-interacting, non-detected particles and nothing else. An experiment[66] to search for this decay mode has been recently performed. A limit of below 6 x 10^{-4} (90% confidence level) has been set on the branching ratio for this decay mode as compared to the 3-γ decay mode.

5. ACKNOWLEDGEMENTS

Positron and positronium research at the University of Michigan has been supported by National Science Foundation Grants No. PHY-8403817 and PHY-8605544, by grants from the office of the Vice President for Research of the University of Michigan, and by Richard Wood and Company.

6. REFERENCES

1. Resnick, L., Sundaresan M.K., and Watson, P.J.S., Phys. Rev. D 8, 172 (1973).
2. Poth, H., *et al.*,, Hyperfine Interactions 44, 259 (1988).
3. Neumann, R., Poth, H., Wolf, W., and Winnacker, A., Zeit. f. Phys. A313, 263 (1983).
4. Alekseev, G.D., *et al.*, Yad. Fiz. 40, 139 (1984) [Sov. J. Nucl. Phys. 40, 87 (1984)].
5. Olsen, H.A., Phys. Rev. D 33, 2033 (1986).
6. Brown, B.L. and Leventhal, M., Phys. Rev. Lett. 57, 1651 (1986).
7. Hegstrom, R.A., Rich, A. and Van House, J., Nature 313, 391 (1985); Gidley, D.W., Rich, A., Van House, J. and Zitzewitz, P.W. Nature 297, 639 (1982).

8. Skalsey, M., Paul, D.A.L. and Rich, A., Phys. Rev. C 39, 986 (1989); Skalsey, M., Girard, T.A., Newman, D. and Rich, A., Phys. Rev. Lett. 49, 708 (1982).
9. Roellig, L.O., *et al.*, in "Atomic Physics with Positrons", 233 (1987).
10. Van House, J., and Rich, A., Phys. Rev. Lett. 60, 169 (1988).
11. Van House, J. and Rich, A., Phys. Rev. Lett. 61, 488 (1988); Rich, A., and Van House, J., Jour. of Elec. Micr. Tech. 9, 209 (1988).
12. Brandes, G.R., *et al.*, Rev. Sci. Instr. 59, 228 (1988).
13. Balantekin, A.B., Nucl. Instr. Meth., B24/25, 273 (1987).
14. Ritter, M.W., Egan, P.O., Hughes, V.W., and Woodle, K.A., Phys. Rev. A 30, 1331 (1984).
15. Danzmann, K., Fee, M.S. and Chu, S., Phys. Rev. A 39, 6072 (1989).
16. Hatamian, S., Conti, R.S. and Rich, A., Phys. Rev. Lett. 58, 1833 (1987).
17. Pirenne. J., Arch. Sci. Phys. Nat. 29, 265 (1947).
18. Fulton, T. and Martin, P. C., Phys. Rev. 95, 811 (1954).
19. Fulton, T., Phys. Rev. A 26, 1794 (1982).
20. Caswell, W. and Lepage, G.P., Phys. Rev. A 20, 36 (1979).
21. Cung, V.K., Devoto, A., Fulton, T. and Repko, W.W., Phys. Rev. A 19, 1886 (1979).
22. Caswell, W.E. and LePage, G.P., Phys. Lett. 167B, 437 (1986).
23. Zhao, P., Lichten, W., Layer H., and Bergquist, J., Phys. Rev. Lett. 58, 1293 (1987).
24. Cohen, V. and Taylor, B., Rev. Mod. Phys. 59, 1121 (1987).
25. Lundeen, S.R., and Pipken, F.M., Metrologia 22, 9 (1986).
26. Boshier, M.G., *et al.*, Nature 330, 463 (1987).
27. Hellwig, H., *et al.*, IEEE Trans. Instrm. IM-19, 200 (1970).
28. Chu, S., Mills, A.P., Jr. and Hall, J.L., Phys. Rev. Lett. 52, 1689 (1984).
29. Mills, A.P., Jr., Phys. Rev. A 27, 262 (1983).
30. Mills, A.P., Jr., Berko, S. and Canter, K.F., Phys. Rev. Lett. 34, 1541 (1975).
31. Westbrook, C.I., Gidley, D.W., Conti, R.S., and Rich, A., Phys. Rev. Lett. 58, 1328 (1987).
32. Westbrook, C.I., Gidley, D.W., Conti, R.S. and Rich, A., to be published in Phys. Rev. A 40, November 1989.
33. Adkins, G.S., Annals of Physics 146, 78 (1983).
34. Lepage, G.P., Mackenzie, P.B., Streng K.H. and Zerwas, P.M., Phys. Rev. A 28, 3090 (1983).
35. Adkins, G.S. and Brown, F.R., Phys. Rev. A 28, 1164 (1983).
36. Lepage, G.P. and Adkins, G.S., private communication (1987). This modification was suggested since bound state corrections of order α^2 will occur as well as radiative corrections of order $(\alpha/\pi)^2$.

37. Cleymans, J. and Ray, P.S., Nuovo Cimento 37, 569 (1983).
38. Samuel, A.L., Mod. Phys. Lett. A 3, 1117 (1987).
39. Wahl, W., Proceedings of the Symposium "Production of Low-Energy Positrons with Accelerators and Applications" Giessen, FRG (1986)(to be published in Appl. Phys. A).
40. Gidley, D.W., Westbrook, C.I., Conti, R.S. and Rich, A., in "Atomic Physics with Positrons, NATO ASI Series B, Volume 169", edited by J.W. Humberston and E.A.G. Armour, (Plenum Press, NY, 1988), p 277.
41. Coleman, P.G., Griffith, T.C., Heyland, G.R. and Killeen, T.L., J. Phys. B 8, 1734 (1975).
42. Nico, J.S., Gidley, D.W., Zitzewitz, P.W., and Rich, A., Bull. Am. Phys. Soc. 32, 1051 (1987).
43. Gidley, D.W. and Zitzewitz, P.W., Phys. Lett. 69A, 97 (1978).
44. Van House, J., Rich, A. and Zitzewitz, P.W., Orig. of Life 14, 413 (1984).
45. Gidley, D.W., Rich, A., Sweetman, E. and West, D., Phys. Rev. Lett. 49, 525 (1982).
46. Muta, T. and Niuya, T., Prog. Theor. Phys. 68, 1735 (1982).
47. Bhatia, A.K. and Drachman, R.J., Phys. Rev. A 35, 4051 (1987).
48. Friedman, P.G., Mills, A.P., Jr. and Zuckerman, D., Bull. Am. Phys. Soc. 33, 953 (1988).
49. Bhatia, A.K. and Drachman, R.J., Phys. Rev. A 28, 2523 (1983).
50. Ho, Y.K., J. Phys. B 16, 1503 (1983).
51. Mills, A.P., Jr., Phys. Rev. Lett. 46, 717 (1981).
52. Mills, A.P., Jr., Phys. Rev. Lett. 50, 671 (1983).
53. Conti, R.S., Hatamian, S. and Rich, A., Phys. Rev. A 33, 3495 (1986).
54. Bernreuther W. and Nachtmann, O., Zeit. f. Phys. C11, 235 (1981).
55. Bernreuther, W., Löw, U., Ma, J.P., and Nachtmann, O., Universität Heidelberg preprint HD-THEP-87-25.
56. Mills, A.P., Jr. and Berko, S., Phys. Rev. Lett. 18, 420 (1967); Mills, A.P., Jr., Ph.D. Thesis, Brandeis University (1967).
57. Marko. K. and Rich, A., Phys. Rev. Lett. 33, 980 (1974); Marko, K. Ph.D. Thesis, University of Michigan (1974).
58. Mani, H.S. and Rich, A., Phys. Rev. D 4, 122 (1971).
59. Kobayashi, M. and Maskawa, T., Prog. Theor. Phys. 49, 652 (1973).
60. Wolfenstein, L., Phys. Rev. Lett. 13, 562 (1964).
61. Arbic, B.K., Hatamian, S., Skalsey, M., Van House, J. and Zheng, W., Phys. Rev. A 37, 3189 (1988).
62. Fayet, P. and Mezard, M., Phys. Lett. 104B, 226 (1981).

63. Carboni, G. and Dahme, W., Phys. Lett. 123B, 349 (1983).
64. Amaldi, U., *et al.*, Phys. Lett. 153B, 444 (1985).
65. Orito, S. *et al.*, Phys. Rev. Lett. 63, 597 (1989).
66. Atoyan, G.S., *et al.*, Phys. Lett. 220B, 317 (1989).

TESTS OF QED WITH HIGH RESOLUTION LASER SPECTROSCOPY OF ATOMIC HYDROGEN

D. H. McIntyre[a] and T. W. Hänsch[b]

[a] Department of Physics, Oregon State University
Corvallis, Oregon 97331
[b] Max-Planck-Institut für Quantenoptik, 8046 Garching and
Sektion Physik, University of Munich, 8000 Munich 40, West Germany

INTRODUCTION

The simple hydrogen atom provides an ideal testing ground for the theory of Quantum Electrodynamics (QED). The energy levels can be precisely calculated with QED, and they can be measured with high precision with Doppler-free laser techniques. Our efforts in this field have been concentrated on the transition from the $1S$ ground state to the $2S$ metastable state of atomic hydrogen. This transition has an incredibly narrow natural linewidth of 1.3 Hz, determined by the two photon decay lifetime of the $2S$ state. This transition can be studied with high precision with the elegant technique of Doppler-free two-photon laser spectroscopy. Measurement of the transition frequency can be used to determine an experimental value for the Lamb shift of the $1S$ ground state.

In this review, we will first discuss some of the QED theory as applied to the hydrogen atom, pointing out its uncertainties and limitations. We will then turn to the experimental side, focusing on experiments of the $1S$-$2S$ transition. We will discuss the current state of the art and speculate on what can be expected in the future.

THEORY

We can express the energy levels of hydrogen (and other hydrogen-like atoms) as

$$E(n, J, L, F, I) = E_{\mathrm{DC}}(n, J) + E_{\mathrm{RM}}(n, J) + E_{\mathrm{LS}}(n, J, L) + E_{\mathrm{HFS}}(n, J, L, F, I), \qquad (1)$$

where n is the principal quantum number, J is the total angular momentum of the electron, L is the orbital angular momentum of the electron, F is the total atomic angular momentum, and I is the spin of the nucleus. The four terms represent the Dirac Coulomb contribution $E_{\mathrm{DC}}(n, J)$,

New Frontiers in Quantum Electrodynamics and Quantum Optics
Edited by A. O. Barut, Plenum Press, New York, 1990

the reduced mass correction $E_{RM}(n, J)$, the Lamb shift $E_{LS}(n, J, L)$, and the hyperfine structure $E_{HFS}(n, J, L, F, I)$. Our discussion generally follows the conventions set forth by Mohr[1] and by Johnson and Soff.[2] We will briefly discuss each of the terms of Eq. (1); for a more thorough presentation, see Refs. 1 and 2. In what follows, all energies are expressed in units of Hz.

The first term in Eq. (1) is obtained by solving the Dirac equation for an electron moving in the Coulomb field of an infinitely massive point nucleus with charge Ze. The well known result for the energy eigenvalues is

$$E_{DC}(n, J) = \frac{2\,Z^2}{N(N + n + \gamma - k)}\, cR_\infty\,, \tag{2}$$

where $hcR_\infty = \alpha^2 m_e c^2/2$, and

$$k = J + \frac{1}{2}\,, \tag{3}$$

$$\gamma = \sqrt{k^2 - (Z\alpha)^2}\,, \tag{4}$$

$$N = \sqrt{n^2 - 2(n - k)(k - \gamma)}\,. \tag{5}$$

If we expand Eq. (2) in powers of $(Z\alpha)^2$, then we obtain

$$E_{DC}(n, J) \approx -\frac{Z^2}{n^2}\, cR_\infty \left[1 + \frac{(Z\alpha)^2}{n}\left\{\frac{1}{J+1/2} - \frac{3}{4n}\right\} + O\{(Z\alpha)^4\} \right], \tag{6}$$

where the first term represents the familiar nonrelativistic Bohr result and the second term is the Pauli fine structure correction.

The Dirac energy eigenvalues must be modified to account for the finite mass m_N of the nucleus. This reduced mass correction is[3]

$$E_{RM}(n, J) = -\frac{m_e}{m_e + m_N}\left[1 - \left\{\frac{m_N}{m_e + m_N}\right\}^2 \frac{(Z\alpha)^2}{4n^2} \right] E_{DC}(n, J)\,. \tag{7}$$

The first term accounts for nuclear motion by simply replacing the electron mass m_e with the reduced mass $\mu_N = m_e m_N/(m_e + m_N)$. The second term accounts for relativistic effects of the nuclear motion and is derived from a relativistic two-body treatment of the atom. It is included in some definitions of the Lamb shift, but we have followed the convention of Johnson and Soff and have included it here as an additional relativistic correction to the reduced mass contribution.[2]

The Lamb shift term represents small corrections to the Dirac energy levels primarily due to QED effects. Though the term "Lamb shift" originally referred to the $2S_{1/2}$-$2P_{1/2}$ energy separation, it is now often used to describe the composite of all corrections (except reduced mass and hyperfine structure) to the Dirac energy for a particular level. As defined by Johnson and Soff,[2] the Lamb shift includes QED radiative and binding corrections, relativistic reduced mass and relativistic recoil corrections, and corrections due to the finite size of the nucleus. Quantum electrodynamic Lamb shift calculations generally represent a perturbative expansion in terms of $Z\alpha$ and can quickly become intractable. Many higher-order terms have yet to be computed.

We can write the total Lamb shift as[1,2]

$$E_{LS}(n, J, L) = \frac{2}{\pi\alpha} \frac{(Z\alpha)^4}{n^3} cR_\infty \; F(Z\alpha) , \tag{8}$$

where $2\alpha^3 cR_\infty/\pi = \alpha^5 m_e c^2/\pi h = 813.862\,88(11)$ MHz, $F(Z\alpha)$ is the slowly varying function

$$\begin{aligned}
F(Z\alpha) &= A_{40} + \ln(Z\alpha)^{-2} A_{41} + (Z\alpha) A_{50} \\
&\quad + (Z\alpha)^2 [A_{60} + \ln(Z\alpha)^{-2} A_{61} + \ln^2(Z\alpha)^{-2} A_{62}] + O(Z^3\alpha^3) \\
&+ \alpha B_{40} + O(Z\alpha^2) \\
&+ (m_e/m_N)[C_{40} + \ln(Z\alpha)^{-2} C_{41} + (Z\alpha) C_{50}] + O(\alpha m_e{}^2/m_N{}^2) \\
&+ (Zm_e/m_N)\{(m_e/Z\alpha m_N) D_{40} + D_{50} + \ln(Z\alpha)^{-2} D_{51} + (Z\alpha)[D_{60} + \ln(Z\alpha)^{-2} D_{61}]\} \\
&\quad + O(Z^3\alpha^2 m_e/m_N) \\
&+ \alpha\,(r_N/r_e)^2 N_0 ,
\end{aligned} \tag{9}$$

r_N is the rms charge radius of the nucleus, and $r_e = \alpha^2 a_0$ is the classical radius of the electron. The coefficients A_{pq} represent the lowest-order contributions from the self-energy of the electron and from vacuum polarization. Contributions from these effects of a higher-order are included in the coefficients B_{pq}, and are referred to as "higher-order QED" terms. The coefficients A_{pq} and B_{pq} can also be thought of as arising from the emission and reabsorption of 1 and 2, respectively, virtual photons by the bound electron. The coefficients C_{pq} arise from relativistic reduced mass terms while the coefficients D_{pq} describe relativistic recoil corrections. The formulas used to calculate most of these coefficients can be found in Ref. 2 Some of the terms have only recently been calculated and can be found in the original publications.[4,5] A complete tabulation of all these terms can be found in the thesis of R. G. Beausoleil[6] and in a forthcoming publication.[7] The values of all the various contributions to the function $F(Z\alpha)$ for the hydrogen $1S_{1/2}$ and $2S_{1/2}$ states are shown in Table I.

Table I. Contributions to $F(Z\alpha)$ for Hydrogen $1S$ and $2S$ States

Term	$1S_{1/2}$	$2S_{1/2}$
Self-Energy	10.31676(9)	10.54687(9)
Vacuum Polarization	-0.26438	-0.26439
Higher-Order QED	0.00125	0.00125
Radiative Recoil	-0.01571	-0.01608
Nonradiative Recoil	0.00293	0.00333
Finite Nuclear Size	0.00125(3)	0.00125(3)
Total $F(Z\alpha)$	10.04210(10)	10.27223(9)
Total Lamb Shift (MHz)	8172.89(8)	1045.02(1)

As is evident from Table I, the largest contribution to the Lamb shift is given by the electron self-energy term, which arises from the interaction of the bound atomic electron with the vacuum fluctuations of the quantized electromagnetic field. Though computation of this effect is quite complicated, the bulk of the effect can be accounted for with a simple argument first presented by Welton[8] soon after the discovery of the Lamb shift in 1947.

The zero-point modes of the quantized radiation field result in nonzero mean-square values of the electric and magnetic fields in a vacuum. These fluctuating fields couple with the charge of an electron and cause a fluctuation in the electron's equilibrium position. For a bound electron, these fluctuations in position are much faster than the orbital motion. The electron can thus be considered to be a small sphere of uniform charge density rather than a point charge. This has no effect on the binding energy with a point nucleus unless the nucleus-electron separation becomes less than the size of the smeared-out electron. Once the point nucleus penetrates the electron cloud, it will detect less of the electron's charge and the binding energy of the two particles will decrease. The size of the cloud due to the fluctuation of the electron's position is much smaller than the extent of the electronic wave function. Hence, only S states of the atom are affected, since only these states have a nonzero value of the electronic wave function at the origin. It is also clear from this argument that the energy shift is positive, corresponding to a weakening of the potential energy. Welton presents a more quantitative analysis and obtains a result that corresponds to that portion of the coefficient A_{41} due to the self energy of the electron.[8]

Vacuum polarization is the second largest effect that contributes to the Lamb shift. It arises from the creation and annihilation of virtual electron-positron pairs, which allow for a polarizability of the vacuum. In the vicinity of a charged nucleus, a virtual electron tends to be attracted toward the nucleus while a virtual positron tends to be repelled, resulting in a screening of the nuclear charge. In the Coulomb potential, the term Ze for the nuclear charge refers to the charge seen by a distant electron and so already includes this screening. Thus, an electron that gets very close to the nucleus will see an effective charge greater than Ze, and be bound more strongly to the nucleus. Again only S states are appreciably affected, as they have the highest probability of approaching the nucleus.

It is important to point out that the contributions to $F(Z\alpha)$ shown in Table I that are due to the finite proton size are derived using $r_p = 0.805(11)$ fm, as measured by Hand *et al.*[9] However, a recent measurement by Simon *et al.*[10] obtained the result $r_p = 0.862(12)$ fm, which increases the S-state contribution to $F(Z\alpha)$ from 0.001 25(3) to 0.001 43(4). This increases the theoretical 1S and 2S Lamb shifts by 149 kHz and 18 kHz, respectively. This discrepancy must be resolved before QED tests with hydrogen can proceed beyond this level of precision. The uncertainty of the QED terms arises mostly from the A_{60} term of the self energy.[1] Other QED uncertainties are below the level of precision shown in Table I. The total uncertainty is 80 kHz for the 1S Lamb shift and 10 kHz for the 2S state.

The hyperfine structure is well known theoretically,[11] but is even better known experimentally.[12] Optical frequency measurements can thus be compared with theory without any additional uncertainty introduced by the hyperfine splitting.

EXPERIMENT

Tests of QED with hydrogen have mostly been done with the microwave $2S_{1/2}$-$2P_{1/2}$ Lamb shift transition. This transition has been measured with 100 kHz precision,[13] but is limited in resolution by the 100 MHz natural linewidth. At Stanford and now in Munich, we have concentrated our efforts on a different transition with potentially much narrower width: the transition from the 1S ground state to the 2S metastable state.[14] The 1/8 sec lifetime of the upper state implies a natural linewidth of only 1 Hz, or an ultimate resolution better than one part in 10^{15}. This transition is not dipole-allowed, but can be observed by two-photon spectroscopy. Doppler-broadening can be elegantly eliminated by excitation with two counterpropagating laser beams whose first-order Doppler shifts cancel.

Excitation of the hydrogen 1S-2S two-photon transition requires ultraviolet light at 243 nm, approximately half the energy of the vacuum ultraviolet 1S-2P Lyman-α transition. Though there are no tunable lasers at 243 nm, tunable light can be generated by frequency doubling the radiation from a 486-nm dye laser in a nonlinear crystal. This was the approach used in the first hydrogen 1S-2S experiment, performed by Hänsch *et al.*[14] This first experiment and several subsequent experiments[15-18] used powerful pulsed laser sources at 486 nm in order to obtain sufficient second-harmonic generation of the 243-nm light necessary to excite the weak two-photon transition.

However, a continuous-wave (cw) source of 243-nm radiation is clearly required to take full advantage of the resolution available with the narrow natural linewidth of the hydrogen 1S-

2S transition. Efforts to produce sufficient (≈ 1 mW) cw 243-nm radiation began at Stanford University soon after the first pulsed experiments. Unfortunately, simple extension of the second-harmonic generation employed with pulsed lasers to the case of cw lasers met several obstacles. In particular, the nonlinear crystals were quickly damaged by the ultraviolet light that was generated. Efforts soon turned to sum-frequency mixing of radiation from two different lasers in a non-linear crystal. This has the advantage that a powerful ion laser can be used to increase the generated power and a dye laser can provide tunability. This approach culminated in the first continuous-wave measurement of the absolute frequency of the hydrogen 1S-2S transition which was completed at Stanford in 1986.[19,20] The approach taken in this experiment was to mix the frequencies of a 790-nm dye laser and a 351-nm argon-ion laser in a crystal of potassium dihydrogen phosphate (KDP). With the dye laser radiation enhanced in an external ring cavity, this scheme produced several milliwatts of stable, tunable 243-nm light.

In the Stanford experiment, the 1S-2S transition was excited in a low pressure hydrogen cell. Atomic hydrogen flowed through the cell after being produced by microwave (2.45 GHz) dissociation of molecular hydrogen. The hydrogen cell was isolated from the microwave cavity with a length of Teflon tubing, thus avoiding problems with the fields and radiation in the discharge. The 243-nm radiation was enhanced in an optical cavity surrounding the interaction region. Excited 2S atoms soon collide with other hydrogen atoms or molecules and are likely to be transferred to the nearby $2P_{1/2}$ state, from which they quickly decay back to the ground state by the emission of a Lyman-α photon. These collisions reduce the lifetime of the 2S state and so limit the resolution of the experiment, but the Lyman-α light can be used to detect the excitation of the transition. The frequency of the hydrogen 1S-2S transition was measured by heterodyne comparison with the second-harmonic of a 486-nm dye laser that was frequency-stabilized to a Doppler-free transition in molecular tellurium ($^{130}Te_2$) vapor. The $^{130}Te_2$ molecule has a rich absorption spectrum in the blue region, and the frequencies of two strategic $^{130}Te_2$ transitions were recently measured by Barr *et al.*[21] with an accuracy of 4 parts in 10^{10} by interferometric comparison with an iodine-stabilized helium-neon laser. We were thus able to measure the hydrogen 1S-2S transition frequency with an accuracy of 7 parts in 10^{10}, limited by pressure broadening of the transition in the hydrogen observation cell and by the accuracy of the $^{130}Te_2$ frequency standard.

The result for the hydrogen 1S-2S energy interval may be used to derive an experimental value for the 1S Lamb shift. By relying on a precise independent measurement of the Rydberg constant,[22] one can accurately compute the well known Dirac and reduced mass contributions to the energy levels. By taking the difference between these theoretical terms and the experimental 1S-2S transition frequency and incorporating the 2S Lamb shift measurement, one arrives at a result for the Lamb shift of the ground state of hydrogen of 8 173.9(1.9) MHz. This result is in good agreement with the theoretical prediction of 8 172.89(8) MHz.[7] A more recent experiment at Oxford with a similar apparatus has obtained a result of 8 172.93(84) MHz, also in good agreement with theory.[23]

The accuracy of these measurements is still more than an order of magnitude behind that of the 2S Lamb shift measurement.[13] However, there is much room for improvement in the case of the 1S-2S transition even though the present experiments have been limited by the optical frequency standards to which they are compared. Studies of the hydrogen 1S-2S transition can be pursued beyond this level by making comparisons with other narrow transitions in hydrogen. The simple $1/n^2$ dependence of hydrogen energy levels leads to harmonic relationships between many of the level separations, permitting simple experimental comparison. In this manner, QED shifts can be determined without an absolute optical frequency measurement and without precise knowledge of the Rydberg constant, which is essentially cancelled in the comparison.

The most obvious comparison to make is between the 1S-2S transition and the Balmer-β transition at 486 nm, as was done in the original 1S-2S experiments at Stanford.[14-16] The Bohr theory predicts that the $n = 1 \rightarrow 2$ interval is exactly four times the $n = 2 \rightarrow 4$ interval, so that both transitions should be observed at the same 486-nm laser frequency. However, relativistic and quantum electrodynamic corrections remove this degeneracy and cause a displacement of the transitions. Measurement of this small displacement allows a determination of the 1S Lamb shift without having to make an absolute optical frequency measurement.

In our laboratory at the Max Planck Institute in Garching we are working towards improving the resolution of the hydrogen 1S-2S transition by a thousandfold, down to the few kHz level. We are planning to excite the transition with 243-nm light generated by frequency doubling in β-barium-borate (BBO). The Oxford group showed that this crystal was capable of generating several milliwatts of second harmonic without the damage suffered by other crystals.[23] In our lab, Zimmermann *et al.* have generated 1.3 mW of 243-nm light from a crystal of BBO placed in an external cavity that was made to be resonant for both the fundamental wave and the generated second harmonic.[24] The fundamental 486-nm laser has been frequency stabilized to the sub-kiloHertz level by Kallenbach *et al.*,[25] using standard FM techniques. The 1S-2S transition will be excited in an atomic hydrogen beam so that collisional line broadening and shifting will be eliminated. In addition the beam will be cooled to liquid helium temperatures, to decrease the transit time broadening and second-order Doppler shift and to increase the transition probability.

For a precision QED test we are working on a comparison of the 1S-2S transition with the 2S-4S transition, which can be excited with two 972-nm photons. This two-photon transition has a linewidth of 690 kHz, narrower than the 13 MHz of the traditional one-photon Balmer-β line. The 972-nm light is generated with a titanium doped sapphire laser pumped by an argon-ion laser. The 1S-2S and 2S-4S transitions will be compared by generating the second harmonic of the 972-nm laser in a potassium niobate crystal ($KNbO_3$) and heterodyning it the the fundamental 486-nm light which is used in the 1S-2S experiment. The beat note between the two 486-nm beams will be 4836 MHz, assuming we are exciting the $F = 1 \rightarrow 1$ transitions.

This frequency difference has a QED contribution of 868.68 MHz, which consists of the 4*S* Lamb shift plus one-quarter of the 1*S* Lamb shift minus five-quarters of the 2*S* Lamb shift. The rest of the 4836 MHz frequency comes from the Dirac Coulomb term, the reduced mass corrections and the hyperfine structure; the uncertainties of each of these terms being below 1 kHz. The QED term uncertainty is 30 kHz, so a 10 kHz measurement of the beat note will already be able to make a test of QED.

FUTURE POSSIBILTIES

As another example of precision comparisons that are possible, consider two-photon transitions from the hydrogen 2*S* state to high Rydberg states. Biraben *et al.*[26] have excited the hydrogen 2*S*-12*D* two-photon transition with radiation at 750 nm in their recent Rydberg constant measurement. As higher *nD* or *nS* states are accessed, the required wavelength for two-photon excitation from the 2*S* state approaches 729 nm, the third harmonic of which is very close to the 243-nm radiation needed to excite the 1*S*-2*S* transition. Thus these two transitions could be compared by observing a heterodyne signal between the 243-nm light and the third harmonic of the 729-nm light. Such a scheme is particularly attractive because the 2*S*-*nS* transitions have a natural linewidth of approximately 1 GHz/n^3, so that the 2*S*-100*S* transition would have a natural linewidth of approximately 1 kHz.

Interesting QED tests are also possible if the hydrogen 1*S*-2*S* transition is compared to the same transitions in positronium[27] and muonium.[28] Since these atoms are purely leptonic, they have no hadronic structure corrections, which we saw earlier to be a limitation for future hydrogen experiments. In the more distant future, trapped anti-protons[29] could be used to make anti-hydrogen. A comparison between the 1*S*-2*S* transitions of hydrogen and anti-hydrogen could test with extreme precision whether there is a spectroscopic difference between matter and anti-matter.

ACKNOWLEDGEMENTS

We would like to thank all of our colleagues who have worked with us on these experiments, especially R. G. Beausoleil, C. Zimmermann, R. Kallenbach, J. Sandberg, R. G. DeVoe, W. Vassen, D. Meschede, and J. Eberlein.

REFERENCES

1. P. J. Mohr, Ann. Phys. (N.Y.) 88, 26, 52 (1974); Phys. Rev. Lett. 34, 1050 (1975); Phys. Rev. A 26, 2338 (1982); and in The Spectrum of Atomic Hydrogen: Advances, edited by G. W. Series (World Scientific, New Jersey, 1988), p. 111.
2. W. R. Johnson and G. Soff, At. Data Nucl. Data Tables, 33, 405 (1985).

3. G. W. Erickson, J. Phys. Chem. Ref. Data 6, 831 (1977).
4. G. Bhatt and H. Grotch, Phys. Rev. A 31, 2794 (1985); Phys. Rev. Lett. 58, 471 (1987); Ann. Phys. 178, 1 (1987); and H. Grotch, Phys. Scr. T21, 8689 (1988).
5. G. W. Erickson and H. Grotch, Phys. Rev. Lett. 60, 2611 (1988).
6. R. G. Beausoleil, Ph. D. Dissertation, Stanford University, 1986 (unpublished).
7. R. G. Beausoleil, D. H. McIntyre, and T. W. Hänsch, to be published.
8. T. A. Welton, Phys. Rev. 74, 1157 (1948).
9. L. N. Hand, D. J. Miller, and R. Wilson, Rev. Mod. Phys. 35, 335 (1963); D. J. Drickey and L. N. Hand, Phys. Rev. Lett. 9, 521 (1962).
10. G. G. Simon, Ch. Schmitt, F. Borkowski, and V. H. Walther, Nucl. Phys. A333, 381 (1980).
11. H. A. Bethe and E. E. Salpeter, Quantum Mechanics of One- and Two-Electron Atoms (Plenum, New York, 1977).
12. P. Petit, M. Desaintfuscien, and C. Audoin, Metrologia 16, 7 (1980); J. W. Heberle, H. A. Reich, and P. Kusch, Phys. Rev. 101, 612 (1956).
13. S. R. Lundeen and F. M. Pipkin, Phys. Rev. Lett. 46, 232 (1981).
14. T. W. Hänsch, S. A. Lee, R. Wallenstein, and C. Wieman, Phys. Rev. Lett. 34, 307 (1975).
15. S. A. Lee, R. Wallenstein, and T. W. Hänsch, Phys. Rev. Lett. 35, 1262 (1975).
16. C. E. Wieman and T. W. Hänsch, Phys. Rev. A 22, 192 (1980).
17. E. A. Hildum, U. Boesl, D. H. McIntyre, R. G. Beausoleil, and T. W. Hänsch, Phys. Rev. Lett. 56, 576 (1986).
18. J. R. M. Barr, J. M. Girkin, J. M. Tolchard, and A. I. Ferguson, Phys. Rev. Lett. 56, 580 (1986).
19. R. G. Beausoleil, D. H. McIntyre, C. J. Foot, E. A. Hildum, B. Couillaud, and T. W. Hänsch, Phys. Rev. A 35, 4878 (1987).
20. D. H. McIntyre, R. G. Beausoleil, C. J. Foot, E. A. Hildum, B. Couillaud, and T. W. Hänsch, Phys. Rev. A 39, 4591 (1989).
21. J. R. M. Barr, J. M. Girkin, A. I. Ferguson, G. P. Barwood, P. Gill, W. R. C. Rowley, and R. C. Thompson, Opt. Commun. 54, 217 (1985).
22. P. Zhao, W. Lichten, H. Layer, and J. Bergquist, Phys. Rev. Lett. 58, 1293 (1987).
23. M. G. Boshier, P. E. G. Baird, C. J. Foot, E. A. Hinds, M. D. Plimmer, D. N. Stacey, J. B. Swan, D. A. Tate, D. M. Warrington, and G. K. Woodgate, Nature 330, 463 (1987).
24. C. Zimmermann, R. Kallenbach, T. W. Hänsch, and J. Sandberg, Opt. Commun. 71, 229 (1989).
25. R. Kallenbach, C. Zimmermann, D. H. McIntyre, T. W. Hänsch, and R. G. DeVoe, Opt. Commun. 70, 56 (1989).
26. F. Biraben, J. C. Garreau, L. Julien, and M. Allegrini, Phys. Rev. Lett. 62, 621 (1989).
27. S. Chu, A. P. Mills, Jr., and J. L. Hall, Phys. Rev. Lett 52, 1689 (1984).

28. S. Chu, A. P. Mills, Jr., A. G. Yodh, K. Nagamine, Y. Miyake, and T. Kuga, Phys. Rev. Lett 60, 101 (1988).
29. G. Gabrielse, X. Fei, K. Helmerson, S. L. Rolston, R. Tjoelker, T. A. Trainor, H. Kalinowsky, J. Haas, and W. Kells, Phys. Rev. Lett. 57, 2504 (1986).

ON HIGH ENERGY TESTS OF QED

Y.Sakurayama, H.Salecker, and F.C.Simm

Sektion Physik der Universität München
Theresienstr. 37, D–8000 München 2
Munich, W.Germany

In this paper we examine the different possibilities for high energy tests of quantum electrodynamics (QED) from a more fundamental standpoint and especially with a view beyond the standard model. First we stress that the photon propagator (together with the contiguous vertices) i.e., Maxwell's equations together with the minimal form of coupling can be tested in a model–independent way. These tests are limited by the Z_0–resonance. This situation is quite different with the electron propagator (i.e., Dirac's equation always together with the minimal coupling). Here Compton scattering and pair annihilation can give valid tests or in the case of deviations valuable limits even in the region of the Z_0 and beyond. With additional interactions tests concerning excited electrons of spin 1/2 or 3/2 could also be performed, where we could also expect some information about composite structures if we stay in the physically meaningful region where unitarity is still (approximately) valid. Finally we consider tests of the nonlinear effects of QED at high energies and give a table of the different tested lengths (as of Nov.89).

1. INTRODUCTION

In the last decade the principles of QED have been applied very successfully to other fields e.g., gauge theories of weak and strong interactions. The resulting standard theory of electro–weak and strong interactions is nowadays still in very good agreement with low and high energy experience. Many scientists have even tried to base a complete theory of all interactions including gravitation on these principles. It is, therefore, very important to test QED as accurately as possible. Especially at high energies we can hope to find some deviations which could give the lacking information about the foundations of the standard theory.

Let us recall how QED is proceeding so successfully. First Maxwell's and Dirac's equations are coupled in the so–called minimal way with the field quantities interpreted as operators. Then perturbation theory is used for solving the equations. In this step, renormalization must be used. Renormalization introduces, however, no new

New Frontiers in Quantum Electrodynamics and Quantum Optics
Edited by A. O. Barut, Plenum Press, New York, 1990

ingredients into the theory, since in QED it only means that all parts of mass and charge must be taken together to give the observable quantities, as is well known. Even if all of these parts were finite, one would have to renormalize in QED. The theory now consists of all Feynman diagrams described by the propagators of the photon, the electron, and the vertex operator. These three quantities are the elementary constituents which build up any matrix element in any order. However, photon and electron propagators are always multiplied with the contiguous vertices. That is, roughly speaking, only two of these three elementary constituents are observable separately. This situation is already inherent in classical electrodynamics. We never can distinguish between a deviation from Coulomb's law and an extended charge distribution. Therefore, in the second chapter we will consider the testing of the photon propagator with e^-e^-, e^+e^- scattering and lepton pair creation. This can be done even in a model-independent way. Nevertheless, the testing is limited near the Z_0 resonance because QED passes over to the combined electro-weak interaction.

In the third chapter we shall consider the testing of the electron propagator that is roughly speaking a test of the Dirac equation with the minimal coupling to the electromagnetic field with Compton scattering and pair annihilation $e^+e^- \rightarrow 2\gamma$. These processes on the one hand allow testing of QED even in the region of the Z_0 resonance but because of gauge invariance there are a number of compensations, which have been discussed already earlier.[1,2,3]

We will consider in the fourth chapter the effect of additional interactions which allow the existence of excited electrons of spin $\frac{1}{2}$ and $\frac{3}{2}$. Within the region of (approximate) validity of unitarity comparison with Compton scattering and pair annihilation could perhaps give some information beyond the standard model.

In the fifth chapter we will consider the nonlinear effects of QED which allow even with the minimal interaction a test of the electron propagator. At high energies these effects are not unmeasurably small and could be performed with $\gamma\gamma$ interaction. Nonlinear effects in gauge theories are especially interesting since deviations with such effects could perhaps help in determining some of the otherwise undetermined many constants in the standard theory.

2. TESTING OF THE PHOTON PROPAGATOR

The photon propagator can be tested experimentally at high energies with processes where one photon is exchanged in the lowest order. There are e^-e^-, e^+e^- scattering[4,5] and lepton pair creation $e^+e^- \to \mu^+\mu^-$, $\tau^+\tau^-$ (Fig.1). Before the comparison with theoretical considerations radiative corrections have to be subtracted from the experimental results because they depend critically on the experimental arrangements. Also weak corrections have to be applied in the same order (Z_0 exchange). Fortunately the last corrections are small for unpolarized scattering up to about 60 GeV per beam.[6] Even for higher energies the terms with mixed γ and Z_0 exchange can mostly be omitted (reduced radiative corrections). For a discussion of radiative corrections see e.g., Ref.7.

The photon propagator D is given in perturbation theory by (apart from constant factors)

$$D(q^2) = 1/q^2 \tag{2.1}$$

with the energy–momentum transfer q.

To make a quantitative estimate of a possible breakdown of QED one of the authors[4] has proposed already a long time ago to replace (2.1) by the photon propagators of generalized field theories, which has proved useful in regularization. Among

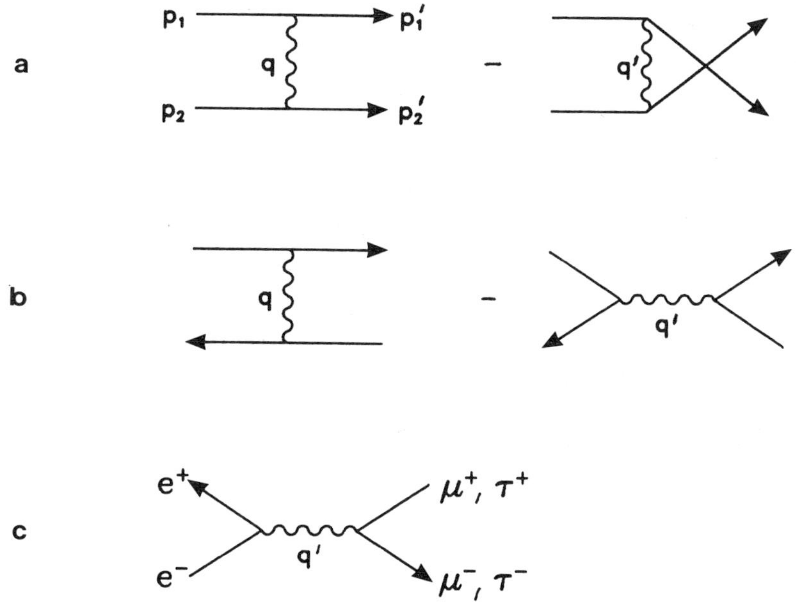

Fig. 1. Lowest–order diagrams for (a) e^-e^-, (b) e^+e^- scattering, and (c) lepton pair creation

other models he used especially the photon propagator

$$D'(q^2) = \frac{1}{q^2} - \frac{1}{q^2 - \Lambda^2} = \frac{-\Lambda^2}{q^2(q^2 - \Lambda^2)} = \frac{1}{q^2}\left[1 - \frac{q^2}{q^2 - \Lambda^2}\right] \tag{2.2}$$

which corresponds to a modified Coulomb potential in classical electrodynamics

$$V' = \frac{1 - e^{-\Lambda r}}{r}. \tag{2.3}$$

However, in quantum theory this kind of modification introduces heavy photons of mass Λ with a negative fine structure constant α as coupling strength. It will, therefore, violate causality which is restored only in the transition to unmodified QED ($\Lambda \longrightarrow \infty$). Nevertheless, this parametrization has been used widely as a measure for an overall breakdown of QED so far.

However, one can do better.[3] In the most general case (provided only that relativistic invariance is still valid) we obtain for the generalized photon propagator

$$D'(q^2) = \frac{F_\gamma(q^2)}{q^2} \tag{2.4}$$

with a scalar function $F_\gamma(q^2)$. Since the $e^+ e^-$ are on the mass shell, $p_1^2 = (p_1')^2 = p_2^2 = (p_2')^2 = m^2$ the vertex operators are in the most general case ($\hbar = c = 1$)

$$\Gamma_\mu(p_1, p_1', q) = e\,[\gamma_\mu F_1(q^2) + (i\kappa/2m)\sigma_{\mu\nu} q^\nu F_2(q^2)]. \tag{2.5}$$

Here κ is the anomalous magnetic moment of the electron, which is known to be very small in QED (about 10^{-3} of the Bohr magneton). If we consider e^-e^- scattering first, we are in the scattering region, q^2 is spacelike and the form factors F_1 and F_2 are likely to decrease apart perhaps from a few possible resonances. Therefore, the second term in Γ_μ can be neglected because of the experimental error, which is typically a few percent. We then obtain in the most general case in the matrix element a form factor[8]

$$F(q^2) = F_1^2(q^2) F_\gamma(q^2) \tag{2.6}$$

which we may expand for small deviations from 1 $[F(0) = 1,\ q^2 < 0]$

$$F(q^2) = 1 \pm q^2/(\Lambda_\gamma^\mp)^2 + \cdots \tag{2.7}$$

(the plus and minus signs denote the upper and lower limits of the experimentally measured cross section).

Under this assumption, which is valid roughly up to a 10% deviation or experimental error, the expansion is independent of the shape of $F(q^2)$, i.e., also independent of the shape of the model we eventually take for testing $F(q^2)$. It depends only upon the slope at $q^2 = 0$, which has the dimension of the square of a length and is equal, up to a numerical factor, to the *rms* radius of $F(q^2)$. (A vanishing or an infinite slope is a very special case; in general this quantity will neither vanish nor become infinite.) That is, comparing the theory with experimental results we can determine Λ_γ in a model–independent way if the error is not too large and if q^2 and $(q')^2$ are known.[8]

The differential cross section with general form factors (2.6) was calculated by Salecker.[4] In the center–of–momentum frame (CMS) we obtain

$$\frac{d\sigma_F^{e^-e^-}}{d\Omega} = \frac{r_0^2}{4\gamma^2}\left[\frac{s^2 + u^2 + 8m^4}{2t^2}F^2(t)\right.$$

$$\left. + \frac{s^2 + 4m^4}{t\,u}F(t)F(u) + \frac{s^2 + t^2 + 8m^4}{2u^2}F^2(u)\right] \tag{2.8}$$

with the Mandelstam variables $s = (p_1 + p_2)^2 = 4E^2$, $t = q^2 = (p_1 - p_1')^2$, $u = (q')^2 = (p_1 - p_2')^2$, where $\gamma = E/m$ in the CMS, and r_0 is the classical electron radius. With a small deviation like the experimental error we can take the expansion (2.7) and obtain (introducing the unmodified Møller cross section $d\sigma_M/d\Omega$)

$$\frac{d\sigma_M}{d\Omega}\left[\frac{d\sigma_F^{e^-e^-}}{d\sigma_M} - 1\right]\left\{\pm\frac{r_0^2}{4\gamma^2}\left[\frac{2s^2 + u^2 + 12m^4}{t} + \frac{2s^2 + t^2 + 12m^4}{u}\right]\right\}^{-1}$$

$$= \frac{1}{(\Lambda_\gamma^{\mp})^2}\,. \tag{2.9}$$

Since $1/\Lambda_\gamma^2$ is the square of a length, the sign on the left side of (2.9) has to be chosen in such a way that $1/\Lambda_\gamma^2$ becomes positive. If we take for $d\sigma_F/d\sigma_M - 1$ the experimental error, we get in general two different lengths $1/\Lambda_\gamma^{\pm}$ according to the upper or lower limit of the experimental error. Under the above assumptions we obtain in this way the tested length $1/\Lambda_\gamma$ independent of any model.

The situation is somewhat different in electron–positron scattering (Fig.1b). Here $(q')^2$ is timelike in the annihilation diagram. Thus e^+e^- scattering offers the possibility of testing the photon propagator also in the timelike region.[4] On the other hand, compensations between eventual deviations in the spacelike and in the timelike regions could occur because the form factors may increase in the timelike region. Since the form factors can only depend upon the energy–momentum transfer squared, their behaviour in the spacelike region could first be tested with e^-e^- scattering and after that with e^+e^- scattering taken for the same spacelike q^2 and $(q')^2$ in the timelike region. In the absence of sufficiently accurate high–energy e^-e^- scattering experiments, this can also be done by measuring the angular distribution of e^+e^- scattering in a sufficiently wide region on both sides of 90° scattering, since $(q')^2$ does not depend upon the angle.

Since we know from earlier (lower energy and/or less accurate) experiments on e^-e^- or e^+e^- scattering that possible deviations can only be small, we may use the expansion (2.7) also in the timelike region since the unphysical region $4m^2 > q^2 > 0$ is very small compared with the presently achieved values of s, t and u. In this case also e^+e^- scattering can give model–independent results. The differential cross section with general form factors $d\sigma_F^{e^+e^-}/d\Omega$ for e^+e^- scattering corresponding to the e^-e^- scattering cross section (2.8) can easily be obtained from (2.8) by the substitution law with $u \leftrightarrow s$,* as has also been calculated by Salecker.[4]

$$\frac{d\sigma_F^{e^+e^-}}{d\Omega} = \frac{r_0^2}{4\gamma^2}\left[\frac{s^2 + u^2 + 8m^4}{2t^2}F^2(t)\right.$$

$$\left. + \frac{u^2 + 4m^4}{ts}F(t)F(s) + \frac{u^2 + t^2 + 8m^4}{2s^2}F^2(s)\right] . \tag{2.10}$$

Under these circumstances we obtain for the square of the tested length corresponding to (2.9)

$$\frac{d\sigma_B}{d\Omega}\left[\frac{d\sigma_F^{e^+e^-}}{d\sigma_B} - 1\right]\left\{\pm\frac{r_0^2}{4\gamma^2}\left[\frac{s^2 + 2u^2 + 12m^4}{t} + \frac{2u^2 + t^2 + 12m^4}{s}\right]\right\}^{-1}$$

$$= \frac{1}{(\Lambda_\gamma^{\mp})^2} , \tag{2.11}$$

where we have introduced the unmodified Bhabha cross section $d\sigma_B/d\Omega$. The remarks made in connection with (2.9) apply here in the same way. Here we have made the tacit assumption that Λ_γ is the same in the spacelike as in the timelike region. Otherwise analyticity (crossing symmetry) would be violated. This could be tested independently by measuring Λ_γ in both regions (there is, however, the unphysical range).

---

* In general, form factors become complex in this case. However, in testing QED we can leave this effect out of consideration.

Another possibility for testing the photon propagator in the timelike region is lepton pair production $e^+e^- \rightarrow \mu^+\mu^-$ or $e^+e^- \rightarrow \tau^+\tau^-$, because only the annihilation diagram occurs. With the assumption of analyticity (crossing symmetry) we can determine Λ_γ in a similar way as in (2.11). However, in this case, there is another tacit assumption included, namely μ–e or τ–e universality. The form factor product similar to (2.6) now also contains the vertex form factor of the μ or τ

$$F_{e\mu(e\tau)}(q^2) = F_{1e}(q^2)F_\gamma(q^2)F_{1\mu(1\tau)}(q^2)$$

with (2.12)

$$\frac{d\sigma^{l^+l^-}}{d\Omega} \Big/ \frac{d\sigma^{e^+e^-}}{d\Omega} = |F_{el}(q^2)|^2, \quad l = \mu, \tau.$$

This experiment could therefore also be used to test μ–e or τ–e universality, which especially at high energy–momentum transfer, is by no means self–evident.

Altogether contrary to often given statements, the photon propagator (together with the contiguous vertices) can be tested in a model–independent way under the above–mentioned assumptions. Then $1/\Lambda_\gamma$ is a model–independent measure of how far the theory has been tested in this respect. Since the photon propagator $D(q^2)$ is also the Green's function of Maxwell's equations, this test is, roughly speaking, a test of Maxwell's equations together with the form of coupling with the electric current, an upper limit for a modification of the electromagnetic field of the electron.

3. TESTING OF THE ELECTRON PROPAGATOR

The next question concerns the testing of the electron propagator. Here two–quantum pair annihilation seems to be the cleanest process[8] (Fig.2), especially if we notice that there are no weak corrections in the lowest order even in the region of the Z_0 resonance.

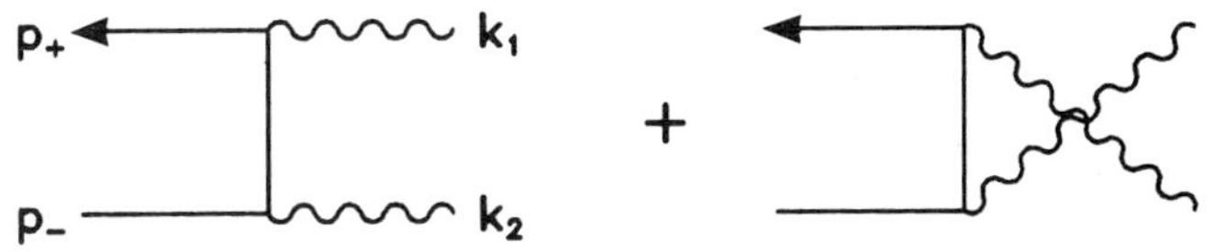

Fig. 2. Diagrams for $e^+e^- \rightarrow \gamma\gamma$ in the lowest order.

The electron propagator contains in the most general case two scalar functions ($\not{p} \equiv \gamma_\mu p^\mu$)

$$S(p) = \frac{G_1(p^2)\not{p} + mG_2(p^2)}{p^2 - m^2} \tag{3.1}$$

compared with the form in lowest order of perturbation theory

$$S(p) = \frac{\not{p} + m}{p^2 - m^2} . \tag{3.2}$$

As is well known, the general vertex $\Gamma_\mu(p, p', k)$ contains up to 12 invariant functions, if all lines are virtual. In the above case with $p_\pm{}^2 = m^2$, $k^2 = 0$ on the mass shell there remain four invariant functions, which are multiplied by the two invariant functions of the electron propagator $S(p)$. Therefore, we obtain in the matrix element a fairly complex expression.

$$\Gamma_\nu(p)S(p)\Gamma_\mu(p) = G_{\mu\nu}(p) \tag{3.3}$$

Hence, the testing of the electron propagator is already much more involved than the testing of the photon propagator. In an earlier work[8] Salecker simplified this expression to get an estimate of the effects to be expected.

The most important difference lies in the fact that the electron line carries charge. To keep the theory self–consistent, charge must be conserved. This requires gauge invariance of the first kind (phase invariance). Besides this, the photon mass is known to be very small.[9] If we assume it to be zero, full gauge invariance is required. This gives already in conventional QED a relation between the vertex and the electron propagator, the well–known Ward–Takahashi equation written symbolically as

$$k \cdot \Gamma_1(q) = S^{-1}(q + k) - S^{-1}(q) . \tag{3.4}$$

With the notation

$$\Gamma_0 \equiv S^{-1} \tag{3.5}$$

we obtain

$$k \cdot \Gamma_1(q) = \Gamma_0(q + k) - \Gamma_0(q) . \tag{3.6}$$

In this form the equation can easily be generalized to n–photon vertices

$$k \cdot \Gamma_n(q) = \Gamma_{n-1}(q+k) - \Gamma_{n-1}(q) \tag{3.7}$$

which has been obtained by many authors.[10] Therefore, if we take a generalized electron propagator S as in (3.1) for testing we get a generalized one–photon vertex Γ_1 as in (3.4), from Γ_1 we get a generalized two–photon vertex Γ_2, etc. In conventional QED Γ_2 and all multiphoton vertices vanish, but this will not necessarily be the case in a generalized theory which we need for testing the electron propagator.

A special solution of this system of equations has been given by Kroll.[1] The general solution is, of course, not unique because a transverse vertex W_n can always be added, i.e., with $k \cdot W_n = 0$. Kroll has especially considered the so–called minimal case where all $W_n = 0$. His solution shows an interesting result, which is of fundamental importance for testing the electron propagator. Any alterations of the electron propagator and the induced vertex modifications [induced through the so–called Chang–Mani equations (3.7)] cancel completely if the electron line is an open one.

Speaking intuitively, charge conservation always requires the same charge on the electron line and this is the observable electron charge because of the open line. This means that the electron propagator cannot be tested with experiments where the diagrams contain only open electron lines in the first order, if we consider the (renormalizable) minimal coupling between current and field.

This result has been extended by Ringhofer[2] to the case where additional generalized vertices W_n are enclosed. The transversality condition $k \cdot W_n = 0$ limits the W_n to such a form that they can only be observed in the lowest (nth) order of perturbation theory if they do not vanish on the mass shell (the anomalous magnetic moment is an example for a W_1 not vanishing on the mass shell). More explicitly, if we split W_n into on– and off–shell parts

$$W_n = W_n^{on} + W_n^{off}, \tag{3.8}$$

it is the effect of W_n^{off} in the matrix elements to change the on–shell parts of the higher order generalized vertices in such a way that no pole terms enter these vertices. It follows, therefore, that in this connection only generalized on–shell vertices can be tested.

As an illustration, consider the simple case of one– and two–photon vertices (Fig.3). We obtain for the operators of the matrix element

$$W_1 S W_1 + W_2 = W_1^{\rm on} S W_1^{\rm on} + \tilde{W}_2 \tag{3.9}$$

with

$$\tilde{W}_2 = W_1^{\rm on} S W_1^{\rm off} + W_1^{\rm off} S W_1^{\rm on} + W_1^{\rm off} S W_1^{\rm off} + W_2 \tag{3.10}$$

Here $k \cdot \tilde{W}_2 = 0$ and the on–shell part of $\tilde{W}_2$ does not contain the electron pole, i.e., it is a true two–photon vertex. Between free electron states only the on–shell part of $\tilde{W}_2$ will contribute. That is, the effect of $W_1^{\rm off}$ in this example is to change the on–shell part of W_2. Therefore, if we examine $e^+e^- \longrightarrow 2\gamma$ experimentally and compare it with theory, we are finally looking for generalized anomalous moments in the seagull diagram, since the anomalous magnetic moment of the electron is much too small to be detected here.

In recent experimental work the so–called $(q/\Lambda)^4$ model is used for testing the electron propagator with reference to the work of Kroll.[1] This is a misunderstanding of Kroll's work. On his page 87, shortly before his equation (89), Kroll introduces this model as an intrinsic vertex modification like our W_n [intrinsic in contrast to the induced vertex modifications Γ_n which are introduced by the Chang–Mani equations (3.7)]. This modification therefore cannot test the electron propagator (as Kroll has already stressed), but only the intrinsic vertex, i.e., in $e^+e^- \longrightarrow 2\gamma$ the magnitude of a possible two–photon vertex (seagull) diagram.

The main assumption for this compensation is the open electron line (if gauge invariance is taken for granted). Therefore, if we consider processes with closed electron loops in the lowest order, the situation changes completely. Scattering processes of this kind are photon–photon and Delbrück scattering as well as photon splitting

Fig. 3. One– and two–photon vertices in a generalized theory.

(Fig. 14). In these cases the electron propagator (with some vertices) can indeed be tested.[11–13] Photon–photon scattering and (perhaps even) photon splitting could also be accomplished with colliding electron–positron beams where two photons are scattered through a photon–photon interaction contribution (Fig.16). We will come back to these interesting processes in chapter 5.

4. SEARCH FOR EXCITED ELECTRONS WITH POLARIZED $e^+e^- \to \gamma\gamma$ AND $e\gamma \to e\gamma$

Although the standard $SU(3)\times SU(2)\times U(1)$ model of strong, electromagnetic, and weak interactions is in fascinating agreement with experiments, in many aspects it seems to be incomplete: It still needs a large number of seemingly unrelated parameters and it does not provide an explanation for such fundamental questions like, e.g., the origin of the repetitive generation pattern of flavors. An appealing approach to some of these questions is the idea of compositeness of the 'elementary' particles.[14] A clear signal of the composite structure is the existence of excited states of leptons and quarks. If the electron, for example, is composite, then its excited states e^*, with masses probably larger than today's available accelerator energies, should also modify the pure QED part of the standard model. QED would become an effective 'low–energy' theory, which has to be corrected for substructure effects at very high energies. In particular, this modifications would affect the propagation function of the electron. Could these substructure effects be tested in simple, pure QED processes like $e^+e^- \to \gamma\gamma$ and $e\gamma \to e\gamma$?

As already stated above, in the minimal modification scheme as introduced by Kroll[1] the electron propagator cannot be tested in processes with open electron lines. However, $e^+e^- \to \gamma\gamma$ and $e\gamma \to e\gamma$ are processes which are described (to lowest order) by Feynman diagrams with open electron lines. One can circumvent the problem only by introducing new intrinsic vertices, as already mentioned. Per definition, vertices of this kind are absent in normal QED, where the electron field is coupled *minimally* to the electromagnetic field.

An intrinsic vertex W_1 (Fig. 4) can be given by, e.g., an *intrinsic* anomalous magnetic moment of the electron leading to an *intrinsic* anomaly $a_i \equiv \frac{1}{2}(g_i-2)$, which has to be clearly distinguished from the anomalous magnetic moment a_{th}, which can be calculated from higher–order perturbation theory in QED including corrections from weak and strong interactions.[15] A usual way to incorporate such intrinsic vertices in the standard theory is to add corresponding effective Lagrangian terms $\mathscr{L}_{eff}$ to the original Lagrangian. The Feynman rules, then, give new vertices represented by

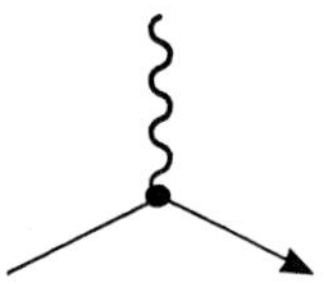

Fig. 4. Intrinsic one–photon vertex W_1.

diagrams as above. To get a_i, for instance, we have to add a term

$$\mathscr{L}_{\text{eff}} = e\frac{1}{\Lambda}\,\bar{\psi}\,\sigma_{\mu\nu}\,F^{\mu\nu}\,\psi \tag{4.1}$$

where e is the electron charge, ψ is the Dirac field of the electron , $F^{\mu\nu}$ is the electromagnetic field tensor, and $\sigma^{\mu\nu} = i/2(\gamma^\mu\gamma^\nu - \gamma^\nu\gamma^\mu)$. The parameter Λ has the dimension of a mass. Since this coupling Lagrangian contains $F^{\mu\nu}$ instead of A^μ it is obvious that W_1 fulfills the transversality requirement $W_1 \cdot k = 0$. Λ is related to a_i by $a_i = 4m/\Lambda$, where m is the mass of the normal electron. If we ascribe a possible difference between the measured anomalous magnetic moment a_{exp} and a_{th} to a_i, we end up with an extremely small value [15]

$$|a_i| \lesssim 2 \times 10^{-10}$$

or, equivalently, $|\Lambda| \gtrsim 10^4$ TeV ! We notice already here that such an interaction does not make sense in this region for some other physical reasons. It is not renormalizable and will violate unitarity there strongly as will be shown in another work.[16]

Another example of an intrinsic vertex is the 2–photon vertex W_2 constructed by Kroll[1] in his $1/\Lambda^4$ model as mentioned above (Fig. 3). It can be derived from an effective Lagrangian proportional to

$$\mathscr{L}_{\text{eff}} = e^2\frac{1}{\Lambda^4}\,F^{\mu\nu}\,F^{\alpha\beta}\{\bar{\psi}\,\sigma_{\mu\nu}\,(i\partial_\tau\gamma^\tau - m)\,\sigma_{\alpha\beta}\,\psi\} + \text{h.c.} \tag{4.2}$$

and was introduced as a QED test model in e^+e^- pair creation experiments. Seagull terms can also be tested in $e^+e^- \to \gamma\gamma$.

In the context of substructure these effective Lagrangians are looked at as low–energy effects of a yet unspecified sub–constituent dynamics. Λ is usually referred to as the 'compositeness scale' or 'substructure scale'. If such effective Lagrangians are added to QED or the standard model for modification, the theory looses some of its most beautiful features: renormalizability and unitarity. One assumes implicitly that these modification terms are valid only for energies $\sqrt{s}$ small compared to the substructure scale. One must be aware that these additional Lagrangians cannot be used for higher–order loop calculations because the resulting infinities cannot be absorbed into physical quantities by renormalization. On the other hand one must take care that the energy $\sqrt{s}$, at which the modifications are used for a parametrization of deviations from the standard model, is small enough compared to the substructure scale, so no unitarity–violating partial waves are present in the test amplitudes[16].

Now, if we add suitable intrinsic vertices, the electron propagator or, equivalently, the structure of the electron can be tested in processes with open electron lines. For instance, we may construct an intrinsic vertex $\hat{W}_1$ which couples the electron ground state via one photon to some excited state e*, which, in turn, modifies the propagation of the electron.

e* Test Models

Intrinsic vertices by which the electron is coupled to an excited state can be described with various effective Lagrangians. For the sake of lucidity we restrict our considerations to point–like couplings of the electron and one photon to excited states of spin $\frac{1}{2}$ and spin $\frac{3}{2}$. The following simple forms were derived from Lorentz invariance principle:

spin–$\frac{1}{2}$ e*:

$$\mathscr{L}_1 = e \frac{\lambda}{2M} \bar{\psi} \sigma_{\mu\nu} F^{\mu\nu} \psi' + \text{h.c.} \tag{4.3}$$

where ψ and ψ' are the Dirac fields of the normal and the excited electron, respectively. M is the mass of e*. We introduce the dimensionless coupling parameter λ in order to allow for a substructure scale M/λ different from the e* mass M.

spin–$\frac{3}{2}$ e*:

$$\mathscr{L}_2 = e \frac{\lambda}{M} \bar{\psi} \gamma_\mu F^{\mu\nu} \psi_\nu + \text{h.c.} \tag{4.4}$$

ψ_ν is the Rarita–Schwinger[17] (RS) vector spinor field of the spin–$\frac{3}{2}$ e*.

These models can be modified to respect a chiral symmetry of the electron field ψ by the substitution

$$\psi \rightarrow (1 \pm \gamma^5)\psi .$$

This is motivated by the fact that, in some models with excited electrons, this chiral coupling suppresses an effect of e* on the magnetic moment of the electron ground state compared to the chiral non–symmetric coupling by a factor of $O(m/M)$ (Ref. 18). This is important, if one wants to preserve the beautiful correspondence of a_{th} and a_{exp} also in these models when the mass of e* is not too large. However, P– and C–invariance is no longer preserved in chiral models of this kind. The implications of this property will be discussed in the context of e* tests with polarized $e^+e^- \rightarrow \gamma\gamma$ and $e\gamma \rightarrow e\gamma$ later on.

For the propagators – which we want to test – we take the expressions that can be derived from relativistic quantum field theory. For spin $\frac{1}{2}$ we get the usual Feynman propagator for a fermion with mass M

$$S(p) = \frac{\not p + M}{p^2 - M^2} \, . \tag{4.5}$$

For the e* field of spin $\frac{3}{2}$ we use the RS–propagator in the form

$$S_{\mu\nu}(p) = \frac{\not p + M}{p^2 - M^2} \left(- g_{\mu\nu} + \frac{1}{3} \gamma_\mu \gamma_\nu + \frac{1}{3M} (\gamma_\mu p_\nu - \gamma_\nu p_\mu) + \frac{2}{3M^2} p_\mu p_\nu \right) . \tag{4.6}$$

We want to mention that the propagator of the RS–field is not unique due to the invariance of the free RS–field under the point transformation[19]

$$\psi_\mu \longrightarrow \psi_\mu + a \, \gamma_\mu \, \gamma^\nu \, \psi_\nu \, , \; a \in \mathbb{R}, \; a \neq -\tfrac{1}{4}.$$

Our specific choice (4.6) results in a relatively simple expression and is also motivated from hadron physics calculations for spin–$\frac{3}{2}$ hyperons (e.g., Ref. 20).

With the new couplings (4.3, 4.4) and propagators (4.5, 4.6) we can construct test amplitudes for $e^+e^- \rightarrow \gamma\gamma$ and $e\gamma \rightarrow e\gamma$ with the help of the usual Feynman rules. These amplitudes, with λ and M as test parameters, have to be added to the amplitudes from the standard model.The lowest–order Feynman diagrams are shown in Fig. 5.

It is noteworthy that in the limit $\sqrt{s} \ll M$ these new diagrams can be reduced to point–like 2–photon contact vertices (seagull vertices) (Fig. 6). Hence, it may be justi-

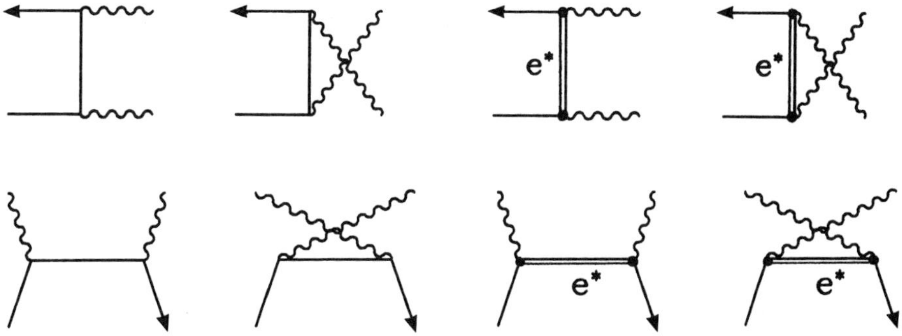

Fig. 5. Lowest–order Feynman diagrams for $e^+e^- \to \gamma\gamma$ and $e\gamma \to e\gamma$ including e* exchange.

fied to relate certain models with 2–photon vertices to propagator tests. But there, one has to distinguish carefully between the various models.

e* Search in $e^+e^- \to \gamma\gamma$ and $e\gamma \to e\gamma$ at High Energy Colliders with Polarized Beams

For future accelerators beam polarization will become an appealing feature, since it permits additional tests of the standard model[21] and of substructure effects[22] in the scattering experiments. At the linear e^+e^- collider at Stanford (SLC) transverse and longitudinal polarization is possible at least for the electron beam.[23] At e^+e^- storage rings (e.g. LEP) polarized beams are available through the 'natural' transverse beam polarization during beam storage.[24] With additional spin rotators transverse polarization can then be transformed into longitudinal polarization.[25]

The possibility of performing direct $e\gamma$ collisions at linear e^+e^- colliders with the help of power lasers was noticed some years ago.[26–28] In short, it is proposed to use the collision of a laser beam ($E_\gamma \simeq 4$ eV) with one $e^\pm$ beam of the accelerator

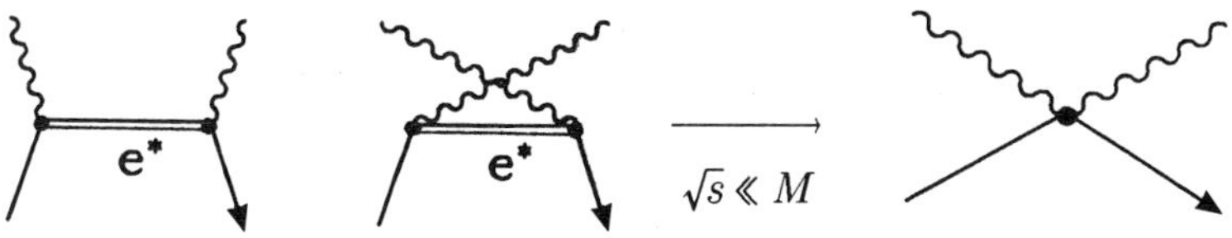

Fig. 6. Reduction of e* exchange diagrams to an intrinsic 2–photon vertex in the limit $\sqrt{s} \ll M$ for Compton scattering.

($E_e \simeq 50$ GeV) to produce an intense high–energy photon beam ($E_\gamma \simeq 40$ GeV) via backward Compton scattering. The ultrahigh–energetic γ beam goes exactly in the direction of the primary $e^\pm$ beam and can then collide head–on with the second $e^\pm$ beam ($\sqrt{s} \simeq 80$–90 GeV). The luminosity for the eγ collision is expected to be slightly below the designed e^+e^- luminosity.[27] An important feature of this γ production mechanism is the possibility of polarizing the high–energy photons just by polarizing the low–energy laser beam.[29] It is not necessary to polarize the $e^\pm$ beam involved in the γ production. At storage rings this method of high–energy photon production is not possible. Here, one must use the Weizsäcker–Williams photons of one beam to perform quasi–real Compton scattering off the particles of the other beam. Since the high energy quasi–real photon will inherit the helicity of the emitting $e^\pm$ almost completely, a longitudinally polarized beam can be an ideal source of circularly polarized high energy Weizsäcker–Williams photons.[30]

These experimental possibilities have motivated us to study the effects of e* according to the test models as introduced above in high–energy $e^+e^- \to \gamma\gamma$ and $e\gamma \to e\gamma$ also for polarized initial states[31]. The kinematical situations at the SLC (or at LEP I at CERN) for these processes are depicted in Fig. 7 and 8. For $e^+e^- \to \gamma\gamma$ the laboratory system is identical to the center–of–mass system (c.m.s.) with $\sqrt{s} \simeq 100$ GeV (Fig. 7). Our numerical examples for $e\gamma \to e\gamma$ were calculated for an 'almost' c.m.s. with $E_\gamma \simeq 36$ GeV and $E_e \simeq 50$ GeV, $\sqrt{s} \simeq 85$ GeV (Fig. 8). The polarization parameters for the incoming particles are related to the scattering plane (xz–plane) and to the helicity frames of the particles[32,33] (the y–axis of the helicity frames are parallel to the y–axis of the laboratory frame). The spin of the electron (positron) is described by the spin components s_{N1}, s_{S1}, and s_{L1} (s_{N2}, s_{S2}, s_{L2}) in the helicity frame of the particle, $-1 \leq s_i \leq 1$. The circular polarization of the photon is characterized by $p_C \in [-1,1]$. The description of the linear polarization of the photon is a little more involved: Let p_L be the degree of linear polarization, $0 \leq p_L \leq 1$. Then we define $p_Q \equiv p_L \sin 2\psi$ and $p_P \equiv p_L \cos 2\psi$ with $p_Q, p_P \in [-1,1]$, where $\psi \in [0,\pi]$ is the angle between the scattering plane (x–axis of the helicity frame) and the polarization plane (Fig. 8).

The polarized differential cross section is most clearly expressed in terms of the unpolarized differential cross section, the polarization asymmetries (PA), and the polarization parameters of the particles.[32] Polarization asymmetries, which in our case refer to the scattering plane and are defined for polarized initial states only, are obtained by relative measurements of the cross section with different beam polarizations when the degree of polarization for every measurement is known. For $e^+e^- \to \gamma\gamma$ and polarized initial states we have, in principle, to calculate 15 different PA. C– and

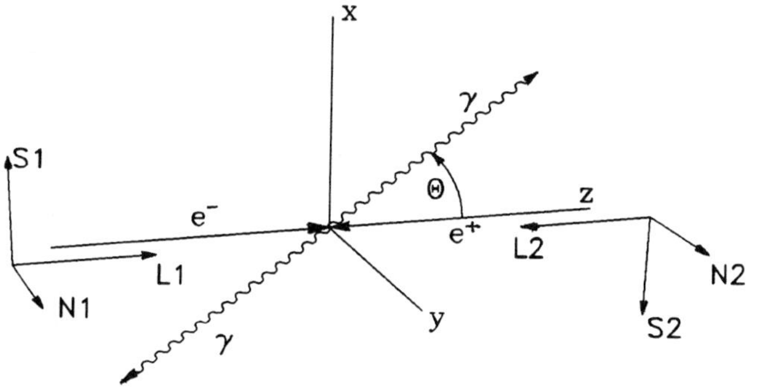

Fig. 7. Kinematics and polarization parameters for $e^+e^- \rightarrow \gamma\gamma$ (SLC or LEP I). The xz–plane is the scattering plane. L.., S.., N.. signify the spin directions in the helicity frames of the fermions.

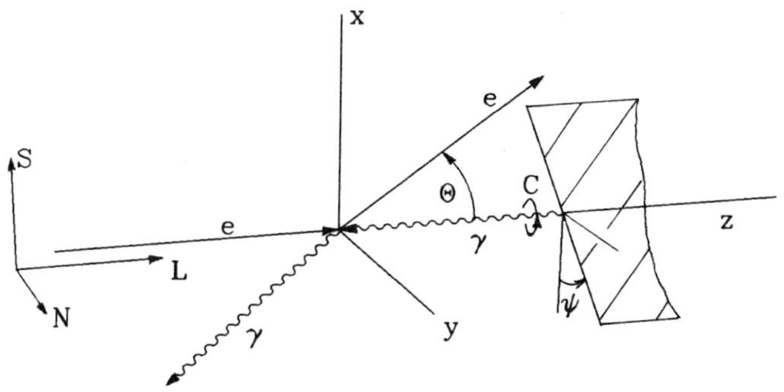

Fig. 8. Kinematics and polarization parameters for $e\gamma \rightarrow e\gamma$ (SLC). The xz–plane is the scattering plane. L, S, N signify the spin directions in the helicity frame of the fermion. C signifies the helicity of the photon. ψ is the angle between the scattering plane and the polarization plane of the photon (hatched).

P– invariance of the scattering process can reduce this number to 6. In lowest–order QED only 3 PA survive in high–energy approximation. This remains true even when we include test amplitudes from our e* models, if they are constructed to obey C– and P– invariance. All of these PA are 2–particle PA, which means that both beams have to be polarized for a measurement. Hence, for $e^+e^- \rightarrow \gamma\gamma$ in lowest–order QED including e* exchange amplitudes the polarized differential cross section in high–energy approximation can be written in the form

$$\frac{d\sigma}{d\Omega}_{pol} = \frac{d\sigma}{d\Omega}_{unpol} \left\{ 1 + s_{N1} s_{N2} A_{NN} + s_{L1} s_{L2} A_{LL} + s_{S1} s_{S2} A_{SS} \right\} . \qquad (4.7)$$

A_{NN}, A_{LL}, and A_{SS} are fairly lenghty expressions, which are most easily derived from the helicity amplitudes of the process.[34,32,33] The detailed form of the unpolarized differential cross section and of A_{NN}, A_{LL}, and A_{SS} in (4.7) can be found elsewhere.[31]

For $e\gamma \to e\gamma$ with polarized initial states, there exist 15 different polarization asymmetries, in principle. P–invariance reduces this number to 7. So, analogously to $e^+e^- \to \gamma\gamma$, we write the polarized differential cross section in the form

$$\frac{d\sigma}{d\Omega}_{pol} = \frac{d\sigma}{d\Omega}_{unpol} \Big\{ 1 + s_N A_N - p_P A_P + s_L p_C A_{LC} + s_S p_C A_{SC} + s_L p_Q A_{LQ} \quad (4.8)$$
$$+ s_N p_P A_{NP} + s_S p_Q A_{SQ} \Big\} .$$

In lowest–order QED only A_{LC} is present in high–energy approximation. The polarization asymmetries and the unpolarized cross section have been calculated elsewhere.[16] There, one can also find the expressions for the helicity amplitudes, which have been calculated with a special method[33] which was very useful for our applications.

<u>Limits on Test Parameters from Future Experiments without Beam Polarization</u>

First, we want to give some idea of the limits on the test parameters λ and M, which can be expected from future $e^+e^- \to \gamma\gamma$ and $e\gamma \to e\gamma$ experiments without beam polarization under the assumption that *no* signal of e^* will be found or, equivalently, if the experiment turns out to be consistent with the standard model within the experimental error. Later on, we shall study the effect of beam polarization, too.

For this purpose it is useful to define a relative deviation δ of the differential cross section in the presence of test models $d\sigma$ from the pure QED value $d\sigma_{QED}$ by

$$\frac{d\sigma}{d\sigma_{QED}} = 1 + \delta , \quad (4.9)$$

where δ is a function of the model parameters λ and M and the scattering angles and depends also on the polarization of the particles. The theoretically predicted deviation δ must be compared to the relative experimental uncertainty δ_{exp}. The experiment is consistent with QED predictions if

$$\delta \lesssim \delta_{exp}.$$

This condition restricts the possible values for λ and M. Assuming $\delta_{exp} = 5\%$, we find for unpolarized $e^+e^- \to \gamma\gamma$ at $\sqrt{s} = 100$ GeV and for $e\gamma \to e\gamma$ at the SLC ($\sqrt{s} \approx 85$ GeV) limits as shown in Fig. 9. Of course, in a realistic experiment a detailed statistical analysis based on specific data of the machine and the detectors will be performed to derive such bounds. But for a rough theoretical estimate our approach is adequate.

Effects of Polarization on e* Search in $e^+e^- \to \gamma\gamma$ and $e\gamma \to e\gamma$

Now, one may ask, whether scattering experiments with polarized beams will be advantageous compared to the unpolarized experiments in the search for e*. For $e^+e^- \to \gamma\gamma$ we have found that polarization of both beams can considerably enhance the sensitivity to e* of spin $\frac{1}{2}$ and $\frac{3}{2}$ (Ref. 31). By this we mean that an experiment with polarized beams gives stronger limits on the e* test parameters, if no deviation from the standard model predictions is found within a certain experimental error. This is

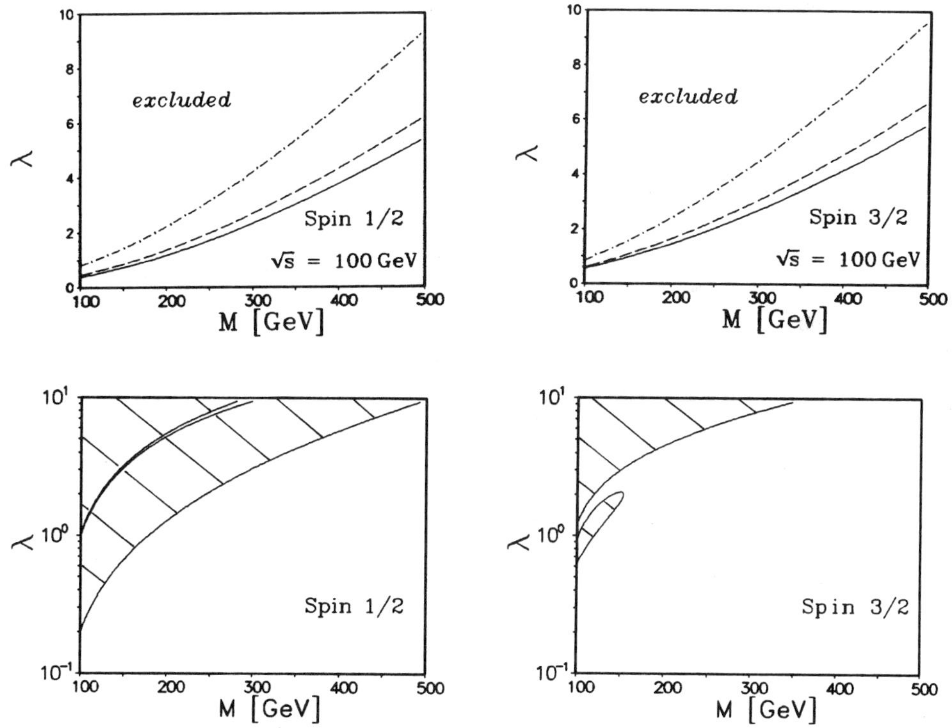

Fig. 9. Boundaries for the coupling strenght λ and the mass M in QED models with excited electrons. A deviation of the unpolarized differential cross section from QED or experimental error of $\delta_{exp} = 5\%$ is assumed. Top: Unpolarized $e^+e^- \to \gamma\gamma$ at a c.m.s. energy of 100 GeV at scattering angles 30° (– · –), 60° (– – –), and 90° (——) (from Ref. 31). Bottom: Unpolarized $e\gamma \to e\gamma$ at the SLC ($\sqrt{s} \approx 85$ GeV) at scattering angles of 90° (excluded region hatched).

shown in Fig. 10, where both beams are transversely polarized. A similar result can be found, if both beams are longitudinally polarized.[6,31] Roughly, this comes from the property of QED to give a very small scattering amplitude for $e^+e^- \to \gamma\gamma$ in the high–energy limit if both fermions have the same helicity. In this way, the background from pure QED can be suppressed compared to the e* test model amplitudes by polarizing the initial state properly. Once signals of e* have been found, polarization asymmetries will provide additional information on the type of coupling. If chiral couplings are present, one–particle asymmetries like A_{L1} and A_{L2}, detectable with longitudinally polarized electron (positron) beams, become non–zero.

For $e\gamma \to e\gamma$ the situation is a little bit more complicated, due to the fact that here the excited electrons can be produced on–shell in the *s*–channel. In contrast to $e^+e^- \to \gamma\gamma$ the suppression of QED background is feasible only at very small scattering angles with a very high degree of polarization of both beams.[16] This restricts the use of polarization to the situation where excited states have already been found in the unpolarized experiment. In particular, near the resonance pole it is possible to measure the imaginary part of the e* test model amplitude. This is important, if additional decay channels different from $e^* \to e\gamma$ are open. Then, the total decay width Γ can exceed the partial width $\Gamma(e^* \to e\gamma)$, which is quite small on the order of $O(\lambda^2\alpha)$ (Ref. 16), and the polarization asymmetry A_{NP} becomes non–zero (Fig. 11). A_{NP} can be measured with transversely polarized electrons and linearly polarized photons (4.8). In the case of chiral couplings one–particle asymmetries like A_L and A_C, which are absent in a P–invariant theory, become non–zero. A_L and A_C can be measured with longitudinally polarized electrons and circularly polarized photons, respectively.

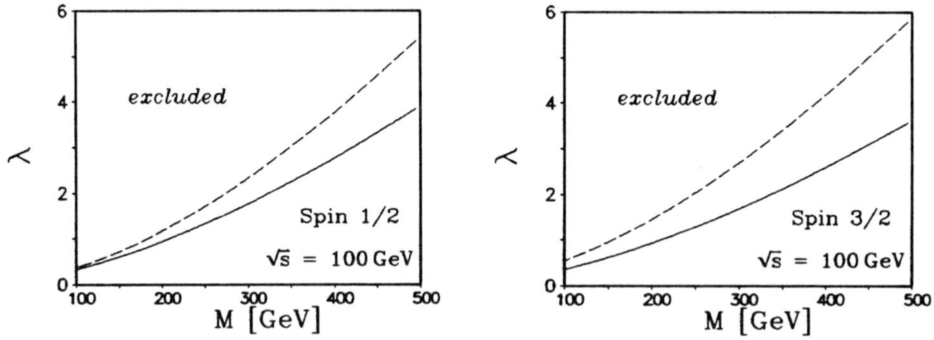

Fig. 10. Effect of transverse polarization on the limits on λ and M in $e^+e^- \to \gamma\gamma$. In 90° scattering a deviation of the polarized and unpolarized differential cross section from QED or, equivalently, an experimental error of $\delta_{exp} = 5\%$ is assumed (from Ref.31). Dashed line: unpolarized beams. Solid line: both beams 90% transversely polarized with $s_{S1}s_{S2} = 0.81$

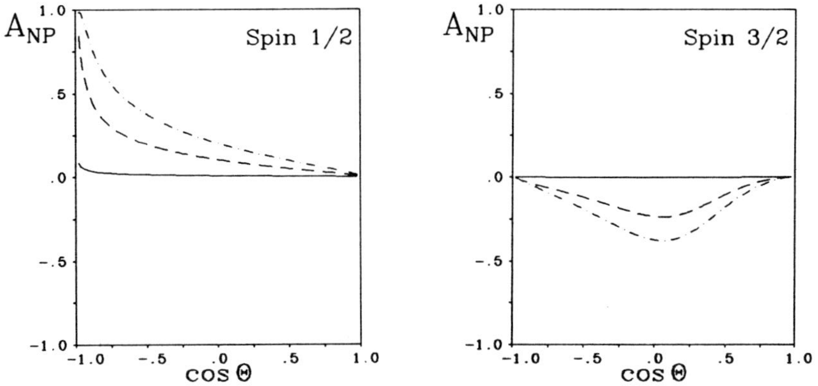

Fig. 11. Two–particle polarization asymmetry A_{NP} in $e\gamma \to e\gamma$ at the resonance of an e* (SLC, $\sqrt{s} = M = 85$ GeV) for different values of Γ. (——) $\Gamma = \Gamma(e^* \to e\gamma)$, (– –) $\Gamma/M = 5\%$, (– · –) $\Gamma/M = 10\%$.

5. HIGH ENERGY TESTS OF THE NONLINEAR EFFECTS IN QED

Vacuum Polarization

The Maxwell equations in vacuum are linear in the field–strength. This linearity of classical electrodynamics has the consequence that light can only interfere with light because of the superposition principle and cannot be scattered by light.

In QED, however, antiparticles must necessarily exist, and the vacuum becomes polarizable through pair productions and annihilations. Therefore, interactions of light with light, for instance scattering of light by light through intermediate vacuum polarizations, is possible. This polarizability of the vacuum implies nonlinear interactions between electromagnetic fields.

The vacuum polarization in the lowest order is of the photon self–energy type with two external real or virtual photons (Fig.12(a)), which yields after renormalization non–negligible contributions to the Lamb–shift, the anomalous magnetic moment of the electron and muon, and radiative corrections of scattering processes.

So far as only electromagnetic interactions are concerned, all vacuum polarization diagrams with odd vertices vanish because of the Furry theorem. So we have in the next order vacuum polarization loops of the photon–photon scattering type with four external real or virtual photons (Fig.12(b)).

For this type we get in the lowest perturbative order after permutations of the photon momenta altogether 6 diagrams each with four vertices, so called box diagrams (Fig.13). If the four photon lines are amputated, we obtain the fourth rank vacuum polarization tensor.

$$G_{\mu\nu\rho\sigma} = T_{\mu\nu\rho\sigma}(1234) + T_{\rho\nu\mu\sigma}(3214) + T_{\nu\mu\rho\sigma}(2134) \ ,$$

$$T_{\mu\nu\rho\sigma} = \frac{1}{i\pi^2} \int d^4p \ \mathrm{tr} \left[\gamma_\mu(\not{p} - m)^{-1}\gamma_\nu(\not{p} - \not{k}_2 - m)^{-1} \right.$$

$$\left. \times \ \gamma_\rho(\not{p} + \not{k}_1 + \not{k}_3 - m)^{-1}\gamma_\sigma(\not{p} + \not{k}_1 - m)^{-1} \right] \ , \qquad (5.1)$$

where m is the mass of the loop particle. Here the usual numerical factors of the Feynman rules are omitted. The three tensors T correspond to the diagrams shown in Fig.13. The degree of divergences of each T is zero, i.e., T is logarithmically divergent. These divergences, however, totally cancel in the sum in $G_{\mu\nu\rho\sigma}$, which thus remains finite. This is just a consequence of gauge invariance of QED.

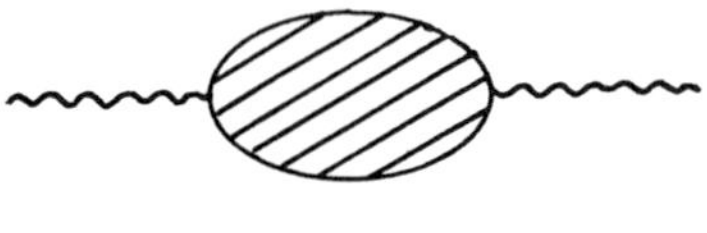

a

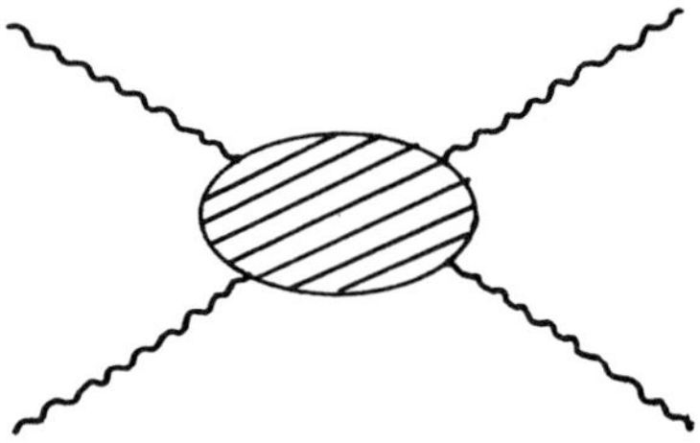

b

Fig. 12. The two simplest types of the vacuum polarization

Main Nonlinear Effects in QED

So far, the tensor $G_{\mu\nu\rho\sigma}$ is not yet analytically calculated in full generality, but analytically exact expressions exist for those cases in which at least two photons are real.[35] These cases provide just some important nonlinear effects in QED. They are Delbrück scattering, photon splitting, and photon–photon scattering (Fig.14). Now we have three main nonlinear effects determined by the closed loop, which hence makeit

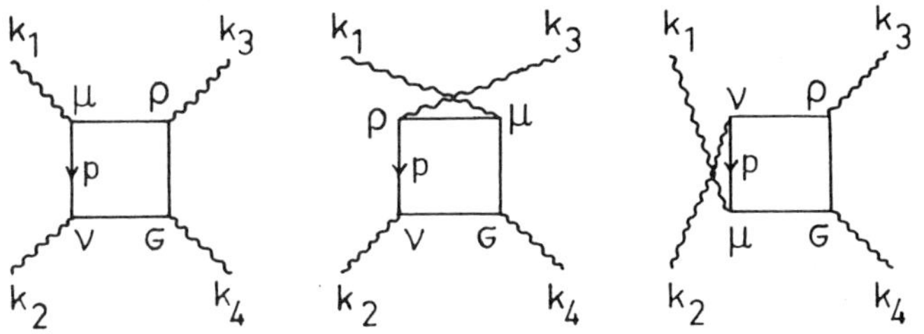

Fig. 13. Vacuum polarization of the photon–photon scattering type in the lowest order. There are 3 further diagrams of this sort which differ only by the arrow directions in the closed loops. All photon momenta are counted as outgoing.

possible to test the electron propagator in a gauge invariant manner, as mentioned previously.

The only one of them which has been experimentally observed so far is Delbrück scattering,[36] while the other effects remain at present unobserved. Delbrück scattering has, however, to be observed in the laboratory system, for which the reaction energy is limited to the energy of the incident beam. Photon splitting is fairly difficult from the experimental point of view.[13] So photon–photon scattering is the most suitable non-linear effect for high energy tests of the electron propagator. Possibilities for high energy photon–photon scattering have recently been proposed by some authors.[26,27] The proposed schemes are based on the converting of the electrons in e^+e^- colliding beams into photons through backward Compton scattering of laser light.

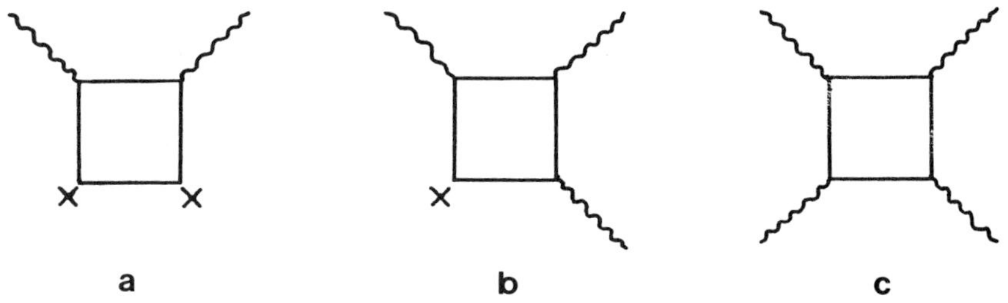

Fig. 14. (a) Delbrück scattering, (b) photon splitting, and (c) photon–photon scattering.

In the next section we shall consider the photon–photon scattering in detail. But the other effects have also been studied theoretically in Refs. 35 and 12 for Delbrück scattering and Refs. 35 and 13 for photon splitting.

Photon–Photon Scattering

The vacuum polarization tensor G was calculated by Costantini et al.[35] for these three cases. For the photon–photon scattering we have to multiply the tensor $G_{\mu\nu\rho\sigma}$ with the polarization vectors of the real photons. If we choose circular polarizations, we have the following helicity amplitudes

$$M_{\lambda_1\lambda_2\lambda_3\lambda_4} = \frac{1}{4}\, G_{\mu\nu\rho\sigma}\, e_1^{\mu}(\lambda_1)\, e_2^{\nu}(\lambda_2)\, e_3^{\rho}(\lambda_3)^*\, e_4^{\sigma}(\lambda_4)^* \,, \tag{5.2}$$

where $\lambda_i = +1$ and $\lambda_i = -1$ are valid for positive and negative helicity, respectively. The differential cross section is now expressed by

$$d\sigma_{\lambda_1\lambda_2\lambda_3\lambda_4}/d\Omega = \frac{\alpha^4}{4\pi^2\omega^2}\, |M_{\lambda_1\lambda_2\lambda_3\lambda_4}|^2 \tag{5.3}$$

for circularly polarized photons and by

$$d\sigma_{\text{unpol}}/d\Omega = \frac{\alpha^4}{4\pi^2\omega^2}\, \frac{1}{4} \sum_{\lambda} |M_{\lambda_1\lambda_2\lambda_3\lambda_4}|^2 \tag{5.4}$$

for unpolarized photons, where ω is the photon energy in the center of mass system of the incoming photons.

Some of these helicity amplitudes are not necessarily different because of symmetry properties. Invariance under parity transformation of QED requires

$$M_{\lambda_1\lambda_2\lambda_3\lambda_4} = M_{-\lambda_1-\lambda_2-\lambda_3-\lambda_4} \,, \tag{5.5}$$

and invariance under time reversal yields

$$M_{\lambda_1\lambda_2\lambda_3\lambda_4} = M_{\lambda_3\lambda_4\lambda_1\lambda_2} \,. \tag{5.6}$$

Finally, since two identical bosons appear in the initial and final states, M must be invariant under the exchange of these photons in each state.

$$M_{\lambda_1\lambda_2\lambda_3\lambda_4} = M_{\lambda_2\lambda_1\lambda_4\lambda_3} \tag{5.7}$$

In this way, from altogether 16 helicity amplitudes we have to calculate only 5. The polarization sum in (5.4) will take, e.g., the following form:

$$\frac{1}{4}\sum_{\lambda} |M_{\lambda_1\lambda_2\lambda_3\lambda_4}|^2 = \frac{1}{2}\Big[\, |M_{++++}|^2 + |M_{++--}|^2$$

$$+ |M_{+-+-}|^2 + |M_{+--+}|^2 + 4|M_{+++-}|^2 \Big]\,. \tag{5.8}$$

The expressions for these 5 basic helicity amplitudes calculated with a set of special representations for the photon polarization[35] are given in the appendix.

In the low energy limit ($\omega \ll m$) they become much simpler (see (A.4)) and yield for the unpolarized differential cross section[35,37]

$$d\sigma_{\text{unpol}}/d\Omega = \frac{139}{8100}\,\frac{\alpha^4}{4\pi^2 m^2}\left[\frac{\omega}{m}\right]^6 (3 + \cos^2\theta)^2\,, \tag{5.9}$$

where θ is the photon scattering angle in the center of mass system. We obtain the total cross section by integrating (5.9) over θ

$$\sigma_{\text{unpol}} = \frac{973}{10125}\,\frac{\alpha^4}{\pi m^2}\left[\frac{\omega}{m}\right]^6 . \tag{5.10}$$

Because of the factor $(\omega/m)^6$ only the electron loop needs to be considered. The magnitude of σ is, for instance, $6\cdot 10^{-64}$ cm^2 for $\omega = 2.1$ eV (D–line of Na). This fact means that the superposition principle of classical electrodynamics is very well valid in the visible region of the electromagnetic spectrum. However, the existence of small deviations from it is conceptually not to be overlooked. Perhaps we may expect in the near future experimental developments to realize the measurement of these small values of σ.

In the high energy limit ($\omega \gg m$) there are two different ω–dependences of the differential cross section with regard to the photon scattering angle. For $\theta = 0^\circ$ we have[11,38]

$$d\sigma_{\text{unpol}}/d\Omega = \frac{\alpha^4}{\pi^2\omega^2}\ln^3\left[\frac{\omega}{m}\right]\cdot\left[\ln\left[\frac{\omega}{m}\right] + 2(2\ln 2 - 1)\right] \tag{5.11}$$

For $\theta \neq 0^o$ the helicity amplitudes take quite different forms given by (A.5) in the appendix, which are independent of the particle mass m in the loop and the photon energy ω. Therefore, in the region of $\omega \gg m$ and $\theta \neq 0^o$ the contributions of all the loop particles with masses m_i much lower than ω must be taken into account by summing the vacuum polarization tensor $G_{\mu\nu\rho\sigma}(m_i)$ over these flavors.

The angular distributions of the unpolarized differential cross section for various values of ω are demonstrated in Fig.15(a), and the total cross section in Fig.15(b). Here only the electron loop is considered. Similar curves have been given in Ref.37. In the high energy region the forward peak becomes narrower with increasing ω because of the different ω–dependence at $\theta = 0^o$ (eq.(11)) and $\theta \neq 0^o$(eqs.(A–5)). The total cross section shows the above mentioned ω–dependence $(\omega/m)^6$ for $\omega \ll m$ and $1/\omega^2$ for $\omega \gg m$. The other prominent feature is a discontinuous slope at $\omega = m$. This is physically quite natural. For $\omega < m$ the photons can only be scattered, i.e., only the dispersive parts of the helicity amplitude are non–zero. For $\omega > m$, however, the photon–photon scattering can also occur via real intermediate e^+e^- pair production and annihilation. Therefore, the absorptive parts will become also non–zero (see (A.3)). So the total cross section has a discontinuity of the slope at the threshold $\omega = m$.

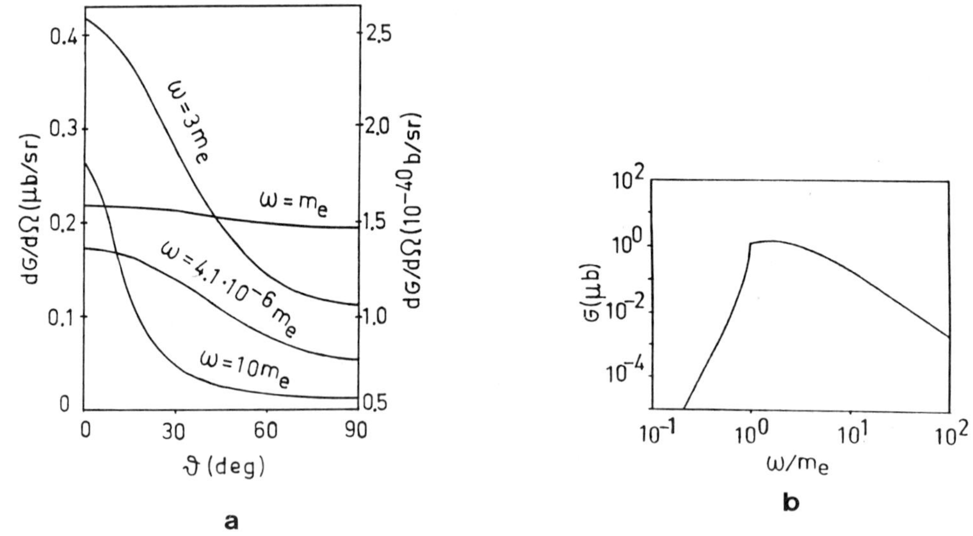

Fig. 15. (a) Unpolarized differential cross sections of photon–photon scattering which are symmetric about $\theta = 0^\circ$ and 180°. For $\omega = 4.1 \cdot 10^{-6}\ m_e$ the right scala is valid. (b) Unpolarized total cross section of the same scattering for which both axis are logarithmic.

Nearly Real Photon–Photon Scattering

At present no high energy photon beams suitable for photon–photon scattering are available. The high energy test of the electron propagator via photon–photon scattering is not yet realizable. There are, however, other possibilities to realize similar processes, using the e^+e^- storage rings. In the two–photon–processes $e^+e^- \longrightarrow e^+e^-\gamma^*\gamma^* \longrightarrow e^+e^-\gamma\gamma$ virtual photon–photon scattering as a part of the whole process can occur as shown in Fig.16(a). $q_1 = p_3 - p_1$ and $q_2 = p_4 - p_2$ are two transversal virtual photons with well defined momentum. If such kinematic conditions as q_1 and q_2 being nearly real (q_1^2, $q_2^2 \simeq 0$) are assumed, we obtain nearly real photon–photon scattering in the intermediate state. Moreover, this part of the process approaches also photon splitting if one of both virtual photons is nearly real (q_1^2 or $q_2^2 \simeq 0$) and Delbrück scattering for both quite virtual photons q_1 and q_2.

Here we choose a symmetrical and coplanar kinematic configuration defined by the relation

$$\vec{p}_1 + \vec{p}_2 = \vec{p}_3 + \vec{p}_4 = \vec{k}_1 + \vec{k}_2 = \vec{q}_1 + \vec{q}_2 = 0\ . \tag{5.12}$$

This is valid in the center of mass system of all the external particles as well as the two virtual photons. In this configuration and in the high energy limit ($p_1^0 = p_2^0 \equiv E$, $p_3^0 = p_4^0 \equiv E\,'$, $k_1^0 = k_2^0 \equiv \omega \gg m_e$), one has

$$q_1^2 = q_2^2 = -\,4EE\,'\,\sin^2(\theta_e/2)\ , \tag{5.13}$$

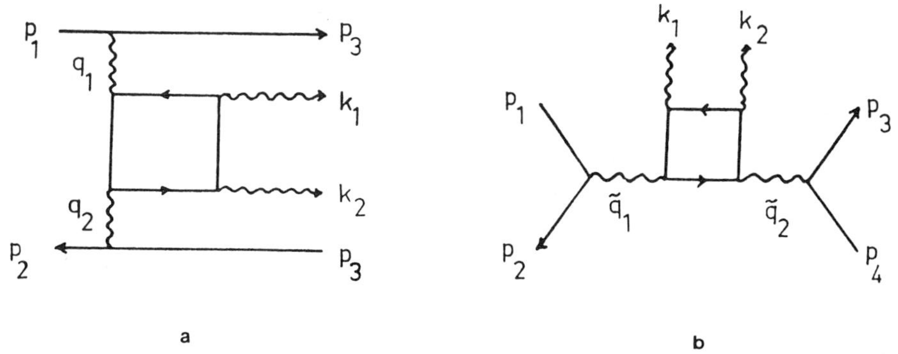

Fig. 16. e^+e^- scattering with intermediate $\gamma\gamma$ scattering. There are 5 further diagrams corresponding to those in Fig. 2 for each type (a) and (b).

where θ_e is the electron scattering angle, and hence nearly real photon–photon scattering in the intermediate state for $\theta_e \ll 1$. For this conditions the cross section will become large because of the photon propagators, and the annihilation diagrams (Fig.16(b)) are negligible.[39]

Of course, the two photons are produced in the lowest order via the double bremsstrahlung shown in Fig.17. Therefore, there are three main contributions to the whole process, i.e., the pure photon–photon scattering contribution $(d\sigma/d\Gamma)_\gamma$, the pure double bremsstrahlung contribution $(d\sigma/d\Gamma)_b$ and their interference terms $(d\sigma/d\Gamma)_{\gamma b}$, where $d\Gamma = d^3p_3 d^3p_4 d\Omega_{k_1}$. It can be shown that the contribution $(d\sigma/d\Gamma)_\gamma$ dominates the other contributions for sufficiently large photon scattering angles θ_γ (Fig. 18), for instance in the region $\theta_\gamma > 8^\circ$ if $E = 20$ GeV, $\omega = 1$ GeV, and $\theta_e = 2$ mrad are assumed, and light leptons e and μ and light quarks u and d in the loop are taken into account.[39]

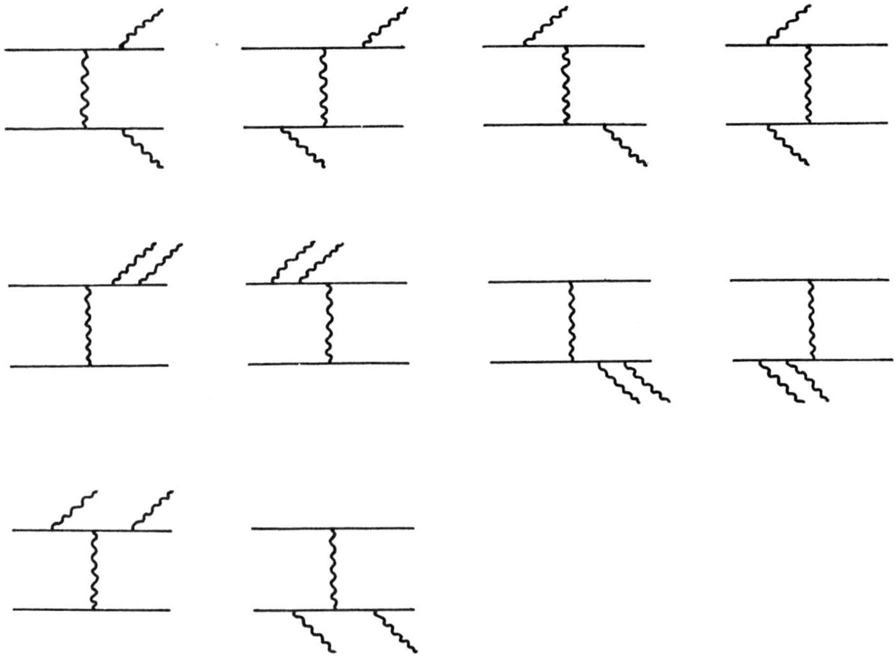

Fig. 17. Double bremsstrahlung. There are altogether 40 diagrams. The other 30 diagrams can be obtained by exchanging the photons and/or the fermions in the final state (annihilation diagrams).

Because of the condition $q_1^2 = q_2^2 \simeq 0$ the well-known equivalent photon approximation[40] can be applied to $(d\sigma/d\Gamma)_\gamma$ and yields (for the exact calculation see Ref.39)

$$\left[\frac{d\sigma}{d\Gamma}\right]_\gamma \simeq \frac{\alpha^2}{(2\pi)^4} \cdot \frac{E'^2}{E^2} \cdot \frac{\omega^2}{\sin^4(\theta_e/2)} \left[\frac{d\sigma_{\gamma\gamma\to\gamma\gamma}}{d\Omega}\right] \left[\frac{1}{|\mathbf{q}_1|^2} + \frac{1}{2EE'}\right] . \tag{5.14}$$

Eq.(14) shows that the θ_γ-dependence of $(d\sigma/d\Gamma)_\gamma$ is approximately the same as that of the differential cross section $d\sigma_{\gamma\gamma\to\gamma\gamma}/d\Omega$ of the real photon-photon scattering. Therefore, in its dominant region real photon-photon scattering can clearly be observed experimentally with the help of this process and used to test the electron propagator of QED in the high energy region.

<u>High Energy Test of the Electron Propagator</u>

For the quantitative test we modify the electron propagator according to Kroll[1] as follows:

$$S(p) = 1/(\not{p} - m) + f/(\not{p} - M) \tag{5.15}$$

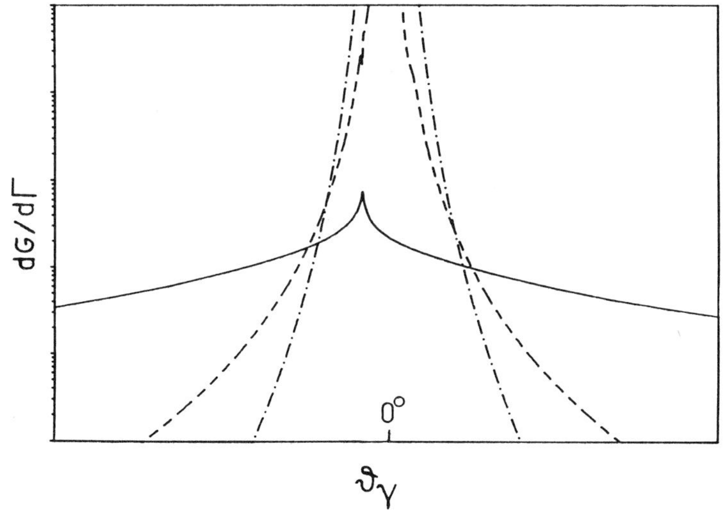

Fig. 18. Qualitative demonstration of the dominance of $(d\sigma/d\Gamma)_\gamma$ in the sideward scattering region: ——— $(d\sigma/d\Gamma)_\gamma$, - - - $(-\,d\sigma/d\Gamma)_{\gamma b}$ —·—·— $(d\sigma/d\Gamma)_b$.

With a special choice of the parameter $f = -1$, one obtains the modified vacuum polarization tensor[1]

$$G^{\text{mod}}_{\mu\nu\rho\sigma} = G_{\mu\nu\rho\sigma}(m) + G_{\mu\nu\rho\sigma}(M) \,. \tag{5.16}$$

Now we introduce again the relative deviation

$$\delta = \frac{(d\sigma/d\Gamma)^{\text{mod}}_{\gamma} - (d\sigma/d\Gamma)^{\text{QED}}_{\gamma}}{(d\sigma/d\Gamma)^{\text{QED}}_{\gamma}} \,, \tag{5.17}$$

where $(d\sigma/d\Gamma)^{\text{mod}}_{\gamma}$ is the differential cross section calculated by using the vacuum polarization tensor (5.16), and $(d\sigma/d\Gamma)^{\text{QED}}_{\gamma}$ the ordinary one of QED. δ is a function of the test mass M, which falls off monotonously with increasing M in the region $M < \omega$ according to the mass dependence of $(d\sigma/d\Gamma)_{\gamma}$ given in (5.9). If no discrepancy between the experiment and the theory is evident, δ has to be put equal to the relative experimental error. In this way, the lower limit of the parameter M will be determined, up to which the electron propagator of QED is tested by the experiment in a gauge invariant manner with the renormalizable interaction given by (5.15).

Some results for this minimal value of M at appropriate photon scattering angles θ_{γ} are shown in Tab.1 for various beam energies E, photon energies ω (here always

Table 1. Minimal values of the testing mass M for $\omega = 5, 15, 60$ GeV.

E (GeV)	ω (GeV)	θ_e (mrad)	flavors	θ_{γ} (deg)	$(d\sigma/d\Gamma)_{\gamma+\gamma b}$ (cm^2/GeV^2sr^3)	M (GeV)
10	5	0.0511	e,μ,τ u,d,s	−25.5 +25.7	0.818 0.809 $\times 10^{-28}$	3.88
30	15	0.0170	e,μ,τ u,d,s,c	−27.5 +27.9	1.288 1.261 $\times 10^{-28}$	11.25
120	60	0.00426	e,μ,τ u,d,s,c,b	−31.0 +31.1	1.128 1.123 $\times 10^{-28}$	45.00

$2\omega = E$), and electron scattering angles θ_e. The cross sections to be measured include all particles (flavors) in the loop which yield non–negligible contributions at the given photon energy. In these examples the value of δ is always limited by $|\delta| \leq 0.1$, which is a typical experimental error of this sort of experiments at present.

6. CONCLUSIONS

We will consider here the approximate numbers which have been given experimentally in testing the different parts of QED. Although, we have tried in the first place to study the physical meaning of the tested limits, it is still interesting to see, how far the agreements go up to date quantitatively. We observe that the electromagnetic field (Maxwell's equations) with its minimal current coupling is in agreement with experience down to about $5 \cdot 10^{-17}$ cm, that is together with macroscopic measurements (e.g., of the magnetic field of the earth) over almost 27 orders of magnitude.

The Dirac equation can be tested, of course, much less precisely since the field quantities are not directly observable.

Table 2. Lower limits (95% confidence) on different QED test parameters

$\Lambda_\gamma^+ \gtrsim$ 0.4 TeV (TASSO Coll.[41])	e^+e^- scattering
$\Lambda_\gamma^- \gtrsim$ 0.6 TeV (TASSO Coll.[41])	
$\Lambda_\gamma^+ \gtrsim$ 0.35 TeV (MARK J Coll.[42])	$e^+e^- \longrightarrow \mu^+\mu^-$
$\Lambda_\gamma^- \gtrsim$ 0.27 TeV (VENUS Coll.[43])	
$\Lambda_\gamma^+ \gtrsim$ 0.28 TeV (JADE Coll.[44])	$e^+e^- \longrightarrow \tau^+\tau^-$
$\Lambda_\gamma^- \gtrsim$ 0.27 TeV (HRS Coll.[45])	
$M^+_{e*} \gtrsim$ 94 GeV* (TOPAZ Coll.[46])	$e^+e^- \longrightarrow \gamma\gamma$
$M^-_{e*} \gtrsim$ 82 GeV (VENUS Coll.[47])	

* Here, a spin–$\frac{1}{2}$ e* with $\lambda = 1$ is assumed (cf.4.3). M^+ and M^- correspond to positive and negative deviations from QED, respectively.

The basic 5 helicity amplitudes appearing in (5.8) for photon–photon scattering have been calculated by Costantini et al.[35] in terms of Mandelstam variables and dilogarithmic functions. It should be noted, however, that the results given below depart from those of these authors in the different notation and metric of the Mandelstam variables and the explicit writing of the loop mass m.

$$M_{++++} = 1 + 2(1+\frac{2t'}{s'})B(t') + 2(1+\frac{2u'}{s'})B(u') + 2(\frac{t'^2+u'^2}{s'^2} - \frac{1}{s'})[T(t') +$$

$$+ T(u')] + (-\frac{1}{t'} + \frac{1}{2s't'})I(s',t') + (-\frac{1}{u'} + \frac{1}{2s't'})I(s',u')$$

$$+ (-2\frac{t'^2+u'^2}{s'^2} + \frac{4}{s'} + \frac{1}{t'} + \frac{1}{u'} + \frac{1}{2t'u'})I(t',u') ,$$

$$M_{+++-} = -1 + (-\frac{1}{s'} - \frac{1}{t'} - \frac{1}{u'})[T(s') + T(t') + T(u')] + (\frac{1}{u'} + \frac{1}{2s't'})I(s',t')$$

$$+ (\frac{1}{t'} + \frac{1}{2s'u'})I(s',u') + (\frac{1}{s'} + \frac{1}{2t'u'})I(t',u') , \qquad \text{(A.1)}$$

$$M_{++--} = -1 + \frac{1}{2s't'}I(s',t') + \frac{1}{2s'u'}I(s',u') + \frac{1}{2t'u'}I(t',u') ,$$

$$M_{+-+-} = M_{++++}(s' \leftrightarrow u') , \quad M_{+--+} = M_{++++}(s' \leftrightarrow t') ,$$

where the Mandelstam variables are here defined by

$$s' = (k_1 + k_2)^2/4m^2 , \; t' = (k_1 + k_3)^2/4m^2 , \; u' = (k_1 + k_4)^2/4m^2 . \qquad \text{(A.2)}$$

In contrast to the early definition of s, t, u in chapter 2, the factor $1/4m^2$ has been introduced in (A.2) in order to simplify the expressions in (A.1). The transcendental functions B, T, I are given by

$$B(x) = \text{Re}\left[-1 + \frac{b(x)}{2}\ln\left[\frac{b(x)+1}{b(x)-1}\right]\right] - i\theta(x-1)\,\frac{\pi}{2}\,b(x) ,$$

$$T(x) = \text{Re}\left[\frac{1}{2}\ln\left[\frac{b(x)+1}{b(x)-1}\right]\right]^2 - i\theta(x-1)\,\pi\,\text{arcosh}\sqrt{x} , \qquad \text{(A.3)}$$

$$I(x,y) = \frac{1}{2a}\,\mathrm{Re}\left[\,\phi\left[\frac{a+1}{a+b(x)}\right] + \phi\left[\frac{a+1}{a-b(x)}\right] - \phi\left[\frac{a-1}{a+b(x)}\right] - \phi\left[\frac{a-1}{a-b(x)}\right]\,\right] \tag{A.3}$$

$$- i\theta(x-1)\,\frac{\pi}{2a}\ln\left[\frac{a+b(x)}{a-b(x)}\right] + (x \leftrightarrow y)\ ,$$

with

$$\mathrm{a} = \sqrt{1-1/x-1/y}\ ,\quad b(x) = \sqrt{1-1/x}\ ,$$

$$\phi(z) = \int_0^z \mathrm{d}t\,\frac{\ln(1-t)}{t} \quad \text{(Spence function)}\ .$$

In the low energy limit ($\omega \ll m$) we have

$$M_{++++} \simeq -(22/45)s'^2\ ,\quad M_{+--+} \simeq -(22/45)t'^2\ ,\quad M_{+-+-} \simeq -(22/45)u'^2\ ,$$

$$M_{++--} \simeq -(2/15)(s'^2 + t'^2 + u'^2)\ ,\quad M_{+++-} \simeq 0\ , \tag{A.4}$$

and in the high energy limit ($\omega \gg m$) for $\theta_\gamma \neq 0^0$

$$M_{++++} \simeq 1 + (2x-1)L_2 + \frac{1}{2}\,[x^2 + (1-x)^2\,](L_2^2 + \pi^2)\ ,$$

$$M_{+-+-} \simeq 1 + (1-\frac{2}{x})(L_1 - i\pi) + \frac{1}{2x^2}\,[1 + (1-x)^2](L_1^2 - 2i\pi L_1)$$

$$M_{+--+} = M_{+-+-}(x \leftrightarrow 1-x)\ ,\quad M_{+++-} = M_{++--} \simeq -1\ , \tag{A.5}$$

where $x = \cos^2(\theta_\gamma/2)$, $L_1 = -\ln(1-x)$, and $L_2 = \ln[\,(1-x)/x\,]$.

REFERENCES

1. N.M. Kroll, Nuovo Cimento **45a** (1966) 65
2. K. Ringhofer, Acta Phys. Austr. **31** (1970) 67
3. K. Ringhofer and H. Salecker, paper presented at the 1975 Int. Symp. on Lepton and Photon Interactions at High Energies, Stanford University, Stanford, California (see the lecture of R. Hofstadter, proceedings of this conference, p. 873); Found. Phys. **10** (1980) 185
4. H. Salecker, Z. Naturf. **8a** (1953) 16, **10a** (1955)349
5. S.D. Drell, Ann. Phys. (NY) **4** (1958) 75
6. T. Anders, W. Jachmann, H. Salecker, and K. Wadan, in: "Old an New Questions in Physics, Cosmology, Philosophy, and Theoretical Biology": Essays in Honour of Wolfgang Yourgrau, A. van der Merwe, ed., Plenum Press, New York (1983) p. 411

7. F.A. Berends and A. Böhm, in "High Energy Electron–Positron Physics", A. Ali and P. Söding, eds., World Scientific, Singapore (1988) p. 28
8. H. Salecker, Z. Phys. **160** (1960) 385
9. L. Davis, Jr., A.S. Goldhaber, and M.M. Nieto, Phys. Rev. Lett. **35** (1975) 1402
10. E. Kazes, Nuovo Cimento **13** (1959) 1226
    K. Nishijima, Phys. Rev. **119** (1960) 485
    N.P. Chang and H.S. Mani, Phys. Rev. **B134** (1964) 896
11. N. Liebsch, thesis, University of Munich (1974)
12. J. Kraus, Nucl. Phys. **B89** (1975) 133
13. H.–D. Steinhöfer, Z. Phys. **C18** (1983) 139
14. for reviews see: M. Peskin, in "Proceedings of the 1981 International Symposium on Lepton and Photon Interactions at High Energies", Bonn, W. Pfeil, ed., Bonn, (1981) p. 880, and "Proceedings of the 1985 International Symposium on Lepton and Photon Interactions at High Energies", Kyoto, M. Konuma and K. Takahashi, eds., Kyoto, (1985) p. 714
15. T. Kinoshita, CERN preprint, CERN–TH.5097/88 (July 1988)
16. H. Salecker and F.C. Simm, to be published
17. W. Rarita and J. Schwinger, Phys. Rev. **60** (1941) 61
18. S.J. Brodsky and S.D. Drell, Phys. Rev. **D22** (1980) 2236
    F.M. Renard, Phys. Lett. **B116** (1982) 264
19. Y. Takahashi, "An Introduction to Field Quantization", Pergamon Press, Oxford (1969)
    C.J.C. Burges and H.J. Schnitzer, Nucl. Phys. **B228** (1983) 464
20. G. von Gehlen, Springer Tracts in Modern Physics **59** (1971) 164
21. G. Altarelli, in: "Polarization at LEP", Vol. 1, CERN 88–06, G. Alexander et al., eds., (1988) p. 14
22. B. Schrempp, F. Schrempp, N. Wermes, and D. Zeppenfeld, Nucl. Phys. **B296** (1988) 1
    F. Schrempp, DESY report, DESY 89–047 (April 1989)
23. C.Y. Prescott, SLAC–PUB–2630 (October 1980) –2649 (November 1980)
    C.Y. Prescott in: "Proceedings of the 1980 International Symposium on High–Energy Physics with Polarized Beams and Polarized Targets", Lausanne, C. Joseph and J. Soffer, eds., Birkhäuser, Basel (1981) p. 34
    K. C. Moffeit, SLAC–PUB–4746 (October 1988)
24. A.A. Sokolov and I.M. Ternov, Sov. Phys. Doklady **8** (1964) 1203
    B. W. Montague, Phys. Reports **113** (1984) 1
25. V.N. Baier, Sov. Phys. Uspekhi **14** (1972) 695
26. C. Akerlof, preprint UM–HE 8159, Ann Arbor, 1981
27. I.F. Ginzburg, G.L. Kotkin, V.G. Serbo, and V.I. Telnov, Nucl. Instr. Methods **205** (1983) 47
28. F.M. Renard, Z. Phys. **C14** (1982) 209
29. I.I. Gol'dman and V.A. Khoze, Sov. Phys. JETP **30** (1970) 501
30. N. Dombey, Rev. Mod. Phys. **41** (1969) 236
31. A. Feldmeier, H. Salecker, and F.C. Simm, Phys. Lett. **223B** (1989) 234
32. C. Bourrely, E. Leader, and J. Soffer, Phys. Reports **59** (1980) 95
33. T.B. Anders and W. Jachmann, J. Math. Phys. **24** (1983) 2847, J. Math. Phys. **28** (1987) 221
34. M. Jacob and G. Wick, Ann. Phys. (NY) **7** (1959) 404
    A.O. Barut, "The Theory of the Scattering Matrix", Macmillan, New York (1967)
35. V. Costantini, B. de Tollis, and G. Pistoni, Nuovo Cimento **A2** (1971) 733
36. see e. g.: G. Jahrlskog, L. Jönsson, S. Prünster, H.D. Schulz, H.J. Willutzki, and G.G. Winter, Phys. Rev. **D8** (1973) 3813
37. B. de Tollis, Nuovo Cimento **35** (1965) 1182
38. A. Achieser, Phys. Zeits. Sowj. **11** (1937) 263
39. Y. Sakurayama and H. Salecker, to be published.
40. see e. g.: H. Terazawa, Rev. of Mod. Phys. **45** (1973) 615
41. TASSO Coll.: W. Braunschweig et al., Z. Phys. **C37** (1988) 171
42. MARK J Coll.: B. Adeva et al., Phys. Lett. **179B** (1986) 177

43. VENUS Coll.: Y. Unno, Proc. XXIV Int. Conf. on High Energy Physics, Munich (1988) 860
44. JADE Coll.: W. Bartel et al., Z. Phys. **C30** (1986) 371
45. HRS Coll.: K.K. Gan et al., Phys. Lett. **153B** (1985) 116
46. TOPAZ Coll.: I. Adachi et al., Phys. Lett. **200B** (1988) 391
47. VENUS Coll.: K. Abe et al., submitted to Z. Phys. **C** (1989)

FINITE QUANTUM ELECTRODYNAMICS

G. Scharf
Institut für Theoretische Physik, Universität Zürich,
8001 Zürich, Switzerland

What *is* certain is that we do not have a good mathematical way to describe the theory of quantum electrodynamics.

R.P.Feynman (1985)[1]

1. Introduction

The word 'finite' in the title means that the approach to QED, we are going to present, is free from the well-known ultraviolet divergences. How can it happen that a theory developes divergences ? There are only two possibilities: 1.) There is an error somewhere. 2.) The theory rests on badly defined basic equations. There is no other possibility. In fact, if the initial equations are well-defined and no error is made then divergences can never appear. That is the strength of mathematics. This trivial remark immediately leads to the question which of the two possibilities is realized in conventional QED. The answer depends on the approach one considers. In the widely used lagrangean approach, one starts from a Lagrangean

$$L = L_0 + e\overline{\psi}(x)\gamma^\mu\psi(x)A_\mu(x). \tag{1.1}$$

Although this is a well-defined expression in classical field theory, it is badly defined if ψ and A are quantized interacting fields. Hence, this approach corresponds to possibility 2.).

There is a better approach which goes back to Stückelberg and Bogoliubov [2]. It consists of directly constructing the S-matrix, starting from the first order of perturbation theory

$$S_1(g) = ie\int d^4x\ :\overline{\psi}_{\rm in}(x)\gamma^\mu\psi_{\rm in}(x):\ A^{\rm in}_\mu(x)g(x). \tag{1.2}$$

Here $\psi_{\rm in}$, $A^{\rm in}$ are incoming free quantized fields such that this is a well-defined equation. $g(x)$ is a C-number switching function. From (1.2) one then constructs the second order $S_2(g)$. It contains contributions from loop graphs (vacuum polarization, self-energy). Nevertheless, if we work carefully enough, there cannot be infinities. In this way, we get a rigorous finite theory which only contains

New Frontiers in Quantum Electrodynamics and Quantum Optics
Edited by A. O. Barut, Plenum Press, New York, 1990

mathematically well-defined quantities. This improves the situation with regard to the motto above.

As an introduction we will describe the method of Bogoliubov in the case of quantum mechanics. We will give two different constructions of the quantum-mechanical S-matrix. The first one is by using the Schrödinger equation

$$i\frac{d}{dt}\psi(t) = (H_0 + V(t))\psi(t). \tag{1.3}$$

The solution ψ is given by a unitary propagator

$$\psi(t) = U(t,s)\psi(s). \tag{1.4}$$

The S-matrix is then determined by the strong time limits

$$S = \lim_{\substack{s\to-\infty \\ t\to+\infty}} e^{iH_0 t}U(t,s)e^{-iH_0 s}$$

$$\stackrel{\text{def}}{=} U_0(0,\infty)U(+\infty,-\infty)U_0(-\infty,0). \tag{1.5}$$

To compute S, we go over to the interaction picture

$$\psi(t) = e^{-iH_0 t}\varphi(t) \tag{1.6}$$

$$i\frac{d}{dt}\varphi(t) = \tilde{V}(t)\varphi(t). \tag{1.7}$$

The quantities with tilde are always sandwiched between free time evolution operators:

$$\tilde{V}(t) \stackrel{\text{def}}{=} e^{iH_0 t}V(t)e^{-iH_0 t} \tag{1.8}$$

$$\tilde{U}(t,s) \stackrel{\text{def}}{=} e^{iH_0 t}U(t,s)e^{-iH_0 s}. \tag{1.9}$$

The S-matrix (1.5) is then given by

$$S = \lim_{\substack{s\to-\infty \\ t\to+\infty}} \tilde{U}(t,s). \tag{1.10}$$

In order to calculate (1.10), we integrate (1.7)

$$\varphi(t) = \varphi(s) - i\int_s^t dt_1\, \tilde{V}(t_1)\varphi(t_1) \tag{1.11}$$

and iterate this integral equation

$$\varphi(t) = \left[1 + \sum_{n=1}^{\infty}(-i)^n \int_s^t dt_1 \int_s^{t_1} \cdots \int_s^{t_{n-1}} dt_n\, \tilde{V}(t_1)\cdots\tilde{V}(t_n)\right]\varphi(s). \tag{1.12}$$

This is the well-known Dyson series. The domain of integration in (1.12) is a simplex in $\mathbb{R}^n$. This integral can be extended to an integral over a cube, maintaining the correct time ordering. Taking the (strong) limits $s \to -\infty$ and $t \to +\infty$, we obtain for the S-matrix (1.10)

$$S=\sum_{n=0}^{\infty}\frac{(-i)^n}{n!}\int_{-\infty}^{+\infty}dt_1\int_{-\infty}^{+\infty}dt_2\cdots\int_{-\infty}^{+\infty}dt_n\,T\Big[\tilde V(t_1)\cdots\tilde V(t_n)\Big]. \tag{1.13}$$

The time ordering symbol T means that the factors are arranged with decreasing time coordinate.

Now we give a second derivation of this result (1.13) *without using the Schrödinger equation.* Since in quantum field theory one does not have well-defined dynamical evolution equations, in general, it is this derivation which will be taken over to QED. In view of this later application, let us multiply the interaction by a switching function $g(t)$ which vanishes rapidly for $t\to\pm\infty$. We want to construct S as a power series in g of the form

$$S(g)=1+\sum_{n=1}^{\infty}\frac{1}{n!}\int dt_1\cdots dt_n\,T_n(t_1,\cdots t_n)g(t_1)\cdots g(t_n). \tag{1.14}$$

By definition, $T_n(t_1,\cdots t_n)$ must be symmetric with respect to permutations of the t_j, otherwise the contribution to (1.14) would be zero.

Instead of the Schrödinger equation, we will use a causality property of the S-matrix which follows from (1.10). Let us consider two switching functions g_1, g_2 with disjoint support in time

$$\operatorname{supp} g_1\subset(-\infty,s)\,,\ \operatorname{supp} g_2\subset(s,+\infty). \tag{1.15}$$

Then, in virtue of (1.10) we have

$$S(g_1+g_2)=U_0(0,\infty)U(+\infty,-\infty)U_0(-\infty,0)=$$

$$=U_0(0,\infty)U(+\infty,s)U_0(s,0)U_0(0,s)U(s,-\infty)U_0(-\infty,0)=S(g_2)S(g_1). \tag{1.16}$$

This equation expresses causality in the sense that, what happens for $t<s$ (described by $S(g_1)$) is not influenced by what happens for $t>s$ (described by $S(g_2)$). This causality condition implies the following conditions for the n-point functions T_n in (1.14)

$$T_n(t_1,\cdots t_n)=T_m(t_1,\cdots t_m)T_{n-m}(t_{m+1},\cdots t_n) \tag{1.17}$$

if all $\{t_1,\cdots t_m\}$ are greater than all $\{t_{m+1},\cdots t_n\}$.

We now claim that all n-point functions T_n in (1.14) can be inductively determined by means of (1.17), provided the first order $T_1(t)$ is given. Let us consider the construction of T_2. We introduce the operator valued function

$$A_2(t_1,t_2)\stackrel{\text{def}}{=}T_2(t_1,t_2)-T_1(t_1)T_2(t_2). \tag{1.18}$$

It follows from the causality condition (1.17) that

$$A_2(t_1,t_2)=0\quad\text{if}\quad t_1>t_2. \tag{1.19}$$

Considering t_2 as reference point, we call this an advanced function. Similarly,

$$R_2(t_1,t_2)\stackrel{\text{def}}{=}T_2(t_2,t_1)-T_1(t_2)T_1(t_1) \tag{1.20}$$

is the retarded function because

$$R_2(t_1, t_2) = 0 \quad \text{if} \quad t_1 < t_2. \tag{1.21}$$

In the difference

$$D_2(t_1, t_2) \overset{\text{def}}{=} R_2 - A_2 \tag{1.22}$$

$$= T_1(t_1)T_1(t_2) - T_1(t_2)T_1(t_1) \tag{1.23}$$

the two-point function T_2 drops out because of symmetry. Consequently, this function D_2 is known from the given T_1. Then R_2 and A_2 can be identified individually since they have disjoint supports (see fig.1). Multiplying by Θ-function, we find

$$R_2(t_1, t_2) = \Theta(t_1 - t_2) \cdot D_2(t_1, t_2). \tag{1.24}$$

The two-point function can then be obtained from (1.20)

$$T_2 = R_2 + T_1(t_2)T_1(t_1)$$

$$= \Theta(t_1 - t_2)T_1(t_1)T_2(t_2) + \Theta(t_2 - t_1)T_1(t_2)T_1(t_1). \tag{1.25}$$

This is just the time ordered product

$$T_2(t_1, t_2) = T\Big[T_1(t_1)T_2(t_2)\Big]. \tag{1.26}$$

The corresponding n-th order result is

$$T_n(t_1, \cdots t_n) = T\Big[T_1(t_1) \cdots T_1(t_n)\Big], \tag{1.27}$$

in agreement with (1.13).

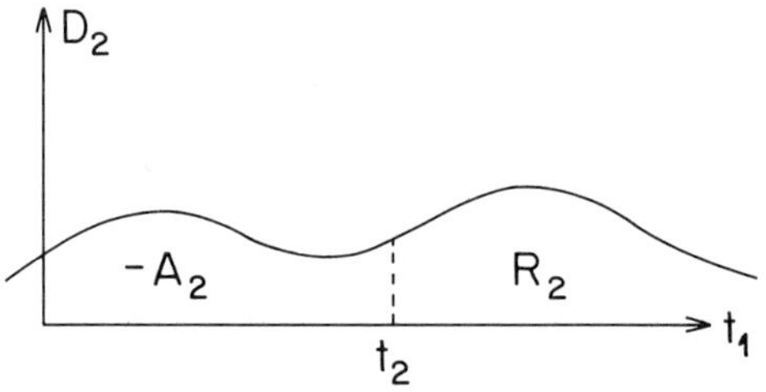

Fig. 1. Splitting of $D_2(t_1, t_2)$ into retarded and advanced parts.

It is well-known that in field theory, the naive time-ordred product (1.27) contains ultraviolet divergences. These are due to the fact that T_1 and therefore all D_n now are operator-valued *distributions* and not just operators as in quantum mechanics, and in general, it is not possible to multiply those distributions by Θ-functions as in (1.24). Nevertheless, we will have to split causal distributions into retarded and advanced parts, and it is exactly this point where we must be careful (see sect.3). Before turning to this subject, we have to generalize the construction to field theory.

2. Causal Perturbation Theory in Quantum Field Theory

Let $g(x)$ now be a switching function on Minkowski space, $x = (t, \boldsymbol{x})$, which we assume to be in Schwartz space $\mathcal{S}(\mathbb{R}^4)$. We look for an expansion of the S-matrix of the following form

$$S(g) = 1 + \sum_{n=1}^{\infty} \frac{1}{n!} \int d^4x_1 \dots d^4x_n \; T_n(x_1, \dots, x_n) g(x_1) \cdots g(x_n). \tag{2.1}$$

The T_n are the operator-valued n-point distributions. Similarly, the inverse S-matrix can be expanded

$$S(g)^{-1} = 1 + \sum_{n=1}^{\infty} \frac{1}{n!} \int d^4x_1 \dots d^4x_n \; \tilde{T}_n(x_1, \dots, x_n) g(x_1) \cdots g(x_n), \tag{2.2}$$

where the distributions $\tilde{T}_n$ can be computed by formal inversion of (2.1). Since, by definition, $T_n(x_1, \dots, x_n)$ is symmetric in $x_1, \dots, x_n$, it is convenient to use a set-theoretical notation $X = \{x_1, \dots, x_n\}$. We then have

$$\tilde{T}_n(X) = \sum_{r=1}^{n} (-)^r \sum_{P_r} T_{n_1}(X_1) \cdots T_{n_r}(X_r), \tag{2.3}$$

where the second sum runs over all partitions P_r of X into r disjoint subsets

$$X = X_1 \cup \dots \cup X_r, \quad X_j \neq \emptyset, \quad |X_j| = n_j. \tag{2.4}$$

All products of distributions in (2.3) are well-defined, because the arguments are disjoint sets of points such that the products are direct products of distributions. Note the special case

$$\tilde{T}_1(x) = -T_1(x). \tag{2.5}$$

In analogy to (1.16), causality of the S-matrix is now expressed by the condition

$$S(g_1 + g_2) = S(g_2) \cdot S(g_1), \tag{2.6}$$

if the support of g_1 in Minkowski space is earlier then the support of g_2 in some Lorentz frame. The latter property we shall denote by

$$\operatorname{supp} g_1 < \operatorname{supp} g_2 \tag{2.7}$$

in the following. As in (1.17), this implies the following causality condition for the n-point distributions

$$T_n(x_1, \dots, x_n) = T_m(x_1, \dots x_m) \cdot T_{n-m}(x_{m+1}, \dots x_n) \tag{2.8}$$

if $\{x_{m+1}, \dots x_n\} < \{x_1, \dots x_m\}$. Similarly, the inverse of (2.6)

$$S(g_1 + g_2)^{-1} = S(g_1)^{-1} S(g_2)^{-1} \tag{2.9}$$

leads to

$$\tilde{T}_n(x_1,\ldots,x_n)=\tilde{T}_m(x_1,\ldots x_m)\tilde{T}_{n-m}(x_{m+1},\ldots x_n) \tag{2.10}$$

if $\{x_1,\ldots x_m\}<\{x_{m+1},\ldots x_n\}$.

Now we are ready to introduce the retarded n-point distributions

$$R_n(x_1,\ldots,x_n)=T_n(x_1,\ldots,x_n)+\sum_{P_2}T_{n-n_1}(Y,x_n)\tilde{T}_{n_1}(X), \tag{2.11}$$

where the sum runs over all partitions

$$P_2\ :\ \{x_1,\ldots x_{n-1}\}=X\cup Y\ ,\quad X\neq\emptyset \tag{2.12}$$

into disjoint subsets with $|X|=n_1\geq 1$, $|Y|\leq n-2$. Note that this sum, which will be abbreviated by R'_n in the following, contains T_j's with $j\leq n-1$ only. The last argument x_n in (2.11) has a distinguished meaning. It is the reference point for the support of R_n. This is expressed in the following

Proposition.

$$\operatorname{supp}R_n(x_1,\ldots,x_n)\subseteq\Gamma^+_{n-1}(x_n), \tag{2.13}$$

where Γ^+_{n-1} is the $(n-1)$-dimensional closed forward cone

$$\Gamma^+_{n-1}(x_n)=\{(x_1,\cdots x_{n-1})\,|\,(x_j-x_n)^2\geq 0,\ x_j^0\geq x_n^0,\ \forall j\}. \tag{2.14}$$

This is a consequence of causality. Because of the fundamental importance of this proposition, we give the idea of the proof (for the full proof see [3][4]). Suppose for simplicity that $x_1^0<x_n^0$ and all other $x_j^0>x_n^0$. We have to show that $R_n=0$. By causality (2.8), we have

$$T_n(x_1,\ldots,x_n)=T_{n-1}(x_2,\ldots x_n)\cdot T_1(x_1). \tag{2.15}$$

Let us consider the sum in (2.11)

$$R'_n(Z,x_n)=\sum_{\substack{X\cup Y=Z\\ X\neq\emptyset}}T_{n-n_1}(Y,x_n)\tilde{T}_{n_1}(X). \tag{2.16}$$

Regarding the point x_1, the following cases occur: 1.) $x_1\in Y$, then the contribution to (2.16) is

$$\sum_{P'_2}T_{n-1-n_1}(Y',x_n)T_1(x_1)\tilde{T}_{n_1}(X') \tag{2.17}$$

by causality (2.8), where P'_2 are the partitions of

$$P'_2\ :\ \{x_2,\ldots x_{n-1}\}=X'\cup Y'. \tag{2.18}$$

2.) $x_1\in X$, then by (2.10) the contribution is

$$\sum_{P'_2}T_{n-1-n_1}(Y',x_n)\tilde{T}_1(x_1)\tilde{T}_{n_1}(X'). \tag{2.19}$$

Taking (2.5) into account, these two contributions cancel. There remains the case 3.) $x_1 \equiv X$. Now the contribution is $T_{n-1}(x_2, \ldots x_n)\tilde{T}_1(x_1)$, and from (2.15) and (2.5) we find that

$$R_n = T_n + R'_n = 0. \tag{2.20}$$

Since this is true in any Lorentz system, we arrive at (2.13).

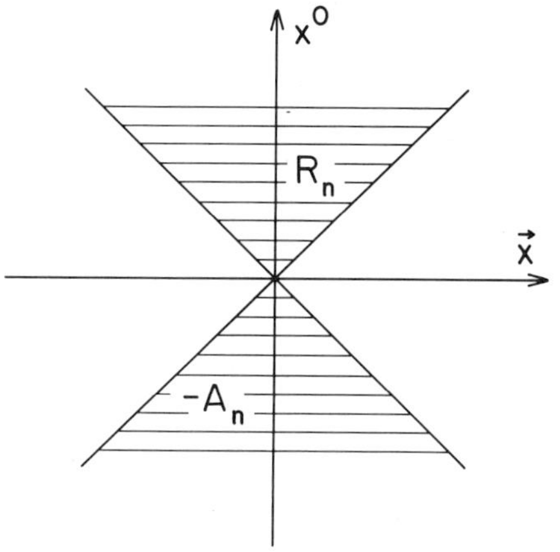

Fig. 2. Retarded and advanced parts of the causal distribution D_n.

Beside the retarded distribution we define the advanced n-point distribution

$$A_n(x_1, \ldots, x_n) = T_n(x_1, \ldots, x_n) + A'_n \quad \text{where} \tag{2.21}$$

$$A'_n = \sum_{P_2} \tilde{T}_{n_1}(X) T_{n-n_1}(Y, x_n). \tag{2.22}$$

Again, A'_n is determined by the lower T_j's, $j < n$, and we have the support property

$$\operatorname{supp} A_n \subseteq \Gamma^-_{n-1}(x_n), \tag{2.23}$$

where Γ^-_{n-1} is the $(n-1)$-dimensional backward cone. In the difference

$$D_n(x_1, \ldots, x_n) \stackrel{\text{def}}{=} R_n - A_n = R'_n - A'_n \tag{2.24}$$

the n-point distribution T_n drops out, such that this quantity is known in terms of $T_1, \ldots T_{n-1}$. D_n has a causal support (fig.2)

$$\operatorname{supp} D_n \subseteq \Gamma^+_{n-1}(x_n) \cup \Gamma^-_{n-1}(x_n). \tag{2.25}$$

The splitting of this causal distribution gives R_n (or $-A_n$) individually. Summing up, the n-point distributions T_n can be constructed inductively in the following way: Suppose $T_1, \ldots T_{n-1}$ are already constructed, then we form R'_n, A'_n and D_n (2.24). The splitting of D_n gives R_n and T_n then follows from (2.11)

$$T_n = R_n - R'_n. \tag{2.26}$$

3. Splitting of Causal Distributions

It was mentioned in the introduction that this is the most subtle step of the whole construction. In fact, it was not before 1973 when Epstein and Glaser treated this problem correctly [3]. The distributions which we have to split in QED are of the following form

$$D_n(x_1,\ldots,x_n) = \sum_k : \prod_j \overline{\psi}(x_j) d_n^k(x_1,\ldots,x_n) \prod_l \psi(x_l) :: \prod_n A(x_n) :, \quad (3.1)$$

where ψ and A are the free electron-positron and radiation fields and the double dots mean normal ordering of these free fields. The numerical distributions d_n^k have causal support (2.25) and are assumed to be tempered $\in \mathcal{S}'(\mathrm{I\!R}^{4m})$, because we will use Fourier transformation. There remains to split these numerical distributions as follows

$$d_n^k = r_n - a_n \quad (3.2)$$

$$\operatorname{supp} r_n \subseteq \Gamma_{n-1}^+(x_n) \quad , \quad \operatorname{supp} a_n \subseteq \Gamma_{n-1}^-. \quad (3.3)$$

Because of translation invariance, it is sufficient to put $x_n = 0$ and to consider

$$d(x) \stackrel{\text{def}}{=} d_n^k(x_1,\ldots x_{n-1},0) \in \mathcal{S}'(\mathrm{I\!R}^m), \quad m = 4n-4. \quad (3.4)$$

Because of the support properties (3.3), the behaviour of $d(x)$ in the neighbourhood of $x = 0$ is essential for the splitting procedure. It can be characterized by means of the following definition due to Epstein and Glaser:

Definition. *The distribution $d(x)$ is called singular of order ω at $x = 0$, if*

$$|\langle d, \varphi\rangle| \le K(\varepsilon) \sum_{|a|\le M} \sup_x (1+|x|)^N |x|^{L(a)} |D^a\varphi(x)| \quad (3.5)$$

for all $\varphi \in \mathcal{S}(\mathrm{I\!R}^m)$, where

$$L(a) = \max(0, |a| - \omega - \varepsilon), \quad (3.6)$$

$$1 > \varepsilon > 0 \quad ; \quad M, N \ge 0 \quad \textit{integers}.$$

Here multi-index notation for partial derivatives is used:

$$D^a \stackrel{\text{def}}{=} \frac{\partial^{a_1+\ldots a_m}}{\partial x_1^{a_1}\ldots \partial x_m^{a_m}} \quad , \quad |a| = a_1 + \ldots + a_m.$$

In the splitting problem we have to distinguish two cases:

a) $\omega \le -1$: In this case trivial splitting by multiplication with a Θ-function is possible. Let $v = (v_1,\ldots v_m)$ with all v_j time-like $v_j \in V^+$. Then, in virtue of (3.3), the splitting can be achieved by multiplication with (see fig.3)

$$\chi(x) = \Theta(v\cdot x) : \quad (3.7)$$

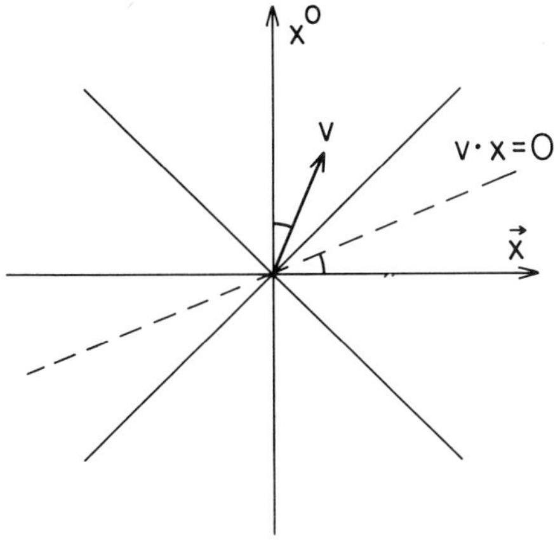

Fig. 3. The cutting hyper-plane in the distribution splitting

$$\langle r(x),\, \varphi(x)\rangle \stackrel{\text{def}}{=} \langle d(x),\, \chi(x)\varphi(x)\rangle,\ \forall\varphi\in\mathcal{S}. \tag{3.8}$$

This is a well-defined distribution because d can be extended to $\varphi \in C^M(\mathbb{R}^m\setminus 0)$ with

$$|D^a\varphi(x)| \leq \text{const}\cdot(1+|x|)^{-N}|x|^{-L(a)}, \quad x\neq 0, \tag{3.9}$$

where

$$L(a) = |a| - \omega - \varepsilon > 0. \tag{3.10}$$

$\chi(x)\varphi(x)$ is such a function on the support of d, because the discontinuity at $x = 0$ is allowed by (3.9).

b) $\omega \geq 0$: In this case trivial splitting is impossible and we must be careful here. Since

$$L(a) < 0 \quad \text{for} \quad |a| \leq \omega, \tag{3.11}$$

it follows from (3.9) that

$$D^a\varphi(x) \to 0 \quad \text{for} \quad x\to 0,\ |a|\leq\omega. \tag{3.12}$$

That means, splitting by multiplication with step function in only possible if

$$(D^a\varphi)(0) = 0 \quad \text{for} \quad |a|\leq\omega. \tag{3.13}$$

To achieve that, we introduce an auxiliary function $w(x)\in\mathcal{S}(\mathbb{R}^m)$ with

$$w(0) = 0\,,\ D^a w(0) = 0 \quad \text{for} \quad 1\leq|a|\leq\omega \tag{3.14}$$

and define

$$(W\varphi)(x) \stackrel{\text{def}}{=} \varphi(x) - w(x)\sum_{|a|=0}^{\omega}\frac{x^a}{a!}(D^a\varphi)(0). \tag{3.15}$$

This is a test function satisfying (3.13). Consequently,

$$\langle r(x),\, \varphi(x)\rangle \stackrel{\text{def}}{=} \langle d(x),\, \chi(x)W\varphi\rangle \tag{3.16}$$

defines a tempered distribution with $\operatorname{supp} r \subseteq \Gamma_m^+(0)$. It agrees with $d(x)$ on $\Gamma_m^+(0)\setminus 0$ *in the sense of distributions*, because a test function $\varphi \in \mathcal{S}$ with $\operatorname{supp}\varphi \subset \Gamma_m^+(0)\setminus 0$ vanishes at $x = 0$, together with all its derivatives, so that the additional terms in (3.15) vanish. But without these terms, the splitting would not make sense for arbitrary $\varphi \in \mathcal{S}$.

In sharp contrast to case a), the splitting b) is not unique. It depends on the auxiliary function $w(x)$ (3.14). If $\tilde{w}(x)$ is another such function, then

$$\langle r - \tilde{r},\, \varphi\rangle = \int dx\, d(x)\chi(x)(\tilde{w}(x) - w(x)) \sum_{|a|=0}^{\omega} \frac{x^a}{a!}(D^a\varphi)(0)$$

$$\stackrel{\text{def}}{=} \sum_{|a|=0}^{\omega} C_a(D^a\varphi)(0). \tag{3.17}$$

This means that r is only determined up to a distribution with point support at $x = 0$

$$\tilde{r} - r = \sum_{|a|=0}^{\omega} \tilde{C}_a D^a\delta(x). \tag{3.18}$$

The undetermined local terms are not fixed by causality, additional physical normalization conditions are necessary to fix them.

All explicit calculations in QED are best done in momentum space. For this reason, we must investigate the splitting procedure in p-space. Let us first compute the inverse Fourier transforms

$$(x^a w)\check{}\,(p) = (iD_p)^a\check{w}(p) \quad \text{and}$$

$$(D^a\varphi)(0) = \langle(-)^a D^a\delta,\, \varphi\rangle = (-)^a\langle\widehat{D^a d},\, \check{\varphi}\rangle$$
$$= (-)^a(2\pi)^{-m/2}\langle(-ip)^a,\, \check{\varphi}\rangle = (2\pi)^{-m/2}\langle(ip)^a,\, \check{\varphi}\rangle.$$

Then we get by the convolution theorem

$$\langle r,\, \varphi\rangle = \langle\chi d,\, W\varphi\rangle$$

$$= (2\pi)^{-m/2}\Big\langle \hat{\chi} * \hat{d},\, \check{\varphi} - \sum_a \frac{1}{a!}(iD_p)^a\check{w}(p)(2\pi)^{-m/2}\langle(ip')^a,\, \check{\varphi}\rangle\Big\rangle$$

$$\stackrel{\text{def}}{=} \langle\hat{r},\, \check{\varphi}\rangle. \tag{3.19}$$

The retarded distribution is therefore given in momentum space by

$$\hat{r}(p) = (2\pi)^{-m/2}\int dk\, \hat{\chi}(k)\Big[\hat{d}(p-k)$$

$$-(2\pi)^{-m/2}\sum_a \frac{(-)^a}{a!}p^a \int dp'\, \hat{d}(p'-k)D^a_{p'}\check{w}(p')\Big]$$

$$= (2\pi)^{-m/2} \int dk\, \hat{\chi}(k) \Big[\hat{d}(p-k)$$

$$-(2\pi)^{-m/2} \sum_a \frac{p^a}{a!} \int dp'\, \check{w}(p') D^a_{p'} \hat{d}(p'-k) \Big]. \qquad (3.20)$$

In QED, where all fermions are massive, it is possible to let

$$w(x) \to 1 \quad \text{i.e.} \quad \check{w}(p') \to (2\pi)^{m/2} \delta(p').$$

This leads to the following unique result

$$\hat{r}(p) = (2\pi)^{-m/2} \int dk\, \hat{\chi}(k) \Big[\hat{d}(p-k) - \sum_{|a|=0}^{\omega} \frac{p^a}{a!} (D^a \hat{d})(-k) \Big]. \qquad (3.21)$$

The subtracted terms are the beginning of the Taylor series at $p = 0$. This is an ultraviolet "regularization" in the usual terminology. It should be stressed, however, that here this is a consequence of the causal distribution splitting and not an ad hoc recipe.

In order to have the Fourier transform of χ (3.7) in a simple form, we choose coordinates such that $v = (v_1^0 = 1, 0, \ldots)$. This leads to

$$\chi(x) = \Theta(x_1^0) \quad \text{and}$$

$$\hat{\chi}(p) = (2\pi)^{m/2-1} \delta(\boldsymbol{p}_1, p_2, \ldots p_m) \frac{i}{p_1^0 + i0}. \qquad (3.22)$$

Let us consider the case where the argument p of $\hat{r}$ (3.21) has all its components in the forward cone. Without loss of generality, it is possible to assume p to be parallel to v. Then (3.21) can be simplified as follows

$$\hat{r}(p_1^0) = \frac{i}{2\pi} \int\limits_{-\infty}^{+\infty} dk_1^0 \frac{1}{k_1^0 + i0} \Big[\hat{d}(p_1^0 - k_1^0, 0, \ldots)$$

$$- \sum_{a=0}^{\omega} \frac{p_1^{0a}}{a!} (-)^a D^a_{k_1^0} \hat{d}(q_1^0 - k_1^0, 0, \ldots) \Big|_{q_1^0 = 0} \Big]. \qquad (3.23)$$

Integrating the last term by parts and using $k_1^0 - p_1^0 = k_0'$ as a new integration variable (omitting the prime) in the first term, we arrive at

$$\hat{r}(p_1^0) = \frac{i}{2\pi} \int\limits_{-\infty}^{+\infty} dk_0'\, \hat{d}(-k_0') \Big[\frac{1}{p_1^0 + k_0' + i0}$$

$$- \sum_{a=0}^{\omega} \frac{p_1^{0a}}{a!} \frac{\partial^a}{\partial k_0'^a} \frac{1}{k_0' + i0} \Big].$$

The square bracket can be easily computed

$$[\ldots] = \left(-\frac{p_1^0}{k_0' + i0}\right)^{\omega+1} \frac{1}{p_1^0 + k_0' + i0}.$$

Changing the variable of integration $k_0 = -k_0'$, we obtain the following result

$$\hat{r}(p_1^0) = \frac{i}{2\pi}(p_1^0)^{\omega+1} \int\limits_{-\infty}^{+\infty} dk_0 \, \frac{\hat{d}(k_0)}{(k_0 - i0)^{\omega+1}(p_1^0 - k_0 + i0)}. \tag{3.24}$$

This is a subtracted dispersion relation. In order to write down the result for arbitrary

$$p = (p_1, \ldots p_{n-1}) \quad , \quad p_j \in V^+, \tag{3.25}$$

we introduce the variable of integration $t = k_0/p_1^0$ which finally gives

$$\hat{r}(p) = \frac{i}{2\pi} \int\limits_{-\infty}^{+\infty} dt \, \frac{\hat{d}(tp)}{(t - i0)^{\omega+1}(1 - t + i0)}. \tag{3.26}$$

The dispersion integral (3.26) is converging at infinity if

$$|\hat{d}(p)| < \text{const} \cdot |p|^{\omega+\delta}, \quad 0 \le \delta < 1 \tag{3.27}$$

for $|p| \to \infty$. It can be shown that a causal distribution which satisfies this condition (3.27) is indeed singular of order ω. We see that it is terribly important to choose the singular order ω correctly. If ω is to small, then ultraviolet divergences appear. The result (3.26) is a special solution of the splitting problem, called the symmetrical solution because it has generally the same symmetry properties as $\hat{d}(p)$. In virtue of (3.18), the general solution is then given by

$$\tilde{r}(p) = \hat{r}(p) + \sum_{|a|=0}^{\omega} C_a p^a. \tag{3.28}$$

The undetermined polynomial in (3.28) must be fixed by other physical conditions. We will return to this point at the end of the next section.

For arbitrary p and general $\hat{d}$, the calculation leading to (3.26) does not go through. Then we may write the dispersion relation in the form

$$\hat{r}(p) = \frac{i}{2\pi} \int\limits_{-\infty}^{+\infty} dt \, \frac{1}{t + i0} \Bigg[\hat{d}(p - tv)$$

$$- \sum_{|a|=0}^{\omega} \frac{p^a}{a!} (D_k^a \hat{d})(k - tv)\Big|_{k=0}\Bigg], \tag{3.29}$$

which follows directly from (3.21).

4. Application to QED

In order to start the inductive construction, the one-point distribution T_1 must be given. In QED, it is of the following well-known form

$$T_1(x) = ie : \overline{\psi}(x)\gamma^\mu\psi(x) : A_\mu(x) = -\tilde{T}_1(x). \tag{4.1}$$

The field operators in (4.1) are ordinary free fields. The normal ordering is necessary to have a stable vacuum. For constructing the two-point distribution, we must first compute A_2' (2.22)

$$A_2'(x_1, x_2) = -T_1(x_1)T_2(x_2)$$

$$= e^2\gamma^\mu_{ab}\gamma^\nu_{cd} : \overline{\psi}_a(x_1)\psi_b(x_1) :: \overline{\psi}_c(x_2)\psi_d(x_2) : A_\mu(x_1)A_\nu(x_2). \tag{4.2}$$

The products in (4.2) are normally ordered by means of Wick's theorem. For this purpose we need the various contractions which are commutators (in case of Bose fields) or anticommutators (in case of fermi fields) between the absorption and emission parts. Remember, all fields are decomposed like $A = A^{(+)} + A^{(-)}$, where $A^{(+)}$ contains emission operators and $A^{(-)}$ absorption operators only. In QED, only the following three contractions appear:

$$\overbrace{\psi_a(x)\overline{\psi}_b(y)} \stackrel{\text{def}}{=} \{\psi_a^{(-)}(x), \overline{\psi}_b^{(+)}(y)\} = \frac{1}{i}S_{ab}^{(+)}(x-y) \tag{4.3}$$

$$\overbrace{\overline{\psi}_a(x)\psi_b(y)} \stackrel{\text{def}}{=} \{\overline{\psi}_a^{(-)}(x), \psi_b^{(+)}(y)\} = \frac{1}{i}S_{ba}^{(-)}(y-x) \tag{4.4}$$

$$\overbrace{A_\mu(x)A_\nu(y)} \stackrel{\text{def}}{=} [A_\mu^{(-)}(x), A_\nu^{(+)}(y)] = g_{\mu\nu}iD_0^{(+)}(x-y). \tag{4.5}$$

The commutation functions are given by

$$S^{(\pm)}(x) = (i\not\partial + m)D^{(\pm)}(x) \tag{4.6}$$

$$D^{(\pm)}(x) = \pm\frac{i}{(2\pi)^3}\int d^4p\,\delta(p^2 - m^2)\Theta(\pm p^0)e^{-ipx}. \tag{4.7}$$

Because of the normal orderings in (4.2), only contractions between ψ and $\overline{\psi}$ with different coordinates must be taken into account:

$$\begin{aligned} A_2'(x_1, x_2) = e^2\gamma^\mu_{ab}\gamma^\nu_{cd}\Big[&: \overline{\psi}_a(x_1)\psi_b(x_1)\overline{\psi}_c(x_2)\psi_d(x_2) : \\ + : \overline{\psi}_a(x_1)\psi_d(x_2) : \frac{1}{i}S_{bc}^{(+)}(x_1 - x_2) &+ : \psi_b(x_1)\overline{\psi}_c(x_2) : \frac{1}{i}S_{da}^{(-)}(x_2 - x_1) \\ &- S_{bc}^{(+)}(x_1 - x_2)S_{da}^{(-)}(x_2 - x_1)\Big] \\ \times \Big[&: A_\mu(x_1)A_\nu(x_2) : + g_{\mu\nu}iD_0^{(+)}(x_1 - x_2)\Big]. \end{aligned} \tag{4.8}$$

The distribution R_2' is obtained by interchanging x_1 and x_2 and by replacing the indeces of summation $\mu \leftrightarrow \nu$, $a \leftrightarrow c$, $b \leftrightarrow d$. Then we compute $D_2 = R_2' - A_2'$, we write down those terms only which we will discuss in the following:

$$D_2(x_1,x_2)=e^2\gamma^{\mu}_{ab}\gamma^{\nu}_{cd}\Big\{$$

$$:\overline{\psi}_a(x_1)\psi_b(x_1)\overline{\psi}_c(x_2)\psi_d(x_2):\, g_{\mu\nu}i\Big[D_0^{(+)}(x_2-x_1)-D_0^{(+)}(x_1-x_2)\Big] \quad (4.9)$$

$$-:\psi_b(x_1)\overline{\psi}_c(x_2):\,\Big[S_{da}^{(+)}(x_2-x_1)D_0^{(+)}(x_2-x_1)$$

$$+S_{da}^{(-)}(x_2-x_1)D_0^{(+)}(x_1-x_2)\Big]g_{\mu\nu} \quad (4.10)$$

$$-:\overline{\psi}_a(x_1)\psi_d(x_2):\,\Big[S_{bc}^{(-)}(x_1-x_2)D_0^{(+)}(x_2-x_1)+$$

$$+S_{bc}^{(+)}(x_1-x_2)D_0^{(+)}(x_1-x_2)\Big]g_{\mu\nu}+\dots\Big\} \quad (4.11)$$

It is straight-forward to associate Feynman graphs to the individual terms: (4.9) describes electron scattering, whereas (4.10) and (4.11) give the self-energy. But in contrast to ordinary Feynman rules, all products of distributions in (4.10-11) are well-defined, we will see this explicitly below.

Let us first briefly consider the tree graph (4.9). The square bracket is just the Pauli-Jordan function

$$D_0^{(+)}(-y)-D_0^{(+)}(y)=-D_0(y),\quad y=x_1-x_2, \quad (4.12)$$

which has a trivial causal splitting ($\omega=-2$)

$$-D_0(y)=-D_0^{\rm ret}(y)+D_0^{\rm av}(y). \quad (4.13)$$

Therefore, the retarded distribution R_2 contains the factor $-D_0^{\rm ret}(y)$. Since R_2' had a factor $D_0^{(+)}(-y)=-D_0^{(-)}(y)$, the two-point distribution contains

$$T_2=R_2-R_2'\sim -D_0^{\rm ret}(y)+D_0^{(-)}(y)=-D^F(y), \quad (4.14)$$

which is just the Feynman propagator. This shows that the causal perturbation theory leads to the naive Feynman rules in the case of tree graphs. This is not true for loop graphs, where the splitting is non-trivial.

To illustrate this important point, we consider the self-energy contributions (4.10-11). Using $y=x_1-x_2$ again, we have to compute

$$D_2(x_1,x_2)=:\overline{\psi}(x_1)d(y)\psi(x_2): \quad (4.15)$$

where

$$d(y)=-e^2\gamma^{\mu}[S^{(-)}(y)D_0^{(+)}(-y)+S^{(+)}(y)D_0^{(+)}(y)]\gamma_{\mu}. \quad (4.16)$$

We will look at the first term in (4.16)

$$d_-(y)\stackrel{\rm def}{=}-S^{(-)}(y)D_0^{(-)}(y) \quad (4.17)$$

in some detail, which is the contribution from R_2'. The Fourier transform of (4.17) is given by

$$\hat{d}_-(p) = (2\pi)^{-4} \int d^4q\, \Theta(q^0 - p^0)\delta((p-q)^2)(\not{q} + m)\Theta(-q^0)\delta(q^2 - m^2). \quad (4.18)$$

Since q lies on the mass shell $q^2 = m^2$, $q^0 < 0$ and $p - q$ on the backward light-cone $(p-q)^2 = 0$, $p^0 - q^0 < 0$, it follows that $p = p - q + q$ is time-like. For the further calculations, it is therefore convenient to use a Lorentz frame such that $p = (p_0, \mathbf{0})$. The calculation of (4.18) is then quite simple [4], due to the presence of the two δ-distributions. We shall obtain

$$\hat{d}(p) = e^2(2\pi)^{-3}\Theta(p^2 - m^2)\operatorname{sgn} p_0$$

$$\times \left(1 - \frac{m^2}{p^2}\right)\left[m - \frac{\not{p}}{4}\left(1 + \frac{m^2}{p^2}\right)\right]. \quad (4.19)$$

Now we have to split the distribution (4.19). Since $\hat{d}(p) \sim |p|$ for large $|p|$, we have $\omega = 1$ according to (3.27). Considering time-like p, the splitting is carried out by the following dispersion integral

$$\hat{r}(p) = \frac{i}{2\pi}\operatorname{sgn} p_0 \int\limits_{-\infty}^{+\infty} dt\, \frac{\hat{d}(tp)}{(t - ip_0 0)^2(1 - t + ip_0 0)}. \quad (4.20)$$

The difference to (3.26) is due to the fact that we have included the possibility of negative p_0. To show what kind of integrals we get, let us consider the contribution of the scalar term $\sim m$ in (4.19) in some detail

$$r_1(p) \stackrel{\text{def}}{=} e^2(2\pi)^{-4} i \operatorname{sgn} p_0$$

$$\times \int\limits_{-\infty}^{+\infty} dt\, \frac{m\Theta(t^2p^2 - m^2)}{t^2(1 - t + ip_0 0)}\operatorname{sgn}(tp_0)\left(1 - \frac{m^2}{t^2p^2}\right). \quad (4.21)$$

The contribution from the 1 in the last bracket is

$$r_{11}(p) \stackrel{\text{def}}{=} e^2(2\pi)^{-4} im \int\limits_{m^2/p^2}^{\infty} \frac{ds}{s(1 - s + ip_0 0)}$$

$$= e^2(2\pi)^{-4} im\left[\mathrm{P} \int\limits_{m^2/p^2}^{\infty} ds\left(\frac{1}{s} + \frac{1}{p_0^2 - s}\right)\right.$$

$$\left. - i\pi \operatorname{sgn} p_0 \int\limits_{m^2/p^2}^{\infty} ds\, \frac{\delta(s-1)}{s}\right]$$

$$= e^2(2\pi)^{-4} im\left[\log\left|\frac{p^2}{m^2} - 1\right| - i\pi \operatorname{sgn} p_0 \Theta\left(\frac{p^2}{m^2} - 1\right)\right]. \quad (4.22)$$

All other integrals are computed similarly, the total result is

$$\hat{r}(p) = e^2(2\pi)^{-4} i \Big\{ \Big(\log |1 - b^2| - i\pi \,\mathrm{sgn}\, p_0 \Theta(p^2 - m^2) \Big)$$

$$\times \Big[m(1 - \frac{1}{b^2}) - \frac{\not{p}}{4}(1 - \frac{1}{b^4}) \Big] + \frac{\not{p}}{4b^2} - m + \frac{\not{p}}{8} \Big\}, \tag{4.23}$$

where we have introduced the dimensionless parameter

$$b^2 = \frac{p^2}{m^2}. \tag{4.24}$$

The result (4.23) has to be combined with the contribution $-\hat{r}'$ from R_2' (4.17) to obtain the two-point function

$$T_2(x_1, x_2) \stackrel{\text{def}}{=} i \, : \overline{\psi}(x_1) \Sigma(x_1 - x_2) \psi(x_2) \, : . \tag{4.25}$$

Then the $\mathrm{sgn}\, p_0$ in (4.23) is changed into a factor 1, so that we end up with the following result for the self-energy

$$\hat{\Sigma}(p) = e^2(2\pi)^{-4} \Big\{ \Big(\log |1 - b^2| - i\pi \Theta(p^2 - m^2) \Big)$$

$$\times \Big[m(1 - \frac{1}{b^2}) - \frac{\not{p}}{4}(1 - \frac{1}{b^4}) \Big] + \frac{\not{p}}{4b^2} - m + \frac{\not{p}}{8} \Big\}. \tag{4.26}$$

It is a remarkable feature of this calculation of the self-energy that the result (4.26) is free of any infrared divergence. There is only a discontinuity on the mass shell $p^2 = m^2$. On the other hand, in the conventional calculation of the self-energy by regularization of a Feynman integral, infrared divergences appear. The two observations lead to the conclusion that these divergences are caused by the ultraviolet regularization, they are not inherent in the self-energy. Obviously, one is punished for a bad treatment of the ultraviolet problem by infrared divergences! Not all infrared singularities in the conventional formalism are of this spurious type, the vertex function, for example, has an inherent infrared singularity.

The distribution (4.19) which we have split has singular order $\omega = 1$. Consequently, the general solution (3.28) of the problem contains an undetermined linear polynomial in p which we write in the following form :

$$\tilde{\Sigma}(p) = \hat{\Sigma}(p) + C_0 + C_1(\not{p} - m). \tag{4.27}$$

The constants C_0, C_1 are not fixed by causality, they must be determined by other physical conditions. One such condition is

$$\tilde{\Sigma}(\not{p} = m) = 0, \tag{4.28}$$

which means that the electron mass is not changed by this radiative correction (no *finite* mass renormalization). According to (4.26) this leads to

$$C_0 = \frac{5}{8} m. \tag{4.29}$$

The constant C_1 is connected with the free constant in the vertex function by the Ward identity (see (5.17)), i.e. with charge normalization.

5. Properties of the S-matrix

Our construction of the S-matrix was perturbative. We cannot make any precise statement about the (probably asymptotic) convergence of the perturbation series (2.1). Accordingly, the properties of the S-matrix, we are going to study here, are properties of all orders of perturbation theory. Consequently, all such properties are proved by induction : One shows that the first order $T_1(x)$ has the property and that the inductive construction does not destroy it.

5.1 Normalizability

This subject is called renormalizability in standard textbooks and it is very complicated because one has to fight with the so-called overlapping divergences. In finite QED there is no ultraviolet divergence problem. The problem here simply consists of showing that the number of undetermined constants C_j does not increase with the order n of perturbation theory. Then, finitely many normalization conditions of the type (4.28) are sufficient to determine S completely. The problem obviously is to determine the singular order of arbitrary graphs. Normalizability of QED is contained in the following

Theorem. The singular order $\omega(g)$ of a graph g is equal to

$$\omega(g) = 4 - \frac{3}{2} f_g - b_g, \tag{5.1}$$

where f_g is the number of external fermions and b_g the number of external photons in g.

The expression (5.1) is well-known from naive power counting in momentum space. The important point is that it is independent of the order n of perturbation theory. Then there exist finitly many classes of graphs with $\omega \geq 0$, only, namely the following 5 classes:

1.) $\omega = 4$: These are vacuum graphs because they have no external line at all.

2.) $\omega = 2$: These graphs with 2 external photons contribute to the photon self-energy (vacuum polarization).

3.) $\omega = 1$: These graphs with 2 external fermions contribute to the electron self-energy.

4.) $\omega = 0$: The graphs with 2 external fermions and one photon contribute to the vertex function. There is a second class with

5.) $\omega = 0$: These graphs have 4 external photons which describes photon-photon scattering.

Further possibilities do not exist as a consequence of Furry's theorem.

The idea of the proof of (5.1) is quite simple. In the inductive construction, we form tensor products of n-point distributions as

$$T^1_{n_1}(x_1, \ldots x_{n_1}) \cdot T^2_{n_2}(y_1, \ldots y_{n_2}). \tag{5.2}$$

Let us assume that the individual factors have singular orders according to (5.1):

$$\omega_1 = 4 - \frac{3}{2} f_1 - b_1 \quad , \quad \omega_2 = 4 - \frac{3}{2} f_2 - b_2. \tag{5.3}$$

The tensor product has singular order $\omega = \omega_1 + \omega_2$. Then we have to perform contractions between the two factors in the course of normal ordering. Let us consider l contractions between photon operators. These give rise to l factors $D_0^{(+)}$ which encreases the singular order as follows

$$\begin{aligned} \omega &= \omega_1 + \omega_2 + 2l - 4 \\ &= 4 - \frac{3}{2}(f_1 + f_2) - (b_1 + b_2 - 2l). \end{aligned} \tag{5.4}$$

Since the l contractions have diminished the number of external photons by $2l$, the new result (5.4) is again in agreement with (5.1). The contractions of Fermi operators are treated similarly. Finally one has to show that ω is not changed by the distribution splitting. Detailed proofs can be found in [3] and [4].

5.2 Gauge Invariance

In the framework of finite QED, gauge invariance is expressed as follows : We consider the n-point distribution T_n normally ordered with respect to the photon operators

$$T_n(x_1, \ldots, x_n) = \sum_{l=0}^{n} \sum_{1 \le k_1 < \ldots < k_l \le n} t_{k_1 \ldots k_l}^{\mu_1 \ldots \mu_l}(x_1, \ldots, x_n)$$

$$\times : A_{\mu_1}(x_{k_1}) \cdot \ldots \cdot A_{\mu_l}(x_{k_l}) : . \tag{5.5}$$

The t's contain Fermi operators: $t_{k_1 \ldots k_l}^{\mu_1 \ldots \mu_l}$ is the sum of all graphs of order n with external photon lines at the vertices $x_{k_1}, \ldots x_{k_l}$ and no other external photon lines, the external fermions being arbitrary. Each t is symmetrical in $(x_{k_1}, \mu_1) \ldots (x_{k_l}, \mu_l)$. A gauge transformation

$$A_\mu(x) \longrightarrow A_\mu(x) + \partial_\mu \Lambda(x) \tag{5.6}$$

must formally give no contribution to the physical S-matrix $S(g = 1)$. Integrating (5.5) over $x_1, \ldots, x_n$ and performing partial integrations in the Λ-terms, we arrive at

$$\frac{\partial}{\partial x_{k_j}^{\mu_j}} t_{k_1 \ldots k_l}^{\mu_1 \ldots \mu_l}(x_1, \ldots, x_n) = 0, \tag{5.7}$$

for all $1 \le l \le n$, all $1 \le j \le l$, all $1 \le k_1 < \ldots < k_l \le n$ and all $(x_1, \ldots, x_n) \in \mathbb{R}^{4n}$. This is perturbative gauge invariance.

To prove (5.7) by induction, let us assume that (5.7) holds for all m-point distributions with $m \le n - 1$. Going from $n - 1$ to n, we have first to form (2.16)

$$R'_n(x_1, \ldots, x_n) = \sum_X T_{n-n_1}(Y, x_n) \tilde{T}_{n_1}(X). \tag{5.8}$$

Each term in (5.8) is a product of lower T's with disjoint arguments and, therefore, gauge invariant. The same is true for A'_n (2.22) and $D_n = R'_n - A'_n$. There remains to prove that gauge invariance is preserved under distribution splitting.

The starting point for this is the divergence relation

$$\partial_\nu d^\nu \stackrel{\text{def}}{=} \frac{\partial}{\partial x_n^\nu} d^{\mu_1 \ldots \mu_{l-1} \nu}_{k_1 \ldots k_l}(x_1, \ldots, x_{n-1}, x_n) = 0, \tag{5.9}$$

where we have introduced a shorthand notation and have taken $x_{k_j} = x_n$, $\mu_j = \nu$ for simplicity. Expanding (5.9) into normal products in the Fermi operators, it is important to note that the derivative can produce a δ-distribution

$$\frac{\partial}{\partial x_n^\nu}\Big(\overline{\psi}(x_n)\gamma^\nu S_{\substack{\text{ret}\\ \text{av}}}(x_n - x)\Big) = i\overline{\psi}(x_n)\delta^4(x_n - x) \tag{5.10}$$

$$\frac{\partial}{\partial x_n^\nu}\Big(S_{\substack{\text{ret}\\ \text{av}}}(x - x_n)\gamma^\nu \psi(x_n)\Big) = -i\delta^4(x - x_n)\psi(x_n), \tag{5.11}$$

taking the free Dirac equation into account. Therefore, combinations of field operators which differ from each other in d^ν may agree in $\partial_\nu d^\nu$. We will show this in detail for the vertex function and self-energy in all orders.

In this case there are three classes of graphs (fig.4) which must be taken together. The total d-distribution can be written as follows

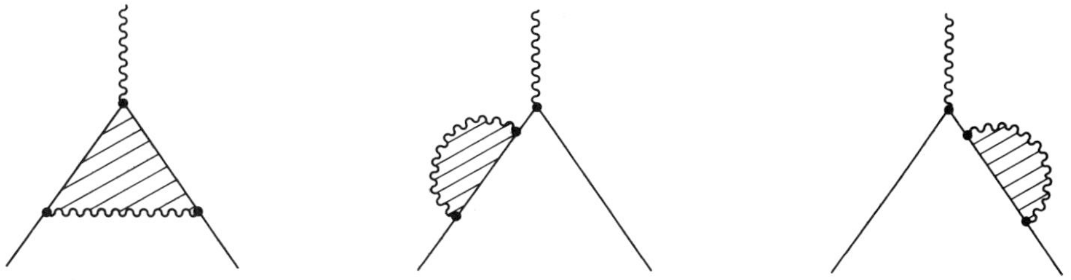

Fig. 4. Vertex and self-energy diagrams which must be taken together in the proof of gauge invariance.

$$d^\nu(x_1, \ldots, x_{n-1}, x_n) \stackrel{\text{def}}{=} :\overline{\psi}(x_1)\Lambda^\nu_d(x_1 - x_n, \ldots x_{n-1} - x_n)\psi(x_2):$$

$$+\Big[:\overline{\psi}(x_n)\gamma^\nu S^{\text{av}}(x_n - x_1)\Sigma^1_{\text{ret}}(x_2 - x_1, \ldots x_{n-1} - x_1)\psi(x_2):$$

$$+ :\overline{\psi}(x_1)\Sigma^2_{\text{ret}}(x_1 - x_2, x_3 - x_2, \ldots x_{n-1} - x_2)S^{\text{ret}}(x_2 - x_n)\gamma^\nu\psi(x_n):\Big] -$$

$$-\Big[:\overline{\psi}(x_n)\gamma^\nu S^{\rm ret}(x_n-x_1)\Sigma^1_{\rm av}(x_2-x_1,\dots x_{n-1}-x_1)\psi(x_2):$$

$$+:\overline{\psi}(x_1)\Sigma^2_{\rm av}(x_1-x_2,x_3-x_2,\dots x_{n-1}-x_2)S^{\rm av}(x_2-x_n)\gamma^\nu\psi(x_n):\Big], \quad (5.12)$$

where $\Sigma^{1,2}$ are self-energy functions which are retarded or advanced with respect to x_1 or x_2. Using (5.10-11) and introducing relative coordinates $y_i = x_i - x_n$, the divergence relation (5.9) assumes the following form

$$0=-\Big(\partial_{y_1\nu}+\dots+\partial_{y_{n-1},\nu}\Big)\Lambda^\nu_d(y_1,\dots y_{n-1})$$

$$-i\Big[-\delta(y_1)\Sigma^1_{\rm ret}(y_2,\dots y_{n-1})+\delta(y_2)\Sigma^2_{\rm ret}(y_1,y_3,\dots y_{n-1})\Big]$$

$$+i\Big[-\delta(y_1)\Sigma^1_{\rm av}(y_2,\dots y_{n-1})+\delta(y_2)\Sigma^2_{\rm av}(y_1,y_3,\dots y_{n-1})\Big]. \quad (5.13)$$

After Fourier transformation in the y_i, $1\le i\le n-1$, we get

$$0=(p_{1\nu}+\dots p_{n-1,\nu})\hat\Lambda^\nu_d(p_1,\dots p_{n-1})+\frac{1}{(2\pi)^2}\Big[\hat\Sigma^1_{\rm ret}(p_2,\dots p_{n-1})$$

$$-\hat\Sigma^2_{\rm ret}(p_1,p_3,\dots p_{n-1})\Big]-\frac{1}{(2\pi)^2}\Big[\hat\Sigma^1_{\rm av}(p_2,\dots p_{n-1})-\hat\Sigma^2_{\rm av}(p_1,p_3,\dots p_{n-1})\Big], \quad (5.14)$$

for all $p\in\mathbb{R}^{4(n-1)}$.

We need the general causal splittings of the various distributions. The vertex function ($\omega=0$) is given by

$$\hat\Lambda^\nu_{\rm ret}(p_1,\dots p_{n-1})=\frac{1}{2\pi i}\int dt\,\frac{\hat\Lambda^\nu_d(tp_1,\dots tp_{n-1})t}{(t+i0)^2(t-1-i0)}+C\gamma^\nu, \quad (5.15)$$

where we have added a factor t in the numerator and denomenator. For the self-energy functions ($\omega=1$) we have

$$\hat\Sigma^{1,2}_{\rm ret}(q_1,\dots q_{n-2})=\frac{1}{2\pi i}\int dt\,\frac{(\hat\Sigma^{1,2}_{\rm ret}-\hat\Sigma^{1,2}_{\rm av})(tq_1,\dots tq_{n-2})}{(t+i0)^2(t-1-i0)}+$$

$$+C^{1,2}_1+C^{1,2}_2\not{q}_1+C^{1,2}_3(\not{q}_2+\not{q}_3+\dots+\not{q}_{n-2}). \quad (5.16)$$

Because of symmetry in the inner vertices, each self-energy function has three free constants. Since the result does not depend on the reference point in the distribution splitting, they are related to each other

$$C^1_1=C^2_1\,,\ C^1_2=-C^2_2\,,\ C^1_3=-C^2_2+C^2_3. \quad (5.17)$$

To prove gauge invariance, we have to show that

$$0=(p_{1\nu}+\dots+p_{n-1,\nu})\hat\Lambda^\nu_{\rm ret}(p_1,\dots p_{n-1})+$$

$$+\frac{1}{(2\pi)^2}\Big[\hat\Sigma^1_{\rm ret}(p_2,\dots p_{n-1})-\hat\Sigma^2_{\rm ret}(p_1,p_3,\dots p_{n-1})\Big]. \quad (5.18)$$

Inserting (5.15-16) and using (5.14) and (5.17), we shall obtain

$$0=\frac{1}{2\pi i}\int dt\,\frac{1}{(t+i0)^2(t-1-i0)}\Big\{\hat{\Lambda}_d^\nu(tp_1,\dots tp_{n-1})t(p_{1\nu}+\dots+p_{n-1,\nu})+$$

$$+\frac{1}{(2\pi)^2}\Big[(\hat{\Sigma}^1_{\rm ret}-\hat{\Sigma}^1_{\rm av})(tp_2,\dots tp_{n-1})-(\hat{\Sigma}^2_{\rm ret}-\hat{\Sigma}^2_{\rm av})(tp_1,tp_3,\dots tp_{n-1})\Big]\Big\}+$$

$$+(p_{1\nu}+\dots+p_{n-1,\nu})C\gamma^\nu+\frac{1}{(2\pi)^2}\Big[C_1^1+C_2^1\not{p}_2+C_3^1(\not{p}_3+\dots+\not{p}_{n-1})$$

$$-C_1^2-C_2^2\not{p}_1-C_2^2(\not{p}_3+\dots+\not{p}_{n-1})\Big]$$

$$=(\not{p}_1+\dots+\not{p}_{n-1})\Big(C+\frac{1}{(2\pi)^2}C_2\Big).$$

Hence, gauge invariance holds if and only if

$$C=-\frac{1}{(2\pi)^2}C_2. \tag{5.19}$$

This is the well-known connection between the normalization constants for the self-energy and vertex functions.

From (5.18) one obtains the corresponding equation for the t-distributions

$$0=(p_{1\nu}+\dots+p_{n-1,\nu})\hat{\Lambda}^\nu(p_1,\dots p_{n-1})+$$

$$+\frac{1}{(2\pi)^2}\Big[\hat{\Sigma}^1(p_2,\dots p_{n-1})-\hat{\Sigma}^2(p_1,p_3,p_4,\dots p_{n-1})\Big]. \tag{5.20}$$

This is the most important one of the Ward-Takahashi identities which are equivalent to gauge invariance. By appropriate choice of the normalization constants in the distribution splittings, all these identities can be satisfied. For the complete proof we refer to [5].

5.3 Unitarity

The most important and most subtle property of the S-matrix is unitarity. The subtlety comes from the fact that, because of the gauge structure, the photon Fock space $\mathcal{F}$ contains more elements than are physically distinguishable. A general one-photon state $\Psi\in\mathcal{F}$ is of the following form

$$\Psi=\sum_{\mu=0}^{3}\int d^3k\,\hat{f}_\mu(\boldsymbol{k})e^{-i\omega t}a_\mu(\boldsymbol{k})^+\Omega$$

$$\stackrel{\rm def}{=}\sum_\mu\Psi_\mu. \tag{5.21}$$

Here a_μ^+ are ordinary creation operators for all 4 components of the electromagnetic potential. We introduce the usual *positive definite* scalar product

$$(\Phi,\Psi)\stackrel{\rm def}{=}\sum_{\mu=0}^{3}\int d^3k\,\hat{f}_\mu(\boldsymbol{k})^*\hat{g}_\mu(\boldsymbol{k}), \tag{5.22}$$

which defines the topology and enables us to apply the usual Hilbert space techniques. But, in order to get the correct commutation relations (4.5), the scalar potential in the field operators must be skew-adjoint

$$A^0(x) = i\,(2\pi)^{-3/2}\int\frac{d^3k}{\sqrt{2\omega}}\Big(a^0(\boldsymbol{k})e^{-ikx} + a^0(\boldsymbol{k})^+e^{ikx}\Big), \tag{5.23}$$

which is accomplished by the factor i.

Since A^0 is not hermitean, scalar photons should be considered as unphysical. We therefore use the radiation gauge

$$\Phi_0 = 0 \quad , \quad \partial_j\Phi_j(x) = 0, \tag{5.24}$$

and we call the space generated by these one-photon states together with the usual fermion states the physical subspace $\mathcal{F}_{\mathrm{phys}}$. Any state $\Psi \in \mathcal{F}$ is obtained from a unique $\Phi \in \mathcal{F}_{\mathrm{phys}}$ by gauge transformation, for example

$$\Psi_\mu(x) = \Phi_\mu(x) + \partial_\mu\Lambda(x) \tag{5.25}$$

in case of a one-photon state. Let us denote the projection operator on the physical subspace $\mathcal{F}_{\mathrm{phys}}$ by P.

Since the scalar potential $A^0(x)$ (5.23) is not hermitean, we cannot expect unitarity to hold on the entire Fock space $\mathcal{F}$. On the other hand, the transition probabilities for all possible final states must add up to one in every scattering process. This is guaranteed by unitarity of the physical S-matrix on the physical subspace $\mathcal{F}_{\mathrm{phys}}$

$$\lim_{g\to 1} PS(g)^+PS(g)P = P. \tag{5.26}$$

It is our aim now to prove a perturbative version of this. The photon sector in $\mathcal{F}_{\mathrm{phys}}$ is generated by transverse photon operators

$$A^\mu_{\mathrm{tr}}(x) = PA^\mu(x)P =$$

$$= (2\pi)^{-3/2}\int\frac{d^3k}{\sqrt{2\omega}}\sum_{n=1}^{2}\varepsilon^\mu_n(\boldsymbol{k})\Big[a_n(\boldsymbol{k})e^{-ikx} + a_n(\boldsymbol{k})^+e^{ikx}\Big]. \tag{5.27}$$

Here

$$\varepsilon^\mu_n(\boldsymbol{k}) = (0,\varepsilon_n) \quad , \quad \boldsymbol{k}\cdot\varepsilon_n(\boldsymbol{k}) = 0 \quad , \quad n = 1,2 \tag{5.28}$$

are two transverse orthogonal and real polarization vectors and

$$a_n(\boldsymbol{k}) = \varepsilon^\nu_n(\boldsymbol{k})a^\nu(\boldsymbol{k}).$$

In order to see what is going on, let us investigate the lowest order terms of the expression

$$PS(g)^+PS(g)P = P + \int P[T_1(x) + T_1(x)^+]Pg(x)dx$$

$$+\int PT_1(x_1)^+PT_1(x_2)Pg(x_1)g(x_2)dx_1dx_2+$$

$$+\frac{1}{2}\int P[T_2(x_1,x_2)^+ + T_2(x_1,x_2)]Pg(x_1)g(x_2)dx_1dx_2+\ldots \tag{5.29}$$

For (5.26) to be valid, all terms on the right side except P must vanish. The first order term vanishes indeed, because

$$T_1(x)=ie\ :\overline{\psi}(x)\gamma^\mu\psi(x):\ A_\mu(x) \tag{5.30}$$

is skew-adjoint on $\mathcal{F}_{\mathrm{phys}}$. Concerning the second order terms, we concentrate on the electron scattering graph (cf. (4.9) (4.14))

$$T_2(x_1,x_2)=-ie^2\ :\overline{\psi}(x_1)\gamma^\mu\psi(x_1)\overline{\psi}(x_2)\gamma_\mu\psi(x_2):\ D_0^F(x_1-x_2). \tag{5.31}$$

It contributes

$$\frac{1}{2}\Big[T_2(x_1,x_2)^+ + T_2(x_1,x_2)\Big]=e^2\ :\overline{\psi}(x_1)\gamma^\mu\psi(x_1)$$

$$\times\,\overline{\psi}(x_2)\gamma_\mu\psi(x_2):\ \mathrm{Im}\,D_0^F(x_1-x_2) \tag{5.32}$$

to (5.29) with

$$\mathrm{Im}\,D_0^F(x)=(2\pi)^{-4}\pi\int d^4k\,e^{-ikx}\delta(k^2). \tag{5.33}$$

The term (5.32) should be compensated by some contribution from the $PT_1^+PT_1P$ term. As usual, the product is normally ordered here by means of Wick's theorem

$$PT_1(x_1)^+PT_1(x_2)P=e^2\ :\overline{\psi}(x_1)\gamma_\mu\psi(x_1)\overline{\psi}(x_2)\gamma_\nu\psi(x_2):$$

$$\times[:\ A^\mu_{\mathrm{tr}}(x_1)A^\nu_{\mathrm{tr}}(x_2)\ :+\overline{A^\mu_{\mathrm{tr}}(x_1)A^\nu_{\mathrm{tr}}(x_2)}]+\ldots \tag{5.34}$$

The pairing of the transverse field operators (5.27) is given by

$$\overline{A^\mu_{\mathrm{tr}}(x_1)A^\nu_{\mathrm{tr}}(x_2)}=(2\pi)^{-3}\int\frac{d^3k}{2\omega}e^{-ik(x_1-x_2)}\sum_{n=1}^{2}\varepsilon_n^\mu(\boldsymbol{k})\varepsilon_n^\nu(\boldsymbol{k}). \tag{5.35}$$

The two polarization vectors (5.28) together with the two vectors

$$\eta^\mu=(1,0,0,0)\quad,\quad\hat{k}^\mu=\frac{k^\mu-(k^\nu\eta_\nu)\eta^\mu}{|\boldsymbol{k}|} \tag{5.36}$$

form a basis in Minkowski space. The completeness relation then implies

$$\sum_{n=1}^{2}\varepsilon_n^\mu(\boldsymbol{k})\varepsilon_n^\nu(\boldsymbol{k})=-g^{\mu\nu}+\eta^\mu\eta^\nu$$

$$-\frac{1}{\boldsymbol{k}^2}\Big[k_0^2\eta^\mu\eta^\nu+k^\mu k^\nu-k_0(k^\mu\eta^\nu+k^\nu\eta^\mu)\Big]. \tag{5.37}$$

From $-g^{\mu\nu}$ on the right side we get the covariant distribution

$$d_0^{(+)}(x)\stackrel{\mathrm{def}}{=}-(2\pi)^{-3}\int\frac{d^3k}{2\omega}\,e^{-ikx}\Big|_{k_0=\omega}$$

$$= -(2\pi)^{-3} \int d^4k\, \delta(k_0^2 - \boldsymbol{k}^2)\Theta(k_0)e^{-ikx}. \tag{5.38}$$

It is important to note that only the part symmetric in x_1, x_2 contributes to (5.29). Therefore, instead of (5.38), we must consider

$$\frac{1}{2}[d_0^{(+)}(x) + d_0^{(+)}(-x)] = -\frac{1}{2}(2\pi)^{-3} \int d^4k\, \delta(k^2)e^{-ikx}. \tag{5.39}$$

This is just the negative of $\operatorname{Im} D_0^F(x)$ (5.33), consequently this term compensates (5.32), as it should. There remains to discuss the remaining non-covariant terms in (5.37).

The distribution multiplied by $\eta^\mu \eta^\nu$ in (5.35) is equal to

$$\frac{1}{2}(2\pi)^{-3} \int d^4k\, \delta(k^2)e^{-ikx}\Big(1 - \frac{k_0^2}{\boldsymbol{k}^2}\Big) = 0. \tag{5.40}$$

The other terms in (5.37) contain at least one k^μ, which leads to a gradient in x-space. Since in (5.34) the distribution is multiplied by conserved current densities, the whole expression is a divergence. Then, by partial integration in (5.29), the derivative is shifted to the test function g. For $g = 1$ there is no contribution. Hence unitarity is satisfied in the lowest orders as a consequence of current conservation. In higher orders unitarity follows in virtue of gauge invariance (5.7). For the complete proof we refer to [4].

References

[1] R.P.Feynman: QED, The Strange Theory of Light and Matter, Princeton, N.J. (1985)

[2] N.N.Bogoliubov, D.V.Shirkov: Introduction to the Theory of Quantized Fields, New York (1959)

[3] H.Epstein, V.Glaser: *Ann.I.H.Poincaré* **29**, 211 (1973)

[4] G.Scharf: Finite Quantum Electrodynamics, Springer-Verlag, Berlin Heidelberg (1989)

[5] M.Dütsch, F.Krahe, G.Scharf: Gauge Invariance in Finite QED, *Nuovo Cimento A* (1990), to appear.

FOUNDATIONS OF SELF-FIELD QUANTUMELECTRODYNAMICS

A.O. Barut

Department of Physics, University of Colorado
Boulder, Colorado, 80309, USA

The program of selffield approach to Quantumelectrodynamics and radiative processes is almost completed. We give here the main principles of this theory which is conceptually much simpler than the standard perturbative QED. In this formulation the electromagnetic and the electron fields are not quantized; it is a classical field theory. A complete relativistic dynamics of two or more particles interacting via the electromagnetic field is developed by virtue of the new approach. We review numerical results obtained.

I. INTRODUCTION

There are many approaches to radiative processes, or more generally, to electromagnetic interactions of charged particles. We should welcome this multitude because different ways of looking at the same physical phenomena can only bring clarity and hopefully enlightenment. I list those different formulations which are definite and more or less complete:

(i) Second quantized quantum field theory, or the perturbative QED[1].

(ii The S matrix theory of electromagnetic interactions, either from unitarity, analyticity and successive pole approximation[2], or from regularization of the product of distributions[3]. Both of these lead to the renormalized perturbation theory with particles on the mass shell.

(iii) Path integral method. Either path integrals of Maxwell–Dirac fields[4], or path integrals directly from the classical particle trajectories[5].

(iv) Source theory[6].

(v) Selffield quantumelectrodynamics.

Of these only the selffield approach is in the long tradition of classical radiation theory and classical electrodynamics and is the subject of these lectures.

It is often stated that a large number of radiative phenomena conclusively show that the electromagnetic field, and further the electron's field, is quantized as a system of infinitely many oscillators with their zero point energies. The radiative phenomena are listed in Table I. We shall show that all these processes can also be understood and calculated in the selffield approach which does not quantize the fields. The quantum properties of the electromagnetic field are reduced here to the quantum properties of the source. One avoids thereby some of the difficulties of the quantized

New Frontiers in Quantum Electrodynamics and Quantum Optics
Edited by A. O. Barut, Plenum Press, New York, 1990

fields, such as the infinite zero point energy and other infinities of the perturbative QED.

TABLE I. RADIATIVE PROCESSES

Spontaneous emission
Lamb shift
Anomalous magnetic moment
Vacuum polarization
Casimir effect between parallel plates
Casimir Polder potentials
Planck-distribution law for blackbody radiation
Unruh effect
QED in cavities
$e^+ - e^-$ system:
positronium spectrum
positronium annihilation
pair production and annihilation
$e^+ - e^-$ scattering
Relativistic many body problem with retardation
Electron - photon system:
photoelectric effect
Compton effect
Bremsstrahlung

This lecture tells the story of the developments of selffield QED and it is good to begin from the beginning, namely the classical electrodynamics.

II. CLASSICAL ELECTRODYNAMICS

The selfconsistent treatment of coupled matter and electromagnetic field goes back to H.A. Lorentz[7]. The electromagnetic field has as its source all the charged particles which in turn move in this total electromagnetic field. We have thus the Maxwell's equations coupled to the equations for matter:

$$\left.\begin{array}{l} 1) \quad F_{\mu\nu}{}^{,\nu} = -j_\mu \\ \\ 2) \text{ Equation of motion of matter in the field } F \end{array}\right\} \qquad (1)$$

These equations, both, can be derived from a single action principle. It has the general from

$$W = \int [\text{Kinetic energy of matter } - j_\mu A^\mu - 1/4\ F_{\mu\nu} F^{\mu\nu}] \tag{2}$$

The last term is the action density of the field and the middle term represents the interaction of the matter current with the field.

We shall keep this general framework throughout also for quantum electrodynamics. The only change will be in the specific form of the current or how we describe the matter, the electron.

Classical electrodynamics *per se* is usually associated with the current of point particles moving along wordlines. But we can have more general extended sources of currents, as we shall see. For a number of point particles the current is given by

$$j_\mu(x) = \sum_i e_i \int ds_i \dot{x}_{i\mu}(s_i) \delta\left(x - x_i(s_i)\right) \tag{3}$$

Hence the fundamental equations are

$$F_{\mu\nu}{}^{,\nu} = -j_\mu = -\sum_i e_i \int ds_i \dot{x}_i(s_i) \delta\left(x - x_i(s_i)\right) \tag{4}$$

Here s_i are invariant time-parameters on the worldlines of the particles, and dots represent differentiation with respect to these times.

The equations of motions of the worldlines are

$$m_i \ddot{x}_{i\mu} = e_i F_{\mu\nu} \dot{x}_i^\nu \quad , i = 1, 2, 3, \ldots \tag{5}$$

It is essential for the selfconsistency of our system that the field F entering the last equation is the field produced by all the particles including the particle i, namely the selffield. Hence we divide F into two parts

$$m_i \ddot{x}_{i\mu} = e_i F_{\mu\nu}^{(\text{other particles})} \dot{x}_i^\nu + e_i F_{\mu\nu}^{\text{self}} \dot{x}_i^\nu \tag{6}$$

The selffield can be obtained from the Lienard-Wiechert potential

$$A_\mu(x) = \int dx_\mu(s) D\left(x - x(s)\right) = e \int ds \dot{x}_\mu(s) D\left(x - x(s)\right) \tag{7}$$

but is formally infinite at the position of the particle. It must be treated properly, for example, by analytic continuation onto the world line[8]. This leads to the final Lorentz-Dirac equation for each particle (in natural units $c = \hbar = 1$)

$$m\ddot{x}_\mu = eF_{\mu\nu}^{\text{ext}} \dot{x}^\nu + \frac{2}{3} e^2 \left(\dddot{x}_\mu + (\ddot{x})^2 \dot{x}_\mu\right) \tag{8}$$

This is the basic nonperturbative equation of classical electrodynamics. Here m is now the renormalized mass. Furthermore we must find solutions of this equation which have the property that whenever the external force is zero the electron moves like a free particle, $m\ddot{x}_\mu = 0$, that is the second term must vanish together with the external field. This is part of the renormalization program. The important feature of this equation is that all radiative effects are now expressed in a closed, we repeat, non-perturbative way. The price we pay for this is that the equation is not only nonlinear but also contains the third derivatives. The selffield approach to quantumelectrodynamics has the goal of finding the analogous nonlinear, nonperturbative equation in the case of quantum currents. It is clear that radiative effects like the Lamb shift, anomalous magnetic moment, spontaneous emission, etc. have their counterparts also in classical electrodynamics.

As a second example of a classical current we consider the classical model of the Dirac electron which describes a spinning and charged relativistic point particle. In this model the worldline of the point particle is a helix, called zitterbewegung, and the orbital angular momentum of the helix in the rest frame of the center of mass accounts for the spin and the magnetic moment of the particle. The generalization of the Lorentz-Dirac equation for this case has recently been given[9]:

$$\dot{\pi}_\mu = eF^{\text{ext}}_{\mu\nu} v^\nu + e^2 \left(g_{\mu\nu} - \frac{v_\mu v_\nu}{v^2} \right) \left[\frac{2}{3} \frac{\ddot{v}^\nu}{v^2} - \frac{9}{4} \frac{(v \cdot \dot{v}) \dot{v}^\nu}{v^4} \right] \tag{9}$$

where

$$\pi_\mu = p_\mu - eA_\mu, \quad v = \dot{x} \text{ and } v^2 \neq 1 \text{ due to spin.}$$

There are other classical models of the electron. A remarkable one is due to Lees[10] amd Dirac[11] in which a charged shell is held stable with a surface tension. In the equilibrium position the surface tension can be expressed in terms of the mass of the electron so that this model has exactly again two parameters, mass and charge, like the point worldline. The Lorentz-Dirac equation for this model to my knowledge has not been worked out yet.

III. SCHRÖDINGER AND DIRAC CURRENTS QUANTUMELECTRODYNAMICS

Quantumelectrodynamics has the same two basic equations (1). Only the form of the current j is different. According to Schrödinger and Dirac the electron is described not by a worldline but by a field $\psi(x,t)$ and the basic coupled equations (1) become

$$F^{;\nu}_{\mu\nu} = -j_\mu \quad , \; F_{\mu\nu} = A_{\nu,\mu} - A_{\mu,\nu}$$

and

$$(\gamma^\mu i\partial_\mu - m)\,\psi(x) = e\gamma^\mu \psi(x) A_\mu(x) \tag{10}$$

for the relativistic Dirac case, and

$$i\frac{\partial\psi}{\partial t} = \left(-\frac{1}{2m}\left[(\vec{p} - e\vec{A})^2 \right] + eA_0 \right) \psi \tag{11}$$

for the nonrelativistic Schrödinger case. The currents for a number of electrons is

$$j_\mu(x) = \sum_i e_i \bar{\psi}_i(x) \overset{(i)}{\gamma}_\mu \psi_i(x) \tag{12}$$

with a similar expression for the Schrödinger current.

Again the field A_μ is the sum of an external and a selffield parts:

$$A_\mu = A^{\text{ext}}_\mu + A^{\text{self}}_\mu \tag{13}$$

With the choice of gauge $A^{\mu}_{,\mu} = 0$ the Maxwell equations become

$$\Box A_\mu = j_\mu(x) = \sum_i e_i \bar{\psi}_i(x) \overset{(i)}{\gamma}_\mu \psi_\mu(x) \tag{14}$$

so that the selffield can be expressed in terms of the current as

$$A_\mu(x) = \int dy D(x-y) j_\mu(x) \tag{15}$$

where $D(x-y)$ is the appropriate Green's function corresponding to initial and boundary conditions. Equation (15) is our generalized Lienard-Wiechert potential. Thus the light emitted by a source depends on the nature and preparation of the current, and also on the nature of the environment determining the Green's function. Furthermore the whole light cone where ψ is different from zero contributes to the field at the field point and not just a single intersection of the worldline with the light cone, as in the case of a point particle.

Thus the selffield can be eliminated from the coupled Maxwell-Dirac equations. Inserting A_μ into the equation of motion we obtain

$$\left\{\gamma^\mu \left(i\partial_\mu - e_k A^{\text{ext}}_\mu\right) - m_k\right\} \psi_k(x) = e_k \gamma^\mu \psi_k(x) \int dy D(x-y) \sum_i e_i \bar{\psi}_i(y) \gamma_\mu \psi_i(y) \tag{16}$$

Here A^{ext} is a fixed external field whose sources are far away and not dynamically relevant. In the next Section we shall treat two or many body systems in which we shall eliminate completely <u>all</u> the fields in favor of the currents. Eq. (16) is a nonlinear integral equation for ψ analogous to the nonlinear equation of the classical electrodynamics. The corresponding equation for the Schrödinger case is ($\hbar = 1$)

$$i\frac{\partial \psi}{\partial t} = \left[-\frac{1}{2m}\left(\vec{p} - e\vec{A}^{\text{ext}} - e\vec{A}^{\text{self}}\right)^2 + e\left(A_0^{\text{ext}} + A_0^{\text{self}}\right)\right]\psi \tag{17}$$

where the selfpotentials are

$$A_0^{\text{self}} = \int dy D(x-y) \sum_k e_k \psi_k^*(y) \psi_k(y), \quad \vec{A}^{\text{self}}(x) = \int dy D(x-y) \sum_k \psi_k^*(y) \frac{\nabla}{i} \psi_k(y) \tag{18}$$

In writing these equations we have assumed that the ψ-current is an actual material charge current, and not just a probability current. Thus we are inevitably led to contemplate the interpretation and foundations of quantum theory. The foundations of quantumelectrodynamics and that of quantum theory must be the same, for quantum mechanics was invented to understand the interactions between light and matter. Not surprisingly, it was Schrödinger who first formulated the selfconsistent coupled Maxwell and matter field equations, i.e., the program of Lorentz, for the new wave mechanics and insisted that for the selfconsistency of the theory the self field of the electron must be included as a nonlinear term. Schrödinger however calculated only the static part of the selfenergy and obtained unacceptable large selfenergies. Subsequently quantum electrodynamics went into a different direction. The selffield was dropped completely. Instead, one introduced a separate quantized radiation field with its own new degrees of freedom and coupled this to the quantized matter field. In the selffield approach the electromagnetic field has no separate degrees of freedom, they are determined by the source's degrees of freedom, but then we must include the full nonlinear selffield term. We shall come to this duality between the two approaches and to the questions of interpretation of quantum theory after the developments of the selffield QED.

IV. RADIATIVE PROCESSES IN AN EXTERNAL (COULOMB) FIELD

The basis of selffield quantumelectrodynamics is conceptually very simple and is completely expressed by the single equation (16). All QED processes in an external field listed in Table I should be derived from this single equation. To perform actual calculations it is much simpler and more direct to work with the action rather than with the equations of motion. The action W can, up to an overall δ-function, be related to the energies of the system for bound state problems, and to the scattering amplitude for scattering problems.

The action for the system (10) is

$$W = \int dx \left[\bar{\psi}(x)(\gamma^\mu i\partial_\mu - m)\psi(x) - e\bar{\psi}(x)\gamma^\mu\psi(x)A_\mu(x) - \frac{1}{4}F_{\mu\nu}F^{\mu\nu} \right] \tag{19}$$

Here we shall express $A_\mu(x)$ in terms of ψ using (15). For bound state problems the action of the electromagnetic field can be reexpressed by a partial integration, using (10), as

$$-\frac{1}{4}\int dx F_{\mu\nu}F^{\mu\nu} = +\frac{1}{2}\int dx j_\mu(x)A^\mu(x) \tag{20}$$

Putting all together we have the action underlying our nonlinear equation (16), namely

$$\begin{aligned} W = \int dx \Big[&\bar{\psi}(x)\left(\gamma^\mu\left(i\partial_\mu - eA_\mu^{\text{ext}}\right) - m\right)\psi(x) \\ &- \frac{e^2}{2}\int dy \bar{\psi}(x)\gamma^\mu(x)\psi(x)D(x-y)\bar{\psi}(y)\gamma_\mu\psi(y) \Big] \end{aligned} \tag{21}$$

We shall consider now the single electron problem in an external field.

We expand the classical field ψ into a Fourier series

$$\psi(x) = \sum\!\!\!\!\!\!\int_n \psi_n(\vec{x})e^{-iE_n t} \tag{22}$$

and shall try to determine the expansion coefficients $\psi_n(\vec{x})$ and the spectrum E_n– discrete and continous. This expansion is quite different than the one used in standard QED and quantumoptics, namely the Coulomb series expansion, for example, in the Coulomb field,

$$\psi(x) = \sum\!\!\!\!\!\!\int_n c_n(t)\psi_n^c(\vec{x})$$

Here one derives equations for the time-dependent coefficients $c_n(t)$. The idea behind is that the system has definite levels and the perturbation will cause transitions between these levels. In our formulation, due to selfenergy, there are no definite (discrete) levels as exact eigenstates of the system to begin with, but the equations will determine the spectrum. In fact it will turn out that only the ground state of the system will be a stable eigenstate followed by a continuum with spectral concentrations around the unperturbed spectrum.

If we insert the Fourier expansion into the action we obtain

$$W = \sum_{nm} \int dx \Bigg\{ \bar{\psi}_n(\vec{x}) e^{iE_n x^0} \left[\gamma^\mu \left(i\partial_\mu - eA_\mu^{\text{ext}}\right) - m\right] \psi_m(\vec{x}) e^{-iE_m x^0} -$$

$$- \frac{e^2}{2} \int dy \bar{\psi}_n(\vec{x}) \gamma^\mu \psi_m(\vec{x}) e^{i(E_n - E_m)x^0} D(x-y) \bar{\psi}_r(\vec{y}) \gamma_\mu \psi_s(\vec{y}) e^{i(E_r - E_s)y^0} \Bigg\} \tag{23}$$

Time integrations can be performed using

$$D(x-y) = -\frac{1}{(2\pi)^4} \int dk \frac{e^{-ik(x-y)}}{k^2} \tag{24}$$

and we can write the interaction part of the action entirely in terms of the Fourier components of the current

$$\begin{aligned} W_{\text{int}} = &+ \frac{e^2}{2} \sum_{nmrs} \delta(E_n - E_m + E_r - E_s) \int \bar{\psi}_n(\vec{x}) \gamma^\mu \psi_m(\vec{x}) \\ &\times \frac{e^{i\vec{k}\cdot(\vec{x}-\vec{y})}}{(E_n - E_m)^2 - \vec{k}^2} \bar{\psi}_r(\vec{y}) \gamma_\mu \psi_s(\vec{y}) d\vec{x} d\vec{y} d\vec{k} \end{aligned} \tag{25}$$

For the exact solutions of our equations the action W will vanish identically. We will now solve the system iteratively.

To lowest order of iteration we take the field to be given by the solutions of the external field problem without the selfenergy terms, and the energies to be shifted by a small amount:

$$\begin{aligned} \psi_n(x) &= \psi_n^{\text{ext}}(x) \\ E_n &= E_n^{\text{ext}} + \Delta E_n \end{aligned} \tag{26}$$

The first term in (22) therefore gives simply, using the orthonormality of ψ_n's,

$$\begin{aligned} W_0 &= \int d\vec{x} \sum_{nm} \bar{\psi}_n \left(\gamma^0 E_n^{\text{ext}} - \vec{\gamma}\cdot\vec{p} - m - eA_\mu^{\text{ext}}\right) \psi_m \delta(E_n - E_m) \\ &\Rightarrow \sum_{nm} \Delta E_n \delta(E_n - E_m) \delta nm \end{aligned}$$

In the second term we separate the terms according to $E_n = E_m$, $E_r = E_s$ and according to $E_n = E_s$, $E_m = E_r$, the two ways of satisfying the overall δ-function. And since $W = 0$ to this order of iteration we can solve for ΔE_n. The action and the total energy of the system are related by a δ-function. Cancelling this δ-function and also the sum over n to obtain the energy shift of a fixed level n, we obtain

$$\begin{aligned} \Delta E_n = &\frac{e^2}{2} \int d\vec{x} \bar{\psi}_n(\vec{x}) \gamma_\mu \psi_n(\vec{x}) P \int \frac{d\vec{k}}{(2\pi)^3} \int d\vec{y} \frac{e^{i\vec{k}\cdot(\vec{x}-\vec{y})}}{k^2} \cdot \sum_s \bar{\psi}_s(\vec{y}) \gamma^\mu \psi_s(\vec{y}) \\ &- \frac{e^2}{2} \sum_s \int d\vec{x} d\vec{y} \bar{\psi}_n(\vec{x}) \gamma_\mu \psi_s(\vec{x}) \int \frac{d\vec{k}}{(2\pi)^3} e^{i\vec{k}\cdot(\vec{x}-\vec{y})} \bar{\psi}_s(\vec{y}) \gamma^\mu \psi_n(\vec{y}) \cdot \\ &\cdot \left[\frac{1}{E_s - E_n - k} - \frac{1}{E_s - E_n + k}\right] \\ &- \frac{e^2}{2} \sum_{\substack{s \\ (s<n)}} \int d\vec{x} d\vec{y} \bar{\psi}_n(\vec{x}) \gamma_\mu \psi_s(\vec{x}) \int \frac{d\vec{k}}{(2\pi)^3} e^{i\vec{k}\cdot(\vec{x}-\vec{y})} \bar{\psi}_s(\vec{y}) \gamma^\mu \psi_n(\vec{y}) \cdot \\ &\cdot \frac{i\pi}{2k} \delta(E_s - E_n - k) \end{aligned} \tag{27}$$

This can be written in the form

$$\Delta E_n = -\frac{e^2}{2}\sum_m\!\!\!\!\!\!\int \frac{d\vec{k}}{(2\pi)^3}\frac{j^\mu_{nn}(\vec{k})j^{mm}_\mu(-\vec{k})}{\vec{k}^2} - \frac{e^2}{2}\sum_{m<n}\!\!\!\!\!\!\int \frac{d\vec{k}}{(2\pi)^3} j^\mu_{nm}(\vec{k})j^{mn}_\mu(-\vec{k})\frac{i\pi}{2}\delta(E_m - E_n - k)$$
$$- \frac{e^2}{2}\sum_m\!\!\!\!\!\!\int \frac{d\vec{k}}{(2\pi)^3} j^\mu_{nm}(\vec{k})j^{mn}_\mu(-\vec{k})\frac{1}{2k}\left[\frac{1}{E_m - E_n - k} - \frac{1}{E_m - E_n + k}\right] \tag{28}$$

Thus the energy shifts are entirely expressed in terms of the integrals over the Fourier spectra of currents of all states. The first term corresponds to vacuum polarization, the second to spontaneous emission, and the third term to the Lamb shift proper. In arriving at these results we have used the causal Green's function and separated the integrals into a principal and a imaginary part according to the formula

$$\frac{1}{x} = P\frac{1}{x} \pm i\pi\delta(x) \tag{29}$$

All the main QED effects are obtained here from a single expression. In fact one can also read off the anomalous magnetic moment $(g-2)$ from this expression as we shall show in Section VI.

The evaluation of these expressions is a rather laborious technical problem. We have to use relativistic Coulomb wave functions for both the discrete and contious spectrum and integrate the products of such functions and sum over the whole spectrum. We shall indicate some of these calculations and give results in Section VIII. The most important feature of the present formulation is that there are no infrared nor ultraviolet divergences.

The spontaneous emission term in Eq. (27) has been exactly evaluated[12]. We have now complete relativistic spontaneous decay rates for all hydrogenic states[13]. Table II shows some of these results.

TABLE II. Decay rates (s^{-1}) in hydrogen and muonium

Transition	Hydrogen	Muonium
$2S_{1/2} \to 1S_{1/2}$	2.4964×10^{-6}	2.3997×10^{-6}
$2S_{1/2} \to 1P_{1/2}$	5.194×10^{-10}	5.172×10^{-10}
$2P_{1/2} \to 1S_{1/2}$	2.0883×10^{8}	2.0794×10^{8}
$2P_{3/2} \to 1S_{1/2}$	4.1766×10^{8}	4.1587×10^{8}
$2P \to 1S_{1/2}$	6.2649×10^{8}	6.2382×10^{8}

The vacuum polarization term has also been evaluated analytically[14] to lowest order term in $\alpha(Z\alpha)^4$. This is the most divergent term in perturbative QED and vanishes in the nonrelativistic limit.

The Lamb shift term which correctly reduces to the standard expressions in the dipole approximation has also been shown to be finite and will be evaluated in closed form[15].

In all these calcuations, since we are using Coulomb wave fuctions instead of the plane waves, the individual integrals are all finite. The summation over all the discrete and continous levels are done by means of the relativistic Coulomb Green's functions.

V. QUANTUMELECTRODYNAMICS OF THE RELATIVISTIC TWO-BODY SYSTEM

One of the most important and perhaps unexpected features of the selffield formulation of quantumelectrodynamics turned out to be a nonperturbative treatment of two and many body systems in closed from. It is well known that bound state problems cannot be treated in perturbative QED starting from first principles. Instead one begins from a Schrödinger or Dirac-like equation obtained from some approximation to the Bethe-Salpeter relations and then calculates the perturbation diagrams to the bound state solutions of these equations. What one really needs is a genuine two-body relativistic equation which includes all the radiative terms as well as all the recoil corrections at once. We shall now discuss the principles of this theory.

In nonrelativistic quantum theory the many body problem is formulated in configuration space by a wave equation with pair potentials $v_{ij}(x_i - x_j)$ of the form

$$\left(\frac{p_1^2}{2m_1} + \frac{p_2^2}{2m_2} + \ldots V_{12} + V_{13} + V_{23} + \ldots\right)\psi(x_1, \ldots, x_n; t) = i\hbar\frac{\partial\psi}{\partial t}$$

This a priori not obvious. We may also think that each particle has its own field $\psi(x)$ and satisfy a wave equation with a potential coming from the charge distribution of the other particles. For two particles, for example, we would have the coupled Hartree-type equations

$$i\hbar\frac{\partial\psi_1(\vec{x}_1, t)}{\partial t} = \left(-\frac{\hbar^2}{2m_1}\Delta + \int\frac{\psi_2^*(x_2, t)\psi_2(x_2, t)}{|\vec{x}_2 - \vec{x}_1|}d\vec{x}_2\right)\psi_1(\vec{x}_1, t)$$
$$i\hbar\frac{\partial\psi_2(\vec{x}_2, t)}{\partial t} = \left(-\frac{\hbar^2}{2m_2}\Delta + \int\frac{\psi_1^*(\vec{x}_1, t)\psi_1(\vec{x}_1, t)}{|\vec{x}_1 - \vec{x}_2|}d\vec{x}_1\right)\psi_2(\vec{x}_2, t)$$

These two formulations are closely related but not identical. We shall see that they correspond to two different types of variational principles and actually describe two different types of physical situations. Quantum theory has a separate new postulate for two or more particles, namely that the state space is the tensor product of one particle state spaces. This leads immediately to the first formulation in configuration space. Such combined systems are called in the axiomatic of quantum theory "nonseparated" systems with all the nonlocal properties of quantum theory. But this postulate does not apply universally. There are other systems, namely the "separated" systems, which are described by the second type of equations. For example, for the system hydrogen molecule the two protons are separated, whereas the two electrons are nonseparated. The superposition principle holds for the nonseparated systems only. We shall now see how all this comes about from two different basic variational princples in the relativistic case (the nonrelativistic case is similar).

Consider a number of matter fields $\psi_1(x), \psi_2(x)\ldots$ The action of these fields interacting via the electromagnetic field is

$$W = \int dx\left\{\sum_k \bar{\psi}_k\left(\gamma^\mu i\partial_\mu - m_k\right)\psi_k - j_\mu(x)A^\mu(x) - \frac{1}{4}F_{\mu\nu}F^{\mu\nu}\right\} \qquad (30)$$

where the current j^μ is the sum of Dirac currents for each field

$$j^\mu(x) = \sum_k e_k\bar{\psi}_k(x)\gamma^\mu\psi_\mu(x) \qquad (31)$$

Again in the gauge $A^\mu{}_{,\mu} = 0$ we obtain the equations for the electromagnetic field as

$$\Box A_\mu = j_\mu = \sum_k j_\mu^{(k)} \tag{32}$$

with the solution

$$A_\mu(x) = \int dy D(x-y) j_\mu(x). \tag{33}$$

If we insert this into the action both in the $j^\mu \cdot A_\mu$ term as well as in the term $-(1/4)F_{\mu\nu}F^{\mu\nu}$, and using the identity (20), we obtain

$$W = \int dx \sum_k \bar{\psi}_k (\gamma^\mu i\partial_\mu - m_k)\psi_k - \sum_{k,\ell} \frac{1}{2} \int dx dy j_\mu^{(k)}(x) D(x-y) j^\mu_{(\ell)}(y) \tag{34}$$

The interaction action is a sum of current-current interactions containing both the mutual interaction terms, e.g.

$$-\frac{e_1 e_2}{2} \int dx dy \bar{\psi}_1(x)\gamma^\mu \psi_1(x) D(x-y) \bar{\psi}_2(y)\gamma_\mu \psi_2(y) - (1 \leftrightarrow 2)$$

and the self interaction terms like

$$-\frac{e_1^2}{2} \int dx dy \bar{\psi}_1(x)\gamma^\mu \psi_1(x) D(x-y) \bar{\psi}_1(y)\gamma_\mu \psi_1(y)$$

If we vary this action with respect to each field ψ_k separately we obtain coupled nonlinear equations. For example for two particles

$$\begin{aligned}
(\gamma^\mu i\partial_\mu - m_1)\psi_1 &= \frac{e_1 e_2}{2}\gamma^\mu \psi_1 \int dy D(x-y)\bar{\psi}_2(y)\gamma_\mu \psi_2(y) \\
&\quad + \frac{e_1^2}{2}\gamma^\mu \psi_1 \int dy D(x-y)\bar{\psi}_1(y)\gamma_\mu \psi_1(y) \\
(\gamma^\mu i\partial_\mu - m_2)\psi_2 &= \frac{e_1 e_2}{2}\gamma^\mu \psi_2 \int dy D(x-y)\bar{\psi}_1(y)\gamma_\mu \psi_1(y) \\
&\quad + \frac{e_2^2}{2}\gamma^\mu \psi_2 \int dy D(x-y)\bar{\psi}_2(y)\gamma_\mu \psi_2(y)
\end{aligned} \tag{35}$$

Next let us define a composite field Φ by

$$\Phi(x_1, x_2) = \psi_1(x_1)\psi_2(x_2) \tag{36}$$

This is a 16-component spinor field. We can rewrite our action (34) entirely in terms of the composite field. This is straightforward in the mutual interaction terms. In the kinetic energy and selfinteraction terms we multiply suitable by normalization factors. For example for the first kinetic energy term we get

$$\int dx_1 \bar{\psi}_1(x_1)(\gamma^\mu i\partial_\mu - m_1)\psi_1(x_1) \cdot \int d\sigma_2 \bar{\psi}_2(x_2)\gamma \cdot n \psi_2(x_2)$$

where $d\sigma_2 n^\mu = d\sigma_2^\mu$ is a 3-dimensional volume element perpendicular to the normal n^μ. Similarly for the other kinetic energy term. The selfenergy terms need two such normalization factors. The resultant action in terms of the composite field is then

$$\begin{aligned} W = \Bigg[& \int dx_1 d\sigma_2 \bar{\Phi}(x_1 x_2)(\gamma^\mu \pi_{1\mu} - m_1) \otimes \gamma \cdot n \Phi(x_1 x_2) \\ & + \int dx_2 d\sigma_1 \bar{\Phi}(x_2 x_1) \gamma \cdot n \otimes (\gamma^\mu \pi_{2\mu} - m_2) \Phi(x_2 x_1) \\ & - e_1 e_2 \int dx_1 dx_2 \bar{\Phi}(x_1 x_2) \gamma^\mu \otimes \gamma_\mu D(x_1 - x_2) \Phi(x_1 x_2) \Bigg] \end{aligned} \tag{37}$$

The generalized canonical momenta $\pi_{i\mu}$ are given further below. Here and through the rest of the paper we shall write spin matrices in the form of tensor products $\otimes$, the first factor always referring to the spin space of particle 1, the second to particle 2. We shall give the selfenergy terms explicitely below.

Now our second variational principle is that the action be stationary not with respect to the variations of the individual fields but with respect to the total composite field only. This is a weaker condition than before and leads to an equation for Φ in configuration space. For bound state problems only the symmetric Green's function contributes and it contains a $\delta(x^2)$-function which we decompose relative to the space-like surface with normal n^μ as follows

$$\delta(r^2) = \frac{\delta[(r \cdot n) - r_\perp) \pm \delta[(r \cdot n) + r_\perp)}{2r_\perp}, \quad r_\perp = [(r \cdot n)^2 - r^2]^{1/2} \tag{38}$$

where $r_\perp$ is a relativistic three dimensional distance which for $n = (1000)$ reduces to the ordinary distance r. All the integrals in the action (37) are 7-dimensional. For covariance purposes it is necessary to have the vector n^μ. It tells us how to choose the time axis. The vector n is also present, in principle, in the one-body Dirac equation but we usally do not write it when discussing the solutions, but automatically choose it to be $n = (1000)$, i.e., the rest frame. The final form of our two-body equation is then

$$\left\{(\gamma^\mu \pi_1^\mu - m_1) \otimes \gamma \cdot n + \gamma \cdot n \otimes (\gamma^\mu \pi_2^\mu - m_2) - e_1 e_2 \frac{\gamma^\mu \otimes \gamma_\mu}{r_\perp}\right\} \Phi(x_1 x_2) = 0 \tag{39}$$

where now the selfpotentials are inside the generalized momenta

$$\pi_i^\mu = p_i^\mu - e_i A_i^{\mu \text{self}} - e_i A_i^{\mu \text{ext}} \tag{40}$$

with

$$\begin{aligned} A_{\mu,1}^{\text{self}}(x) &= \frac{e_1}{2} \int dz d\sigma_u D(x - z) \bar{\Phi}(z, u) \gamma_\mu \otimes \gamma \cdot n \Phi(z, u) \\ A_{\mu,2}^{\text{self}}(x) &= \frac{e_2}{2} \int d\sigma_z du D(x - u) \bar{\Phi}(z, u) \gamma \cdot n \otimes \gamma_\mu \Phi(z, u) \end{aligned} \tag{41}$$

We note that the last term in (39) can also be put into the potential A_μ, one half for each particle; $\frac{1}{2} e_2 \frac{1 \otimes \gamma_\mu}{r}$ and $\frac{1}{2} e_1 \frac{\gamma_\mu \otimes 1}{r}$, respectively.

The self potentials are nonlinear integral expressions. The arguments of Φ consist of seven variables because $\Phi(x_1, x_2)$ is different from zero only if $(x_1 - x_2)$ is lightlike; only then there is a communication between the particles. This means that we have one time-variable and three space variables for each particle. We see this more cleary if we introduce center of mass and relative variables according to

$$\begin{aligned} \Pi &= \pi_1 + \pi_2 \,, \qquad \pi = \pi_1 - \pi_2 \\ x &= x_1 - x_2 \,, \qquad X = x_1 + x_2 \end{aligned} \tag{42}$$

Then equation (39), without the selffield terms for simplicity, becomes

$$\left\{ \Gamma^\mu \Pi_\mu + k^\mu \pi_\mu - \frac{e_1 e_2}{r_\perp} \gamma^\mu \otimes \gamma_\mu - m_1 I \otimes \gamma \cdot n - m_2 \gamma \cdot n \otimes I \right\} \Phi = 0 \tag{43}$$

where we have introduced

$$\Gamma^\mu = \frac{1}{2}(\gamma^\mu \otimes \gamma \cdot n + \gamma \cdot n \otimes \gamma^\mu)$$

and

$$k^\mu = \frac{1}{2}(\gamma^\mu \otimes \gamma \cdot n - \gamma \cdot n \otimes \gamma^\mu) \tag{44}$$

We see now that $k \cdot n = 0$, i.e., the component of k^μ parallel to n^μ vanishes which means that the component of the relative momentum π_μ parallel to n_μ drops out of the equation automatically. For $n = (1000)$, in particular, we have

$$\left\{ \Gamma^0 \Pi_0 - \vec{\gamma} \cdot \vec{\Pi} - \vec{k} \cdot \vec{\pi} - \frac{e_1 e_2}{r} \gamma^\mu \otimes \gamma_\mu - m_1 I \otimes \gamma_0 \quad m_2 \gamma_0 \otimes I \right\} \Phi = 0 \tag{45}$$

Thus we have only one time variable conjugate to the center of mass energy Π_0 and three degrees of freedom for the center of mass momentum $\vec{\Pi}$ and three degrees of freedom for the relative momentum $\vec{\pi}^0$; π_0 does not enter, as it should be so on physical grounds. In contrast the Bethe-Salpeter equation has two time coordinates. Since Π_0 is the "Hamiltonian" of the system we obtain, by multiplying (45) by Γ_0^{-1} the Hamiltonian form of the two-body equation

$$\Pi_0 \Phi = \left\{ \vec{\alpha} \cdot \vec{\Pi} + (\vec{\alpha}_1 - \vec{\alpha}_2) \cdot \vec{\pi} + \frac{e_1 e_2}{r} (1 - \vec{\alpha}_1 \cdot \vec{\alpha}_2) + m_1 \beta_1 \cdot I + m_2 I \cdot \beta_2 \right\} \Phi$$

where we have defined

$$\vec{\alpha} \equiv \frac{1}{2}(\vec{\alpha}_1 + \vec{\alpha}_2); \; \vec{\alpha}_i = \gamma_i^0 \vec{\gamma}_i, \beta_i = \gamma_{0i}, \; i = 1, 2 \tag{46}$$

Our two-body equation has the form of a generalized Dirac equation, now a 16-component wave equation. In fact it reduces to the one-body Dirac equation in the limit when one of the particles is heavy.

The above developments are completely relativistic and covariant. The physical results are independent of the vector n although a vector n must appear for manifest covariance. Thus recoil corrections are included to all orders. Further interesting properties of the equation, beside being a one-time relativistic equation, are that relative and center of mass terms in the Hamiltonian are additive, and radial and angular parts of the relative equation are exactly separable. It has also a non-relativistic limit to the two-body Schrödinger equation. We shall discuss numerical results in Section VIII.

VI. THE INTERPRETATION OF NEGATIVE ENERGY STATES

It is often stated that only in second quantized field theory can one have an adequate description of antiparticles and negative energy solutions where one changes the roles of the creation and annihilation operators for the negative energy solutions. We shall now show that there is also a consistent way of dealing with the negative energy solutions and antiparticles in the Dirac equation as a classical field theory and elaborate how we obtain the annihilation potential in positronium, for example.

There are actually not one but two Diract equations

$$\begin{aligned}(\gamma \cdot p - m)\psi_I &= 0 \\ (\gamma \cdot p + m)\psi_{II} &= 0\end{aligned} \tag{47}$$

obtained from the factorization of the Klein-Gordon operator, for example. By convention we just peak one and work with the complete set of solutions of this equation. Now the negative energy solutions of ψ_I coincide with the positive energy solutions of ψ_{II}. Furthermore in the presence of the electromagnetic field with minimal coupling we have the two equations

$$\begin{aligned}(\gamma \cdot (p - eA) - m)\,\psi_I &= 0 \\ (\gamma \cdot (p - eA) + m)\,\psi_{II} &= 0\end{aligned} \tag{48}$$

and we can easily prove that

$$\psi_I(-p, -e) = \psi_{II}(p, e) \tag{49}$$

that is the negative energy momentum solutions of ψ_I coincide with the positive energy solutions of ψ_{II} of opposite charge. Therefore we should consider positive energy solutions of both equations as physical particles. The total number of such physical solutions is the same as the total number of both positive and negative energy solutions of a single Dirac equation.

With this interpretation we obtain quite naturally the annihilation diagrams and annihilation potentials between particles and antiparticles. Consider our interaction action

$$\int dx dy \bar{\psi}_1(x)\gamma^\mu \psi_1(x) D(x - y)\bar{\psi}_2(y)\gamma_\mu \psi_2(y)$$

Here the classical fields $\psi_i(x)$ contain all positive and negative energy solutions according to our general expansion (22). Separating positive and negative energy solutions as

$$\psi_n(x) = \psi_{E_N > o} + \psi_{E_N < o} \equiv \psi_n^+ + \psi^-$$

and inserting it into the action we get 16 terms. In the limiting case of the lowest order scattering, where we replace the fields by plane wave solutions, we have essentially two distinct types of vertices at each point x or y, namely

$$\bar{\psi}^+(x)\gamma_\mu \psi^+(x)$$

and (50)

$$\bar{\psi}^+(x)\gamma_\mu \psi^-(x)$$

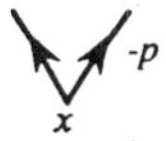

In the second case we have used our interpretation of the negative energy solutions as the antiparticles with reversed energy-momentum p_μ. The complete interaction action to this order consists of all combinations of these two vertices located at x and

y for particles 1 and 2 multiplied with the Green's function $D(x-y)$. Of these 16 terms some cannot be realized because of the overall energy-momentum conserving δ-function, $\delta(p_1+p_2-p_3-p_4)$, and we are left with two disctinct types of terms

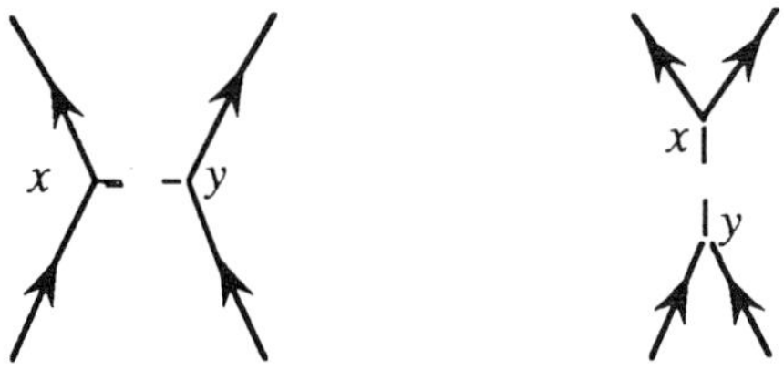

(51)

plus the same terms with particles and antiparticles interchanged. This result agrees with the standard QED. But we shall go a bit further and apply it to bound state problems in Section VII after a discussion of the case of identical particles.

Identical Particles

For two identical particles we use the postulate of the first quantized quantum theory that the field is symmetric or antisymmetric under the interchange of all dynamical variables of identical particles. In our formulation we go back to the original action principle and assume that the current j_μ is antisymmetric in the two fields

$$j_\mu = \frac{1}{2} e \left(\bar{\psi}_1 \gamma_\mu \psi_2 - \bar{\psi}_2 \gamma_\mu \psi_1\right), \quad e_1 = e_2 = e \tag{52}$$

This implies in the interaction action

$$W_{\text{int}} = \frac{1}{4} e^2 \left[\int dx dy \bar{\psi}_1(x) \gamma_\mu \psi_2(x) D(x-y) \bar{\psi}_1(y) \gamma^\mu \psi_2(y) \right. \\ \left. - \int dx dy \bar{\psi}_1 \gamma_\mu \psi_2 D(x-y) \bar{\psi}_2 \gamma^\mu \psi_1 + (1 \leftrightarrow 2) \right] \tag{53}$$

and again when the fields are expanded we see that identical particles with exactly the same wave functions i.e., the same quantum numbers or the same state, will not interact and that in the lowest approximation we will get besides the direct interaction also an exchange term as shown in the following diagrams

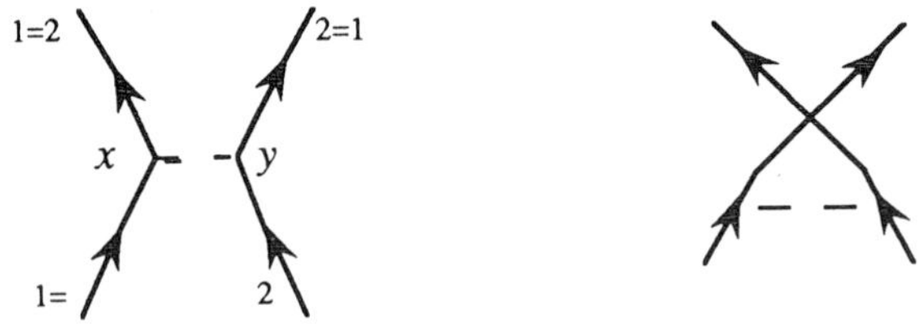

Finally we combine the two effects, identical particles and particle-antiparticle properties, to discuss systems like electron-positron complex and positronium. According to our discussion this system is just a part of the larger electron-electron system taking into account the interpretation of the negative energy levels and the identicity of the particles.

VI. CALCULATION OF THE ANOMALOUS MAGNETIC MOMENT $(g-2)$

We show now that our basic interaction action also contains besides Lamb shift, spontaneous emission and vacuum polarization also the anomalous magnetic moment in the same single expression. We shall also introduce at this occasion the more general four-dimensional energy-momentum Fourier expansion instead of the energy Fourier expansion (22) which was appropriate for the fixed external field problem.

The interaction action is given by

$$W_{\text{int}} = -\frac{e^2}{2}\int dxdyj^{\mu}(x)D(x-y)j_{\mu}(y) \tag{54}$$

We expand the fields as four dimensional Fourier integrals

$$\psi(x) = \int dpe^{-ipx}\psi(p) \tag{55}$$

and insert it into the action

$$W_{\text{int}} = -\frac{e^2}{2}\frac{1}{(2\pi)^4}\int dxdydkdpdqdrds\bar{\psi}(p)\gamma^{\mu}\psi(p)\frac{e^{-ik(x-y)}}{k^2+i\varepsilon}\bar{\psi}(r)\gamma_{\mu}\psi(s)$$
$$\times\, e^{i(p-q)x+(r-s)y}$$

which can be written as

$$W_{\text{int}} = -\frac{e^2}{2}(2\pi)^4\int dpdqdrdsj^{\mu}(p,q)\frac{1}{(r-s)^2+i\varepsilon}j_{\mu}(r,s)\delta(p-q+r-s) \tag{56}$$

where $j^{\mu}(p,q)$ stand for the double Fourier transform

$$j^{\mu}(p,q) = (2\pi)^8\int dxdy\bar{\psi}(x)\gamma^{\mu}\psi(y)e^{-ipx}e^{iqy} \tag{57}$$

The δ-function arises from the x,y and k-integrations. We separate the action into two terms to satisfy the δ-function

$$\begin{aligned} &(i) \quad p=q, \text{ hence } r=s \\ &(ii) \quad p=s, \text{ hence } q=r \end{aligned} \tag{58}$$

Again as before the first corresponds to vacuum polarization; the second term contains Lamb shift and spontaneous emission as real and imaginary parts of the energy shift ΔE. It also contains the anomalous magnetic moment as the coefficient of the magnetic part of the Lamb shift for any external field. For the calculation of $(g-2)$ it is thus not necessary to solve a problem with an external magnetic field or to solve any external problem for that matter.

The second term with (ii) can be written as

$$W_{\text{int}}^{(ii)} = -\frac{e^2}{2}\int dxdydpdqD(x-y)j^{\mu}(q,p)j_{\mu}(q,p)e^{ipx}e^{i(y-x)}q \tag{59}$$

Here we recognize the c-number electron propagator-function $S(x-y)$

$$\int dq\psi(q)\bar{\psi}(q)e^{i(y-x)q} \equiv S(x-y) \tag{60}$$

which satisfy the inhomogeneous wave equation

$$[\gamma^\mu(p_\mu - eA_\mu) - m] = i\delta(x-y) \tag{61}$$

Let us also take the Fourier transform of S

$$S(x-y) = \frac{1}{(2\pi)^4}\int dpe^{-ip(x-y)}S(p) \tag{62}$$

Then the action (59) becomes

$$W^{(ii)}_{\text{int}} = -\frac{e^2}{2}i\int dpdP\bar{\psi}(p)\frac{\gamma^\mu S(p)\gamma_\mu}{(p-P)^2+i\varepsilon}\psi(p) \tag{63}$$

It is related to the energy, more precisely to a mass shift by an overall δ-function and we can write

$$\Delta E = \frac{W^{(ii)}_{\text{int}}}{(2\pi)^4} = \int dp\bar{\psi}(p)\Delta M(p)\psi(p) \tag{64}$$

where we have introduced an effective *mass matrix* by

$$\Delta M(p) = \frac{e^2}{2}\frac{i}{(2\pi)^4}\int ds\frac{\gamma^\mu S(p-s)\gamma_\mu}{s^2 - i\varepsilon} \tag{65}$$

It remains now to evaluate the mass matrix ΔM. First we expand the Green's function or propagator in an external field as follows

$$\frac{1}{\gamma^\mu(p_\mu - eA_\mu) - m} = \frac{\not{p} - e + m}{p^2 - m^2} + 2e\frac{p\cdot A(\not{p}+m)}{(p^2-m^2)^2} - ie\frac{(\not{p}+m)\gamma^\mu\gamma^\nu F_{\mu\nu}}{(p^2-m^2)^2} + \ldots$$

where

$$\not{p} \equiv \gamma^\mu p_\mu,\ \not{A} = \gamma^\mu A_\mu,\ p\cdot A = p^\mu A_\mu. \tag{66}$$

It turns out that only the third term which is gauge invariant gives a nonvanishing contribution to lowest order in α: the terms containing A_μ give vanishing contributions. The mass operator becomes

$$\Delta M(p) = \frac{e^2}{2}\frac{i}{(2\pi)^4}(-ie)\int ds\frac{\gamma^\mu\left((p-s)-m\right)\gamma^\lambda\gamma^\varphi F_{\lambda\varphi}\gamma_\mu}{s^2\left((p-s)^2-m^2\right)^2} \tag{67}$$

The integrals can be performed giving[16]

$$\Delta M(p) = \frac{e^2}{2}\frac{i}{(2\pi)^4}(-ie)\frac{2}{m}\int_0^1 dy(1-y)\sigma^{\mu\nu}F_{\mu\nu}$$

and finally

$$\Delta M(p) = -\frac{\alpha}{2\pi}\frac{e}{2m}\sigma_{\mu\nu}F_{\mu\nu} = -\frac{\alpha}{2\pi}\frac{e}{2m}(\vec{\sigma}\cdot\vec{B} + i\vec{\alpha}\cdot\vec{E}) \tag{68}$$

Thus we recognize the anomalous magnetic moment to this order in front of the $\vec{\sigma}\cdot\vec{B}$ term for any field as we have mentioned.

If we insert the mass operator into the energy shift formula (64) we can evaluate it as the expectation value of the operator $\vec{\sigma}\cdot\vec{B}+i\vec{\alpha}\cdot\vec{E}$. For relativistic Coulomb problem for example the magnetic part of the Lamb shift can be analytically evaluated exactly.[17]

This way of calculating the anomalous magnetic moment also shows now how to calculate higher order terms. We must take more terms in the Green's function expansion (66). This may be much simpler than the diagrammatic method of perturbative QED where there are already 891 Feynman diagrams in the order $(\alpha/\pi)^4$.

VII. COVARIANT ANALYSIS OF RADIATIVE PROCESSES FOR TWO-BODY SYSTEMS

In this section we discuss how to treat radiative processes, like Lamb shift, etc., for a system like positronium or muonium beyond the naive reduced mass method. As mentioned above the action formalism is more convenient than the equations of motion.

We go back to our covariant 2-body action (37) and separate center of mass and relative coordinates and momenta according to

$$\begin{array}{lll} r = x_1 - x_2 & x_1 = R + \frac{1}{2}r & P = p_1 + p_2 \\ R = \frac{1}{2}(x_1 + x_2) & x_2 = R - \frac{1}{2}r & p = \frac{1}{2}(p_1 - p_2) \\ q = z - u & z = Q + \frac{1}{2}q & p_1 = \frac{1}{2}P + p \\ Q = \frac{1}{2}(z + u) & u = Q - \frac{1}{2}q & p_2 = \frac{1}{2}P - p \end{array} \tag{69}$$

All quantities here are four-vectors. Then the action becomes

$$\begin{aligned} W = \int dRdq\bar{\Phi}(R,r)\Big\{ &\left[\gamma\cdot\left(\frac{1}{2}P+p\right)-m_1\right]\otimes\gamma\cdot n+\gamma\cdot n\otimes\left[\gamma\cdot\left(\frac{1}{2}P-p\right)-m_2\right] \\ &- e_1e_2D(r) - \frac{e_1^2}{2}\int dQdq\gamma_\mu\otimes\gamma\cdot nD\left(R-Q+\frac{1}{2}(r-q)\right)\bar{\Phi}(Q,q)\gamma^\mu\otimes\gamma\cdot n\Phi(Q,q) \\ &- \frac{e_2^2}{2}\int dQdq\gamma\cdot n\otimes\gamma_\mu D\left(R-Q-\frac{1}{2}(r-q)\right)\bar{\Phi}(Q,q)\gamma\cdot n\otimes\gamma^\mu\Phi(Q,q)\Big\}\Phi(R,r) \end{aligned} \tag{70}$$

In the absence of a fixed external field the system is translationally invariant and the generalization of the Fourier expansion (22) is the four dimensional Fourier transform of the composite field $\Phi(R,r)$ which has actually one time variable, $\Phi(R,r_\perp)$, the relative coordinates is a 3-vector $r_\perp$ perpendicular to n.

$$\Phi(R,r_\perp) = \int\frac{d^4P}{(2\pi)^4}e^{iPR}\psi(P,r_\perp)$$

We insert this expansion everywhere in our action and obtain

$$W = \int dR dr_\perp \int \frac{dP_n}{(2\pi)^4}\frac{dP_m}{(2\pi)^4}\bar{\psi}(P_n, r_\perp)e^{-iP_nR}\Bigg\{ [\Gamma_\mu P^\mu + \mathcal{L}_{rel}(r_\perp, p)]\, e^{iP_mR}$$

$$-\frac{1}{2}\int \frac{dP_r}{(2\pi)^4}\frac{dP_s}{(2\pi)^4} dQ dq_\perp dk \Bigg[\frac{e_1^2}{2}\frac{e^{-ik[R-Q+\frac{1}{2}(r_\perp - q_\perp)]}}{k^2}\gamma^\mu \otimes \gamma\cdot n\bar{\psi}(P_r, q_\perp)e^{-iP_rQ}\gamma_\mu$$

$$\otimes \gamma \cdot n e^{iP_sQ}\psi(P_s, q_\perp)$$

$$+\frac{e_2^2}{2}\frac{e^{-ik[R-Q-\frac{1}{2}(r_\perp - q_\perp)]}}{k^2}\gamma\cdot n\otimes\gamma^\mu\bar{\psi}(P_r, q_\perp)e^{-iP_rQ}\gamma\cdot n\otimes\gamma_\mu e^{iP_sQ}$$

$$\times\psi(P_s, q_\perp)\Bigg]\Bigg\}\psi(P_m, r_\perp) \tag{71}$$

where $\mathcal{L}_{rel}$ is the Lagrangian of the relative motion and is given by

$$\mathcal{L}_{rel}(r_\perp, p) = (\gamma^\mu p_\mu - m_1)\otimes\gamma\cdot n + \gamma\cdot n\otimes(-\gamma^\mu p_\mu - m_2) - \frac{e_1 e_2}{r_\perp}\gamma^\mu\otimes\gamma_\mu \tag{72}$$

and

$$\Gamma_\mu = \frac{1}{2}(\gamma_\mu\otimes\gamma\cdot n + \gamma\cdot n\otimes\gamma_\mu) \tag{73}$$

The result of performing the R and Q-integrations, letting

$$k_\mu = \gamma_\mu\otimes\gamma\cdot n - \gamma\cdot n\otimes\gamma_\mu \tag{74}$$

is

$$W = \int \frac{dP_n}{(2\pi)^4} dP_m dr_\perp \bar{\psi}(p_n, r_\perp)\Bigg\{\Bigg[\Gamma_\mu P^\mu + k^\mu p_\mu - m_1 I\otimes\gamma\cdot n - m_2\gamma\cdot n\otimes I - \frac{e_1e_2}{r_\perp}\gamma^\mu\otimes\gamma_\mu\Bigg]$$

$$\times\delta(P_n - P_m) - \frac{1}{2}\int\frac{dP_r}{(2\pi)^4}dP_s\frac{dk}{(2\pi)^4}dr_\perp dq_\perp\delta(P_n - P_m - k)\delta(P_r - P_s - k)\Big[e_1^2 e^{-i\frac{1}{2}k(r_\perp - q_\perp)}$$

$$\times\gamma^\mu\otimes\gamma\cdot n\bar{\psi}(P_r, q_\perp)\gamma_\mu\otimes\gamma\cdot n + e_2^2 e^{i\frac{1}{2}k(r_\perp - q_\perp)}\gamma\cdot n\otimes\gamma^\mu\bar{\psi}(P_r, q_\perp)\gamma\cdot n\otimes\gamma^\mu\Big]\psi(P_S, q_\perp\Bigg\}$$

$$\psi(P_m, r_\perp) \tag{75}$$

Introducing the form factors

$$\overset{(1)}{T^\mu}{}_{nm}(k) \equiv \int dr_\perp\bar{\psi}(P_n, r_\perp)e^{\frac{i}{2}kr_\perp}\gamma^\mu\otimes\gamma\cdot n\psi(P_m, r_\perp)$$

and (76)

$$\overset{(2)}{T^\mu}{}_{nm} = \int dr_\perp\bar{\psi}(P_n, r_\perp)e^{-\frac{i}{2}kr_\perp}\gamma\cdot n\otimes\gamma^\mu\psi(P_m, r_\perp)$$

for the two particles, we can write the action in the compact form

$$W = \sum_{nm} dr_\perp\bar{\psi}(P_n, r_\perp)\,[\Gamma_\mu P^\mu + \mathcal{L}_{rel}]\,\psi(P_m, r_\perp)\delta(P_n - P_m)$$

$$-\frac{1}{2}\sum_{nm}\sum_{rs}\frac{dk}{(2\pi)^4}\frac{1}{k^2}\left[e_1^2\overset{(1)}{T^k}{}_{nm}\overset{(1)}{T^{rs}}{}_\mu + e_2^2\overset{(2)}{T^\mu}{}_{nm}\overset{(2)}{T^{rs}}{}_\mu\right]\delta(P_n - P_m + k)\delta(P_r - P_s - k) \tag{77}$$

where

$$\mathcal{L}_{rel}(r_\perp) = k^\mu p_\mu - m_1 I \otimes \gamma \cdot n - m_2 \gamma \cdot n \otimes I - \frac{e_1 e_2}{r_\perp} \gamma^\mu \otimes \gamma_\mu$$

Note that $k^\mu n_\mu = 0$.

Now we can perform the k^0-integration, and without loss of generality set $n = (1000)$, and obtain

$$\begin{aligned} W = & \sum\!\!\!\!\!\!\int_{nm} \delta(E_n - E_m)\delta(\vec{P}_n - \vec{P}_m) d\vec{r} \bar{\psi}_n(\vec{P}_n, \vec{r})[\Gamma_0 P^0 - \vec{\Gamma} \cdot \vec{P} + \mathcal{L}_{rel}]\psi_m(\vec{P}_m, \vec{r}) \\ & - \frac{1}{2} \sum\!\!\!\!\!\!\int_{nm} \sum\!\!\!\!\!\!\int_{rs} \frac{d\vec{k}}{(2\pi)^4} \frac{1}{\omega_{nm}^2 - \vec{k}^2} \Big[e_1^2 \overset{(1)}{T^\mu}{}_{nm}(\omega_{nm}, \vec{k}) \overset{(1)}{T^{rs}}{}_\mu(\omega_{rs}, \vec{k}) \\ & + e_2^2 \overset{(2)}{T^\mu}{}_{nm}(\omega_{nm}, \vec{k}) \overset{(2)}{T^{rs}}{}_\mu(\omega_{rs}, \vec{k}) \Big] \delta(\omega_{nm} + \omega_{rs}) \delta(\vec{P}_n - \vec{P}_m + \vec{k}) \delta(\vec{P}_r - \vec{P}_s - \vec{k}) \end{aligned} \tag{78}$$

Now we look at the selfinteraction terms only and expand the denominator

$$\begin{aligned} W^{\text{self}} = & \frac{1}{2} \sum\!\!\!\!\!\!\int_{nm} \sum\!\!\!\!\!\!\int_{rs} \frac{d\vec{k}}{(2\pi)^4} \Big[e_1^2 \overset{(1)}{T^\mu}{}_{nm}(\omega_{nm}, \vec{k}) \overset{(1)}{T^{rs}}{}_\mu(-\omega_{nm}, -\vec{k}) + e_2^2 \overset{(2)}{T^\mu}{}_{nm}(\omega_{nm}, \vec{k}) \overset{(2)}{T^{rs}}{}_\mu(-\omega_{nm}, -\vec{k}) \Big] \\ & \times \delta(\omega_{nm} + \omega_{rs}) \left\{ (\delta(\omega_{rs} - k) + \delta(\omega_{rs} + k)) \frac{2\pi i}{2k} + P \frac{1}{ik} \left(\frac{1}{\omega_{rs} - k} - \frac{1}{\omega_{rs} + k} \right) \right\} \\ & \delta(\vec{P}_n - \vec{P}_m + \vec{k}) \delta(\vec{P}_n - \vec{P}_m + \vec{P}_r - \vec{P}_s) \end{aligned} \tag{79}$$

where P stands for the principal value of the integral and $\sum\!\!\!\!\!\!\int$ means a summation over discrete states and an integration over the continuum states. As in the case of the Coulomb problem and $(g-2)$-calculation, we separate the two terms corresponding to

$$\text{(a) } n = m, \text{ hence } r = s$$

and

$$\text{(b) } n = s, \text{ hence } m = r$$

and dictated by the δ-functions and obtain finally

$$\begin{aligned} W^{\text{self}} = & -\frac{1}{2} \sum\!\!\!\!\!\!\int_{ns} \frac{d\vec{k}}{(2\pi)^4} \left\{ e_1^2 \overset{(1)}{T^\mu}{}_{nn}(0, \vec{k}) \overset{(1)}{T^{ss}}{}_\mu(0, -\vec{k}) + e_2^2 \overset{(2)}{T^\mu}{}_{nn}(0, \vec{k}) \overset{(2)}{T^{ss}}{}_\mu(0, -\vec{k}) \right\} \\ & \times \left\{ \frac{i\pi}{k} (\delta(k) + \delta(-k)) + \frac{1}{2k} P \left(-\frac{2}{k} \right) \right\} \delta(\vec{P}_n - \vec{P}_m + \vec{k}) \delta(\vec{P}_n - \vec{P}_m + \vec{P}_r - \vec{P}_s) \\ & - \frac{1}{2} \sum\!\!\!\!\!\!\int_{nm} \frac{d\vec{k}}{(2\pi)^4} \left\{ e_1^2 \overset{(1)}{T^\mu}{}_{nm}(\omega_{nm}, \vec{k}) \overset{(1)}{T^{rs}}{}_\mu(-\omega_{nm}, -\vec{k}) + e_2^2 \overset{(2)}{T^\mu}{}_{nm}(\omega_{nm}, \vec{k}) \overset{(2)}{T^{nm}}{}_\mu(-\omega_{nm}, -\vec{k}) \right\} \\ & \times \left\{ \delta(\omega_{nm} - k) + \delta(\omega_{nm} + k) \frac{i\pi}{k} + P \frac{1}{2k} \left(\frac{1}{\omega_{nm} - k} - \frac{1}{\omega_{nm} + k} \right) \right\} \end{aligned} \tag{80}$$

We recognize again the following terms:

(i) Term containing $\delta(k) + \delta(-k)$. The contribution of this term to the dk-integral vanishes.

(ii) The term $P1/k$: This term corresponds to vacuum polarization.

(iii) The term with $(i\pi/k)[\delta(\omega_{nm}-k)+\delta(\omega_{nm}+k)]$. This term gives the spontaneous emission or absorption from level n to m or vice versa.

(iv) The term with $P\frac{1}{2k}\left(\frac{1}{\omega_{nm}-k}-\frac{1}{\omega_{nm}+k}\right)$ gives the Lamb shift.

These are our formulas for the radiative processes of the two-fermion system[18]. In the limit they go over to the fixed center Coulomb problem on the one hand, and for free particles to perturbative QED results.

For identical particles and particle-antiparticle system like positronium we have to antisymmetrize our currents as discussed in Sec. VI. Thus the mutual interaction action has two terms. The first is the usual direct interaction term

$$\begin{aligned} W^{\rm int}_{e^-e^+}(1) &= -e^2\int dxdy\bar{\Psi}_I(x)\overset{(1)}{\gamma^\mu}\psi_I(x)D(x-y)\bar{\psi}_{II}(x)\overset{(2)}{\gamma_\mu}\psi_{II}(y) \\ &= -e^2\int dxdy\bar{\phi}_I(x,y)\overset{(1)}{\gamma^\mu}D(x-y)\overset{(2)}{\gamma_\mu}\phi(x,y) \end{aligned} \tag{81}$$

corresponding to the potential

$$V=-e^2\overset{(1)}{\gamma^\mu}\frac{1}{r}\overset{(2)}{\gamma_\mu} \tag{82}$$

The second term is

$$\begin{aligned} W^{\rm int}_{e^-e^+}(2) &= e^2\int dxdy\bar{\phi}(x,y)\overset{(1)}{\gamma^\mu}D(x-y)\overset{(2)}{\gamma_\mu}\phi(y,x) \\ &= \frac{e^2}{2\pi}\int d\vec{r}\bar{\phi}_E(p_0,-\vec{r})\overset{(1)}{\gamma^\mu}\frac{e^{i\vec{k}\cdot\vec{r}}}{(p_0+p_0')^2-\vec{k}^2}\overset{(2)}{\gamma_\mu}\phi_E(p_0,-\vec{r})\frac{d\vec{k}}{(2\pi)^3} \end{aligned} \tag{83}$$

where p_0 and p_0' are the initial and final state energies and E is total conserved center of mass energy of the whole system. In the positronium the relative momentum is approximately zero so that we can set

$$p_\mu \cong (m,\vec{o})$$

and the action becomes

$$W^{\rm int}_{e^-e^+}(\text{annihilation}) \cong \frac{2\alpha}{\mathrm{m}^2}\int d\vec{r}\bar{\phi}(\vec{r})\overset{(1)}{\gamma^\mu}\delta(\vec{r})\overset{(2)}{\gamma_\mu}\phi(-\vec{r}) \tag{84}$$

Now we show that this term gives correctly the annihilation contribution to the hyperfine splitting in the $n=1$ state of positronium, for example. The effective potential above (84), when inserted into our wave equations gives an energy shift only for the levels $j=l\cong 0$ and for $j-1=l\cong 0$.

$$\delta E(j=\ell=0)\cong\frac{m\alpha^4}{2n^3}$$

and

$$\delta E(j-1=\ell=0)\cong\frac{m\alpha^4}{4n^3}$$

The difference is the annihilation contribution in the hyperfinesplitting

$$\delta E_{Hfs}(\text{annihilation}) \cong \frac{3m\alpha^4}{12n^3} \tag{85}$$

To this order it agrees with perturbative QED. It is however obtained here in first quantized QED with selffields.

VIII. FURTHER RESULTS

There are still discrepancies between theory and experiment in almost all tests of QED. The Table III summarizes all measured levels in positronium, muonium and Hydrogen, positronium lifetimes, the anomalous magnetic moments a_e and a_μ, and some theoretical values in parenthesis. In reviewing some of these discrepancies, W-Lichten[18] writes "It seems likely that the problem lies in the difficulty of QED calculations which have not been carried out to a high order enough, perhaps a totally new type of calculation is needed". The self field approach to QED provides a new type of calculation. It's important to have a complementary or alternate method to perturbative QED, for a theory is tested not only against experiment but also against other theories in order to clarify the basic assumptions and concepts, specially in view of recent results that perturbative QED might be inconsistent or a trivial theory. Selffield QED modifies our notion of the quantized radiation field and the interpretation of quantum theory. The emphasis is shifted from the field to the source of radiation, an electronic charge distribution which objectively and deterministically evolves as a classical field and produces a selffield which acts back on the charge itself. Quantized properties of the light reflect the discrete frequencies of the oscillating charge distribution.

The two-body relativistic equation discussed in Sections V and VII gives us a possibility to make improved calculations for positronium and muonium, in particular. In positronium, the experiments seem to be more accurate than the theory and the perturbative calculations remain incomplete[19]. Considerable analytical work has been done on the study of the two-body equation (39)ff: separation of radial and angular parts and further reduction of the radial equations[20]. It turns out that the two-body equation, when the electromagnetic potentials are kept to order α^4, is exactly soluble with an energy spectrum

$$E^2 = \frac{M^2+\Delta m^2}{2} + \frac{M^2-\Delta m^2}{2}\left[1+\frac{\alpha^2}{(n_r+\ell)^2}\right]^{-1/2} \tag{86}$$

with $M = m_1 + m_2$, $\Delta m = m_1 - m_2$, generalizing the Dirac spectrum. We have treated the remaining potentials of order α^5 and higher as perturbations. But having tested the equation in this way, one can now make direct nonperturbative numerical calculations. The treatment of the negative energy states and the covariance of the equation has also been discussed[22] according to the methods outlined in Section VI. We give here some of the results[20].

1) For parapositronium, eq. (86), is exact including terms of the order α^4 since normal and anomalous magnetic moment terms do not contribute to this order. It gives

$$E^{\text{para ps}} = sm - \frac{m\alpha^2}{4n^2} - \frac{m\alpha^4}{2n^3(2j+1)} + \frac{11}{64}\frac{m\alpha^4}{n^4} + 0(\alpha^6) \tag{87}$$

TABLE III. BOUND STATE TESTS OF QED

POSITRONIUM

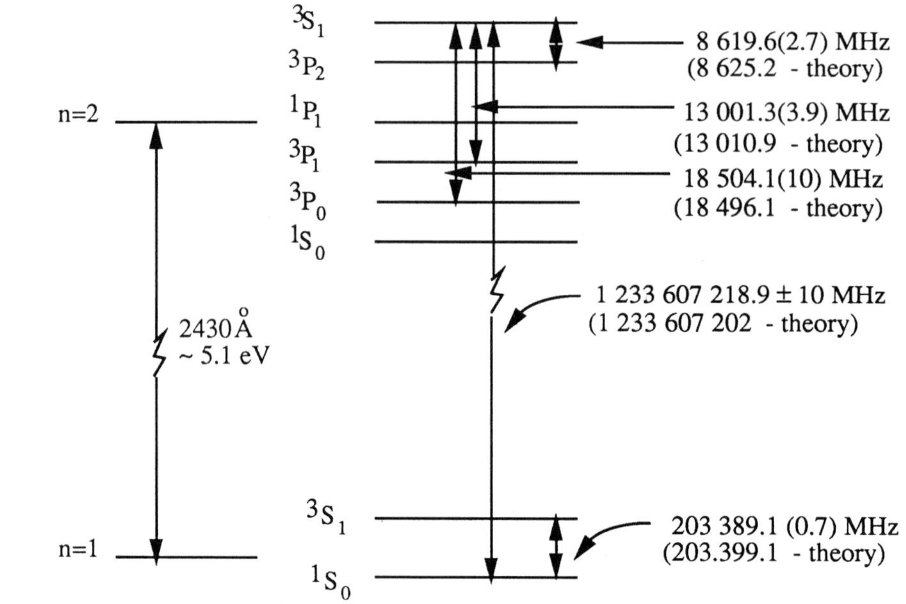

$\tau\ ^3S_1 \rightarrow 3\gamma = 7.0514\ \mu s^{-1}$
(7.0383 - theory)

$\tau\ ^1S_0 \rightarrow 2\gamma = 7.994\ ns^{-1}$
(7.9866 - theory)

$a_e = 1\ 159\ 652\ 188.4\ (4.3) \times 10^{-12}$
(1 159 652 192 (108)) - theory)

MUONIUM (AND H)

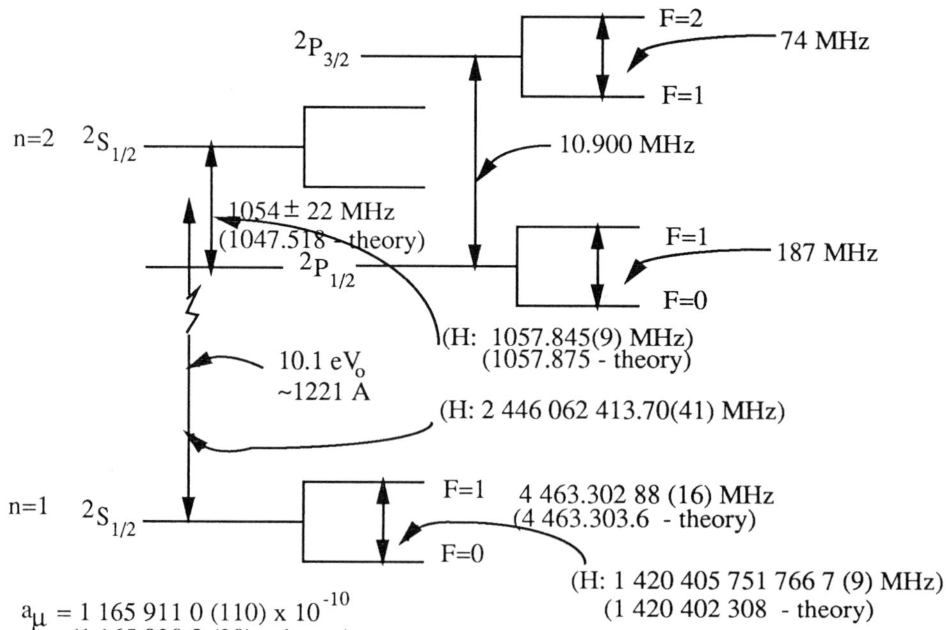

$a_\mu = 1\ 165\ 911\ 0\ (110) \times 10^{-10}$
(1 165 920 3 (20) - theory)

2) Introducing the anomalous magnetic moments a_1, a_2, which are in the selfenergy term, as a Pauli-coupling we obtain the ground-state hyperfine splitting

$$\Delta E^{\mathrm{Hfs}} = \frac{8}{3}\frac{\zeta}{(1+\zeta)^2} m\alpha^4 \left[(1+a_1)(1+a_2) - \frac{3}{4}\zeta a_2^2 + \frac{3}{4}a_1\frac{a_2}{(1+\zeta)^2}\right]$$

with $\zeta = m_1/m_2$. Numerically this gives 1420.348 MHz for H, and 4.463.060 MHz for muonium, compared to the experimental values 1420.405752 and 4.463302, respectively.

3) Positronium hypofinesplitting including the annihilation term, eq. (85), gives

$$\Delta E^{\mathrm{Hfs}} = \frac{7}{12}m\alpha^4 + \frac{5}{12}m\alpha^4\left(\frac{\alpha}{2\pi}\right).$$

This "Lambshift" term $-\frac{\alpha}{2\pi}(\frac{16}{9}+\ell n 2)m\alpha^4$ has to be added perturbatively, but we hope to calculate these terms and more eventually numerically.

4) Positronium ($n = 2, n = 1$) splitting, including annihilation and anomalous magnetic moment contributions

$$\Delta E_{21} = \frac{3}{8}Ry - 0.468098\alpha^2 Ry - \frac{\alpha^2 Ry}{2\pi}\frac{35}{96}.$$

5) Positronioum fine structure

$$\Delta E(2^3S_1 - 2^3P_2) = -\underset{(\mathrm{recoil})}{\frac{1}{12}\alpha^2 Ry} + \underset{(\mathrm{annihilation})}{\frac{7}{48}\alpha^2 Ry} - \underset{\left(\substack{\mathrm{normal\ magnetic}\\ \mathrm{moment}}\right)}{\frac{7}{480}\alpha^2 Ry} + 0(\alpha^5)$$

$$= \frac{23}{480}\alpha^2 Ry + 0(\alpha^5)$$

6) H or muonium ($n = 2$, $n = 1$) splitting

$$\begin{aligned}\Delta E_{21} = &\frac{3}{8}\mu\alpha^2 + \frac{8}{128}\mu\alpha^4 + \frac{15}{128}\mu\alpha^4\frac{\zeta}{1+\zeta}\\ &-\frac{7}{16}\mu\alpha^4\left(1+\frac{2(a_1+\zeta^2 a_2)}{(1+\zeta)^2} - 2a_1a_2\frac{\zeta}{1+\zeta}\right)\\ &-\frac{7}{12}\frac{\mu^2}{M}\alpha^4(1+a_1+a_2+a_1a_2)\end{aligned}$$

where $M = m_1 + m_2$, $\zeta = m_1/m_2$, $\mu = \frac{m_1 m_2}{M}$.

For other details and applications of self-field QED we refer to the literature listed in the Appendix.

References

1. See e.g. W. Thirring, *Principles of Quantumelectrodynamics* (Academic Press, N.Y. 1958).
   J.D. Bjorken and S.D. Drell, *Quantum Fields* (McGraw Hill, N.Y. 1965).
   R.P. Feynman, *Quantum electrodynamics* (Benjamin, N.Y. 1961).

2. A.O. Barut, in *Quantumelectrodynamics*, ed. P. Urban, Suppl. Acta Physica Austriaca, Vol. 2 (Springer, Berlin 1965); and *The Theory of the Scattering Matrix*, (Macmillan, N.Y. 1962), Ch. 13.

3. G. Scharf, *Finite Quantumelectrodynamics*, Springer 1989, and These Proceedings.

4. See e.g. C. Itzykson and J.-B. Zuber, *Quantum Field Theory* (McGraw Hill, N.Y. 1964).

5. A.O. Barut and I.H. Duru, *Phys. Reports*, **172**, 1 (1989).

6. J. Schwinger, *Particles, Sources and Fields*, Vol. II (Addison Wesley, Reading, MA 1973).

7. H.A. Lorentz, *The Theory of Electrons*, (Dover, N.Y. 1952).

8. A.O. Barut, *Phys. Rev.*, **D10**, 3335 (1974).

9. A.O. Barut and N. Ünal, *Phys. Rev.*, **A40**, 5404 (1989).

10. A. Lees, *Philos. Mag.*, **28**, 385 (1939).

11. D.A.M. Dirac, *Proc. Roy. Soc.*, (London), **A268**, 57 (1962).

12. A.O. Barut and Y. Salamin, *Phys. Rev.*, **A37**, 2284 (1988).

13. A.O. Barut and Y. Salamin, Quantum Optics (in press).

14. A.O. Barut and N. Ünal, A Regularized Analytic Evaluation of Vacuum polarization in Coulomb field (to be published), and see also N. Ünal, These Proceedings.

15. A.O. Barut, J. Kraus, Y. Salamin and N. Ünal (to be published).

16. For more details on $(g-2)$ calculations see A.O. Barut and J.P. Dowling, *Zeits. f. Naturf.*, **44a**, 1051 (1989) and A.O. Barut, J.P. Dowling and J.F. van Huele, *Phys. Rev.*, **A38**, 4405 (1988).

17. For more details see A.O. Barut and N. Ünal, Complete QED of the electron-positron system and positronium (to be published).

18. W. Lichten, in *The Hydrogen Atom*, edit. by G.F. Bassani *et al* (Springer-1989), p. 39.

19. A. Rich *et al*, These proceedings.

20. A.O. Barut and N. Ünal, *J. Math. Phys.*, **27**, 3055 (1986), *Physica*, **142A**, 457 and 488 (1987).

21. A.O. Barut, *Physica Scripta*, **36**, 493 (1987), and in *"Constraint Theory and Relativistic Dynamics"*, L. Lusanna *et al*, edit., (World Scientific, 1987), p. 122.

Appendix: References on Selffield QED

1. Nonperturbative QED: The Lamb shift, A.O. Barut and J. Kraus, *Found. of Physics*, **13**, 189 (1983).

2. QED based on selfenergy, A.O.Barut and J.F. van Huele, *Phys. Rev. A*, **32**, 3887 (1985)

3. QED based on selfenergy vs. quantization of fields: Illustration by a simple model. A.O. Barut, *Phys. Rev. A*, **34**, 3502 (1986)

4. An exactly soluble relativistic quantum two fermion problem. A.O. Barut and N. Ünal, *J. Math. Phys.*, **27**, 3055 (1986)

5. A new approach to bound state QED. I. Theory, *Physica Scripta*, **142A**, 457 (1987), II. Spectra of positronium, muonium and Hydrogen. A.O. Barut and N.Ünal, *Physica*, **142A**, 488 (1987)

6. An approach to finite non-perturbative QED. A.O. Barut in "Proc. 2nd Intern. Symposium on Foundations of Quantum Mech.", *Phys. Soc. of Japan*, 1986, p. 323.

7. On the treatment of Møller and Breit potentials and the covariant 2-body equation for positronium and muonium, A.O. Barut, *Physica Scripta*, **36**, 493 (1987)

8. On the covariance of two-fermion equation, A.O. Barut, in "Constraint theory and Relativistic dynamics", (L. Lusanna *et al*, edit.) World Scientific, 1987; p. 122

9. Formulation of nonperturbative QED as a nonlinear first quantized classical field theory, A.O. Barut in "Differential Geometric Methods in Theoretical Physics", (H. Doebner *et al*, edit's), World Scientific 1987; p. 51

10. QED based on selfenergy: Spontaneous emission in cavities, A.O. Barut and J.P. Dowling, *Phys. Rev. A*, **36**, 649 (1987)

11. QED based on self energy: The Lamb shift and long-range Casimir-Polder forces near boundaries, A.O. Barut and J.P. Dowling, *Phys. Rev. A*, **36**, 2550 (1987)

12. QED based on selfenergy, A.O. Barut, *Physica Scripta*, **T21**, 18 (1988)

13. Relativistic spontaneous emission, A.O. Barut and Y. Salamin, *Phys. Rev. A*, **37**, 2284 (1988)

14. QED based on selfenergy: A nonrelativistic calculation of $(g-2)$, A.O. Barut, J.P. Dowling and J.F. van Huele, *Phys. Rev. A*, **38**, 4405 (1988)

15. QED based on selfenergy: Cavity dependent contributions of $(g-2)$, A.O. Barut and J.P. Dowling, *Phys. Rev. A*, **38**, 2796 (1989)

16. QED based on self fields, A Relativistic calculation of $(g-2)$, A.O. Barut and J.P. Dowling, *Zeits. f. Naturf*, **44A**, 1051 (1989)

17. Path integral formulation of QED from classical particle trajectories, A.O. Barut and I.H. Duru, *Phys. Reports*, **172**, 1–32 (1989)

18. Problème relativiste á deux corps en électyrodynamique quantique, *Heelv. Phys. Acta*, **62**, 436 (1989)

19. Selffield QED: The two-level atom, A.O. Barut and J.P. Dowling, *Phys. Rev. A*, (1 March 1990 issue)

20. QED based on selffields: On the origin of thermal radiation detected by an accelerated observer, A.O. Barut and J.P. Dowling, *Phys. Rev. A*, (March 1990)

21. Relativistic Spontaneous emission in Heisenberg representation. A.O. Barut and Y. Salamin, Quantum Optics, (1990)

22. Relativistic $2S_{1/2} \rightarrow 1S_{1/2} + \gamma$ decay rates of H-like atoms for all Z, A.O. Barut and Y. Salamin, *Phys. Rev. A*,

23. QED-The unfinished business, A.O. Barut in "Proc. III. Conf. Math. Physics" (World Scientific, 1990)

24. A regularized analytic evaluation of vacuum polarization in Coulomb field, A.O. Barut and Ünal, *Phys. Rev. D*,

25. The Foundations of Self-field Quantumelectrodynamics, A.O. Barut in "New Frontiers of Quantumelectrodynamics and Quantumoptics", (edit. by A.O. Barut), Plenum Press, NY 1990

26. QED Based on Selffields: Cavity effects, J.P. Dowling, *ibid*,

27. Complete QED of the electron-positron system and positronium, A.O. Barut and N. Ünal,

QED BASED ON SELF-FIELDS: CAVITY EFFECTS

Jonathan P. Dowling

Max-Planck-Institute für Quantenoptik
D-8046 Garching
Federal Republic of Germany

I. INTRODUCTION

The second quantization of the electromagnetic field was performed for one of the first times in a seminal paper by Dirac in 1927;[1] this step is usually considered to have been the dawn of QED. Although the second quantization hypothesis allowed Dirac to derive the Einstein A coefficient of spontaneous emission—further results were not forthcoming until approximately 20 years later when regularization and renormalization schemes were invented to treat the various singular expressions which arose in the theory.[2] One particular singularity, however, seems to resist being swept under the rug and continually keeps crawling back out again. It is well known that the second quantization procedure predicts that the electromagnetic vacuum contains a zeropoint energy corresponding to $\frac{1}{2}\hbar\omega_k$ per normal mode k of the field. This gives rise to a divergent vacuum energy density,[3] which one may theoretically renormalize away by demanding that the photon destruction and creation operators a and $a^\dagger$ be normally ordered, i.e. with the destruction operators all to the right. From the standpoint of general relativity this seems unsatisfactory, however. The stress energy tensor $T^{\mu\nu}$ has a physical connection to the spacetime metric $g^{\mu\nu}$ via the Einstein field equations. A divergent vacuum $T^{\mu\nu}$ would imply an infinite curvature for the universe, and such a curvature can not be removed simply by performing some sort of transfinite shift of the energy scale. An infinite vacuum energy corresponds to an infinite cosmological constant $\Lambda_{\text{thy}} = \infty$. Yet observations of the motion of distant galaxies puts an upper limit on the cosmological constant of $\Lambda_{\text{exp}} < 10^{-56}\,\text{cm}^{-1}$. This, the famous *Cosmological Constant Problem*,[4] casts doubt on the physical reality of vacuum field fluctuations. Nevertheless, many of us apparently would like to have our vacuum and eat it too. There is a longstanding tradition in QED to take the existence of the zeropoint fluctuations as real things, and to use them to carry out calculations of radiative effects. By coupling an electron to the vacuum fluctuations one may obtain a satisfactory account of the Lamb shift and spontaneous emission—although one can not get easily a sensible value for $g-2$ using this method.[5—8] This process of coupling the electron to the zeropoint field as a method of calculation we shall call nonrelativistic QED. Casimir forces and apparatus contributions to such things as spontaneous emission, the Lamb shift, and $g-2$ can also be calculated in this fashion.[9—11] Can nonrelativistic QED be trusted? If taken seriously it gives good results for for spontaneous emission and the nonrelativistic Lamb shift, cryptic or ambiguous results for $g-2$, and complete nonsense for the cosmological constant. How can we be comfortable with any prediction of this theory in the absence of a fully relativistic calculation to back it up? In standard QED one second quantizes the free electromagnetic field separately from the electronic field, and only then does one couple the two entities. Is this procedure valid? As Einstein has warned us:

> I feel that it is a delusion to think of the electrons and the fields as two physically different, independent entities. Since neither can exist without the other, there is only *one* reality to be described, which happens to have two different aspects; and the theory ought to recognize this from the start instead of doing things twice.[12]

Perhaps the vacuum field is not physically real after all—in fact some argue that it can not be—perhaps it is only a mathematical artifact of the second quantization procedure.[12,13] General relativity also presents us with a second troubling problem inherit in the quantum field theoretic notion of the vacuum. The normal mode decomposition of the electromagnetic field is unique only in Minkowski space. In curved spacetime this is not so, and hence different observers will see different vacua. This conclusion has as its consequence such phenomena as Hawking radiation from a black hole, and the Unruh effect in which an accelerating detector registers a thermal bath of photons. This is quite distressing—if an inertially moving detector and a uniformly accelerating detector are near each other in spacetime, the inertial one sees nothing, while the accelerating one sees a Planck distribution of photons. If these photons are real, why doesn't the inertial detector see them too? Such paradoxes have led P. C. W. Davies to conclude that the concept of 'particle' (in this case 'photon') breaks down in curved spacetime.[14] This is pretty strong stuff! The acceptance of standard quantum field theory implies that the particle notion is nonsense in curved space. Can such a conclusion be avoided? Yes, it is possible to rescue the notion of 'the photon' if we abandon the quantum field notion of 'the vacuum.'

It is standard folklore to believe that radiative effects such as spontaneous emission and the Lamb shift are caused by the interaction of the electron with zeropoint fluctuations. If we dispose of the vacuum fluctuations, what then is the causative agent behind these radiative corrections? There exist perfectly respectable classical analogs of spontaneous emission and the Lamb shift. A harmonically bound charge will exhibit a line broadening and a level shift if the equation of motion includes radiation reaction—and no field fluctuations are needed to explain this result. Is it somehow possible to take the classical theory of radiation reaction and generalize it to a quantum mechanical setting? Schrödinger was one of the first to point out that the back reaction of the electron's own field on itself must be added to the Schrödinger equation in order to have an equation of motion which could be considered complete.[15] Fermi also tried something along this line.[16] By inserting a classical-like radiation reaction term into the Schrödinger equation, he arrived at what essentially was the neoclassical theory of Crisp and Jaynes.[17] This approach yields the correct Einstein A coefficient—but a nonexponential "chirped" decay profile. (Recent work seems to indicate that the nonphysical decay law of the neoclassical theory is a mathematical error arising from an invalid application of the superposition principle in a nonlinear theory.[18]) The neoclassical approach seems a bit *ad hoc*, and it turns out that there is a more natural and complete way to include radiation reaction in quantum mechanics.

In 1938 Dirac was able to derive the classical, covariant Abraham-Lorentz equation of motion for a charge which includes radiation reaction.[19] In particular, he had to assume that the electromagnetic potential A_μ surrounding a charge is symmetric in the retarded and advanced fields to arrive at his result. In 1945, Wheeler and Feynman elaborated on the work of Fokker, Tetrode, and Schwarzschild to produce the *Absorber Theory* or *Action at a Distance Electrodynamics.*[20—22] The idea here is that one can produced all of electrodynamics—Maxwell's equations and the Abraham-Lorentz-Dirac (ALD) equation of motion—from a single action principle *if* one assumes that the action density is symmetric with respect to future and past, or, equivalently, in the retarded and advanced fields. It is well known that Wheeler and Feynman never produce a quantum version of this theory. Süssman has produced a fully second quantized version of action at a distance electrodynamics, from which he was able to arrive at the A coefficient.[21] Barut and his coworkers, however, have gotten this coefficient—and more—with intermediate versions of the theory which are not second quantized, but rather which extend the action principle of Wheeler and Feynman to include Schrödinger, Pauli, and Dirac action principles, rather than just the classical.[18,23—25] The contention is that the covariant inclusion of radiation reaction is to be done instead of—not in addition to—second quantization. This approach to QED has led to correct results for relativistic accounts of spontaneous emission,[26] the Lamb shift,[27] the $g-2$ anomaly,[28] and vacuum polarization.[29] In the nonrelativistic approximation, spontaneous emission and the Lamb shift,[25] $g-2$ of the electron,[8] cavity QED effects,[30—32] and the Unruh effect[33] have been calculated. In this paper we shall summarize some of the cavity results as well as the Unruh effect calculation. It should be mentioned that Casimir forces can be equally well derived in the self-field approach, even though there are no vacuum fluctuations.[31,34] All phenomena which hitherto were thought to be caused by zeropoint energy can apparently be explained in terms of self-fields. In fact Jaynes has shown[12] that the radiation reaction spectrum over the linewidth of an atom is equal to the vacuum fluctuation spectrum. In the self-field approach the vacuum field is assumed to be zero for all moments of the correlation functions. For example, to trigger spontaneous emission, an atom produces a radiation reaction field on itself in just the right amount to cause a decay. Compare this to the quantum field philosophy in which one must fill the entire universe with an infinite density zeropoint energy in order to get spontaneous emission for a single atom. Since in self-field QED the vacuum field is identically zero, there is no longer a cosmological constant problem. Self-field theory *predicts* a cosmological constant of zero—in excellent agreement with experiment. The self-field approach also solves the paradox that usual QED leads to in curved space. It has been shown that the Unruh effect can be calculated in self-field theory, and the result is precisely the same as in standard QED.[33] An accelerating detector responds *as if* bathed with thermal photons, whereas the inertially moving detector sees nothing. But now the conclusion is different: The thermal photons are not real, but rather the accelerating agent directly stimulates the self-field of the detector—the causative agent of spontaneous emission—and forces the atom into a superposition of states which corresponds to a thermal distribution. This neatly accounts for the fact that a nearby inertial detector sees no photons, and hence rescues the concept of photon as a particle. The cost of saving the photon is the loss of a dynamic, interactive vacuum. Zeropoint fluctuations in empty space are perhaps only a useful fiction, from the self-field frame of mind. Even the nonrelativistic calculation of $g-2$ in the self-field calculation unambiguously gives the correct sign, in contradistinction to a standard QED vacuum field calculation.[5,7,8]

It should be mentioned that in the context of standard QED there seems to be a dual relation between the vacuum fluctuation and the radiation reaction interpretations.[6,12,35—39] The consensus here appears to be that, *within the framework of standard QED*, both interpretations are required for a cogent theory of spontaneous emission. There remains the possiblity that a modified version of QED might not contain zeropoint fluctuations at all—that they might only be a mathematical subterfuge introduced by the second quantization prescription. It is proposed that self-field QED, so far at least to order α, is such a theory.

II. CAVITY EFFECTS IN QED

In fully relativistic QED, the freespace Feynman propagator is a globally defined Green's function for the electromagnetic field. As such, its structure will depend on the environment and the external boundary conditions imposed on the field. Thus the presence of conducting surfaces, for instance, in the neighborhood of an atom will alter such radiative effects as spontaneous emission, the Lamb shift, $g-2$, etc. whose calculations depend explicitly on the form of the propagator. In nonrelativistic QED one demands that the vacuum field obeys some appropriate boundary conditions, and then one couples the atom to the modified vacuum in order to calculate for apparatus induced changes to the usual freespace radiative corrections. In self-field QED the self-field A_μ^{self} of the electron is eliminated from the total action through use of the *same* Feynman's Green function used in standard QED. Hence, one expects similar boundary corrections as in standard QED, but now the understanding is that it is the radiation reaction field of the atom—and not the zeropoint field—which is adapting itself to a new environment.

For example, it is experimentally well verified that the spontaneous emission rate of an atom between parallel conducting plates can be suppressed nearly completely.[40,41] In standard QED the interpretation goes something as follows. Consider a two level atom of frequency ω_o. In freespace the atom finds all modes of the vacuum available to it, including that which also has frequency ω_o which is capable of stimulating spontaneous emission. Suppose now the atom is placed between parallel plates whose spacing L is too small to support the vacuum mode corresponding to ω_o. This occurs when $L < \lambda_o/2 := \pi/\omega_o$. In this case, even the zeroth harmonic corresponding to one half of a wavelength of a standing wave can not fit between the plates, and so the ω_o mode of the vacuum vanishes and spontaneous emission turns off. Those accustomed to this account of the phenomenon may find it difficult to believe that a theory without zeropoint fluctuations can produce the same result. Let us see how. Reconsider the two level atom in freespace. The atom is exposed to its own radiation reaction field, which when Fourier analyzed, contains all the same frequency components found in the vacuum field before. Hence there is a Fourier component of the self-field of frequency ω_o at hand to trigger spontaneous emission of our atom. Now between parallel mirrors, each Fourier component of the reaction field must separately obey the new boundary conditions. The condition $L < \lambda_o/2$ will completely wipe out all Fourier components of the self-field with frequency of ω_o or lower—and spontaneous emission will cease.

III. THE SELF-FIELD ACTION FORMALISM

In analogy to classical, action at a distance electrodynamics we wish to specify an action $W := \int dx\, w(x)$ which has as its Euler-Lagrange equations of motion Maxwell's equations for the electromagnetic (EM) field, and for the particle, an equation which includes radiative effects. We assume that the action density $w(x)$ consists of a free particle term w_o, a free field term w_f, and an interaction term w_i. Hence, the general form of w is, with the convention $\hbar = c = 1$,

$$\begin{aligned} w &= w_o + eA_\mu j^\mu + \frac{1}{4}F_{\mu\nu}F^{\mu\nu} \\ &=: w_o + w_i + w_f \end{aligned} \tag{1}$$

This expression at first appears to be the usual semiclassical action density. In standard semiclassical quantum theory, however, it is assumed that the A_μ which appears here is an external field from sources assumed to be at infinity. In particular A_μ would not include the self-field of the charge. But, according to Schrödinger,[15] this self-field *must* be included if one hopes to have a complete description of radiation reaction. To account for the self-field in a covariant fashion we make the *ansatz* that the EM potential surrounding the charge can be split into an external and self-field contribution as

$$A_\mu = A_\mu^{\text{ext}} + A_\mu^{\text{self}} \tag{2}$$

where A_μ^{ext} and A_μ^{self} obey the homogeneous and inhomogeneous Maxwell's equations, respectively, in a localized interaction region surrounding the charge. These equations are

$$\partial_\nu F_{\text{ext}}^{\mu\nu} = 0 \tag{3a}$$

$$\partial_\nu F_{\text{self}}^{\mu\nu} = e\, j^\mu \tag{3b}$$

The general solution to the nonhomogeneous equation (3*b*) can be written with a Green's function $D_{\mu\nu}(x-y)$ as[20]

$$A_\mu^{\text{self}}(x) = e \int dy\; D_{\mu\nu}(x-y)\, j^\nu(y) \tag{4}$$

where $x := x_\mu$, $y := y_\mu$, and $dy := d^4y$. Thus the integral is carried out over all Minkowski space. Equation (4) is the single most important feature of the self-field approach to QED, and hence bears a brief discussion. Expression (4) allows one to eliminate the self-field from the action, and hence from the equations of motion, in a covariant fashion. If we allowed our interaction region to include the entire universe, then $A_\mu = A_\mu^{\text{self}}$ alone and there would no longer be any A_μ^{ext}. Hence, as is apparent from the form of this equation, all electromagnetic potentials A_μ have their origin in some source current. As a consequence, electromagnetic fields do not exist independent of the sources that produce them. Considerations such as these have led to revival of the *Schrödinger Interpretation of Quantum Mechanics* in which the electron wave function is viewed as an actual distribution of electronic charge, as opposed to the usual probabilistic interpretation.[42—44] There is no such thing as a 'free' EM field, and consequently no such thing as a vacuum EM field. The self-field approach predicts that the vacuum is empty of electromagnetic energy, and that the EM vacuum contributes zero to the cosmological constant. A common criticism of self-field QED is that since the EM field is treated classically, one can not hope to obtain a complete theory of QED, since the field is not quantized. But from equation (4) it is clear that A_μ^{self} will be classical only if the source current j_μ is classical. On the other hand, if j_μ corresponds to a Schrödinger, Pauli, or Dirac current, then clearly A_μ^{self} will have quantum mechanical properties also, which it inherits from the quantum mechanical source. It is our position that the second quantization of the EM field is perhaps an unnecessary duplication of what is already contained in this expression (4). Maybe one does not have to second quantize A_μ if it is *already* quantized, inasmuch as it always exhibits quantum properties due to the quantized source which produced it. As a longstanding critic of the second quantization procedure, E. T. Jaynes tells us:

> One can hardly imagine a better way to generate infinites in physical predictions than by having a mathematical formalism with $(\infty)^2$ more degrees of freedom than are actually used by Nature.[12]

In the action density w of equation (1) we have not specified the form of $w_\circ$ or j_μ. Actually, after specifying $w_\circ$, the requirement that the variation of the total action

W with respect to A_μ yields an extremum give us Maxwell's equations and identifies the form of j_μ. The four cases of interest to us here are

1. *Classical Action Density and Current*

$$w_i = m\dot{z}^2 - eA_\mu \dot{z}^\mu \tag{5a}$$

$$j^\mu = \int d\tau \; e\dot{z}^\mu \; \delta(x - z(\tau)) \tag{5b}$$

2. *Schrödinger Action Density and Current*

$$w_i = \psi^* \Big[\frac{1}{2m}(\overleftarrow{\nabla} + ie\mathbf{A}) \cdot (\overrightarrow{\nabla} - ie\mathbf{A}) + eA_\circ - i\frac{\partial}{\partial t}\Big]\psi \tag{6a}$$

$$j^\mu = \psi^* \Big[1, \frac{1}{2mi}\overleftrightarrow{\nabla} - \frac{e}{m}\mathbf{A}\Big]\psi \tag{6b}$$

3. *Pauli Action Density and Current*

$$w_i = \phi^* \Big\{\frac{1}{2m}\big[(\overleftarrow{\nabla} + ie\mathbf{A})\cdot\sigma\big]\big[\sigma\cdot(\overrightarrow{\nabla} - ie\mathbf{A})\big] + eA_\circ - i\frac{\partial}{\partial t}\Big\}\phi \tag{7a}$$

$$j^\mu = \phi^* \Big[1, \frac{1}{2mi}\overleftrightarrow{\nabla} + \frac{1}{2m}(\overleftarrow{\nabla}\times\sigma - \sigma\times\overrightarrow{\nabla}) - \frac{e}{m}\mathbf{A}\Big]\phi \tag{7b}$$

4. *Dirac Action Density and Current*

$$w_i = \overline{\Psi}\big[\gamma^\mu(i\partial_\mu - eA_\mu) - m\big]\Psi \tag{8a}$$

$$j^\mu = \overline{\Psi}\gamma^\mu\Psi \tag{8b}$$

Variation of W with respect to z_μ, ψ, ϕ, or Ψ yields, respectively, the ALD, Schrödinger, Pauli, or Dirac equations of motion. The classical version is essentially the absorber theory of Wheeler and Feynman. The Schrödinger and Pauli actions produce nonrelativistic versions of self-field QED—without and with spin respectively. Finally, in the Dirac case, we have a fully covariant theory of electronic motion which includes radiation reaction. One can then treat this final relativistic version as a possible candidate for a complete theory of QED—perhaps equivalent or dual to the usual second quantized version.

IV. SPONTANEOUS EMISSION BETWEEN MIRRORS

To illustrate the self-field methodology, we now sketch a calculation of how the Einstein A coefficient for spontaneous emission changes between parallel, perfectly conducting mirrors. For this problem it is sufficient to consider the Schrödinger action density and corresponding current found in equation (6). The total action density w of equation (1) then reduces to

$$\begin{aligned} w =& \psi^*\Big[\Big(\frac{1}{2m}\nabla^2 - i\frac{\partial}{\partial t} + eA_\circ^{\text{ext}}\Big) + \Big(\frac{ie}{2m}\mathbf{A}^{\text{self}}.\nabla\Big) + \Big(\frac{e}{2}A_\circ^{\text{self}}\Big)\Big]\psi \\ =:& \psi^*\Big[(H_1) + (H_2) + (H_3)\Big]\psi \\ =:& w_1 + w_2 + w_3 \end{aligned} \tag{9}$$

To arrive at this relation we have used the Coulomb gauge $\nabla \cdot \mathbf{A} = 0$ and the weak field approximation $|\mathbf{A}^{\text{ext}}| \approx 0$. We have also neglected terms of order α^2 or higher, where the fine structure constant $\alpha := e^2/4\pi$ in our units. Finally, to reproduce a hydrogen-like atom we took $\mathbf{A}^{\text{ext}} = 0$ and $A_\circ^{\text{ext}} = -Ze/r$. If we were to set $A_\mu^{\text{self}} \equiv 0$

in equation (9) we would arrive at just the usual Schrödinger equation for hydrogen. However, since $A_\mu^{\rm self}$ is proportional to the electron current via equation (4), we can not consistently set it equal to zero unless $j_\mu \equiv 0$ over all spacetime—in which case we have no electron! Hence, by *reductio ad absurdum* it is clear that the inclusion of $A_\mu^{\rm self}$ in the Schrödinger equation is *required*. Consequently, the Schrödinger action contains nonlinear, nonlocal terms of the generic form

$$W^{\rm self} = \frac{e^2}{2} \int\int dx\; dy\; j^\mu(x)\; D_{\mu\nu}(x-y)\; j^\nu(y) \tag{10}$$

implying a nonlinear, nonlocal, integro-differential Schrödinger equation. In freespace, the Green's function $D_{\mu\nu}$ in the Coulomb gauge has the form.

$$D_{ij}(x-y) = \frac{1}{(2\pi)^4} \int dk\; \frac{e^{-ik\cdot(x-y)}}{k^2 + i\epsilon} \left(\delta_{ij} + \hat{k}_i \hat{k}_j\right) \tag{11a}$$

$$D_{00}(x-y) = \frac{1}{(2\pi)^4} \int dk\; \frac{e^{-ik\cdot(x-y)}}{\lambda^2 + i\epsilon} \tag{11b}$$

$$D_{i0}(x-y) = D_{0i}(x-y) = 0 \tag{11c}$$

where $\lambda := |\mathbf{k}|^2$, $k^2 := k^\mu k_\mu$, and the $+i\epsilon$ in the denominator insures that the correct symmetry between retarded and advanced solutions to Maxwell's equations are obtained—a choice which is required if the action is to have an extremum.[22,—20] With this choice of Green's function, equation (4) can be written as

$$\mathbf{A}^{\rm self}(x) = -\frac{e}{(2\pi)^4} \int\int dy\; dk\; \frac{e^{-ik\cdot(x-y)}}{k^2 + i\epsilon} \left[\mathbf{j}(y) - \hat{\mathbf{k}}(\hat{\mathbf{k}}\cdot\mathbf{j}(y))\right] \tag{12a}$$

$$A_0^{\rm self}(x) = \frac{e}{(2\pi)^4} \int\int dy\; dk\; \frac{e^{-ik\cdot(x-y)}}{\lambda^2 + i\epsilon} \rho(y) \tag{12b}$$

where ρ and $\mathbf{j}$ are the time and space components of the current j_μ as given in equation (6*b*). In our notation above we use $dy := d^4y$, $dk := d^4k$, and $\hat{\mathbf{k}} := \mathbf{k}/|\mathbf{k}|$. If this expression for $\mathbf{A}^{\rm self}$ is inserted into the action density w as given in expression (9), one can extract from the H_1 term a complex energy shift to level n which is given by

$$\begin{aligned} \mathcal{E}_1^{(n)} &= \frac{W_1^{(n)}}{2\pi} \\ &= \frac{2\alpha}{3\pi} \sum_m \omega_{nm}^3 |\mathbf{r}_{nm}|^2 \int_0^\infty \frac{d\lambda}{\omega_{nm} - \lambda} - \frac{2\alpha i}{3} \sum_{\substack{m \\ m<n}} \omega_{nm}^3 |\mathbf{r}_{nm}|^2 \\ &:= \delta E_n - i A_n \end{aligned} \tag{13}$$

where $\omega_{nm} := E_n - E_m$ and the $\mathbf{r}_{nm}$ are the usual matrix elements of the atomic position operator $\mathbf{r}$. This result is just Bethe's nonrelativistic Lamb shift formula, and also the Einstein A coefficient, which appears here as a damping term. A complex energy shift is interpreted as a line broadening in the usual fashion, giving rise to spontaneous emission.[18,25]

The self-field $A_\mu^{\rm self}$ we see is a globally defined quantity whose form depends on the Green's function which in turn depends on the environment. Clearly, if the Green's function is changed, the Lamb shift and spontaneous emission rates—and in fact all radiative corrections—must change. The self-field must meet the newly imposed boundary conditions. It is easy to show that the Green's function between parallel mirrors can be constructed by the method of images. Suppose we place two perfectly

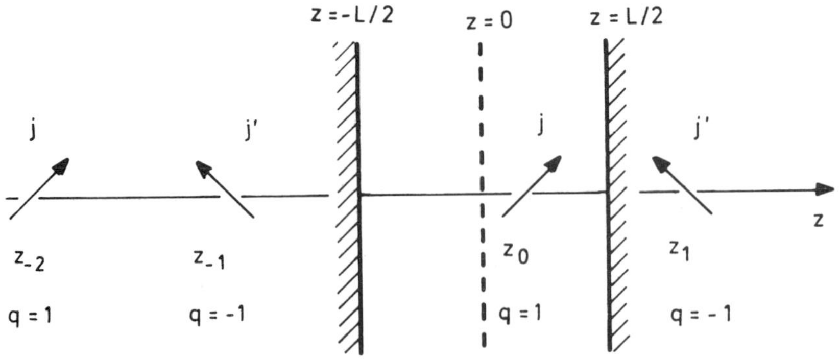

Fig. 1. A unit charge between parallel plates, located at z_o, and the resultant series of image charges of charge $(-1)^p$ located a $z_p := pL + (-1)^p z_o$, for $p = \pm 1, \pm 2, \pm 3 \ldots$

conducting, plane mirrors normal to the z axis at $z = \pm L/2$. A unit charge placed on the axis at $z = z_o$ will have image charges located at $z_p := pL + (-1)^p z_o$, where $p = \pm 1, \pm 2, \ldots$ (See Fig. 1) The total Green's function is then seen to be the infinite sum of translated freespace Green's functions, and it has the form

$$D^{\text{mirrors}}_{\mu\nu}(x - y) = \sum_{p=-\infty}^{\infty} D^{\text{freespace}}_{\mu\nu}(x - y'_p - z_p) \tag{14}$$

where

$$y'_p := \begin{cases} (y_o, y_1, y_2, y_3), & \text{if } p \text{ even} \\ (y_o, y_1, y_2, -y_3), & \text{if } p \text{ odd} \end{cases}$$

and $z_p := z_p^\mu := (z_p, 0, 0, z_p)$ is the image location in Minkowski space evaluated at the retarded time $t = z_p/c$.

After some work, we can show that the use of this Green's function gives rise to a complex energy shift similar to that seen in expression (13), but now in the limit that the plate spacing $L \ll c/A_o$,where A_o is the freespace value the A coefficient becomes the following expression[30]

$$A_n = \alpha \sum_{\substack{m \\ m<n}} \omega_{nm}^2 |\mathbf{r}_{nm}|^2 \sum_{p=1}^{[[\sigma_{nm}]]} \left\{ \left[(1 + \zeta_{nm}) + (1 - 3\zeta_{nm}) \left(\frac{p}{\sigma_{nm}} \right)^2 \right] \right.$$
$$\left. - \left[(1 - 3\zeta_{nm}) + (1 + \zeta_{nm}) \left(\frac{p}{\sigma_{nm}} \right)^2 \right] \cos \left[\pi p \left(\frac{2z_o}{L} - 1 \right) \right] \right\} \tag{15}$$

where $\sigma_{nm} := L\omega_{nm}/\pi$ and $\zeta_{nm} := |z_{nm}|^2/|\mathbf{r}_{nm}|^2$. The notation $[[x]]$ stands for the 'greatest integer less than x' function. This formula (15) agrees with the standard QED calculations of Barton,[45] Milonni and Knight,[46] Philpott,[47] and an experiment done by Hulet, Hilfer and Kleppner.[41] In particular—if we average formula (14) over the plate separation, assuming a uniform distribution of atoms between the plates, we obtain the A coefficient as a function of the plate spacing L as indicated in Fig. 2. As was mentioned earlier in this paper, for a plate spacing of $L < \lambda_o/2$, where λ_o is the wavelength of a particular two level transition we wish to suppress, spontaneous decay does not occur. This is because a Fourier component of A_μ^{self} with frequency less than ω_o can not meet the parallel mirror boundary conditions and hence it is no longer available in the radiation reaction field surrounding the atom to trigger spontaneous decay. The qualitative structure of the graph is in good agreement with a similar experimental graph,[41] even up to the prediction of an enhancement factor of $3A^\circ/2$ when $L = \lambda_\circ/2$. This enhanced spontaneous emission rate can be viewed as a cooperative Dicke superradiance phenomenon between the atom and its images.

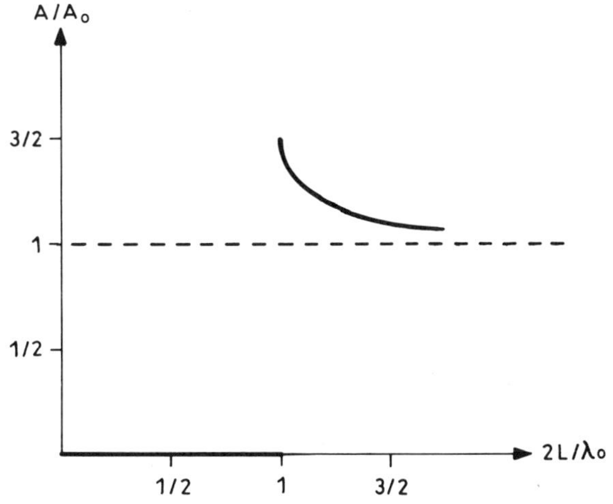

Fig. 2. The Einstein A coefficient between parallel mirrors as a function of the plate spacing L. The freespace value is $A_\circ$ *and* $\lambda_\circ$ *is the wavelength of the emitted photon. Below the critical plate separation* $L = \lambda_\circ/2$ *spontaneous emission is suppressed. At the criticla separation, the rate is enhanced by a factor of* $3/2$*, but then approaches asymptotically the freespace value as* $L \to \infty$.

V. LAMB SHIFT NEAR A SINGLE PLATE

Just as the imaginary energy shift found in equation (13) changes with changing boundary conditions—so does the real part of this shift. Hence one would expect a boundary induced change in the Lamb shift as well as the spontaneous emission rate. Here we give the results of a sample calculation for the apparatus correction to the energy level n of a hydrogen atom which is located a distance R from a single conducting plane. Only one image is needed to construct the necessary Green's function here, in contrast to the double mirror case.

Schematically, the Green's function construction can be viewed as the sum of two Green's functions, where the image Green's function is evaluated at a retarded

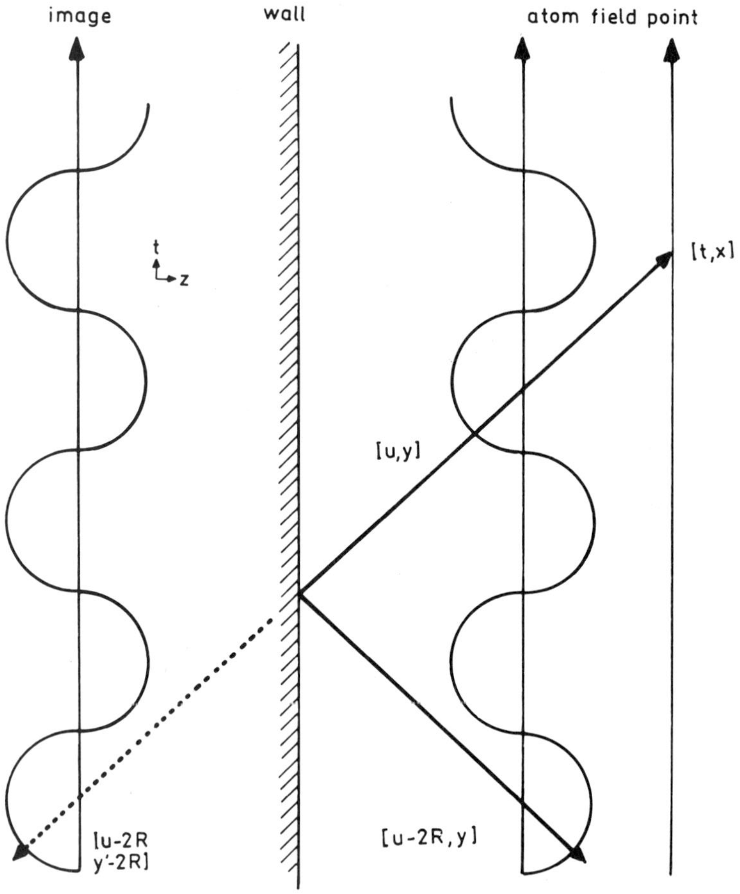

Fig. 3. The value of the electromagnetic Green's function at the point $(t, \mathbf{x})$ is determined by the current at the source point at $(u, \mathbf{y})$ and by the image current at the image source point $(u - 2R, \mathbf{y}' - 2\mathbf{R}$, where we define $\mathbf{y}' := (y_1, y_2, -y_3)$ and $\mathbf{R} := (0, 0, R)$. Notice that the image source point is retarded in both space and time.

spacetime point. (See Fig. 3.) The energy shift turns out to be given by[31]

$$\Delta E_n = \frac{\alpha}{\pi} \sum_m \omega_{nm}^3 |\mathbf{r}_{nm}|^2 \Big\{ (1 - \zeta_{nm}) \Big[\frac{1}{\mu_{nm}^2} - \frac{f(\mu_{nm})}{\mu_{nm}} + \pi \Theta_{nm} \frac{\cos\mu_{\rm nm}}{\mu_{\rm nm}} \Big] - (1 + \zeta_{nm}) \Big[\frac{g(\mu_{nm})}{\mu_{nm}^2} + \frac{f(\mu_{nm})}{\mu_{nm}^3} + \pi \Theta_{nm} \Big(\frac{\sin\mu_{\rm nm}}{\mu_{\rm nm}^2} + \frac{\cos\mu_{\rm nm}}{\mu_{\rm nm}^3} \Big) \Big] \Big\} \quad (16)$$

where ζ_{nm} is defined as before, and $\mu_{nm} := 2\omega_{nm}R$. We have also defined a step function as $\Theta_{nm} := \Theta(n - m)$ which turns on only if the level n is an excited state.

The functions f and g are defined as

$$f(z) := \int_0^\infty \frac{\mathrm{sint}}{\mathrm{t+z}} dt \qquad \text{and} \qquad g(z) := \int_0^\infty \frac{\mathrm{cost}}{\mathrm{t+z}} dt$$

The expression (16) has two limiting cases of interest. For an atom far from the plate we have ($R \to 0$)

$$\Delta E_n \sim \frac{\alpha}{\pi} \sum_{\substack{m \\ m \neq n}} \omega_{nm}^3 |\mathbf{r}_{nm}|^2 \left\{ \frac{4}{\mu_{nm}^4} + \pi \Theta_{nm} \right.$$
$$\left. \times \left[(1-\zeta_{nm}) \frac{\cos\mu_{\mathrm{nm}}}{\mu_{\mathrm{nm}}} - (1+\zeta_{nm}) \left(\frac{\sin\mu_{\mathrm{nm}}}{\mu_{\mathrm{nm}}^2} + \frac{\cos\mu_{\mathrm{nm}}}{\mu_{\mathrm{nm}}^3} \right) \right] \right\} \tag{17}$$

The first term in the curly braces of (17) is the usual Casimir-Polder, long range, Van der Waals energy between the atom and its image in the wall. From this energy one obtains the a force via

$$F = -\frac{\partial(\Delta E)}{\partial R}$$

The additional terms proportional to Θ_{nm} are corrections for excited states. These terms have also been found by Barton.[48] In the opposite extreme of the limit $R \to \infty$ we have

$$\Delta E_n \sim -\frac{\alpha}{16R^2} \langle n | r^2 + z^2 | n \rangle + \frac{\alpha}{4Rm^2} \langle n | p_x^2 + p_y^2 - 2p_z^2 | n \rangle$$
$$+ \frac{2\alpha}{3\pi} \sum_m \left(|\mathbf{r}_{nm}|^2 - 2|\zeta_{nm}|^2 \right) \ln 2R|\omega_{nm}| + \frac{\alpha}{4\pi m R^2} \tag{18}$$

The first term is the standard London energy for an atom interacting nonretardedly with its image in the wall. The second term, since it contains components of p^2, corresponds to an anisotropic change in the electron mass. In general, the mass of an electron in a cavity is a tensor quantity; the inertial response of an electron to a given force will depend on the direction of the force. (One way to visualize this is to consider that the electromagnetic part of the electron's mass is bound up in the electric field lines, and that some of these lines are now cut short by the conductor, hence changing the mass.) The third term in equation (18) is something like an anisotropic contribution to the Lamb shift, and the final term is a level independent contribution to the Van der Waals energy. These results also agree with those of Barton,[48] and the reader is refered thither for more detailed discussion of these terms.

VI. $g-2$ OF AN ELECTRON NEAR A MIRROR

In 1989, Hans Dehmelt shared the Nobel Prize in physics for his work with the Penning trap. His group has made to-date the most accurate measurements of the electron g factor, by comparing the cyclotron and spin precession frequencies, ω_{cyc} and ω_{spin} respectively, of an electron electromagnetically bond in such a trap. The electron anomaly factor a can be written as

$$a := \frac{g-2}{2} \equiv \frac{\frac{g}{2}-1}{1} \equiv \frac{\omega_{\mathrm{spin}} - \omega_{\mathrm{cyc}}}{\omega_{\mathrm{cyc}}} \tag{19}$$

so long as one is in freespace. Dehmelt's result is[49]

$$a_{\mathrm{exp}} = 1\ 159\ 652\ 193\ (4) \times 10^{-12} \tag{20}$$

Unfortunately the experiment is not carried out in freespace, but rather inside a conducting cavity whose dimension is of the order of 1cm in diameter. Since a_{exp} is compared to a theoretical calculation of $a_{\rm thy}$ using *freespace* Feynman integrals—it becomes of crucial importance to know at what stage of accuracy does the apparatus introduce a systematic error into $a_{\rm exp}$. The trap alters the radiation reaction field, in the self-field picture, and so we would expect both $\omega_{\rm spin}$ and $\omega_{\rm cyc}$ to change, altering the value of $a_{\rm exp}$. There has been quite a confusion in the literature over whether or not there was a first order correction to $\omega_{\rm spin}$ as well as $\omega_{\rm cyc}$.[32,45,49—58] The consensus seems to now be that there is a correction to $\omega_{\rm cyc}$ but no first order boundary induced change in $\omega_{\rm spin}$. This fact is due to a subtle cancellation between apparatus corrections to the electron magnetic moment and to those of the mass.[32,58] To see how this occurs, let us take the simple case of an electron undergoing a cyclotron orbit in a plane parallel to, and a distance R from, a plane conducting mirror. Instead of the Schrödinger action of equation (6) we now use the Pauli action of (7) so that we may include the electron spin in an elementary fashion. With the same approximations as we made in the Schrödinger action to arrive at the action density (9), we now obtain from the Pauli action density (7) the simplified density

$$\begin{aligned} w =&\phi^*\Big[-\frac{1}{2m}\nabla^2 - i\frac{\partial}{\partial t} + \frac{ie}{m}\mathbf{A}^{\rm ext}\cdot\nabla - \frac{e}{2m}\sigma\cdot\mathbf{B}^{\rm ext} \\ &+\frac{e}{2}A_{\rm o}^{\rm self} + \frac{ie}{2m}\mathbf{A}^{\rm self}\cdot\nabla - \frac{e}{4m}\sigma\cdot\mathbf{B}^{\rm self}\Big]\phi \end{aligned} \tag{21}$$

where we have now set $A_{\rm o}^{\rm ext} \equiv 0$ but $\mathbf{A}^{\rm ext} = (\mathbf{B}^{\rm ext}\times\mathbf{r})/2$. Without the terms labeled with the superscript *self*, equation (21) would be the usual semiclassical expression for an electron with spin, executing Landau orbits in a homogeneous magnetic field. The $A_{\rm o}^{\rm self}$ and $\mathbf{A}^{\rm self}\cdot\nabla$ terms, similar as to before, give rise to mass renormalization, Lamb shift and spontaneous emission effects, while that proportional to $\sigma\cdot\mathbf{B}^{\rm self}$ is responsible for the nonzero value of $g-2$ in freespace.[8,28] Apparatus dependent shifts in the $\sigma\cdot\mathbf{B}^{\rm self}$ energy would alter $\omega_{\rm spin}$ and hence $g-2$. Changes in the cyclotron frequency come about through changes in the $\mathbf{A}^{\rm self}\cdot\nabla$ term.

The boundary condition that $\mathbf{E}_{\|}$ and $\mathbf{B}_{\perp}$ vanish identically on the surface of a conductor S can be covariantly written as[56—58]

$$F_{\mu\nu}\, n_\alpha\, \epsilon^{\mu\nu\alpha\beta}|_S \equiv 0, \qquad \forall\beta \tag{22}$$

where $n_\alpha := (0,\hat{\mathbf{n}})$ and $\hat{\mathbf{n}}$ is everywhere normal to the surface S in the restframe of the conductor. This constraint (22) can be met by the axial gauge condition

$$n_\mu\, A^\mu|_S \equiv 0 \tag{23}$$

which gives rise to the Green's function:

$$D_{\mu\nu}(x-y) = -\frac{1}{(2\pi)^4}\int dk\,\frac{e^{ik\cdot(x-y)}}{k^2+i\epsilon}\left(g_{\mu\nu} - \frac{n_\mu\, k_\nu + n_\nu\, k_\mu}{n\cdot k} + \frac{n^2\, k_\mu\, k_\nu}{(n\cdot k)^2}\right) \tag{24}$$

Of course the choice of the Green's function can not affect our answer, since the theory is gauge invariant, but it can greatly simplify our calculations—which is why we use this special Green's function here. Restricting ourselves to the case where we have a single plane surface, we can set $\hat{\mathbf{n}} := \hat{\mathbf{z}}$ and write equation (4) with the help of (24) as

$$\mathbf{A}^{\rm self}(x) = -\frac{e}{(2\pi)^4}\int\!\!\int dy\, dk\,\frac{e^{ik\cdot(x-y)}}{k^2+i\epsilon}\left[\mathbf{j}(\mathbf{y}) - \frac{(\mathbf{k}\cdot\mathbf{j})\hat{\mathbf{z}} + (\hat{\mathbf{z}}\cdot\mathbf{j})\mathbf{k}}{\hat{\mathbf{z}}\cdot\mathbf{k}} + \frac{(\mathbf{k}\cdot\mathbf{j})\mathbf{k}}{(\hat{\mathbf{z}}\cdot\mathbf{k})^2}\right] \tag{25a}$$

$$A_{\rm o}^{\rm self}(x) = -\frac{e}{(2\pi)^4}\int\!\!\int dy\, dk\,\frac{e^{ik\cdot(x-y)}}{k^2+i\epsilon}\left[1 - \frac{\omega^2}{(\hat{\mathbf{z}}\cdot\mathbf{k})^2}\right]\rho(y) \tag{25b}$$

where ρ and $\mathbf{j}$ are the space and time components of the Pauli current j_μ, which are given in (7*b*). Using the fact that $\mathbf{B}^{\text{self}} = \nabla \times \mathbf{A}^{\text{self}}$, we may now compute the boundary induced effect on the $\sigma \cdot \mathbf{B}^{\text{self}}$ energy term using the method of images as before. After isolating the the electron's magnetic moment μ we can extract the plate correction $\Delta\mu$ as

$$\frac{\Delta\mu}{\mu} = -\frac{\alpha}{4Rm} \tag{26}$$

However, there is a plate induced mass correction similar to the tensorial mass change we saw before in the apparatus dependent Lamb shift expression (18). So long as the electron is constrained to move in a plane parallel to the conducting mirror, this mass correction is given by

$$\frac{\Delta m}{m} = -\frac{\alpha}{4Rm} \tag{27}$$

(Since translation invariance is broken in the $\hat{\mathbf{z}}$ direction, momentum of the electron is not conserved in this direction—which is part of the reason we restrict the motion to a plane parallel to the conductor. The breaking of translation invariance in the z direction also manifests itself in the tensorial character of the electron mass.) The total correction to the magnetic spin energy is, to first order in the unitless parameter α/Rm, given by

$$\begin{aligned}\Delta E^{\text{spin}} &= -\frac{e}{2m}\sigma\cdot\mathbf{B}^{\text{ext}}\left(1+\frac{\Delta\mu}{\mu}\right)\\ &= -\frac{e}{2m_\circ}\sigma\cdot\mathbf{B}^{\text{ext}}\left(1-\frac{\Delta m}{m}+\frac{\Delta\mu}{\mu}\right)\\ &= -\frac{e}{2m_\circ}\sigma\cdot\mathbf{B}^{\text{ext}}\end{aligned} \tag{28}$$

where the first order correction vanishes if we express the mass m near the plate in terms of the observed freespace mass $m_\circ$. Hence there is not first order change of the spin precession frequency ω_{spin}. One can now calculate the cyclotron frequency shift. This turns out to be essentially a classical effect, and it *does* remain intact at order α/Rm, and hence is the dominant effect:

$$\omega_{\text{cyc}} \sim \frac{\alpha}{Rm} \tag{29}$$

For the presently used cavities, the systematic error introduced by the cavity wall is about 1 part in 10^{12}, which is precisely the accuracy of the current $g-2$ experiments. It would seem that apparatus contributions to ω_{cyc} are now a limiting factor on the accuracy of this type of measurement of $g-2$.

VII. THE UNRUH EFFECT AND HAWKING RADIATION

We now move along to a more arcane boundary condition in QED which gives rise to the Unruh and the related Hawking radiation. Consider a general quantum mechanical detector. For our purposes, we demand that the detector have a complete set of energy levels, and that it couple to the electromagnetic field. (A hydrogen atom would serve quite well.) At a temperature of $T = 0°K$ an inertially moving detector sees no photons—although some might argue that the detector 'sees' the vacuum fluctuations since they trigger such effects as the freespace spontaneous emission and Lamb shift. From the self-field point of view, however, such radiative effects are an integral part of a complete Dirac equation, say, which incorporates the radiation reaction field. So from our point of view an inertially moving detector registers no photons in a vacuum.

Shortly after Hawking showed that the event horizon of a black hole appears to emit thermal radiation,[59] Unruh proved that a uniformly accelerating detector in a vacuum appears to register a thermal bath of photons at a temperature proportional to the acceleration.[60] It turns out that the two phenomena are related via a conformal transformation. The Unruh radiation can be thought of as being emitted from the event horizon of Rindler space. (Rindler space is a coordinate system used to describe a uniformly accelerating reference frame.) In particular, a detector accelerating uniformly through a vacuum with a constant acceleration a responds *as if* it is immersed in a Planckian distribution of photons at a temperature T given by

$$T = \frac{\hbar a}{2\pi k c} \tag{30}$$

where k is Boltzmann's constant.

This result leads to a rather paradoxical conclusion: An inertial detector near an accelerating detector will see nothing, while the other sees a thermal flux of photons. If the photons are really real and exist as physical entities in the surrounding space—then how come the inertial detector can't see them? To understand this, let's consider the quantum field theoretic, plane wave mode decomposition of the electromagnetic potential, namely

$$A_\mu(x) = \frac{1}{(2\pi)} \int dk \left[a_\mu(k)\, e^{-ik\cdot\mathbf{x}} + \text{h.c.} \right] \tag{31}$$

This decomposition is unique *only* in Minkowski space—hence *only* in Minkowski space is 'the vacuum' uniquely defined. In curved spacetime this decomposition is not unique and in general different observers will see different vacua. Considerations such as these have lead P. C. W. Davies to conclude that the notion of particle—'photon' in this case—breaks down in curved spacetime.[14] From the self-field point of view, the problem is not with the notion of 'particle', but rather with the quantum field notion of 'vacuum'. Here is just one more example of how the idea of a dynamic, fluctuating zeropoint field leads to extreme difficulties and apparent contradictions in other areas of physics.

The Unruh effect can also be calculated from the self-field point of view, but now the interpretation is entirely different. The detector, by the very fact that it is assumed to couple to the EM field, must contain a self-field of its own, given by equation (4). This self-field becomes modified by an accelerating agent in such a way so that it acts back on the detector and drives it into a superposition of excited states. When thermodynamically analyzed, an accelerating detector appears *as if* it is subjected to a bath of thermal photons. But this is just an illusion. The radiation reaction field of the detector is merely being perturbed by the acceleration. There are no real photons surrounding the detector, and hence the concept of 'photon' can be saved, since there is nothing for a neighboring inertial detector to detect. There is no paradox. In general relativity, as Davies has repeatedly emphasized, one may not discuss 'the vacuum' independent of the worldline of the detector which is being used to observe deviations away from this vacuum. This point highlights why differing detectors apparently see different vacua. In self-field theory all the vacua are identical and empty; the differing detectors are only responding to *themselves* in differing fashions.

To carry out the calculation of the Uhruh effect[60] we use the fully covariant version of the self-field theory embodied in the Dirac action of equation (8). The self-field contribution to the total action is then given exactly by equation (10). We use the usual QED causal Green's function $D_{\mu\nu} := -\eta_{\mu\nu} D$ in the Feynman gauge, where

$D(x-y)$ can be written in any of the equivalent forms

$$\begin{aligned} D(x-y) &= \frac{i}{4\pi^2}\left\{\frac{1}{(x-y)^2} - i\pi\delta\left[(x-y)^2\right]\right\} \\ &= \frac{i}{4\pi^2}\frac{1}{(x-y)^2+i\epsilon} \\ &= \frac{1}{(2\pi)^4}\int\frac{e^{-ik\cdot(\mathbf{x}-\mathbf{y})}}{k^2+i\epsilon} \end{aligned} \tag{32}$$

We wish now to boost this Green's function into the accelerating frame of Rindler coordinates. If we take the acceleration vector $\mathbf{a} := (0,0,a)$ in the z direction, then in the zt plane, these coordinates can be written as:

$$x_o = \sinh(\tau) \qquad y_o = \sinh(\tau') \tag{33a}$$

$$x_3 = \cosh(\tau) \qquad y_3 = \cosh(\tau') \tag{33b}$$

with all other components zero. Here τ and τ' are the propertimes associated with the source point x_μ and the field point y_μ, respectively, of the Green's function $D(x_\mu - y_\mu)$ which is comoving with the detector. We have set $\hbar = c = a = 1$. In these coordinates the Green's function (32) becomes, in the dipole approximation where $\mathbf{x} \approx \mathbf{y}$,

$$\begin{aligned} D(x-y) &\approx \frac{1}{16\pi^2}\operatorname{csch}\left(\frac{i\Delta\tau}{2}+i\epsilon\right) \\ &= \frac{1}{4\pi^2}\sum_{p=-\infty}^{\infty}\left(\Delta\tau + 2\pi i p + i\epsilon\right) \end{aligned} \tag{34}$$

where $\Delta\tau := \tau - \tau'$, and we have expanded the hyperbolic cosecant in an infinite partial fraction expansion. If one now makes the Fourier expansion

$$\psi(x) = \sum_n \psi_n(\mathbf{x})\, e^{-iE_n t} \tag{35}$$

for each ψ appearing in equation (10), one arrives at an expression for the self-field contribution to the total action, given by

$$W^{\text{self}} = -\frac{\alpha}{2\pi}\sum_{n,m}\langle n|\gamma^\mu|m\rangle\langle m|\gamma_\mu|n\rangle \sum_{p=-\infty}^{\infty}\int\!\!\int d\tau\, d\tau' \frac{e^{i\omega_{nm}\Delta\tau}}{\left(\Delta\tau + 2\pi i p\right)^2} \tag{36}$$

This may be converted to a transition probability G per unit time by replacing the double integral with a single integral via the prescription $\int\int d\tau d\tau' \to \int d(\Delta\tau)$. Carrying out the integration, and then summing a remaining geometric series yields

$$G = \frac{\alpha}{2\pi}\sum_{\substack{m \\ m<n}}\left[\frac{\hbar\omega_{nm}}{2} + \frac{\hbar\omega_{nm}}{\exp\left(\frac{2\pi c\omega_{nm}}{a}\right)-1}\right] \tag{37}$$

From the form of this equation, we see that by taking $a \to 0$ we recover only the term $\hbar\omega_{nm}/2$ in the curly braces. This is suggestive of the zeropoint field of standard QED, but here we have only the radiation reaction field of an inertial detector. This observation supports the suggestion of Jaynes, Milonni and others that the zeropoint fluctuations are perhaps a mathematical subterfuge which mimic the physical radiation reaction field.[12,13] When $a > 0$ we see that there is a Planckian contribution to the transition rate, which corresponds precisely to the Unruh temperature given in equation (30). But there are no real thermal photons—just as in the inertial case the detector is merely responding to itself. Only now this self-response has been modified by the force required to maintain the detector in an accelerating frame. As we mentioned earlier, a detector outside of a black hole sees a similar thermal radiation. Indeed, a detector accelerating at a rate a is by the equivalence principle indistinguishable from a detector at rest in a gravitational field $\mathbf{g}$ of strength $g = a$. From the self-field view, then, Hawking radiation is in some sense just as fictitious as the Unruh radiation. It would be our contention that the black hole is not emitting radiation in the usual sense, but rather it is perturbing the metric around itself such that a nearby detector responds to the curvature as if it were bathed with thermal photons.

VIII. CONCLUSION

We have summarized here how the self-field theory of QED can be used to account satisfactorily, at least to order α, for an array of boundary induced changes in the radiative corrections found in QED. We have emphasized that it is not necessary to invoke the notion of zeropoint fluctuations to construct an interpretive framework of a boundary modified radiative effects. One can view the whole process for the point of view of a quantum analog to classical radiation reaction theory. This is consistent with the self-field point of view, which looks upon all radiative effects as arising from the correct inclusion of the back reaction upon a charge of its own self field. It is our position that an electromagnetic field does not exist independent of the source which produced it—in the spirit of the quote from Einstein which appeared at the beginning of this work.

ACKNOWLEDGEMENT

I would like to thank Professor A. O. Barut for all of his help and guidance in the execution of the calculations upon which this this work has been based, and for all the work he did in organizing such a great conference in İstanbul. I would also like to thank G. Barton and M. Kreuzer for some interesting discussions and correspondences concerning cavity QED.

REFERENCES

1. P. A. M. Dirac, Proc. R. Soc. London, Ser. **A**, 243 (1927).
2. *Selected Papers on Quantum Electrodynamics*, edited by J. Schwinger (Dover, New York, 1958).
3. E. A. Power, *Introductory Quantum Electrodyamics* (Longmans, London, 1964).
4. S. Weinberg, Rev. Mod. Phys **61**, 1–24 (1989).
5. T. A. Welton, Phys. Rev. **74**, 1157 (1948).
6. H. B. Callen and T. A. Welton, Phys. Rev. **83**, 34 (1951).
7. H. Grotch and E. Kazes, Phys. Rev. **D 13**, 2851 (1976).
8. A. O. Barut, J. P. Dowling, and J. F. van Huele, Phys. Rev. **A 38**, 4405 (1988).
9. H. G. B. Casimir, Proc. K Ned. Akad. Wet. **51**, 793 (1948).
10. H. G. B. Casimir and D. Polder, Phys. Rev. **73**, 360 (1948).
11. G. Barton and N. S. J. Fawcett, Phys. Rep. **170**, 1–95 (1988).
12. E. T. Jaynes, "Probability in Quantum Theory," to appear in *Proceedings of the 1989 Workshop on Complexity, Entropy, and the Physics of Information* (Addison-Wesley Publishing Co.).
13. P. W. Milonni, Phys. Script. **T 21**, 1020 (1988).

14. N. D. Birrell and P. C. W. Davies, *Quantum Fields in Curved Space* (Cambridge University Press, Cambridge, U.K., 1982).

15. E. Schrödinger, Ann. Phys. **82**, 265 (1926).

16. E. Fermi, Rend. Lincei **5**, 795 (1927).

17. M. D. Crisp and E. T. Jaynes, Phys. Rev. **179**, 1253 (1969).

18. A. O. Barut and J. P. Dowling, "Self-field QED: The two level atom," (submitted to Phys. Rev. A).

19. P. A. M. Dirac, Proc. R. Soc. London, Ser. **167 A**, 148 (1938).

20. A. O. Barut, *Electrodynamics and Classical Theory of Fields and Particles* (Dover, New York, 1980), Chapter VI.

21. G. Süssman, Seit. f. Phys. **131**, 629 (1952).

22. J. A. Wheeler and R. P. Feynman, Rev. Mod. Phys **17**, 157 (1945).

23. A. O. Barut and J. Kraus, Found. Phys. **13**, 189 (1983).

24. A. O. Barutand J. Kraus, Phys. Rev. **D 16**, 161 (1977).

25. A. O. Barut and J. F. van Huele, Phys. Rev. **A 32**, 3187 (1985)

26. A. O. Barut and Y. I. Salamin, Phys. Rev. **A 37**, 2284 (1988).

27. A. O. Barut and Y. I. Salimin (submitted to J. Math. Phys.).

28. A. O. Barut and J. P. Dowling, Z. Naturforch. **44**a, (November, 1989).

29. A. O. Barut and N. Unal (unpublished).

30. A. O. Barut and J. P. Dowling, Phys. Rev. **A 36**, 649 (1987).

31. A. O. Barut and J. P. Dowling, Phys. Rev. **A 36**, 2550 (1987).

32. A. O. Barut and J. P. Dowling, Phys. Rev. **A 39**, 2796 (1989).

33. A. O. Barut and J. P. Dowling, "QED based on self-fields: On the origin of the thermal radiation detected by an accelerating observer," (submitted to Phys. Rev. A).

34. P. W. Milonni, Phys. Rev. **A 25**, 1315 (1982).

35. J. R. Ackerhalt, P. L. Knight, and J. H. Eberly, Phys. Rev. Lett. **30**, 456 (1973).

36. J. R. Senitzky, Phys. Rev. Lett. **31**, 955 (1973).

37. P. W. Milonni, J. R. Ackerhalt, and W. A. Smith, Phys. Rev. Lett. **31**, 958 (1975).

38. P. W. Milonni and W. A. Smith, Phys. Rev. **A 11**, 814 (1975).

39. J. Dalibard, J. Dupont-Roc, and C. Cohen-Tannoudji, *J. Phys.* (Paris) **43**, 1617(1982).

40. P. Goy, J. M. Raymund, M. Goss and S. Haroche, Phys. Rev. Lett. **50**, 1903 (1983).

41. R. G. Hulet,E. S. Hilfer, and D. Kleppner, Phys. Rev. Lett. **55**, 2137 (1985).

42. A. O. Barut, Found. Phys. **17**, 549 (1987).

43. A. O. Barut, Found. Phys. **18**, 95 (1987).

44. A. O. Barut, Ann. Phys. **45**, 31 (1988).

45. G. Barton, Proc. R. Soc. London, Ser. **320 A**, 251 (1970).

46. P. W. Milonni and P. L. Knight, Opt. Commun. **9**, 119 (1973).

47. M. R. Philpott, Chem. Phys. **53**, 101 (1976).
48. G. Barton, J. Phys. **7 B**, 2134 (1974).
49. H. Dehmelt, Phys. Rev. **D 34**, 722 (1986).
50. M. Babiker and G. Barton, Proc. R. Soc. London, Ser. **326 A**, 255,277 (1972).
51. G. Barton and H. Grotch, J. Phys. **10 A**, 1201 (1977).
52. E. Fischbach and N. Nakagawa, Phys. Rev. **D 30**, 2356 (1984).
53. L. S. Brown, G. Gabrielse, K. Helmerson, and J. Tan, Phys. Rev. **A 32**, 3204 (1985).
54. L. S. Brown, G. Gabrielse, K. Helmerson, and J. Tan, Phys. Rev. Lett. **55**, 44 (1985).
55. D. G. Boulware, L. S. Brown, T. Lee, Phys. Rev. **D 32**, 729 (1985).
56. K. Svozil, Phys. Rev. Lett. **54**, 742 (1985).
57. M. Kreuzer and K. Svozil, Phys. Rev. **D 34**, 1429 (1986).
58. M. Kreuzer, J. Phys. **21 A**, 3285 (1988).
59. S. Hawking, Commun. Math. Phys. **43**, 199 (1975).
60. W. G. Unruh, Phys. Rev. **D 14**, 870 (1976).

FINITE VACUUM POLARIZATION
IN SELF ENERGY FORMULATION OF QED

Nuri Unal

Dicle University, Physics Department
Diyarbakir, Turkey

ABSTRACT

The vacuum polarization term is evaluated in the QED based on the self energy. A finite result is obtained for the first time and it agrees with the standard QED at the lowest order.

INTRODUCTION

Electrodynamics is the interaction of the charged particles with the electromagnetic field. This interaction is described by the coupled field-matter action and we derive the basic equations of the electrodynamics from this action. These are coupled matter-field equations. The electromagnetic field is described by the Maxwell equations with the matter source. In the classical electrodynamics, the matter source and the radiation field are described by c-number variables and in QED both are described by q-number variables.

In the classical electrodynamics radiative processes are discussed non-perturbatively. Recently, a new formulation of the QED has been developed and a general formula for the radiative processes of the QED has been obtained. The Lamb shift, anomalous magnetic moment, spontaneous emission and the Compton scattering all appear in the same, single expression.(1)–(3)

Main technical problem of the QED is the evaluation of this general expression. In principle it can be solved non-perturbatively. In perturbation theory, the radiative processes are evaluated order by order by using the plane waves. Here we evaluate them by using the well defined wave functions which depend on the external fields of the particular problem (e.g., the relativistic Coulomb wave functions or the wave functions of the Landau orbits).

The fundamental problem of the QED is the regularization and the renormalization. We have the mass renormalization also in the classical electrodynamics.(4) The most divergent term of the perturbation theory is the vacuum polarization. The vacuum polarization energy shift of the bound states was studied by Wichmann and Kroll in their classic paper.(5) They calculated the energy shift of the hydrogenic atoms by using the Dirac-Coulomb wave functions.

The aim of this work is to reevaluate the vacuum polarization term by using the Dirac-Coulomb Wave functions. We will show for the first time that this can be

New Frontiers in Quantum Electrodynamics and Quantum Optics
Edited by A. O. Barut, Plenum Press, New York, 1990

done without infinities and renormalization. The paper is organized as follows: In Section two we review the derivation of the vacuum polarization energy shift term. In Section three we discuss the density of the energy eigenstates of the Dirac-Coulumb problem and their sum over all positive and negative energies. In Section four we discuss the Mellin transformation of the energy shift and in Section five we do the summation over all positive and negative energies and over all quantum numbers κ. Finally we compare these results with the standard, renormalized quantumelectrodynamics.

DERIVATION OF THE ENERGY SHIFT

The most convenient way to formulate the theory is to use the action which is related to the physical observables, namely the energy and the invariant S-matrix in the case of bound and scattering problems, respectively. The Maxwell-Dirac action of the electromagnetic field $A_\mu(x)$ and the matter field $\psi(x)$ is

$$A = \int dx \left\{ \bar{\psi}(x) \left[\gamma^\mu (p_\mu - eA_\mu) - m\right] \psi(x) - \frac{1}{4} F_{\mu\nu} F^{\mu\nu} \right\} \tag{1}$$

where we have used the natural units $\hbar = c = 1$. This action gives linear equations for $A_\mu(x)$ which can be solved formally in terms of the source field $\bar{\psi}(x)\gamma_\mu\psi(x)$:

$$A_\mu(x) = e \int dy D(x-y) \bar{\psi}(y)\gamma_\mu\psi(y). \tag{2}$$

We substitute this into eq. (1) and do partial integration. The result is

$$\begin{aligned} A = \int dx \Big\{ \bar{\psi}(x) \left[\gamma^\mu (p_\mu - eA_\mu^{ext}) - m\right] \psi(x) - e^2 \bar{\psi}(x)\gamma_\mu\psi(x) \int dy D(x-y) \cdot \\ \cdot \bar{\psi}(y)\gamma^\mu\psi(y) \Big\} - \frac{e^2}{2} \int d\sigma_\mu(x) dy dz D(x-y) \bar{\psi}(y)\gamma_\nu\psi(y) \cdot \\ \partial^{[\mu}_{(x)} D(x-z)\bar{\psi}(z)\gamma^{\nu]}\psi(z) \end{aligned} \tag{3}$$

Here we neglect the surface terms and expand the matter field $\psi(x)$ into its Fourier components,

$$\psi(x) = \sum\!\!\!\!\!\!\int_n \psi_n(\vec{x}) e^{-iE_n t} \tag{4}$$

where $\psi_n(\vec{x})$ are the unknown Fourier components. We perform all time integrations with the causal Green's function $D_c(x-y)$. Then the total energy of the system is

$$\begin{aligned} \mathcal{E} = & \sum\!\!\!\!\!\!\int_n \int d\vec{x}\, \bar{\psi}_n(\vec{x}) \left(\gamma^\mu \left(p_\mu - eA_\mu^{ext}\right) - m\right) \psi_n(\vec{x}) \\ & - \frac{e^2}{2} \sum\!\!\!\!\!\!\int_n \int d\vec{x}\, \bar{\psi}_n(\vec{x})\gamma_\mu\psi_n(\vec{x}) \sum\!\!\!\!\!\!\int_s d\vec{y}\, \bar{\psi}_s(\vec{y})\gamma^\mu\psi_s(\vec{y}) P \int \frac{d\vec{k}}{(2\pi)^3} \frac{e^{i\vec{k}\cdot(\vec{x}-\vec{y})}}{k^2} \\ & + \frac{e^2}{2} \sum\!\!\!\!\!\!\int_{ns} \int d\vec{x}\, d\vec{y}\, \bar{\psi}_n(\vec{x})\gamma_\mu\psi_s(\vec{x}) \int \frac{d\vec{k}}{(2\pi)^3} e^{i\vec{k}\cdot(\vec{x}-\vec{y})} \bar{\psi}_s(\vec{y})\gamma^\mu\psi_n(\vec{y}) \cdot \\ & \left\{ \left[\frac{1}{E_s - E_n - k} - \frac{1}{E_s - E_n + k} \right] + \frac{i\pi}{2k} \left[\delta(E_s - E_n - k) + \delta(E_s - E_n + k)\right] \right\} \end{aligned} \tag{5}$$

Each term of eq. (5) represents, respectively, the total action of the particle in external field, the vacuum polarization, the self energy and the spontaneous decay.

The expression eq. (5) is the fundamental equation of the self-energy method and it is in principle, non-perturbative. In practice, we solve it by iteration. For the lowest order iteration we choose the Fourier components ψ_n as to be the relativistic Coulomb wave functions ψ_n^c. Then we write E_n as $E_n^c + \Delta E_n$. The total action is zero for solution of the equation of motion and this gives the energy shift of any fixed level n as

$$\Delta E_n = \frac{e^2}{2}\int d\vec{x}\,\bar{\psi}_n(\vec{x})\gamma_\mu\psi_n(\vec{x})P\int\frac{d\vec{k}}{(2\pi)^3}\int d\vec{y}\,\frac{e^{i\vec{k}\cdot(\vec{x}-\vec{y})}}{k^2}\cdot\sum\!\!\!\!\!\!\int_s\bar{\psi}_s(\vec{y})\gamma^\mu\psi_s(\vec{y})$$

$$-\frac{e^2}{2}\sum\!\!\!\!\!\!\int_s\int d\vec{x}\,d\vec{y}\,\bar{\psi}_n(\vec{x})\gamma_\mu\psi_s(\vec{x})\int\frac{d\vec{k}}{(2\pi)^3}e^{i\vec{k}\cdot(\vec{x}-\vec{y})}\bar{\psi}_s(\vec{y})\gamma^\mu\psi_n(\vec{y})\cdot$$

$$\cdot\left[\frac{1}{E_s-E_n-k}-\frac{1}{E_s-E_n+k}\right]$$

$$-\frac{e^2}{2}\sum\!\!\!\!\!\!\int_{\substack{s\\(s<n)}}\int d\vec{x}\,d\vec{y}\,\bar{\psi}_n(\vec{x})\gamma_\mu\psi_s(\vec{x})\int\frac{d\vec{k}}{(2\pi)^3}e^{i\vec{k}\cdot(\vec{x}-\vec{y})}\bar{\psi}_s(\vec{y})\gamma^\mu\psi_n(\vec{y})\cdot$$

$$\cdot\frac{i\pi}{2k}\delta(E_s-E_n-k) \tag{6}$$

where each term of eq. (6) corresponds to the energy shift of the vacuum polarization, self energy and the spontaneous emission of the level n. Here we are interested in the evaluation of the energy shift of the vacuum polarization. It is given by

$$\Delta E_n^{\mathrm{V.P.}} = \frac{e^2}{2}\int d\vec{x}\,d\vec{y}\,\bar{\psi}_n(\vec{x})\gamma_\mu\psi_n(\vec{x})\bar{\psi}_s(\vec{y})\gamma^\mu\psi_s(\vec{y})P\int\frac{d\vec{k}}{(2\pi)^3}\frac{e^{i\vec{k}\cdot(\vec{x}-\vec{y})}}{k^2} \tag{7}$$

DENSITY OF STATES

We expand $\exp[i\,\vec{k}\cdot(\vec{x}-\vec{y})]$ into spherical harmonics, perform the $\vec{k}$ integration and $\hat{x}$, $\hat{y}$ angular integrations. Then eq. (7) becomes

$$\Delta E_n^{\mathrm{V.P.}} = -\frac{e^2}{4\pi}(2J_n+1)\sum_{lm}\sum\!\!\!\!\!\!\int_s(2J_s+1)\int dr\,dr'\,V_\ell(r,r')\Big\{W_{nn}^{\ell m}W_{ss}^{\ell m}\Big[|fn|^2+|g_n|^2\Big]\Big[|fs|^2+|g_s|^2\Big]-\Big[f_n^\dagger(r)\,\vec{K}_{nn}^{\ell m}\,g_n(r)+h.c.\Big]\Big[f_n^\dagger(r')\,\vec{K}_{ss}^{\ell m}\,g_s(r')+h.c.\Big]\Big\} \tag{8}$$

where $V_\ell(r,r')$ and $W_{nn}^{\ell m}, \vec{K}_{nn}^{\ell m}$ are

$$V_\ell(r,r') = \frac{r^2 r'^2}{2\ell+1}\cdot\frac{r_<^\ell}{r_>^{\ell+1}} \tag{9}$$

$$W_{nn}^{\ell m} = (-1^{\frac{1}{2}-m_n}\left(1+(-1)^{\ell}\right)\sqrt{2\ell+1}\begin{pmatrix} j_n & j_n & \ell \\ -1/2 & 1/2 & 0 \end{pmatrix}\begin{pmatrix} j_n & j_n & \ell \\ -m_n & m_n' & m \end{pmatrix}$$

$$\vec{K}_{nn}^{\ell m} = (-1)^{m_n}\sqrt{2\ell+1}\sum_{j_n'} 2(2j_n'+1)\left(1+(-1)^{\ell+1}\right)\begin{pmatrix} j_n & j_n' & \ell \\ -1/2 & 1/2 & 0 \end{pmatrix}$$

$$\left[\begin{pmatrix} 1 & j_n & j_n' \\ 0 & 1/2 & -1/2 \end{pmatrix}\left(\theta(\kappa_n)+\theta(-\kappa_n)\right) - \left(\theta(\kappa_n)-\theta(-\kappa_n)\right)\begin{pmatrix} 1 & j_n & j_n' \\ 1 & -1/2 & -1/2 \end{pmatrix}\right]$$

$$\left[\sqrt{2}\hat{e}_+\begin{pmatrix} 1 & j_n & j_n' \\ 1 & m_n' & -1-m_n' \end{pmatrix}\begin{pmatrix} j_n & j_n' & \ell \\ -m_n' & 1+m_n' & m \end{pmatrix} + h.c. - \right.$$

$$\left.\hat{e}_3\begin{pmatrix} 1 & j_n & j_n' \\ 0 & m_n' & -m_n \end{pmatrix}\begin{pmatrix} j_n & j_n' & \ell \\ -m_n' & m_n & m \end{pmatrix}\right] \tag{10}$$

In eq. (8) f and g are the radial Dirac-Coulomb wave functions. Discrete states are given by

$$\begin{pmatrix} f_n(r) \\ g_n(r) \end{pmatrix} = U_n(r)\begin{pmatrix} \sqrt{1+\varepsilon_n} & (F_n(r)+G_n(r)) \\ \sqrt{1-\varepsilon_n} & (F_n(r)-G_n(r)) \end{pmatrix} \tag{11}$$

where

$$U_n(r) = \left[\frac{\Gamma(2\gamma_n+n_r+1)}{4N_n(N_n-\kappa_n)n_r!}\right]^{1/2}\frac{(2p_N)^{3/2}(2p_N r)^{\gamma_n-1}e^{-p_N r}}{\Gamma(2\gamma_n+1)} \tag{12}$$

with

$$\begin{aligned} F_n(r) &= n_r\phi(1-n_r,\ 2\gamma_n+1;\ 2p_N r) \\ G_n(r) &= (N_n-\kappa_n)\phi(-n_r,\ 2\gamma_n+1;\ 2p_N r) \end{aligned} \tag{13}$$

with

$$\begin{aligned} &p_N = \frac{Z\alpha}{N_n}\ ,\quad \varepsilon_n = \frac{E_n}{m} = (1+p_N^2)^{1/2},\quad N = \left[n^2-2n_r(|\kappa_n|-\gamma_n)\right]^{1/2} \\ &n_r = n-|\kappa_n|\ ,\quad \kappa_n = \pm(j_n+1/2),\quad \gamma_n = \sqrt{\kappa_n^2-(Z\alpha)^2} \end{aligned} \tag{14}$$

In eq. (13) ϕ is the confluent hypergeometric function. By using the power series expansion of it we write $F_n(r)$ and $G_n(r)$ as

$$\begin{aligned} F_n(r) &= \textstyle\sum_{n_1} A_{n_1}(2p_N r)^{n_1} \\ G_n(r) &= \textstyle\sum_{n_1} B_{n_1}(2p_N r)^{n_1} \end{aligned} \tag{15}$$

where

$$\begin{pmatrix} A_{n_1} \\ B_{n_1} \end{pmatrix} = \frac{1}{(2\gamma_n+1)_{n_1}n_1!}\begin{pmatrix} n_r(1-n_r)_{n_1} \\ (N_n-\kappa_n)(-n_r)_{n_1} \end{pmatrix} \tag{16}$$

Then the density of the bound state n is

$$|f_n|^2+|g_n|^2 = \textstyle\sum_{n_1n_2} a_{n_1n_2}\lambda_{n_1}(p_N r)\lambda_{n_2}(p_N r) \tag{17}$$

where $a_{n_1n_2}$ and $\lambda_{n_i}(p_N r)$ are

$$a_{n_1n_2} = \left[\frac{\Gamma(2\gamma_n+n_r+1}{4N_n(N_n-\kappa_n)n_r!\Gamma^2(2\gamma_n+1)}\right](A_{n_1}A_{n_2}+B_{n_1}B_{n_2}-2\varepsilon_n A_{n_1}B_{n_2}) \tag{18}$$

and

$$\lambda_{n_i}(p_N r) = (2p_N)^{3/2}(2p_N r)^{\gamma_n+n_i-1}e^{-p_N r},\ i = 1,2 \tag{19}$$

In eq. (8) we have the sum over all states. This sum can be written in terms of the Green's function of the Dirac-Coulomb problem. Then the sum of the charge density over all positive and negative energies is given by

$$\begin{aligned}&\frac{1}{2}\left[\sum\!\!\!\!\!\!\int_{E_s>0} - \sum\!\!\!\!\!\!\int_{E_s<0}\right](|f_s|^2 + |g_s|^2)\\ &= \frac{1}{2\pi i}\left(\int_{C+} + \int_{C-}\right) dz \text{ trace } K(r;r';z)\end{aligned} \tag{20}$$

where the contour C_+ and C_- are shown in figure 1.

Green's function of the Dirac-Coulomb problem is

$$K(r,r';z) = \frac{1}{k(z)}\begin{pmatrix} W_1^{(2)} & (r_>,z) \\ W_2^{(2)} & (r_>,z)\end{pmatrix}\left(W_1^{(1)}(r_<,z), W_2^{(1)}(r_<,z)\right) \tag{21}$$

where $W^{(1)}(r_<,z)$ and $W^{(2)}(r_>,z)$ are the regular solutions of the system at the origin and infinity, respectively. They are given by

$$\begin{aligned}W_{\frac{1}{2}}^{(1)}(r_<,z) &= \left(2r_<\sqrt{z^2-1}\right)^{\gamma} e^{ir_<\sqrt{z^2-1}}\left[\left(\kappa - \frac{iZ\alpha}{\sqrt{z^2-1}}\right)\begin{pmatrix} i\sqrt{z+1} \\ \sqrt{z-1}\end{pmatrix}\right.\\ &\phi\left(\gamma - i\nu, 2\gamma+1, -2i\sqrt{z^2-1}r_<\right)\\ &+\begin{pmatrix} i\sqrt{z+1} \\ -\sqrt{z-1}\end{pmatrix}(\gamma - i\nu)\phi\left(\gamma+1-i\nu, 2\gamma+1, -2i\sqrt{z^2-1}r_<\right)\end{aligned} \tag{22}$$

and to obtain $W^{(2)}$ we replace $r_<$ by $r_>$ and ϕ by χ. The quantity $k(z)$ is a normalization factor given by

$$k(z) = -2\sqrt{z^2-1}\left(\kappa - \frac{iZ\alpha}{\sqrt{z^2-1}}\right)\frac{\Gamma(-\gamma-i\nu)\Gamma(2\gamma+1)}{\Gamma(-2\gamma)\Gamma(\gamma-i\nu)}\exp\left[\frac{i\pi}{2}(2\gamma+1)\right] \tag{23}$$

In eq. (22) ϕ and χ are the solutions of the confluent hypergeometric differential equation regular at the origin and infinity, respectively, and their integral representations are

$$\begin{aligned}\phi(\alpha,\gamma;z) &= \frac{\Gamma(\gamma)}{\Gamma(\alpha)\Gamma(\gamma-\alpha)}\int_0^1 dt\ e^{zt}t^{\alpha-1}(1-t)^{\gamma-\alpha-1}\\ \chi(\alpha,\gamma;z) &= \frac{\Gamma(\alpha+1-\gamma)}{\Gamma(\alpha)\Gamma(1-\gamma)}\int_0^\infty dt\ e^{-zt}t^{\alpha-1}(1+t)^{\gamma-\alpha-1}\end{aligned} \tag{24}$$

Then we write the sum over all states s as

$$\begin{aligned}&\sum\!\!\!\!\!\!\int_s (2J_s+1)W_{ss}^{\ell m}\left[|f_s|^2 + |g_s|^2\right] = -2\sum_{\kappa=1}^{\infty}|\kappa|W_{ss}^{\ell m}\left(\int_{C+} + \int_{C-}\right)\frac{dz}{2\pi i}\cdot\\ &\cdot\int_0^1 dt\ t^{\alpha-1}(1-t)^{2\gamma-\alpha}\int_0^\infty dt' t'^{\alpha'-1}(1+t')^{2\gamma-\alpha'}T_{\alpha\alpha'}\cdot\\ &\cdot\frac{2i}{\sqrt{z^2-1}\ r'^2}\left(-2ir'\sqrt{z^2-1}\right)^{2\gamma}\cdot e^{2ir'\sqrt{1-z^2}(1-t+t')}\end{aligned} \tag{25}$$

where we have used the symmetry of $|f_s|^2 + |g_s|^2$ with respect to $\pm\kappa$ and $T_{\alpha\alpha'}$ is given by

$$T_{\alpha\alpha'} = \left\{ \frac{iZ\alpha}{\sqrt{z^2-1}} \left[\frac{\delta_{\alpha,\gamma-i\nu}\, \delta_{\alpha',\gamma-i\nu} + \delta_{\alpha,\gamma+1-i\nu}\, \delta_{\alpha',\gamma+1-i\nu}}{\Gamma(\alpha)\Gamma(2\gamma+1-\alpha')} \right] \right.$$
$$\left. -z \left[\frac{\delta_{\alpha,\gamma-i\nu}\, \delta_{\alpha',\gamma+1-i\nu}}{\Gamma(\alpha)\Gamma(2\gamma+1-\alpha')} - \frac{\delta_{\alpha,\gamma+1-i\nu}\, \delta_{\alpha',\gamma-i\nu}}{\Gamma(\alpha')\Gamma(2\gamma+1-\alpha)} \right] \right\} \tag{26}$$

We substitute eq. (17) and (25) into eq. (8). Then $\Delta E_n^{V.P.}$ becomes

$$\Delta E_n^{V.P.} = \alpha(2J_n+1) \sum_{\ell m} W_{nn}^{\ell m} \sum_{\kappa=1}^{\infty} 2\kappa W_{ss}^{\ell m} \sum\nolimits_{n_1 n_2} a_{n_1 n_2} \left(\int_{C+} + \int_{C-} \right) \frac{dz}{2\pi i}$$
$$\cdot \frac{2i}{\sqrt{z^2-1}} T_{\alpha,\alpha'} \cdot \int_0^1 dt t^{\alpha-1}(1-t)^{2\gamma-\alpha} \int_0^\infty dt' t'^{\alpha'-1}(1+t')^{2\gamma-\alpha'} .$$
$$\cdot \int_0^\infty r^2\,dr \int_0^\infty r'^2\,dr' \lambda_{n_1}(p_N r)\lambda_{n_2}(p_N r) \left(-2i\sqrt{z^2-1}r'\right)^{2\gamma} e^{+2i\sqrt{z^2-1}(1-t+t')r'}$$
$$\frac{r_<^\ell}{(2\ell+1) r_>^{\ell+1}} \tag{27}$$

Next we perform r and r' integrations:

$$R = \int_0^\infty r^2\,dr \int_0^\infty r'^2\,dr' \frac{r_<^\ell}{(2\ell+1)r_>^{\ell+1}} (2p_N)^3 \left(-2i\sqrt{z^2-1}r'\right)^{2\gamma} (2p_N r)^{2\gamma_n+n_1+n_2-2} .$$
$$\cdot\, e^{-2p_N + 2i\sqrt{z^2-1}(1-t+t')r'}$$
$$= \frac{1}{2\ell+1} \frac{\Gamma(b)(-iy)^{2\gamma}}{(1+ay)^b} \left\{ c_1^{-1}\ {}_2F_1\left[1,\ b; c_1+1;\ (1+ay)^{-1}\right] \right.$$
$$\left. + c_2^{-1}\ {}_2F_1\left[1,\ b; c_2+1;\ ay(1+ay)^{-1}\right] \right\} \tag{28}$$

where

$$b = 2\gamma + 2\gamma_n + n_1 + n_2 + 1$$
$$c_1 = 2\gamma_n + n_1 + n_2 + \ell + 1 \tag{29}$$
$$c_2 = 2\gamma + \ell + 1$$

and

$$y = \sqrt{z^2-1}/p_N$$
$$a = -i(1-t+t') \tag{30}$$

THE MELLIN TRANSFORMATION OF $\Delta E; t-t'$ INTEGRATIONS

The Mellin transformation of $f(x)$ is defined by

$$\mathcal{F}(w) = M[f] \equiv \int_0^\infty dx x^{w-1} f(x) \tag{31}$$

The inverse transformation is

$$f(x) = M^{-1}[\bar{f}] \equiv \int_{C_w - i\infty}^{C_w + i\infty} \mathcal{F}(w) x^{-w}\, dw/2\pi i \tag{32}$$

To obtain the Mellin transformation of $R(y)$ with respect to y we use the series representation of ${}_2F_1$, transform term by term and then resum. The result is

$$\mathcal{R}(w) = a^{-(2\gamma+w)} \frac{\Gamma(2\gamma+w)\Gamma(b-2\gamma-w)}{(\ell+w)(\ell+1-w)} \tag{33}$$

We write the $\Delta E_n^{V.P.}$ as the inverse Mellin transformation of $\bar{R}(w)$. It is

$$\begin{aligned} \Delta E_n^{V.P.} &= \gamma(2J_n+1)\sum_{\ell m} W_{nn}^{\ell m} \sum_{\kappa=1}^{\infty} 2\kappa W_{ss}^{\ell m} \sum_{n_1 n_2} a_{n_1 n_2} \left\{ \int_{C+} + \int_{C-} \right\} \frac{dz}{2\pi i} \\ &\cdot \frac{2i}{\sqrt{z^2-1}} \int_{C_w - i\infty}^{C_w + i\infty} \frac{dw}{2\pi i} \frac{\Gamma(2\gamma+w)\Gamma(b-2\gamma-w)y^{-w}}{(\ell+w)(\ell+1-w)((-i)^{2\gamma+w}} J_{\alpha\alpha'}(w) T_{\alpha\alpha'} \end{aligned} \tag{34}$$

where $J_{\alpha\alpha'(w)}$ is

$$J_{\alpha\alpha'}(w) = \int_0^1 dt \int_0^\infty dt' t^{\alpha-1}(1-t)^{2\gamma-\alpha} t'^{\alpha'-1}(1+t')^{2\gamma-\alpha'}(1-t+t')^{-(2\gamma+w)} \tag{35}$$

We calculate $J_{\alpha\alpha'}(w)$ for the complex values of w as follows: t'-integration gives

$$\begin{aligned} J_{\alpha\alpha'}(w) &= \int_0^1 dt t^{\alpha-1}(1-t)^{2\gamma-\alpha} B(w,\alpha')(1-t)^{-2\gamma-w} \\ &\cdot {}_2F_1\left(2\gamma+w, \alpha'; \alpha'+w; \frac{t}{t-1}\right) \end{aligned} \tag{36}$$

It can also be rewritten as

$$J_{\alpha\alpha'}(w) = \Gamma(w) \int \frac{dv}{2\pi i} \Gamma(-v) \frac{\Gamma(\alpha+v)\Gamma(\alpha'+v)\Gamma(2\gamma+w+v)(-1)^{\gamma+v}}{\Gamma(\alpha+v+w)\Gamma(\alpha'+w+v)\Gamma(2\gamma+w)}$$

or [5]

$$\begin{aligned} J_{\alpha\alpha'}(w) = \Gamma(w) \int \frac{dv}{2\pi i} \Gamma(-v) \frac{\Gamma(\alpha'+v)\Gamma(2\gamma+w+v)(-1)^{\alpha+v}}{\Gamma(\alpha'+v+w)\Gamma(2\gamma+w)(\alpha+v)} \times \\ \times {}_2F_1(1-w, \alpha+v; \alpha+v+1; 1) \end{aligned} \tag{37}$$

where the hypergeometric function ${}_2F_1$ is convergent for the positive values of $Re(w)$. In the w-plane we have the poles at $w = b - 2\gamma + r$ with $r = 0, 1, 2, \ldots$ and this gives a $(Z\alpha)$ expansion of $\Delta E_n^{V.P}$. The first pole is at

$$w_0 = b - 2\gamma = 2\gamma_n + n_1 + n_2 + 1 \cong 3 \tag{38}$$

It is obtained for $n_1 = n_2 = 0$ and $\gamma_n \cong 1$ which corresponds to $j_n = 1/2$. Thus the coefficient of the lowest order vacuum polarization is proportional to $\alpha(Z\alpha)^4$. Here we calculate $J_{\alpha\alpha'}(w)$ only for $j_n = 1/2$ and $\ell_n = 0$ or S-waves. In this case ℓ is zero or one.

In $W_{nn}^{\ell m}\left(\overset{\rightharpoonup}{K}{}_{nn}^{\ell m}\right)$ it is zero (one) and

$$W_{nn}^{00} = W_{ss}^{00} = 1 \tag{39}$$

Then $J_{\alpha\alpha'}(w_0)$ is

$$J_{\alpha\alpha'}(w_0 \approx 3) = \Gamma(3) \int \frac{dv}{2\pi i} \Gamma(-v) \frac{\Gamma(2\gamma+3+v)}{\Gamma(2\gamma+3)} \cdot$$

$$\cdot \sum_{p=0}^{2} \frac{(-1)^{\alpha+v+p}\,\Gamma(3)}{\Gamma(3-p)p!(\alpha'+v)(\alpha'+v+1)(\alpha'+v+2)(\alpha+v+p)} \tag{40}$$

In the v-integration we choose the contour such that the power of (-1), or $\alpha+v+p$ will be non-negative. The poles and the integration contours are shown in figure 2.

The results of the v-integrations are

$$J_{\alpha\alpha'}(w_0 \approx 3) = \frac{\Gamma(3)}{\Gamma(2\gamma+3)} \left\{ \frac{\Gamma(\alpha)\Gamma(2\gamma+3-\alpha)}{2} [-\psi(\alpha) + \psi(2\gamma+3-\alpha) - 3 + \log(-1)] \right.$$

$$+ 2\Gamma(\alpha+1)\Gamma(2\gamma+2-\alpha)[\psi(\alpha+1) - \psi(2\gamma+2-\alpha) + \frac{1}{2} - \log(-1)]$$

$$\left. + \frac{\Gamma(\alpha+2)\Gamma(2\gamma+1-\alpha)}{2} [-\psi(\alpha+2) + \psi(2\gamma+1-\alpha) + \frac{3}{2} + \log(-1)] \right\}$$

$$J_{\alpha,\alpha+1}(w_0 \approx 3) = \frac{\Gamma(3)}{\Gamma(2\gamma+3)} \left\{ \frac{\Gamma(\alpha)\Gamma(2\gamma+3-\alpha)}{3!} + \Gamma(\alpha+1)\Gamma(2\gamma+2-\alpha) \right.$$

$$[-\psi(\alpha+1) + \psi(2\gamma+2-\alpha) - 2 + \log(-1)]$$

$$\left. - \Gamma(\alpha+2)\Gamma(2\gamma+1-\alpha)[-\psi(\alpha+2) + \psi(2\gamma+1-\alpha) + \log(-1)] \right\}$$

$$J_{\alpha,\alpha-1}(w_0 \approx 3) = \frac{\Gamma(3)}{\Gamma(2\gamma+3)} \left\{ -\frac{\Gamma(\alpha-1)\Gamma(2\gamma+4-\alpha)}{3!} + \Gamma(\alpha)\Gamma(2\gamma+3-\alpha) \right.$$

$$[\psi(\alpha) - \psi(2\gamma+3-\alpha) - \frac{3}{2} - \log(-1)] + \Gamma(\alpha+1)\Gamma(2\gamma+2-\alpha)$$

$$\left. [-\psi(\alpha+1) + \psi(2\gamma+2-\alpha) + 1 + \log(-1)] \right\} - \Gamma(\alpha+2)\Gamma(2\gamma+1-\alpha)/3! \Big\} \tag{41}$$

Then we substitute $J_{\gamma-i\nu,\gamma-i\nu} J_{\gamma+1-i\nu,\gamma+1-i\nu}$, $J_{\gamma-i\nu,\gamma+1-i\nu}$ and $J_{\gamma+1-i\nu,\gamma-i\nu}$ into eq. (34).

The result is

$$\Delta E_n^{V.P.} = -\alpha \sum_{\kappa=1}^{\infty} 2\kappa \Big[\int_{c+} + \int_{c-} \Big] \frac{dz}{2\pi i} \cdot \frac{4i(-i)^{-2\gamma-3}}{6(z^2-1)^2} \cdot \left(\frac{Z\alpha}{N_n}\right)^3 \cdot$$

$$\cdot \Bigg\{ \frac{iZ\alpha}{\sqrt{z^2-1}} \Bigg[\frac{\Gamma(\gamma+3+i\nu)}{\Gamma(\gamma+1+i\nu)} [-\psi(\gamma-i\nu)+\psi(\gamma+3+i\nu)-3+\log(-1)]$$

$$+\frac{2\Gamma(\gamma+1-i\nu)\Gamma(\gamma+2+i\nu)}{\Gamma(\gamma-i\nu)\Gamma(\gamma+1+i\nu)}[-\psi(\gamma+2+i\nu)+\psi(\gamma+1-i\nu)+\frac{1}{2}-\log(-1)]$$

$$+\frac{\Gamma(\gamma+2-i\nu)}{2\Gamma(\gamma-i\nu)}[\psi(\gamma+1+i\nu)-\psi(\gamma+2-i\nu)+\frac{3}{2}+\log(-1)]$$

$$+\frac{\Gamma(\gamma+2+i\nu)}{\Gamma(\gamma+i\nu)}[\psi(\gamma+2+i\nu)-\psi(\gamma+1-i\nu)-3+\log(-1)]$$

$$+\frac{2\Gamma(\gamma+2-i\nu)\Gamma(\gamma+1+i\nu)}{\Gamma(\gamma+1-i\nu)\Gamma(\gamma+i\nu)}[\psi(\gamma+2-i\nu)-\psi(\gamma+1+i\nu)+\frac{1}{2}-\log(-1)]$$

$$+\frac{\Gamma(\gamma+3-i\nu)}{2\Gamma(\gamma+1-i\nu)}[-\psi(\gamma+3-i\nu)+\psi(\gamma+i\nu)+\frac{3}{2}+\log(-1)]\Bigg]$$

$$-z\Bigg[-\frac{\Gamma(\gamma+3+i\nu)}{3!\Gamma(\gamma+i\nu)}+\frac{\Gamma(\gamma+1-i\nu)\Gamma(\gamma+2+i\nu)}{|\Gamma(\gamma-i\nu)|^2}$$

$$[-\psi(\gamma+1-i\nu)+\psi(\gamma+2+i\nu)-2+\log(-1)]-\frac{\Gamma(\gamma+2-i\nu)\Gamma(\gamma+1+i\nu)}{|\Gamma(\gamma-i\nu)|^2}$$

$$[-\psi(\gamma+2-i\nu)+\Gamma(\gamma+1+i\nu)+\log(-1)]+\frac{\Gamma(\gamma+3+i\nu)}{3!}$$

$$-\frac{\Gamma(\gamma+1-i\nu)\Gamma(\gamma+2+i\nu)}{|\Gamma(\gamma-i\nu)|^2}[\psi(\gamma+1-i\nu)-\psi(\gamma+2+i\nu)+\frac{3}{2}-\log(-1)]$$

$$+\frac{\Gamma(\gamma+3-i\nu)}{3!\Gamma(\gamma-i\nu)}-\frac{\Gamma(\gamma+2-i\nu)\Gamma(\gamma+1+i\nu)}{|\Gamma(\gamma-i\nu)|^2}$$

$$[-\psi(\gamma+2-i\nu)+\psi(\gamma+1+i\nu)+1+\log(-1)]\Bigg]\Bigg\} \tag{42}$$

In eq. (42) we have only z-integration and κ-summation. All Γ and ψ functions are analytic functions of $i\nu$. So the integrand is analytic except the branch cuts at $|z| \geq 1$. We deform the contour of z-integration as in figure 1. Then

$$\left(\int_{C+} + \int_{C-}\right) \frac{dz}{2\pi i} I(z) = \left(\int_{C_1} + \int_{C_2} + \int_{C_3} + \int_{C_4} + 2\int_I\right) \frac{dz}{2\pi i} I(z) \tag{43}$$

where C_1, C_2, C_3 and C_4 are the segments of the circle with the radius R and I is from $-iR$ to $+iR$. When R goes to infinity the contribution of C_1, C_2, C_3 and C_4 to the integral is zero and I gives a finite contribution.

To do the z-integration and κ-summation we assume $Z\alpha$ to be small. Then we expand all the Γ and ψ functions into the series of $Z\alpha$ or ν and take only the terms up to $Z\alpha$ in the curly bracket of eq. (42). We also approximate γ to $|\kappa|$. Then we get the lowest order contribution to the vacuum polarization of the s-waves. It is

$$\Delta E_n^{V.P.} = 4\alpha(Z\alpha)\left(\frac{z\alpha}{N_n}\right)^3 \sum_{\kappa=1}^{\infty}(-1)^\kappa \kappa \int_{-i\infty}^{-i\infty} \frac{dz}{2\pi i(z^2-1)^{5/2}}$$
$$\Big\{\Big[-\frac{1}{4}(\kappa+1)(3(\kappa+2)-5\kappa)\Big] \tag{44}$$
$$-z^2\left[\frac{1}{6}\left(\kappa(\kappa+1)+\kappa(\kappa+2)+(\kappa+1)(\kappa+2)\right)-\frac{5}{2}\kappa^2\right]+O(Z\alpha)\Big\}$$

The κ-summation and z-integration are done as follows:

$$\sum_{k=1}^{\infty}(-1)^k k(k+1)(k+2) = -\frac{3}{8}$$
$$\sum_{k=1}^{\infty}(-1)^k k^2(k+1) = \sum_{k=1}^{\infty}(-1)^k k^2(k+2) = \sum_{k=1}^{\infty}(-1)^k k^3 = \frac{1}{8} \tag{45}$$

$$\int_{-i\infty}^{i\infty} \frac{dz}{2\pi i(z^2-1)^{5/2}} = \frac{2i}{3\pi}$$
$$\int_{-i\infty}^{i\infty} \frac{dz}{2\pi i}\frac{z^2}{(z^2-1)^{3/2}} = -\frac{i}{3\pi} \tag{46}$$

Then we substitute all these values into eq. (44). It gives

$$\Delta E_n^{V.P.} = -\frac{4\alpha}{3\pi}\cdot\frac{29}{144}\cdot(Z\alpha)\cdot\left(\frac{Z\alpha}{N_n}\right)^3, \text{ for the S}-\text{states.} \tag{47}$$

In the standard QED, the energy shift is calculated as the expectation value of the Uehling potential between the S-states[7] and it is given by

$$\Delta E_n^{V.P.} = \frac{4\alpha}{3\pi}\cdot\frac{29}{145}\cdot(Z\alpha)\cdot\left(\frac{Z\alpha}{n}\right)^3, \text{ for the S}-\text{states.} \tag{48}$$

The difference between these two results is of order $\alpha(Z\alpha)^5$. The result in eq. (47) is obtained from the QED based on self energy without any renormalizations, for the first time. Eq. (48) is obtained by the renormalization process.

ACKNOWLEDGEMENTS

This talk is based on work done in collaboration with Professor A.O. Barut[8].

REFERENCES

1. A.O. Barut and J. Kraus, *Foundations of Physics*, **13**, 189 (1983).

2. A.O. Barut and J.F. Van Heule, *Phys. Rev.*, **A 32**, 3187 (1985).

3. A.O. Barut in "Foundations of Radiation Theory and Quantumelectrodynamics", Plenum Publ. Co., N.Y., (1980), (A.O. Barut ed.).

4. A.O. Barut "Electrodynamics and Classical Theory of Fields and Particles", (1964, second edition, Dover 1980).

5. E. Wichmann and N. Kroll, *Phys. Rev.*, **101**, 83 (1956).

6. I.S. Gradshteyn and I.M. Ryzhik, "Tables of Integrals. Series and Products", (Academic Press. N.Y., 1965).

7. E.A. Uehling, *Phys. Rev.*, **48**, 55 (1935).

8. A.O. Barut and N. Ünal, to be published.

FINITE FORMULATION OF THE SCHWINGER-DYSON EQUATIONS IN Q.E.D.

J. F. Van Huele and M. Berrondo

Department of Physics & Astronomy
Brigham Young University
Provo, UT 84602.

INTRODUCTION

The standard view on Quantum Electrodynamics (Q.E.D.) can be summarized quoting Pais[1]:

"The agreement between experiment and theory" .... "the highest point in precision reached anywhere in the domain of particles and fields, ranks among the highest achievements in twentieth-century physics. Meanwhile the battle with the infinite continues."

Most physicists have learned to live with this ambiguous aspect of Q.E.D. as the wrong theory that gives the right numbers. One can hardly call this situation scientifically sound or aesthetically pleasing. In addition it should be remarked that this "success" of Q.E.D. has led to the development of other gauge field theories and has determined to a large extent how physicists construct particle models today. This is true independently of the fact that no theory to date has come up with numbers that are remotely as good as the ones provided by Q.E.D.

While it is true that agreement between theory and experiment is generally excellent a few comments are in order :

1) Considering the fundamental nature of Q.E.D. as the theory of the interaction between matter and radiation, agreement, no matter how excellent, can only be considered perfect if the difference between experimental and theoretical numbers can be made arbitraririly small (at least in principle). As such and as a laboratory to measure fundamental constants in nature, Q.E.D. is certainly not a closed subject.

2) Such improvement requires a higher precision on the numbers given by the experiment and by the theory. While the experiments have made substantial progress and can be expected to continue to do so, increasing the accuracy of the theoretical numbers has been a slower process[2].

3) Discrepancies between experiment and theory have appeared (and disappeared) in the past. Presently the matter is not unambiguously settled[3,4].

4) The level of precision is such that effects arising from outside what is traditionally considered Q.E.D. become important. As an example, at present the uncertainty on the theoretical value of the Lamb-shift (2s-2p shift in Hydrogen) due to the uncertainty on the proton radius is believed to be larger than the contribution of the next term to be evaluated in the perturbation series. The question then becomes one of how much one should reasonably

New Frontiers in Quantum Electrodynamics and Quantum Optics
Edited by A. O. Barut, Plenum Press, New York, 1990

expect from Q.E.D. and of when one does call an effect a "non-Q.E.D. effect" (e.g. does a discrepancy in the life-time of positronium[5] qualify?).

These considerations by themselves justify a continued study of Q.E.D. Even more importantly the shortcomings of the formal aspects of the theory need to be addressed.

It has been remarked that the procedure of renormalization in Q.E.D consists in subtracting the nonsense from the sense[6]. Renormalizability then appears as the very attractive feature of being able to remove the nonsense in a consistent way. Unfortunately, consistency does not imply limpidity and a lot of effort goes into carefully turning infinite quantities into finite ones (regularization) and assigning the correct experimental value to the physical parameters (renormalization).

It appears that the effort can be better spent in constructing a theory where the inconsistencies do not have to be removed *a posteriori* because they have never been introduced in the theory to begin with.

In this work we present an approach to the quantum theory of electrodynamics where regularization and renormalization have no place. Indeed infinities do not appear at any stage of the calculation. Parameters do not have to be redefined. There is no subtraction mechanism. Renormalization conditions do not have to be imposed.

Of course, besides renormalization there are other problematic aspects in Q.E.D. such as its perturbative character and the lack of solid knowledge concerning the behaviour of the series, the infra-red divergences, as well as the many questions we might hope that Q.E.D. would answer for us whereas it does not. (Q.E.D. does not provide the masses of electromagnetic particles[7–9] or a value[10] for α.)

To tackle the first problem Dyson[11], and Schwinger[12] developed equations for the propagators of Q.E.D. The original program to solve Q.E.D. exactly with the help of the Green's function equations was never achieved however because of a fundamental difficulty: in their original (unrenormalized) form, these equations contain divergences, whereas in their renormalized form, the equations are multiplied by infinite constants. Only when one expands the equations in powers of α does one obtain a reasonable result: precisely the series from perturbation theory!

We give here exact and finite expressions for the Schwinger-Dyson equations. Here again there is no need for regularization or subtractions since all the quantities we work with are finite from the very beginning.

FINITE ELECTRODYNAMICS AND SCHWINGER-DYSON EQUATIONS

We now proceed to construct a version of Quantum Electrodynamics[13] that is free from the ambiguities and infinities from the canonical formalism[14]. To this end we choose to avoid introducing ill-defined quantities such as products of fields evaluated at the same space-time point. In particular, the equal-time commutation relations for interacting Heisenberg fields are ill-defined as is well known[15]. On the other hand, the S-operator and n-point propagators are always well-defined. So are the free fields, their equations of motion, and their equal-time commutation relations.

We also replace the canonical Lagrangian (variational) approach by the constructive approach developed in axiomatic field theories[15–17]. We therefore assume general properties such as causality, locality, uniqueness of the vacuum state and asymptotic completeness, along with the canonical dynamical equations for the free fields.

The dynamical equations for the interacting electron field $\psi(x)$ and photon field $A_\mu(x)$ are given by:

$$(i\gamma \cdot \partial - m)\psi(x) = ef(x) \tag{1}$$

and

$$\partial^2 A_\mu = e j_\mu(x) \tag{2}$$

in the Lorentz gauge, with the source j_μ, and interaction term f given in terms of the S-operator as:

$$e j_\mu(x) = S^{-1}\partial^2 T[a_\mu(x)S] \tag{3}$$

and

$$ef(x) = S^{-1}(i\gamma \cdot \partial - m)T[\phi(x)S], \tag{4}$$

where $a_\mu(x) = A_{in,\ \mu}(x)$ and $\phi(x) = \psi_{in}(x)$ are the asymptotic free photon and electron fields[14,16] respectively, and T is the time-ordering operator. In order to obtain the photon propagator, we first rewrite Eq. (2) in integral form:

$$A_\mu(x) = a_\mu(x) - e\int D(x-x')j_\mu(x')dx' \tag{5}$$

where $D(x-x')$ is the free photon propagator. From the definition of the exact (scalar) photon propagator[18] :

$$\mathcal{D}(x-x') = \frac{i}{3}\langle TA_\mu(x)A^\mu(x')\rangle, \tag{6}$$

and using the stability of the one-photon states[14,16], we obtain[13] the following expression for $\mathcal{D}$:

$$\begin{aligned}\mathcal{D}(x-x') &= D(x-x') + \\ &+ i\frac{e^2}{3}\{\theta(t-t')\int D(x-y)\langle j_\mu(y)j^\mu(y')\rangle D(y'-x')dydy' + \\ &+ (x \longleftrightarrow x')\},\end{aligned} \tag{7}$$

in terms of the current-current correlation function, whose Fourier transform has the form[14] :

$$\int \langle j_\mu(y)j^\mu(y')\rangle e^{ik(y-y')} = 3\theta(k^0)J(k^2). \tag{8}$$

In momentum space, Eq. (8) turns out to be the spectral representation of $\mathcal{D}(k)$, with the *correct* spectral density[13] $J(\lambda)/\lambda^2$:

$$\mathcal{D}(k) = D(k) + \frac{e^2}{2\pi}\int_0^\infty \frac{J(\lambda)}{\lambda^2}\frac{1}{\lambda - k^2 - i\epsilon}d\lambda. \tag{9}$$

A similar expression can be obtained for the electron propagator[13]. Eq. (9) is the normalized form of the photon propagator and requires no renormalization.

The first order term in the S-operator defines the interaction as:

$$S^{(1)} = -ie\int :\bar{\phi}(x)\gamma_\mu\phi(x)a^\mu(x): dx, \tag{10}$$

which, in turn gives the current $j_\mu^{(1)}(x) =: \bar{\phi}(x)\gamma_\mu\phi(x):$ to first order, and the correction to the spectral density $J^{(2)}$ to second order:

$$J^{(2)}(k^2) = -\frac{1}{3}\mathrm{tr}\int \gamma_\mu S^-(p-k)\gamma^\mu S^+(p)\frac{d^4p}{(2\pi)^4}, \qquad \text{for } k^0 > 0. \tag{11}$$

where:

$$S^{\pm}(p) = \pm 2\pi\, i\, \theta(\pm k^0)(\gamma \cdot p + m)\delta(p^2 - m^2). \tag{12}$$

The generalization of this expression for the *irreducible* (scalar) polarization operator $\mathcal{P}$ is obtained by replacing the internal electron lines $S^{\pm}$ by the exact ones $\mathcal{G}^{\pm}$, and one of the elementary vertices γ_μ by the exact normalized vertex Γ_μ. The exact spectral density $Q(k^2)$ in the *irreducible* case is thus given as[19] :

$$Q(k^2) = -\frac{1}{3}\mathrm{tr} \int \gamma_\mu \mathcal{G}^-(p-k)\Gamma^\mu(p-k,p;k)\mathcal{G}^+(p)\frac{d^4p}{(2\pi)^4}, \qquad \text{for } k^0 > 0, \tag{13}$$

where $\mathcal{G}^{\pm}$ are *defined* as[19]:

$$\mathcal{G}^{\pm}(p) \equiv S^{\pm}(p) + S^{\pm}(p)\mathcal{M}(p)\mathcal{G}(p), \tag{14}$$

in terms of the exact electron propagator $\mathcal{G}$, and the irreducible electron mass operator $\mathcal{M}$.

The normalized vertex Γ_μ is obtained as the Fourier transform of the functional derivative[12,20]:

$$e\,\Gamma_\mu(x,y;z) = -\left.\frac{\delta \mathcal{G}^{-1}[x,y|a_\mu]}{\delta a^\mu(z)}\right|_{a_\mu=0}. \tag{15}$$

Finally the exact *irreducible* polarization operator's spectral representation is:

$$\mathcal{P}(k^2) = \frac{e^2}{2\pi}k^4 \int_0^\infty \frac{Q(\lambda)}{\lambda^2}\frac{1}{\lambda - k^2 - i\epsilon}\,d\lambda. \tag{16}$$

and the exact photon propagator[18,19] :

$$\mathcal{D}(k) = D(k) + D(k)\mathcal{P}(k)\mathcal{D}(k). \tag{17}$$

The electron propagator can be obtained in an analogous way[19], thus closing the set of normalized Dyson equations.

This set of new equations differs from the usual Dyson equations[11,12] in the sense that they do not need any renormalization. Indeed all quantities appearing in the equations are finite by construction. The equations obviously do not contain renormalization constants. They constitute the appropriate starting point to find nonperturbative solutions to Q.E.D.

CONCLUSIONS

To conclude we stress that we have achieved two main points :

We have constructed a finite theory of Quantum Electrodynamics which, if solved perturbatively, leads to the same development as standard renormalized perturbative Q.E.D. In the low-energy regime, we do not expect new results, but we may be able to obtain numbers in a more straightforward way since we do not have to deal with the counting and the regularization of the Feynman graphs.

Secondly, we have obtained a closed set of finite equations involving the two-point propagators in Q.E.D.(electron propagator and photon propagator) and the vertex function. These equations are exact, and the hope exists that they can be solved at least in some approximate way. Approximation schemes to solve the unrenormalized Schwinger-Dyson equations for Q.E.D. have been developed in the literature[21–23] and can hopefully be successfully applied or generalized to the normalized equations. In the high-energy domain we can hope that these new, nonperturbative solutions will help us understand some new and intriguing experimental phenomena[24] and theoretical considerations[25–30] that have surfaced in the study of Quantum Electrodynamics.

REFERENCES

1. A. Pais, "Inward Bound," Oxford University Press, New York (1986).
2. R.S. Van Dyck Jr., P.B. Schwinberg, and H.G. Dehmelt, "New High-Precision Comparison of Electron and Positron g Factors", Phys. Rev. Lett., 59:26 (1987).
3. M.A. Samuel, "Is There a Breakdown of Quantum Electrodynamics?", Phys. Rev. Lett., 57:3133 (1986).
4. T. Kinoshita, Comment on Ref.3., Phys. Rev. Lett., 61:2898 (1988); M.A. Samuel, Reply to the Comment, ibid., 61:2899 (1988).
5. C.I.Westbrook, D.W. Gidley, R.S. Conti and A. Rich, "New Precision Measurements of the Orthopositronium Decay Rate : A discrepancy with Theory", Phys. Rev. Lett., 58:1328 (1987)
6. R. Glauber, remarks at this conference.
7. Y. Nambu, "An Empirical Mass Spectrum of Elementary Particles", Prog. Theoret. Phys., 7:595 (1952).
8. G. Rosen, "Radiative Reaction and a Possible Theory for the Muon", Nuovo Cimento, 32:1037 (1964).
9. A.O. Barut, "Lepton Mass Formula", Phys. Rev. Lett., 42:1251 (1979).
10. D.J. Gross, "On the calculation of the fine-structure constant", Phys. Today, 42(12):9 (Dec 1989)
11. F.J. Dyson, "The Radiation Theories of Tomonaga, Schwinger, and Feynman", Phys. Rev., 75:486 (1949); "The S-Matrix in Quantum Electrodynamics", Phys. Rev., 75:1736 (1949).
12. J. Schwinger, "The Theory of Quantized Fields", Proc. Natl. Acad. Sci. (N.Y.), 37:452 (1951).
13. M. Berrondo and R. Jáuregui, "Minimal Theory of Quantum Electrodynamics", Phys. Rev., D33:455 (1986).
14. N.N. Bogoliubov and D.V. Shirkov, "Introduction to the Theory of Quantized Fields", Interscience, New York (1959).
15. P. Roman, "Introduction to Quantum Field Theory", John Wiley & Sons, New York (1969).
16. H. Lehmann, K. Symanzik, and W. Zimmermann, "Zur Formulierung quantisierter Feldtheorien", Nuovo Cimento, 1:205 (1955); "The Formulation of Quantized Field Theories. II", Nuovo Cimento, 6:319 (1957).
17. N.N. Bogoliubov, A.A. Logunov, and I.T. Todorov, "Introduction to Axiomatic Quantum Field Theory", Benjamin, New York (1975).
18. V.B. Berestetskii, E.M. Lifshitz and Pitaevskii, "Quantum Electrodynamics", Pergamon, Oxford (1982), Chaps. 11 and 12.
19. M. Berrondo and J.F. Van Huele, "Normalized Dyson Equations in Quantum Electrodynamics", submitted for publication.
20. C. Itzykson and J.B. Zuber, "Quantum Field Theory", McGraw-Hill, New York (1980).
21. H.S. Green, J.F. Cartier, and A.A. Broyles, "Electron Propagator without Renormalization", Phys. Rev., D18:1102 (1978).
22. J.F. Cartier, A.A. Broyles, R.M. Placido, and H.S. Green, "Finite, Unrenormalized, Nonperturbative Solution to the Schwinger Dyson Equations of QED", Phys. Rev., D30:1742 (1984).
23. P.J. Rembiesa "Consistent Method of truncating the electron self-energy in nonperturbative QED", Phys.Rev., D33:2333 (1986) therein.
24. W. Greiner, Ed., "Physics of Strong Fields", NATO Advanced Study Institute 153B, Plenum, New York (1986) and these proceedings.
25. M. Gell-Mann and F.E. Low, "Quantum Electrodynamics at Small Distances", Phys. Rev., 95:1300 (1954).
26. K. Johnson, M. Baker, and R. Willey, "Self-Energy of the Electron" Phys. Rev., 136:B1111 (1964).

27. K. Johnson and M. Baker, "Some Speculations on High-Energy Quantum Electrodynamics", Phys. Rev., D8:1110 (1973) and references therein.
28. R. Fukuda and T. Kugo, "Schwinger Dyson Equation for Massless Vector Theory and the Absence of a Fermion Pole", Nucl. Phys., B117:250 (1976).
29. K. Kondo, Y. Kikukawa, and H. Mino, "Phase Structure of QED in the Framework of the Schwinger Dyson Equation", Phys. Lett., B220:270 (1989).
30. E.B. Manoukian, "Modification of the "Coulomb" Interaction at Small Distances in Finite Quantum Electrodynamics", Phys. Rev., D25:3420 (1982) and references therein.

VACUUM-CONFINEMENT QED PROCESSES IN THE OPTICAL MICROSCOPIC CAVITY

F. De Martini

Dipartimento di Fisica, Universita' "La Sapienza"
Roma, 00185 Italy

INTRODUCTION

The problem of the interaction of atoms and molecules with the radiation field in its quantum ground-state and in the presence of electromagnetic boundaries has attracted in the past a great deal of attention both on the theoretical (1,2) and experimental sides (3-6). In recent times spontaneous-emission (SpE) enhancement-inhibition processes in macroscopic cavity structures have been investigated in the microwave, infrared and optical zones of the spectrum (3,4,5). In the present work several experimental schemes for QED vacuum-confinement are investigated together with the corresponding resonant processes of molecular or atomic dynamics related to the interaction of atoms with the electromagnetic (e.m.) field, namely SpE and stimulated-emission (StE). Furthermore, we shall investigate experimentally a new process of photon emission in "extreme confinement" condition and when relativistic retardation becomes important. We refer to this process as: "photon-localization" in optical-emission (7). We shall see that in these cases the presence of e.m. boundaries affects in a dramatic and sometimes unexpected way the manifestation of well established fundamental processes that are generally taken as unlikely sources of surprises. For instance, by the combined action of a single-boundary and of a well collimated laser pump beam, the field modal structure of SpE may be drastically modified in a controlled way ("periodic optical pumping") while the temporal structure of the same process, the atomic lifetime T, can be changed by the use of a novel optical device we first introduced: the Fabry-Perot (FP) "Microscopic cavity" (or microcavity). The peculiar resonant quasi-single-mode behaviour of the microcavity is found to have large effects on the "phase-transition" and laser-gain in StE. In this respect the physical interpretation of the "anomalies" in the near-to-boundaries photon-emission processes has led to an interesting debate in recent times. In fact two entirely different physical models, vacuum-field and atom self-reaction, account correctly for the same set of phenomena. For which modern QED expresses a correct formal theory. In this respect we comment on several points that appear significant:

1. The two fields possess quite different locality properties. Furthermore the ubiquitous vacuum-field gives rise to a pre-existing "external" force that is not generated by the charge itself.

2. Two formal devices have been found that somewhat relate the two models: (a) The apparent interchangeability of self-reaction and zero-point field in SpE may be considered to be but one example of a very general fluctuation-dissipation theorem (6). (b) Different operator-ordering procedures adopted within the Heisenberg field-atom coupling theory lead formally to excluding either one of the models.

New Frontiers in Quantum Electrodynamics and Quantum Optics
Edited by A. O. Barut, Plenum Press, New York, 1990

3. In this and similar cases of ambiguity the attribution of a specific model to a physical process is often determined by the historical context and by the scientific "paradigms" existing at the time the effect has been first analyzed, by the authority of the proponent the first interpretation and, possibly, by lack of a timely criticism or discussion. Examples of this are the Casimir-Polder effect (8), The Welton's model for the Lamb shift (9), The Purcell's effect in microwave spectroscopy (1). In these cases the model involving the zero-point field is invoked with no alternatives while the ambiguity involving self-reaction and other models is almost exclusively taken up with similar but more recently discovered quantum-optical effects.

4. The Casimir effect is a fundamental concept of modern topology and quantum-field theory with relevance in a number of areas including the MIT bag model for hadrons in QCD and quantum effects in modern cosmology (10). In the context of elementary particle physics and cosmology it is related to to the confinement of the vacuum field by proper boundaries or in adequate topological structures. However, it has been shown recently in the context of quantum-optics that the long-range Van der Waals attraction forces (sometimes called "Casimir forces") between a single atom and an e.m. boundary may be attributed alternatively to charge self-reaction, to charge-image interaction and to a few other classical models. As a consequence, a number of scientists active in that cultural context are drawn to believe that all possible effects involving the zero-point field can in fact be explained by different models, either classical (11) or quantum-mechanical (5). This seems at least premature. For, while we should generally attribute a great heuristic value to alternative simpler models for any microscopic phenomenon, it is nevertheless not difficult to prove that, at least in this case, that claim presently lacks of an adequate general demonstration. For instance, in the case of the Casimir effect no quantum models alternative to the zero-point picture seem to have been found which explain correctly the attractive force acting between two macroscopic boundaries made either of metal or of multilayer-coated nonconductive plates. In fact, from a theoretical point of view, these objects can hardly be taken as mere collections of single atoms. In addition the theory underlying some of the classical models possesses aspects that appear somewhat misleading on logical grounds. For instance, the "image" concept, which is a shortcut for solving boundary value problems, should not be taken seriously whenever the physical boundaries themselves are essential parts of the problem, as for the Casimir effect.

5. Among other effects in physics and cosmology no models alternative to the one based on the quantum 0-point field have been found to explain the relativistic Unruh radiation process for an accelerated observer (13).

In the present paper we are not going to speculate further on the possible models apt to "explain" the experimental quantum-optical results we intend to present. While we are aware of the existence of alternative pictures, we take here the viewpoint we find more appealing and that is likely the right one: the one which assumes as basic the reality of the vacuum-fluctuations and of their dynamical effects.

SPONTANEOUS EMISSION

1. Controlled change of the SpE modal structure

A sizable electromagnetic-mode confinement is achieved in the space adjacent the boundaries, i.e., conducting or nonconducting, multilayer-coated, plates. This space extends over a distance of the order of the wavelength λ of the radiation involved in the process. According to QED, this perturbation of the modal structure of the field generally acts on

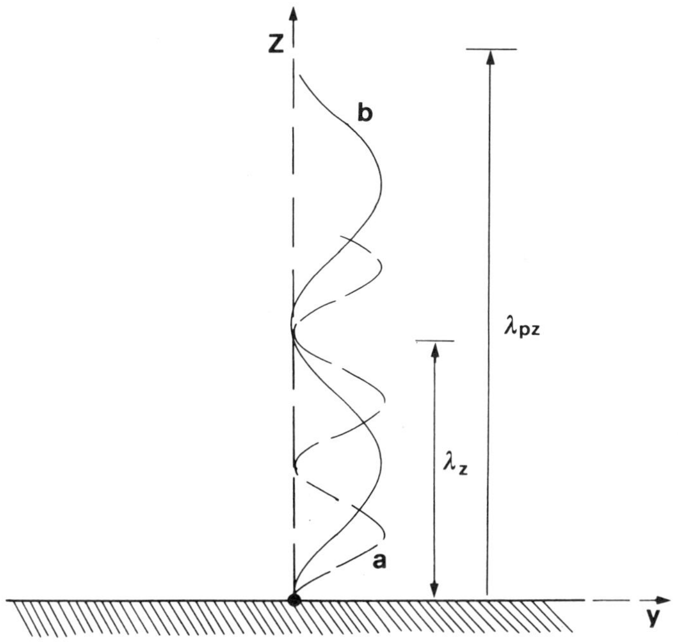

FIG.1-Wiener-fringe pattern due to mirror self-reflection of 0-point field

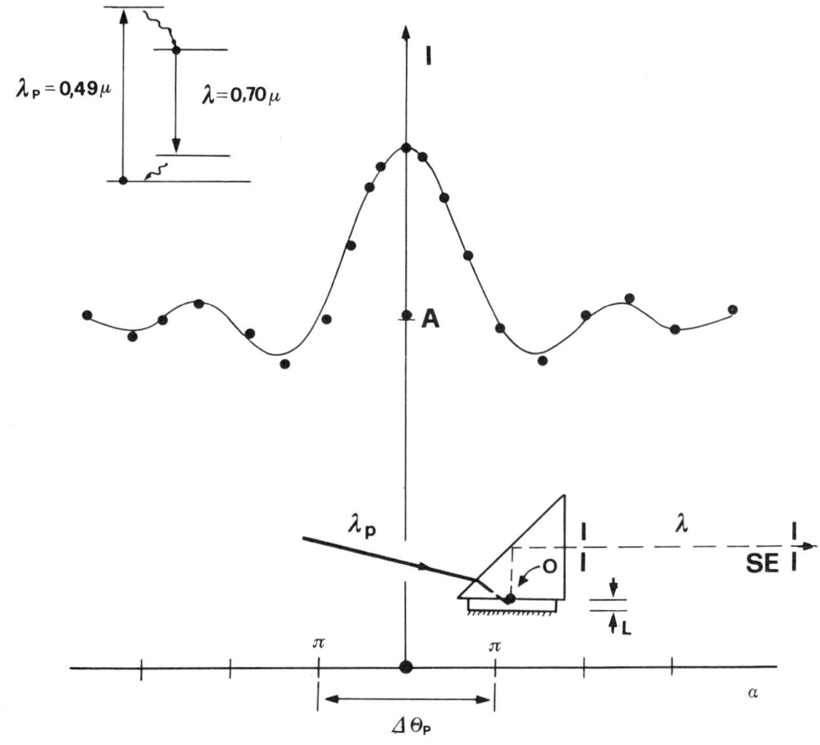

FIG.2-SpE intensity as function of the "Phase-matching" angle $\alpha(\Theta_P)$.

the dynamics of atomic processes as SpE and Lamb's shift. This has been first demonstrated by the Goettingen chemical-physics group (14) by observation of an intensity angular redistribution effect of the SpE-light emitted by Eu-complex-dye monolayers taken at fixed distances $d \approx \lambda$ from the reflecting surface by fatty-acid layers. A different, far easier technique called "periodic optical pumping" has been recently introduced by our laboratory (15). The method is based on a process of "phase-matching" established between a set of standing-wave fringes (Wiener fringes) obtained by mirror self-interference of a collimated laser pump-beam, with wavelength λ_p, and the "vacuum-field self-interference" pattern resulting from the quantum description of SpE (with emission wavelength λ) in front of the mirror surface. Physically, the SpE intensity detected over a particular e.m. h-mode emerging from the mirror is strongly determined by the amount of superposition of the mentioned SpE "quantum-interference" and of the fringing spatial pattern of excited molecules existing in front of the mirror. This last one is in turn proportional to to the e.m. intensity pattern of the pump-beam close to the mirror itself. For instance, if SpE is observed in z-direction orthogonal to the mirror, the maximum SpE intensity is detected if the following "phase-matching" condition is satisfied: $k_z = k_{pz}$, where $k_z = 2\pi n/\lambda$, $k_{pz} = k_p \cos\Theta_p$, $k_p = 2\pi n_p/\lambda_p$ and Θ_p is the angle made by the direction of the incoming pump-beam and the z-axis (Fig.1). The angular light-intensity re-distribution of SpE from a thin slab ($L=8\mu$) of ruby crystal (flatness$\approx\lambda/10$) is reported in Ref.15. As shown by the inset of Fig.2, the slab was glued with a transparent adhesive to a SF10 glass prism held on a mount rotatable around his axis, trace 0. A thick layer of silver was deposited on the other surface of the slab. The SpE observational h-mode has been chosen orthogonal to the mirror (z-axis) and the SpE intensity was detected by a photomultiplier, after suitable collimation and focalization on a 1mm diameter fiber-optic end. The pump was provided by a CW Ar-laser operating at λ_p=4880A. In Fig.2 the SpE intensity at the wavelength λ=6943A is reported as function of the phase-matching angle $\alpha = (k_z - k_{pz})/k_z$ together with the corresponding theoretical curve evaluated by theory (15). In our experiment the combined action of two independent interference effects is able to create a sharply defined radiation pattern which is not determined by simple diffraction. This novel effect allows for the first time to re-direct over a narrow beam the SpE radiation by altering in a controllable way the spatial properties of a spontaneous quantum process. This effect must not be confused with the one investigated by K.Drexhage: this one leads to a much broader angular distribution of the SpE pattern (14). To our knowledge the effect reported in (15) has not previously investigated in connection with noise fields in the context of optics, radio or microwave spectroscopy.

2. Microscopic-Cavity, Optical Casimir Effect

Adding a second plane mirror parallel to the first one at a distance $d \approx \lambda$ along the normal to the surface, i.e. making a FP "microscopic-cavity", leads to more pronounced effects on all physical processes involving atom-radiation coupling as SpE, Lamb's shifts, etc. Owing to multiple field interference, a marked effect of mode-confinement and a strong selection of radiation modes is determined, in a controlled way, through d-variations, with a large perturbation of the mode density along the allowed k-directions (16). Note that, at $d < \bar{d} \equiv (\lambda/2)$ the cavity allows propagation over a single atom-field resonance frequency, thus realizing a novel condition of quantum-statistical-electrodynamics. In the original experiment the piezoelectrically (PZT)-tuned FP mirocavity (Fig.3) consisted of two plane dielectric-coated mirrors (cavity "finesse"$\approx$170 (20,34)). A steady flow of 10^{-4} ethanol solution of tetraphenylnaphtacene dye (free-space $T_o \equiv (1/A) = 13$ nsec.) was kept between mirrors. The optical pumping of the active medium was provided by SHG at $\lambda_p = 0.53\mu$ by an ultrastable self-injected Nd-Yag laser with pulses of 2ns duration (17). During the experiment, the intensity of the pumping beam was taken well below the StE threshold, for $d<\bar{d}$. The pump beam was also well collimated and injected into the cavity through one mirror (A) with a selectable angle, taken respect to the cavity axis, Θ_p in order to take advantage of the already described "periodic optical pumping" technique. By SpE intensity measurement we found that $\Theta_p=48°$ was the angle corresponding to a selective molecular excitation located at $z=\lambda/4$ from mirror (B). This last mirror was 96% reflecting the SpE and the pump wavelenghts (λ,λ_p) while mirror (A) was 98% reflection-coated at λ and anti-reflection-coated at λ_p. The SpE light emitted from the cavity was filtered by an interference filter centered at λ=6328A (10A passband), focussed on the end of an optical fiber and then recorded by a large quantum-efficiency RCA C31034A photomultiplier (2ns risetime). The SpE waveform data were computer processed by interfacing with a LeCroy 8013A fast Waveform Digitizer (1.3 GHz bandwidth). The cavity aligment was maintained by means of two small disk-shaped (PZT) transducers while the cavity d-spacing was controlled by a Micro-Controle positioner (by 1000 A steps) and, on a finer scale, by a large cylindrical (PZT). We found that the cavity alignment, for $d\approx\lambda$ was kept particularly stable by the surface tension of the dye solution. The mirror spacing d could be varied over a distance ranging from about one millimeter down to a very small fraction of λ thus realizing, for $d<\bar{d}$, the remarquable quantum-mechanical effects related to the field-mode elimination and the topological configuration of the Casimir effect (8). The rigorous quantum theory of microcavity SpE has been given recently by F.De Martini, M.Marrocco and R.Loudon (18). There a strong microcavity SpE anysotropy and a striking effect of non-exponential SpE decay are reported. In the present paper, mostly dealing with experiments and physical ideas, we shall limit ourselves to give an approximate quantum theory.

In Heiseinberg's representation, the dynamics of the single atom-field coupling is expressed, as usually, by a Dicke hamiltonian written in terms of the atomic displacement (π^+,π) and field-operators (a_k^+,a_k) belonging to the mode-k (10). By a Markov approximation we obtain the evolution equations for the atom and field belonging to the (single) mode-h which is probed out of the cavity, for instance along the cavity axis:

$$d(\pi^+\pi)/dt \;=-d/dt\left[\sum\nolimits_k(a_k^+a_k)\right] \qquad (\omega\approx\omega_k) \qquad (1a)$$

$$d(a_h^+a_h)/dt=-(g_h\; f_h)\;(\pi\; E^\circ{}_h+h.c.)\;+\;\delta_h(\pi^+\pi) \qquad (1b)$$

Equations similar to (1b) can be written for any cavity k-mode. By solving the set of dynamical equations we get the SpE decay-rate for an atom placed at a distance z from one mirror and with dipole-moment μ orthogonal and parallel to mirrors, respectively.

$\Gamma\;(z)\equiv3A(1-R)^{-2}*\;(I_1-I_2)$; $\qquad\Gamma\;(z)\equiv(3/2)A(1-R)^{-2}\;*\;(I_1+I_2)$ $\qquad$ and:

$I_1(z)\equiv\int G(z,\Theta)d\Theta$ $\qquad$; $\qquad I_2(z)\equiv\int G(z,\Theta)*\cos^2\Theta\;\; d\Theta$ $\qquad$ where:

the integrals are limited between angles 0 and $\pi/2$; A is the free-space SpE rate, Θ the angle made by k with the cavity axis; $E^\circ{}_h\equiv a_h^+(0)*\exp(i\omega_h t)+h.c.$; $f_k\equiv2\sin(kz\cos\Theta)$; $g_h\equiv|\mu|[4\pi^2\omega_k Y/(hV)]^{½}\cos\alpha$; α≡angle between μ and field-polarization; $Y\equiv[(1-R)^2(1+F\sin^2(kd\cos\Theta)]^{-1}$; ω≡atomic resonant-frequency; $F\equiv4R/(1-R)^2$; $R^2\equiv R_1*R_2$, R_i≡mirror reflectivities; $\delta_k\equiv(2g_k f_k)^2[\delta(\omega-\omega_k)+\delta(\omega+\omega_k)]$; $G(z,\Theta)\equiv\sin\Theta\,\sin^2(kz\cos\Theta)/[1+F\sin^2(kd\cos\Theta)]$. Equation (1a), expressing energy-conservation, shows that the time evolution of the atomic energy operator depends on the full set of k-modes available to atoms for spontaneous decay. Then the atomic decay-rate Γ may be substantially affected by the cavity resonance only if all, or a large portion of, k-modes are confined within the cavity itself. Furthermore, by averaging over the vacuum state, eq.(1b) shows that the SpE decay-time of the radiated energy, $\langle0|a_h^+a_h|0\rangle$, detected under pulse-excitation, coincides with the decay-time of the molecular excitation, $T=\Gamma^{-1}$.

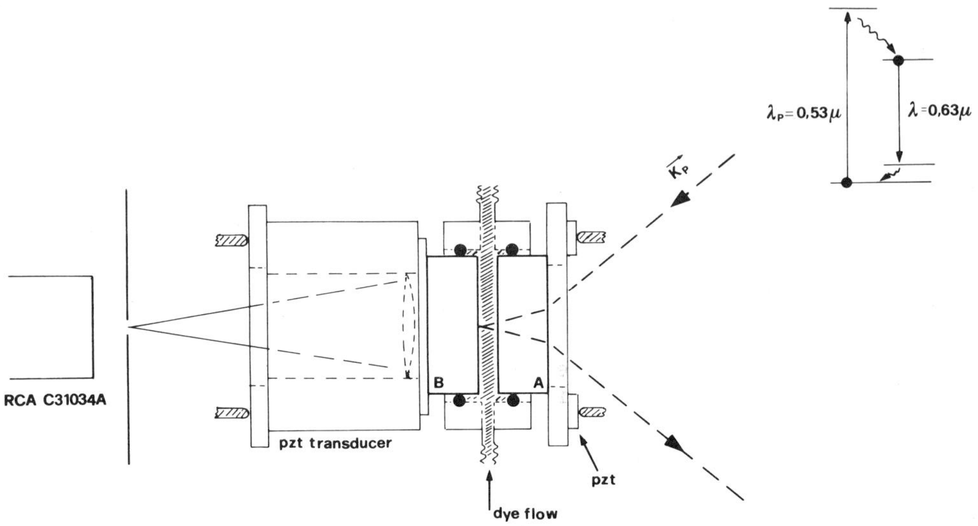

FIG.3-Microscopic-cavity.

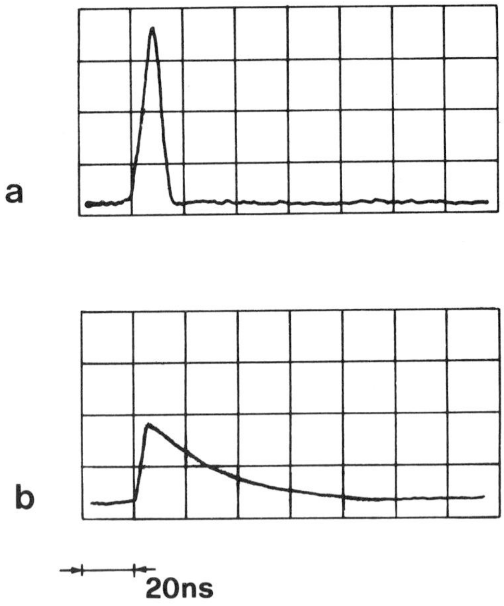

FIG.4-Detector signals showing SpE-enhancement (a) and SpE-inhibition (b).

On the other hand, eq.(1b) shows also that if a CW field-intensity measurement is made on the observational mode-h, the cavity transfer-function (Airy-function) Y plays a large role in determining the resonant behaviour of the detected intensity, even in absence of any sizable resonant change of Γ. Thus realizing, in this last case a mere spatial intensity-redistribution effect. The experimental evidence of the variation of the SpE lifetime T upon cavity tuning is reported in Fig.4 (20). Trace (a) showing SpE-enhancement is obtained for $d=\bar{d}$, while SpE-inhibition shown by (b) corresponds to $d \approx \bar{d}/4$. These data were taken by setting the "periodic optical pumping" angle at $\Theta_p=48°$. A 20% decrease of the effect was found by setting Θ_p close to its extreme values 0°,90°. In summary, we have given a most direct demonstration that the fundamental time of a quantum phenomenon, the SpE time, is altered by a process of non-local space symmetry-breaking, thus realizing a new type of "Casimir-effect".

3. Anomalous spontaneous-stimulated phase-transition

In an active "microcavity" the condition of maximum SpE-enhancement corresponds to resonant coupling of atoms with a single spatial field mode. In the present section we investigate briefly the condition according to which the SpE-enhancement $d \equiv \bar{d}$ is actually nonexistent in a microcavity as SpE is overcome and merges into StE at very low levels of molecular excitation, thus realizing for the first time a virtually "zero-threshold laser" action. This is a direct consequence of the mentioned peculiar single-mode condition in quantum-statistical-electrodynamics.

Assume that, under pulsed- (or CW-) excitation an atom or molecule placed in a microcavity with $d=\bar{d}$ emits a fluorescence photon with a wavelenght λ at time $t=t_o$ with a homogeneously-broadening linewidth $D\lambda=\lambda\, D\nu/\nu$, and suppose further that this is the only photon present in the cavity at that time. Since there is only one spatial mode available, k, any other photon emitted by any atom existing in the cavity within the time interval $\delta t=1/D\nu=t_c$ ("coherence time") is a StE photon, i.e. emitted "in phase" with the first one and then belonging to a photon coherent-state. In other words, owing to the complete elimination of the statistical radiation-mode "reservoir" in a single-mode microcavity, if more than one atom, out of a large number $\bar{n}$ of excited, StE-cooperating ones, undergoes a radiative decay within δt, a collective state is established in the medium owing to the quantum stimulation process or, if appropriate conditions are met, to a "superradiant" process. Of course, the transformation of any SpE decay in a collective StE process, and then in the absence of any fluorescence loss, leads to a large value of the StE gain. So that, in practice, the only way to achieve the SpE process (i.e., photon emission over the vacuum state $|0\rangle$) in presence of any lowest nonzero degree of excitation, is to provide the presence of only one atom in the microcavity i.e., by associating it with an atomic "trap". This behaviour results from the theory reported in Sect.2: Eqs.(1 a,b) solved in closed form lead to the expression of the laser "gain" for $d=\bar{d}$, μ randomly oriented: $G=3\pi^2 k^3 F^{1/2} |\mu|^2 \bar{n}/(h\omega)$. This peculiar effect has been studied in our laboratory with a microcavity similar to the one used for SpE study. The experimental results are reported in (21). This peculiar StE dynamics belongs to the context of second-order phase-transition theory: in our cooperative system the "ordering" process is so largely overwhelming "disorder" (here provided by mirror absorption) that, once one photon is stored in the cavity, any additional SpE process provides the symmetry-breaking field to establish the phase-transition to the state of nonzero average field $\langle E\rangle$, or to the Glauber coherent-state. As a final remark, note that the strikingly anomalous behaviour of atoms when confined in a microcavity (anomalous SpE-time, anomalous StE-phase-transition, anomalous StE-gain) is not obtainable by d-dimension scaling the ususal active macro-cavity effects. As already said (7), it is a fact referable to a quantum-mechanical effect of the same class of non-local space symmetry-breaking particle confinement processes taking place over the scale of the De-Broglie wave-length of the test-particle, λ_D (7). Examples of these processes are: the dynamical-electromagnetic Casimir effect, all kinds of particle-diffraction, the electron-energy quantization in semiconductor multiple-quantum-wells. This kind of behaviour is of course quite general and the substance of the above statements could be extended to include other space dimensions as Compton's and Planck's lenghts. In the latter case relevant "confinement" effects are, for istance the QCD bag model for hadrons (10), the Hawking's theory of black-hole evaporation (22), several boundary problems in non-abelian-gauge theory (23,24) and in the Klein-Kaluza theories of unified interactions (25). As remarked, these processes are sometimes properly grouped under the broad class of the generalized Casimir-effects (8).

TRANSVERSE QUANTUM-CORRELATIONS IN THE ACTIVE MICROCAVITY

In the present Section we take into consideration, we believe for the first time, the process of transverse Bose-Einstein (BE) inter-atom coupling taking place after any photon emission process, SpE or StE, according to the plane-wave QED photon-delocalization model (7). In fact, consider that before the emission of a photon, the corresponding energy-quantum "stored" in the excited atom is localized within a dimension $r<\lambda$. After emission, according to the generally adopted QED model, the photon is fully delocalized over an infinite plane-wave (18,26, 27). In the present experiment, the transversal "extent" (i.e., in the direction orthogonal to the average mode propagation **k**-vector) of the field-mode associated with the emitted photon is investigated via the inter-atom StE BE-correlations in extreme confinement conditions, viz., within space-time distances $d \leq \bar{d} \equiv \lambda/2$. This problem, involving a peculiar transverse relativistic-retardation process, should be taken as the basic one of laser physics.

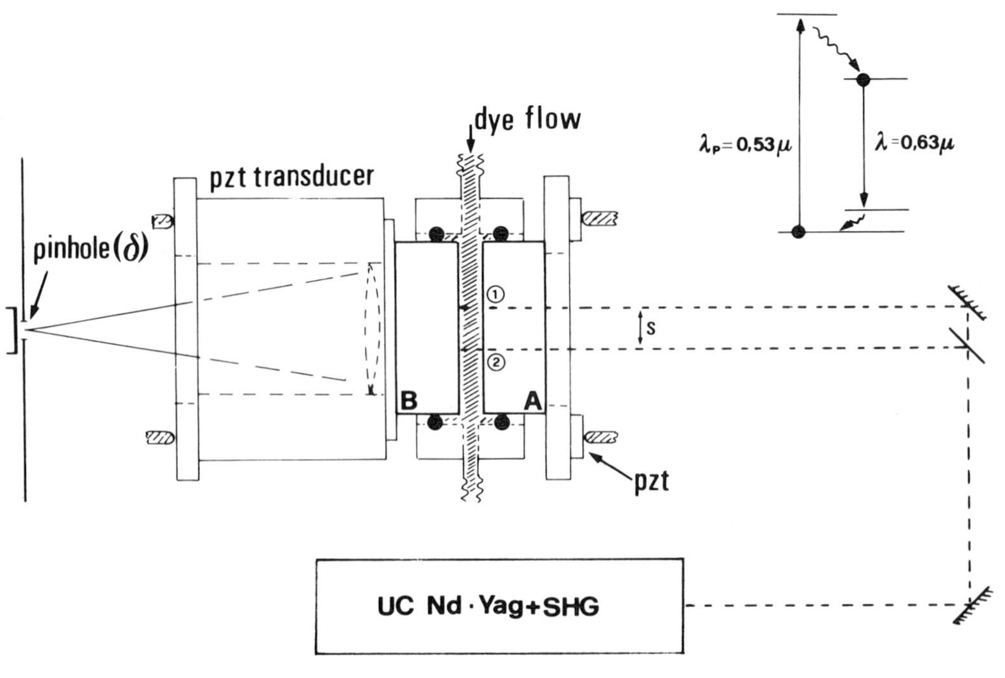

FIG.5-Schematic diagram of the experimental apparatus.

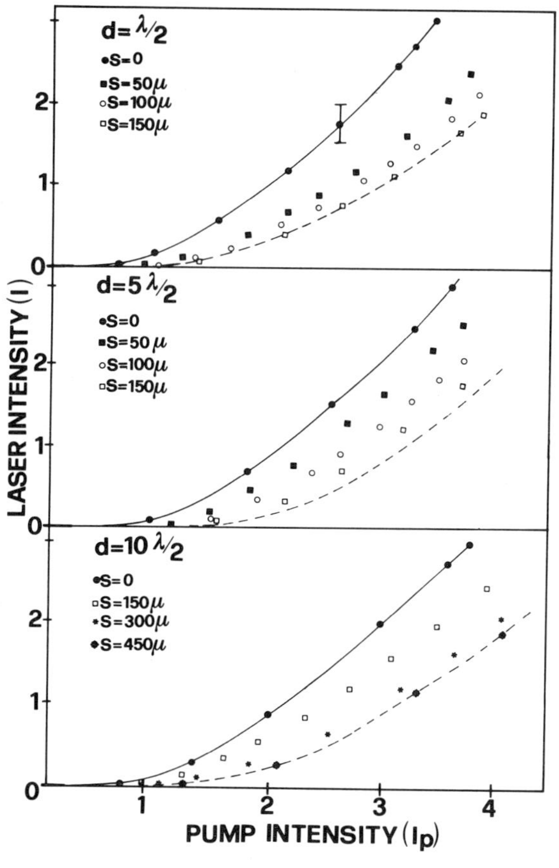

FIG.6-Gain curves showing the loss of StE coupling between two microlasers

Take two equal and dynamically equivalent cylindrical "microlasers" (1)-(2), with diameter δ , placed at an adjustable mutual distance s in the microcavity active-plane, and coupled to the same cavity-allowed mode as for istance to the quasi plane-wave **k**-mode orthogonal to the cavity mirrors (A,B) (Fig.5). To a large approximation this is the only allowed resonant mode for microcavity spacing $d=\bar{d}$ (16,21). In our experiment this condition was obtained by focussing on two $\delta\approx30\mu m$ spots on the active-plane and by a common lens (f.l.=25 cm) two TEM00 single-mode pump beams ($\lambda_P=0.53\mu m$) whose non-collinearity before focusing was micrometrically controlled. The problem tackled in this work may be reformulated by the following question: "To what extent a StE photon emitted over the common **k**-mode by one microlaser determines the "gain" of the other microlaser,in spite of a macroscopic distance s,taken orthogonally to **k** and externally imposed on the two lasers?". According to QED,since a photon delocalization is expected after emission over the transversal extension of the mode, full inter-laser correlation is also expected in spite of the peculiar topoogical configuration of the experiment (7). Let's analyze the problem by considering the "single-mode" case $d=\bar{d}$. Let m_1, m_2 the numbers of photons emitted by microlasers (1-2) respectively within the coherence time $t_c=[\lambda/(\nu\ \Delta\lambda)]$ over the common k-mode ($\Delta\lambda\equiv$band-width of the detected radiation). The time evolution equation for m_1 is given in the form: $dm_1/dt=G(1+m_1+\alpha m_2)$, where the "degree of correlation", $\alpha=\alpha(s,d)$, $0\leq\alpha\leq1$, represents interlaser coupling. The extreme values taken by α in its existence range correspond to full laser independence and to full correlation, respectively. An identical equation holds for m_2 by interchanging indexes. At last, the overall emitted photon-number $m\equiv(m_1+m_2)$, relative to condition $\alpha=1$, is related to $\alpha(s)$ through: :

$$f(s,d)\equiv(m(s,d)/m(0,d))= 2\sinh[G(\alpha+1)]*\exp[G(\alpha-1)/2]/[(\alpha+1)\sinh(2G)] \quad (1)$$

where $G\equiv(g*d*I_P)$ is the low-signal gain proportional to the microlaser pump-intensity, I_P, and to d, in first approximation (26). Note that the overall output "gain" is strongly dependent on α, as it almost <u>doubles</u> in the case of $\alpha=1$. The measure of gain as function of s and d is precisely the method we adopt to investigate the quantum correlation process. Before reporting the experimental results, let us give some details on the apparatus. A negative branch,unstable-cavity,Q-switched neodymium-doped yttrium-aluminum-garnet laser equipped with a second-harmonic-generator, provided TEM00 single-mode pulses at λ_P with τ=10nsec duration to pump the two microlasers upon focussing in the microcavity active-plane (17,31). The mutual polarizations of the pump beams were taken at 90° in order to avoid field interference effects within the pumping process. The pump intensity was kept at a level such that no laser-saturation was detected,as shown by the slope of the "gain" curves, Fig.6 (32). The cavity was similar to the one already reported. Mirror A was transparent to λ_P and reflecting R_1= 99,9% of the microlaser light at λ=6328 A, the detected wavelength. Mirror B was $R_2\approx98\%$ reflecting λ and λ_P. A flow of a 10^{-2} ethanol solution of DCM dye was kept between the mirrors. The cavity zeroth-order **k**-mode, was focussed by a 30cm f.l.lens, with lens aperture =10 mm, onto a pinhole of diameter δ'=10μm. The StE light reaching the phototube (RCA C31034A-02) was filtered by a $\Delta\lambda$=8A interference-filter centered at λ. The signals were processed by a computer-interfaced Le-Croy 9400 digital oscilloscope. The plots of the output-radiation intensity, I ,emitted from the microcavity Vs the pump intensity for $d=\bar{d},5\bar{d},10\bar{d}$ are reported in Fig.6, for various s-values. The I_P scales for the three cases are renormalized to compensate for the effect of different d in determining $G(d*I_P)$. A progressive loss of inter-microlaser correlation for increasing distance s is shown in Fig.6 by the progressive departure of the gain curves from the "full-coupling curves", obtained with s=0, toward the curves of complete decoupling, the dotted ones in Fig 6. These ones are obtained by doubling the values of I, I_P related by the full-coupling curves. Fig.6 shows also that the effect of correlation loss is increasingly less pronounced for increasing $d>\bar{d}$, approaching asymtotically (i.e.,for a macroscopic cavity,$d>>\bar{d}$) the general behavior $\alpha(s)\approx 1$ expected according to standard theory. The measurement of $I(I_P,s)$ leads to the determination of $\alpha(s)$, through Eq.2. The α-curves for $d=\bar{d},5\bar{d},10\bar{d}$ shown in Fig.7 together with related best-fit gaussian plots, reproduce the relevant correlation-increasing behaviour for increasing $d=N\bar{d}$. An approximate, simplified explanation of this behaviour may be given as follows. Consider the microcavity forward-mode,i.e.,with average **k**-vector orthogonal to mirrors,and corresponding to $N\geq 1$. For an active FP-cavity with "finesse", $f=\pi R^{1/2}/(1-R);R^2\equiv R_1 R_2$, this mode may be considered as a superposition of plane-waves with a k-space distribution assigned by the Airy-function $Y\propto[1+(2f\sin\psi/\pi)^2]^{-1}$ and by the expression of the Fabry-Perot (FP) interference phase $\psi=\pi N\cos\Theta$, as function of the emission-angle Θ. The distribution FWHM is found:$\Delta\Theta_N=2(fN)^{-1/2}$ (5,10). The superposition of plane-waves, over each of which the forward-emitted photons are delocalized, determines according to obvious Feynman-path-interference considerations,a limitation of the transversal coherence-length over which BE-ycorrelations (e.g., StE) can effectively take place. By an equivalent, more transparent approach, the extent of the photon-gas Gibbs phase-space corresponding to the cavity forward-mode is expressed, for a rectangular spatial cross-section $\Delta x*\Delta y$, by: $\Delta x*\Delta y*\Delta z*\Delta p_x*\Delta p_y*\Delta p_z=h^3$, where $\Delta z,\Delta p_z$ refer to the time-coordinate (33). By expressing $\Delta p_x*\Delta p_y$ in terms of $\Delta\Theta_N$, we finally obtain for cylindrical symmetry the expression of the: "*Transverse quantum-correlation length*" , $\ell_N=2\lambda(fN)^{1/2}$. Then, two cylindrical microlasers, with diameter δ,sharing a common FP cavity-mode can be coupled by STE if $\ell_N>\delta$. This relation leads to a further interesting insight of the dynamics of the process. In fact, if $\delta<\ell_N$, the microcavity-allowed k-space distribution is *sharper* that the one requested, in non-confinement condition, by Fresnel diffraction from the circular ends of each active microlaser , or: $\Delta\Theta_N<\Delta\Theta_D\approx\lambda/\delta$. In other words, in these conditions, the microcavity *inhibits diffraction*. The back-reaction of the field to this anomalous conditions consists of a kind of a "coupling halo"

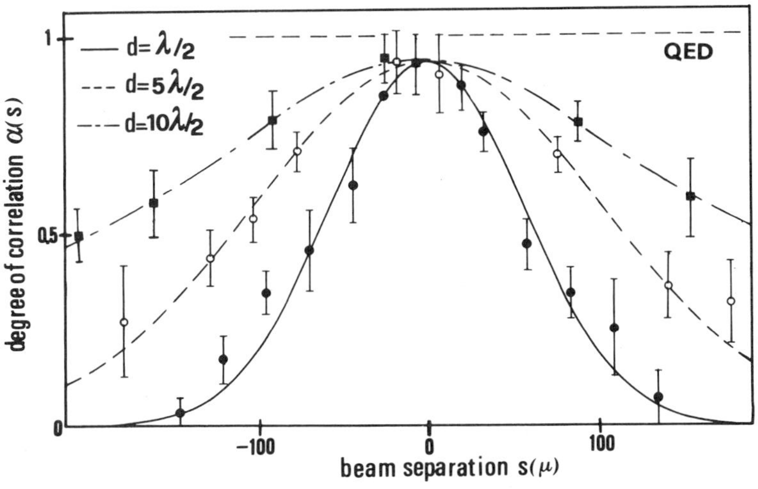

FIG.7-"Degree of Bose-Einstein correlation" in Stimulated-Emission.

i.e., a cylindrical region of thickness ½(ℓ_N-δ) sorrounding the microlasers,in which the quantum-correlations can take place. We refer to this condition as to the: *Microcavity-Regime*. On the other hand,if $\ell_N < \delta$ no StE correlation in transverse-direction is possible outside the active regions: here diffraction inhibits external correlations. This identifies the: *Diffraction-Regime*. Since in this regime no transversal quantum-correlation takes place over radial-distances $\ell > \ell_N$, the maximum number of StE-interacting atoms is $\bar{n}=\frac{1}{4}\pi\eta d\ell_N{}^2=\frac{1}{2}\pi\eta f N^2\lambda^3$, being η the volume-density of excited atoms. The above considerations lead to conclude that in open-space, i.e.,outside a high-Q strongly mode-selective cavity, it is impossible to detect transverse field-delocalization effects using our technique of StE correlation between active spots apart by $\ell > \lambda$: there in fact the diffraction-regime is always realized. In addition to that, and very important, since the "transverse quantum-correlation length" identifies a region of spatial-coherence for the quantum-field, including the vacuum-field, ℓ_N identifies the transverse extent of the modes (and then, of the microcavity mirrors) that are effectively involved in the dynamics of Spontaneous-Emission from one excited atom (15). The microcavityregime is demonstrated experimentally in Fig.7 for increasing $N=2d/\lambda$. The value of f ($f\approx170$) has been determined by measurement of the angular-distribution $\Delta\Theta_N$, according to our theory (34). This one has been further substantiated by the results of a similar experiment involving a long microcavity, $N=10^3$, terminated by simple glass-windows, $f\approx0.3$. In summary, we have investigated by a new technique the rather unexplored condition of transverse-nonlocality of the electromagnetic-field. Consequently our work may stimulate future endeavour of fundamental scientific relevance. For instance, it may suggest the realization of transverse-EPR-type experiments opening then a new field of quantum-mechanical speculation (35).

REFERENCES

1. E.M.Purcell, Phys.Rev. 69, 681, 1946.
2. C.H.Townes and A.Schawlow, "Microwave Spectroscopy", McGraw Hill, N.Y. 1955, p.336; J.R.Ackeralt, P.L.Knight and J.H.Eberly, Phys.Rev.Lett. 30, 456, 1973; J.R.Ackeralt and J.Eberly, Phys.Rev. D10, 3350, 1974; I.R.Senitzky, Phys.Rev.Lett. 31, 955, 1973; P.W.Milonni, J.Ackeralt, W.A.Smith, Phys.Rev.Lett. 31, 958, 1973; D.Kleppner, Phys.Rev.Lett., 47, 233, 1981; J.Dalibard, J.Dupont-Roc and C.Cohen-Tanndouji, J.de Physique, 43, 1617, 1982; P.W.Milonni, Phys.Rev. A25, 1315, 1982.
3. P.Goy, J.Raymond, M.Gross, S.Haroche, Phys.Rev.Lett. 50, 1903, 1983; R.G.Hulet, E.S.Hilfer and D.Kleppner, Phys.Rev.Lett. 55, 2137, 1985; D.Meschede, H.Walther, G.Muller, Phys.Rev.Lett. 54, 551, 1985, reported SpE enhancement-inihibition in microwave transitions between Rydberg levels of alkali atoms; P.Filipovicz, P.Meystre, G.Rempe, H. Walther, Opt.Acta, 32, 1105, 1985; G.Gabrielse, H.Dehmelt, Phys.Rev. Lett. 55, 67, 1985; W.Jhe, A.Anderson, E.A.Hinds, D.Meschede, L.Moi, S.Haroche, Phys.Rev.Lett. 58, 666, 1987 reported similar effects with Cs atoms in the infrared (λ=3.5μ) by CW-intensity detection.
4. D.Heinzen, J.Childs, J.Thomas, M.Feld, Phys.Rev.Lett. 58, 1320, 1987.
5. P.W.Milonni and W.A. Smith, Phys.Rev. A11, 814, 1975.
6. E.T.Jaynes, in "Coherence and Quantum Optics", L.Mandel (ed.) Plenum, N.Y. 1978.
7. F.De Martini, M.Marrocco, D.Murra, "Coherence and Quantum Optics VI", J.H.Eberly, L.Mandel, E.Wolf (eds.), Plenum, N.Y. 1990. "Extreme-confinement" implies field-trapping within dimension $d\approx\lambda_D$, the De-Broglie λ of the corresponding particle. For photons, $\lambda_D=\lambda$.
8. H.Casimir, Proc.Ned.Aka.Wet., Amsterdam 60, 793, 1948; Phys Rev. 73, 360, 1948 (with D.Polder); J.Chim.Phys. 46, 407, 1949; G.Plunien, B. Muller, W. Greiner, "The Casimir Effect", Phys.Rep. 134, 89, 1986.

9. T.A.Welton, Phys.Rev. 74, 1157, 1948.
10.C.Itzykson, J.Zuber, "Quantum Field Theory", Ch.3, MacGraw Hill, N.Y., 1980; E.A.Power "Introductory Quantum Electrodynamics", Elsevier, N.Y., 1965; P.Candelas, Ann.of Phys., 143, Academic-Press, N.Y. 1982.
11.T.H.Boyer, Phys.Rev. A7, 1832, 1973; Phys.Rev. A9, 2078, 1974.
12.H.Morawitz, Phys.Rev. 187, 1792, 1969.
13.W.Unruh, Phys.Rev. D14, 870, 1976; P.C.Davies, J.Phys. A8, 609, 1975; T.H.Boyer, Phys.Rev. D29, 1089, 1984; P.Milonni, private comm.
14.K.H.Drexhage, "Progress in Optics", E.Wolf (ed.), Vol.12, North-Holland Amsterdam 1974.
15.F.De Martini, Phys.Lett. A115, 421, 1986; F.De Martini and G.Innocenti "Quantum Optics IV", J.Harvey, F.Walls, (eds.) Springer-Verlag 1986. In these papers a heuristic interpretation of the "period-optical-pumping" effect based on a classical-stochastic description of the vacuum field is given (11). The same paper reports the first realization of the microcavity with application to a SpE process. Cfr.also: F.De Martini, G.Innocenti, International Quantum Electronics Conference, (I.Q.E.C.), San Francisco, 1986, and Baltimore, 1987.
16.M.Born and E.Wolf, "Principles of Optics", Ch.8, MacMillan, N.Y. 180. In a plane FP-microcavity additional allowed "radial" radiation-modes are present. They correspond to waveguide-type field propagation in directions parallel to mirrors with polarization orthogonal to them. These modes are virtually unaffected by cavity resonance and have been found to be weakly excited in our experimental conditions.
17.C.Brito Cruz, E.Palange, F.De Martini, Opt.Com. 39, 331, 1981; S.Giacomini, G.Innocenti, P.Mataloni, F.De Martini, App.Opt. 26,3179,1987.
18.R.Loudon, "The Quantum Theory of Light", Ch.5, Clarendon Press, Oxford 1983. Equations (1) are found to be independent of the operator-ordering procedure adopted in the Heisenberg'equation. Note that when only one coupling radiation-mode is involed the Heisenberg equations are solvable in closed form. The exact microcavity-SpE quantum-theory including results of a rigorous computer study on vacuum-confinement by a multilayer-dielectric bounded microcavity is due to F.De Martini M.Marrocco and R.Loudon, I.Q.E.C.'90, Anaheim, Cal., 1990.
19.The molecular fluorescence taking place in our work over a broad-band continuum of SpE wavelenghts suggests a further nice comparison with the Casimir effect where an analogous continuum of radiation modes are selected by the cavity. Note also in this connection that SpE-inhibition condition tested experimentally at λ=6328A actually extends over the entire set of SpE wavelenghts $\lambda'>\lambda$, thus giving rise to a metastable state of statistical non-equilibrium in the medium.
20.F.De Martini, G.Innocenti, G.R.Jacobovitz, P.Mataloni, Phys.Rev.Lett. 59, 2955, 1987.
21.F.De Martini and G.R.Jacobovitz, Phys.Rev.Lett. 60, 1711, 1988.
22.S.Hawking, Nature 284, 30, 1974; Cfr W.G.Unruh, Ref. (13).
23.P.C.Davies and S.D.Unwin, Phys.Lett. 98B, 274, 1981; P.C.Davies and D. J.Toms, Phys.Rev.D31, 1336, 1985.
24.C.Quigg, "Gauge theories of the Strong, Weak and E.M. Interactions", Benjamin, London, 1983.
25.T.Applequist and A.Chodos, Phys.Rev.Lett. 50, 141, 1983.
26.M.Sargent, M.O.Scully, "Laser Physics", Addison-Wesley, N.Y. 1974; A. Yariv, "Quantum Electronics", Wiley, N.Y. 1967.
27.W.Heitler, "The quantum Theory of Radiation", Carendon, Oxford, 1960.
28.A detailed mean-field theory based on the solution of Heisenberg equations for a superradiant-medium with rigorous account of diffraction will be reported elsewhere. Cfr: Ref.18; N.Rehler, J.H.Eberly, Phys. Rev.A, 3, 1735, 1971; J.A.Stratton, "Electromagnetic Theory", Mc Graw Hill, N.Y., 1942, p.256.
29.In our experiment the coherence-time of emitted radiation was $t_c \approx 1$psec. The retardation-time among microlasers s=100μ apart, was t_r=0.5 psec.

Since the pump pulse duration $\tau \gg t_r$, trivial retardation in transverse microcavity radial-directions was irrelevant.

30. Note the peculiar field-retardation picture taking place in our experiment. In an idealized situation, a SpE photon emitted by microlaser (1) cannot interact with (2) in a time shorter than t_r as a StE correlation correspons to an exchange of information between the active points (1-2). An information transfer is certainly provided by photons belonging to the microcavity radial-modes. However, StE effects taking place on radial-modes cannot be detected by our lens-pinhole mode-selective detector which is only sensitive to StE on the forward mode. For retardation in atom-radiation processes cfr: E.Fermi, Rev. Mod.Phys. 4, 7, 1932; S.Kikuchi, Zeit.Phys. 66, 8, 1930; M.Fierz, Helv.Phys.Acta, 23, 731, 1950; P.Milonni and P.Knight, Phys.Rev.A,10, 4,1974; V.Bykov, Sov.Phys.Usp. 27,8, 1984.

31. A.E.Siegman, IEEE J.Quantum Electronics , QE12-35, 1976.

32. Any StE-saturation process would tend to cancel the evidence of photon localization by flattening down the $\alpha(s)$ curves of Fig.7.

33. R.Feynman, A.Hibbs, "Quantum Mechanics and Path Integrals", Mc Graw Hill N.Y., 1965, Ch.1; K.Huang, "Statistical Mechanics", Wiley, N.Y., 1963, Ch.9; With usual macroscopic cavities, $N \gg 1$ and $\ell_N \gg \lambda$: e.g., with f=100, d=10cm, λ=0.7μm is $\ell_N \approx$1cm.

34. Over short radial distances $\approx \ell_N$ the f-values, measured by emitted-intensity angular-distribution, can be very high as determined mostly by R. The small f-values reported in works 1,2 were obtained by measuring the d-dependence of microcavity transmission using a large diameter beam (>10mm). There large f-limiting effects due to imperfect mirror-planeity are overwhelming.

35. A.Einstein, B.Podolsky and N.Rosen, Phys.Rev.47,777,1935. The problem investigated in this paper is similar to the space-time transverse-localization argument given by Einstein at the 1927 Solvay Conference and dealing with causality in particle-detection on a diffracted wavefront: cfr. N.Bohr, in P.Shlipp, "Albert Einstein Philosopher Scientist", Cambridge University Press, 1982; F.De Martini, in: *N.Bohr Symposium*, La Rivista di Storia della Scienza, 2, 557, 1985.

STOCHASTIC ELECTRODYNAMICS AND HYDROGEN ATOM*

Armelle Denis**

Département de Physique Théorique
Université de Genève
1211 Genève 4, Switzerland

Stochastic electrodynamics is a classical theory of particles and fields, in fact it is a particular version of classical electrodynamics, the departure with usual classical picture being the introduction of an universal stochastic electromagnetic field (classical "zero point" field) whose spectral density is, namely, Lorentz invariant. This theory has been developed by several authors, see, for instance, in the review "Brief survey of stochastic electrodynamics" by T.M. Boyer [1].

Since the early and extensive works of Marshall during the 1960-1966 years, there has been made efforts to understand connections between stochastic and quantum theories, and extensive calculations have been performed for harmonic oscillator systems (see ref. therein [1]). But, in view to decide if stochastic electrodynamics is a physical theory, some physical phenomena which are well described by quantum theories should be recovered within that classical context of stochastic electrodynamics. One of these physical phenomena is that of the sharp atomic spectra.

The calculation of the atomic spectra in the framework of stochastic electrodynamics has been performed until the end, for the hydrogen atom, from a linear response formalism for stochastic multiperiodic systems [2], but the results are completely incompatible with experiments.

BASIC ELEMENTS OF STOCHASTIC ELECTRODYNAMICS

In this theory, see [1] and [2], one considers a charged particle under the influence of a deterministic force (usually deriving from a potential), and submitted to the Lorentz-Dirac damping force of classical electrodynamics $\vec{F}_d = (2e^2/3c^3)\,\dddot{\vec{r}}$, *moreover* this particle is submitted to a random electromagnetic force $\vec{F}_s = e(\vec{E} + \dot{\vec{r}}/c \wedge \vec{B})$ due to a random electromagnetic field with spectral density $S(\omega) = (h/3\pi c^3) \mid \omega \mid^3$. Thus one are dealing with a kind of Brownian motion. The stochastic differential equation this particle is submitted to a random electromagnetic force $\vec{F}_s = e(\vec{E} + \vec{r}/c \wedge \vec{B})$ due to a random electromagnetic field with spectral density $S(\omega) = (h/3\pi c^3) \mid \omega \mid^3$. Thus one are dealing with a kind of Brownian motion. The stochastic differential equation associated with this motion is the so-called "Brafford-Marshall" equation:

$$m\ddot{\vec{r}} = -\vec{grad}\ V(r) + \vec{F}_d(t) + \vec{F}_s(t)$$

Since the stochastic force has *not* a constant spectrum the phase-space trajectories do not correspond to a Markov process and it is very difficult to solve the problem in general. However, damping and stochastic forces being small with respect to the deterministic force, convenient Markovian approximation may be used and it corresponds to a Fokker-Planck type equation . Moreover, from the Haken's procedure , this equation may be expressed in a reduced space of "relevant" constants of motion corresponding to the unperturbed deterministic motion. For a general multiperiodic system, three actions variables may be taken as "relevant" constants of motion for expressing the reduced Fokker-Planck equation. A direct derivation, of this reduced Fokker-Planck equation, from the equations of the motion, was also performed .

For an isotropic potential V(r), the three action variables

$$\xi_1 = J_r + J_\theta + J_\varphi \ , \quad \xi_2 = J_\theta + J_\varphi \ , \quad \xi_3 = J_\varphi \ ,$$

with $J_k = \frac{1}{2\pi} \oint p_k dq_k$, may be introduced, moreover for a central potential these three preceding variables reduce to the two rotation-invariant variables ξ_1 and ξ_2.

Then, the reduced Fokker-Planck equation for the probability density $\Pi(\xi_1, \xi_2)$ was derived as

$$\xi_2 \frac{\partial \Pi}{\partial t} = \sum_{n=-\infty}^{+\infty} \left(n\frac{\partial}{\partial \xi_1} + \frac{\partial}{\partial \xi_2} \right) \left[f_n \Pi + g_n \left(n\frac{\partial \Pi}{\partial \xi_1} + \frac{\partial \Pi}{\partial \xi_2} \right) \right]$$

where

$$f_n(\xi_1, \xi_2) = \frac{2e^2}{3c^2} \mid c_n(\xi_1, \xi_2) \mid^2 (n\omega_1 + \omega_2)^3$$

$$g_n(\xi_1, \xi_2) = \pi e^2 \xi_2 \mid c_n(\xi_1, \xi_2) \mid^2 S(n\omega_1 + \omega_2)$$

$S(\omega)$ being the spectral density of the stochastic force, c_n being the Fourier components of the orbit in a plane ($x + iy = \sum_{n=-\infty}^{+\infty} c_n(\xi_1, \xi_2) exp\ i(n\omega_1 + \omega_2)$, $z = 0$) and c_n depending on the specific potential considered.

In the hydrogen atom case, a single charged particle (electron) is moving in a central field of force deriving from the Coulomb potential $V(r) = -\frac{e^2}{r}$. In this case

$$\omega_1 = \frac{me^4}{\xi_1^3} \ , \quad \omega_2 = 0$$

and

$$c_n = \frac{\xi_1^2}{me^2 n} \left[J_n' \left(\frac{n\sqrt{\xi_1^2 - \xi_2^2}}{\xi_1} \right) + \frac{\xi_2}{\sqrt{\xi_1^2 - \xi_2^2}} J_n \left(\frac{n\sqrt{\xi_1^2 - \xi_2^2)}}{\xi_1} \right) \right]$$

where $J_n(x)$ denotes the Bessel functions of order n.

Moreover, in the hydrogen atom case, instead of the (ξ_1, ξ_2) variables, it is possible to introduce other variables, for instance the ($L = \xi_1$ Delaunay action, $\eta = \sqrt{1-\epsilon^2}$ (with $\xi_2 = L\eta$, ϵ being the eccentricity of the orbit)) variables or the (E energy, M momentum) variables.

Later, in order to compute the absorption coefficient of the hydrogen atom, whose maxima should be correspond to the line of atomic spectra, it will be necessary to derive, from the preceding reduced Fokker-Planck equation, a stationary probability density Π_{st} corresponding to some stationary stochastic motion.

HYDROGEN ATOM ABSORPTION COEFFICIENT IN STOCHASTIC ELECTRODYNAMICS

In order to write down a formula, for the absorption coefficient of the hydrogen atom [2], it a linear response formalism for stochastic multiply periodic systems (these systems being obtains by letting some stochastic perturbation act on deterministic multiperiodic systems) it was first developed from the linear response theory, in the classical case, due to Kubo.

In this case

a)- the unperturbated system is described by a phase-space stationary probability density $\Pi_{st}(\vec{q}, \vec{p})$,

b)- the external perturbation is corresponding to the interaction $H_1(t) = -\vec{q}e\vec{E}(t)$ (perturbation by an *exterior* electromagnetic field in the dipolar approximation).

Then, the formula for the mean power absorbed by the system, from the perturbating field, in the frequency range $[\omega, \omega + d\omega]$, was obtained as

$$P = \frac{1}{2}\omega \sum_i \sum_j \chi''_{q_i q_j} e^2 \delta_{ij} \frac{8\pi}{3} u(\omega) d\omega$$

where $u(\omega)$ is the spectral energy density, χ'' the imaginary part of the susceptibility tensor and δ_{ij} is corresponding to the isotropy assumption on the non-monochromatic perturbation.

The absorption coefficient $a_u(\omega)$ is now defined from the relation $P = a_u(\omega)u(\omega)d\omega$ and writes

$$a_u(\omega) = \frac{4\pi}{3} e^2 \omega \sum_j \chi''_{q_j q_j}(\omega)$$

which, from

$$\chi''_{q_j q_j} = \frac{1}{2i} \int_{-\infty}^{+\infty} \left[\int q_j^{(\theta)} \frac{\partial \Pi_{st}}{\partial p_j} \, d\vec{q} d\vec{p} \, \right] e^{-i\omega\theta} d\theta$$

writes :

$$a_u(\omega) = \frac{4\pi}{3} e^2 \frac{\omega}{2i} \int_{-\infty}^{+\infty} \left[\int \sum_j q_j^{(\theta)} \frac{\partial \Pi_{st}}{\partial p_j} \, d\vec{q} d\vec{p} \, \right] e^{-i\omega\theta} d\theta \qquad (*)$$

* where $(\vec{q}^{\,(\theta)}, \vec{p}^{\,(\theta)})$ denotes the phase-point at the time θ when starting from the point $(\vec{q}, \vec{p})$ at time 0.

In the case of *multiperiodic systems*, instead of the canonical variables $(\vec{q}, \vec{p})$, the new action and angle variables $(\tilde{J}, \tilde{w})$ may be introduced, and from the fact that each q_k, due to the multiply periodic expansion theorem, may be expressed as a Fourier series in the form

$$q_k(t) = \sum_{\tilde{n}=-\infty}^{+\infty} q_{\tilde{n}}^{(k)}(\tilde{J}) e^{i(\tilde{n}.\tilde{w})}$$

(with $\tilde{J}$ for $(J_1, ..., J_f)$,$\tilde{n}$ for $(n_1, ..., n_f)$ and $\tilde{n}.\tilde{w}$ for $(n_1 w_1 + ... + n_f w_f)$), the expression of the absorption coefficient may be reexpressed, in terms of the $(\tilde{J}, \tilde{w})$ variables and of the Fourier series of $q_j^{(\theta)}$, as

$$a_u(\omega) = -\frac{4\pi^2 e^2}{3} \omega \int \sum_{\tilde{n}} | q_{\tilde{n}} |^2 \tilde{n}.\tilde{\omega} \ \delta(\omega - \tilde{n}.\tilde{\omega}) \sum_k n_k \frac{\partial P}{\partial J_k} d\tilde{J}$$

where $| q_{\tilde{n}} |^2 = \sum_j | q_{\tilde{n}}^j |^2$ and where $P(\tilde{J}) = \Pi_{st}(\tilde{J})(2\pi)^f$ may be considered as the renormalized density probability in the reduced space $\tilde{J} = (J_1, ..., J_f)$.

In view to obtain the absorption coefficient for the *hydrogen atom*, from the preceding more general formula, a sufficient condition must be fulfilled : the exterior electromagnetic field, from which the hydrogen atom absorbs, must be weaker than the random field of stochastic electrodynamics. Then, considering that the energy density per unit frequency of the random electrodynamics field is of order of $30.10^6 J/m^3$ and that of ordinary light sources, as used in spectroscopy, is of order $(\frac{10^{-8}}{3})10^4 J/m^3$ the condition is fulfilled and the preceding formula may be used for the hydrogen atom's absorption coefficient calculation in the framework of stochastic electrodynamics.

In the hydrogen atom case the various quantities, implied in the preceding formula, are

$$\tilde{J} = (\xi_1, \xi_2, \xi_3) \ , \ \omega_1 = \omega_2 = 0 \ , \ \tilde{n}\tilde{\omega} = n\omega_1 \ , \ \omega_1 = \frac{me^4}{\xi_1^3}$$

$$q_{\tilde{n}} = (x_{n,\pm1,\pm1} \ , \ y_{n,\pm1,\pm1} \ , \ z_{n,\pm1,0})$$

and $x_{n,\pm1,\pm1}, y_{n,\pm1,\pm1}, z_{n,\pm1,0}$ being functions of c_n , ξ_2 , ξ_3.

The final expression for the hydrogen atom absorption coefficient, in the case of a non-specified stationary probability density $P(\tilde{J})$, after integration over the ξ_3 variable and variables change (ξ_1, ξ_2) into (L, η) (for details see [2]), takes the form

$$a_u(\omega) = -\frac{4\pi^2}{9} \frac{e^6}{\omega^2} \sum_{n=0}^{\infty} \left[\left(\frac{mne^4}{\omega} \right)^{1/3} n \int_0^1 \eta \left. \frac{\partial P}{\partial L} \right|_{L_0} F_n(\eta) \, d\eta + \int_0^1 \eta \left. \frac{\partial P}{\partial \eta} \right|_{L_0} G_n(\eta) \, d\eta \right]$$

* Errata : in [2] eq. (2.6) $a_u(\omega)=.....$ is written in an erroneous form and must be read as the preceding expression (*) and there are also some other little misprints in this article which have been corrected here.

where

$$L_0 = \left(\frac{mne^4}{\omega}\right)^{1/3}$$

$$F_n(\eta) = 2\left[J_n'^2(n\epsilon) + \frac{\eta^2}{\epsilon^2} J_n^2(n\epsilon)\right]$$

$$G_n(\eta) = 4\left[J_n'(n\epsilon) J_n(n\epsilon)\right] \frac{\eta}{(1-\eta^2)^{1/2}}$$

Now, in order to perform the computer calculation of $a_u(\omega)$, the $P(\eta, L)$ density probability must be specified.

But, exact stationary density corresponding to the reduced Fokker-Planck equation of the hydrogen atom would be very difficult to obtain because the diffusion coefficients are very complicated functions and the detailed balance is not exactly fulfilled. Moreover, it was established that the process is nonrecurrent, and this property excludes the existence of "finite" (i.e. with finite integral) stationary density. *

However, in order to investigate what the absorption coefficient of the hydrogen atom would be in the case of a reasonable stationary density, with a correct value for the energy, a Fokker-Planck equation with diffusion coefficients slightly modified, in such a way as to fulfilled the detailed balance, was considered. Then, the stationary probability density of this modified reduced Fokker-Planck equation takes the form

$$P(\eta, L) = exp - \left(\frac{g(\eta)L}{h}\right)$$

with $g(\eta) = 2\eta$, but this is still a non-integrable density in similarity with the exact reduced Fokker-Planck equation. Therefore, in order to get an integrable density the $g(\eta)$ function was modified as to make it different from zero at $\eta = 0$. Three kinds of modified function were used one of them being

$$g(\eta) = 2\eta \quad if \quad b \leq \eta \leq 1, \qquad g(\eta) = 2b \quad if \quad 0 \leq \eta \leq b \qquad (**)$$

b being a constant determined from the condition that the mean energy must be finite.

The three choices of $g(\eta)$ gave rise to three similar curves concerning the values of the absorption coefficient arising from computer calculations. And, as it is illustrated on figure 1, the absorption curve does not exhibit absorption at the very specific frequencies of the hydrogen atom but, behaves in a continuous form, exhibiting only one very broad maximum which is completely incompatible with experiments.

* Actually, reference (14) therein [2] must be completed by 23, 753 (1982).

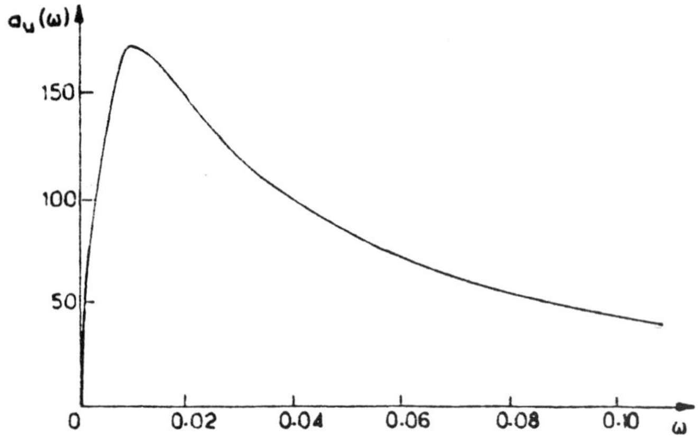

Figure 1. Absorption coefficient $a_u(\omega)$. $a_u(\omega)$ and ω are expressed in atomic units. These values of $a_u(\omega)$ were computed from the function (**) with b=0.630.

REFERENCES

[1] T.M. Boyer , in Foundation of Radiation Theory and Quantum Electrodynamics, ed. A.O. Barut , (Plenum Press, 1980) p.49-63.

[2] A. Denis , L. Pesquera , P. Claverie , Physica 109A, 178 (1981).

NON-LOCAL EFFECTS IN QED

Antony Valentini*

Institute for Theoretical Physics
Technical University Vienna
Karlsplatz 13, A-1040 Vienna, Austria

INTRODUCTION

In this article I shall review some recent work[1] which began with an attempt to clarify the physical meaning of the non-vanishing of the Feynman photon propagator $D_F(x)$ for spacelike separations. The unphysical case of the interaction between two strictly bare atoms, separated by a distance R, is first considered, with attention focused on interaction times $t < R/c$. (For a discussion of the electron-electron case, see Ref.1). It is found that, while the total probability distribution for the bare energy state of each atom is strictly local (in accordance with relativistic causality), the joint probability distribution for the atomic states shows a non-local correlation at $t < R/c$, which was not present in the initial state at t=0. This result would seem to violate any local hidden-variables theory.

The non-local correlations arise not only from the non-vanishing of $D_F(x)$ for spacelike x, but also from interference between the two indistinguishable ways for the pair of atoms to emit a pair of photons. This latter effect has some similarities to the celebrated Brown-Twiss effect[2], and to the non-local and non-classical two-photon interference effect recently observed by Ghosh and Mandel[3].

The bare-state calculation of the joint photon emission interference effect is easily generalised to physical dressed atoms in free space, with the resulting prediction that two such atoms, separated by R and both excited at t=0, should show at $t < R/c$ a non-local correlation in their joint decay probability. For example, the atoms could have a tendency to decay together, even though they are causally disconnected. This result is analogous to the Brown-Twiss effect, where photons tend to arrive at the detectors in pairs. The crucial difference, of course, is that here the photons tend to be emitted in pairs, leaving behind a non-locally correlated pair of source atoms, this correlation developing in a time $t < R/c$. For appropriate single-atom emission spectra, a negatively correlated joint decay is also possible (which is analogous to the Ghosh-Mandel case). As well as

*Present address: International School for Advanced Studies, Strada Costiera 11, I-34014 Trieste, Italy.

New Frontiers in Quantum Electrodynamics and Quantum Optics
Edited by A. O. Barut, Plenum Press, New York, 1990

discussing the effects of atomic recoil, the possibility is considered that only emitted photon pairs which occupy the same "cell in phase space" will contribute to the non-local effect. An experimental test of the predicted non-locally correlated joint spontaneous emission seems feasible.

It is also conceivable that a finite physical analogue of the bare-state case could be constructed by "partially undressing" atoms by confining them in conducting cavities. However, this proposal requires further study.

Finally, we find that the origin of the non-local effects may be understood physically in two complementary ways: (i) In terms of non-local photon propagation, and (ii) In terms of non-locally correlated vacuum-field fluctuations. This latter interpretation is based on a somewhat related paper by Unruh and Wald[4]. Clearly these alternative explanations may be seen as "particle" versus "field" interpretations respectively.

Before discussing the details, some historical background seems appropriate.

BACKGROUND

Ever since the formulation of covariant QED, the Feynman photon propagator $D_F(x)$ has played a central role. While it is well-known[5] that $D_F(x)$ does not vanish for spacelike-separations x in configuration space, the physical significance of this fact has so far not been clarified. In the early days of QED it was argued that the non-locality of $D_F(x)$ would be unobservable, essentially on grounds of the Uncertainty Principle[6]. A related problem was first formulated by Fermi[7] in 1932: Two atoms A and B are separated by a fixed distance R. At t=0, A is in the ground state, B is excited, and no photons are present. One may ask about the probability for atom A to be found excited at later times $t < R/c$. Fermi, using various approximations, calculated the amplitude at $t < R/c$ to find atom A excited, with B in the ground state, and no photons present. He obtained a vanishing result, which has since been re-obtained by several other authors[8], who all use similar approximations to Fermi's.

It was first pointed out by Shirokov[9] that the locality of Fermi's result is an artifact of an approximation in evaluating integrals over photon momentum. This approximation consists of artificially extending frequency integrals over $(0,\infty)$ to $(-\infty,\infty)$, where the integrand contains dominant resonant terms, and then dropping the non-resonant terms from the integrand. However, an exact calculation[9] gives a <u>non-zero</u> result at $t < R/c$. More recently, Rubin[10] has shown that it is the non-resonant part of the amplitude which is non-local. We note that even if one makes the commonly-used rotating-wave approximation, one only obtains a local result if the non-resonant part is also dropped. Inclusion of photon frequencies $\omega \neq \omega_o$ inevitably leads to non-locality. Finally, a recent paper[11] shows that attempts to dismiss the effect by appeal to the Uncertainty Principle, as done in the early days of relativistic QED, are inadequate.

Two points must be stressed before proceeding: (i) All of the above-mentioned treatments are based on calculations using <u>bare</u> states, which are then claimed to apply to the interaction between real physical dressed atoms over a <u>finite</u> time interval. (ii) The non-local amplitude for photon exchange via $D_F(x)$ refers only to events where no photons are emitted into the final state[12]. For these reasons, the physical interpretation and relevance of previous treatments has been at best obscure.

CASE OF TWO BARE TWO-LEVEL ATOMS

Consider a pair of strictly bare atoms A and B, located at $\vec{0}$ and $\vec{R}$ respectively. Assume that, at t=0, A is in the bare ground state $|E_0\rangle$, B is in the bare excited state $|E_1\rangle$, and no photons are present. We will discuss the evolution of the atom-field system for times $t < R/c$. We consider transitions from bare states at t=0 to bare states at $t > 0$. Observation of these transitions formally requires measurement of the bare operators at t=0 and at $t > 0$. For the purposes of clarity, we assume that measurement of the bare (Hermitian) operators is formally possible. Clearly, energy "non-conserving" processes $E_0 \rightleftarrows E_1 + \gamma$ are possible for finite times. However the total energy H is of course strictly conserved at all times. It is the bare energy alone which is not conserved, just as the kinetic energy alone of a classical particle cannot be conserved in a region of non-uniform potential.

We assume that each atom is confined, throughout the time of interest, to a spatial region which is small compared to emitted photon wavelengths, so that the dipole approximation may be applied. This condition also implies an uncertainty in total momentum for each atom which is large compared to emitted photon momenta, which ensures that recoil effects may be ignored (see below). The Hamiltonian is then taken to be

$$H = H_{OA} + H_{OB} + H_{OEM} + e\vec{X}_A\cdot\vec{E}(\vec{0},t) + e\vec{X}_B\cdot\vec{E}(\vec{R},t) \tag{1}$$

where $\vec{X}_i$ is the displacement of electron i from its nucleus. This describes a gauge-independent retarded electric-dipole interaction between the atoms, and is of course extensively used in quantum optics. We say retarded because it generates, in the Heisenberg picture, a causal retarded evolution for the electric field operator $\vec{E}$. Note that (1) includes the contribution from the inter-Coulomb interaction[13].

It is straightforward to show[1], using perturbation theory, that the second-order (in e) transition amplitude $\langle E_1E_00|S^{(2)}(t,0)|E_0E_10\rangle$ for pure photon exchange is

$$8\pi\alpha c l_i l_j \int_0^t dt_1 \int_0^{t_1} dt_2 \cos\omega_o(t_1-t_2)\cdot\frac{\partial^2}{\partial t_1^2}\Delta^{tr}_{+ij}(t_1-t_2,\vec{R})$$

where the relevance of the transverse propagator function

$$\Delta^{tr}_{+ij} = \frac{1}{2(2\pi)^3}\int\frac{d^3\vec{k}}{\omega}\, e^{-i\omega t}e^{i\vec{k}\cdot\vec{x}}(\delta_{ij} - \hat{k}_i\hat{k}_j)$$

is made clear. For $t < R/c$, calculation of the $\vec{k}$-integral gives the result

$$\frac{8\alpha c^2}{\pi R^4}\int_0^t dt_1\int_0^{t_1} dt_2\cdot\frac{\vec{l}^2(1+\lambda^2)-2(\vec{l}\cdot\hat{R})^2}{(1-\lambda^2)^3}\cdot\cos\omega_o(t_1-t_2) \tag{2}$$

Here $\vec{l} = \langle E_1|\vec{X}|E_0\rangle$, which may be assumed real[14], ω_o is the resonant frequency $(E_1-E_0)/\hbar$, and $\lambda = c(t_1-t_2)/R$. Clearly, the amplitude does not vanish for $t < R/c$.

At this point one may be concerned that relativistic causality is being violated. However, it is easy to show that this cannot be the case, simply because of the causal propagation of the electric field $\vec{E}$ in the Heisenberg picture: If one writes down the Heisenberg-picture equations of motion for the operators $\vec{X}_A$, $\vec{P}_A$ of atom A, only the value of $\vec{E}$ at the location of atom A makes an appearance, and this value has a retarded dependence on the operators $\vec{X}_B$, $\vec{P}_B$ of atom B. Thus, for $t < R/c$, the operators of atom A are independent of those of atom B, and so they act only in the Hilbert space of states of atom A. This implies that all expectation values of A operators are independent of atom B for all $t < R/c$, so that, for example, the total inclusive probability distribution $P_A(E_A)$ for the energy state of atom A is precisely local.

Consistency of the locality of $P_A(E_A)$ with the non-locality of (2) requires that there exist further non-local processes, contributing to $P_A(E_A)$, which cancel the non-locality present in (2). To find out what these processes are, it is convenient to consider the limit of very short times $t \longrightarrow 0$. We may then write, where $H_I(t)$ is the interaction Hamiltonian,

$$S(t,0) = \sum_{n=0}^{\infty} (-i/\hbar)^n (t^n/n!) H_I^n(0)$$

which may be regarded as an expansion in powers of et. It is found[1] that the processes of Fig.1 contribute, to order α^2, to the total inclusive excitation probability of atom A. We introduce the quantity

$$\Sigma(\vec{R}) = \frac{1}{V} \sum_{\vec{k}r} \omega(\vec{1}\cdot\vec{\varepsilon}_{\vec{k}r})^2 e^{i\vec{k}\cdot\vec{R}}$$

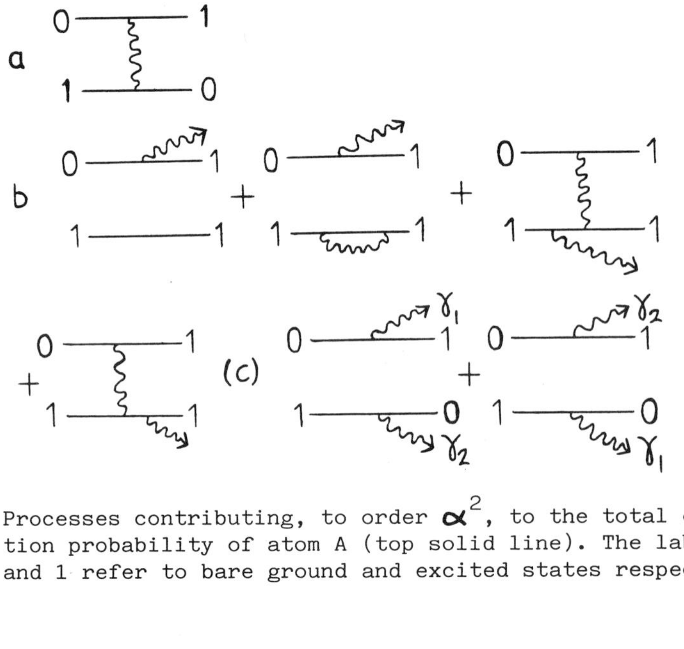

Fig. 1. Processes contributing, to order α^2, to the total excitation probability of atom A (top solid line). The labels 0 and 1 refer to bare ground and excited states respectively.

(where V is an infinite normalisation volume) which may be evaluated to give

$$\Sigma(\vec{R}) = -(2c/\pi^2 R^4)(\vec{I}^2 - 2(\vec{I}\cdot\hat{\vec{R}})^2) \tag{4}$$

for $R \neq 0$ and

$$\Sigma(\vec{0}) = (\Omega^4/12\pi^2 c^3)\vec{I}^2 \tag{5}$$

where Ω is an infinite high-frequency cut-off. The amplitudes shown in Fig.1 (a)-(c) are then respectively found to be (where $|\phi_0\rangle = |E_0 E_1 0\rangle$)

$$\langle E_1 E_0 0|S^{(2)}|\phi_0\rangle = -(2\alpha\pi c/V)t^2\Sigma(\vec{R})$$

$$\langle E_1 E_1 \vec{k}r|S^{(1)} + S^{(3)}|\phi_0\rangle = -(2\alpha\pi c/V)^{1/2} t\sqrt{\omega}(\vec{I}\cdot\vec{\varepsilon}_{\vec{k}r})$$

$$+ (2\alpha\pi c/V)^{3/2} t^3 V\sqrt{\omega}(\vec{I}\cdot\vec{\varepsilon}_{\vec{k}r})\left[\Sigma(\vec{0}) + e^{-i\vec{k}\cdot\vec{R}}\Sigma(\vec{R})\right]$$

$$\langle E_1 E_0 \vec{k}_1 r_1 \vec{k}_2 r_2|S^{(2)}|\phi_0\rangle = (2\alpha\pi c/V)t^2\sqrt{\omega_1\omega_2}(\vec{I}\cdot\vec{\varepsilon}_{\vec{k}_1 r_1})(\vec{I}\cdot\vec{\varepsilon}_{\vec{k}_2 r_2})$$

$$\times(e^{-i\vec{k}_1\cdot\vec{R}} + e^{-i\vec{k}_2\cdot\vec{R}})$$

One then obtains the following probabilities (after summing over emitted photon states)

$$P(E_1 E_0 0) = (2\alpha\pi c)^2 t^4(\Sigma(\vec{R}))^2 \tag{6}$$

$$P(E_1 E_1 1\gamma) = 2\alpha\pi c t^2\Sigma(\vec{0}) - 2(2\alpha\pi c)^2 t^4[(\Sigma(\vec{0}))^2 + (\Sigma(\vec{R}))^2] \tag{7}$$

$$P(E_1 E_0 2\gamma) = (2\alpha\pi c)^2 t^4[(\Sigma(\vec{0}))^2 + (\Sigma(\vec{R}))^2] \tag{8}$$

where the final states contain none, one, and two photons respectively.

Each of (6)-(8) depends on R, despite $t < R/c$. Further, in the context of the bare theory, each is in principle separately observable, by measuring the bare atomic states together with simultaneous measurement of the final photon content. Since measurement of the final photon content requires observation of the whole system, these results do not lead to any superluminal signalling. It may be shown[1] that inclusion of undetected soft photons does not affect the observability of, say, the pure photon exchange probability (6).

The non-locality of (6) arises, of course, from photon exchange via the Feynman propagator (Fig.1(a)). The non-locality of (8) has a different origin: It arises from interference between the two ways that the pair of atoms can emit a pair of photons (Fig.1(c)). The atoms have a relative phase $e^{-i\vec{k}\cdot\vec{R}}$ between their amplitudes for emission of a photon of momentum $\vec{k}$, which gives a factor $\cos(\vec{k}_1 - \vec{k}_2)\cdot\vec{R}$ in the interference term. This factor gives, upon summation over final-photon momenta, an R-dependent result for the interference term, which is just the second term of (8). The non-locality of (7) arises from both of these effects (there is an interference between the amplitudes for the single final-state photon to be emitted from A or from B - see Fig.1(b)).

To obtain the total excitation probability $P_A(E_1)$ of atom A, one must sum over all possible final states of atom B and also over the final photon content of the system. Thus $P_A(E_1)$ is given by the sum of (6)-(8), so that

$$P_A(E_1) = 2\alpha\pi ct^2 \Sigma(\vec{0}) - (2\alpha\pi c)^2 t^4 (\Sigma(\vec{0}))^2 \tag{9}$$

The dependence on R, contained in the factors $\Sigma(\vec{R})$, cancels exactly. Further, calculation of the excitation probability of atom A in the absence of atom B exactly reproduces the result (9) (this probability being non-zero because an isolated bare atom can excite itself by emitting a photon). Thus $P_A(E_1)$ is completely independent of the presence of atom B, and so relativistic causality is strictly preserved. Our earlier argument based on the local evolution of $\vec{E}$ implies that the cancellation of the non-local terms must occur also for $t \not\rightarrow 0$, and to all orders in α.

It may be noted that to work in the dipole-approximation one should, strictly speaking, for consistency, introduce a finite cut-off $\Omega \sim c/\Delta x$, where Δx is the accuracy of spatial localisation of the atoms (and of course $\Delta x \geq$ atomic diameter). This artificial cut-off, which is introduced merely to give a simplified dipole-approximation model, leads to[1] a small acausality of the electric field operator for times within $\sim \Delta x/c$ of the light cone. This should lead to a small acausality in $P_A(E_1)$ near the light cone. (The cancellation of the non-local terms in (9) does actually still occur for finite Ω, but this is because we have considered times $t \rightarrow 0$, far from the light cone). The acausality near the light cone is clearly just an artifact of the simplified dipole-approximation model with finite Ω. This is evident from the precise causal evolution of $\vec{E}$ in the exact theory, which ensures precise causality for $P_A(E_1)$.

Of considerable interest is the joint probability distribution $P_{AB}(E_A E_B)$ of the atomic energy states, where all the final-photon states are summed over. From (6)-(8) we have

$$P_{AB}(E_1 E_0) = (2\alpha\pi c)^2 t^4 [(\Sigma(\vec{0}))^2 + 2(\Sigma(\vec{R}))^2] \tag{10}$$

$$P_{AB}(E_1 E_1) = 2\alpha\pi ct^2 \Sigma(\vec{0}) - 2(2\alpha\pi c)^2 t^4 [(\Sigma(\vec{0}))^2 + (\Sigma(\vec{R}))^2] \tag{11}$$

Further calculation, similar to the above, shows that

$$P_{AB}(E_0 E_0) = P_{AB}(E_1 E_1) \tag{12}$$

$$P_{AB}(E_0 E_1) = 1 - 4\alpha\pi ct^2 \Sigma(\vec{0}) + 2(2\alpha\pi c)^2 t^4 [(3/2)(\Sigma(\vec{0}))^2 + (\Sigma(\vec{R}))^2] \tag{13}$$

The joint-atom distribution (10)-(13) has the following striking properties: The total distribution for any one atom is local (independent of R), while the joint distribution contains non-local R-dependent terms. Clearly, the joint distribution is not equal to the product of the individual distributions, $P_{AB}(E_A E_B) \neq P_A(E_A) P_B(E_B)$, and we have a correlation between the atomic energy states, at $t < R/c$, which was not present at t=0. (At t=0 we do have $P_{AB} = P_A P_B$ and so the states are initially uncorrelated). Since measurement of $P_{AB}(E_A E_B)$ requires observation of both atoms, and since $P_A(E_A)$ and $P_B(E_B)$ are both local, there is clearly no possibility of any superluminal signalling, and relativistic causality is preserved. However, some sort of non-locality is clearly present. We may say that evolution of the bare atom-field system via the Schrödinger equation has generated non-local correlations out of an uncorrelated initial joint-atom state.

PARTIALLY UNDRESSED ATOMS

The above calculations do not prove the existence of non-local correlations for real non-bare atoms. They apply only to the unphysical case of strictly bare atoms. In the next section, the above non-local effect arising from two-photon interference is generalised to the case of physical dressed atoms in free space. However, such atoms are qualitatively different to bare ones, since self-excitation of the ground state by photon emission does not occur. In the present section, we suggest the possibility that a finite physical analogue of the bare case could be constructed by "partially undressing" physical atoms by means of conducting cavities.

It is well-known that an atom inside a conducting cavity interacts with the discrete photon cavity-modes only. The contribution of non-discrete modes to, for example, its self-energy, is "missing", and this missing contribution is finite. (For electrons, the effect has rigorously been shown, in relativistic QED, to give a finite cutoff-independent decrease in the renormalised mass[15]). We could call such an atom "partially undressed".

Consider an atom, confined to a cavity, which is in the partially undressed stationary ground state. If the atom were suddenly released from the cavity, we expect that after a time $\sim|\text{cavity size}|/c$, the atomic state would no longer be stationary, and would begin to evolve in time. We expect that photons would then be created, which is analogous to photon emission from an initially unexcited bare atom. The difference, of course, is that in the bare case the atom interacts with the full free-space photon-mode spectrum, and the emitted photon energy diverges, while for the partially undressed case, where the newly released atom will interact only with the "missing" non-discrete modes, the radiated energy will be finite.

We propose that a pair of partially undressed atoms could be released from a pair of separate cavities (separated by R) and allowed to interact for a time $t < R/c$, after which they are reconfined. The initial and final atom-cavity states could then be measured, giving a close analogue to the bare case (where bare states are measured both initially and finally). We suggest that non-local correlations could be present, just as in the bare case.

A precise analysis of this proposal would be difficult. As a rough model, one could replace the set of free-space photon modes in the Hamiltonian (1) by the set of non-discrete modes only, i.e. one could assume that the discrete modes "dress" the atoms, while their mutual interaction is carried by the non-discrete modes. The results for the case $t \longrightarrow 0$ would then be the same as the bare case, except that the sum over free-space modes in the quantity $\Sigma(\vec{R})$(Eqn.(3)) would be replaced by a sum over non-discrete modes only. Note that such a replacement does not affect the cancellation of the non-local terms in each individual probability distribution.

While a lot of work remains to be done to clarify this proposal, a significant step in this direction has been taken by Rubin[16], who has calculated the probability for non-local photon exchange between essentially a pair of partially undressed atoms as above, and finds a non-vanishing result. Rubin notes that the dressing effects for an atom in a cavity depend on the cavity volume, and therefore they cannot eliminate the non-local photon exchange between two such atoms, since this depends on the larger quantisation volume V of free space.

FREE FULLY-DRESSED ATOMS: NON-LOCALLY CORRELATED JOINT SPONTANEOUS EMISSION

A physical atom, alone in empty space, will eventually settle into its stable fully-dressed ground state. A precise theoretical treatment of such states is lacking. In particular one does not know how to rigorously treat the interaction between a pair of physical dressed atoms over finite times. Use of bare states cannot be correct, of course, since the bare ground state is not stationary (and emits photons with divergent energy). One might then suggest using the rotating-wave approximation (RWA), as is commonly done in practice, which makes the ground state stable. However, (i) The RWA is indeed just an approximation, and (ii) The RWA leads to non-causal propagation for the energy density of the electromagnetic field[17], and thus violates relativistic causality. Finally, there are serious ambiguities even in the definition of dressed excited bound states, which are related to the problem of meaningfully separating an atom's virtual photon cloud (self-field) from "real emitted photons"[18]. For these reasons, we proceed on the basis of general arguments.

We consider a pair of fully-dressed two-level atoms A and B, in otherwise empty space, denoting the dressed ground and excited state of each atom by $|g\rangle$ and $|e\rangle$ respectively (assuming that these can be meaningfully defined). The definition of $|e\rangle$ may have to include details of the experimental excitation process[18]. Consider first the case of atom A in state $|g\rangle$ at $\vec{0}$, with B in state $|e\rangle$ at $\vec{R}$, at t=0. We assume that one can meaningfully say that no "real photons" are present at t=0. Now, if B were absent, the state of A would not change with time. If we assume that, in the presence of atom B, the total excitation probability of atom A is independent of atom B for $t<R/c$, then this probability must vanish for all $t<R/c$. But then, if atom A remains in the ground state for all $t<R/c$, there is no possibility of any non-local correlation between the atoms, for this initial joint-atom state. Further, this tentative argument implies that for fully dressed states, the non-local part of the photon exchange amplitude vanishes. Our treatment is perhaps oversimplified: It could be that a careful discussion, including details of the process of excitation of atom B (in a realistic experimental setting), would alter the picture. For example, it might be that atom A will inevitably have a non-zero total excitation probability (perhaps due to the inevitable presence of "real photons"), which would invalidate the above argument, and thus open up the possibility that non-local correlations, and a non-local photon exchange amplitude, could in fact be present. Being unable to treat this case more precisely, we turn to a different process which we are able to calculate.

Non-locally Correlated Joint Spontaneous Emission

We now consider the case where the atoms A and B are both excited, in the state $|e\rangle$, at t=0 (and no real photons are present). Each atom has some non-zero amplitude to have decayed at $t>0$, and we want to consider the joint probability $P(gg2\gamma)$, at $t<R/c$, that both atoms have decayed to the ground state $|g\rangle$ with emission of a photon. Now, when alone in otherwise empty space, each initially excited atom has some time-dependent amplitude, denoted $f(\vec{k}r,t)$, to have decayed into the ground state with emission of a photon of momentum $\vec{k}$ and polarisation r. If A and B are both present and separated by $\vec{R}$, experience with the bare case suggests that their respective amplitudes for decay with photon emission will be $f(\vec{k}r,t)$ and $f(\vec{k}r,t)e^{-i\vec{k}\cdot\vec{R}}$ (with a relative phase $e^{-i\vec{k}\cdot\vec{R}}$ arising from the spatial dependence of the $\vec{k}$-mode of the electric field operator). With this assumption, we can calculate the probability for joint spontaneous emission, at $t<R/c$, in terms of the experimentally-measureable emission characteristics of a single atom.

The amplitude at time $t<R/c$ to find both atoms in the ground state, with two emitted photons $\vec{k}_1 r_1$ and $\vec{k}_2 r_2$ present, will be

$$f(\vec{k}_1 r_1, t) f(\vec{k}_2 r_2, t)(e^{-i\vec{k}_2\cdot\vec{R}} + e^{-i\vec{k}_1\cdot\vec{R}}) \tag{14}$$

where we have assumed that the two possible ways of emitting a pair of photons from the two atoms are indistinguishable, and add coherently (this assumption is examined below). The total probability at time $t<R/c$ that both atoms are found in the ground state, with two emitted photons, is then

$$P(gg2\gamma) = \int d^3\vec{k}_1 \int d^3\vec{k}_2 \; p(\vec{k}_1,t)p(\vec{k}_2,t)(1 + \cos(\vec{k}_1-\vec{k}_2)\cdot\vec{R}) \tag{15}$$

where $p(\vec{k},t)d^3\vec{k}$ is the experimentally measureable total probability at time t for a single (isolated) initially excited atom to have decayed with emission of a photon into $d^3\vec{k}$. The first term in (15) is just the square of the total single-atom decay probability with photon emission, which is the total result expected for $t<R/c$ in the absence of non-local effects. However, the second term in (15) is non-zero. For example, if $p(\vec{k},t)$ is spatially isotropic, the second term may be rewritten

$$\left| \int_0^\infty d\omega \, I(\omega,t) \, \frac{\sin(\omega R/c)}{(\omega R/c)} \right|^2$$

where $I(\omega,t)d\omega$ is the total single-atom decay probability at time t by photon emission into the frequency range $d\omega$. For a narrow atomic linewidth $\Delta\omega$ such that

$$\Delta\omega << \pi c/2R \tag{16}$$

we have, approximately,

$$P(gg2\gamma) = P_0(t)\left(1 + \frac{\sin^2(\omega_0 R/c)}{(\omega_0 R/c)^2}\right) \tag{17}$$

where ω_0 is the resonant frequency, and $P_0(t)$ is the square of the total single-atom decay probability with photon emission. If instead the photons are emitted into a sufficiently narrow range of angles only, one may obtain[1]

$$P(gg2\gamma) = 2P_0(t) \tag{18}$$

which is the usual doubling of the classically expected probability for two identical bosons to enter the same final state.

According to (15), then, the probability at $t<R/c$ for both atoms to have decayed with photon emission depends on their separation R, and is not equal to the product of their individual decay probabilities. (And of course the bare case shows a similar effect). For our fully-dressed atoms, we assume that a transition $|e\rangle \longrightarrow |g\rangle$ is always accompanied by the emission of a single photon (whose frequency can range over the linewidth $\Delta\omega$). Then $P(gg2\gamma)$ is equal to the total joint decay probability $P_{AB}(gg)$, and we have a non-local correlation between the pair of atomic energy states, just as in the bare case. We note that $P_{AB}(gg)$, which is given by (15), could be observed by measurement of the pair of atomic states only (which may be done, for example, by measuring the ionisation time in a strong external electric field). Measurement of the emitted photons is not necessary.

Locality of each probability distribution. If the total distributions

$P_A(E_A)$, $P_B(E_B)$ for each atom are local, as we showed for the bare case, then there must exist additional non-local processes, to cancel the non-locality present in $P_{AB}(gg)$, as occurs in the bare case. In the absence of a precise theory of dressed-state interactions, we are unable to discuss these other processes in any detail. However, let us assume that the state at $t>0$ is a superposition of just $|ee0\rangle$, $|eg\vec{k}r\rangle$, $|ge\vec{k}r\rangle$ and $|gg\vec{k}_1 r_1 \vec{k}_2 r_2\rangle$, where $P(gg2\gamma)$ depends on R and $P_A(E_A)$, $P_B(E_B)$ do not. It is then easy to deduce that $P(ee0)$, $P(eg1\gamma)$ and $P(ge1\gamma)$ all depend on R also. This seems to be in accord with the analogous bare case $|\phi_o\rangle = |E_1 E_1 0\rangle$, where the survival amplitude contains a non-local term due to the exchange of two photons.

Validity of the factor $e^{-i\vec{k}\cdot\vec{R}}$. Clearly this factor arises from the dipole approximation, and is only valid if the atoms are confined to spatial regions which are small compared to emitted photon wavelengths. In addition, inspection of the bare case shows[1] that use of this simple relative phase factor is correct only if $e^{i(\omega-\omega_o)t} \approx 1$ for all emitted ω, i.e. if $t\Delta\omega \ll 1$, or $t \ll \tau$ where τ is the single-atom lifetime.

Inclusion of atomic recoil. Our dipole-approximation calculations apply when the atoms are fixed, throughout the duration of the experiment, at $\vec{0}$ and $\vec{R}$ to within an uncertainty small compared to emitted photon wavelengths. The uncertainty in total momentum for each atom is then large compared to emitted photon momenta, which ensures that a measurement of the atomic recoils cannot reveal which photon was emitted by which atom. Thus the recoil of the atoms will not destroy the coherence of the two routes for joint photon emission. More formally[1], the centre-of-mass momentum wavefunctions of the atoms are almost unchanged by the photon emission, so that the final wavefunction corresponding to emission of $\vec{k}_1$ is almost identical to that corresponding to emission of $\vec{k}_2$. This implies that the coherence of the alternative joint-emission routes is destroyed by a negligible amount, and we recover the result (15).

Indistinguishability of the Emitted Photons

We have assumed that the alternative routes for joint photon emission are indistinguishable in principle. However, this assumption is called into question if the emitted photons are actually detected. For imagine a photodetector placed very close to each initially excited atom. If, at $t<R/c$, the detector near atom A is triggered, one might argue that this photon cannot possibly have come from the other atom, since it would have had to travel at a speed greater than c.

This problem can be avoided by choosing the single-atom lifetime τ to be large compared to R/c. This ensures that the spread $\Delta|\vec{k}| = \Delta\omega/c$ of emitted photon momenta is small compared to $\hbar/R$. Then, if the momenta $\vec{k}_1, \vec{k}_2$ of the emitted photons are measured to an accuracy better than $\Delta|\vec{k}|$, which would allow the two momenta to be distinguished, then the uncertainty in position of the detected photons would be large compared to the atomic separation R, making it impossible to deduce which photon came from which atom. On the other hand, if a detector were placed close to each atom, in such a way that the uncertainty in each detected photon position were small compared to R, then the assumption of photon paths with speed c would enable one to identify the source of each photon. However, the uncertainty in photon momentum would then be large compared to $\Delta|\vec{k}|$, and one could not distinguish $\vec{k}_1$ from $\vec{k}_2$. Thus for $\tau \gg R/c$, we can safely assume indistinguishability (and thus coherence) for the alternative routes for joint photon emission.

Essentially we are saying that for the case $\tau \gg R/c$, the emitted photons occupy the same "cell in phase space" (i.e. $\Delta x \Delta p \lesssim \hbar$ where $\Delta x \sim R$ and

$\Delta p \sim \hbar \Delta\omega/c$), making them indistinguishable, and therefore they must interfere. As mentioned in the Introduction, our effect is somewhat related to the Brown-Twiss effect[2], and to the recent Ghosh-Mandel experiment[3] which demonstrates interference between the two indistinguishable amplitudes for a pair of photons to travel from two sources to two detectors. It is well-established experimentally that such interference does take place when the detected photons occupy the same cell in phase space[19]. We may reasonably expect the same phenomenon to take place here.

The more general case, where τ is not large compared to R/c, needs further consideration. If, as in (15), all emitted $\vec{k}_1, \vec{k}_2$ are assumed to be indistinguishable (which certainly occurs naturally in the bare-state calculation), then the photon wavepackets must have an overlap in configuration space, as well as in momentum space, at $t < R/c$, despite the fact that emission of the packets by the atoms begins only at t=0. Clearly the configuration space photon wavepackets are propagating at superluminal speed. It is not clear whether this is really a problem or not. Firstly, for the bare case, this phenomenon certainly occurs, together with non-local photon exchange, and yet the atomic states behave locally, and relativistic causality is preserved. Secondly, the notion of a localised photon wavepacket travelling in space at the speed c is a naive one, since it is well-known[20] that no local position operator can be defined for photons. It is rather the electromagnetic field operators, and the electromagnetic energy-density, which propagate in space at the speed of light. We may note also that our bare-state calculation makes use of delocalised photon modes as a basis set. However, since all photon variables are summed over in $P_{AB}(E_A E_B)$, the result should not depend on the basis set used.

It may seem, then, that the result (15) will also be valid for the case where τ is not large compared to R/c. But one may then be concerned that a single atom might be able to excite a distant photodetector at superluminal speed: If, in the two-atom case, a detector near A has an "alternative" amplitude to be excited by atom B, at times $t < R/c$, then perhaps the detector could be excited by B even in the absence of atom A? Clearly, to resolve this issue, one would have to include the photodetectors as part of the system. Such an analysis, for the case of real non-bare states, would be difficult, noting that there is as yet no really precise theory of photodetection.

We suggest that the end result of a rigorous dressed-state analysis might be that the probability for joint spontaneous emission (or for subsequent joint photodetection) is effectively given by coherent summation within each cell in phase space, followed by incoherent summation over the different cell contributions. In that case we would have

$$P(gg2\gamma) = \frac{1}{2} \sum_{\vec{k}_1 r_1 \vec{k}_2 r_2} P^{eff}(gg\vec{k}_1 r_1 \vec{k}_2 r_2) \tag{19}$$

where, approximately,

$$P^{eff}(gg\vec{k}_1 r_1 \vec{k}_2 r_2) = |f(\vec{k}_1 r_1, t) f(\vec{k}_2 r_2, t)(e^{-i\vec{k}_1 \cdot \vec{R}} + e^{-i\vec{k}_2 \cdot \vec{R}})|^2 \tag{20}$$

for $|\vec{k}_1 - \vec{k}_2| \leq \Delta$, while for $|\vec{k}_1 - \vec{k}_2| \geq \Delta$

$$P^{eff}(gg\vec{k}_1 r_1 \vec{k}_2 r_2) = 2|f(\vec{k}_1 r_1, t) f(\vec{k}_2 r_2, t)|^2 \tag{21}$$

where Δ is roughly $\sim \hbar/R$. In (19)-(21), only photon momenta within Δ of eachother are interfering. This ensures that photon paths with superluminal "speed" make no contribution (by which we mean that an approximate simultaneous measurement, at $t<R/c$, of the emitted photons' positions and momenta, would not enable one to rule out, on the assumption of photon propagation at speed c, the coherent contribution of these two photons to the non-locality of $P_{AB}(E_A E_B)$).

For the case $\tau >> R/c$, most of $\vec{k}_1, \vec{k}_2$ satisfy $|\vec{k}_1 - \vec{k}_2| \leq \Delta$, and the prediction (19)-(21) reduces to (15). It may be hoped that a careful experimental test could distinguish which is the correct general result, (15) or (19)-(21). Either way, our prediction for the case $\tau >> R/c$ is the same.

Experimental Test

We require a pair of initially excited atoms, at t=0, whose position wavefunctions are essentially confined to $\vec{0}$ and $\vec{R}$ with an accuracy better than the magnitude of emitted photon wavelengths. If the experiment is performed over a time T, this accuracy of confinement has to be maintained, in order for our dipole-approximation calculations to be valid, and to avoid destruction of the non-local effect by atomic recoil. The spreading of the atomic wavepackets in time T does not lead to any problems if the emitted wavelengths are sufficiently large[1]. In any case it should be possible to avoid this point by means of modern trapping techniques.

To experimentally define the excited state requires a time $\Delta t \gtrsim 1/\omega_o$ (otherwise the energy "uncertainty" is greater than the energy difference between excited and ground states). We must work with timescales large compared to Δt, and so we require

$$T >> \lambda_o/c \tag{22}$$

where λ_o is the resonance wavelength. Since $T<R/c$, we then have

$$\lambda_o << R \tag{23}$$

To maximise the magnitude of the non-local effect, it may be necessary to impose the condition

$$\tau >> R/c \tag{24}$$

on the lifetime τ of the atomic states, as discussed above. We are free to choose R to satisfy (23). For a given λ_o, there is no upper limit in principle for τ, so that (24) can also be satisfied (if necessary), whatever the value of R.

With such a set-up, it should be possible to measure the probability at $t<R/c$ for joint spontaneous decay, either by measurent of the emitted photons, or by direct measurement of the atomic states (say by measuring the ionisation time in a strong external electric field). We remark on the magnitude of the expected non-local correlation: For a spatially isotropic decay, the condition (24) is equivalent to (16), so that , if (24) is necessary, the result (17) will be applicable. But then (23) implies that the second term of (17) will be small compared to the first (with ratio $\sim(\lambda_o/2\pi R)^2$), and the magnitude of the non-local correlation will be small. However, the magnitude can be increased by using a spatially well-directed decay, for which the joint decay may be enhanced by a factor of two[1], as in (18).

PHYSICAL INTERPRETATION

So far we have interpreted the non-local effects in terms of non-local photon propagation. This corresponds to a "particle" viewpoint. However, a "field" interpretation is also possible, this interpretation being suggested by related work due to Unruh and Wald[4]. In this work, Unruh and Wald consider a uniformly accelerated two-level detector (essentially a two-level atom), in the presence of a quantised massless real scalar field. Such a detector may become excited, with a thermal spectrum - the celebrated Unruh effect. Now Unruh and Wald have analysed the appearance of this process from the viewpoint of an inertial observer. If the path of the detector is confined to the right-hand-wedge of the light-cone, they find that the state of the quantum scalar field in the spacelike-separated left-hand wedge shows a correlation with the state of the detector. At the same time, causality is preserved since each total probability is completely local. This result is analogous to the non-local atom-atom correlations considered here. Now Unruh and Wald give an interesting explanation for the origin of the correlation found by them: Consider the free scalar field values $\phi_o(\vec{x}_1,0+)$, $\phi_o(\vec{x}_2,0-)$ at spacelike-separated points. It is well-known[21] that the mean value

$$\langle 0|\phi_o(\vec{x}_1,0+)\phi_o(\vec{x}_2,0-)|0\rangle = i\Delta_F(\vec{x}_1-\vec{x}_2,0) = 1/(4\pi^2|\vec{x}_1-\vec{x}_2|^2) \qquad (25)$$

where Δ_F is the massless scalar Feynman propagator. Now, if the values of ϕ_o at $\vec{x}_1$ and $\vec{x}_2$ were statistically independent, the left-hand-side of (25) would factorise as a product of mean values

$$\langle 0|\phi_o(\vec{x}_1,0+)|0\rangle\langle 0|\phi_o(\vec{x}_2,0-)|0\rangle$$

and this product would be <u>independent</u> of the separation $|\vec{x}_1-\vec{x}_2|$. But this contradicts (25), so clearly the values of ϕ_o at spacelike-separated points are statistically correlated. Thus Unruh and Wald explain their correlation as originating from correlations of the free field in the initial vacuum state at spacelike-separated points.

This leads us to a physically appealing explanation for our non-local effects, as being due to non-local correlations in the vacuum fluctuations of the zero-point electromagnetic field. It is easy to show that the vacuum expectation value

$$\langle 0|E_i^o(\vec{x}_1,0)E_j^o(\vec{x}_2,0)|0\rangle = -(\hbar c/\pi^2\varepsilon_o R^4)(\delta_{ij} - 2\hat{R}_i\hat{R}_j) \qquad (26)$$

where $\vec{E}^o(\vec{x},t)$ is the free operator electric field. Now, at least for the bare case, one may regard all interactions as being due to perturbation of the atomic states by interaction with the free zero-point field (this is clear in the interaction picture). And the electric field operator is a physically observable quantity, which interacts (locally) with the atoms via their dipole moments. It is then clear that the spacelike-separated atoms may "pick up" the initial correlation (26) in the electric field. If one considers "photon exchange", with "emission" at x_1 and "absorption" at x_2, one may regard this as a joint local interaction, at x_1 and x_2, of the pair of atoms with the vacuum fluctuations of the field. Since these fluctuations are non-locally correlated, so are the interactions. This point of view gives a natural explanation for our non-locally correlated joint spontaneous emission: If we regard the emission from each atom as being caused by local vacuum fluctuations, then non-locally correlated emission becomes a rather obvious result, in view of the non-locally correlated character of these fluctuations.

One may wonder to what extent the simple bare vacuum state $|0\rangle$ provides an accurate description of real empty space. Certainly the physical vacuum is more complex than that. Nevertheless the physical picture of vacuum-field fluctuations, based on the free electromagnetic vacuum state $|0\rangle$, is generally accepted by many physicists, and gives an intuitive explanation of phenomena such as spontaneous emission and the Lamb shift. It seems reasonable to suppose, from this success, that the picture of the physical vacuum based on the state $|0\rangle$ does have some bearing on the physical properties of real empty space. It is perhaps more convincing to note that the real physical presence of vacuum fluctuations may in a sense be "derived" by considering the quantum-mechanical limits to measurement of the electromagnetic field, as done by Bohr and Rosenfeld[22]. If one accepts this physical picture, then one must also accept the non-local correlations present in such fluctuations, which follow from the free-field commutation relations.

CONCLUSION

The theoretical evidence presented above seems sufficiently firm to make an experimental test worthwhile. While we were able to give a rigorous treatment for bare atoms, our treatment of real non-bare atoms necessarily relies on (apparently very reasonable) general arguments. In this connection we note that very clean technical evidence for the reality of the effects comes from the above-mentioned paper of Unruh and Wald[4]: Their correlation of the accelerated detector state with the state of the quantum field in the other spacelike-separated wedge of the light-cone is calculated for infinite time intervals, for which the use of bare states as a description of real physical states is rigorously justified by adiabatic switching of the interaction. Their calculation effectively proves that the effects cannot be an unphysical artifact of the finite-time bare-state theory.

If the effects were observed, this would be evidence that the dynamics of the Schrödinger equation for the atom-field system can generate non-local correlations from an initially uncorrelated joint-atom state. This would certainly be a severe challenge for the class of local hidden-variables theories. If the effects were not observed, then apart from the possibility that QED is incorrect, one would have to consider that the standard methods and concepts presently in use for making QED calculations may give rise to spurious non-local effects which are not really present in the theory. If one wished to have some theoretical understanding of QED-processes over finite time intervals, these methods and concepts would then need to be revised. We conclude that an experimental test would be very interesting, whatever the outcome.

REFERENCES

1. A. Valentini, Non-local correlations in quantum electrodynamics; Non-local correlations in quantum electrodynamics. I. Bare states; Non-local correlations in quantum electrodynamics. II. Physical dressed states (to be published)
2. R.H. Brown and R.Q. Twiss, Nature 177:27 (1956)
3. R. Ghosh and L. Mandel, Phys.Rev.Lett. 59:1903 (1987)
4. W.G. Unruh and R.M. Wald, Phys.Rev.D 29:1047 (1984)
5. C. Itzykson and J. Zuber, "Quantum Field Theory," McGraw-Hill, New York (1980)
6. For such arguments in the case of the photon propagator see M. Fierz, Helv.Phys.Acta 23:731 (1950). For a presentation of Fierz's argument in English see G. Källén, "Quantum Electrodynamics," Springer, Berlin (1972) section 23. For criticism of this argument see M.I. Shirokov, Usp.Fiz.Nauk. 124:697 (1978) [Sov.Phys.Usp. 21:345 (1978)] and also Ref.11 below. It was claimed by Pauli that non-local effects from the Feynman electron propagator would be masked by pair creation. See R.P. Feynman, "The Theory of Fundamental Processes,"

Benjamin, New York (1962)
7. E. Fermi, Rev.Mod.Phys. 4:87 (1932)
8. E.A. Power and S. Zienau, Phil.Trans.Roy.Soc.London Ser.A 251:427 (1959); P.W. Milonni and P.L. Knight, Phys.Rev.A 10:1096 (1974);see also the review by Shirokov (Ref.6 above)
9. M.I. Shirokov, Yad.Fiz. 4:1077 (1966)[Sov.J.Nucl.Phys. 4:774 (1967)]
10. M.H. Rubin, Phys.Rev.D 35:3836 (1987)
11. A. Valentini, Phys.Lett.A 135:63 (1989)
12. B. Ferretti, in: "Old and New Problems in Elementary Particles," Academic Press, New York (1968)
13. E.A. Power and T. Thirunamachandran, Phys.Rev.A 28:2649 (1983); 28:2663 (1983); 28:2671 (1983)
14. R. Loudon, "The Quantum Theory of Light," Clarendon, Oxford (1983)
15. K. Svozil, Phys.Rev.Lett. 54:742 (1985); M. Kreuzer and K. Svozil, Phys.Rev.D 34:1429 (1986)
16. M.H. Rubin, The effect of renormalisation on noncausal effects in quantum field theory, to be published
17. G. Compagno, G.M. Palma, R. Passante, and F. Persico, Europhys.Lett. 9:215 (1989)
18. W. Heitler, "The Quantum Theory of Radiation," Dover, New York (1984). See section 16.
19. For an interesting discussion see P.W. Milonni, in: "The Wave-Particle Dualism," S. Diner, D. Fargue, G. Lochak, and F. Selleri, eds., Reidel, Dordrecht (1984)
20. T.D. Newton and E.P. Wigner, Rev.Mod.Phys. 21:400 (1949)
21. See, for example, Ref.5 above.
22. N. Bohr and L. Rosenfeld, Kgl.Dan.Vid.Sel.Mat.Fys.Medd. 12, number 8 (1933) (English translation in R.S. Cohen and J.J. Stachel, "Selected Papers of Léon Rosenfeld," Reidel, Dordrecht (1979)). A discussion is also given in the book by Heitler (Ref.18 above).

THE ROLE OF PLANCK'S CONSTANT IN THE LAMB SHIFT STANDARD FORMULAS

Roger Boudet

Université de Provence
Pl. Hugo, 13331 Marseille , France

INTRODUCTION

The most fundamental question in Quantum Mechanics is the meaning of the Planck constant h. Certainly, the key to the answer depends on clarifying the relations that connect this constant to the euclidean structure of the Minkowski spacetime.

Is it possible to make these relations explicit?

a) In the relativistic Dirac theory of the spin 1/2 particles?
The answer is positive and has been discussed in detail in an other Conference[1].

b) In the quantum theory of fields (QFT) applied to the electromagnetic field?
The answer depends on the preliminary question:

Is REALLY the constant h TO BE OR NOT TO BE associated with the electromagnetic field?

The aim of this lecture is to show that, at least in the Lamb shift calculation, Planck's constant does not appear as an intrinsic property of the field, but is introduced into the field by means of the source - only the source - of the field. Futhermore we will give a very short and simple way to predict the Lamb shift formulas.

1. THE PLANCK CONSTANT AND THE THEORY OF QUANTUM FIELDS

The reduced Planck constant $\hbar = h/2\pi$ appears in QFT by means of the fundamental commutation rules

(1) $$[q_j, p_k] = i\hbar\ \delta_{jk} .$$

But two features lead to think that these rules are insubstantial :

a) An algebraic contradiction with the Schrödinger electron theory.

Though the Schrödinger theory has been a model for the theory of quantum fields, the "number i " which appears to be associated with $\hbar$ in both theories, has the former a geometrical interpretation[2] which cannot be the same as in the latter[1]. But because the electromagnetic field is real (each quantity considered in QFT being corrected by the addition of its complex conjugate) the interpretation given to "i" in this last theory is irrelevant. As well, one can say that the role played by i in QFT is insubstantial.

b) An insubstantial introduction of h in the QFT Lamb shift calculation.
Surely, as i enters only in combination with $\hbar$ in the rules (1), and if the role played by i is

New Frontiers in Quantum Electrodynamics and Quantum Optics
Edited by A. O. Barut, Plenum Press, New York, 1990

insubstantial, one may find, somewhere in the theory, a place where $\hbar$ is introduced in perfect insubstantiality. Indeed such a place exits! And moreover this place is a crucial point of the calculation which is considered as an "outstanding triumph " of QFT: the Lamb shift!

One cand find, in the passage which concerns the operation of renormalization, the following formula (W. Heitler, " The quantum theory of radiation ", third edition, eq. (4') p.341.)

(2)
$$\frac{\Psi_o^*(\vec{r})\,\Psi_n(\vec{r})\,\Psi_n^*(\vec{r'})\,\Psi_o(\vec{r'})}{|\vec{r}-\vec{r}|}$$

$$= \frac{1}{2\pi^2\,\hbar c}\int (\Psi_o^*(\vec{r})\,e^{i(\vec{k}.\vec{r})/\hbar c}\,\Psi_n(\vec{r}))\,(\Psi_n^*(\vec{r'})\,e^{-i(\vec{k}.\vec{r'})/\hbar c}\,\Psi_o(\vec{r'}))\,\frac{\vec{dk}}{k^2}\,.$$

One observes that $\hbar$ appears in the right hand side of the equation . By in what way ? One has used the " Fourier transform " of $1/R$ $(R = |\vec{r}-\vec{r'}|)$:

(3)
$$\frac{1}{R} = \frac{1}{2\pi^2}\int e^{\,i(\vec{k}.\vec{R})}\,\frac{\vec{dk}}{k^2}\,,\quad (k = |\vec{k}|),$$

where $\vec{dk}/k^2 = d\Omega\,dk$, $d\Omega = \sin\theta\,d\theta\,d\varphi$, deduced from the relations

$$\frac{\pi}{2} = \int_0^{\infty}\frac{\sin x}{x}\,dx\,,\quad \frac{\sin(kR)}{kR} = \frac{1}{2}\int_0^{\pi} e^{ikR\cos\theta}\sin\theta\,d\theta = \frac{1}{4\pi}\int e^{i(\vec{k}.\vec{R})}\,d\Omega.$$

Then one has made the change of variable $\vec{k}\rightarrow\vec{k}/\hbar c$, in such a way that the formula looks like a nice decomposition into plane waves of the static Coulomb field, in accordance with the decomposition of the total field into plane waves (" the photons $\vec{k}$"), that the rules (1) allows one to define.

But what has been done in reality ? One has written

$$\frac{1}{R} = \frac{1}{R}\ \text{(exact !)},\qquad \frac{1}{R} = \frac{1}{R}\times\frac{2}{\pi}\times\frac{\pi}{2}\ \text{(right !)}.$$

The Planck constant is here ! I repeat, the Planck constant is here, inside $\pi/2$ (not inside $2/\pi$!) !! Indeed , one has

(4)
$$\frac{\pi}{2} = \int_0^{\infty}\frac{\sin(kR)}{k}\,dk = \int_0^{\infty}\frac{\sin\left(\frac{kR}{\hbar c}\right)}{\frac{k}{\hbar c}}\cdot\frac{dk}{\hbar c}\,.$$

It is not false , but what does it mean ? It means that one has achieved the quantum theory of the number π !!

Without the above introduction of $\hbar$ inside $\pi/2$, there is no QFT operation of renormalization possible , and so no QFT Lamb shift calculation.

But in itself , such an operation makes no sense ! You can put whatever you want instead of $\hbar$ in the relation (4) : the Boltzmann constant , the value of the dollar etc ....

2. THE LAMB SHIFT IN FINITE ELECTRODYNAMICS

Recently, Barut and his collaborators have developed[3,4] a method of calculation of the Lamb shift, quite different from the one used in QFT, in which it appears that the Planck constant is introduced into the field only by means of the source of the field. The Barut Finite Quantum electrodynamics model takes into account both quantum properties of the electron and the unquantified properties of its own electromagnetic field. An advantage of this model is that, at no time, infinite quantities are to be considered, in contrary to what is done in the QFT Lamb shift calculation, Futhermore the QFT standard Lamb shift formulas appear, as a very good approximation of the Barut's ones (see n°5) .

We propose here the analysis of the general structure of the Lamb shift formulas, by using a Finite Electrodynamics model. We define Finite Electrodynamics as a theory in which the electromagnetic field obeys the pure Maxwell laws. Moroever, one supposes that the charge current which is the source of the field is represented by a function whose properties allow the potential of the field to be defined by a convergent integral. The nature, quantum or classical, of the source does not play an essential role in the theory, only the mathematical properties of the current are to be taken into account (if the source is to be associated to a quantum particle this theory becomes the Finite Quantumelectrodynamics Theory).

For example the radiation of Lorentz's electron does not belong to this theory, because the current is a distribution, not a function. But the pure Maxwell field associated with the currents (as defined in First Quantization) of a Schrödinger or Dirac electron bound in an hydrogenlike atom, is in accordance with the assumptions of the theory. In contrast, because in Second Quantization such currents are each considered as the sum of an infinity of plane waves, QFT is outside this theory.

For achieving our model, using only the pure Maxwell theory (in which Planck's constant does not intervene) we will establish a general formula associated with the self-energy of a general time-periodic system of charges S. This formula is convenient whatever the nature, quantum or classical, of the system S may be. For finding the Barut Lamb shift formulas and so the standard formulas, it will be sufficient to replace, just at the last line of the calculation, the charge currents which are the sources of the self field, by the Schrödinger or the Dirac currents associated with an electron bound in an atom. So, it will appear in an indisputable way, that Planck's constant is introduced in the Lamb shift calculation by means not of the field, but of the source - only the source -of the field. Futhermore this process precisely explains how this introduction is done.

A corollary of this result is that, at least for the Lamb shift calculation the presence of $\hbar$ inside the rules (1) is abusive, and so has not to be considered as expressing a law of Nature.

We want to emphasize that our model of construction of the Lamb shift formulas provides a prediction of these formulas. But it is only a prediction not a proof, and it is convenient only for the Lamb shift in free space. A complete proof needs the coupling of the Schrödinger or Dirac equations with the Maxwell ones, as it is done in the works of Barut et al. However, because of its shortness and its mathematical simplicity (the quantum theory of the electron is just introduced by means of the form taken by the current, and just at the end of the calculation) , this model constitutes a highly credible way of verification of the calculation made by Barut and his collaborators.

3 THE INTERACTION ENERGY OF A TIME PERIODIC SYSTEM OF CHARGES WITH ITS OWN FIELD

We consider some general system S of charges whose current is time-periodic :

$$(5) \qquad J_\mu(x^o,\vec{r}) = \cos(\omega x^o)\, \overset{I}{j}_\mu(\vec{r}) + \sin(\omega x^o)\, \overset{II}{j}_\mu(\vec{r}),$$

where $x^o = ct$, and $\partial^\mu J_\mu = 0$. For simplicity we will suppose that, for the calculation of the energy, J_μ is reduced to one term. Let

$$(6) \qquad W_I = -\frac{1}{T}\int_0^T \int J_\mu(x^o,\vec{r})\, A^\mu(x^o,\vec{r})\, dx^o\, d\vec{r} \ , \qquad (T=\pi/\omega)$$

be the time average of the intereaction energy of S with its own field, where A^μ is the potential of the field created by J_μ.

For defining A^μ, one considers the solutions of Maxwell's equations, in the Lorentz gauge and in free space :

$$(7)\quad A^{\mu}_{\pm}(x^o,\vec{r}) = \int \frac{J^{\mu}(x^o \pm |\vec{r}-\vec{r}'|,\vec{r}')}{|\vec{r}-\vec{r}'|}\, d\vec{r}', \quad A^{\mu}_{o} = \frac{1}{2}(A^{\mu}_{-} + A^{\mu}_{+}).$$

Because J_{μ} is time-periodic, one can take A^{μ} equal to A^{μ}_{o} , A^{μ}_{-} or A^{μ}_{+} indifferently for the calculation of W_I (see Ref[5] App.A) .

One obtains immediately

$$(8)\quad W_I = -\frac{1}{2}\int\int \frac{\cos(\omega|\vec{r}-\vec{r}'|)}{|\vec{r}-\vec{r}'|}\, j_{\mu}(\vec{r})\, j^{\mu}(\vec{r}')\, d\vec{r}\, d\vec{r} \quad (j_{\mu} = j^{I}_{\mu} \text{ or } j^{II}_{\mu}).$$

Writing $\vec{R} = \vec{r} - \vec{r}'$, $R = |\vec{R}|$, we achieve the following decomposition

$$(9)\quad -\frac{\cos(\omega R)}{R} = \frac{1-\cos(\omega R)}{R} - \frac{1}{R},$$

from which we deduce the following decomposition of W_I :

$$(10)\quad W_I = W^H_I + W^S_I .$$

W^H_I is the part of the energy which corresponds to the part of the potential giving the long range electromagnetic field. W^S_I corresponds to the static Coulomb energy.

This decomposition of W_I corresponds to a kind of " renormalization ". But all the integrals considered here are convergent, provided that $j_{\mu}(\vec{r})$ is a continuous function which decreases when $|\vec{r}|$ increases, in a way that makes the integral defining A_{μ} convergent .

One has, using the residues theorem (see Réf[5] App. A)

$$(11)\quad 1-\cos(\omega R) = \int_0^{\infty} \frac{\sin(kR)}{\pi k} \cdot \frac{|\omega|}{(|\omega|+k)}\, dk + P\int_0^{\infty} \frac{\sin(kR)}{\pi k} \cdot \frac{|\omega|}{(|\omega|-k)}\, dk ,$$

From $\dfrac{\sin(kR)}{kR} = \dfrac{1}{4\pi}\displaystyle\int e^{i(\vec{k}.\vec{R})}\, d\Omega$, and denoting

$$(12)\quad T^{\mu}(\vec{k}) = \int e^{i(\vec{k}.\vec{R})}\, j^{\mu}(\vec{r})\, d\vec{r},$$ one can write

$$(13)\quad W^H_I = w^H_I(\omega) + w^H_I(-\omega) ,$$ where

$$(14)\quad w^H_I(\omega) = \frac{\omega}{8\pi^2}\int \frac{T_{\mu}(\vec{k})\, T^{\mu}(-\vec{k})}{\omega - k}\, \frac{d\vec{k}}{k^2}$$

is to be considered in principal value if $\omega > 0$.

If the Coulomb gauge is used, one has to write (see Ref.[5] App. C)

$$(15) \qquad w_I^H(\omega) = \frac{\omega}{8\pi^2} \int \frac{\vec{T}^{\perp}(\vec{k}) \cdot \vec{T}^{\perp}(-\vec{k})}{(\omega - k)} \frac{d\vec{k}}{k^2}$$

where $\vec{T}^{\perp}$ is the component of $\vec{T} = (T^1, T^2, T^3)$ orthogonal to the vector $\vec{k}$.

All the integrals considered converge, provided that the integrals

$$(16) \qquad U^{\mu} = \int j^{\mu}(\vec{r})\, d\vec{r} \quad \text{converge}$$

Concerning the complexity of the formalism of the creation and annihilation operators and of the use of Fock spaces , the mathematical tools used here seem very simple. Why is the calculation so simple ? Perhaps because the fundamental laws of Nature are not so complicated.

4. THE SELF-ENERGY OF A SYSTEM OF CHARGES WITH STATIONARY STATES AND TIME PERIODIC TRANSITION CURRENTS

We considerer a general system S of charges, which can have stationary states, in which the charge current is time-independent, each state n being associated with a constant energy level E_n. We suppose furthermore that, during the passage from a state n to another state p, a time periodic charge current $J^{\mu}_{np} = J^{\mu}_{pn}$ whose frequency $\omega_{np} = -\omega_{pn}$ is well defined, is created.

Now, we consider the self-energy W(n,p) of S, corresponding to a transition (n,p). We are more precisely interested in the part $W^H(n,p)$ of this energy which corresponds to the long range part of the electromagnetic field, created by the current J^{μ}_{np}.

Because the constant energy level E_n is a time-averaged energy, it is interesting to consider $W^H(n,p)$ as being time-averaged, and then $W^H(n,p)$ is both independent:
a) of the duration of the transition (because the current is of a time periodic type)
b) of the effective number of transitions (n,p),

We will define $W^H(n,p)$ as

$$(17) \qquad W^H(n,p) = \frac{1}{2} W_I^H(n,p).$$

$W_I^H(n,p)$ is defined as in (8) and (10) , J^{μ} and ω being replaced by J^{μ}_{np} and ω_{np}.

A justification of the above factor 1/2 is given, when S is an electron, in Ref[6] and [4] (see also Ref [7] App. A). We can notice that, at least if the Coulomb gauge is used, $W^H(n,p)$ corresponds to the time average of the energy due to the self Lorentz force of the system S, during the transition (n,p), and thus may be considered as a contribution to the modification of the levels of energy E_n and E_p.

Then we decompose $W^H(n,p)$ into a sum of two terms , w^H_{np} and w^H_{pn} that we associate with the state n and the state p respectively, by using (13):

$$(18) \qquad W^H(n,p) = w^H(\omega_{np}) + w^H(\omega_{pn}), \qquad w^H_{np} = w^H(\omega_{np}).$$

The relation (14) shows that, if $\omega_{n,p}$ is related to the energy levels as

$$(19) \qquad \omega_{np} = \frac{E_n - E_p}{c\,\tau_{np}} ,$$

where $\tau_{np} = \tau_{pn}$ is some positive constant having the dimension of an action, the integral associated with the upper level is the one which has to be considered in principal value.

There is a profound reason for this association, related to the relative stability of the levels E_n and E , which appears in the Barut method [3,4] of calculation of the Lamb shift. But this reason in our opinion goes beyong the particular case of the electron and concerns all the systems S defined as above, if a long lifetime is to be associated to each transition.

Finally , we take the sum W_n^H of the self-energies w_{np}^H for all the transitions $p \rightarrow n$

$$(20) \quad W_n^H = \sum_{p\neq n} w_{np}^H = \sum_{p\neq n} \frac{\omega_{np}}{16\,\pi^2} \int \frac{T^{\mu}_{np}(\vec{k})\, T_{np\mu}(-\vec{k})}{(\omega_{np} - k)} \cdot \frac{\vec{dk}}{k^2} .$$

A point must be emphasized : what we do is not at all an energy balance depending on some conservation law. Rather, what we are doing is a simple inventory of the time-averaged self-energies of the system S, and a classification, state by state, of these energies.

If one uses the Coulomb gauge and the relation (19), and the total charge of the system S is q, then setting

$$(21) \quad \vec{j}_{np}(\vec{r}) = q\,\vec{j'}_{np}(\vec{r}) , \quad \frac{\vec{u}_{np}}{c} = \int \vec{j'}_{np}(\vec{r})\, \vec{dr} ,$$

eq. (20) becomes

$$(22) \quad W_n^H = \sum_{p\neq n} \frac{1}{16\,\pi^2} \cdot \frac{q^2}{c\tau_{np}} (E_n - E_p) \int \frac{\vec{T'}^{\perp}_{np}(\vec{k}) . \vec{T'}^{\perp}_{np}(-\vec{k})}{E_n - E_p - c\tau_{np} k}\, c\tau_{np} \frac{\vec{dk}}{k^2} .$$

If one makes the dipole approximation, i.e. if one replaces in (22) , exp $(i(\vec{k},\vec{r}))$ by unity, (12) becomes (16), and we have, in spherical coordinates θ, φ, of the direction of the vector $\vec{u}_{np}$,

$$\vec{T'}^{\perp}_{np}(\vec{k}) = \vec{T'}^{\perp}_{np}(-k) = \sin^2\theta \; \vec{u}^{\,2}_{np}/c^2 ,$$

because the angle of the vector $\vec{T'}^{\perp}_{np}(\vec{k}) = \vec{T'}_{np}(-k) = \vec{u}^{\perp}_{np}$ with this direction is equal to $\theta - \pi/2$.

So, after integration over θ and φ , a factor $\int_0^{2\pi} \int_0^{\pi} (\sin^2\theta) \sin\theta \, d\varphi \, d\theta = 8\pi/3$ is to be introduced.

Futhermore, the integrals over k diverge for the upper limit of integration. After having made the change of variable $K = c\,\tau_{np}\,k$, and chosen some " cut-off" K_{np}

for each integral, one obtains the relation

$$(23)\quad W_n^H = \sum_{p\neq n} \frac{2}{3\pi} \cdot \frac{q^2}{c\tau_{np}} (E_n - E_p) \frac{\vec{u}_{np}^{\,2}}{4c^2} \int_0^{K_{np}} \frac{dK}{E_n - E_p - K},$$

that we will call the generalized Bethe formula.

5. THE INTRODUCTION OF PLANCK'S CONSTANT: THE LAMB SHIFT

Nothing has been said up to now about the nature of the charge currents considered. At no time the Planck constant has been used. Nervertheless (what a coincidence !) the relation (23) is exactly the Bethe formula[8] giving the Lamb shift in the nonrelativistic approximation if q is taken to be equal to the charge e of the electron; if one writes

$$(24)\qquad \tau_{np} = \hbar ,$$

(so that $q^2/c\tau_{np} = \alpha$ the fine structure constant) and

$$(25)\qquad \vec{u}_{np} = 2\vec{v}_{np} , \quad |\vec{v}_{np}| = \left| \int \psi_n^*(\vec{r}) \frac{\hbar}{im} \nabla\psi_p(\vec{r})\, d\vec{r} \right| ,$$

where the $\psi_p(\vec{r})$ are the solutions of the Schrödinger equation for an electron bound in an hydrogen-like atom; and if each K_{np} is a convenient " cut-off ". The charge current associated with the transition $p \rightarrow n$ is (if $\psi_n , \psi_p \in \mathbb{R}$)

$$(26)\quad \vec{J}_{np}(t,\vec{r}) = \frac{c\hbar}{m} \sin(\Omega_{np} t)\, (\psi_n(\vec{r})\nabla\psi_p(\vec{r}) - \psi_p(\vec{r})\nabla\psi_n(\vec{r})),$$

where $\Omega_{np} = (E_n - E_p)/\hbar$. Furthermore one has

$$(27)\quad \left[\int e^{i(\vec{k}\cdot\vec{r})} (\psi_n\nabla\psi_p d\vec{r}\right]^{\perp} = -\left[\int e^{i(\vec{k}\cdot\vec{r})} (\psi_p\nabla\psi_n d\vec{r}\right]^{\perp},$$

(see Ref 7, App. B) which explains the factor 2 in the relation (25) (See also Ref[9] p.248 and Ref[10], p.3). The relation (22) gives exactly the Barut and Van Huele[4] non relativistic formula of the Lamb shift whose each term is a convergent integral.

By calculating these integrals, we have in compared (Réf 7), term by term, this formula with the Bethe formula, in the case where the state n is the 1S state and where the states p are the kP_0 (k = 2,3,...) states. It is a partial comparison because all the states and the continuum are to be taken into account for the shift of a state. But this partial result shows that the numerical values of the two formulas must be very close, provided that the " cut-off" is taken equal to αmc^2 (a "good "cut-off from the QFT point of view).

The relation (20) corresponds to the relativistic Barut and Kraus[3] formula and, so, to the standard QFT relativistic formula (Ref[11]).

We can see now how Planck's constant is introduced in the Lamb shift formulas. It is introduced by means of the relation (11) and the relation (19) in which τ_{np} is taken equal to $\hbar$. The "photon k " appears simply as a point in the field of integration of the integral (11) expressing cos(ωr).

6. CONCLUSION

Not only is the Theory of Quantum Fields placed by the Lamb shift outside the laws of Mathematics, but moreover it is placed outside the laws of Reason : one has seen, as a necessity of the QFT renormalization, that Planck's constant is to be put inside $\pi/2$ and not inside $2/\pi$! So the Lamb shift calculation which has been achieved by Barut and his collaborators is not to be presented as an option to the QFT one. There is no alternative. If one concedes some meaning to the word " rationality " in Physics, only the Barut calculation is reasonable.

Nevertheless :

a) though the Lamb shift appears as a refutation of the constitutive rules of the quantization of fields, QFT remains a good tool for studying Electrodynamics. It is not surprising that some of the beautiful experiments of the "Institut d'Optique d'Orsay" or the "Laboratoire de Spectroscopie Hertzienne de l'Ecole Normale Supérieure "(France) for example , are presented as a confirmation of QFT. Whenever an electromagnetic wave may be considered as a plane wave with periodic boundary conditions, the use of QFT (i.e. the theory of the harmonic oscillators applied to Maxwell's equations) is in perfect accordance with the laws of Maxwell.

b) both the pure Maxwell theory and QFT are unable to explain what is a real photon, i.e. the mysterious behaviour in spacetime of the energy which may be associated to the transition from a upper to a lower state of an electron bound in an atom.

But, in our opinion, if the mystery is some day removed, that will be done by the use of, perhaps subtle but surely rigourous, laws of Mathematics.

REFERENCES

1. **R. Boudet,** "The Planck constant and the Clifford algebra of the Minkovski spacetime " , 2nd Workshop on Clifford -Algebras and their Applications to Mathematical Physics Montpellier , France, Sept. 17 - Sept. 30 , 1989 .
2. **D. Hestenes ,** J. Math. Phys., 8, 798 (1967) .
3. **A.O. Barut, J. Kraus.** Found .Phys., 13, 189 (1983) .
4. **A.O. Barut J.F. Van Huele,** Phys. Rev. A, 32 , 3187 (1985)
5. **R.Boudet,** Ann. Fond. L. de Broglie (Paris) , 14, 119 (1989) .
6. **V.F. Weisskopf** , Phys. Rev. 56, 72 (1939) .
7. **B. Blaive, R. Boudet,** Ann. Fond. L. de Broglie (Paris) 14, 147 (1989).
8. **H. Bethe,** Phys. Rev., 72, 339 (1947).
9. **L. Schiff,** " Quantum Mechanics ", N.Y. (1955) .
10. **P.W. Milonni** , Phys. Rep. , 25, 1 (1976) .
11. **B. Davis,** Am. J. Phys., 50. 331 (1982) .

AHARONOV-BOHM EFFECT

AND CASIMIR INTERACTIONS

I.H. Duru

Research Institute for Basic Sciences
TUBITAK
P.K.74, 41470 Gebze
TURKEY

INTRODUCTION

Vacuum polarizations of the fields when the quantization volumes are bounded or the space acquires non-trivial topological structures lead to interesting phenomena which are known as the Casimir effects. A well known example is the existence of the attractive force between two uncharged parallel conductive planes which is proportional to the inverse fourth power of the separation of the plates[1]. In this effect the boundary planes constrain the electromagnetic field modes to form a discrete set in the perpendicular direction; thus, the effective topology in momentum space is not Euclidean, but $R^2 \times S^1$[2]. There is a long list of Casimir interactions involving several other geometrical boundaries[3].

There exist Casimir interactions associated with even a simpler geometry, namely a multiply connected space which is obtained by excluding x_3-axis[4]. Such a space can be realized by placing an "impenetrable conducting wire" or an "Aharonov-Bohm solenoid" parallel to x_3-axis. At this point we emphasize that the realization of the boundaries is dependent on the choice of the fields for which we study the vacuum structure. For example, a "perfectly conducting wire" provides a multiply connected space for the photon field, while for charged particles an A-B solenoid is required.

In this work we will employ the second realization, namely the Aharonov-Bohm effect[5]. When an infinitely long and tightly wound solenoid, with a magnetic field $\vec{B}$ confined in it, is placed in the space it is possible to create an outside region in which there is a vector potential $\vec{A}$, but the field $\vec{B} = \vec{\nabla} \times \vec{A}$ is zero. An electron moving in the outside region picks up an extra phase $e\Phi$ which is determined by the line integral of $\vec{A}$ along the electron path

$$e\Phi = e \int d\,\vec{\ell} \cdot \vec{A} \tag{1}$$

For a closed path, this phase can be expressed in terms of the enclosed magnetic flux

$$e\Phi = e \oint d\,\vec{\ell} \cdot \vec{A} = e \int\!\!\int d\,\vec{\sigma} \cdot \vec{B} \tag{2}$$

New Frontiers in Quantum Electrodynamics and Quantum Optics
Edited by A. O. Barut, Plenum Press, New York, 1990

The phase in (1) is observable in the interference experiments.

In sections II and III we show that a Casimir force exists between two parallel A-B solenoids which confine the fluxes n_1 and n_2 within them, respectively. Following the customary method employed in studying the interactions of parallel plates[2], one can calculate the vacuum expectation value of the energy for the massless field by taking the required coincidence derivatives of the Green function. The vacuum energy includes the "self interaction" terms which involves n_1^2 and n_2^2, and a "mutual interaction" term with $(n_1 n_2)^2$. The "self interaction" terms have some divergent parts, which are to be eliminated using some suitable renormalization techniques. However to investigate the force between two flux tubes it is sufficient to consider the "mutual interaction" term which is finite. This term depends on the distance between the fluxes, and by virtue of this fact, the tubes attract each other by a force which is inversely proportional to the third power of their separation.

In section IV we briefly present a result on the changes occuring in the radiation reactions of the H-atom when it is placed near an A-B solenoid.

Finally we give the Casimir force between a conducting plane and a line singularity.

THE GREEN FUNCTION

The Casimir energy between the parallel plates is calculated by using the Green function for the electromagnetic field which vanishes at the perfectly conducting boundaries. In our case, since the space around the A-B solenoid is characterized by a vector potential, we have to use the Green function of a charged field. For an infinitely long and very tightly wound cylindrical solenoid of radius R, which has its centre at the origin and its axis in the x_3 direction, the potential in the outside region is given by

$$\vec{A} = \frac{\Phi}{2\pi}\frac{1}{\rho}\left(\sin\theta\ \hat{x}_1 - \cos\theta\ \hat{x}_3\right), \tag{3}$$

where ρ, θ are the polar coordinates in the $x_1 x_2$ - plane and Φ is the constant flux confined in the solenoid. To calculate the expectation value of the energy-momentum tensor one needs the symmetric Green function. For a charged particle of mass μ it can be represented by the path integral (with $h = c = 1$)

$$G(x, x') = \frac{1}{2}\left(K(x, x') + K(x', x)\right), \tag{4}$$

where

$$K(x, x') = \int_0^\infty dW\ e^{-i\mu^2 W} \int D^4 x \exp\left[\frac{i}{4}\int_0^W dw\left(-\dot{t}^2 + \dot{\vec{x}}^2 + 4eA\cdot\dot{x}\right)\right] \tag{5}$$

Here overdots stand for the derivative with respect to the parameter time w. In this work we do not need the explicit form of the above path integral which can be found in the literature[6].

Inserting the potential (3) into (5) we observe that the interaction term can be integrated explicitly

$$\frac{i}{4}\int_0^W dw(4eA\cdot\dot{x}) = -i\int_0^W dw\ e\frac{\Phi}{2\pi}\dot{\theta} = -ie\frac{\Phi}{2\pi}\left(\theta - \theta' \pm 2\pi n\right), \tag{6}$$

with $\theta(w=0)=\theta'$, $\theta(w=W)=\theta$. The integer n is the winding number which distinguishes the different homotopy classes. In order to take into account all the paths connecting x' and x we have to sum over n, which by virtue of the Poisson formula

$$\sum_{n=-\infty}^{\infty} e^{2\pi i n \frac{e\Phi}{2\pi}} = \sum_{n=-\infty}^{\infty} \delta\left(n - \frac{e\Phi}{2\pi}\right),$$

implies the flux quantization.

$$\Phi = \frac{2\pi}{e} n.$$

The above quantization condition has appeared as a consequence of our assumption of single valuedness for the particle wave function under 2π rotations. However we emphasize that the foregoing discussion about the interaction of two solenoids is valid whether the fluxes are quantized or not. We also remark that for spinning particles $2\pi n$ in (6) would be replaced by $4\pi n$, thus the flux would be quantized to half of the value given by (7).

After renormalizing by subtracting the Minkowski space Green function $G_M(x,x')$, the final form of the Green function becomes

$$G_n(x,x') = \cos\, n(\theta-\theta')G_M(x,x') - G_M(x,x'), \tag{8}$$

where,

$$G_M(x,x') = \int_0^\infty dW e^{-i\mu^2 W} \int D^4x \exp\left[\frac{i}{4}\int_0^W dw\left(-\dot{t}^2 + \dot{\vec{x}}^2\right)\right]. \tag{9}$$

Using the formal resemblance with the well-known form of the free particle kernel for the non-relativistic particle motion[7], eq. (7) can be written as

$$G_M(x,x') = \frac{i}{8}\int_0^\infty \frac{dW}{W^2} e^{-i\mu^2 W} e^{\frac{i}{4W}\left[-(t-t')^2 + (\vec{x}-\vec{x}')^2\right]}. \tag{10}$$

For the massless fields, it becomes[2]

$$G_M^{\mu=0}(x,x') = D(x,x') = \frac{1}{(2\pi)^2\left[-(t-t')^2 + (\vec{x}-\vec{x}')^2\right]}. \tag{11}$$

The formula (8) can easily be generalized to the case of more than one solenoid. If we have a second solenoid which is parallel to the first one the Green function is

$$G_{n_1 n_2}(x,x') = [\cos\, n_1(\theta-\theta')\cos\, n_2(\theta_2-\theta_2') - 1]\, G_M(x,x'). \tag{12}$$

Here n_1 and n_2 are the fluxes confined within the first and second solenoids and the angle θ_2 is given by

$$\sin\theta_2 = \frac{x_2}{\rho_2},\ \rho_2 \equiv \sqrt{(x_1-a)^2 + x_2^2}. \tag{13}$$

To study the interaction of two solenoids, we calculate the vacuum expectation value of the energy for a massless field given by

$$\langle 0|T_{tt}|0\rangle = \lim_{x \to x'} \frac{1}{4}\left(\partial_t \partial_{t'} + \partial_{x_1}\partial_{x_1'} + \partial_{x_2}\partial_{x_2'} + \partial_{x_3}\partial_{x_3'}\right) D_{n_1 n_2}(x, x') \tag{14}$$

where

$$D_{n_1 n_2}(x, x') = \frac{1}{2\pi^2} \frac{\cos n_1(\theta - \theta')\cos n_2(\theta_2 - \theta_2') - 1}{-(t - t')^2 + (\vec{x} - \vec{x}')^2}. \tag{15}$$

Since we are interested in the interaction energy, it is sufficient to consider only the terms involving the product $n_1 n_2$. In this limit $x \to x'$ the first two terms in (14) cancel each other. Then only the derivatives in $x_1 x_2$-plane contribute, and we can write

$$\langle 0|T_{tt}|0\rangle_{n_1 n_2} = \lim_{\substack{\rho \to \rho' \\ \alpha \to 0}} \frac{1}{8\rho^2(1 - \cos\alpha)}\left[\left(\partial_{x_1}\partial_{x_1'} + \partial_{x_2}\partial_{x_2'}\right)\cos n_1\alpha \cos n_2\alpha_2 - \frac{8(\cos n_1\alpha \cos n_2\alpha_2 - 1)}{2\rho^2(1 - \cos\alpha)}\right], \tag{16}$$

where

$$\alpha = \theta - \theta', \ \alpha_2 = \theta_2 - \theta_2'. \tag{17}$$

In writing (16) we have calculated the coincidence limit derivatives of the denominator of $D_{n_1 n_2}$ in the Cartesian coordinates. In order to develop the remaining derivatives in (16) it is convenient to use the differentiations with respect to α. For that purpose we first express α_2 in terms of α. In the limit α, $\alpha_2 \to 0$, and using (13), we can write

$$\begin{aligned}
\sin n_2\alpha_2 &\cong n_2\alpha_2 \cong n_2(\sin\theta_2\cos\theta_2' - \sin\theta_2'\cos\theta_2) \\
&\cong \frac{n_2}{\rho_2\rho_2'}\left[x_2(x_1' - a) - x_2'(x_1 - a)\right], \\
&\cong \frac{n_2}{\rho_2\rho_2'}\left[\rho\sin\theta(\rho\cos\theta' - a) - \rho\sin\theta'(\rho\cos\theta - a)\right], \\
&\cong \frac{n_2}{\rho_2\rho_2'}(\rho^2 - ax)a,
\end{aligned}$$

which implies

$$\alpha_2 \cong \frac{1}{\rho_2^2}(\rho^2 - ax)\alpha. \tag{18}$$

We can then write (for $\alpha \to 0$)

$$\partial_{x_1} = -\partial_{x_1'} \cong \frac{x_2}{\rho}\partial_\alpha, \ \partial_{x_2} = -\partial_{x_2'} \cong -\frac{x_1}{\rho_2^2}\partial_\alpha, \tag{19}$$

Using (19), (18) and (16) we get

$$\langle 0|T_{tt}|0\rangle_{n_1 n_2} = \frac{n_1^2 n_2^2}{16\pi^2}\frac{1}{\rho^2\rho_2^2}\left(-1 + \frac{(\rho^2 - ax_1)^2}{\rho^2\rho_2^2}\right). \tag{20}$$

To have a more symmetric formula we shift the x_1 coordinate by $x_1 \to x_1 + a/2$ and obtain the mutual interaction term of the energy density vaccum expectation value for two fluxes at $x_1 = \pm a/2$:

$$\langle T_{tt} \rangle_{n_1 n_2} = \frac{n_1^2 n_2^2}{16\pi^2} \frac{a^2 \rho^2 \sin^2\theta}{\left((\rho^2 + a^2/4)^2 - a^2\rho^2 \cos^2\theta \right)^2}. \tag{21}$$

By integrating this expression over $x_1 x_2$-plane we obtain the total interaction energy in the slice of space with unit thickness having the normal in x_2-direction

$$E_{n_1 n_2} = \int_0^{2\pi} d\theta \int_0^\infty d\rho\rho \langle T_{tt} \rangle_{n_1 n_2}. \tag{22}$$

Introducing the dimensionless variable

$$\nu = 4\rho^2/a^2 \tag{23}$$

we have

$$\begin{aligned} E_{n_1 n_2} &= -\frac{n_1^2 n_2^2}{4\pi^2} \frac{1}{a^2} \int_0^\infty d\nu \int_0^\pi d\theta \frac{4\nu \sin^2\theta}{\left((\nu - 1)^2 + 4\nu \sin^2\theta \right)^2} \\ &\equiv -\frac{n_1^2 n_2^2}{4\pi^2} \frac{1}{a^2} I, \end{aligned} \tag{24}$$

where by virtue of symmetry around $\theta = \pi$ we have written the range of θ as from 0 to π. From (24) we see that $E_{n_1 n_2}$ is inversely proportional to a^2. To calculate the proportionality intergral I, we have to exclude the cross sections of the flux tubes which are located at $\nu = 1$, $\theta = 0, \pi$. We divide I into two parts

$$I = I_1 + I_2 \tag{25}$$

The range of the first term is defined as

$$I_1 : \left(\int_0^{1-\delta} d\nu + \int_{1+\delta}^\infty d\nu \right) \int_0^\pi d\theta,$$

with $\delta = \frac{4R^2}{a^2} << 1$, $R =$ radius of flux tubes, and covers the $x_1 x_2$-plane excluding the ring of radius 1 and thickness 2δ. We perform $d\theta$ integration[8] by simplifying it as

$$\int_0^\pi d\theta \frac{4\nu \sin^2\theta}{\left((\nu - 1)^2 + 4\nu \sin^2\theta \right)^2} = -\lim_{\beta \to 0} \int_0^\pi d\theta \frac{1}{(\nu - 1)^2 + 4\nu\beta \sin^2\theta},$$

and arrive at

$$I_1 = \left(\int_0^{1-\delta} d\nu + \int_{1+\delta}^\infty \right) \frac{2\nu}{(\nu - 1)(\nu + 1)^3}.$$

Inserting the indefinite integral

$$\int \frac{d\nu\ \nu}{(\nu - 1)(\nu + 1)^3} = \frac{1}{4} \left(-\frac{1}{(\nu + 1)^2} + \frac{1}{(\nu + 1)} + \frac{1}{2} \ell n \frac{\nu - 1}{\nu + 1} \right),$$

we obtain

$$I_1 = \frac{\pi}{2} \left(\frac{1 - \delta}{(2 - \delta)^2} - \frac{1 + \delta}{(2 + \delta)^2} + \ell n \frac{1 + \delta/2}{1 - \delta/2} \right),$$

which in the limit $\delta << 1$ can be written as (with $u = \nu - 1$):

$$I_2 \cong \frac{1}{4} \int_{-\delta}^{\delta} du(1-u) \int_{\delta}^{\pi-\delta} \frac{d\theta}{\sin^2\theta} \cong 1 + 0(\delta^2). \tag{27}$$

Combining (27), (26), (25) and (24) we obtain (up to order δ)

$$E_{n_1 n_2} = -\frac{n_1^2 n_2^2}{4\pi^2} \frac{1}{a^2}. \tag{28}$$

The presence of this interaction energy term in the region around the solenoids, implies an attractive force per unit length given by

$$F_{n_1 n_2} = -\frac{\partial}{\partial a} E_{n_1 n_2} = -\frac{n_1^2 n_2^2}{2\pi^2} \frac{1}{a^3}. \tag{29}$$

This result is to be compared to the usual Casimir force between the parallel plates which per unit area is given by[2]

$$F_{\text{plate}} = -\frac{\pi^2}{240} \frac{1}{a^4}. \tag{30}$$

For the typical experimental distance $a = 0.5$ μm, this force is $\cong 0.2$ dyn/cm^2 [9]. To have a reasonable guess of the magnitude of this force, expecting that in practice only the small enough fluxes would be totally confined within the solenoids, we can assume $n_1 \cong n_2 \cong 1$. Then for $a = 0.5$ μm we obtain

$$F_{n_1 n_2} \cong 0.5 \times 10^{-5} \text{dyn/cm}, \tag{31}$$

which is very small.

RADIATION REACTIONS AND A-B EFFECT

The experimental verifications of the Casimir interactions are usually achieved by measuring the changes in the radiation reaction energy levels of the atoms caused by the boundaries. Motivated by this fact, the shift occuring in the Lamb shift of the H-atom when it is placed before a solenoid can be calculated[10]. For that purpose it is enough to replace the free space Green function $D(x, x')$ of the self radiation field of the atom by

$$D_\Lambda(x, x') = \cos\Lambda(\theta - \theta')D(x, x'), \tag{32}$$

where Λ is the flux confined in the solenoid. If the distance between a solenoid and the atom is large compared to the Bohr radius, the Lamb shift of the nth energy level is changed by the amount

$$\Delta E_n = \frac{e^2\pi^2\Lambda^2}{M^2 a^2} \sum_{N'} (\delta_{mm'} + \delta_{m,m'-1})\, \omega_{NN'}\, \vec{\zeta}_{NN'} \cdot \vec{\zeta}_{N'N}, \tag{33}$$

with $\omega_{NN'} = E_n - E'_n$, $M =$ reduced mass, and

$$\vec{\zeta}_{NN} = \int_0^\infty dr r^3 \int_{-1}^{1} dx\; x \mathcal{N}_N\, \mathcal{N}_{N'} P_\ell^m(x) R_{n\ell}(r) \frac{\vec{\nabla}}{i} P_\ell^{m'}(x) R_{n\ell}(r),$$

Here N stands for the set of quantum numbers $n\ell m$ and $\mathcal{N}$, $R_{n\ell}$ are the normalization factor and the radial wave functions of the H-atom while $P_\ell^m(x)$ is the Legendre Polynomial.

Note that $1/a^2$ dependence in (33) is common to all Casimir energies involving linear boundaries.

where Λ is the flux confined in the solenoid. If the distance between a solenoid and the atom is large compared to the Bohr radius, the Lamb shift of the nth energy level is changed by the amount

$$\Delta E_n = \frac{e^2\pi^2\Lambda^2}{M^2a^2} \sum_{N'} (\delta_{mm'} + \delta_{m,m'-1})\,\omega_{NN'}\,\vec{\zeta}_{NN'} \cdot \vec{\zeta}_{N'N}, \tag{33}$$

with $\omega_{NN'} = E_n - E'_n$, M = reduced mass, and

$$\vec{\zeta}_{NN} = \int_0^\infty drr^3 \int_{-1}^{1} dx\; x\mathcal{N}_N\, \mathcal{N}_{N'} P_\ell^m(x) R_{n\ell}(r) \frac{\vec{\nabla}}{i} P_\ell^{m'}(x) R_{n\ell}(r),$$

Here N stands for the set of quantum numbers $n\ell m$ and $\mathcal{N}$, $R_{n\ell}$ are the normalization factor and the radial wave functions of the H-atom while $P_\ell^m(x)$ is the Legendre Polynomial.

Note that $1/a^2$ dependence in (33) is common to all Casimir energies involving linear boundaries.

SOLENOID AND PLANE BOUNDARY

In this final section we briefly present the case of a solenoid and a plane boundary. For an impenetrable plane at $x_1 = 0$ and a line parallel to x_3-axis with flux Λ placed at $x_1 = a$, the renormalized Green function is written by using the method of images as

$$G(x, x') = \frac{\cos\Lambda(\theta - \theta')}{2\pi^2} \left[\frac{1}{-(t-t')^2 + (\vec{x} - \vec{x}')^2} - \frac{1}{-(t-t')^2 + (\vec{x} - \vec{x}')_a^2} \right] - \frac{1}{2\pi^2} \frac{1}{-(t-t')^2 + (\vec{x} - \vec{x}')^2}, \tag{34}$$

where

$$(\vec{x} - \vec{x}')_a^2 = (x_1 + x_1' - 2a)^2 + (x_2 - x_2')^2 + (x_3 - x_3')^2.$$

Repeating the procedure presented in section III we obtain a Casimir energy depending on the inverse square of the distance which leads to an attractive force between the plane and the unit length of the flux line:

$$F = -\frac{\Lambda^2}{16\pi^2}(1 - \ell n2)\frac{1}{a^3} \tag{33}$$

REFERENCES

1. H.B.G.. Casimir, *Proc. K. Ned. Akad. Wet.*, **51**, 793 (1948); *Physica*, **19**, 846 (1956).

2. N.D. Birrell and P.C.W. Davies, Quantum Fields in Curved Space (Cambridge Univ. Press, Cambridge, 1982).

3. G. Plumien, B. Müller and W. Greiner, *Phys. Rep.*, **C134**, 87 (1986).

4. I.H. Duru, ICTP, Trieste, Report No. IC/89/128.

5. W. Ehrenberg and R. E. Siday, *Proc. Phys. Soc.*, **62B**, 8 (1949); Y. Aharonov and D. Bohm, *Phys. Rev.*, **115**, 485 (1959).

6. See for example I.H. Duru and N. Ünal, *Phys. Rev.*, **D34**, 959 (1986).

7. R.P. Feynman and A.R. Hibbs, Quantum Mechanics and Path Integrals (McGraw Hill, New York 1965).

8. W. Gröbner and N. Hofreiter, Integral Tafeln; Zweiter Teil, (Springer Verlag, Wien, New York 1965).

9. V.M. Mostepanenko and N.N. Trunov, *Sov. Phys. Usp.*, **31**, 965 (1988).

10. A.O. Barut and I.H.Duru, submitted for publication.

MACROSCOPIC COHERENT STATES OF THE QUANTIZED ELECTROMAGNETIC FIELD

Alfred Rieckers
Institut für Theoretische Physik
Tübingen, West Germany

We formulate a theoretical framework for studying coherent states independently of a special representation of the photon Weyl algebra. A complete classification of all non-Fock coherent states (with a non-zero photon density in infinite space) is given. Their GNS-representation is shown to exhibit a phase operator and a classical part of the represented field operator. The discussion of the photon counting distribution and of the dynamic generation by means of a Dicke model indicates that these states describe the coherent light of a laser for high intensities, without neglecting its quantum properties.

In Glauber's[1] coherence condition

$$
\begin{aligned}
&< E^{(-)}_{\alpha_1}(x_1)\ldots E^{(-)}_{\alpha_n}(x_n)E^{(+)}_{\alpha_{n+1}}(x_{n+1})\ldots E^{(+)}_{\alpha_{2n}}(x_{2n}) >= \\
&tr[\rho E^{(-)}_{\alpha_1}(x_1)\ldots E^{(-)}_{\alpha_n}(x_n)E^{(+)}_{\alpha_{n+1}}(x_{n+1})\ldots E^{(+)}_{\alpha_{2n}}(x_{2n})] = \\
&\prod_{i=1}^{n} \mathcal{E}_{\alpha_i}(x_i)\mathcal{E}_{\alpha_{n+i}}(x_{n+i})
\end{aligned}
\tag{1}
$$

ρ is a density operator in the Fock space, but the parts $E^{(\pm)}_\alpha(x)$, of the quantized electric field at the point $x \in \mathbb{R}^4$ are not well defined operators and the c-number expressions $\mathcal{E}_\alpha(x)$ may also be singular. A connection with the rigorous formalism is gained by "smearing the field with a test function" as, e.g., for the creation operator

$$
a^*(f) := \sum_\alpha \int E^{(-)}_\alpha(x) f_\alpha(x) d^4x \; . \tag{2}
$$

Here f is an element of the linear space E spanned by the eigenmodes in the cavity. E has a scalar product $(\cdot|\cdot)$ but is not complete, i.e., E is a pre-Hilbert space. To avoid mathematical difficulties from the unboundedness of $a^*(f)$ and its Hermitian conjugate $a(f)$ one works with the unitary *Weyl operators*

$$W(f) := exp[\frac{1}{\sqrt{2}}(a^*(f) + a(f))], \quad f \in E. \tag{3}$$

These satisfy the relations

$$W^*(f) = W(-f), f, g \in E \tag{4a}$$

$$W(f)W(g) = exp\,(-\frac{i}{2}Im(f|g))W(f+g). \tag{4b}$$

From (4) follows that the set of all finite linear combinations

$$\left\{\sum_{i=1}^{n} c_i W(f_i); f_i \in E, c_i \in \mathbb{C}\right\} \tag{5}$$

constitutes an algebra which is invariant under the *-operation. Its norm closure gives the abstract C^*-*Weyl-algebra* $\mathcal{W}$(E) which is uniquely associated with E^2. As is well-known there are many inequivalent representations of $\mathcal{W}$(E), but a great deal of the formalism can be done without a representation at all. Especially the notion of a state can be introduced independently of a representation dependent density operator by considering the expectation values as a sufficient characterization. We supplement the bracket symbol in the first line of (1) by a letter φ for the state and write for the expectations

$$< \varphi; A >, \quad A \in \mathcal{W}(E) . \tag{6}$$

In this way φ is a linear functional on $\mathcal{W}(E)$ which is positive and normalized (to give $W(0)$ the value unity). The set $\mathcal{S}$ of all these states is *convex* (the mixture of two states is on the line segment between these states and has to be in $\mathcal{S}$) and compact in the $w^\star - topology$, where the latter is given by the convergence of the expectation values for all observables $A \in \mathcal{W}(E)$.

With every $\varphi \in \mathcal{S}$ is associated the function

$$C_\varphi : E \to \mathbb{C}, \; C_\varphi(f) :=< \varphi; W(f) > , \tag{7}$$

the *characteristic function.* It satisfies

$$C_\varphi(0) = 1 \tag{8a}$$

and

$$\sum_{i,j=1}^{n} \bar{c}_i c_j exp[\frac{i}{2}Im(f_i|f_j)]C_\varphi(f_j - f_i) \geq 0 \tag{8b}$$

for all $n \in \mathbb{N}$ and all $c_i \in \mathbb{C}$ and $f_i \in E$. On the other hand, if $\mathcal{C}(E)$ denotes the set of functions fulfilling (8), then every $C \in \mathcal{C}(E)$ determines a unique state φ (such that $C_\varphi = C$).

For every $\varphi \in \mathcal{S}$ there exists a Hilbert space $\mathcal{H}_\varphi$, a *-homomorphic mapping

$$\pi_\varphi : \mathcal{W}(E) \to \mathcal{B}(\mathcal{H}_\varphi) \quad (\equiv \textit{all bounded operators in } \mathcal{H}_\varphi)$$

and a cyclic vector $\Omega_\varphi \in \mathcal{H}_\varphi$, such that

$$< \varphi; A >= (\Omega_\varphi | \pi_\varphi(A)\Omega_\varphi) \quad \forall A \in \mathcal{W}(E). \tag{9}$$

Not in all such representations (each one being called the *GNS-representation* of φ) there exist the field operators. For simplicity let us take into account only *analytic* states for which $C_\varphi(tf)$ is extensible to an analytic function $C_\varphi(z; f)$ around $z = 0$ for every $f \in E$. Then the unbounded field operator

$$\phi_\varphi(f) := d\pi_\varphi(W(tf))/dti|_{t=0} \tag{10}$$

and all its powers $\phi^n_\varphi(f), n \in \mathbb{N}$, are self-adjoint with a common dense domain containing Ω_φ. The creation operators $a^\star_\varphi(f) := (\phi_\varphi(f) - i\phi_\varphi(if))/\sqrt{2}$ and their Hermitian conjugates are closed operators which together with their powers have again Ω_φ in their domain. For analytic states we have now the pre-requisites to give the correlation functions and normally ordered correlation functions a well-defined meaning.

For the Fock vacuum state φ_F we have $C_F(f) = exp[-\|f\|^2/4]$ and $C_F(z; f) = exp[-z^2\|f\|^2/4]$, showing that φ_F is entire analytic. The vacuum vector Ω_F is in fact in the domain of all powers of the Fock-represented field operators.

For an arbitrary analytic $\varphi \in \mathcal{S}$ we have (for a z-neighbourhood of 0)

$$\begin{aligned} C_\varphi(z; f) &= \sum_{n=0}^{\infty} \frac{(iz)^n}{n!} (\Omega_\varphi | \phi^n_\varphi(f)\Omega_\varphi) \\ &= C_F(z; f) N_\varphi(z; f) \end{aligned} \tag{11}$$

where the *normally ordered characteristic* function is

$$N_\varphi(z; f) = \sum_{k,l=0}^{\infty} \left(\frac{iz}{\sqrt{2}}\right)^{k+l} \frac{1}{k!}\frac{1}{l!} (\Omega_\varphi | a^\star_\varphi(f)^k a_\varphi(f)^l \Omega_\varphi) \, . \tag{12}$$

An analytic state φ is called *classical*, if $N_\varphi(1; f)$ fulfills the classical positive-definiteness condition, that is (8b) without the exponential.

In the smeared coherence condition (1) we have at the right-hand side a product over factors of the form $\sum_\alpha \int \mathcal{E}_\alpha(x) f_\alpha(x) d^4x =: L(f)$ and their complex conjugates. Here L is a linear form on E, the set of which we denote by E'. Depending on the state, L may be continuous in the norm of E - and then is called *bounded* - or not.

1. *Definition*: A state $\varphi \in \mathcal{S}$ is called (all order) *coherent*, if it is analytic and if there is an $L \in E'$, such that (in its GNS-representation)

$$\begin{aligned} &(\Omega_\varphi | a^\star_\varphi(f_1) \ldots a^\star_\varphi(f_k) a_\varphi(g_1) \ldots a_\varphi(g_l) \Omega_\varphi) \\ &\quad = L(f_1) \ldots L(f_k) \bar{L}(g_1) \ldots \bar{L}(g_l) \end{aligned} \tag{13}$$

for all $k = l \in \mathbb{N}_0$ and all $f_i, g_i \in E$. The set of all coherent states with the same $L \in E'$ - the *L-coherent states* - is denoted by $\mathcal{S}_L$.

It is clear that $\mathcal{S}_L$ has in general more than one element, since (13) does not specify the normally ordered correlations for $k \neq l$. The following propositions are based on results of [3].

2. *Proposition*: $\mathcal{S}_L$ is convex and $w^\star$ -compact.

The compactness of a convex set $\mathcal{K}$ guarantees the existence of states in the *extremal boundary* $\partial_e\mathcal{K}$ (which consists of points in $\mathcal{K}$ not being in the interior of a line segment between two points of $\mathcal{K}$). More specifically, it can be shown[4], that every point of such a $\mathcal{K}$ has a decomposition into points of $\partial_e\mathcal{K}$ by means of a probability measure (supported by $\partial_e\mathcal{K}$). If $\mathcal{K} = \mathcal{S}$, the extremal boundary $\partial_e\mathcal{S}$ constitutes the pure states, and the mentioned extremal decomposition is not unique. If $\mathcal{K} = \mathcal{S}_L$, the extremal decomposition into states of $\partial_e\mathcal{S}_L$ decomposes L-coherent states into less mixed L-coherent states. But it is not clear from the outset that the states in $\partial_e\mathcal{S}_L$ are pure and whether the extremal decomposition for states in $\mathcal{S}_L$ is unique or not. In fact, the most popular coherent states are in $\partial_e\mathcal{S}_L$.

3. *Proposition*: For an arbitrary $L \in E'$ it holds:

a) the function

$$C_L(f) := C_F(f)exp[i\sqrt{2}Re\ L(f)],\ \ f \in E, \tag{14}$$

is in $\mathcal{C}$(E) and defines a state $\varphi_L \in \mathcal{S}$, which is entire analytic, extremal coherent, classical, and pure;

b) the following conditions on a state $\varphi \in \mathcal{S}$ are equivalent:
(i) $\varphi = \varphi_L$ (i.e. φ has the characteristic function C_L),
(ii) φ is analytic and satisfies (13) for all $k, l \in N_0$,
(iii) φ is analytic and satisfies (13) for $(k,l) = (0,1)$ and $(k,l) = (1,1)$,
(iv) φ is analytic and it holds in its GNS-representation

$$a_\varphi(f)\Omega_\varphi = \bar{L}(f)\Omega_\varphi, \quad for\ all\ f \in E.$$

The general coherent states may be characterized as follows.

4. *Proposition*: For an arbitrary $L \in E'$ an analytic state ω is in $\mathcal{S}_L$, iff its normally ordered characteristic function has the form

$$N_\omega(z;f) = \sum_{k,l=0}^{\infty} \left(\frac{iz}{\sqrt{2}}\right)^{k+l} \frac{1}{k!}\frac{1}{l!}L(f)^k\bar{L}(f)^l c(k,l), \tag{15}$$

where $c(k,l)$ is a positive definite matrix with $c(k,k) = 1, \quad \forall\ k \in \mathbb{N}_0$ (implying $|c(k,l)| \le 1, \quad \forall\ k,l \in \mathbb{N}_0$).

From (15) one can deduce

5. *Proposition*: For an L-coherent state ω the following conditions are equivalent:
(i) L is bounded,
(ii) the total number expectation value $(\Omega_\omega|\sum_{n=1}^{\infty} a_\omega^\star(e_n)a_\omega(e_n)\Omega_\omega)$ has the same finite value for all orthonormal basis systems of $\bar{E}$ (the norm closure of E),
(iii) ω is realizable by means of a density operator in the Fock representation.

In spite of there being in some sense more unbounded linear forms in E' than bounded ones, the non-Fock coherent states can be classified in a more complete manner.

6. *Proposition*:

a) If $L \in E'$ is *unbounded* then the following is true:

(i) $\omega \in \mathcal{S}_L$, iff in the series (15) $c(k,l) = d(k-l)$, for all $k,l \in \mathbb{N}_0$, where $d : \mathbb{Z} \to \mathbb{C}$ is a positive-definite function with $d(0) = 1$;

(ii) $\omega \in \mathcal{S}_L$, iff there is a probability measure μ on $[0, 2\pi)$ such that

$$\omega = \int_0^{2\pi} \varphi_{(e^{i\vartheta} L)} d\mu(\vartheta), \tag{16}$$

where then

$$d(k-l) = \int_0^{2\pi} e^{i(k-l)\vartheta} d\mu(\vartheta);$$

(iii) for $\omega \in \mathcal{S}_L$ (16) is the only decomposition into states in $\partial_e \mathcal{S}_L$ (so that $\mathcal{S}_L$ is a so-called *simplex*);

(iv) for $\omega \in \mathcal{S}_L$ (16) is the only decomposition into pure states (in $\partial_e \mathcal{S}$) ;

(v) for $\omega \in \mathcal{S}_L$ any decomposition into other states by means of a discrete probability measure is automatically a decomposition into L-coherent states; i.e. $\mathcal{S}_L$ is a *face*;

(vi) every $\omega \in \mathcal{S}_L$ is classical.

(b) If $L \in E'$ is *bounded* (with norm $\|L\| = sup\{|L(f)|; f \in E, \|f\| = 1\}$) it holds:

(i) for every sequence $(\vartheta_r \in \mathbb{R}; r \in \mathbb{N}_0)$ the coefficient matrix

$$c(k,l) = exp\,(-\|L\|^2 \sum_{r=0}^{\infty} exp\,(-i\vartheta_{r+k} + i\vartheta_{r+l}) \frac{\|L\|^{2r}}{r!} \tag{17}$$

gives via Prop. 4 an L-coherent state which is pure and thus in $\partial_e \mathcal{S}_L$. Iff $\vartheta_r = r\vartheta$ for all $r \in \mathbb{N}_0$ and some $\vartheta \in \mathbb{R}$ it holds $c(k,l) = exp(-i\vartheta(k-l))$ and only then $c(k,l)$ is of the form $d(k-l)$;

(ii) the states of i) with $c(k,l) \neq d(k-l)$ do not have a decomposition of the form (16);

(iii) if φ is a state of (i), then $\int_0^{2\pi} C_\varphi(e^{i\vartheta} f) d\vartheta / 2\pi$ is the characteristic function of an L-coherent state which is invariant under gauge transformations of the first kind and has also a decomposition of the form (16), so that the decomposition into extremal L-coherent states is not unique ($\mathcal{S}_L$ is not a simplex);

(iv) since every $\omega \in \mathcal{S}_L$ can be realized by a density operator in Fock space (according to Prop. 5) the gauge invariant states of (iii) have a decomposition into pure number eigenstates (and not only into pure coherent states), so that the decomposition into pure states is not unique;

(v) the decomposition into number states mentioned in (iv) is discrete and not supported in $\mathcal{S}_L$: thus $\mathcal{S}_L$ is not a face;

(vi) the pure coherent states of (i) with $\vartheta_r \neq r \cdot \vartheta$ are not classical.

Since the L-coherent states with unbounded L have an infinite particle number (cf. Prop. 5), they will be called *macroscopic*. In order to analyze their physical properties in more detail it is useful not only to look at their correlation functions but also to investigate the observables in their GNS-representation. The Fock representation will be designated by an index F.

7. *Proposition*: Let be ω a macroscopic L-coherent state. Then its GNS-representation may be realized in the Hilbert space

$$\mathcal{H}_\omega = \mathcal{H}_F \otimes \mathcal{L}^2([0, 2\pi), \mu) \tag{18}$$

with the cyclic vector

$$\Omega_\omega = \Omega_F \otimes 1 \ , \tag{19}$$

where μ is the measure of (16). The closure $\mathcal{M}_\omega$ of $\pi_\omega(\mathcal{W}(E))$ in the weak operator topology (a so-called *von Neumann-algebra*) contains the phase operator

$$\Theta_\omega(\psi \otimes \chi(\vartheta)) := \psi \otimes \vartheta\chi(\vartheta) \tag{20}$$

where $\psi \in \mathcal{H}_F$ and $\chi \in \mathcal{L}^2([0, 2\pi), \mu)$. The creation operator of this representation has the form

$$a^\star_\omega(f) = a^\star_F(f) \otimes 1 + 1 \otimes L(f) e^{i\Theta_\omega} \ . \tag{21}$$

Iff μ is absolutely continuous with respect to the Lebesque measure $d\vartheta$, the prescription

$$N_\omega[\psi \otimes \chi(\vartheta)] := (N_F\psi) \otimes \chi(\vartheta) + \psi \otimes (\frac{d}{id\vartheta})\chi(\vartheta) \tag{22}$$

gives rise to a self-adjoint particle number operator with

$$[\Theta_\omega, N_\omega]_- \subset i\,1 \ . \tag{23}$$

While the proof of Prop. 7 requires only conventional operator algebraic techniques the physical implications are remarkable. The phase operator (20) is not a c-number, if μ has more than one point in its support. It commutes with any operator of $\mathcal{M}_\omega$ and thus lies in the center of $\mathcal{M}_\omega$. That means that Θ_ω is compatible with all observables of the photon Weyl algebra and such an observable will be called *classical.* Since all classical (central) observables of $\mathcal{M}_\omega$ are bounded functions of Θ_ω, the decomposition (16) makes all classical observables (relevant for ω) fluctuation free, since the measure μ of $\varphi_{(e^{i\vartheta}L)}$ is the point measure at ϑ. According to Prop. 6 a)(iv) a macroscopic L-coherent state is made pure by specifying its (macroscopic) phase, and this is the only way to do it. By means of this macroscopic phase the field operator (21) acquires a classical part which is a c-number, iff the phase is fixed. Only if the phase distribution is purely continuous the classical field parts can be counted and the number operator (22) for both the quantum mechanical and classical field excitations exists. This number operator is not affiliated with the original field algebra and does not commute with Θ_ω. Just the part which counts the classical field excitations gives rise to the canonical commutation relations (23). By the built-in particle number renormalization ω for itself obtains the number eigenvalue zero.

To understand the classical features of the photon field we should compare it with the condensed particle field operators for superfluid bosons[2,5]. In both cases the order phenomenon is the phase correlation (giving rise to the macroscopic phase observable). The phase correlated bosons (giving the super current in the material case) may be counted experimentally but behave otherwise classically in virtue of the reduced degrees of freedom due to the correlations. Our classical photon field looks a little bit simple since our system is spatially homogeneous. Then all classical observables have no dependence on the position variable. In a macroscopically inhomogeneous system[7] the linear form L(f) may depend on a $q \in R^3$.

The point we want to emphasize is that the classical part of the field in (21) gives countable photons but a c-number intensity. This is just the situation, in which a *Poissonian* photon counting distribution may be derived[6,8] by so-called semi-classical arguments. Here this situation is the collective part of a purely quantum mechanically founded theory.

Let us discuss finally, how these macroscopic coherent states may be generated.

8. *Proposition*: Let $L \in E'$ be an unbounded linear form for which there exists a sequence $\{g_N \in E;\ N \in \mathbb{N}\}$ with

$$\|g_N\|^2 = (N/\rho) + o(N), lim_N (g_N|f) = L(f), \forall f \in E \ , \tag{24}$$

for some $\rho > 0$. Denoting $\hat{g}_N := g_N/\|g_N\|$ it holds

$$(a_F^{\star N}(\hat{g}_N)\Omega_F|\pi_F(W(f))a_F^{\star N}(\hat{g}_N)\Omega_F)/N! =: C_N(f)$$

$$= C_F(f)L_N(\frac{\rho}{2}|(\hat{g}_N|f)|^2) \ , \tag{25}$$

where L_N is a Laguerre polynomial. This sequence of number eigenstates converges for $N \to \infty$ to the non-Fock, gauge invariant $(\rho/2)^{1/2}L$-coherent state ω with the characteristic function

$$C_\omega(f) = lim_N C_N(f) = C_F(f)J_0(\rho^{1/2}|L(f)|)$$

$$= C_F(f)\int_0^{2\pi} exp[i\ Re(e^{i\vartheta}\rho^{1/2}L(f))]\frac{d\vartheta}{2\pi} \ , \tag{26}$$

where J_0 is a Bessel function, and its integral decomposition gives a special case of (16).

If g_N is a plane wave in the volume $V_N = N/\rho$ then (24) is valid with L the δ-function for a sharp momentum. But (24) is also satisfied for L a δ-function in position space. The first case may be viewed as a stationary, monochromatic laser beam spread over a macroscopic volume with finite photon density ρ. The second case may be interpreted as an ideally localized laser pulse with extremely high photon density. Both are well defined macroscopic coherent states on the photon Weyl algebra. But there are many other unbounded linear forms L, which can be approximated in the form (24), where the first condition may be achieved sometimes by renumbering the elements of the sequence. Since the pure coherent states φ_L of Prop. 3 and the gauge invariant coherent states with a characteristic function of the form (26) are completely determined by L, the macroscopic ones may be approximated by the corresponding Fock states in terms of a sequence $\{g_N\} \subset E$, which has to fulfill only the second condition of (24). Observe that the arise of a macroscopic phase and number operator is completely independent of the form of L and its approximating g_N's.

In a simple coupling model of Dicke type we have the total Hamiltonian

$$H = H_m + H_r + \lambda(R\, a^\star(g) + R^\star a(g)) \; , \tag{27}$$

where H_r is the free radiation part, H_m a Hamiltonian for a macroscopic number of two level atoms with frequency ω and $R^\star$ the sum of the rising operators of the atoms, whereas $g \in E$ gives the relative weight of the coupling strength for the various modes. In an appropriate representation[9] which is determined by the inversion and cooperation of the atomic states, any cycle averaged Schrödinger state of the total system converges for $t \to \infty$ to a state, the restriction of which to the photon algebra is L-coherent with

$$L(f) = \sum_\alpha \int_{\mathbb{R}^3} \tilde{\bar{g}}_\alpha(k)[c_1 \delta(\omega - |k|) + c_2 \frac{1}{\omega - |k|}] \tilde{f}_\alpha(k) d^3k \tag{28}$$

for all $f \in E$ and some $c_{1/2} \in \mathbb{C}$. If the coupling $\tilde{g}_\alpha(k)$ - the momentum representative restricted to the light cone - does not vanish strongly at the resonance surface $|k| = \omega$, the linear form L is unbounded. This demonstrates that a macroscopic atomic system radiates a macroscopic coherent state of the electromagnetic field by resonant coupling.

REFERENCES

1. R.J. Glauber, in: "Quantum Optics and Electronics", C. De Witt et al., ed., Gordon & Breach, New York (1965)
2. O. Bratteli and D.W.. Robinson, "Operator Algebras and Quantum Statistical Mechanics" I, II, Springer (1979, 1981)
3. R. Honegger and A. Rieckers, to be published in Publ. RIMS Kyoto Univ. (1990)
4. E.M. Alfsen, "Compact Convex Sets and Boundary Integrals", Springer, Berlin (1971)
5. J.T. Lewis and J.V. Pule, Commun. Math. Phys. 36:1 (1974)
6. R. Loudon, "The Quantum Theory of Light", Clarendon Press, Oxford (1985)
7. R. Brendle, Z. Naturforsch. 40a:1189 (1985)
8. J.R. Klauder and E.C.G. Sudarshan, "Fundamentals of Quantum Optics", W.A. Benjamin, New York (1968)
9. R. Honegger and A. Rieckers, in preparation

ATOM OPTICS

David W. Keith and David E. Pritchard

Physics department and Research Laboratory of Electronics
Massachusetts Institute of Technology
Cambridge, MA 02139

INTRODUCTION

By atom optics we mean the rich collection of emerging techniques by which atoms may be manipulated in the manner of light in classical optics. Existing atom optical elements include mirrors, lenses, and diffractive optics including beam splitters as well as dissipative elements such as slowers, 'coolers', and traps which have no analogue in classical optics. To date, these atom optical elements have been realized as demonstrations of principal, we hope that we will soon see some of them used as tools in real experiments. We must caution the reader that this paper is intended as a introduction and enticement to atom optics, not as an exhaustive survey. Most of the paper will be devoted to atom interferometers; first general comments on beam splitters and interferometer geometries, then a detailed look at the one we are currently constructing, and finally a discussion of a few possible experiments with atom interferometers. The final section of the paper will describe an assortment of atom optical elements, concluding with a return to nearer term experimental realities — the need for the rapid development of atom sources that are both slow *and* bright.

ATOM INTERFEROMETERS

Gratings

The key component necessary for the construction of an atom interferometer is a coherent beam splitter. Therefore we will first discuss the available atom beam splitters with special regard to their suitability for constructing an atom interferometer.

Due to the large potential energy of atoms in solids the tunnelling depth of a free atom with thermal energy is less than atomic dimensions; thus, beam splitters based on partial transmission appear impossible. We now list three general classes of beam splitters for atoms.

1) Reflective diffraction gratings. Although it was not perceived as such, the first atomic beam splitter was demonstrated in 1929[1]; it was the diffraction of atoms from the surface of ionic crystals. Because the interatomic spacing in a crystal surface is of the same order as the de Broglie wavelength (λ_{dB}) of typical atomic beams, the angular separation of the diffracted beams is of order unity (i.e. ~1 rad). Atoms may be specularly reflected by surfaces when the de Broglie wavelength corresponding to the momentum perpendicular to the surface is much lager than the surface roughness. It should be possible to use this effect to diffract atoms by grazing incidence reflection from a high quality laminar grating. We are currently trying to demonstrate this type of atom diffraction grating.

2) Transmission diffraction gratings. In 1983 our group demonstrated the Kapitza-Dirac effect in which atoms are diffracted from a standing wave of near resonant light[2].

The grating period in the standing wave is 1/2 the optical wavelength, thus the angular separation of the diffracted orders is $2\lambda_{dB}/\lambda_{light}$ which is ~60 µrad for a thermal sodium beam. In 1988 we demonstrated the diffraction of atoms by transmission through a fabricated periodic structure[3]. The transmission gratings are arrays of slits with a spatial period of 0.2 µm in a 0.5 µm-thick gold membrane.

3) Conventional beam splitters. If one could make a transmission grating micro structure with a surface sufficiently smooth to reflect atoms incident at some grazing angle while still transmitting atoms through the slits then one would have a near analog to the half silver beam splitter used in conventional optics. Unless the grating period is sufficiently small, such a device will still waste about 1/2 of the flux by scattering atoms into orders other than the desired 0th order reflected and transmitted beams.

Interferometer Geometries

Interferometers have different geometries and properties depending on what class of grating is used. Irrespective of the class of grating used, the poor velocity width of existing atom beam sources ($\Delta v/v \sim 1\text{-}10^{-3}$) force one to design a white fringe interferometer in which an achromatic central fringe is assured by using equal path lengths on either side of the interferometer. We now define various quantities needed for the discussion of interferometer properties; the atom de Broglie wavelength (λ_{dB}), the angle of incidence of the atom beam on the grating measured with respect to the grating surface (θ), and the grating period p. The height (h) (measured along the grating lines) and width (w) of the beam are also needed to determine the requirements on flatness and alignment. The various requirements on relative alignment of the gratings are of two types. The first is on the flatness of the gratings and the relative alignment of the grating surfaces, that is the collinearity of the vectors normal to the grating surfaces. The second is the alignment of the grating lines, that is the relative alignment of the gratings with respect to rotations about the surface normals.

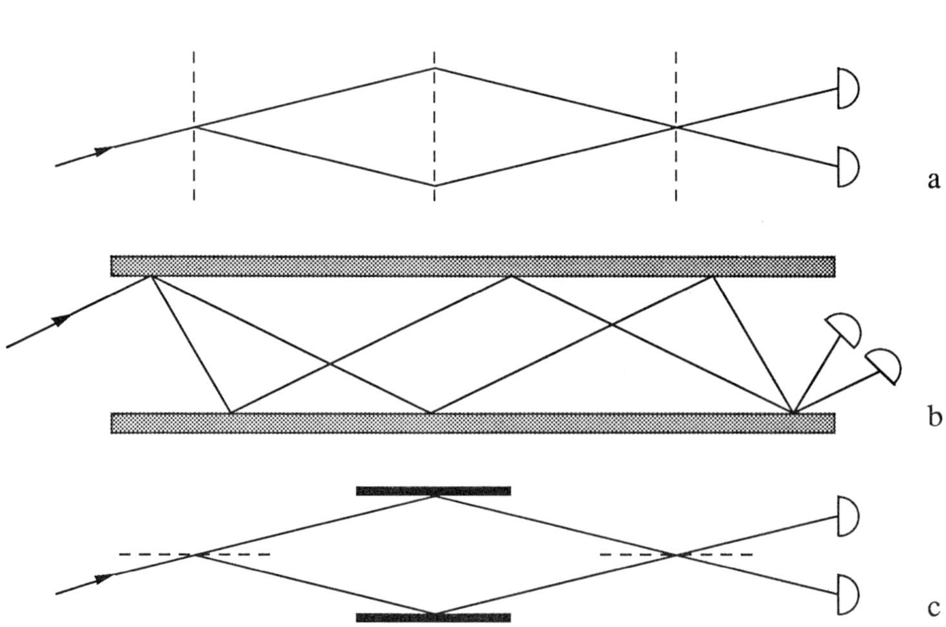

Figure 1. Three different interferometer geometries, transmission gratings (a), reflection gratings (b), and thin reflective gratings (c). In all cases angles are greatly exaggerated and diffracted beams that do not end at one of the detectors are not shown.

For transmission gratings it is well known that an arrangement of three equally spaced gratings has may desirable properties[4]. This (Fig. 1a) is the same geometry as is used for neutron interferometers. This type of interferometer is completely insensitive to the incident angle and is achromatic, all requirements on relative alignment are on the scale of the grating period — independent of , and much smaller than, λ_{dB}. The grating lines must be parallel to $\sim p\sin(\theta)/h$, and the requirement on grating surface alignment expressed as a requirement on $\Delta\theta$ is that $\Delta\theta < p\sin^2(\theta)/w\cos(\theta)$.

For reflection gratings the dependance of the separation of the 0th and 1st order diffracted beams on the angle of the incident beam make it harder to find geometries which are white fringe[5]. Figure 1b shows an example of a white fringe geometry for reflection gratings. The requirement on surface alignment is λ_{dB}/d where d is the larger of h and w. The conditions on line alignment are the same as for transmission gratings, $\sim p\sin(\theta)/h$.

The properties of an atom interferometer made with 'conventional' beam splitters (fig. 1c) are different in several important respects. The requirement on grating surface alignment is λ_{dB}/d where d is as above, this is independent of the grating period. There is no requirement on the grating line alignment. Unlike the two previous cases the area of a conventional beam splitter interferometer is independent of λ_{dB}, which is important when one considers using atom interferometers as rotation sensors.

Our Interferometer

We are currently constructing a three transmission grating interferometer for sodium atoms. We now turn to a detailed description of this interferometer with the hope that the problems involved have some general interest. The present (late October 1989) state of this experiment is that all of its components have worked once, and we are hard at work. Our interferometer differs from the design described above only in that the the interference is detected as a spatial variation of particle density at the third grating, rather than by the variation in intensity in two beams with different directions of propagation in the far field. This detection scheme is of course only possible with amplitude gratings, it has the advantage that it requires only 2/3 the length of the separated beam method which gives us 3/2 greater separation of the beams in the interferometer for the fixed length of our beam tube.

The interferometer is built with a grating spacing of 60 cm giving us a 60 µm beam separation at the middle grating. This allows us to completely separate a 30 µm wide beam which would have an intensity of $\sim 10^6$ sec^{-1} using our existing apparatus with no gratings in place. A realistic estimate of our anticipated final signal strength may therefor be obtained from the properties of the individual gratings. Attenuation caused by the primary grating and the grating support structure gives an intensity in the 0th order of 1/8 of the incident intensity, and of 1/16 in each of the $\pm 1^{st}$ orders. These factors combine to give an intensity at the maximum of a fringe after transmission through all three gratings of only 0.005 of the incident intensity. The near field detection scheme limits the theoretical fringe contrast to 4:1, resulting in a final interference signal of ~0.004 of the incident intensity. Thus, the final interference signal through the interferometer is anticipated to be at most $\sim 4 \times 10^3$ sec^{-1}: this signal will be reduced by any misalignment of the gratings. This signal greatly exceeds the noise of the detector ~ 10 sec^{-1}, allowing us in principal to see the fringes with a S/N of ~4 after a 0.01 sec averaging time.

There are a variety of experimental complications not mentioned in this description of our interferometer. We will now discuss the two of these which appear the most problematical and which are likely to be problems in any atom interferometer: vibration isolation and grating alignment.

We begin our discussion of vibration isolation with a review of the vibration problems relevant to our interferometer. There are two requirements, the first is that the three gratings are stationary relative to each other to within ~1/4 period (50 nm) during the time the final grating integrates the intensity at a given position. Thus, the rms amplitude of relative vibrations integrated over all frequencies greater than the reciprocal of the integration time must be less than ~50 nm. The second requirement is on motion of the gratings as a unit due to acceleration of the center of mass of the grating system during the time it takes for the atoms to traverse the interferometer, the motion due to this acceleration must also be less than ~1/4 period. In our interferometer the transit time is 1.3 msec which implies that the rms acceleration below ~900 Hz must be less than 10^{-2} ms^{-2}.

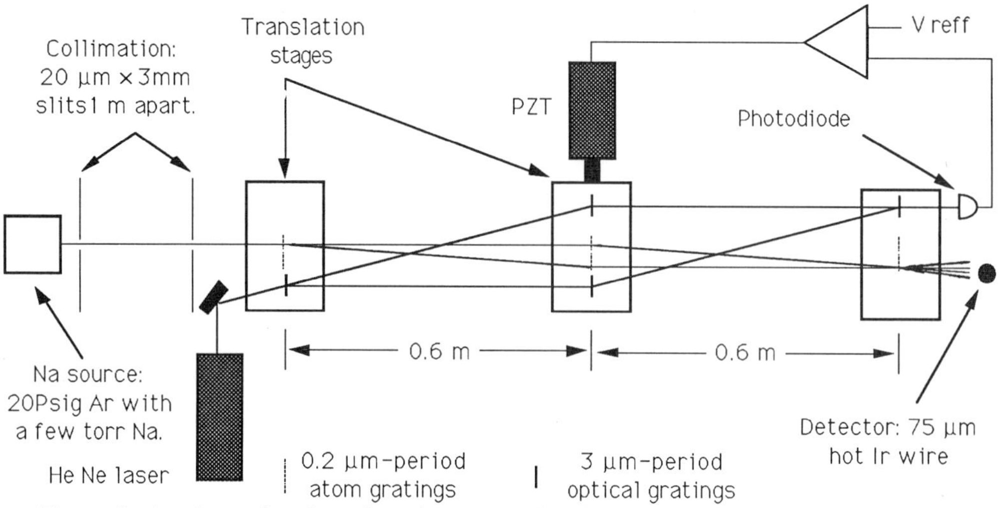

Figure 2. A schematic of our interferometer showing the active vibration isolation system. Not to scale.

We have attacked our vibration problem using a combination of passive isolation and active feedback. The passive isolation system consists of small pneumatic feet which support the apparatus and act like damped springs with a 2 Hz resonant frequency. This simple isolation system reduced the rms motion due mainly to building noise by an order of magnitude to ~ 0.5 µm. The active feedback system is used to stabilize the relative positions of the three gratings at frequencies below ~150 Hz. This system works best at low frequencies (< 10 Hz) where the passive system is least effective. The reduction of relative motion provided by the active system will allow us to use much longer integration times when we are looking for the interference signal. The active feedback system uses a laser interferometer which has the same transmission grating geometry as the atom interferometer. The gratings for the optical interferometer are mounted on the same three translation stages as the matter wave gratings in order to record the exact relative alignment of the matter wave interferometer. The error signal from the optical interferometer provides a measure of the relative alignment of the three grating platforms, it is applied to a Peazo-electric translator (PZT) through a feedback network in order to stabilize the platforms. Using this system we have reduced the relative rms motion of the gratings from ~1500 to 40 nm.

In order that all points along the height (3 mm) of our ribbon shaped beam have the same phase of interference signal it is necessary that the gratings be aligned with respect to rotations about the beam axis to an angle of $\sim 10^{-5}$ rad. We have accomplished this by using a technique based on the optical polarizing properties of the gratings. The 0.2 µm-period grating lines act as wire grid polarizers for light. In principal, it would be possible to align two gratings by rotating them so as to maximize the amount of light transmitted through the pair. This is not practical because the transmitted intensity is proportional to the square of the relative angle between the gratings (for small angles), requiring intensity comparisons to a part in 10^{10}. However, if the polarization of the incident light is modulated about some center angle at frequency f, the amount of light transmitted with modulation frequency 2f, is linearly proportional to the angle between the grating and the center angle. We have used this technique[6] to align the gratings inside our machine to better than 10^{-4} rad, which will be sufficient for our purposes since we can afford to search through the final range of ~10 possible angles.

ATOM INTERFEROMETER EXPERIMENTS

We expect the atom interferometers will one day prove useful in the study of a number of problems in precision metrology, fundamental quantum mechanics, and atomic physics.

Metrology, especially General Relativity

In principal atom interferometers could be used in the manner of optical interferometers to measure fundamental quantities such as acceleration, length, and angular velocity. In practice, atom interferometers are very unlikely to be useful in the measurement of length or acceleration. This is because their advantages over optical interferometers are only due to the ratio of optical to atom de Broglie wavelength in the case of reflective interferometers (not likely to work for small λ_{db}), or on the ratio of optical wavelength to grating period in the case of diffractive interferometers. In either case these advantages will be outweighed by the superior fringe resolution and response time available from optical interferometers. In the area of basic metrology the promise of atom interferometers is in the sensing of inertial rotations, in this case both the low speed (compared to light) and the short wavelength of atoms are advantageous. The Sagnac effect sensitivity measured in radians of interferometer phase shift per unit of angular rotation frequency is $4\pi mA/h$ for a matter wave interferometer of area A, whereas it is $4\pi A/c\lambda$ for an optical interferometer operating at wavelength λ. For example, in order for rotation at one earth rate $\Omega_e \approx 10^{-5}$ sec^{-1} to cause a shift of one fringe in an interferometer using Xe atoms, it would need to have an enclosed area of 10^{-4} m^2, to achieve the same sensitivity in an interferometer using 0.5 μm light would require an area of 10^6 m^2. Of course, optical interferometers have the advantage that is is easy to fold the beam path so that the light makes many trips around the enclosed area effectively multiplying the sensitivity (and decreasing the frequency response) by the number of round trips. However, even if it were possible to build an optical ring cavity that had decay times equal to the millisecond transit times typical of atom interferometers, it would still be less sensitive by the wavelength ratio. It is worth noting that for interferometers using diffractive beam splitters at small incident angles (the simplest technology), the fact that $A \propto \lambda_{dB}$ means that the rotation sensitivity of the interferometer is inversely proportional to the atoms velocity; independent of the mass.

The obvious use for such precise rotation sensors is for tests of general relativity such as the search for the relativistic frame drag. The relativistic effects which might be observable with these techniques are as follows[7]: new limits on the preferred frame parameter in the PPN formalism ($\sim 10^{-8}\,\Omega_e$)[8], the velocity dependant frame drag ($\sim 10^{-9}\,\Omega_e$)[9], and the true Lense Thirring effect ($\sim 10^{-10}\,\Omega_e$)[9]. The second two of these effects are most easily measured by comparing an orbiting gyroscope to the position of the fixed stars as measured from a platform fixed to the gyro. The difficulty of measuring the frame drag can be appreciated when one considers that Everitt et al at Stanford[10] have been developing an experiment of this type (which employes a magnetically levitated spinning superconducting sphere as the gyro) for the last twenty years.

Fundamental Tests of Quantum Mechanics

Most of the experiments in fundamental quantum mechanics that have been performed using neutron interferometers could be improved by using atom interferometers. This is due both to the range of atomic properties potentially available and to the high brightness of atom sources as compared to neutron sources. We will consider two experiments that have not been performed with neutrons; an atom Hanbury Brown and Twiss experiment and a Berry's phase experiment with electric fields and integer spin particles.

Although not an interferometer in the same sense as described above, a conceptually simple application of atom beam splitters is the possibility of experimentally measuring the atom atom correlation functions in atom beams. The general picture of such experiments is shown in Figure 3, it is closely analogous the the Hanbury Brown and Twiss[11] experiment that measured second order correlations in photon counting. When performed using a 'classical' light source this experiment gives a coincidence rate at t=0 which is twice the rate at $t \to \pm\infty$. This may be interpreted as photon bunching due to the Bose statistics of the electromagnetic field. There has been much recent interest in this phenomena which has centered around the production of anti-bunched states of the of the electromagnetic field in which the coincidence rate goes to zero at t=0. Correlation experiments with atoms would give access to quantum counting statistics in a fundamentally different regime: unlike photons, atoms are either bosons or fermions and is possible to define a positional wave function for atoms. One expect that given Δt (defined below) small enough the coincidence rate at t=0 will be zero for a beam of fermions and will be twice the rate at $t \to \pm\infty$ for a beam of Bosons from a thermal source.

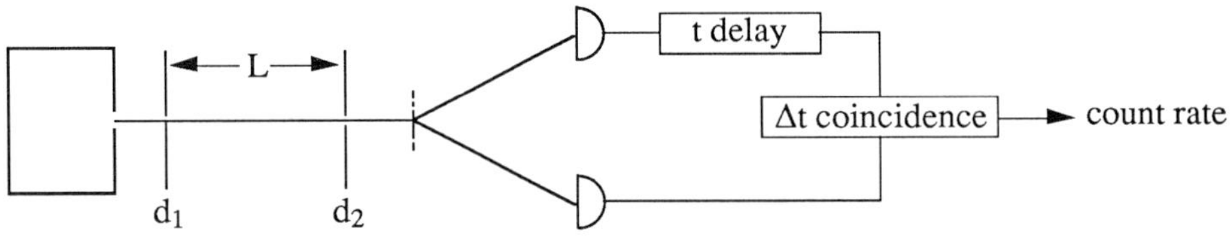

Figure 3. Schematic of an atom Hanbury Brown and Twiss experiment. d_1 and d_2 are the diameters of the collimation pin-holes.

The practical possibility of performing such an atom correlation experiment depends on the expected counting rate. We will calculate the coincidence counting rate as a function of the following beam parameters. At first we will assume that the experimental parameter Δt, the time window within which counts are registered as coincident can be chosen as small as is necessary.

Quantity	Symbol	Typical value
source brightness	B	10^{17}-10^{21} sr^{-1} sec^{-1} cm^{-2}
speed ratio	$s=v/\Delta v$	1-10^{-3}
mass	m	1-100 AMU
mean velocity	v	10^{2}-10^{5} cm sec^{-1}

We want the detectors to sample the same single transverse mode of the atom wave function. Therefore, the second aperture must fit within the diffraction pattern of the first i.e. $d_2 = L\,\lambda_{dB}/d_1$, which implies that the flux through the second slit is given by (ignoring factors of $\pi/4$)

$$f = B\,d_1^2\left(\frac{d_2^2}{L^2}\right) = B\lambda_{dB}^2$$

In addition to requiring that the detectors sample the same transverse mode we will also choose Δt to be equal to the atom coherence time τ so that the detectors sample a single longitudinal mode.

$$\tau = \frac{\lambda_{dB}}{v}s$$

If we assume that the counting rate is low, i.e. that the number of counts in Δt is small;

$$f\Delta t = \frac{\lambda_{dB}^3 s}{v} \ll 1$$

then the coincidence rate given by Poisson statistics is

$$f^2\Delta t = \frac{B^2 S h^5}{m^5 v^6} \to \frac{10^{-12} B^2 S}{m^5 v^6} \quad \text{(cgs with m in AMU)}$$

We believe that it will soon become experimentally feasible to measure the effects of atom-atom correlation in beams. An ideal system would be metastable He for which laser slowing techniques and fast detectors are readily available. For example; a laser slowed He* source with a final velocity of 100 cm sec^{-1} and a brightness of 10^{13} sr^{-1} sec^{-1} cm^{-2} seems experimentally realizable, and would give a coincidence counting rate of 1 Hz. In this case the coherence time is $\tau = 10^{-7}$ sec, and one can make Δt equal to (or even less than) τ by counting the He* directly on electron multipliers.

Atom interferometers are an ideal system in which to investigate the predictions by M. V. Berry[12], Aharonov, and Anandan[13] regarding modifications to the adiabatic theorem. Despite the numerous recent tests of Berry's phase, atom interferometers allow a test of the theory which is novel in several respects. A Berry phase experiment involving the effect of electric fields on Na would be the first such experiment involving non-zero mass bosons and the first where the perturbing field appeared quadratically in the hamiltonian. It should also be possible to test the Aharonov-Anandan geometrical phase in the case of non-adiabatic change.

Atomic Physics

An atom interferometer measures any interaction which differentially affects the energies of particles traveling along separate paths through the interferometer. Thus, atom interferometers could be used to measure such quantities as the electric polarizability or magnetic susceptibility of atomic ground states, or to measure a basic null effect such as the charge neutrality of atoms. In order to determine what problems atom interferometers are most suitable for we must consider the factors which limit the precision of interferometric measurements. The relative precision with which a white fringe interferometer can be used to measure a differential energy shift ΔE is limited by the number of visible fringes, which is approximately given by the speed ratio (S) of the atom beam. The relative precision of energy measurement, $\Delta E/E$ is limited to $\Delta f/S$ where Δf is the fractional accuracy of fringe resolution. If the interferometer is shot noise limited $\Delta f \propto 1/\sqrt{n}$ where n is the total number of atoms counted to determine the phase of the interferometer fringe. These considerations suggest that atom interferometers may be most profitably employed as balance (null) meters i.e., when used to balance the effects of two different interactions applied to opposite sides of the interferometer. For example, an interferometer only slightly more advanced than our first device should be able to measure the ground state polarizability of sodium to $\sim 10^{-2}$ but, its ability to measure the ratio of polarizability to magnetic susceptibility would be limited only by the precision with which the strength of the individual fields could be controlled — perhaps two orders of magnitude better.

ATOM OPTICAL ELEMENTS

Most of the recent work in atom optics has involved the use of light pressure forces to manipulate the atom beams. Mirrors[14], lenses[15], and gratings for atoms have been demonstrated using the stimulated gradient forces on atoms in near-resonant optical radiation. We will not discuss these developments, instead we will review atom optical elements that do not involve light forces. we ignore light force atom optics both to contain the discussion and because of the obvious advantages of developing atom optical elements that are independent of laser technology. It is interesting to note that all the grating types (except the Kapitza-Dirac effect and the reflection/transmission grating) described above have been realized for neutrons and for X-rays. It is fruitful to look for alternative atomic optical elements based on the technology developed for x-ray optics. It should be possible to adopt grazing incidence x-ray mirrors, lenses, and diffraction gratings for use with atom beams. These techniques are based on the specular reflection of atoms from smooth surfaces, which may occur when the surface roughness is much less than the wavelength corresponding to the momentum of the atom perpendicular to the surface. For example, efficient specular reflection of reactive alkali atoms with thermal velocity at angles of up to 40 mrad has recently been reported by Haroche et al[16]. At least two reflective lenses for atoms have recently been demonstrated. Doak has made a cylindrical lens for a He beam by reflecting it off an Au coating on a bowed mica wafer at angles of about 30 degrees[17]. A most favorable system for demonstrating atomic reflection is the reflection of H off of films of He at cryogenic temperatures. Berkhout et al[18] have made a spherical mirror coated with liquid He that focuses an 18 mm diameter beam of H-atoms down to 0.5 mm. Further progress in reflective atom optics is hampered by the deficiency of theoretical or empirical knowledge of the necessary conditions for the reflection of atoms, especially slow atoms, from surfaces.

Another class of x-ray optical elements that could be adapted for use with atoms is based on transmission through micro-fabricated structures. Atom optical elements based on transmission have the advantage that they work for any atomic species independent of surface physics or laser technology. Since our demonstration of transmission diffraction gratings for atoms we have used similar methods[19] to produce 200 nm-period gratings as thin as 5 nm. These gratings can be tilted so as to increase their effective dispersive power, in addition they are a first step towards the reflection/transmission gratings described above. Fabrication methods similar to ours have been used to produce free standing zone-plates which should work as lenses for atom beams.

Slow sources

A key barrier to practical use of most of the atom optical devices discussed above is the poor brightness of existing slow atom sources. A number of radiation pressure atom slowers[20] have been demonstrated. They all work by arranging that an atom decelerating in the slower is continually exposed to radiation that is tuned slightly to the red of the atomic resonance and is directed opposite to the atomic velocity. This Doppler tuning condition may be met either by frequency chirping the laser or by Zeeman tuning the atom's resonance. In either case the atoms accumulate random transverse momentum due to the scattering of the incident photons, the rms transverse momentum is proportional to the square root of the number of photons needed to slow the atom times the total change in atom momentum.

The tools necessary to increase the brightness of slowed beams are available, but they have not as yet been assembled into a bright slow source. The simplest way to increase brightness is to apply transverse cooling in the form of 'red molasses' to the atoms emerging from the end of the slower. A more powerful general method for increasing brightness is to first apply transverse cooling followed by a lens (which alone, increases flux but not brightness), followed by a second region of transverse cooling at the focus of the lens. Another possibility would be to replace the cooler-lens-cooler combination with a single two dimensional spontaneous force optical trap. It is clear that there are no theoretical barriers to the development of laser slowed and intensified atom sources — the development of such sources is a worth while challenge for experimentalists in atom optics.The work on beam splitters was funded by the National Science Foundation (PHY86-05893) with help from the Joint Services Electronics Program (DAAL03-86-K-0002) which supports the M.I.T. Submicron Structures Laboratory. Work on the Atom interferometer is supported by O.N.R. (N0001489-J-1207) and A.R.O. (DAA L03-89-K-0082).

REFERENCES

[1] I. Estermann and O. Stern, Z. *Physik* **61**, 95 (1930).

[2] P. E. Moskowitz, P.L. Gould, S. R. Atlas, and D.E. Pritchard, *Phys. Rev. Lett.* **51**, 370 (1983); P. J. Martin, B. G. Oldaker, A. H. Miklich, and D. E. Pritchard, *Phys. Rev. Lett.* **60**, 515 (1988).

[3] D. W. Keith, M. L. Schattenburg, Henry I. Smith, and D. E. Pritchard, *Phys. Rev. Lett.* **61**, 1580 (1988).

[4] B. J. Chang, R. Alferness, and E. N. Leith, *Appl. Optics,* **14**, 1592 (1975). For the specific case of atom interferometers see V.P. Chebotayev et al., *J. Opt. Soc. Am. B* , **2**(11), 1791 (1985).

[5] Steven J. Wark, William A. Hamilton and Goffrey I. Opat, *J. Modern Optics.*, **34**, 1375 (1987); D. E. Pritchard and D. W. Keith, U.S. Patent pending.

[6] E. H. Anderson, A. M. Levine, and M. L. Schattenburg, *Appl. Optics Lett.*, **27**, 3522 (1988).

[7] L. E. Stodolsky, Gen. Rel. and Gravitation. **11**, 391 (1979).

[8] M. O. Scully, M. S. Zubairy, and M. P. Haugan, *Phys. Rev. A*, **24**, 2009 (1981).

[9] C. W. Misner, K. S. Thorne, and J. A. Wheeler, *Gravitation* (Freeman, San Francisco, 1973), p. 1117.

[10] J. D. Fairbank, B. S. Deaver Jr, C. F. W. Everitt, and P.F. Michelson eds., *Near Zero* (W. H. Freeman and company, New York, 1988) VI.3.

[11] R. Hanbury Brown and R. Q. Twiss, *Nature*, **177**, 27 (1956).

[12] M. V. Berry, *Proc R. Soc. Lond. A*, **392**, 45 (1984).

[13] Y. Aharonov and J. Anandan, *Phys. Rev. Lett.*, **58**, 1593 (1987).

[14] V. I. Balykin, V. S. Letokhov, et al., *JETP Lett.* **45**, 353 (1987); V. I. Balykin, V. S. Letokhov, et al., Phys. Rev. Lett. **60**, 2137 (1988).

[15] J.E. Bjorkholm, R.R. Freeman, A. Ashkin, and D.B. Pearson, *Phys. Rev. Lett.* **41**, 1361 (1978); V.I. Balykin and V.S. Letokhov, Opt. Com. **64**(2),151 (1987)

[16] A. Anderson, S. Haroche, E. A. Hinds, W. Jhe, D. Mexchede, and L. Moi, *Phys. Rev. A* , **34**, 3513 (1986).

[17] Bruce Doak (AT&T), personal communication.

[18] J. J. Berkhout, O. J. Luiten, I. D. Setija, T. W. Hijmans, T. Mizusaki, and J. T. M. Walraven, *Phys. Rev. Lett.*, **63**, 1689 (1989).

[19] A. M. Hawryluk, N. M. Ceglio, R. H. Price, J. Melngailis, and H. I. Smith, *J. Vac. Sci. Technol.*, **19**(4), 897

(1981); N.M. Ceglio, A.M. Hawryluk, and R.H. Price., Proc. S.P.I.E. **316** *(High Resolution Soft X-ray Optics)*, 134 (1981); E. H. Anderson, C. M. Horwitz, and H. I.Smith, *Appl. Phys. Lett.*, **49**, 874 (1983); H. I. Smith, E. H. Anderson, A. M. Hawryluk, and M. L. Schattenburg, in *X-Ray Microscopy*, (Springer Series in Optical Sciences, vol 43), eds. D. Rudolph and G. Schmahl,
(Springer-Verlag), Berlin, Heidelberg, 1984.
[20] William D. Phillips, John V. Prodan, and Harold J. Metcalf, *JOSA-B*, **2**, 1751 (1985).

NEUTRON OPTICS

Mirjana Božić

Institute of Physics, P.O.Box 57
YU-11001 Beograd, Yugoslavia

I. INTRODUCTION

From this large field called "Neutron optics" in this lecture I will speak about the topic which became important during last fifteen years and to which we could give the title: "Physical reality in quantum domain seen through neutron interferometer". Accordingly, the study of Schrödinger and of Bohr-Heisenberg-Jordan-von Neuman meaning of ψ function is given in Chapter II. Neutron interferometer is described in Chapter III. Three groups of experiments with neutron interferometer are reviewed and analyzed in Chapter IV.

II. TWO APPROACHES IN QUANTUM THEORY

A. Schrödinger real ψ field

Schrödinger discovery[1,2] of wave mechanics represents one specific synthesis of mechanics, wave theory and Planck quantum hypothesis.

Basic elements from mechanics are the classical Hamilton and Lagrange functions for a particle

$$H = \frac{\vec{p}^2}{2m} + V(\vec{r}) = T + V \tag{1}$$

$$L = \frac{\vec{p}^2}{2m} - V(\vec{r}) = T - V \tag{2}$$

New Frontiers in Quantum Electrodynamics and Quantum Optics
Edited by A. O. Barut, Plenum Press, New York, 1990

and the principle of least action

$$\delta \int_{t_0}^{t} 2T\,dt = \delta \int_{A}^{B} [2m(E-V)]^{1/2} ds = 0 \tag{3}$$

Basic elements from wave theory are the wave function $\psi(\vec{r}, t)$ required to satisfy the wave (d'Alembert) equation

$$\Delta\psi - \ddot{\psi}/u^2 = 0 \tag{4}$$

and Fermat's principle

$$\delta \int_{A}^{B} \frac{ds}{u} = 0 \tag{5}$$

where u is the wave (phase) velocity.

The synthesis of mechanical and wave elements is now realized with the aid of Planck hypothesis and analogy between the principle of least action and Fermat's principle. The analogy leads to[3]

$$u = C/[2m(E-V)]^{1/2}. \tag{6}$$

Then one identifies the group velocity of the wave of frequency ν given by

$$1/v_g = \frac{d(\nu/u)}{d\nu} \tag{7}$$

with particle velocity $v = p/m$ and find the relationship between the phase velocity and the kinetic and total energies of the particle

$$u = E/(2mT)^{1/2}. \tag{8}$$

Schrödinger then looked for those solutions of the wave equation (4) whose time dependence is of the form

$$\psi(\vec{r}, t) = \phi(\vec{r}) \cdot \exp(-iEt/K) \tag{9}$$

where the constant K must have the physical dimension of action (energy time). Now since the frequency of the wave is obviously

$$\nu = E/2\pi K \tag{10}$$

Schrödinger could not "resist the temptation"[2] to use Planck's relation

$$E = h \cdot \nu \tag{11}$$

i.e. to put K equal to $h/2\pi = \hbar$.

By combining the latter relation with (8) one finds for the wavelength

$$\lambda = h/[2m(E - V)]^{1/2} = h/p \tag{12}$$

which is the de Broglie relation originally derived[4] relativistically, Schrödinger's derivation is nonrelativistic.

By introducing equations (8), (9) and (11) into d'Alembert equation (4) Schrödinger found the equation for amplitude $\phi(\vec{r}\,)$

$$\nabla^2\phi(\vec{r}\,) + \frac{2m}{\hbar^2}(E - V)\cdot\phi(\vec{r}\,) = 0. \tag{13}$$

In order to get the equation for a function $\psi(\vec{r}, t)$ whose time dependence is not restricted to the dependence of the form $exp(-iEt/\hbar)$ Schrödinger later, in the fourth paper of Ref.1, used the property of the stationary solution

$$E\psi(\vec{r}, t) = i\hbar\frac{\partial\psi(\vec{r}, t)}{\partial t} \tag{14}$$

to eliminate the term $E\phi(\vec{r}\,)$ from (13) and arrived at

$$-(\hbar^2/2m)\nabla^2\psi + V\psi = i\hbar\partial\psi/\partial t. \tag{15}$$

Therefore, in Schrödinger theory ψ is a wave function of the real field. The derivation of the equation is ***independent*** of the nature of this field. That means, the question of the nature of the field is a separate one. In answering this question Schrödinger made the hypothesis that the charge of the electron is not concentrated in a point, but is spread out through the whole space, proportional to the quantity $\psi\psi^*$. In the case of more than one, say of N, electrons the real continuous partition of the charge is a sort of mean of the continuous multitude of all possible configurations of the corresponding point-charge model, the mean being taken with the quantity $\psi\psi^*$ as a sort of weight-function in the configuration space. The physical meaning given to ψ by Schrödinger has not been accepted by Bohr, Born, Heisenberg, Jordan and von Neumann who developed the scheme in which ψ is probability amplitude and $\psi\psi^*$ is probability density. Schrödinger original interpretation of ψ has been revived[5] and elaborated[6] by Barut and colaborators.

B. Statistical interpretation of $|\psi|^2$

According to the statistical interpretation[7,8], $\psi(\vec{r}, t)$ is a probability amplitude of the particle's presence. Since the possible positions of the particle form a continuum, the probability $dP(\vec{r}, t)$ of the particle being, at time t, in a volume element

$d^3\vec{r} = dxdydz$ situated at the point $\vec{r}$ must be proportional to $d^3\vec{r}$. $|\psi(\vec{r},t)|^2$ is then interpreted as the corresponding probability density, with

$$dP(\vec{r},t) = C\ |\psi(\vec{r},t)|^2 d^3\vec{r} \tag{16}$$

where C is a normalization constant.

Quantum mechanics associates with the probability amplitude the superposition principle or/and the principle of spectral decomposition. This "or/and" reflects the fact that the two principles are mutually connected but not completly equivalent.

Superposition principle for probability amplitudes is equivalent to the superposition principle for wave fields, which states that the resultant field ψ in a medium with many sources of the field is the sum or linear superposition of fields (φ_1, φ_2) from different sources.

$$\psi(\vec{r},t) = C_1\varphi_1(\vec{r},t) + C_2\varphi_2(\vec{r},t) \tag{17}$$

In the case of material particles the amplitudes $\varphi_i(\vec{r},t)$ are associated with different routes,"trajectories" of the particle.

The principle of spectral decomposition applies to the measurement of an arbitrary physical quantity and states: *i*) The result found must belong to a set of eigen results $\{a_i\}$, *ii*) With each eigenvalue a_i is associated an eigenstate, that is, an eigenfunction $\psi_i(\vec{r})$. This function is such that, if $\psi(\vec{r},t_0) = \psi_i(\vec{r})$ (where t_0 is the time at which the measurement is performed), the measurement will always yield a_i. *iii*) For any $\psi(\vec{r},t)$, the probability P_i of finding the eigenvalue a_i for a measurement at time t_0 is found by decomposing $\psi(\vec{r},t_0)$ in terms of the functions $\psi_i(\vec{r})$:

$$\psi(\vec{r},t_0) = \sum_i C_i\psi_i(\vec{r}). \tag{18}$$

Then:

$$P_i = |C_i|^2 / \sum_j |C_j|^2 \tag{19}$$

If we square left and right hand side of the superposition principle we obtain the equality:

$$|\psi(\vec{r},t)|^2 = \sum_i |C_i|^2|\psi_i(\vec{r})|^2 + \sum_{\substack{i,j \\ i\neq j}} C_i^* C_j \psi_i^*(\vec{r})\psi_j(\vec{r}) \tag{20}$$

which is in the heart of our difficulties to understand physical reality in quantum domain, or more specifically, to understand the content of the notion "wave-particle duality". Namely, probability density of the particle presence on the left hand side of the above equality is expressed through two kinds of quantities: ones $(|C_i|^2, |\psi_i|^2)$ which are probabilities and probability densities and therefore have physical reality and meaning and the other ones ($C_i = |C_i|e^{i\gamma_i}, C_i^* = |C_i|e^{-i\gamma_i}, \psi_i = |\psi_i|e^{i\beta_i}, \psi_i^* = |\psi_i| \cdot e^{-i\beta_i}$ which are complex numbers with phases γ_i, β_i which according to the statistical interpretation have no physical reality and no interpretation.)

Shortly, probability density on the left hand side is equal to the sum of two terms, first term is the sum of products of probabilities and probability densities and therefore is the probability density whereas second term contains quantities which are neither probabilities neither probability densities. This is a consequence of the fact that probabilities (19) are introduced on the basis of the superposition law (18) for probability amplitudes and not on the basis of the expression (20) for probability density. Consequently the latter term has no name in classical theory of probability whereas in statistical interpretation of quantum mechanics it has been named[9] the "interference term". But the notion of interference comes from wave theory so that the right hand side of the relation (20) is the sum of terms representing probabilities being concepts appropriate for particle picture and of interference terms which are associated with wave picture. This relation expresses in the most direct way the content of the notion "wave-particle duality" as understood by de Broglie[10] and de Broglie school[11,12], that means that quantum objects (quantons[13]) are "particles and waves" but not "particles or waves" as claimed by Bohr[14] and Copenhagen interpretation.

The differences in the understanding of two schools have been expressed recently in an original way by Rauch[15]: "All the results of the neutron interferometric experiments are well described by the formalism of quantum mechanics. According to the complementarity principle of the Copenhagen interpretation, the wave picture has to be used to describe the observed phenomena. The question how the well-defined particle properties of the neutron are transferred through the interferometer, is not a meaningful within this interpretation, but it should be an allowed one from the physical point of view. Therefore, other interpretations should also be included in the discussion of such experiments.... As an experimentalist, one appreciates the pioneering work of the founders of quantum mechanics, who created this basic theory with so little experimental evidence. Now we have much more direct evidence, even on a macroscopic scale but, nevertheless, one notices that the interpretation of quantum mechanics goes beyond human intuition in certain cases."

Similar questions and views are expressed in quantum optics by Loudon[16] and Glauber[17]. Loudon writes[16]:" For Young's experiment each photon must be capable of interfering with itself in such a way that its probability of striking the second screen at a particular point is proportional to the calculated intensity at that point. This can be achieved only if each photon passes partly through both pinholes, so that it can have a knowledge of the entire pinhole geometry as it strikes the screen. There is indeed no way in which one can simultaneously assign a photon to a particular pinhole and record the interference pattern. If a phototube is placed behind one of the pinholes to detect photons passing through, then it is not possible to avoid obscuring that pinhole, with consequent destruction of the interference pattern.

These remarks are in agreement with the principles of quantum mechanics. Photons do not interact with each other, and interference effects must be sought in the process by which each single photon passes from the source to the second screen. Quantum-mechanically, the interference occurs between the probability amplitudes for passage from source to screen via the two different paths corresponding to the two pinholes. The intensity on the second screen is proportional to the square modulus of the sum of the two probability amplitudes".

Instead of Young experiment, with two detectors Glauber have proposed and analyzed[17a] Young experiment with an amplifier behind each pinhole and came to the similar conclusion[17b] that each amplifier should get an information from both slits.

Recently, Barut[5b] advanced the thesis that the controversies about the meaning of the function ψ could be solved by making clear distinction between two different notions of the wave function ψ: one (denoted by a small ψ) describing a single quantum system in a single event, the other (denoted by a capital Ψ) describing the typical behaviour of a single quantum system in repeated experiments. The two wave functions correspond to different types of experiments and answer to different type of questions.

III. NEUTRON INTERFEROMETER

Neutron interferometer consists of three identical plates, each having the role to split each incident neutron beam (wave) in two beams.

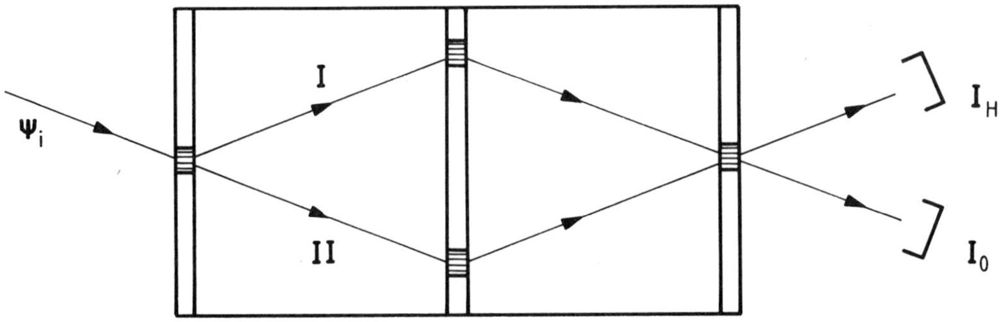

Fig.1. Horisontal cross section of neutron interferometer

Splitting of the beam has been nicely explained[19] in the theory of dynamical neutron diffraction (Appendix). The result of splittings on three plates is that wave functions in the directions I_0 and I_H behind the interferometer are composed of wave functions coming from paths I and II

$$\psi_0 = \psi_0^I + \psi_0^{II}$$
$$\psi_H = \psi_H^I + \psi_H^{II} \tag{21}$$

Moreover, in forward direction I_0 the relation

$$\psi_0^I = \psi_0^{II} = \psi_0 \tag{22}$$

is fulfilled.

The series of experiments with neutron interferometer consist in the application of a device along one of two paths. The effect of this device on the wave function $\psi_0^{II} = \psi_0^I$ is described with the aid of quantum mechanical operator $\widehat{U}$ as follows:

$$\psi_0^{II'} = \widehat{U}\psi_0^{II} = \widehat{U}\psi_0^I. \tag{23}$$

Consequently, wave function and intensity in the detector I_0 in case that the device is inside the interferometer are:

$$\psi_0 = \psi_0^I + \psi_0^{II'} = \psi_0^I + \widehat{U}\psi_0^I \tag{24}$$

$$I_0 = |\psi_0^I + \widehat{U}\psi_0^I|^2 \tag{25}$$

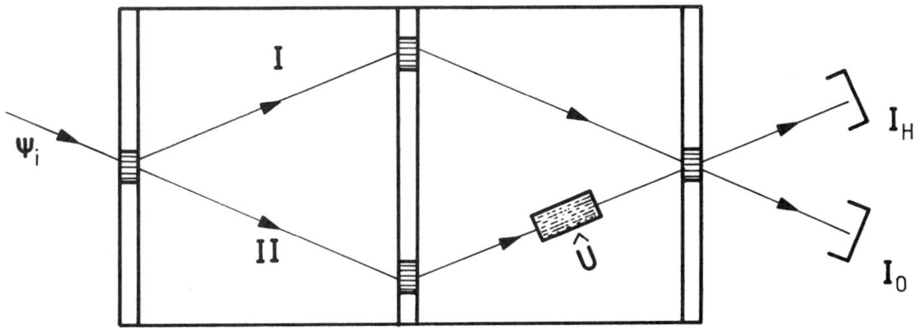

Fig.2. Neutron interferometer supplied with quantum mechanical operator.

IV. NEW INSIGHTS BROUGHT BY NEUTRON OPTICS

Here, we will review and analize three groups of experiments, each group associated with a particular device and particular operator $\widehat{U}$:

$$\text{a)}\quad \widehat{U} = \exp i\chi \quad \text{: specific crystal plate-phase shifter} \tag{26a}$$

$$\text{b)}\quad \widehat{U} = \exp -i\frac{\alpha}{2}\widehat{\sigma}_z \quad \text{: magnetic field-spin rotation} \tag{26b}$$

$$\text{c)}\quad \widehat{U} = \exp -i\frac{\pi}{2}\widehat{\sigma}_y \cdot \exp i\chi \quad \text{: Mezei coil combined with crystal plate-spin flip combined with phase shifter} \tag{26c}$$

In the experiments a) and b) the dependence of intensity I_0 on χ and $\alpha/2$, respectively, has been investigated[20,21]. In the experiment c) the dependence of polarization on χ is studied[22,23]. All experimental results are in perfect agreement with following simple theoretical expressions:

$$\text{a)}\quad I_0 = 2|\psi_0^I|^2(1+\cos\chi) \tag{27a}$$

$$\text{b)}\quad \frac{I_0(\alpha)}{I_0(0)} = \frac{1}{2}(1+\cos\frac{\alpha}{2}) \tag{27b}$$

$$\text{c)}\quad I_0 = \text{const}; \qquad \vec{P} \equiv \psi_0^+\widehat{\sigma}\psi_0 = \cos\chi\vec{i} + \sin\chi\vec{j} \tag{27c}$$

A. Experiments with phase shifter stress the importance of relative phases

The big progress made by neutron interferometry[24] is in stressing and demonstrating in many different ways the importance and physical reality of relative phases. In the experiment with crystal plate (phase shifter, see Fig. A1) the number of neutrons in the detector I_0 changes with the change of the width of the plate (being proportional to relative phase) despite the fact that the crystal plate does not change the probability that neutron takes one of two paths inside the interferometer.

We know mathematically from the expression (27a), that with the change of the width of the plate intensity I_0 changes because of the change of the relative phase of two beams, but in the statistical interpretation we do not know the physical mechanism of this change because we do not know the physical meaning of χ. So the puzzle is how the plate influences *each* neutron, independently of which way neutron chooses. This puzzle can not be solved by arguing, as Bohr would argue, that it is impossible simultaneously to register the interference and to know which path neutron takes. In order to understand how the plate influences the behavior of *each* neutron at the exit from the interferometer, we do not need to know "which path neutron takes" because we see that phase shifter influences each neutron independently of which path it has taken. This ingenious experiment which demonstrates directly the dependence of I_0 on χ shows that the information on the path of the particular neutron is neither relevant nor necessary for the (non)understanding of the physical mechanism of the dependence of I_0 on χ.

B. On the interpretation of the spinor phase change in a magnetic field in neutron interferometry

The standard explanation of the experiment[25] with static magnetic field along path II is based on the fact that the spin dependent part of the time evolution operator

$$U_S(t) = \exp(-iH_S t/\hbar) = \exp(i\mu\hat{\sigma}_z Bt/\hbar) \tag{28}$$

has the same form as the operator of rotations around the z-axis acting on the spinors

$$R_Z(\alpha) = \exp(-i(\alpha/2)\sigma_z) \tag{29}$$

with

$$\alpha = -(2\mu B/\hbar)t = 2|\mu|Bt/\hbar \tag{30}$$

where $\mu = (g/2)(e\hbar/2Mc) = -6.031 \cdot 10^{-8}(eV/T)$ is the magnetic dipol moment of the neutron ($g = -3.82$).

Neglecting the reflection on the field boundaries and assuming the relation $L = tv$ between the length of the field L, the time of passage t and the velocity of neutrons v, the following basic relations between spinors associated with paths (I) and (II) have been utilized

$$\psi_0^{II} = \psi_0^I e^{-i\alpha/2} \quad \text{if} \quad \psi_0^I = \phi_0^I(\vec{r})\begin{pmatrix}1\\0\end{pmatrix} \tag{31a}$$

$$\psi_0^{II} = \psi_0^I e^{i\alpha/2} \quad \text{if} \quad \psi_0^I = \phi_0^I(\vec{r})\begin{pmatrix}0\\1\end{pmatrix} \tag{31b}$$

where

$$\alpha = 2|\mu|BL/\hbar v = 2|\mu|BmL/\hbar^2 k \tag{32}$$

with $v = \hbar k/m$.

Experimental results[21] for unpolarized neutron beam confirm almost ideally theoretical dependence:

$$\frac{I(\alpha)}{I(0)} = \frac{|{}^+\psi_0^I + e^{-i\alpha/2}\cdot{}^+\psi_0^I|^2 + |{}^-\psi_0^I + e^{i\alpha/2}\cdot{}^-\psi_0^I|^2}{|{}^+\psi_0^I + {}^+\psi_0^I|^2 + |{}^-\psi_0^I + {}^-\psi_0^I|^2} = \frac{1}{2}[1 + \cos(\alpha/2)] \tag{33}$$

A rather different explanation has been advanced by Mezei[26,27] according which the phase shift $\mp\alpha/2$ between ψ_0^{II} and ψ_0^I is in fact the "accumulated" phase shift

$$\beta_\pm = \int \delta k_\pm(r)dr = \mp\int (m|\mu|B(r)/\hbar^2 k)dr \tag{34}$$

due to the longitudinal Stern-Gerlach effect[28]. If the field is homogeneous this phase shift is equal to:

$$\beta_{\pm} = \mp m|\mu|BL/\hbar^2 k, \qquad \text{(hence equal to } \mp\alpha/2) \tag{35}$$

Mezei correctly argues[26] that magnetic field acts as a conservative potential and that momentum of the neutron inside the field has two values, $p' = \hbar k'$ and $p'' = \hbar k''$, determined by the relations

$$\hbar^2 k^2/2m = \hbar^2 k'^2/2m + |\mu|B = \hbar^2 k''^2/2m - |\mu|B \equiv \hbar\omega' + |\mu|B \equiv \hbar\omega'' - |\mu|B \tag{36}$$

where $p = \hbar k$ and $E_k = \hbar^2 k^2/2m$ are momentum and kinetic energy of the free neutron.

But, can we associate the phase difference between ψ_0^{II} and ψ_0^{I} with the above changes of momenta? Evidently, Mezei's argument uses the analogy with optics. In optics, the phase difference of two plane waves φ_f and φ_0, one passing through the medium and the other not, in the region $x > L$, is equal to $\Delta\varphi = [kx + (k' - k)L - \omega t] - [kx - \omega t] = (k' - k)L$, because the frequency of light ν satisfying $E = h\nu = \hbar\omega$ does not change in the medium. Only the phase velocity $c = \lambda\nu = 2\pi\nu/k$ of light changes and takes, let us say, the value c'. Consequently the wave vector k changes into $k' = 2\pi\nu/c'$.

Neutron wave vector (as well as photon wave vector) takes new value inside the medium. But this fact alone is not sufficient to conclude that the phase shift induced by magnetic field is "accumulated" phase shift due to longitudinal Stern-Gerlach effect, and not, due to the change of phase associated with spin rotations[29] By factorising appropriatly two basic time-dependent neutron state functions

$$\begin{aligned} \psi'(x,t) &= e^{ik'x}\begin{pmatrix}1\\0\end{pmatrix} \cdot e^{-i[(\hbar^2 k'^2/2m)+|\mu|B]t/\hbar} \\ &\equiv e^{i(k'x-\omega' t)} \cdot e^{-i(\omega-\omega')t}\begin{pmatrix}1\\0\end{pmatrix} \\ \psi''(x,t) &= e^{ik''x}\begin{pmatrix}0\\1\end{pmatrix} e^{-i[(\hbar^2 k''^2/2m)-|\mu|B]t/\hbar} \\ &\equiv e^{i(k''x-\omega'' t)} \cdot e^{-i(\omega-\omega'')t}\begin{pmatrix}0\\1\end{pmatrix} \end{aligned} \tag{37}$$

we found that the phases of coordinate and spin parts of the states ψ' and ψ'' are:

$$\varphi_c' = k'x - \omega' t, \qquad \varphi_c'' = k''x - \omega'' t \tag{38}$$

$$\varphi_s' = -(\omega - \omega')t = |\mu|Bt/\hbar, \qquad \varphi_s'' = -(\omega - \omega'')t = -|\mu|Bt/\hbar \tag{39}$$

It follows now that magnetic field of length L (such that $|\mu|B \ll \hbar^2 k^2/2m$) in one arm of the interferometer induces the following phase difference in the coordinate part of the neutron wave function

$$\Delta\varphi_c' = (k' - k)L - (\omega' - \omega)t, \qquad \Delta\varphi_c'' = (k'' - k)L - (\omega'' - \omega)t. \tag{40}$$

The rest is the phase change in the spin part of ψ:

$$\Delta\varphi_s' = (\omega' - \omega)t, \qquad \Delta\varphi_s'' = (\omega'' - \omega)t. \tag{41}$$

In the usual approximation of weak fields we have the approximative relations

$$k' - k = -\Delta k = \mu Bm/\hbar^2 k = -|\mu|Bm/\hbar^2 k = k - k'' \tag{42}$$

and

$$(\omega' - \omega)t = \hbar t(k'^2 - k^2)/2m \simeq -\Delta k \cdot L \qquad \text{using} \qquad \hbar tk/m = L$$

$$(\omega'' - \omega)t = \hbar t(k''^2 - k^2)/2m \simeq \Delta k \cdot L \tag{43}$$

Hence,

$$\Delta\varphi_c' \simeq \Delta\varphi_c'' \simeq 0$$

$$\Delta\varphi' = \Delta\varphi_s' + \Delta\varphi_c' \simeq \Delta\varphi_s' \simeq -\Delta k \cdot L, \qquad \Delta\varphi'' = \Delta\varphi_s'' + \Delta\varphi_c'' \simeq \Delta\varphi_s'' \simeq \Delta k \cdot L \tag{44}$$

Therefore, the phase change of the whole wave function is identical with the spinor phase change. We conclude that the spinor phase change, of a particle moving through a magnetic field exists independently of the change of its momentum (and follows for weak fields the law of transformation of spinors under rotation), in the same way as it exists for the spinor at rest in the magnetic field.

C. Verification of spin state superposition law

The experiments[22,23] with spin flipper and phase shifter verify directly from right to left the spin state superposition law

$$|\frac{\pi}{2}, \chi\rangle = \left[\begin{pmatrix}1\\0\end{pmatrix} + \exp i\chi \begin{pmatrix}0\\1\end{pmatrix}\right] / \sqrt{2} \tag{45}$$

according which the superposition at the right hand side is the eigenstate of the operator

$$\widehat{S}_{xy} = \widehat{S}_x \cos\chi\vec{i} + \widehat{S}_y \sin\chi\vec{j} \tag{46}$$

Experiment confirms that the beam I_0 has no component of polarization parallel to that of interfering constituents, in accordance with the expression (27c). The importance of the experiment comes also from the fact that it is considered to be the realization of Wigner's gedanken experiment[30] and that it resolves the dilemma in the theory of measurement in favour of the orthodox interpretation. (With small change[31,32] in the experimental set up this experiment could be closer to the realization of Wigner gedanken experiment).

According to the orthodox interpretation of quantum measurement initial state vector of the system—object-plus-apparatus is a simple product of the state φ_a of apparatus and of the state of the object $\sum_\nu C_\nu \varphi_0^\nu$. The measurement is described as the following change of the state vector of object-plus-apparatus

$$\varphi_a \times \sum_\nu C_\nu \varphi_0^\nu \quad \rightarrow \quad \sum_\nu C_\nu \varphi_a^\nu \times \varphi_0^\nu \tag{ξ}$$

The alternative view presupposes that the result of measurement is not a state vector, such as (ξ) but a so-called mixture, namely, *one* of the state vectors

$$\varphi_a^{(\mu)} \times \varphi_0^{(\mu)} \tag{η}$$

and that this particular state will emerge from the interaction between object and apparatus with the probability $|C_\mu|^2$. So, Wigner formulated the problem of measurement to be to answer to the question whether process (η) or process (ξ) is valid in quantum measurement. From the experiment with spin flipper and phase shifter it was concluded that (ξ) was valid and therefore that orthodox interpretation of quantum measurement was correct. We are not satisfied[31,32] with this conclusion because position coordinate of the particle is treated as apparatus, so that the conclusion has been derived on the basis of particular object-plus-apparatus system which is in fact spin-and-position coordinates of the particle. On the other hand, in dynamical theory of diffraction, positional coordinate of the neutron inside neutron interferometer is quantum-mechanical. Therefore, the assumption that positional coordinate may play the role of classical macroscopic apparatus is not in our oppinion consistent with the theory of neutron interferometer.

We would like more to see here the analysis of dilemma in quantum measurement theory in which neutron interferometer would be treated as apparatus and neutron (it's positional and spin coordinate) as object . Then, we would conclude from the results of the experiments[22,23] that the pure state of the object evolves inside the apparatus into a new pure state.

CONCLUSION

Neutron optics provided a variety of ingenious experiments for the study of the physical reality in quantum domain, in particular, for the study of the content of the notion wave-particle duality. Most of those experiments verify directly the dependence of the neutron intensity at the exit from the interferometer on the relative phase of two wave functions associated with two paths inside the interferometer. So, those experiments emphasize the role of phases and of relative phases–quantities which do not have interpretation in statistical interpretation of quantum mechanics. It is usually said that the results of those experiments agree with statistical predictions of the behaviour of the quantum ensemble. But, it is hard to understand this statistical behaviour because we do not have proper description of individual particles, what comes from the fact that we do not know the physical meaning of phases and of relative phases.

By arguing that the expression for intensity (probability density) derived from the superposition principle expresses in the most direct way the content of the notion wave–particle duality, as understood by de Broglie and de Broglie school, we conclude that neutron interferometry encourages the investigations based on the "particle and wave" concept of quantum objects.

APPENDIX: DYNAMICAL NEUTRON DIFFRACTION

One starts with Schrödinger equation assuming a strictly periodic potential $V(\vec{r}) = V(\vec{r} + \vec{R}_n)$ between the nuclei at positions $\vec{R}_n$ and the neutron of the energy E and mass m. With the aid of Bloch ansatz

$$\psi(\vec{r}) = u(\vec{r})e^{i\vec{k}\vec{r}} \tag{A.1}$$

and Fourier series for the function $u(\vec{r})$ and periodic potential $V(\vec{r})$

$$u(\vec{r}) = \sum_{\vec{G}} u(\vec{G})\exp(i\vec{G}\vec{r}), \qquad V(\vec{r}) = \sum_{\vec{G}} V(\vec{G})\exp(i\vec{G}\vec{r}) \tag{A.2}$$

Schrödinger equation leads to the homogeneous system of linear equations for $u(\vec{G})$.

$$\left[\frac{\hbar^2}{2m}(\vec{K} + \vec{G})^2 - E\right] u(\vec{G}) = -\sum_{\vec{G}'} V(\vec{G} - \vec{G}')u(\vec{G}') \tag{A.3}$$

In the case of thermal neutrons Fermi potential describes well neutron-nucleus interaction

$$V(\vec{r}) = \frac{2\pi\hbar^2}{m}\sum_i b_c\delta(\vec{r} - \vec{R}_i) \tag{A.4}$$

where b_c is the bound coherent scattering length, so that coefficients $V(\vec{G})$ can be written in the form:

$$V(\vec{G}) = \frac{1}{V}\int V(\vec{r})\exp(-i\vec{G}\vec{r})d\vec{r} = \frac{2\pi\hbar^2 b_c F_g}{mV_c} \tag{A.5}$$

V_c is the volume of the unit cell and F_g is the geometrical structure factor of the crystal. One-beam and two-beam approximations of equations (A.3) are important for neutron interferometer, first one for phase shifters and second one for beam splitters.

One-beam approximation is applicable if $|(K+G)^2 - k^2| \gg |2mV(0)/\hbar^2|, \forall G \neq 0$, and reads

$$\left[\frac{\hbar^2}{2m}K^2 - E\right] u(0) = -V(0)u(0) \tag{A.6}$$

The latter equation gives the relation between K inside the crystal and initial k of the neutron beam

$$K^2 = k^2 \left[1 - \frac{V(0)}{E}\right] \tag{A.7}$$

Now, in an analogy with optics, index of refraction has been introduced:

$$n \equiv \frac{K}{k} \approx 1 - \frac{V(0)}{2E} = 1 - \lambda^2 \frac{Nb_c}{2\pi}$$

Consequently, when conditions for one beam approximation are satisfied the whole effect of the crystal of length D reduces to the change of the phase of the neutron state by

$$\chi = (K - k)D = (n - 1)kD \tag{A.8}$$

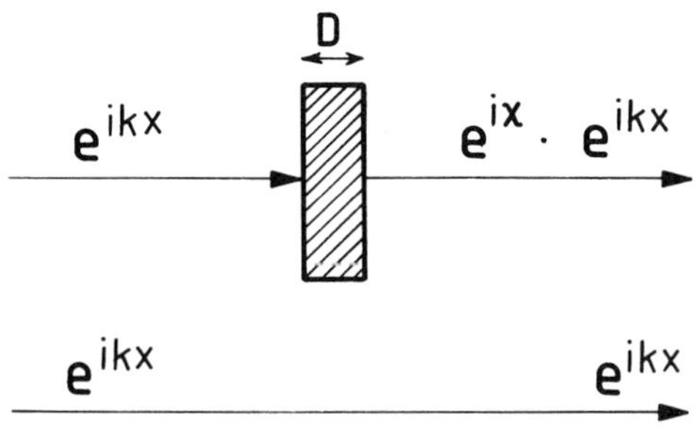

Fig.A1. Phase shifting with the aid of phase shifter

Two beam approximation is applicable if $|(K + G)^2 - k^2| \gg |2mV(0)/\hbar^2|$ for $\forall G$ except $\vec{G} = 0$ and $\vec{G} = \vec{G}_0$. Therefore, only $u(0)$ and $u(\vec{G}_0)$ are assumed to be different from zero so that system (A.3) reduces to:

$$\begin{aligned} \left[\frac{\hbar^2}{2m}K^2 - E + V(0)\right] u(0) + V(-\vec{G}_0)u(\vec{G}_0) &= 0 \\ V(\vec{G}_0)u(0) + \left[\frac{\hbar^2}{2m}(\vec{K} + G_0)^2 - E + V(0)\right] u(G_0) &= 0 \end{aligned} \tag{A.9}$$

By combining the secular determinant of the latter equation and boundary conditions for wave fields, one finds that within the crystal exist four excited waves having wave vectors $\vec{K}_1, \vec{K}_2, (\vec{K}_1 + \vec{G}_0)$, and $(\vec{K}_2 + \vec{G}_0)$ given by:

$$\begin{aligned} K_i^2 &= k^2(1 + 2\epsilon_i) \qquad i = 1, 2 \\ \vec{K}_i &= \vec{k} + (k\epsilon_i/\cos\gamma)\vec{n} \\ (\vec{K}_i + \vec{G}_0)^2 &= k^2(1 + 2\epsilon_i/b + \alpha). \end{aligned} \tag{A.10}$$

$$\epsilon_{1,2} = \frac{1}{4}\left\{-\alpha b - (1+b)\frac{V(0)}{E} \pm \sqrt{\left[\alpha b - (1-b)\frac{V(0)}{E}\right]^2 + 4b\frac{V(\vec{G}_0)V(-\vec{G}_0)}{E^2}}\right\} \quad (A.11)$$

where

$$\begin{aligned} \alpha &= (G_0^2 + 2\vec{k}\cdot\vec{G}_0)/k^2 \\ b &= \cos\gamma/\cos\gamma_{G_0} \end{aligned} \quad (A.12)$$

and γ and γ_{G_0} are the cosines of the angles between the normal on the surface and the incident and reflected beam, respectively. Shortly, whole wave function consists of two wave fields (1 and 2) such that

$$\psi^{1,2} = \psi_0^{1,2} + \psi_{G_0}^{1,2} \quad (A.13)$$

so that the effect of the crystal on the initial beam is as represented on Fig.A2.

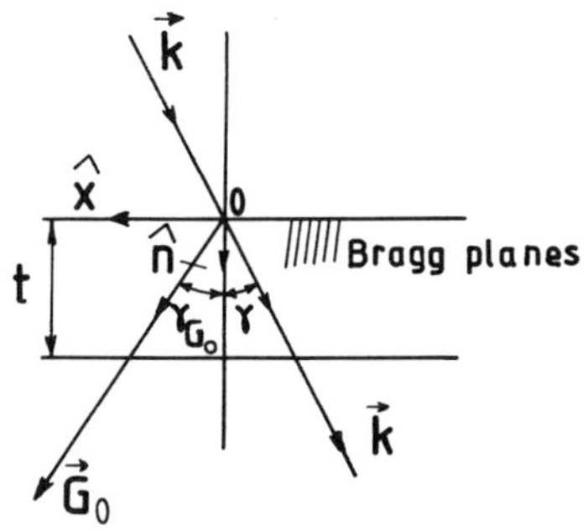

Fig.A2. Splitting of the incident beam in two beams by reflection on Bragg planes normal to the crystal surface.

REFERENCES

1. E. Schrödinger, Quantisierung als Eigenwertproblem, Ann. Physik. 79: 361; 79: 489; 80: 437; 81: 109 (1926).
2. E. Schrödinger, An Undulatory Theory of the Mechanics of Atoms and Molecules, Phys. Rev. 28: 1049 (1926).
3. E. Schrödinger, "Four Lectures on Wave mechanics", Blackie & Son, London (1929).
4. L. de Broglie, Sur la possibilité de relier les phénomènes d' interférences et de diffraction à la théorie des quanta de lumière , C. R. Acad. Sci. Paris. 183: 447 (1926); Sur la possibilité de mettre en accord la théorie électromagnétique avec la nouvelle Mécanique ondulatoire, C. R. Acad. Sci. Paris 184: 81 (1927).
5. A. O. Barut, a) Schrödinger's Interpretation of ψ as a Continuous Charge Distribution, Ann. der Physik. 45: 31 (1988); b) The Revival of Schrödinger's

Interpretation of Quantum Theory, Found. Phys. Lett.1: (1988).
6. A. O. Barut and J. F. Van Huele, Quantumelectrodynamics Based on Self-Energy, Phys. Rev. A 32: 3887 (1985); A. O. Barut and J. P. Dowling, Quantumelectrodynamics Based on Self-Energy: Lamb Shift in Cavities, Phys. Rev. A 36: 2550 (1987).
7. M. Born, On the quantum mechanics of collissions, Z. für Physik. 37: 863 (1926), translation into English reprinted in "Quantum Measurement Theory", J. A. Wheeler and W. H. Zurek, eds., Princeton Univ. Press, Princeton -New Jersey (1983).
8. C. C. Tannoudji, Bernard Dik, Franck Laloë, "Mécanique quantique", Hermann, Paris (1977).
9. C. I. Tomonaga, "Quantum Mechanics", Volume II, North-Holland, Amsterdam (1966) p. 268.
10. L. de Broglie, "Une tentative d'interpretation causale et non-lineaire de la mechanique ondulatoire (La théorie de double solution)", Gauthiers-Villars, Paris (1956).
11. J. P. Vigier, Louis de Broglie – Physicist and Thinker, Found. Phys. 12: 923 (1982): J. P. Vigier, Causal non-local interpretation of neutron interferometry experiments, EPR correlations and quantum statistics, in: "Symposium on the Foundations of Modern Physics", P. Lahti and P. Mittelstaedt eds., World Scientific Publishing Co. Singapore (1985).
12. C. Dewdney, Particle trajectories and interference in a time dependent model of neutron single crystal interferometer, Phys. Lett. A 109: 377 (1985); C. Dewdney, P. R. Holland and A. Kyprianidis, A quantum potential approach to spin superposition in neutron interferometry, Phys. Lett.A 121: 105 (1987).
13. J. M. Levy-Leblond, "Quantique-Rudiments", Inter Editions, Paris (1984).
14. N. Bohr, "Atomic physics and human knowledge", Wiley, New York (1958).
15. H. Rauch, "Proc. of 2nd International Sympozium on Quantum Mechanics", Tokyo (1986).
16. R. Loudon, "The quantum theory of light", Clarendon Press, Oxford (1986).
17. R. J. Glauber, a) in "New Techniques and Ideas on Quantum Measurement Theory", D. M. Greenberger, ed. Ann. New York Acad. Sciences 480: 336 (1986); b) Lecture given on this school.
18. H. Rauch, W. Treimer, and U. Bonse, Test of a single crystal neutron interferometer, Phys. Lett. A 47: 369 (1974).
19. H. Rauch and D. Petraschek, "Dynamical Neutron Diffraction and its Application", in Topics in Current Physics, ed. H. Dachs, Springer-Verlag, Berlin 6 (1978).
20. H. Rauch, Scope of neutron interferometry, in "Neutron interferometry", U. Bonse and H. Rauch eds. Clarendon Press, Oxford (1979).
21. H. Rauch, A. Zeilinger, G. Badurek, A. Wilfing, W. Bauspiess and U. Bonse,

Verification of coherent spinor rotations of fermions, Phys. Lett. A 54: 425 (1975).
22. G. Badurek, H. Rauch, J. Summhammer, U. Kischko and A. Zeilinger, Direct verification of the quantum spin-state superposition law, J.Phys. A 16: 1133 (1983).
23. G. Badurek, H. Rauch and J. Summhammer, Time-Dependent Superposition of Spinors, Phys. Rev. Lett. 51: 1015 (1983).
24. G. Badurek, H. Rauch and J. Summhammer, Polarized neutron interferometry: a survey, Physica B 151: 82 (1988).
25. A. Zeilinger, General Formulation of Spin Rotations in Neutron Interferometry, Z. Physik B, 25: 97 (1976).
26. F. Mezei, Neutron spin interference effects, in "Neutron interferometry", U. Bonse and H. Rauch eds. Clarendon Press, Oxford (1979).
27. F. Mezei, Zeeman energy, interference and neutron spin echo: a minimal theory, Physica B 151: 74 (1988).
28. B. Alefeld, G. Badurek, H. Rauch, Longitudinal Stern-Gerlach effect with slow neutrons, Phys. Lett. A 83: 32 (1981).
29. A. O. Barut and M. Božić, On the interpretation of the spinor phase change in a Magnetic field in Neutron Interferometry, submitted for publication.
30. E. P. Wigner, The Problem of Measurement, Am. J. Phys. 31: 6 (1963).
31. M. Božić , A proposal for an additional verification of the spin-state superposition law, Phys. Lett.A 115: 417 (1986).
32. M. Božić , A Proposal for the Completion of the Realization of Wigner's Gedanken-Experiment, Ann. New York Acad. Sciences 480: 565 (1986).

TIME, RELATIVITY AND QUANTUM THEORY*

Constantin Piron**

Département de Physique Théorique
Université de Genève
1211 Genève 4, Switzerland

QUANTUM MECHANICS

As we explain in [1], during these twelve last years Quantum Mechanics has developed deeply and, as a result, a really new theory has emerged. We want to give here a summary on the basic concepts, interpretation and rules of such new theory.

The modern description of physics is based on two notions, the notion of property and the notion of state. Given a system, a part of reality conceived as existing in space and time, we associate to each experimental project a particular property of the system. By experimental project we means an experiment that you can perform on the system and where you have exactly defined in advance what is for you a positive result. We say that the corresponding property is actual for the system as it has been prepared or given if in the event of the experiment the positive result would be certainly obtained. Other properties which can be defined and which are not actual are said to be potential.

The state of the system is nothing else that the full subset of all actual properties. Of course in general the system changes, in other words its state changes with time. Some actual properties becoming potential while others, potential at the start, appear in actuality.

On $\mathcal{L}$, the set of all properties we define an order relation. A property a is said to be stronger than a property b (i.e. $a < b$) if whenever a is actual b is actual as well.

On Σ, the set of all possible states we define an orthogonality relation. Two states $\mathcal{E}_1$ and $\mathcal{E}_2$ are said to be orthogonal (i.e. $\mathcal{E}_1 \perp \mathcal{E}_2$) if there is an experimental project such that the positive result (in the event you would perform the experiment) is certain in $\mathcal{E}_1$ but impossible in $\mathcal{E}_2$.

With very simple axioms it is possible to characterize the mathematical structures of $\mathcal{L}$ and Σ. To do this it is very convenient to define first the states and then the properties as a secondary notion. Let us suppose given as abstract set the set Σ of all possible states. Each property $a \in \mathcal{L}$ will be now represented by a subset A of Σ. By definition a state $\mathcal{E}$ will be in A if and only if the property a is actual in $\mathcal{E}$. Define for any subset A of Σ, $A^{\perp}$ the subset of all states orthogonal to each one of A and

New Frontiers in Quantum Electrodynamics and Quantum Optics
Edited by A. O. Barut, Plenum Press, New York, 1990

also $A^{\perp\perp}$ the subset of all states orthogonal to each one of $A^{\perp}$. Evidently $A \subset A^{\perp\perp}$ since orthogonality is a symmetric relation. In such framework it is not difficult to prove that A represents a property if and only if A is bi-orthogonal (i.e. $A = A^{\perp\perp}$). Moreover a singleton, a subset containing only one state, represents always a property, in other words we have $\{\mathcal{E}\}^{\perp\perp} = \{\mathcal{E}\}$ for each $\mathcal{E} \in \Sigma$. In the following we will identify $\mathcal{L}$ with the lattice of all bi-orthogonal subsets of Σ.

Since $A^{\perp} = A^{\perp\perp\perp}$, $A^{\perp}$ is a property and the map $A \mapsto A^{\perp}$ is well defined on $\mathcal{L}$. $A^{\perp}$ is called the orthocomplement of A and the map itself the orthocomplementation.

A property A is said to be classical if for each state $\mathcal{E} \in \Sigma$ either $\mathcal{E} \in A$ or $\mathcal{E} \in A^{\perp}$. If A is classical $A^{\perp}$ is classical and for any collection of classical properties their upper limit and their lower limit are also classical. This means that the set of all classical properties forms a sublattice of $\mathcal{L}$, called the classical property lattice. Two states $\mathcal{E}_1$ and $\mathcal{E}_2$ are said to be classically equivalent if for any classical $A \in \mathcal{L}$, $\mathcal{E}_1 \in A$ if and only if $\mathcal{E}_2 \in A$. An equivalence class of states is called a macrostate and the classical property lattice turns out to be given by all the subsets in the set of macrostates. The set of macrostates is also called the set of superselection variables. From this analysis we see that in "first approximation" any physical system appears as a classical one !

When the physical system is one entity making a whole, we can impose more axioms and thus justify the Hilbert space structure [2]. But when the system is composed of two or more separated entities such new axioms fail and cannot be maintained without contradictions [3]. In fact two separated entities taken together must be described in the following way. Given Σ_1 and Σ_2 the sets of states of each entity and $\perp_1$ and $\perp_2$ the corresponding relations of orthogonality, the whole system has a set of states Σ given by the direct product $\Sigma_1 \times \Sigma_2$. Two global states $(\mathcal{E}_1, \mathcal{E}_2)$ and $(\mathcal{E}'_1, \mathcal{E}'_2)$ are orthogonal if and only if $\mathcal{E}_1 \perp_1 \mathcal{E}'_1$ or $\mathcal{E}_2 \perp_2 \mathcal{E}'_2$. It can also be proved that the corresponding property lattice is as before the lattice of bi-orthogonal subset of Σ. One important point is that if Σ_1 is the set of rays of a Hilbert space and Σ_2 the set of rays of an other one the property lattice of the global system cannot be embedded in the linear closed subspace lattice of the tensor product of these two Hilbert spaces. It is then not possible to describe two separated entities just by the closed subpaces of the tensor product even if you restrict the states to the ones in $\Sigma_1 \times \Sigma_2$.

Two important ingredients of the theory are symmetries and observables. A symmetry is an automorphism i.e. a permutation of the states of Σ which conserves the orthogonality relation in the two directions. Any symmetry generates an automorphism of the orthocomplemented lattice structure of $\mathcal{L}$. Any symmetry generates a permutation of the macrostates and an automorphism of the boolean lattice structure of the classical properties. If Σ is the set of rays of a Hilbert space any symmetry on Σ is generated by a unique either unitary or an antiunitary transformation, but defined up to a phase i.e. up to a complex number of module one [2]. Finally let us remark that many symmetries of Σ are just mathematical objects which have no direct interpretation.

An observable is a map from some complete boolean lattice $\mathcal{B}$ in the property lattice $\mathcal{L}$ which conserves the lattice structure and the orthocomplementation. In the particular case where $\mathcal{L}$ is a classical property lattice, i.e. the set of subsets of a set Σ, it turns out that $\mathcal{B}$ must also be the set of subset of some set (say E) and the observables can be realized as the inverse images of maps from Σ to E. On the opposite case when Σ is the set of rays of some separable Hilbert space any observable can be interpreted as the spectral family of some self-adjoint operator [2]. As for the symmetries many observables are just mathematical objects without direct physical interpretation. However each experimental project which has a complete set of well-defined outcomes defines an observable.

To conclude this short introduction to the new Quantum Mechanics let us discuss equations of motion. If the set of states Σ follows a motion which is deterministic, orthogonal states at the end must arise from orthogonal states at the beginning. This follows directly from the definition of orthogonality. In fact two states are orthogonal if and only if there is some experimental project where the positive result is certain in one of the states and is impossible in the other one. If such a particular project can be performed at the final time, the following project can be imagined at the initial time : let the system make its motion until the final time and then perform the experiment. Since the motion is deterministic this new experiment is able to demonstrate the orthogonality of the two corresponding initial states. Then, if moreover the motion is also reversible the orthogonality relation is preserved.

For an entity such kind of motion is then induced by an unitary transformation and then obeys a Schrödinger equation. But if the motion is only deterministic the corresponding equation is of order 3 in ψ [4] and then non-linear.

For two separated entities the reversible motion obeys an equation of the type :

$$i\partial_t(\psi_1\psi_2)_t = \hbar^{-1}\left(\frac{p_1^2}{2m_1} + \frac{p_2^2}{2m_2} + V(q_1, \psi_2) + V(q_2, \psi_1)\right)(\psi_1\psi_2)_t$$

This is also a non-linear equation.

PARTICLE MODELS

A particle, as other entity, is described in general by a family of Hilbert spaces $\mathcal{H}_\alpha$ indexed by $\alpha \in \Omega$ where Ω is the set of superselection variables. To sum up, we have the following representation :

Property : by a family of projectors, $P = \{P_\alpha\}$ $\quad P_\alpha^2 = P_\alpha = P_\alpha^\dagger$

State : by one ray, $\alpha_0 \in \Omega$ and $\psi_{\alpha_0} \in \mathcal{H}_{\alpha_0}$

Observable : by a family of self-adjoint operators, $A = \{A_\alpha\}$ $\quad A_\alpha = A_\alpha^\dagger$

Symmetry : by a family of unitary operators, $U_{\alpha\beta}$: $\quad \mathcal{H}_\beta \to \mathcal{H}_\alpha \quad \beta \mapsto \alpha$

Let us also recall the following statistical formula :

$$< A >= \int_\Omega tr(\rho_\alpha A_\alpha) d\mu(\alpha)$$

where $\{\rho_\alpha\}$ is a family of density matrixes and $d\mu(\alpha)$ is a density of probability on Ω.

As in [5] we will define a particle as an entity which admits observables like momentum, position and time but no other ones. But each observable obeys particular covariance relations, characteristic of its nature. And in our case the following three imprimitivity systems must be satisfied :

1) for momentum $\underline{p}$:

$$\begin{array}{ccc} P(\mathbb{R}^3) & \xrightarrow{\underline{p}} & \mathcal{L} \\ \downarrow g & & \downarrow S(g) \\ P(\mathbb{R}^3) & \xrightarrow{\underline{p}} & \mathcal{L} \end{array}$$

2) for position $\underline{q}$:

$$\begin{array}{ccc} P(\mathbb{R}^3) & \xrightarrow{\underline{q}} & \mathcal{L} \\ \downarrow g & & \downarrow S(g) \\ P(\mathbb{R}^3) & \xrightarrow{\underline{q}} & \mathcal{L} \end{array}$$

3) for time $\underline{t}$:

$$\begin{array}{ccc} P(\mathbb{R}) & \xrightarrow{t} & \mathcal{L} \\ \downarrow g & & \downarrow S(g) \\ P(\mathbb{R}) & \xrightarrow{\underline{t}} & \mathcal{L} \end{array}$$

The notation is the usual one : $P(\mathbb{R}^3)$ means the set of subsets of $\mathbb{R}^3$ and $P(\mathbb{R})$ the set of subsets of $\mathbb{R}$. On the other hand, the group G is defined by all possible choices of the zeros of the apparatus scales and so $g \in G$ is either a momentum translation $\vec{\pi}$ acting on momentum only, a space translation $\vec{a}$ acting on space only, a time translation τ acting on time only or also a rotation R acting both on momentum and space. Finally $S(g)$ are symmetries of the property lattice $\mathcal{L}$ defining a representation of G As we have said each element $g \in G$ acts on the corresponding apparatus scales and then its action is passive. But on the contrary $S(g)$ acting on the states of the system is taken in the active sense and in fact we suppose that it is possible to rebuild the "same" system but with more momentum, in another place, in another time and finally also, turned in another direction. With this interpretation it is not possible to add to the group G other elements like dilatations, space inversion or time inversion which cannot act on the system in an active way. The same is true for the change of inertial frame (Galileo or Lorentz transformations) which in contrast are impossible to define in a passive way on the apparatus scales.

Given these three imprimitivity systems it is very remarkable that there are, and there are only, two solutions. The classical model and the quantum one. The classical model is well known :

The state space Σ is $\mathbb{R}^7$ and the property lattice is $P(\mathbb{R}^7)$ the lattice of the subsets of $\mathbb{R}^7$. The group G acts canonically on $(\vec{p}, \vec{q}, t) \in \Sigma$ as you imagine and the three observables are defined by the following morphisms :

$$
\begin{aligned}
\underline{p}: \quad & \Delta \mapsto \underline{p}(\Delta) = \{(\vec{p},\vec{q},t) \mid \vec{p} \in \Delta\} \\
\underline{q}: \quad & \Delta \mapsto \underline{q}(\Delta) = \{(\vec{p},\vec{q},t) \mid \vec{q} \in \Delta\} \\
\underline{t}: \quad & \Delta \mapsto \underline{t}(\Delta) = \{(\vec{p},\vec{q},t) \mid t \in \Delta\}
\end{aligned}
$$

which are nothing other than the inverse images of the following maps :

$$
\begin{aligned}
\vec{p}: \quad & (\vec{p},\vec{q},t) \mapsto \vec{p} \\
\vec{q}: \quad & (\vec{p},\vec{q},t) \mapsto \vec{q} \\
t: \quad & (\vec{p},\vec{q},t) \mapsto t
\end{aligned}
$$

The quantum model is also well known but not really completely as we will see :

It is defined via a family of complex Hilbert spaces $\{\mathcal{H}_t\}$, $t \in \mathbb{R}$, each one identical to $L^2(\mathbb{R}^3, dv)$. The state space Σ is in this case the set of all rays of all these Hilbert spaces and the property lattice is $\bigvee_t P(\mathcal{H}_t)$ the direct union of the $P(\mathcal{H}_t)$ the lattices of projectors on $\mathcal{H}_t$. The representation S of the group G is induced by the following unitary transformations :

$$
\begin{aligned}
(U(\vec{\pi})\psi)_t(\vec{x}) &= e^{i/\hbar\ \vec{\pi}\cdot\vec{x}}\psi_t(\vec{x}) \\
(U(\vec{a})\psi)_t(\vec{x}) &= \psi_t(\vec{x}-\vec{a}) \\
(U(\tau)\psi)_t(\vec{x}) &= \psi_{t-\tau}(\vec{x}) \\
(U(R)\psi)_t(\vec{x}) &= \psi_t(R^{-1}\vec{x})
\end{aligned}
$$

The three observables are defined by the following morphisms:

$$
\begin{aligned}
&\Delta \longmapsto \underline{p}(\Delta) \ \text{ with } \ (\underline{p}(\Delta)\tilde{\psi})_t(\vec{k}) = \chi_\Delta(\vec{k})\tilde{\psi}_t(\vec{k}) \\
&\Delta \longmapsto \underline{q}(\Delta) \ \text{ with } \ (\underline{q}(\Delta)\psi)_t(\vec{x}) = \chi_\Delta(\vec{x})\psi_t(\vec{x}) \\
&\Delta \longmapsto \underline{t}(\Delta) \ \text{ with } \quad (\underline{t}(\Delta)\psi)_t = \begin{cases} 0 & \text{if } t \notin \Delta \\ \psi_t & \text{if } t \in \Delta \end{cases}
\end{aligned}
$$

where

$$
\tilde{\psi}_t(\vec{k}) = (2\pi\hbar)^{-3/2} \int_{-\infty}^{\infty} dV\ e^{-i/\hbar\ \vec{k}\cdot\vec{x}}\psi_t(\vec{x})
$$

and χ_Δ is the characteristic function which is 1 on Δ and 0 outside.

As everybody knows, such morphisms can also be defined via the spectral families of the following families of operators :

$$
\begin{aligned}
p_t^{(i)}\psi_t(\vec{x}) &= -\,i\hbar\frac{\partial}{\partial x^{(i)}}\psi_t(\vec{x}) \qquad && i = 1,2,3 \\
q_t^{(i)}\psi_t(\vec{x}) &= x^{(i)}\psi_t(\vec{x}) && i = 1,2,3 \\
t_t\psi_t(\vec{x}) &= t\psi_t(\vec{x})
\end{aligned}
$$

Apparently this is like Quantum Mechanics, but here the observable time exists, it is well defined and it is given by the family of operators $t_t = t\mathbb{1}$. Since this observables is compatible (commutes) with all other observables, there is no Heisenberg uncertainty relation between time and energy. Such conclusion is also valid for a relativistic quantum particle which must be described with the same model since it is unique in the quantum case.

We have found only two models just because we have imposed the absence of other independent observables, internal or not. If one relaxes this condition there are more solutions, in particular models with intrinsic angular momentum i.e. with spin. There is the quantum particle with spin 1/2, which turns out to be the only one with spin and no internal variables. In this model each one of the Hilbert spaces $\mathcal{H}_t$ is identical to $\mathbb{C}^2 \otimes L^2(\mathbb{R}^3, dV)$. In other words each vector of $\mathcal{H}_t$ is of the form $\psi = \binom{\psi_1}{\psi_2}$ with $\psi_i \in L^2(\mathbb{R}^3, dV)$.

The representation of G is identical to the preceding one except for rotations where now :

$$\left(U(R)\psi\right)_t(\vec{x}) = e^{i/2\ \vec{\sigma}\cdot\vec{n}\vartheta}\psi_t(R^{-1}\vec{x})$$

σ are the well known Pauli matrices, $\vec{n}$ is the unit vector defined by the direction of the rotation axis and ϑ the rotation angle.

There are new independent observables which are $\frac{1}{2}\sigma_1, \frac{1}{2}\sigma_2$ et $\frac{1}{2}\sigma_3$.

MAXWELL EQUATIONS

Let us first recall the Maxwell theory, in the four-dimensional formalism. First we have to consider as given :

a one-form,

$$A = A_\mu dx^\mu$$

a two-form,

$$B = \frac{1}{2} B_{\mu\nu} dx^\mu \wedge dx^\nu$$

and an equation,

$$dA = B$$

from which follows the well known first part of Maxwell Equations :

$$dB = 0$$

Secondly we have also to consider as given :

a two-form of second kind,

$$H = \frac{1}{4} H^{\mu\nu} \varepsilon_{\mu\nu\rho\lambda} dx^\rho \wedge dx^\lambda$$

a three-form of second kind,

$$J = \frac{1}{6} J^\mu \varepsilon_{\mu\nu\rho\lambda} dx^\nu \wedge dx^\rho \wedge dx^\lambda$$

and an equation

$$dH = J$$

which is the second part of Maxwell Equations and from which follows the well known conservation law,

$$dJ = 0$$

As they are written, such equations are obviously invariant for all diffeomorphisms. But this is no more the case when we impose in addition the invariance of the phenomenological equation,

$$B_{\mu\nu} = \frac{1}{2}\mu_{\mu\nu\rho\lambda}H^{\rho\lambda}$$

which describes the relations between $\vec{B}$ and $\vec{H}$ and $\vec{E}$ and $\vec{D}$.To give more inside on such four-index tensor let us first remark that by very definition it is antisymmetric in the index pair (μ,ν) and also in the index pair (ρ,λ). By just changing the notation we can rewrite this four-index tensor as a two-index tensor μ_{ab} where a and b go from 1 to 6 and label the pairs (1,2), (2,3), (3,1), (1,4), (2,4) and (3,4). In particular for the vacuum, we find :

$$\mu_{ab} = \begin{cases} 0 & \text{if } a \neq b \\ \mu_0 & \text{if } a = b = 1,2,3 \\ -\varepsilon_0^{-1} & \text{if } a = b = 4,5,6 \end{cases}$$

We want to prove that the only diffeomorphisms which preserve the phenomenological equations for the vacuum are the ones which act on the tangent space as dilatations and/or Lorentz transformations. If α is the corresponding jacobean matrix acting covariantly on the tangent space, it is easy to check the following :

$B_{\mu\nu}$ is transformed by :

$$\alpha \otimes \alpha$$

$\varepsilon_{\mu\nu\rho\lambda}$ is transformed by :

$$|\,\alpha\,|^{-1}\;\alpha \otimes \alpha \otimes \alpha \otimes \alpha$$

i.e. like scalar, since ε_{1234} takes the value 1 in any bases.

$H^{\mu\nu}$ is transformed by :

$$\|\alpha\|\;\alpha^{-1\sim} \otimes \alpha^{-1\sim}$$

And, in conclusion, $\mu_{\mu\nu\rho\lambda}$ is transformed by :

$$\|\alpha\|^{-1}\alpha \otimes \alpha \otimes \alpha \otimes \alpha$$

Lemma : For the vacuum, the 36 equations

$$\mu_{\mu\nu\rho\lambda} = \hat{g}_{\mu\rho}\hat{g}_{\nu\lambda} - \hat{g}_{\mu\lambda}\hat{g}_{\nu\rho}$$

with $\hat{g}_{\mu\nu}$ a symmetric two index tensor have only two solutions :

$$\hat{g}_{\mu\nu} = \pm\mu_0^{\frac{1}{2}}(1,1,1,-c^2)$$

where $c^2 = \varepsilon_0^{-1}\mu_0^{-1}$

The proof of this lemma is easy and can be done by hand (or read in [6]).

Finally, if we impose to $\hat{g}_{\mu\nu}$ to be transformed by

$$\|\alpha\|^{-\frac{1}{2}}\alpha \otimes \alpha$$

the new $\hat{g}_{\mu\nu}$ will satisfy the new equations and for this reason, $\mu_{\mu\nu\rho\lambda}$ will be invariant if and only if $\hat{g}_{\mu\nu}$ is also invariant since you cannot change the signe of all the $\hat{g}_{\mu\nu}$ by linear transformations. The rest of the proof is trivial.

If μ_0 and ε_0 are constant in space and time the group of covariance is Poincaré and dilatations. If you accept some singularities, you will find each one of the possible conformal groups.

TAI or INTERNATIONAL ATOMIC TIME

1) The SI, the unit of time or the second is the duration of

9192 631 770

periods of the radiation corresponding to the transition between the two hyperfine levels of the ground state of the caesium-133 atom, as adopted in 1967 by the Conférence Générale des Poids et Mesures, and as it is realized on the surface of the rotational geoid (according to the declaration in 1980 by the CCDS, "Comité Consultatif pour la Definition de la Seconde").

The integration of such frequency gives a measure of the time, needed in order to specify when an event actually did occur. The accuracy of the two best caesium clocks is (in 1987) :

$$\sim 2 \times 10^{-14}$$

2) The Atomic International Time, called TAI, is the coordinate time defined in a geocentric reference frame with a second SI [7]. Special relativity predicts that a moving clock will record less time compared with a clock at rest in an inertial reference frame. For low velocities $v^2 \ll c^2$, the ratio of times recorded by the moving clock and the one at rest reduces to $1 - \frac{v^2}{2c^2}$. The general relativity predicts the same kind of effect and for weak field the corresponding ratio is $1 - \frac{u}{c^2}$ where u is the gravity potential (this can be derived from the Einstein equivalence principle). In conclusion the TAI has been defined (after average on different clocks) as the t^* given by the following formula :

$$\begin{aligned} t^* = \tau &+ \frac{1}{c^2}\int_{t_0}^{t} (u^*(t) - u_o^*)dt \\ &+ \frac{1}{c^2}\int_{t_0}^{t} \frac{1}{2} v_g^2 dt \\ &+ \frac{1}{c^2}\int_{t_0}^{t} \omega \dot{L} r^2 \cos^2 \varphi dt \end{aligned}$$

where :

τ is the proper time given by the clock,
u^* is the apparent gravity potential (earth gravity and centrifugal potential),
u_0^* is the same at the sea level,
v_g is the relative velocity,
ω is the angular velocity,

$\dot{L}$ is the longitudinal velocity and
φ is the latitude.

In the neighborhood of the earth we have :

$$u^*(t) - u_o^* = c^2 gh(t) \sim 1 \times 10^{-13}\text{s/Km} .$$

On one day and for all the neighborhood of the earth TAI is actually defined up to ± 5 ns.This means that in spite of the arbitrary imposed to the motions of the clocks, all the events like the ones defined by clocks and incoming electromagnetic signals can be consistently labelled by one parameter t^*. In other words for a given value of t^* all such events are confined in the corresponding three-dimensional flat subspace of $\mathbb{R}^4$ and so no dispersion in time direction is observed. This experimental fact is in complete contradiction with any fully covariant relativistic theory included the author's theory itself [8].

In conclusion we must accept to live in a three dimensional space without possibility to go in a fourth dimension. The interpretation of relativity proposed by H. Minkowski [9] and usually accepted must be reconsidered. As we will see in the following the relativistic covariance is dynamical and the Poincaré group acts on motions but not on individual states. The physics is really the same in each Galilean reference frame, but to interpret a result one must take account of the frame chosen at the very beginning when one has constructed $\mathbb{R}^4$ cartesian product of $\mathbb{R}^3$ by $\mathbb{R}$.

DIRAC EQUATION

According to our quantum model of particle, whatever state one chooses the position observable takes no definite value and so when the particle is located in some part of the space $\mathbb{R}^3$ it is located only potentially at each one of these points. But in contrast the time observable has in each state a definite value. This situation creates difficulty in the Minkowski interpretation of relativity where only points (event) in $\mathbb{R}^4$ are relevant. This difficulty disappears completely if one accepts the conclusions of the preceding section and considers Minkowski space as a formal construction in one given Galilean frame, construction which is useful to manifest explicitly the dynamical covariance of the motions.

As example of such covariance let us consider the Schrödinger equation of a particle :

$$i\partial_t \psi_t(\vec{x}) = H\psi_t(\vec{x})$$

In non-relativistic mechanics we have :

$$H = \hbar^{-1}\Big(\frac{1}{2m}[\vec{p} - \vec{A}(\vec{q},t)]^2 + V(\vec{q},t)\Big)$$

The solutions of this equation are particular functions on $\vec{x}$ and t which are called motions. The Galileo group acts on this set of functions changing a motion $\psi_t(\vec{x})$ in $\psi'_t(\vec{x}) = \psi_t(\vec{x} + \vec{v}t)$. This new function is also a motion but for the new operator :

$$H' = \hbar^{-1}\Big[\frac{1}{2m}(\vec{p} - \vec{A}(\vec{q} + \vec{v}t, t))^2 + V(\vec{q} + \vec{v}t, t) - \vec{v}\vec{p}\Big]$$

which can be written as

$$H' = \hbar^{-1}\Big[\frac{1}{2m}(\vec{p} - \vec{A}(\vec{q}+\vec{v}t,t) - m\vec{v})^2 + V(\vec{q}+\vec{v}t,t) - \vec{v}\vec{A}(\vec{q}+\vec{v}t,t) - \frac{1}{2}m\vec{v}^2\Big]$$

The Galileo dynamical covariance of this theory means that H' must be interpreted as the Schrödinger operator in a new frame for a particle of the same mass, interacting with a new four-potential :

$$A'_i = A_i$$
$$-V' = -V + \vec{v}\vec{A}$$

But one has also to take account of the gauge transformation :

$$\vec{p} \mapsto \vec{p} - m\vec{v}$$
$$H \mapsto H - \frac{1}{2}m\vec{v}^{\,2}$$

corresponding to the changes of zero scales for momentum and energy.

Moreover the four-momentum, $(\vec{p}, -H)$, are transformed according to the covariant action of the Galileo group, in the same way as the four-potential $(\vec{A}, -V)$. This can be easily checked since in fact we have imposed at the beginning of this calculation :

$$p'_i = p_i$$
$$-H' = -H + \vec{v}\vec{p}$$

Finally, the change of the old initial condition $\psi_{t_0}(\vec{x})$ in a new one $\psi'_{t'_0}(\vec{x}) = \psi_{t'_0}(\vec{x}+\vec{v}t)$ defines a unitary transformation which induces a symmetry on the property lattice. In conclusion, the laws of physics in the new reference frame are the same as they are in the old one.

After this example we are able to consider the relativistic case and discuss the Dirac equation. This equation, written by Dirac in 1928 is the following :

$$i\partial_t\psi_t(\vec{x}) = \hbar^{-1}\big[c\alpha^k(p_k - A_k(\vec{x},t)) + \beta mc^2 + V(\vec{x},t)\big]\psi_t(\vec{x})$$

where α^k and β are the well known 4 by 4 Dirac matrixes. Such equation is nothing else that a Schrödinger equation for a $\frac{1}{2}$-spin particle, but with double number of components.

The Lorentz group acts on the motions, spinor functions solutions of this equation, changing

$$\psi_t(\vec{x}) = \psi(x^\mu)$$

in a new spinor function,

$$\psi'_t(\vec{x}) = \psi'(x^\mu) = S(\Lambda)\psi(\Lambda^{-1\,\mu}{}_\nu x^\nu)$$

where $S(\Lambda)$ is a 4 by 4 matrix depending only of the Lorentz transformation Λ.

As it is well known this new spinor function is solution of a new Dirac equation

$$i\partial_t\psi'_t(\vec{x}) = \hbar^{-1}\big[c\alpha^k(p_k - A'_k(\vec{x},t)) + \beta mc^2 + V'(\vec{x},t)\big]\psi'_t(\vec{x})$$

defined for the new four-potentiel

$$A'_\mu(x^\rho) = \Lambda^{-1\,\nu}{}_\mu A_\nu(\Lambda^{-1\,\rho}{}_\lambda x^\lambda)$$

Here there is no gauge transformation since the zero of the four-momentum is defined in an absolute way without reference to an exterior reference frame.

Choosing $\psi_{t_0}(\vec{x})$ as initial state for the Dirac equation, this determines one solution, the spinor function $\psi(x^\mu)$. Applying to this spinor function the action of a Lorentz transformation we can determine $\psi'(x^\mu)$. Then by choosing for the variable t a new value t_0' and puting these altogether we define the following map :

$$\psi_t(\vec{x}) \quad \mapsto \quad \psi'_{t'}(\vec{x}) \;=\; \psi'(\vec{x}, t')$$

This map is linear inversible and moreover, since the Dirac vector current

$$j^\mu(\vec{x}, t) \;=\; \bar{\psi}(\vec{x}, t)\gamma^\mu \psi(\vec{x}, t)$$

obeys the equation

$$\partial_\mu j^\mu(\vec{x}, t) \;=\; 0 \; ,$$

the scalar product in $\mathbb{C}^4 \otimes L^2(\mathbb{R}^3, dV)$ is conserved.

Such transformation induces a symmetry on the property lattice and, as in the Galilean case, the laws of physics in relativity are the same in the new reference frame.

In résumé, in these lectures, we have see that the very accurate synchronization after corrections of the clocks moving all around the earth imposes the physical reality of our three-dimensional space. Our quantum particle model (i.e. the usual one) turns out to be perfectly adapted to this situation and the laws of relativity derived from Maxwell equations apply here without difficulty but only at the level of the dynamics.

REFERENCES

[1] C. Piron : "Recent Developments in Quantum Mechanics", Helv. Phys. Acta 62, 82 (1989)

[2] C. Piron : "Foundations of Quantum Physics", W.A. Benjamin, Inc. London, Amsterdam... (1976).

[3] D. Aerts : "Description of Many Physical Entities without the Paradoxes Encountered in Quantum Mechanics", Found. Phys. 12, 1131 (1982).

[4] W. Daniel : "Axiomatic Description of Irreversible and Reversible Evolution of a Physical System", to appear in Helv. Phys. Acta.

[5] C. Piron : "New Dialogue on a New Science between F. Salviati, G. Sagredo, and Simplicio", Foundations of Physics 19, 1017 (1989).

[6] C. Piron :"Electrodynamique et optique", 88, Département de physique théorique, CH-1211 Genève 4 (1975).

[7] B. Guinot : "Le temps coordonné",Journées relativistes 1988, 72, Département de physique théorique, CH-1211 Genève 4 (1989).

[8] L. P. Horwitz and C. Piron : "Relativistic Dynamics", Helv. Phys. Acta. 46, 316 (1973).

[9] H. Minkowski : "Space and Time" in "The Principle of Relativity ", Dover Publication, Inc.(1923).

GOOD AND BAD WELCHER WEG DETECTORS

Berthold-Georg Englert[†] and Marlan O. Scully[‡]

[†‡]Center for Advanced Studies and
Department of Physics and Astronomy
University of New Mexico, Albuquerque, New Mexico 87131, USA
and
[†]Sektion Physik
Universität München
D-8046 Garching, West Germany
and
[‡] Max-Planck Institut für Quantenoptik
D-8046 Garching bei München, West Germany

INTRODUCTION

We return to the Stern-Gerlach interferometer (SGI) with micromaser welcher Weg detectors in the spin-up arm, as depicted in Figs. 1 and 2 of the second lecture by Scully, Fearn, and Atherton in these Proceedings. It is our objective to supplement the detailed treatment given in Ref. 1 by reporting a criterion that characterizes good welcher Weg detectors.

To set the stage let us recall the essentials of the physical situation being considered. The cavities are prepared with large average photon numbers $N_j = \langle a_j^\dagger a_j \rangle$, $j = 1, 2$ and the uncertainties δN_j are small compared to these average values. Under these conditions (and pretending that the effect of the SGI on the atom's center-of-mass motion can be ignored) the unitary operator that relates the initial variables of the spin and the photon degrees of freedom to the final ones is [this is Eq. (42) of Ref. 1]

$$U = \exp\left(-i\omega(t_f - t_i)(a_1^\dagger a_1 + a_2^\dagger a_2) - \frac{i}{2}\Phi\sigma_z\right) \times \left(-\frac{1+\sigma_z}{2}\frac{a_2 a_1^\dagger}{\sqrt{N_1 N_2}} + \frac{1-\sigma_z}{2}\right). \tag{1}$$

Here ω is the natural frequency of the cavity modes, t_i and t_f denote the initial and the final time, and Φ is a phase contributing to the net Larmor precession angle. At the initial time, the various degrees of freedom are not correlated, so that the joint density operator is a product:

$$\rho_{\text{spin-photon}}(t_i) = \rho_{\text{spin}}(\sigma_z, \sigma_+, \sigma_-)\rho_1(a_1, a_1^\dagger)\rho_2(a_2, a_2^\dagger). \tag{2}$$

New Frontiers in Quantum Electrodynamics and Quantum Optics
Edited by A. O. Barut, Plenum Press, New York, 1990

The unitary evolution operator (1) turns this into

$$\rho_{\text{spin-photon}}(t_f) = \rho_{\text{spin}}\left(\sigma_z, -e^{-i\Phi}a_1^\dagger a_2\sigma_+/\sqrt{N_1N_2}, -e^{i\Phi}a_1a_2^\dagger\sigma_-/\sqrt{N_1N_2}\right)$$
$$\times\left[\frac{1+\sigma_z}{2}\frac{a_1^\dagger a_2}{\sqrt{N_1N_2}}\tilde{\rho}_1\tilde{\rho}_2\frac{a_1a_2^\dagger}{\sqrt{N_1N_2}} + \frac{1-\sigma_z}{2}\tilde{\rho}_1\tilde{\rho}_2\right], \quad (3)$$

where $\tilde{\rho}_j \equiv \rho_j(e^{i\beta}a_j, e^{-i\beta}a_j^\dagger)$ with $\beta \equiv \omega(t_f - t_i)$. For an incoming x-polarized beam we have

$$\rho_{\text{spin}} = \frac{1}{2}(1+\sigma_x) = \frac{1}{2}\begin{pmatrix} 1 & 1 \\ 1 & 1 \end{pmatrix}, \quad (4)$$

where the 2x2 matrix refers to measurements of σ_z. The corresponding spin-matrix version of the final joint density operator (3) is then

$$\rho_{\text{spin-photon}}(t_f) = \frac{1}{2}\begin{pmatrix} \frac{a_1^\dagger a_2}{\sqrt{N_1N_2}}\tilde{\rho}_1\tilde{\rho}_2\frac{a_1a_2^\dagger}{\sqrt{N_1N_2}} & -e^{-i\Phi}\frac{a_1^\dagger a_2}{\sqrt{N_1N_2}}\tilde{\rho}_1\tilde{\rho}_2 \\ -e^{i\Phi}\tilde{\rho}_1\tilde{\rho}_2\frac{a_1a_2^\dagger}{\sqrt{N_1N_2}} & \tilde{\rho}_1\tilde{\rho}_2 \end{pmatrix}. \quad (5)$$

This is the starting point for our analysis of what characterizes a good welcher Weg detector.

QUALITY OF A WELCHER WEG DETECTOR

It suffices to have which-path information available in one of the cavities, say the first one. We therefore simplify matters by choosing $\rho_2(a_2, a_2^\dagger)$ to project to a coherent state, the eigenvalue of a_2 being $\sqrt{N_2}e^{i\theta_2}$. Consequently,

$$\frac{a_2}{\sqrt{N_2}}\tilde{\rho}_2 = e^{i(\theta_2-\beta)}\tilde{\rho}_2,$$
$$\tilde{\rho}_2\frac{a_2^\dagger}{\sqrt{N_2}} = e^{-i(\theta_2-\beta)}\tilde{\rho}_2, \quad (6)$$

so that (5) becomes

$$\rho_{\text{spin-photon}}(t_f) = \frac{1}{2}\begin{pmatrix} a_1^\dagger(a_1a_1^\dagger)^{-1/2}\tilde{\rho}_1(a_1a_1^\dagger)^{-1/2}a_1 & e^{-i\alpha}a_1^\dagger(a_1a_1^\dagger)^{-1/2}\tilde{\rho}_1 \\ e^{i\alpha}\tilde{\rho}_1(a_1a_1^\dagger)^{-1/2}a_1 & \tilde{\rho}_1 \end{pmatrix}\tilde{\rho}_2, \quad (7)$$

where $\alpha \equiv \pi + \Phi + \beta - \theta_2$ and we replaced the average photon number N_1 by the positive operator $a_1a_1^\dagger$ which is permissible since the spread δN_1 is known to be very small. This replacement has the advantage that the r.h.s. of (7) is explicitly linear in ρ_1 and represents a positive operator of unit trace, that is: a density operator, for arbitrary initial density operators ρ_1 and ρ_2. The factorization in (7) signifies that the photon degree of freedom of the second cavity is not correlated to that of the first one and to the spin degree of freedom. Indeed, the second cavity does not contain which-path information.

The spin-matrix in (7) tells us that spin-coherence is maintained only if $a_1^\dagger(a_1a_1^\dagger)^{-1/2}\tilde{\rho}_1$ differs from $\tilde{\rho}_1$ by nothing more than a predetermined phase factor, which similarly to (6) requires that ρ_1 projects to a highly excited coherent state.

Further, we learn from Eq. (7) that which-path information is available if the photon field described by[2]

$$\rho_1^{\uparrow} \equiv a_1^{\dagger}(a_1 a_1^{\dagger})^{-1/2}\, \tilde{\rho}_1\, (a_1 a_1^{\dagger})^{-1/2} a_1, \tag{8}$$

can be distinguished from the one corresponding to

$$\rho_1^{\downarrow} \equiv \tilde{\rho}_1 . \tag{9}$$

As a first step toward a quantitative criterion for the property "distinguishable," consider the initial photon field to be in a pure state, which is to say that ρ_1 is a projection operator,

$$\rho_1 = |\psi_1\rangle\,\langle\psi_1|. \tag{10}$$

Then

$$\rho_1^{\downarrow} = |\psi_1^{\downarrow}\rangle\,\langle\psi_1^{\downarrow}| \quad \text{with} \quad |\psi_1^{\downarrow}\rangle = \exp(-i\beta a_1^{\dagger} a_1)|\psi_1\rangle, \tag{11}$$

and

$$\rho_1^{\uparrow} = |\psi_1^{\uparrow}\rangle\,\langle\psi_1^{\uparrow}| \quad \text{with} \quad |\psi_1^{\uparrow}\rangle = a_1^{\dagger}(a_1 a_1^{\dagger})^{-1/2}|\psi_1^{\downarrow}\rangle. \tag{12}$$

Now, a measure of how similar or how different $|\psi^{\uparrow}\rangle$ and $|\psi^{\downarrow}\rangle$ are is their squared overlap,

$$|\langle\psi_1^{\downarrow}|\psi_1^{\uparrow}\rangle|^2 = \mathrm{tr}(\rho_1^{\downarrow}\rho_1^{\uparrow}). \tag{13}$$

This is zero for no overlap (indicating full which-path information) and unity for perfect overlap (no which-path information). We are thus led to introducing

$$Q \equiv \frac{\mathrm{tr}(\rho_1^{\downarrow}\rho_1^{\uparrow})}{\mathrm{tr}(\tilde{\rho}^{\,2})} = \mathrm{tr}\left(\rho_1 a_1^{\dagger}(a_1 a_1^{\dagger})^{-1/2}\rho_1 (a_1 a_1^{\dagger})^{-1/2} a\right)/\mathrm{tr}(\rho_1^2) \tag{14}$$

as the quality measure of the welcher Weg detector. The denominator equals one for ρ_1 of the form (10) and normalizes Q for more general density operators ρ_1. The extreme values of Q and their significances are

$$\begin{aligned} Q = 1 :&\ \text{no which-path information available,}\\ &\ \text{lousy welcher Weg detector;}\\[1ex] Q = 0 :&\ \text{full which-path information available,}\\ &\ \text{perfect welcher Weg detector;} \end{aligned} \tag{15}$$

and intermediate values of Q mean that partial which-path information is available.

COMPLEMENTARITY

The principle of complementarity requires that, if we are able to tell along which path the atom traversed the interferometer, spin coherence must be lost. Consequently, $Q = 0$ must imply

$$\mathrm{tr}\left(a_1^{\dagger}(a_1 a_1^{\dagger})^{-1/2}\rho_1\right) = 0. \tag{16}$$

To demonstrate that this is indeed so, we evaluate the numerator in (14) in terms of eigenstates $|\nu\rangle$ and eigenvalues μ_ν of ρ_1:

$$\rho_1|\nu\rangle = |\nu\rangle\mu_\nu, \qquad \mu_\nu \geq 0, \qquad \sum_\nu \mu_\nu = \mathrm{tr}\rho_1 = 1, \tag{17}$$

which yields

$$\sum_{\nu,\nu'} \mu_\nu \mu_{\nu'} \left|\langle\nu|a_1^{\dagger}(a_1 a_1^{\dagger})^{-1/2}|\nu'\rangle\right|^2 = 0. \tag{18}$$

In this double sum of nonnegative terms each term must vanish individually, so that for $\nu = \nu'$ we find in particular

$$\mu_\nu \langle \nu | a_1^\dagger (a_1 a_1^\dagger)^{-1/2} | \nu \rangle = 0, \tag{19}$$

which we sum over ν to establish (16). Thus, $Q = 0$ *does* imply total loss of spin coherence, as required by the principle of complementarity. The reverse, of course, is not true: loss of spin coherence does not indicate the availability of which-path information.*

EXAMPLES

For illustration we evaluate both Q and the spin-coherence measure[1]

$$C = |\langle \sigma_+ (t_f) \rangle| = \left| \mathrm{tr} \left(a_1^\dagger (a_1 a_1^\dagger)^{-1/2} \rho_1 \right) \right| \tag{20}$$

for a few ρ_1's of various kinds, all of which are, of course, subject to the conditions mentioned in the Introduction, namely: $N_1 = \langle a_1^\dagger a_1 \rangle$ has to be a large number with a relatively small uncertainty δN_1.

Number State. If ρ_1 projects to a number state, the eigenvalue of $a_1^\dagger a_1$ being N_1, then $\delta N_1 = 0$ and $Q = 0$ as well as $C = 0$. One has total loss of spin-coherence, and full which-path information is available.

Coherent State. If ρ_1 projects to a coherent state, the eigenvalue of a_1 being $\sqrt{N_1} e^{i\theta_1}$, then $1 \ll \delta N_1 = \sqrt{N_1} \ll N_1$, and

$$Q = C^2 = 1 - \frac{1}{N_1} + \ldots \cong 1. \tag{21}$$

One has spin-coherence maintained and no which-path information. This situation and the previous one are dealt with in Ref. 1.

Thermalized Number State. If one fills the cavity with photons starting form a thermal equilibrium distribution, one could end up with

$$\rho_1 = \left(a_1^\dagger (a_1 a_1^\dagger)^{-1/2} \right)^n (1 - \lambda) \lambda^{a_1^\dagger a_1} \left((a_1 a_1^\dagger)^{-1/2} a_1 \right)^n, \tag{22}$$

where n is a positive integer and the parameter λ is related to the corresponding temperature T by

$$\lambda = \exp\left(-\frac{\hbar\omega}{k_B T} \right), \quad 0 < \lambda < 1. \tag{23}$$

Here one has

$$N_1 = n + \frac{\lambda}{1-\lambda}, \qquad \delta N_1 = \frac{\sqrt{\lambda}}{1-\lambda}, \tag{24}$$

and the constraint $\delta N_1 \ll N_1$ reads

$$n \gg \frac{\sqrt{\lambda}}{1+\sqrt{\lambda}}. \tag{25}$$

---

* See the discussion by one of us (BGE) of the Stern-Gerlach interferometer in these Proceedings.

[The situation of a pure number state is, of course, recovered in the zero-temperature limit $\lambda \to 0$.] With (22) we find

$$C = 0,$$

$$Q = \lambda = \frac{\sqrt{1+4(\delta N_1)^2} - 1}{\sqrt{1+4(\delta N_1)^2} + 1}, \tag{26}$$

so that

$$Q \cong \begin{cases} (\delta N_1)^2 \cong 0 & \text{for } \delta N_1 \ll 1, \\ 1 - 1/\delta N_1 \cong 1 & \text{for } \delta N_1 \gg 1 \, . \end{cases} \tag{27}$$

One has total loss of spin-coherence, but which-path information is only available for sufficiently small δN_1.

This example and the preceding one illustrate the notion that one must be able to distinguish between $N_1 \pm \delta N_1$ photons and $N_1 + 1 \pm \delta N_1$ photons in order to have a functioning welcher Weg detector. Although this sounds plausible it is not true, as demonstrated by the next, and last, example.

Parity States. If ρ_1 has a definite parity in the sense of

$$\mathrm{tr}\left[(-1)^{a_1^\dagger a_1}\rho_1\right] = +1 \qquad \text{or} \qquad -1, \tag{28}$$

then $Q = 0$ and $C = 0$, so that one has total loss of spin-coherence and full which-path information available. It does not matter here whether δN_1 is a large number or not.

Projectors to number states obey (28), naturally. Another example is provided by a mixture of states all of which contain an even number of photons. Such a ρ_1 could be produced by filling the cavity, starting from the vacuum, with the aid of resonant two-photon transitions. If this is done using a standard atomic beam with a Poissonian statistic for the atoms, then one gets

$$\rho_1 = \sum_{n=0}^{\infty} |2n\rangle \frac{(N_1/2)^n}{n!} e^{-N_1/2} \langle 2n|, \tag{29}$$

where the spread $\delta N_1 = \sqrt{2N_1}$ is even larger than for the corresponding coherent state with the same average photon number.

FINAL REMARKS

We have found a criterion, formulated in (15), that enables one to decide whether which-path information is *stored* in (the first cavity of) the welcher Weg detector. How to *retrieve* this information is a different question. If the cavity is prepared in a number state, one needs to determine the final photon number reliably, which is not done easily. Is it less difficult to measure the parity (28) of the photon field? Perhaps.

ACKNOWLEDGMENT

We gratefully acknowledge the support by the U. S. Office of Naval Research, Department of the Navy (ONR).

REFERENCES

1. M. O. Scully, B.-G. Englert, and J. Schwinger, Phys. Rev. A **40**, 1775 (1989). Owing to absent-minded proof reading, Eqs. (57) and (58) in this article are incorrect and should be replaced by the present Eqs. (3) and (5), respectively.
2. Incidentally, the normalized ladder operators that sandwich $\tilde{\rho}_1$ have been used by L. Susskind and J. Glogower, Physics **1**, 49 (1964) as quantum analogs of the classical oscillator phase factors $\exp(\pm i\phi)$.

CENTER-OF-MASS MOTION OF MASING ATOMS

Berthold-Georg Englert,[1] Julian Schwinger,[2] and Marlan O. Scully[3]

Center for Advanced Studies[1,2,3] and
Department of Physics and Astronomy
University of New Mexico, Albuquerque, New Mexico, 87131, USA
and
Sektion Physik[1]
Universität München
D-8046 Garching, West Germany
and
Department of Physics, University of California[2]
Los Angeles, CA 90024, USA
and
Max-Planck Institut für Quantenoptik[3]
D-8046 Garching bei München, West Germany

INTRODUCTION

In a recent paper[1] we studied a Stern-Gerlach interferometer with a which-path detector in one arm. This detector consists of a pair of maser cavities with suitably prepared photon fields. The interferometer can only function properly if the interaction with the maser fields does not affect the center-of-mass (CM) motion of the atoms significantly. In Ref. 1 the reader was assured that this condition is met. It is the purpose of the present contribution to supply a detailed discussion which justifies that assertion.

We consider this physical situation: an atom of a mass m moves along the y-axis (momentum $p_y \equiv p$) through a high-Q maser cavity; the coupling between an internal atomic transition (two states characterized by $\sigma_z' = \pm 1$) and the one relevant photon mode (circularly polarized, photon energy $\hbar\omega$, annihilation and creation operators $a, a^\dagger$) is resonant; the dynamics is governed by the Hamilton operator

$$H = \frac{1}{2m}p^2 + \hbar\omega\left(a^\dagger a + \frac{1}{2}\sigma_z\right) - \frac{1}{2}\hbar g(y)(a\sigma_+ + a^\dagger\sigma_-), \tag{1}$$

where $\sigma_\pm = \sigma_x \pm i\sigma_y$ effect the transition between $\sigma_z' = 1$ and $\sigma_z' = -1$, and the spatial dependence of the coupling strength $g(y)$ originates in the mode function associated with the selected photon mode, as discussed in Ref. 1; consequently, $g(y)$ is large only inside the cavity, and leaks out through the cavity openings to some small extent. It is assumed that nothing of importance happens in the perpendicular x, z-plane, which requires that during the period of interest the atom remains well localized around $x = 0$, $z = 0$. In a typical experiment, the kinetic energy $p^2/(2m)$ is about 10^{-1} eV. If we deal with a magnetic coupling as in Ref. 1, one has $\hbar g \sim 10^{-16}$ eV and needs $\sim 10^{10}$ photons to ensure the atomic transition ("spin flip", see below after Eq. (37)), so that the interaction energy is $\sim 10^{-16}$ eV $\times(10^{10})^{1/2} = 10^{-11}$ eV; if instead the photon field is coupled to an electric dipole transition between Rydberg levels, the coupling strength is increased by 10^5 and the required number

New Frontiers in Quantum Electrodynamics and Quantum Optics
Edited by A. O. Barut, Plenum Press, New York, 1990

of photons is reduced to ~ 1, so that again the interaction energy is of the order of 10^{-11} eV. Thus the ratio interaction energy/kinetic energy is *very* small in any situation. Therefore, the effect of the interaction upon the CM motion of the atom needs to be taken into account to first order, at most.

We take for granted that, in a reasonable experiment, the following requirements are obeyed: (i) the atoms move with well defined velocity and (ii) the spreading of the CM wave function is negligible during the time the interaction takes place. If we denote the initial momentum expectation value by p_0 and the spreads in position and momentum by δy and δp, this means that the atom is prepared such that

$$(i)\quad \frac{\delta p}{p_0} \ll 1, \qquad (ii)\quad \frac{\delta y}{\delta p/m} \gtrsim T, \tag{2}$$

where T is the duration of the interaction, to be specified more precisely below.

ZEROTH ORDER TREATMENT

We being with a zeroth order treatment that disregards all effects on the CM motion. This is the approximation

$$\frac{d}{dt}p(t) = -\frac{\partial H}{\partial y} = \frac{1}{2}\hbar\frac{dg}{dy}(y(t))(a(t)\sigma_+(t) + a^\dagger(t)\sigma_-(t)) \cong 0 \tag{3}$$

which has the consequences

$$p(t) \cong p, \tag{4}$$

$$y(t) = y + \int_0^t dt' \frac{1}{m}p(t') \cong y + \frac{t}{m}p, \tag{5}$$

where $p \equiv p(t=0)$ and $y \equiv y(t=0)$. An implied approximation is

$$\begin{aligned} \frac{d}{dt}a(t) &= \frac{1}{i\hbar}\frac{\partial H}{\partial a^\dagger} = -i\omega a(t) + \frac{i}{2}g(y(t))\sigma_-(t) \\ &\cong -i\omega a(t) + \frac{i}{2}g\left(y + \frac{t}{m}p\right)\sigma_-(t). \end{aligned} \tag{6}$$

The approximate $y(t)$ commutes with $a(t)$ and $\sigma_-(t)$ even for unequal times. Consequently, tracing over the CM coordinates in (6) produces

$$\frac{d}{dt}a(t) \cong -i\omega a(t) + \frac{i}{2}\,\tilde{g}(t)\sigma_-(t), \tag{7}$$

where the numerical function $\tilde{g}(t)$ is the spatial expectation value of the operator $g(y(t))$,

$$\tilde{g}(t) = \langle g(y(t))\rangle = \int dy'\psi^*(y',t)g(y')\psi(y',t). \tag{8}$$

The CM wave function $\psi(y',t)$ is supposed to not be spreading significantly during the interaction, so that

$$|\psi(y',t)|^2 \cong |\psi_0(y'-v_0t)|^2. \tag{9}$$

where $v_0 = p_0/m$ is the well defined velocity of the atom. Thus

$$\tilde{g}(t) \cong \int dy'|\psi_0(y')|^2 g(y'+v_0t), \tag{10}$$

which exhibits the overlap of the atom's CM probability density and, essentially, the cavity mode function. The duration of the interaction, T, is then specified by stating that $\tilde{g}(t)$ is nonzero only for $0 < t < T$, or colloquially: $t = 0$ is *before* the atom enters the cavity and $t = T$ is *after* it has left.

For the spin $(\sigma_z, \sigma_\pm)$ and photon $(a, a^\dagger)$ degrees of freedom we have now the effective (and approximate) Hamilton operator

$$\begin{aligned} H_{\text{eff}} &= \hbar\omega\left(a^\dagger a + \frac{1}{2}\sigma_z\right) - \frac{1}{2}\hbar\,\tilde{g}(t)(a\sigma_+ + a^\dagger\sigma_-) \\ &= \hbar\omega\left(\gamma^2 - \frac{1}{2}\right) - \hbar\,\tilde{g}(t)\gamma. \end{aligned} \tag{11}$$

which introduces

$$\gamma \equiv \frac{1}{2}(a\sigma_+ + a^\dagger\sigma_-) \tag{12}$$

and makes use of

$$\gamma^2 = a^\dagger a + \frac{1}{2}(1 + \sigma_z). \tag{13}$$

The eigenvalues of γ are $\gamma' = 0, \pm\sqrt{1}, \pm\sqrt{2}, \pm\sqrt{3}, \ldots$. Expressed in terms of the common eigenstates $|(a^\dagger a)', \sigma_z'\rangle$ of $a^\dagger a$ and σ_z, the eigenstates of γ are

$$|0, -1\rangle \quad \text{for} \quad \gamma' = 0 \tag{14}$$

and

$$2^{-1/2}(|n, -1\rangle \pm |n-1, +1\rangle) \qquad \text{for} \qquad \gamma' = \pm\sqrt{n}, \tag{15}$$

with $n = 1,2,3,\ldots$.

In the transition from the actual Hamilton operator (1) to the effective one (11) all changes of the CM motion are disregarded. In other words, the CM motion is treated classically. As a consequence, the spatial dependence of the coupling strength, $g(y)$, is translated into a temporal dependence, $\tilde{g}(t)$. In this simplified description, the interaction between the photon degree of freedom $(a, a^\dagger)$ and the internal atomic degree of freedom $(\sigma_z, \sigma_\pm)$ is turned on/off when the atom enters/leaves the cavity. The effective evolution operator, transforming from time t' to time t is then (note that γ is constant in time)

$$U_{\text{eff}}(t', t) = \exp\left[-i\omega(t - t')\left(\gamma^2 - \frac{1}{2}\right) + i\gamma\int_{t'}^{t} dt''\,\tilde{g}(t'')\right]. \tag{16}$$

We are interested in the changes form "before" to "after," which draws our attention to

$$\begin{aligned} \int_0^T dt''\,\tilde{g}(t'') &= \int_{-\infty}^{\infty} dt''\,\tilde{g}(t'') = \int_{-\infty}^{\infty} dt'' \int_{-\infty}^{\infty} dy'\,|\psi_0(y')|^2 g(y' + v_0 t'') \\ &= \int_{-\infty}^{\infty} dy'\,|\psi_0(y')|^2 \int_{-\infty}^{\infty} dy''\,g(y'')/v_0 = \frac{1}{v_0}\int dy''\,g(y''). \end{aligned} \tag{17}$$

Note that the detailed form of the normalized initial CM wave function $\psi_0(y')$ does not matter; only the spatial integral of the coupling strength $g(y)$, which is to say: the spatial integral of the cavity mode function, enters. For a cavity of length L, we introduce the average coupling strength $\bar{g}$ by means of

$$\int dy' g(y') = \bar{g}L. \tag{18}$$

The final form for the overall evolution operator is then

$$U_{\text{eff}} \equiv U_{\text{eff}}(0,T) = \exp\left[-i\omega T\left(\gamma^2 - \frac{1}{2}\right) + i\gamma\bar{g}L/v_0\right]. \tag{19}$$

The ratio L/v_0 appearing here is the time that a classical object would need to traverse through the cavity at speed v_0. If $|\psi_0(y')|^2$ is well localized on the scale set by L, $\delta y \ll L$, as is the normal experimental situation, this classical time is practically equal to T. If, however, $|\psi_0(y')|^2$ has macroscopic extensions, $\delta y \sim L$ or larger, the time T can be much larger than L/v_0.

It should be remarked that H_{eff} with $\tilde{g}(t) = \text{const.}$ is the Hamilton operator of the Jaynes-Cummings model[2] (at resonance and for a circularly polarized photon mode). The possibility of summarizing the effect of the interaction in terms of an average coupling constant $\bar{g}$ and an effective interaction time L/v_0 enables one to apply it to real experiments, although detailed knowledge about the mode function may not be available. There are, indeed, precision experiments that demonstrate agreement between the predictions of the Jaynes-Cummings model and the observed phenomena.[3]

FIRST ORDER TREATMENT

We now return to the full Hamilton operator

$$H = \frac{1}{2m}p^2 + \hbar\omega\left(\gamma^2 - \frac{1}{2}\right) - \hbar g(y)\gamma \tag{20}$$

and present a first order treatment of the effects on the CM motion. Quite generally, the time transformation function from a p'', γ''-state at time t'' to a y', γ'-state at time t' is given by

$$\langle y', \gamma', t' | p'', \gamma'', t'' \rangle = \langle y', \gamma', t'' | \exp(-iH(t' - t'')/\hbar) | p'', \gamma'', t'' \rangle. \tag{21}$$

For $t'' = 0$ ("before") and $t' = T$ ("after") this is

$$\begin{aligned} \langle y', \gamma', T | p'', \gamma'', 0 \rangle = \delta_{\gamma', \gamma''} \exp\left[-i\omega T\left(\gamma'^2 - \frac{1}{2}\right)\right] \\ \times \langle y' | \exp\left[-\frac{i}{\hbar}\frac{p^2}{2m}T + ig(y)\gamma' T\right] | p'' \rangle, \end{aligned} \tag{22}$$

where the fact that $\gamma(t') = \gamma(t'')$ is a constant of motion has been used. We are going to apply this only to CM wave functions which have a small momentum spread around $p'' = p_0$ and for which y' is to the right of the cavity ("after") whereas $y' - v_0 T = y' - p_0 T/m$ is to the left ("before"). With this in mind, and recalling that the interaction energy $\hbar g(y)\gamma'$ causes only a very small momentum change, we have to first order in $g(y)$

$$\frac{p^2}{2m} - \hbar g(y)\gamma' \cong \frac{1}{2m}\left(p - \frac{\hbar}{v_0}g(y)\gamma'\right)^2. \tag{23}$$

We combine this with

$$p - \hbar\frac{\partial f(y)}{\partial y} = \exp(if(y))\; p\; \exp(-if(y)), \tag{24}$$

here for $\partial f(y)/\partial y = g(y)\gamma'/v_0$, to establish

$$\exp\left(-\frac{i}{\hbar}\frac{p^2}{2m}T + ig(y)\gamma' T\right) \cong \exp(if(y))\exp\left(-\frac{i}{\hbar}\frac{p^2}{2m}T\right)\exp(-if(y)) \tag{25}$$
$$= \exp(if(y))\exp(-if(y-pT/m))\exp\left(-\frac{i}{\hbar}\frac{p^2}{2m}T\right),$$

where the last step is based upon

$$\exp\left(-\frac{i}{\hbar}\frac{p^2}{2m}T\right) f(y)\exp\left(\frac{i}{\hbar}\frac{p^2}{2m}T\right) = f(y-pT/m). \tag{26}$$

In $f(y-pT/m)$ it is consistent to replace p by $p_0 = mv_0$, which produces

$$\exp(if(y))\exp(-if(y-pT/m)) \cong \exp(if(y) - if(y-v_0T))$$
$$= \exp\left(i\frac{\gamma'}{v_0}\int_{y-v_0T}^{y} d\tilde{y}'\, g(\tilde{y}')\right). \tag{27}$$

The time transformation function under consideration is then easily evaluated:

$$\langle y', \gamma', T | p'', \gamma'', 0\rangle \cong \delta_{\gamma',\gamma''}\langle y'|p''\rangle$$
$$\times \exp\left[-i\omega T\left(\gamma'^2 - \frac{1}{2}\right) + i\bar{g}L/v_0 - \frac{i}{\hbar}\frac{p''^2}{2m}T\right], \tag{28}$$

where we utilize the physical meaning of $y' - v_0T$ and y' (being to the left and right of the interaction region) and recall the definition (18) of $\bar{g}$.

Since the y'-dependence is only in the $\langle y'|p''\rangle$ factor, a permissible initial wave function $\langle p', \gamma', 0|\rangle = \psi(p')\phi_{\gamma'}$ evolves into the final one given by ($v_0 \to p'/m$)

$$\langle p', \gamma', T|\rangle = \psi(p')\phi_{\gamma'}\exp\left[-i\omega T\left(\gamma'^2 - \frac{1}{2}\right) + i\gamma'\bar{g}L\frac{m}{p'} - \frac{i}{\hbar}\frac{p'^2}{2m}T\right]. \tag{29}$$

In order to make contact with the previous zeroth order result, we consider final measurements of spin-and-photon operators $F(a(T), a^\dagger(T), \vec{\sigma}(T))$:

$$\langle F(a, a^\dagger, \vec{\sigma})\rangle_T = \sum_{\gamma',\gamma''}\int dp' \langle|\gamma', p', T\rangle\langle\gamma'|F(a, a^\dagger, \vec{\sigma})|\gamma''\rangle\langle\gamma'', p', T|\rangle$$
$$\cong \int dp' |\psi(p')|^2 \sum_{\gamma',\gamma''}\phi^*_{\gamma'}\exp\left[i\omega T\left(\gamma'^2 - \frac{1}{2}\right) - i\gamma'\bar{g}L\frac{m}{p'}\right] \tag{30}$$
$$\times \langle\gamma'|F(a, a^\dagger, \vec{\sigma})|\gamma''\rangle\exp\left[-i\omega T\left(\gamma''^2 - \frac{1}{2}\right) + i\gamma''\bar{g}L\frac{m}{p'}\right]\phi_{\gamma''}$$

In the exponentials $p' \cong mv_0$ is, once more, consistent, allowing to replace the remaining p'-integral by its unit value, so that we arrive at

$$\langle F(a, a^\dagger, \vec{\sigma})\rangle_T \cong \langle U_{\text{eff}}^{-1} F(a, a^\dagger, \vec{\sigma}) U_{\text{eff}}\rangle, \tag{31}$$

where the effective evolution operator is the one found by the zeroth order calculation. Note that in both treatments details of the CM wave function do not enter.

Quite analogously, we find that final measurements of CM operators $F(y(T), p(T))$ are given by

$$\langle F(y,p)\rangle_T = \left\langle \tilde{U}_{\text{eff}}^{-1} F(y,p)\, \tilde{U}_{\text{eff}} \right\rangle \tag{32}$$

with

$$\tilde{U}_{\text{eff}} = \exp\left(-\frac{i}{\hbar}\frac{p^2}{2m}T + i\gamma m\bar{g}L/p\right), \tag{33}$$

which depends on the spin-photon operator γ. In view of

$$\tilde{U}_{\text{eff}}^{-1}\, p\, \tilde{U}_{\text{eff}} = p, \tag{34}$$

$$\tilde{U}_{\text{eff}}^{-1}\, y\, \tilde{U}_{\text{eff}} = y + \frac{T}{m}p + \frac{\hbar m\bar{g}L}{p^2}\gamma, \tag{35}$$

we obtain

$$\langle F(y(T),p(T))\rangle = \left\langle F\left(y + \frac{T}{m}p + \frac{\hbar}{p}\frac{m\bar{g}L}{p}\gamma, p\right)\right\rangle . \tag{36}$$

This shows that, whereas there is no net momentum transfer to the atom, there is a net displacement

$$\Delta y = \frac{\hbar}{p}\frac{m\bar{g}L}{p}\gamma \simeq \frac{\hbar}{p_0}\frac{\bar{g}L}{v_0}\gamma \tag{37}$$

resulting from the interaction. In the context of Ref. 1, the photon field in the cavity is such that $\bar{g}L\gamma/v_0 = \pi/2$ ("spin-flip condition"), so that Δy is a quarter of the de Broglie wavelength. With the aid of the uncertainty relation, we thus conclude that Δy is very small compared to the spread δy of the CM wave function:

$$\frac{\Delta y}{\delta y} = \frac{\pi}{2}\frac{\hbar}{\delta y \delta p}\frac{\delta p}{p_0} \le \pi\frac{\delta p}{p_0} \ll 1. \tag{38}$$

Therefore, according to Ref. 4, this net displacement is quite innocuous and does not cause a significant loss of spin coherence in the Stern-Gerlach interferometer.

We close by pointing out that Δy is, not surprisingly, identical with the change in the distance covered by a classical object traversing the interaction region, which change originates in the alteration of the kinetic energy. To see this, we solve the equations of motion

$$\frac{d}{dt}p(t) = \hbar\gamma\frac{\partial g}{\partial y}(y(t)), \qquad \frac{d}{dt}y(t) = \frac{1}{m}p(t) \tag{39}$$

to first order in $g(y)$, with the outcome

$$p(t) = p + \hbar\gamma\frac{m}{p}\left[g\left(y + \frac{t}{m}p\right) - g(y)\right], \tag{40}$$

$$y(t) = y + \frac{t}{m}p + \frac{\hbar\gamma}{p}\left(\frac{m}{p}\int_y^{y+(t/m)p} dy' g(y') - tg(y)\right). \tag{41}$$

If here y and $y + (t/m)p$ are outside the interaction region to the left and right, respectively, we recover

$$y(t) = y + \frac{t}{m}p + \Delta y, \tag{42}$$

with the net displacement Δy found above.

SUMMARY

We consider the change in the center-of-mass motion of a masing atom resulting from the interaction with the photon field. Whereas there is no net momentum transfer to the atom, we find a net displacement which is, however, small and can be understood classically. We demonstrate that the evolution of the photon field involves the spatial photon mode function only in terms of an average coupling constant and is independent of the shape of the atom's center-of-mass wave function. We infer that the success of the Jaynes-Cummings model originates in this lack of sensitivity to particulars.

ACKNOWLEDGMENT

This work was partially supported by the Office of Naval Research.

REFERENCES

1. M. O. Scully, B.-G. Englert, and J. Schwinger, Phys. Rev. A **40**, 1775 (1989), where the present contribution is announced in Ref. 9; see also the second lecture by Scully, Fearn, and Atherton in these Proceedings.
2. E. T. Jaynes and F. W. Cummings, Proc. IEEE **51**, 89 (1963).
3. G. Rempe, H. Walther, and N. Klein, Phys. Rev. Lett. **58**, 353 (1987).
4. J. Schwinger, M. O. Scully, and B.-G. Englert, Z. Phys. D **10**, 135 (1988).

SPIN COHERENCE IN STERN-GERLACH INTERFEROMETERS

Berthold-Georg Englert

Center for Advanced Studies and
Deptartment of Physics and Astronomy
University of New Mexico, Albuquerque, New Mexico 87131, USA
and
Sektion Physik
Universität München
D-8046 Garching, West Germany

INTRODUCTION

Most of the well-known paradoxa of quantum mechanics concern the loss of interference properties resulting from the measurement of one observable when some knowledge about a complimentary one was available before the measurement took place. This phenomenon needs thorough understanding, and it does not suffice to simply state, as von Neumann did, that quantum mechanics consists of two parts, namely (i) the temporal evolution of the state of an isolated system, as described by the Schrödinger equation; and (ii) the "reduction" of that state by the process of measurement. This deus ex machina is a *mathematicians* answer; it only increases the confusion, and as a result one can now read about elaborate interpretation schemes invented to explain this alleged state reduction. In contrast, ***physicists*** know that measurements take time and are performed with the aid of interactions, so that (i) and (ii) are both aspects of one quantum dynamics.

To appreciate this, one must analyze simple but realistic experiments with sufficient accuracy to see, in detail, how measurement emerges from interaction. This is the objective of a collaboration between Julian Schwinger, Marlan Scully and myself, results being partly published in a series of papers[1] entitled "Is spin coherence like Humpty-Dumpty?"* In the present talk I shall report the answer we give to the question

> Is it possible to reunite the two partial beams of a Stern-Gerlach apparatus with such precision that the original spin state is recovered? (1)

The detailed dynamical analysis of this problem requires to go well beyond the extreme idealizations that are so common in the so-called theories of quantum measurements. Rather than contribute to these ambitious developments, let us be content, for a start, with a good quantum theory of *one* measurement.

---

* Humpty-Dumpty is an egg that, in a medieval nursery rhyme (actually a riddle), breaks and cannot be put together again.

New Frontiers in Quantum Electrodynamics and Quantum Optics
Edited by A. O. Barut, Plenum Press, New York, 1990

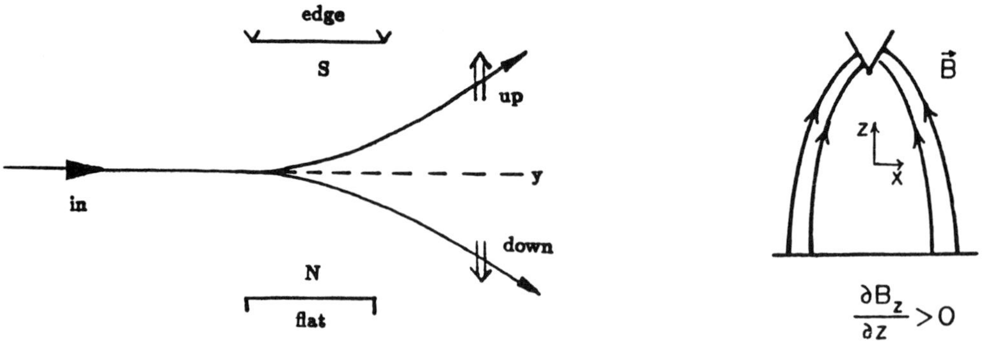

Fig. 1. Side view and cross section of a Stern-Gerlach apparatus.

STERN-GERLACH APPARATUS

Let me begin by reminding you of the essentials of a Stern-Gerlach apparatus (SGA), as sketched in Fig. 1. A beam of spin-1/2 atoms, flying along the y-axis, passes through a symmetric magnetic field whose large B_z inhomogeneity acts upon the atoms' magnetic moments and splits the beam in two: "spin up" and "spin down." Here the spatial properties of the atoms act classically, but there are quantum effects.* The forces $\pm F\vec{e}_z$, exerted on the atoms for the duration T, transfer momenta $\Delta p_z = \pm FT$, which must be large compared to δp_z, the spread in momentum prior to entering the SGA,

$$FT \gg \delta p_z \tag{2}$$

in order to split the beam macroscopically. The force F originates in a spatially dependent potential energy E, $F = -\partial E/\partial z$, which energy contributes to the phase of the wave function, the total change being $-ET/\hbar$. Different parts of the wave function probe the potential energy at different points in space, and as a consequence this phase varies within the wave function because there is a spread δz in position. The phase has therefore an uncertainty given by

$$\delta(-ET/\hbar) = -\delta ET/\hbar = -\frac{\partial E}{\partial z}\delta z T/\hbar = FT\delta z/\hbar \gg \delta p_z \delta z/\hbar \geq \frac{1}{2},$$

or

$$\delta(-ET/\hbar) \gg 1, \tag{3}$$

where the condition (2) and Heisenberg's uncertainty principle have been used. It must be emphasized that the large phase variations (3) are the dispersion of phases in each of the two partial beams, not the relative phases between them. It is this phase dispersion that must be reversed if one wants to reunite the beams and recover the original spin state. Please note, in particular, that (3) is an implication of (2): the price for being able to distinguish the two spin states macroscopically is paid in the hard quantum currency of enormous phase variations.

STERN-GERLACH INTERFEROMETER

When leaving the SGA of Fig. 1, the atoms in the two partial beams not only have different z-values but also different momenta p_z. In other words: the beam is split both in position and in momentum [see (2)]. A single second SGA cannot undo both. Therefore we need altogether four SGAs to construct a Stern-Gerlach interferometer (SGI), as illustrated in Fig. 2. The first SGA is like the one in

* The essential features of the following qualitative argument are due to Heisenberg.[2]

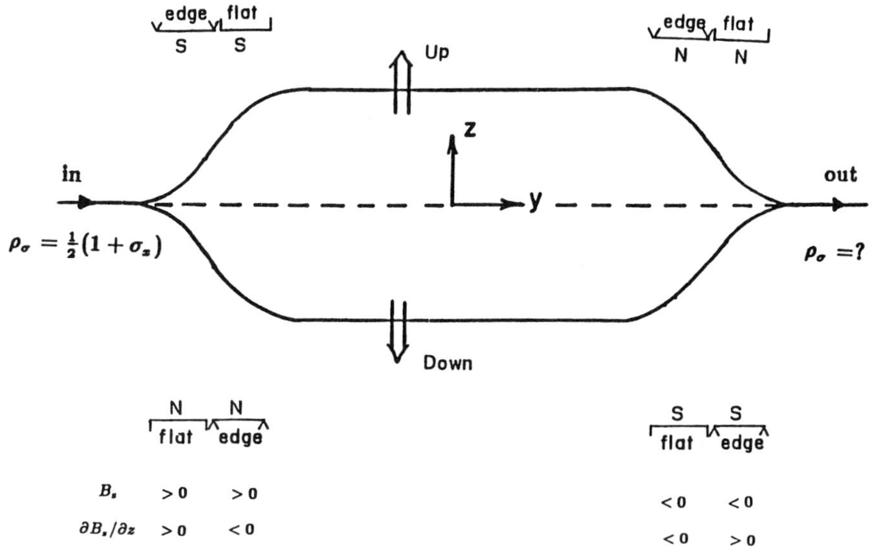

Fig. 2. Side view of the Stern-Gerlach interferometer.

Fig. 1. The second one has the $\partial B_z/\partial z$ field gradient reversed, so that it undoes the splitting in momentum. In the middle of the SGI the atoms travel parallel to the y-axis, the net effect being a displacement along z in either direction. The third and fourth SGAs are symmetrical to the first and second, they are intended to reunite the beams. Ideally, then, one beam leaves the interferometer.

Consider the entering atoms all being polarized in the x-direction, that is

$$\langle\sigma_x\rangle_{\text{in}} = 1,$$

or expressed as a spin density operator,

$$\rho_{\sigma,\text{in}} = \frac{1}{2}(1+\sigma_x)$$

where as always we associate Pauli matrices $\vec{\sigma} = (\sigma_x, \sigma_y, \sigma_z)$ with the 1/2-spin of the atoms. Now imagine that there is yet another SGA, oriented along the x-axis, probing the atoms emerging from the SGI.* If these atoms are again x-polarized,

$$\rho_{\sigma,\text{out}} = \frac{1}{2}(1+\sigma_x), \quad \langle\sigma_x\rangle_{\text{out}} = 1, \tag{4}$$

then they will all be deflected in the $+x$-direction. In contrast, if the emerging beam is unpolarized,

$$\rho_{\sigma,\text{out}} = \frac{1}{2}, \quad \langle\sigma_x\rangle_{\text{out}} = 0, \tag{5}$$

then half of the atoms will be deflected in the $+x$ and half in the $-x$-direction. In principle, then, we can distinguish the two possibilities.

Question (1) thus asks whether the SGI output is unavoidably (5) or conceivably (4). There are two extreme, yet popular positions taken by thoughtful physicist when asked to answer question (1). Some say: "Yea, the emerging beam will be spin coherent since the different parts of the SGI involve magnetic fields which are

* See Fig. 1 in the second lecture by Scully, Fearn, and Atherton.

reversed in subsequent sections of the interferometer, and this is equivalent to time reversal." Others say: "Nay, the output of the SGI will be totally incoherent since the wave functions describing the center-of-mass motion are of finite extent, so that the atom probes different magnetic fields at different points in the wave function; thus there must be a 'scrambling' of the phase of the spin-1/2 atom."* A decision, who has the better arguments, can only be made on the basis of a careful analysis of the SGI.

SIMPLIFIED THEORY OF THE SGI

The dynamical variables are the position $\vec{r}(t)$, the momentum $\vec{p}(t)$, and the magnetic momentum $\vec{\mu}(t) = \mu\vec{\sigma}(t)$ of the atom. Denoting the atomic mass by m, the Hamilton operator is

$$H = \frac{1}{2m}\vec{p}^{\,2} - \vec{\mu}\cdot\vec{B}(\vec{r}), \tag{6}$$

where $\vec{B}(\vec{r})$ is the magnetic field of the SGI. The equations of motion

$$\begin{aligned} &\frac{d}{dt}\vec{r} = \vec{p}/m, \\ &\frac{d}{dt}\vec{p} = \left[\vec{\nabla}\vec{B}(\vec{r})\right]\cdot\vec{\mu}, \\ &\frac{d}{dt}\vec{\mu} = \frac{2\mu}{\hbar}\vec{\mu}\times\vec{B}(\vec{r}) \quad \text{or} \quad \frac{d}{dt}\vec{\sigma} = \frac{2}{\hbar}\vec{\sigma}\times\mu\vec{B}(\vec{r}), \end{aligned} \tag{7}$$

cannot be solved explicitly for a spatially dependent magnetic field, as is the situation in the SGI. We can, however, preliminarily adopt natural simplifications based on the symmetry of the set-up.

The beam runs down the center of the SGI and stays in the y, z-plane, consult Fig. 1. There the dominant field component is B_z, which we shall assume to vary linearly with z. We thus approximate the magnetic field by

$$\vec{B} \cong \vec{e}_z\left(B_z(y) + z\frac{\partial B_z}{\partial z}(y)\right). \tag{8}$$

Here we have already picked $x = 0$, taking for granted that the beam probes the SGI only in the vicinity of the y, z-plane. Further, we consider an experimental arrangement in which the beam has a well-defined velocity in the y-direction. This velocity v will not be affected significantly by the interaction with the magnetic field, so that

$$y \cong vt \tag{9}$$

is justifiable. This turns the spatial dependences of $B_z(y)$ and $\partial B_z/\partial z(y)$ into time dependences,

$$\begin{aligned} \mu B_z(y) &\cong \mu B_z(vt) \equiv \mathcal{E}(t), \\ \mu\frac{\partial B_z}{\partial z}(y) &\cong \mu\frac{\partial B_z}{\partial z}(vt) \equiv F(t), \end{aligned}$$

where the energy parameter $\mathcal{E}(t)$ and the force parameter $F(t)$ are numerical functions of t. We are now left with an effective Hamilton operator

$$H_{\text{eff}} = \frac{1}{2m}p_z^2 - \mathcal{E}(t)\sigma_z - F(t)\sigma_z \tag{10}$$

---

* A poll taken during the lecture in Istanbul identified 15 yea-sayers vs. 25 nay-sayers, and about half the audience had no public opinion.

for the evolution of z, p_z and $\vec{\sigma}$. The irrelevant kinetic energy $(m/2)v^2$ of the now classical y-motion has been omitted.

The corresponding equations of motion

$$\frac{d}{dt}z = p_z/m, \qquad \frac{d}{dt}p_z = F(t)\sigma_z,$$

$$\frac{d}{dt}\sigma_z = 0, \quad \frac{d}{dt}(\sigma_x + i\sigma_y) = \frac{d}{dt}\sigma_+ = \frac{2}{i\hbar}(\mathcal{E}(t) + F(t)z)\sigma_+,$$

are simpler than the set (7) and can be solved explicitly. This results in

$$\begin{aligned} z(t) &= z(0) + \frac{t}{m}p_z(0) + \left(\Delta z(t) + \frac{t}{m}\Delta p(t)\right)\sigma_z(0), \\ p_z(t) &= p_z(0) + \Delta p(t)\sigma_z(0), \\ \sigma_z(t) &= \sigma_z(0), \\ \sigma_+(t) &= e^{-i\Phi(t)} \exp\left(\frac{2i}{\hbar}(p_z(0)\Delta z(t) - z(0)\Delta p(t))\right)\sigma_+(0). \end{aligned} \tag{11}$$

The various numerical functions appearing here are

$$\begin{aligned} \Delta p(t) &= \int_0^t dt' F(t'), \\ \Delta z(t) &= -\int_0^t dt' t' F(t')/m, \\ \Phi(t) &= \frac{2}{\hbar}\int_0^t dt' \mathcal{E}(t'). \end{aligned} \tag{12}$$

The latter is the accumulated Larmor precession angle for the field strength at $z = 0$ whereas Δp and $\Delta z + (t/m)\Delta p$ measure the displacements of the partial beams in momentum and position. Typical time dependences of the force $F(t)$ and these displacements are plotted in Fig. 3. As indicated there, the atom enters the SGI at time $t = 0$ and leaves it at time $t = T$, so that $\mathcal{E}(t)$ and $F(t)$ are nonzero only for $0 < t < T$.

MACROSCOPIC SEPARATION

A functioning SGA enables one to select one partial beam. Therefore the splitting must be macroscopic when the separation is maximal (and $\Delta p = 0$), which ideally happens at time $t = T/2$. In other words:

$$\Delta z(T/2) \gg \delta z(T/2) \cong \delta z, \tag{13}$$

where $\delta z(t)$ is the spread in position of either one of the partial beams at time t and δz is the initial spread. The latter statement of (13) reflects the requirement that in any practical experiment the natural spreading of a wave function must not be too significant during the time T. Since the Hamilton operator (10) implies a spreading identical to that of a free particle, this requirement tells us that

$$T \cong m\delta z/\delta p_z$$

is the order of magnitude of T. The spread in momentum, δp_z, is time independent because (10) is linear in z. With $\Delta p(T/2) = 0$ we have then

$$\left(\Delta z + \frac{t}{m}\Delta p\right)(T/2) = \Delta z(T/2) \gg \delta z \cong \frac{T}{m}\delta p_z \tag{14}$$

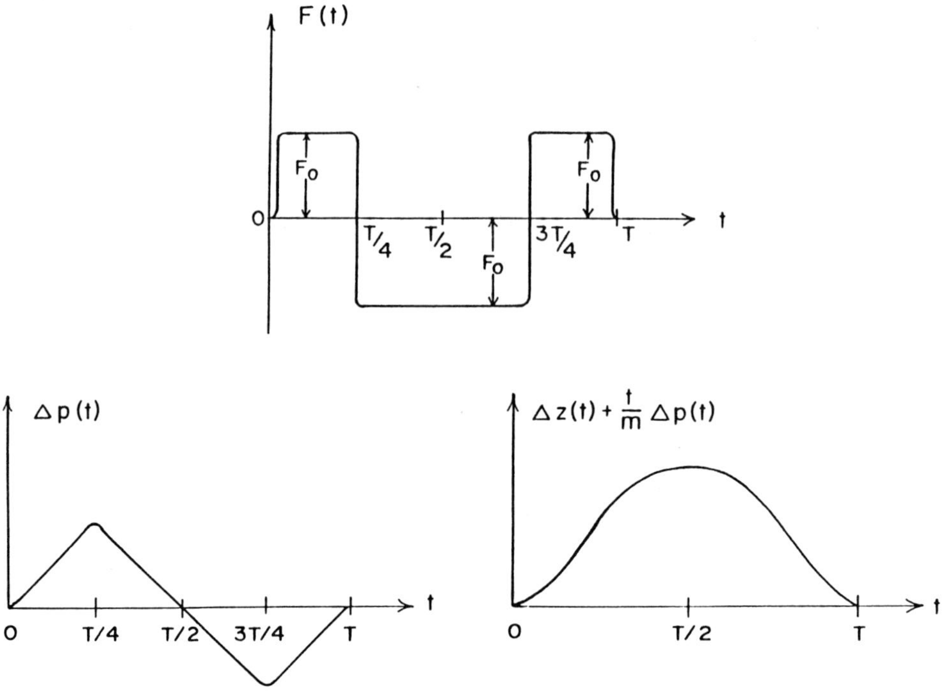

Fig. 3. Typical time dependences of the force $F(t)$, the displacements Δp in momentum and $\Delta z + (t/m)\Delta p$ in position.

and for a force of the type sketched in Fig. 3,

$$\Delta z(T/2) = \alpha F_0 T^2/m,$$

with α of the order 1/20. Combining this with (14) gives

$$\alpha F_0 T \gg \delta p_z,$$

and we are back at the condition (2).

SPIN COHERENCE

The most reasonable measure for spin coherence is the length of the vector $\langle\vec{\sigma}\rangle$,

$$C \equiv |\langle\vec{\sigma}\rangle| = \sqrt{\langle\sigma_z\rangle^2 + |\langle\sigma_+\rangle|^2}.$$

The two extreme situations are

(1) $C = 1$, beam totally polarized, perfect spin coherence;
(2) $C = 0$, beam unpolarized, no spin coherence;

and partially polarized beams, $0 < C < 1$, exhibit partial spin coherence.

As in Fig. 2, consider $\langle\vec{\sigma}\rangle_{\text{in}} = (1,0,0)$, implying $C(t=0) = 1$. Then we have $\langle\sigma_z(t)\rangle = 0$ for all times, indicating that the SGI splits the beam symmetrically, and find with (11)

$$C(t) = |\langle\sigma_+(t)\rangle| = \left|\left\langle \exp\left(\frac{2i}{\hbar}(p_z(0)\Delta z(t) - z(0)\Delta p(t))\right)\right\rangle\right|,$$

the latter expectation value is purely spatial,

$$C(t) = \left| \mathrm{tr} \left\{ \rho_o(z, p_z) \exp\left(\frac{2i}{\hbar} (p_z(0)\Delta z(t) - z(0)\Delta p(t)) \right) \right\} \right|, \tag{15}$$

where $\rho_o(z, p_z)$ is the spatial* density operator of the entering atoms (for the z degree of freedom). It is instructive to write the trace in (15) in terms of integrals over z' or p'_z,

$$\begin{aligned} C(t) &= \left| \int dz' \langle z' + \Delta z(t) | \rho_o | z' - \Delta z(t) \rangle \exp(-2iz'\Delta p(t)/\hbar) \right| \\ &= \left| \int dp'_z \langle p'_z + \Delta p(t) | \rho_o | p'_z - \Delta p(t) \rangle \exp(2ip'_z \Delta z(t)/\hbar) \right|. \end{aligned}$$

Here we observe, for instance, that the momentum displacement $\Delta p(t)$ produces a phase variation in the z' description, which illustrates (3).

The expectation value (15) will vanish, unless both $\Delta z(t)$ and $\Delta p(t)$ are small on the scales set by the initial spreads, δz and δp_z. In particular, half way through the SGI when $\Delta z \gg \delta z$, we have $C = 0$. Spin coherence is lost as soon as the beam is split.

To illustrate (15) let us assume that we know the values of δz and δp_z, in addition to $\langle z \rangle = 0$ and $\langle p_z \rangle = 0$ at $t = 0$, but lack further information. In this situation, we have**

$$\rho_o(z, p_z) = 2 \sinh \mu \, \exp\left(-\frac{\mu}{2} \coth\mu \left[\left(\frac{z}{\delta z} \right)^2 + \left(\frac{p_z}{\delta p_z} \right)^2 \right] \right) \tag{16}$$

where the parameter μ is related to the uncertainty product $\delta z \delta p_z$ by

$$\coth\mu = \frac{\delta z \delta p_z}{\hbar/2}. \tag{17}$$

Here we get

$$C(t) = \exp\left(-2 \left[\left(\frac{\delta p_z \Delta z(t)}{\hbar} \right)^2 + \left(\frac{\delta z \Delta p(t)}{\hbar} \right)^2 \right] \right), \tag{18}$$

which is indeed small, unless

$$|\Delta z| \ll \frac{\hbar}{\delta p_z} \quad \text{and} \quad |\Delta p| \ll \frac{\hbar}{\delta z}. \tag{19}$$

Note that these are stronger requirements than $|\Delta z| \ll \delta z$ and $|\Delta p| \ll \delta p_z$ if $\delta z \delta p_z / \hbar$ is large, as is the situation in a typical atomic beam experiment where this number is about 10^3 [the hyperbolic sine in (16) can then be replaced by μ and the cotangent in (16) and (17) by $1/\mu$].

To achieve $C(T) \lesssim 1$, the final values of Δz and Δp must be small. How small? How large a deviation form the ideal values $\Delta z(T) = 0$, $\Delta p(T) = 0$ can be tolerated? Since a small hermitian operator A gives rise to

$$|\langle e^{iA} \rangle| = 1 - \frac{1}{2}(\delta A)^2 + O(A^4),$$

* The subscript o = orbital is used in place of s = spatial to avoid confusion with s = spin.

** Of all the ρ_o consistent with our knowledge, this one has maximal entropy.

Eq. (15) implies ($t = T$ in C, Δz, and Δp)

$$C \cong 1 - 2\left[\left(\frac{\delta p_z \Delta z}{\hbar}\right)^2 + \left(\frac{\delta z \Delta p}{\hbar}\right)^2\right] + \frac{2}{\hbar}(\langle z p_z + p_z z\rangle - \langle z\rangle\langle p_z\rangle)\frac{\Delta z \Delta p}{\hbar}.$$

This differs from what (18) would produce by the last term, which term, however, is proportional to $\Delta z \Delta p$, the product of two uncontrollable small numbers of arbitrary sign, so that the ensemble average of $\Delta z \Delta p$ over the beam vanishes. Consequently, the requirements (19) are more generally true than the use of (16) suggests.

Taking their product establishes

$$\left|\frac{\Delta z \Delta p}{\hbar}\right| (\ll)^2 \frac{\hbar}{\delta z \delta p_z}. \tag{20}$$

On the other hand, the SGI splits the beam only if the maximal vaues of $\Delta z(t)$ and $\Delta p(t)$ obey

$$|\Delta z|_{\max} |\Delta p|_{\max} (\gg)^2 \delta z \delta p_z,$$

so that we find

$$\frac{|\Delta z \Delta p|_{\text{final}}}{|\Delta z|_{\max} |\Delta p|_{\max}} (\ll)^4 \left(\frac{\hbar}{\delta z \delta p_z}\right)^2.$$

But even if the beam is not split by the SGI, the requirement (20) states that to ensure $C(T) \lesssim 1$, that is

> to ensure a spin coherent output beam, the *macroscopic* magnetic field must be controlled with *submicroscopic* precision. (21)

Is that achievable? Can the yea-sayers lean back and call it a mere technical problem, not a principle limitation? Or is it high time for the nay-sayers who point out that in the end magnets are made of atoms with their own natural quantum uncertainties? I leave it to the reader to decide.

In order to be able to use the SGI for interferometric measurements, one does not need $C \lesssim 1$, it suffices to have a detectable signal. For a marginal signal, $0 < C \ll 1$, the requirements are less stringent. Instead of (20) one has

$$\left|\frac{\Delta z \Delta p}{\hbar}\right| \sim \frac{\hbar}{\delta z \delta p_z}.$$

Yet, a state close to that of minimum uncertainty would still be required if the control is to be on the *microscopic* level.

CORRECTIONS TO THE SIMPLIFIED THEORY

The answer (21) to the question (1) is preliminary. We still have to find out if the approximations that turned the actual Hamilton operator (6) into the effective one (10) are as harmless as they appear.

If, improving upon (9), we take into account the operator nature of y and p_y then we find that when evaluating $\langle \sigma_+(T)\rangle$, the Larmor precession factor $\exp(-i\Phi(T))$ is replaced by

$$\exp(-i\Phi)\left\langle \exp\left(i\Phi\frac{p_y - \langle p_y\rangle}{\langle p_y\rangle}\right)\right\rangle \cong \left\langle \exp\left(-i\Phi\frac{\langle p_y\rangle}{p_y}\right)\right\rangle. \tag{22}$$

This is as it should be. The time integral in Φ, see (12), is a replacment for a y-spatial integration, divided by $v = p_y/m$. The expectation value (22) represents the contribution of what is called the longitudinal Stern-Gerlach effect to C. It arises because the two partial beams have slightly different potential energies, owing to the opposite orientation of the magnetic moments, and therefore also slightly different kinetic energies, and slightly different velocities. As a consequence there is a small relative displacement along the y-axis, which leads to a decrease of the spin coherence.

We now have, for $C \lesssim 1$,

$$C \cong 1 - (\text{as before}) - \frac{1}{2}\left(\Phi\frac{\delta p_y}{\langle p_y\rangle}\right)^2 .$$

So we need

$$|\Phi|\frac{\delta p_y}{\langle p_y\rangle} \ll 1 \text{ for } C \lesssim 1; \quad \sim 1 \text{ for } 0 < C \ll 1. \tag{23}$$

We get an idea of the order of magnitude of the value of Φ that corresponds to a single SGA by first noting that the time-independent field and field inhomogeneity within the magnet define a length

$$\ell_z = B_z \Big/ \frac{\partial B_z}{\partial z}$$

that is assuredly macroscopic:

$$\ell_z \gg \delta z.$$

Then on writing, for one SGA,

$$|\Delta p| = \left|\int dt F(t)\right| = \left|\int dt \mu \frac{\partial B_z}{\partial z}\right| \gg \delta p_z,$$

$$|\Phi| = \left|\frac{2}{\hbar}\int dt \mathcal{E}(t)\right| = \left|\frac{2}{\hbar}\int dt \mu B_z\right| \cong \ell_z|\Delta p|/\hbar$$

we learn that

$$|\Phi| \gg \ell_z \delta p_z/\hbar \gg \frac{\delta z \delta p_z}{\hbar}.$$

Consequently, $|\Phi| \gg 1$ must be quite a large number, involving as it does the product of two ratios between macroscopic and microscopic quantities. This is the contribution to the total Φ from one of the four SGAs. Now looking back at Fig. 2 we note that B_z has different signs in the two halves of the SGI so that by matching the SGAs with high precision, the overall Larmor precession angle can be made small and the condition (23) can possibly be met.

Next, if we take the x-spatial motion into account, we recognize that the atoms have a spread δx and the beams probe the magnetic fields also outside the y, z-plane of symmetry. Therefore, the approximation (8) must be replaced by a realistic field. Maxwell's equations $\vec{\nabla} \cdot \vec{B} = 0$, $\vec{\nabla} \times \vec{B} = 0$ imply a nonvanishing B_x component (mainly inside) and a nonvanishing B_y component (mainly in the fringing fields). Thus,

$$\frac{d}{dt}\sigma_z(t) = \frac{2\mu}{\hbar}(\sigma_x B_y - \sigma_y B_x) \neq 0, \tag{24}$$

the z-component of the magentic moment is not really conserved in a SGA. However, because of the Larmor precession the right-hand side in (24) oscillates very rapidly so that a sizeable change of σ_z cannot accumulate. Nevertheless, the extension of the beam in the x-direction, where it is exposed to an average $B_x \sim \delta x \partial B_z / \partial z$ leads to a loss of coherence, both for σ_z and σ_x, σ_y, that is of the order

$$\left(\delta x \frac{\partial B_z}{\partial z} \Big/ B_z \right)^2 = \left(\frac{\delta x}{\ell_z} \right)^2 .$$

As the squared ratio of microscopic and macroscopic lengths this is small, but it is not zero. A similar contribution arises from the fringing fields.

In view of this, in principle, unavoidable loss of spin coherence we give our answer to question (1):

No! Some loss of spin cohernce must always occur.

The yea-sayers are wrong.

ACKNOWLEDGMENT

I would like to express my gratitude for the hospitality experienced at the Center for Advanced Studies where part of the work reported in this lecture was performed. Thanks to Marlan Scully, each one of my stays in Albuquerque was as pleasant as it possibly could be. This work was partially supported by the Office of Naval Research.

REFERENCES

1. B.-G. Englert, J. Schwinger, and M. O. Scully, Found. Phys. **18**, 1045 (1988); J. Schwinger, M. O. Scully, and B.-G. Englert, Z. Phys. D **10**, 135 (1988); M. O. Scully, B.-G. Englert, and J. Schwinger, Phys. Rev. A **40**, 1775 (1989).
2. W. Heisenberg, Die physikalischen Prinzipien der Quantentheorie (Hirzel, Leipzig, 1930), pp. 33 and 34.

COLLAPSE AND RECREATION OF THE STATE VECTOR IN QUANTUM MECHANICS

Julio Gea-Banacloche

Instituto de Optica "Daza de Valdés"
Serrano 121
28006-Madrid (Spain)

1. INTRODUCTION

There is little question that the formalism of quantum mechanics is correct as a prescription to calculate quantities amenable to experimental observation (at least, it is correct to the present level of experimental accuracy). On the other hand, the interpretation of the formalism has been the subject of controversy almost from the beginning. Does the state vector represent an objective property of an object, or only our knowledge about (the possible outcomes of measurements on) that object? Does it really describe an individual object, or only an ensemble of identical objects? These questions arise in the context of the quantum theory of measurement, and depending on how they are answered the various paradoxes associated with the so-called "collapse of the wavefunction" may appear as trivial or as real difficulties.

If one chooses to think of the state vector as something which exists only in our minds, not directly related to any objective property of an individual system, the "paradoxes" vanish. This solution, however, is not altogether acceptable. The main argument against it is that we would like to think of quantum mechanics as a "universal theory," which should somehow include the classical world as a special limiting case. Since the classical viewpoint is based on the assumption of the existence of well-defined, objective properties for individual objects, we cannot avoid the question of how (or even whether) such concepts may be at all accomodated within the quantum theory.

This, then, is the question with which the present paper will concern itself: under which circumstances is it legitimate to interpret the formalism of quantum mechanics as predicting the objective existence of well-defined properties for individual systems? An answer will be outlined here, and it will be illustrated with an example from quantum optics which may be especially significant as a bridge from the quantum to the "classical" world.

Note that a straightforward attempt to answer this question by choosing the first alternative to the questions asked in the first paragraph—i.e., assuming that state vectors do indeed describe objective properties of individual objects—immediately runs into trouble, for then the "collapse of the state vector" supposed to take place upon the measurement of any observable becomes an incomprehensible event. Indeed, if the measuring apparatus is treated quantum-mechanically, no collapse is apparent. This is the paradox of Schrödinger's cat, which may be given an extra layer of complexity by the introduction of "Wigner's friend." I shall not review here some of the best-known proposed solutions, such as the assumption of a special role played by the observer's consciousness, or the "many-worlds interpretation" according to which no reduction ever takes place, and instead the universe constantly branches out into an inconceivable infinitude of universes where all the possible alternatives are actualized. A fairly complete discussion of these and other ideas may be found in Ref. 1.

It is, however, instructive to show here how the difficulty arises even in what may be the best theory of quantum measurement which we have today, which is the one due to W. H. Zurek [2] based on what he has called "environment-induced superselection rules." His treatment provides some crucial insights into the "collapse of the state vector" process, but even this description is not entirely free from logical difficulties, if one assumes—as is usually done—that one can always write a state vector for every system.

For maximum simplification, assume a quantum object O having two possible states $|a\rangle$ and $|b\rangle$, coupled to a meter M having "pointer states" $|A\rangle$ and $|B\rangle$, coupled to an environment E having many degrees of freedom. If the object O is initially in the state

$$|\psi\rangle_O = \alpha|a\rangle + \beta|b\rangle \tag{1}$$

then the joint evolution will yield a final state vector

$$|\psi\rangle_{OME} = \alpha|a, A, f\rangle + \beta|b, B, g\rangle \tag{2}$$

where f and g represent environment degrees of freedom. For a good meter, the evolution to a state such as (2) is (by hypothesis) guaranteed for a certain basis—the so-called pointer basis—of states $\{|A\rangle, |B\rangle\}$, and, further, the coupling with the environment is such that the overlap between the states $|f\rangle$ and $|g\rangle$ approaches zero very rapidly:

$$\langle f|g\rangle \to 0 \tag{3}$$

Under these conditions, any further measurement carried only on the object and the meter is determined only by the diagonal reduced density matrix

$$\rho_{OM} = |\alpha|^2|a, A\rangle\langle a, A| + |\beta|^2|b, B\rangle\langle b, B| \tag{4}$$

Such a density matrix may describe a statistical ensemble made up of two subensembles, one of systems in pure state $|a, A\rangle$ with statistical weight $|\alpha|^2$, and one of systems in state $|b, B\rangle$ with statistical weight $|\beta|^2$. We may then say that as far as any observation of the object and the meter alone is concerned, an ensemble of object/meters prepared in an identical state would evolve so as to become indistinguishable from the collection of two subensembles just described.

But can we conclude from this that the state vector of each one of the individual systems has collapsed to either $|a, A\rangle$ or $|b, B\rangle$—that is, that a definite outcome has been recorded by the meter in every case? There are actually two obstacles to this interpretation. First, one must note that the decomposition of the density matrix (4) is in no way unique. For instance, introducing the states

$$|\pm\rangle_{OM} = \alpha|a, A\rangle \pm \beta|b, B\rangle \tag{5}$$

we can rewrite Eq. (4) as

$$\rho_{OM} = \frac{1}{2}|+\rangle\langle+| + \frac{1}{2}|-\rangle\langle-| \tag{6}$$

which suggests two distinct subensembles again, but this time one in state $|+\rangle$ and the other one in state $|-\rangle$ (both with the same statistical weight), and in neither of these states is the position of the pointer well defined—as if no measurement had actually taken place.

This first difficulty may actually be resolved using Zurek's assumptions about the pointer basis, and to show how will constitute an important part of this article (Section 3). But there is another difficulty, this one of a logical nature. If we take Eq. (2) to represent really the state of each one of the possible realizations of the total system S made up by the object O, the meter M, and the environment E, we cannot at the same time assume that the individual subsystem (O, M) is, in every realization, *either* in state $|a, A\rangle$ *or* in state $|b, B\rangle$. The reason is that the ket (2) represents an individual total system $S = (O, M, E)$ for which the pointer variable can have either value A or value B—i.e., either value could be observed for that particular realization of S—whereas state $|a, A\rangle$ can only yield value A and state $|b, B\rangle$ can only yield value B. Thus, the individual realizations of the subsystem (O, M) cannot *logically* be in either state $|a, A\rangle$ or (exclusive or) state $|b, B\rangle$ if the the state of the total system is truly given by (2). I emphasize that this is only a *logical* difficulty: it pertains to what we mean by a "state" in quantum mechanics. It is not a *practical* difficulty, since to detect the fact that (2) contains the values A and B in a superposition it is not enough to observe the subsystem (O, M): one would also have to include in the observation some combination of "environment variables," because of (3) and its consequence, (4). For a good measurement, essentially by Zurek's assumptions, such an observation should be impossible in practice (this is what is meant by "environment-induced superselection rules").

Yet the difficulty in principle remains. We even know now from Bell's theorem [3], experimentally verified [4], that the values of, e.g., the pointer variables for an individual total system prepared in state (2) cannot be regarded as having any well-defined values. This flies in the face of our attempts to assign to the individual subsystems (O, M) *either* state $|a, A\rangle$ *or* state $|b, B\rangle$ after the measurement.

There are really only three possible ways out. One is the many-worlds alternative, which would take (2) as the true state of the individual system S (the "universe") which has branched out into two different worlds, in each one of which one of the two alternatives is realized and thus, in that branch, the *relative* state of the (O, M) subsystem is either $|a, A\rangle$ or $|b, B\rangle$ accordingly. Another solution is to retreat to the position mentioned earlier, that state vectors do not really describe individual systems, but only ensembles, and moreover only the state of our knowledge about these ensembles. This also avoids the logical contradictions, but at the cost of rendering the description of individual objects—and thus of the classical world—hopeless. We are left with no clue as to how the observable properties of what we recognize as "individual systems" come to be.

The purpose of this paper is to outline a third solution which preserves the intuitive association between state vectors and definite, objective properties of individual systems. For this, however, one must acknowledge that such a correspondence cannot always be possible: that is, that one cannot always describe every conceivable system by *some* state vector. In this case, in particular, it is reasonable to assume that no state vector description of the environment was possible to begin with (nor, of course, at the end of the process). Then there is really no state vector for the total system S, and equation (2) is to be interpreted as a piece of a density operator which would describe an ensemble of systems. Yet we would like to be able to associate a state vector (namely, $|a, A\rangle$ or $|b, B\rangle$) with the (O, M) part of S. Clearly what we require is some criterion to determine when a state vector description is thus impossible (or meaningless), and also when it is possible and meaningful. Such a criterion is presented here (it may be anticipated that Zurek's concept of a "pointer basis" will play a crucial role), and illustrated with one example from quantum optics which shows precisely the desired transition from the classical to the quantum world: more precisely, how a classical property for an individual system emerges from the quantum-mechanical formalism.

The first step will be a critical review of the concept of state vector and of the applicability of the state vector description: this is done in the following section. The alternative, namely, the density matrix description, and how out of the density matrix formalism one can infer the validity of a state vector description for a reduced part of a larger system, are discussed then in Section 3. The picture of the measurement process which emerges is discussed, along with some conclusions, in Section 4.

2. THE STATE VECTOR AND ITS LIMITATIONS

While it is true that quantum mechanics cannot always make definite predictions for individual systems, it certainly can in some cases, and it is in these situations when the state vector concept finds its most natural applications. For instance, when we know (because of the outcome of a measurement) that a certain property has a certain value (say, a) for a given system, the association of the state vector $|a\rangle$ with that specific system (meaning by this the corresponding eigenstate of the appropriate operator) is quite natural, and in fact much of the formal structure of quantum mechanics can be derived from this correspondence between actual properties for individual systems and state vectors.

The point of view adopted here is, therefore, that a state vector refers primarily to a given, individual system, endowed with a series of well-defined properties; and only secondarily to a possible statistical ensemble of identically prepared systems. The next thing that must be realized is that there are times when such a description in terms of well-defined individual properties—the "state vector description"—is not possible for what we would regard as an "individual system"—although in fact we might as well say that in such cases it is the very individuality of the system that is in question.

This is the case for any "entangled state" of two systems, i.e., a state of the form

$$|\psi\rangle_{12} = \sum |i\rangle_1 |j\rangle_2 \tag{7}$$

of which the most often-quoted example is the singlet state of two spin 1/2 objects:

$$|\psi\rangle_{12} = \frac{1}{\sqrt{2}}(|+-\rangle - |-+\rangle) \tag{8}$$

In such a state *no* component of the spin is defined for *either* one of the "two" subsystems. In fact, the Bell-inequality experiments on states analogous to (8) clearly show that such indefiniteness is truly fundamental (i.e., it is not just that we do not know what the value of these components is, but the system itself does not "know" it) [5]. Under these conditions we cannot think of either of the two subsystems as "having a state vector'. No state vector description is possible for the individual subsystems, and perhaps this is just another way to say that there are *no individual subsystems* anymore in this picture: their individuality has been lost, and however far apart in space they may be, from the point of view of their spins they truly are "one" single entity.

In is interesting to note here that in an analysis of the "Einstein-Podolsky-Rosen paradox" by Cantrell and Scully a few years ago [6] they also conclude "that EPR are correct in pointing out that a *wave function* description of certain phenomena is impossible" (their italics). They go on to argue that the correct characterization of these phenomena is contained in the reduced density matrix formulation of the problem, a point to which I shall return below. The EPR paradox arises in fact because of this entanglement phenomenon where the systems lose their individuality and can no longer be described by a (separate, for each one) state vector.

Since entanglement will almost inevitably occcur whenever two quantum systems interact in anything but the most trivial form, we reach the somewhat startling conclusion that the state vector description for an "individual" system should almost never be possible—not if it has ever interacted with another quantum system. This conclusion is inescapable (except in the "many worlds" picture, of course), but must be qualified. First, sometimes a system may retain much of its identity and be approximately describable by a state vector, even when it is interacting with another quantum system, if the latter is "almost classical"—e.g., a harmonic oscillator in a coherent state with a high average number of quanta. In this case the reduced density matrix for the first system may remain very close to a "pure state," indicating the (approximate) validity of the state vector description.

Secondly, sometimes the dynamics may be such that the reduced density matrix for the subsystem evolves unambiguously towards a definite pure state, in spite of the interaction being "on" all the time; an interesting example of this in quantum optics has recently been discussed [7].

Thirdly, for the state measurement/state preparation concept (fundamental to quantum mechanics) to have any meaning, it must be possible to set up dynamical situations where as a result of the interaction the individual subsystems are indeed left in *some* pure state (that is, the state vector description becomes possible again)—although in general *which* pure state will not be known (unlike in the situation contemplated in the previous paragraph). To describe such "re-creation of the state vector" in a fully quantum-mechanical way we need an objective criterion to know *from the dynamics* that a certain subsystem property must have taken a well-defined value but one unknown to us. In the language of the Schrödinger cat paradox, this would amount to an objective criterion to be able to know with certainty that the cat in the box is definitely alive or definitely dead (not some linear superposition) "without looking" in the box. The following Section concerns itself with possible such criteria.

To summarize the results of this Section, we could say that a state vector description may not always be possible, that when possible it is bound to be only an approximation, and that instead of speaking about the "collapse of the state vector" we could speak of the "collapse of the *state vector description*"—which, for a quantum system, takes place essentially as soon as it interacts appreciably with another quantum system. At this point we lose the individual quantum system—because it loses its individuality. How it may recover it (and how to do physics even in the absence of state vectors) is the subject of the following Section.

3. THE DENSITY MATRIX AND THE EMERGENCE OF WELL-DEFINED PROPERTIES

The way the density operator was initially introduced in quantum mechanics it appeared to be subordinate to the state vector: it expressed a lack of knowledge about the state vector which described a system. But such an assumption—that underlying every density operator there is some state vector for every one of the individual systems making up the "statistical ensemble"—is not really necessary, nor, in view of the foregoing, tenable. We may conclude with Cantrell and Scully that the density operator concept is indeed more fundamental—more general, more widely applicable—than the state vector concept. The price to pay, however, is that the density operator in these more general instances clearly does not describe a single, individual system: rather it contains information on the average values one would obtain by measuring the same properties over an ensemble of "identically prepared " (or identically unprepared!) systems.

Consider now the discussion in the Introduction from this perspective. When the object O is coupled to the meter, the state vector description for O "collapses," because of the interaction. In addition to this, if the environment was not initially describable by a state vector, the state vector description does not even apply to the whole system S. Instead of Eq. (2) we must have a density matrix of the form

$$\rho_{OME} = \sum_n p_n |\psi(n)\rangle\langle\psi(n)| \tag{9}$$

where the states $\{|n\rangle\}$ are any basis that diagonalizes the initial density operator for the environment (with eigenvalues p_n) and which has only a formal significance, and the $|\psi(n)\rangle$ are like Eq. (2), but calculated for different initial environment states $|n\rangle$:

$$|\psi(n)\rangle_{OME} = \alpha|a, A, f(n)\rangle + \beta|b, B, g(n)\rangle \tag{10}$$

Because of the use of a (mixed, in general) density operator, it is clear that we are, in principle, not describing the evolution of an individual system, but of an ensemble of systems. Thus we may not in general expect to obtain a definite prediction for "the" measurement's outcome, because we are dealing, conceptually, not with one, but with many measurements (note, however, that when $\alpha = 0$ or $\beta = 0$ the reduced density matrix ρ_{OM} for the (O, M) subsystem is in a pure state!). Still, we may be able to tell that there is a definite outcome for each measurement, even though we do not know what it is.

This follows, in fact, from Zurek's hypothesis about the "pointer basis." This is a basis of states of (O, M), $\{|a, A\rangle, |b, B\rangle\}$, defined by the property that, if the reduced density matrix ρ_{OM} for the (O, M) subsystem is calculated (by tracing over the environment, as usual) in this basis, it has the following properties:

(i) The diagonal elements ($(\rho_{OM})_{aA,aA}$ and $(\rho_{OM})_{bB,bB}$) are long-lived.

(ii) The off-diagonal elements are very short-lived.

This turns out to be enough to enable us to assert that the collection of identically prepared systems splits into two subsets such that the (O, M) subsystem is, in one of the subsets, in state $|a, A\rangle$, and in the other subset, in state $|b, B\rangle$. The reason why we can make such an assertion is because by (i) above such subsets are long-lived; thus the (mental!) adscription of each individual system to one of the two categories is "time-invariant"—it is preserved by the dynamics. In turn, (ii) above tells us that other possible ways to assign states to the two partitions (e.g., the states $|+\rangle$ and $|-\rangle$ of Eq. (5), which allow one to write the reduced density matrix (4) as (6)) are *not* preserved by the interaction, and are therefore meaningless.

The following toy example will illustrate the last point. Imagine a system with a two-dimensional space of states, with the two alternative bases

$$\{|\phi_1\rangle, |\phi_2\rangle\} \tag{11}$$

and

$$\{|+\rangle, |-\rangle\}, \tag{12}$$

where

$$|\pm\rangle = \frac{1}{\sqrt{2}}\left(|\phi_1\rangle \pm |\phi_2\rangle\right) \tag{13}$$

and an evolution equation for the density matrix given, in the basis (11), by

$$\frac{d\rho_{11}}{dt} = -\frac{d\rho_{22}}{dt} = -\Gamma_{11}(\rho_{11} - \rho_{22}) \tag{14a}$$

$$\frac{d\rho_{12}}{dt} = -\Gamma_{12}\rho_{12}. \tag{14b}$$

We shall assume the decay rates Γ_{11} and Γ_{12} to have very different orders of magnitude. The steady-state solution approached by the system after a sufficiently long time is just $\rho_{s.s.} = 1/2$, which may be written in either of the two bases as

$$\rho_{s.s.} = \frac{1}{2}|\phi_1\rangle\langle\phi_1| + \frac{1}{2}|\phi_2\rangle\langle\phi_2| \tag{15a}$$

$$= \frac{1}{2}|+\rangle\langle+| + \frac{1}{2}|-\rangle\langle-| \tag{15b}$$

The question of interpretation now arises. Given an ensemble of systems described by the dynamical equations (14), which we may assume has had the time to reach the steady state, should we think of each of the individual systems as being in one of the states $|\phi_1\rangle, |\phi_2\rangle$, as Eq. (15a) suggests, or as being in one of the states $|+\rangle, |-\rangle$, as suggested by Eq. (15b)?

The hypothesis of widely different decay rates actually gives us a criterion to choose. Note that Eqs. (12)–(14) immediately imply the dynamical equation

$$\frac{d\rho_{++}}{dt} = -\frac{d\rho_{--}}{dt} = -\Gamma_{12}(\rho_{++} - \rho_{--}) \tag{16}$$

for the diagonal elements of ρ in the basis (12). Now assume that $\Gamma_{12} \gg \Gamma_{11}$ and imagine a subensemble of systems all in state $|+\rangle$: this means that for this subensemble, at some instant (say, $t = 0$), we have $\rho_{++} = 1, \rho_{22} = \rho_{12} = \rho_{21} = 0$. But then, by Eq. (16), a very short time later (say at $t = \Delta t$, with $\Delta t > 1/\Gamma_{12}$ but $\Delta t \ll 1/\Gamma_{11}$) we can no longer say that any of the individual systems in the subensemble is, with any certainty, in the state $|+\rangle$: the probabilities ρ_{++} and ρ_{--} have, indeed, become equal in this short time. In the limit when Γ_{12} is extremely large, the statement that any particular system might be in state $|+\rangle$ (or state $|-\rangle$) at any time becomes, for practical purposes, meaningless, since its state an instant later might be completely different.

On the other hand, if the state vector for the subensemble at $t = 0$ is assumed to be $|\phi_1\rangle$ (or $|\phi_2\rangle$), it follows from Eq. (14a) that after the time $\Delta t \ll 1/\Gamma_{11}$ one still has $\rho_{11} \simeq 1$ (or $\rho_{22} \simeq 1$, respectively). In other words, the mental division of the ensemble described by $\rho_{s.s.}$ into subensembles characterized by $|\phi_1\rangle$ and $|\phi_2\rangle$ is preserved by the interaction of the system with its environment (as reflected in the evolution equation (14)): if an individual system is in state $|\phi_1\rangle$ (or $|\phi_2\rangle$) at some time, a short time later we may be sure that it will still be in that state.

We may say then that the dynamics has rendered possible again (at least for times $\Delta t \ll 1/\Gamma_{11}$) a state vector description of the individual systems—the state vector has been "recreated"—provided that the state in question is chosen to be one of the "pointer basis" (11). We may also say, since a state vector is associated with some well-defined value for a dynamical variable, that the dynamics itself has brought into existence definite properties for the individual system. This happens through the interaction of the system with the environment, because the dissipative nature of Eqs. (14) (which can only arise from a trace over unobserved degrees of freedom) is essential in providing the criterion to select the pointer basis. (The essential role of the environment in this context has been emphasized by Zurek.)

The fact remains, of course, that we do not know, for each individual system, which of $|\phi_1\rangle$ or $|\phi_2\rangle$ is the right state to be assigned. This reflects the fundamental indeterminism of the theory, its intrinsic probabilistic nature. Still, by the argument above, we may have the "moral certainty" that it is one of those two and no other; just as if we flip a coin we may be sure, even if we do not watch it fall, that it has landed heads or tails, and not on the edge. Note that in the case of the coin our certainty also is due to the fact that the probability to find the coin standing on edge goes to zero very quickly with time, in the presence of a perturbing "environment."

One might say that the criterion is to assume that a system which we do not observe must nevertheless be in some definite long-lived state—*if* such long-lived states actually exist. The assumption is that through the system's interaction with the environment most possible states become very short-lived and only a few (the "pointer basis") need be considered.

The need for very different decay rates is essential. In the example above, if $\Gamma_{12} = 2\Gamma_{11}$ we have absolutely no criterion left to decide whether an individual system might be in an eigenstate of $A = a_1|\phi_1\rangle\langle\phi_1| + a_2|\phi_2\rangle\langle\phi_2|$ (such as $|\phi_1\rangle$ or $|\phi_2\rangle$) or in an eigenstate of $B =$

$b_1|+\rangle\langle+| + b_2|-\rangle\langle-|$ (such as $|+\rangle$ or $|-\rangle$). In the first case, the property associated to the operator A would have a well-defined value, whereas in the second case it would be the property associated to B. Clearly A and B would not, in general, commute, so we could not think of both kinds of properties being simultaneously defined for the system. Since we also have no way to decide for one or the other, we might be justified in assuming that neither property really exists for the system. Instead, for a system described by Eq. (14) with $\Gamma_{12} \gg \Gamma_{11}$, we may say that the interaction with the environment causes properties of type A to have well-defined values (at least for times $1/\Gamma_{12} < t < 1/\Gamma_{11}$); whereas if $\Gamma_{11} \gg \Gamma_{12}$, it is the properties of type B which would be well defined.

I shall leave to the reader's imagination to decide whether one might expect a resolution of the Schrödinger cat paradox along these lines. Instead, I shall consider in the remainder of this section an example of a macroscopic system from quantum optics: namely, the field in a laser cavity.

To a reasonably good approximation, for an ensemble of identical lasers high above threshold, the field inside the cavity may be described by the following (steady-state) density operator [8]:

$$\rho = \frac{1}{2\pi}\int |r_0 e^{i\phi}\rangle\langle r_0 e^{i\phi}|\,d\phi \tag{17}$$

in the basis of coherent states of amplitude r and phase ϕ. This suggests that the field in every laser is in some coherent state, with amplitude r_0 and unknown phase ϕ, which varies from one member of the ensemble to the next. However, the same density operator may be written in the number-state basis as

$$\rho = \sum_0^\infty e^{-r_0^2}\frac{r_0^{2n}}{n!}|n\rangle\langle n| \tag{18}$$

which suggests that one might as well say of any given laser that it is in a number state $|n\rangle$ with probability $e^{-r_0^2}r_0^{2n}/n!$.

Which one of these interpretations is more sensible? Should we think of the photon number n or the phase ϕ as being well-defined (though unknown to us) properties of a laser operating high above threshold? We may choose between these two options using the criterion given above. If we look at the dynamical equation of motion for the laser field density matrix we can see that an initial number state would be destroyed (become a mixed state) much faster than an initial coherent state, which remains approximately pure for a long time (the laser's phase-diffusion time, actually) and is indeed the longest-lived kind of initial state [9]. (These statements will be proved in detail in a forthcoming paper.)

Thus, the statement that a laser is at any time in a state with a well-defined number of photons is meaningless, since the number of photons after an extremely small time would certainly not be the same; but the statement that it is, for an appreciable time, in a coherent state with some well-defined phase is perfectly justified. It is interesting to note that this assumption is actually made all the time in quantum optics! The present criterion, based on the lifetime of the different kinds of states under the various dissipative processes going on in a laser, appears to be the only way to justify it.

Once we have satisfied ourselves in this way that we may think of an individual laser as being in a coherent state, we may conclude immediately, among other things, that two independent lasers may yield an interference pattern. This is not at all an obvious conclusion from the density matrix (17) or (18) for the ensemble of lasers, which in fact yields no fringes at all in the average intensity, since when the ensemble average over all the phases is taken, the fringes are wiped out. The existence of fringes in each individual realization of the experiment is usually inferred from a study of the intensity-correlation functions [10]. Our criterion, however, by allowing us to write a certain type of state vector for each individual laser, allows us to predict easily what each individual realization of the interference experiment will look like: not, of course, the position of the fringes, but that there *should* be fringes. It also provides other useful information, such as the stability in time of the fringe pattern, which is determined, naturally, by the lifetime of the coherent state.

The laser, being a simple harmonic oscillator coupled to two reservoirs (the atoms and the cavity losses) is likely to be a paradigm for how macroscopic properties (in this case, the phase of the field) arise for individual quantum systems. Our criterion shows how the emergence of such properties may be inferred from the formalism of ordinary quantum mechanics.

It is worth noting that as more and more experiments with individual quantum systems become possible, it becomes more important to have some criterion to predict what the individual realizations may look like: one may recall, for instance, how predictions differed about what one might observe in "quantum jumps" experiments on a single atomic system, only a few years ago [11]. The present formalism appears to be a natural way to handle such problems.

4. CONCLUSIONS

The solution outlined here to the problem of the "reduction of the wave packet" may be summarized as follows:

(1) We may assume that an ensemble of closed systems may be described by a density operator obeying Schrödinger's equation, and any ensemble of subsystems of the larger system by the corresponding reduced density matrix equation.

(2) It does not follow that each individual system (or subsystem) may be described by a state vector. In particular, a state vector description of an individual system becomes often impossible (or meaningless) when it is allowed to interact strongly with another quantum system. One might call this the (dynamical) collapse of the state vector *description*. One may also view this result as an indication that the system has, in a sense, lost its individuality.

(3) The state vector description for a subsystem may become possible and meaningful again as a result of the interaction of the system with an environment which selects a certain "pointer basis" of privileged states (recreation of the state vector, or of the state vector description). The essential point is the extremely short lifetime of all the states which might diagonalize the density matrix, other than those in the pointer basis (or those very close to them in the Hilbert space). This may be turned into a criterion to assert with confidence that the value of a certain property, corresponding to the pointer basis states, is "an element of reality"—i.e., it exists though it is unknown to us.

(4) It follows that the concept of state vector, and the state-vector description, provides only an approximate description of reality. The state-vector description is meaningful only approximately, for certain systems (or rather for certain degrees of freedom of the given system), and over a finite length of time (the "lifetime" of the state). There is thus no logical need to embrace a many-worlds theory, which is explicitly predicated on the absolute reality of state vectors under any circumstances.

Such an approach to the foundations of quantum mechanics may be said to be objective in that it does not involve any kind of conscious observer: the dynamics alone determines the collapse and subsequent recreation of the state vector picture for each individual system. It is also necessarily probabilistic in nature (the outcome of the individual measurement may not be predicted, but one may be confident that there will be an outcome). Between collapse and recreation one really does not know what happens to the individual system; in a sense, there is *no* individual system for a while. Here is where all of Einstein's "spooky actions at a distance" may freely come in. We have not resolved the EPR paradox, in the sense that both collapse and recreation are potentially nonlocal events.

The approach also has the power to lead to predictions for individual quantum systems, i.e., to determine what a single realization of an experiment may look like. It offers the possibility to describe both quantum and classical systems, even the emergence of classical macroscopic properties. Several instances of the application of these concepts to systems in quantum optics will be presented elsewhere.

REFERENCES

1. *Quantum theory and measurement*, ed. by J. A. Wheeler and W. H. Zurek, Princeton University Press (Princeton, NJ, 1983)
2. W. H. Zurek, Phys. Rev. D **26**, 1862 (1982).
3. J. S. Bell, Physics, **1**, 195 (1964). (Reprinted in Ref. 1 above.)
4. For the most recent experiments, see A. Aspect, J. Dalibard and G. Roger, Phys. Rev. Lett. **49**, 1804 (1982); see also references therein.
5. See, e.g., the discussions by D. Mermin, Physics Today, **38**, no. 4, p. 38 (April 1985); B. d'Espagnat, Sci. Am. **241**, no. 5, p. 128 (November 1979); J. S. Bell, J. Physique (Paris) **42**, C2-41 (1981).
6. C. D. Cantrell and M. O. Scully, Phys. Rep. **43**, 500 (1978).
7. J. J. Slosser, P. Meystre and S. L. Braunstein, Pyhs. Rev. Lett., **63**, 934 (1989).
8. M. O. Scully and W. E. Lamb, Jr., Phys. Rev. **159**, 208 (1967).
9. Y. K. Wang and W. E. Lamb, Jr., Phys. Rev. A **8**, 866 (1973).
10. L. Mandel, Phys. Rev. **134**, A10 (1964).
11. See, e.g., R. J. Cook and H. J. Kimble, Phys. Rev. Lett. **54**, 1023 (1985); D. T. Pegg, R. Loudon and P. L. Knight, Phys. Rev. A **33**, 4085 (1986); A. Schentzle, R. G. DeVoe and R. G. Brewer, Phys. Rev. A **33**, 2127 (1986); J. Javanainen, Phys. Rev. A **33**, 2121 (1986).

QUANTUM NONDEMOLITION MEASUREMENTS AND TESTS OF COMPLEMENTARITY

B.C. Sanders and G.J. Milburn

Department of Physics
The University of Queensland
St Lucia, Queensland, 4067
Australia

PRINCIPLES OF MEASUREMENT

Measurements are performed on dynamical systems to obtain information about some physical quantity. The quantity is measured by causing the system to interact with a detector for some length of time and the value of the quantity is inferred by reading a meter coupled to the detector. In classical mechanics a virtually noiseless detector can be constructed, in principle, which ensures that no noise is transferred to the system during the system–detector interaction. However, a quantum detector possesses quantum noise which cannot be eliminated. During the measurement process the quantum fluctuations of the detector feed into the system and affect the state of the system.

Given a system with the Hermitian density operator ρ, a physical quantity $\mathcal{A}$ is to be measured. The quantity $\mathcal{A}$ is associated with the quantum operator A. Measurement of $\mathcal{A}$ produces a value a and measurements of a collection of identical systems produce a collection of values $\{ a \}$. For $\{ |\alpha >\}$ the set of eigenvalues of A such that $A|\alpha> = \alpha|\alpha>$, the distribution of meter readings $\{\alpha\}$, for an ideal, precise detector, converge in probability to the distribution $< \alpha|\rho|\alpha >$. Ideal, precise measurements of $\mathcal{A}$ thus provide information about the diagonal elements of ρ in the A eigenbasis.

The quantum fluctuations of the quantized detector feed into the system. In general the fluctuations affect the variable of interest over time, so future measurements are subject to quantum uncertainty. However, a class of measurements exist by which the fluctuations do not couple to the variable of interest. The quantum nondemolition (QND) measurements possess the property that a measurement of $\mathcal{A}$ at time t_1 produces a set of values $\{ a \}_1$ such that the set of values $\{ a \}_2$ produced by a measurement of $\mathcal{A}$

at time t_2 can be exactly predicted. Moreover, for precise QND measurements, the value a_1 obtained at time t_1 for one event allows one to predict with certainty what value a_2 will be obtained by a measurement of $\mathcal{A}$ at time t_2. For imprecise QND measurements, as stated implicitly above, the statistics of a reading at time t_1 allow a precise prediction of the statistics at time t_2.

COMPLEMENTARY MEASUREMENTS OF TWO–LEVEL SYSTEMS

We restrict our attention to ideal measurements of two–level systems and introduce plausible experimental schemes by which complementarity can be tested. The two–level system is described by the Pauli Hermitian operators $\underline{\sigma} = (\sigma_x, \sigma_y, \sigma_z)$ which obey the commutator relation $\underline{\sigma} \wedge \underline{\sigma} = i\underline{\sigma}$. We are interested in measuring the complementary physical quantities given by the level occupation number and the degree of coherence between levels. The level occupation number is the physical quantity $\mathcal{N}$ which corresponds to the operator $N = \sigma_z + \frac{3}{2}$. The eigenbasis of N is $\{|1>, |2>\}$ with eigenvalues $\{1,2\}$, respectively.

Here we introduce a general procedure for performing a QND measurement of $\mathcal{N}$ followed by a measurement of the coherence between the levels. The QND measurement is performed by causing the system to interact with a QND detector for some length of time. The resultant value for occupation level number allows one to predict with certainty, for perfectly precise QND measurements, the result of measuring $\mathcal{N}$ at a later time. In fact, for the schemes we present, a precise measurement of $\mathcal{N}$ which produces a value n could be followed by a later measurement of $\mathcal{N}$ and the same value of n would be measured.

It would be convenient to follow the measurement of $\mathcal{N}$ by a phase measurement, but, although phase operators in angular momentum have been studied[2], a scheme for directly measuring phase has not been devised. Instead a statistical process is needed to determine the coherence between levels as we discuss below.

The general pure state in a two–level system is given by

$$|\theta,\phi> = \cos\theta\ |1> + e^{i\phi} \sin\theta\ |2>. \qquad (1)$$

Suppose that we have a large collection of identically prepared states given by (1). An ideal precise QND measurement of $\mathcal{N}$ produces the result that each element of the collection is either in level 1 or in level 2. The probability of being in level 1 is given by $\cos^2\theta$ and is inferred by the statistics obtained from measuring all elements of the collection. However, the precise measurement of $\mathcal{N}$ destroys the coherences between the levels. In the level basis, the measurement process transforms the density matrix to

$$\mathcal{N}\ \text{measurement:}\quad \begin{bmatrix} \cos^2\theta & \tfrac{1}{2}e^{-i\phi}\sin 2\theta \\ \tfrac{1}{2}e^{i\phi}\sin 2\theta & \sin^2\theta \end{bmatrix} \rightarrow \begin{bmatrix} \cos^2\theta & 0 \\ 0 & \sin^2\theta \end{bmatrix};$$

that is, quantum fluctuations in the detector cause a destruction of coherence between the levels and a pure superposition state is transformed to an incoherent mixture of the two levels.

Some coherence between the levels can be preserved by performing an imprecise measurement of $\mathcal{N}$. In this way the determination of $\cos^2\theta$ is not completely accurate, but the coherence between the levels is not completely destroyed. We assume that the detector state can be varied to allow a complete variation of the $\mathcal{N}$ measurement precision from the worst precision case $\eta = 0$ to the perfectly precise case $\eta = 1$. In the experimental schemes presented below, such a variation is possible.

The level distribution is obtained by measuring the quantity $\mathcal{N}$ corresponding to the operator N, or equivalently σ_z. In order to obtain information about the coherence between the levels, the natural quantity to measure is S_ψ corresponding to the operator

$$\sigma_\psi = \sigma_x \cos\psi + \sigma_y \sin\psi \qquad (2)$$

where ψ is a phase determined by the coherence detector. For each value of ψ such that ψ ranges over $0 < \psi \leq 2\pi$, many single–element events are recorded and the statistical mean S_ψ obtained.

From S_ψ the quantum mean $< \sigma_\psi >$ is inferred as

$$< \sigma_\psi > = \mathrm{Re}\,(\rho_{12} e^{i\psi})$$
$$= \tfrac{1}{2}\sin 2\theta \cos(\phi - \psi) \qquad \text{(for superposition state);} \qquad (3)$$

the statistics allow one to infer the off–diagonal element. In fact we are more interested in the degree of coherence between the two levels. By repeating the experiment for many values of ψ, we obtain

$$|< \sigma_\psi >|_{max} = \max\{\,|< \sigma_\psi >| : 0 < \psi \leq 2\pi\} = \tfrac{1}{2}\sin 2\theta$$

and

$$|<\sigma_\psi>|_{min} = \min\{|<\sigma_\psi>| : 0 < \psi \le 2\pi\} = 0.$$

The coherence can be characterized by a visibility (the terminology becomes clear when the experimental schemes are discussed below) which is given by $V = 2\,|<\sigma_\psi>|_{max}$. For a pure state $V = \sin 2\theta$; for a mixture $V = 0$; and a state with some coherence between the levels produces a value of V such that $0 \le V \le 1$. As we shall see, the visibility is a useful experimental parameter.

The logical procedure for performing a test of complementarity in a two–level system is depicted in Fig. 1 as a flow chart. A collection of two–level systems is prepared and each element of the collection is measured separately. The parameter θ is fixed for all elements of the collection, but the collection is partitioned into subsets, each with N elements, and each subset corresponding to some value of ϕ. (In practice ϕ can be identical for all subsets and the parameter ψ in the detector can be varied, but we discuss changing ϕ for clarity). For θ fixed, the experiment is repeated for a range of precision, $0 \le \eta \le 1$. Each time η changes, a new collection of elements is prepared as described above.

For a given precision η, each element undergoes a QND measurement of $\mathcal{N}$ followed by a coherence measurement. The phase ϕ is set to an initial value $\Delta\phi$ and N single–element events proceed. Each element is subjected to a QND level occupation number measurement and the element is inferred to be in either level 1 or 2 by decision process δ. The appropriate counter $P_i(\theta,\eta)$ is incremented. A coherence measurement $I_n(\theta,\eta,\phi)$ is also performed which is related to S_ψ but includes an additive constant that ensures positivity. We use the term I_n because, in the schemes to be discussed, the coherence measurements are invariably intensity measurements of the electromagnetic field. Repetitions of the single–element events provide a mean

$$I(\theta,\eta,\phi) = \sum_{n=1}^{N} I_n(\theta,\eta,\phi)/N. \tag{4}$$

The maximum and minimum with respect to ϕ are determined and the visibility is given by

$$V(\theta,\eta) = \frac{I_{max}(\theta,\eta) - I_{min}(\theta,\eta)}{I_{max}(\theta,\eta) + I_{min}(\theta,\eta)}. \tag{5}$$

The visibility can be plotted as a function of the precision η.

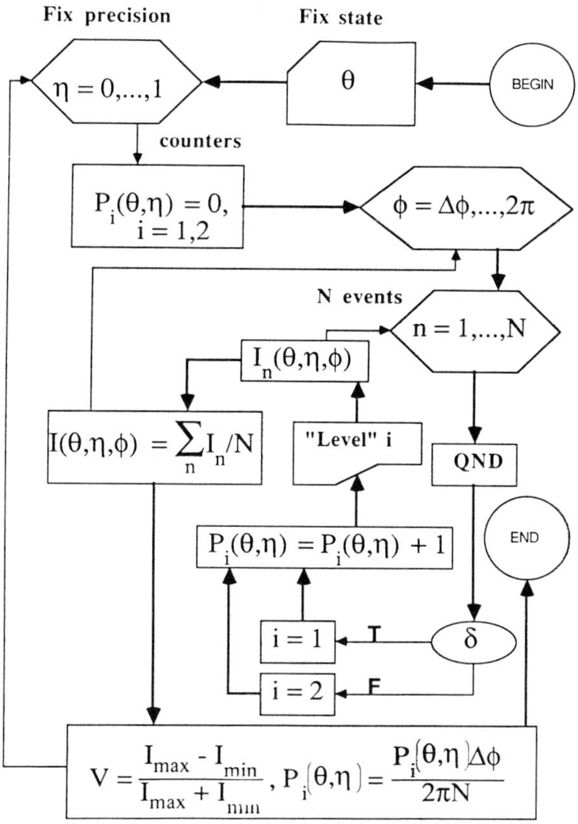

Fig. 1. Flow chart representation for an experimental test of complementarity in a two–level system. Each element of the collection is separately subjected to two complementary measurements providing values for visibility $V(\theta,\eta)$ and level occupation information $P_i(\theta,\eta)$.

At the conclusion of the procedure for given η, the visibility V and proportion of elements in each level $P_i(\theta,\eta)$ is calculated. From many events the value of $\cos^2\theta$ is inferred. As the precision decreases, the conditional probability of observing an element prepared in level 1 (the level being detected) decreases also. Thus, the mistakes increase and an incorrect value of $\cos^2\theta$ is inferred. The deviation of the inferred value of $\cos^2\theta$ from the true value can be seen by plotting the proportion of elements in level 1 as a function of time. In this way we can test the predictions of both visibility and accuracy of determining $\cos^2\theta$ for a continuous range of η. Such an experiment provides the means for testing the complementarity relation.

TWO SCHEMES FOR TESTING COMPLEMENTARITY

We must emphasize at the outset that a careful quantitative analysis of testing complementarity in an experiment has been discussed by Wootters and Zurek for a path vs interference fringes measurement in a double–slit apparatus[3]. We do not discuss the experiment in detail here as we feel that the experiment is out–of–reach with today's technology, but many concepts introduced by Wootters and Zurek are employed here. Rather we discuss two schemes which we hope are closer to realization.

Single–Photon Interference in a Mach–Zehnder Interferometer

Recent experimental advances in single–photon interference measurements[4] and photon number QND measurements[5] could allow for a path determination vs single–photon interference measurement in a Mach–Zehnder interferometer[6]. Such a scheme is closely related to that of Wootters and Zurek[3].

Single–photon interference has been produced by Grangier et al[4] in a Mach–Zehnder interferometer where the single–photon state is produced by a conditional measurement of a two–photon radiative process. In their discussion, they suggest that a photon number QND device could be introduced; we modify their experimental scheme to include such a device. This allows a test of complementarity as described above.

A schematic of the proposal is presented in Fig 2. The vacuum $|0>$ and the one–photon state $|1>$ are fed into the two input ports of the interferometer. Within the interferometer, there are two modes associated with each arm and the mode in arm n is associated with the annihilation operator a_n. A medium with an intensity dependent refractive index (a Kerr medium) is placed in arm 1 and the probe field B interacts with the field a_1 via the medium. For χ a parameter which is proportional to the nonlinearity of the medium and the interaction time, the two fields are transformed to

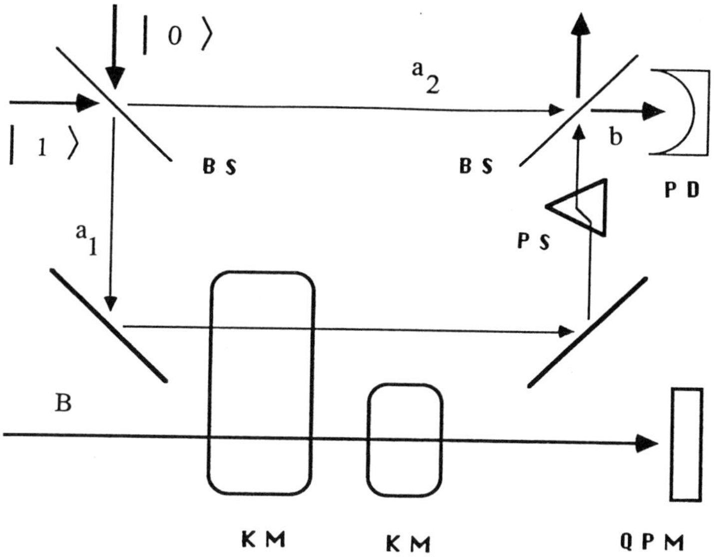

Fig. 2. A Mach–Zehnder interferometer (MZI) with a photon number QND measurement in arm 1. The beamsplitters (BS) are 50/50 and a Kerr medium (KM) and a phase shifter (PS) are introduced in the MZI. The modes in each arm are a_i and the output b is measured by a photon detector (PD). The probe field B interacts with a_1 in the KM, encounters a second KM and undergoes an ideal quadrature–phase measurement (QPM).

$$a_1' = \exp(-i\chi B^\dagger B) a_1 \text{ and } B' = \exp(-i\chi a_1^\dagger a_1) B.$$

A self–action in the medium also induces a nonlinear phase shift on the probe field, but this can be removed by introducing a second Kerr medium for the field[6].

As $a_1^\dagger a_1$ is unchanged by the interaction and is a constant of motion within arm 1, a phase–shift measurement of B provides a QND measurement of $a_1^\dagger a_1$. A quadrature–phase measurement of B (performed, say, by an ideal homodyne detector) measures the quantity corresponding to the operator

$$Y(\Phi) = \tfrac{1}{2}\left[\exp(-i\,[\Phi + \chi\, a_1^\dagger a_1])\, B + \exp(i[\Phi + \chi\, a_1^\dagger a_1]) B^\dagger\right]$$

for Φ the phase of the local oscillator. The probe field is henceforth assumed to be in a (quadrature–phase minimum uncertainty) single–mode coherent state

$$|\alpha> = \exp(\alpha a^{\dagger} - \alpha^{*} a)\ |0>$$

with mean photon number $|\alpha|^2$.

The path of the photon in the interferometer is inferred from the measured phase shift of the probe field. Each event produces a quadrature phase reading y. The readings are normalized with respect to the probe field strength: $c = y/|< B >|$. The distribution of normalized quadrature–phase readings converges to the quadrature–phase marginal distribution

$$P(c) = \tfrac{1}{2}\, P_0(c) + \tfrac{1}{2}\, P_1(c)$$

where $P_n(c)$ is the Gaussian distribution conditional on n photons being in arm 1. From each reading c, one decides whether the value c is a member of distribution $P_0(c)$ or $P_1(c)$. The mean reading conditioned on no photons in arm 1 is $< c >_0$ and on one photon in arm 1 is $< c >_1$. The effective signal is $< c >_0 - < c >_1$ and the signal–to–noise ratio (SNR) for the path determination is

$$\sigma = \frac{|< c >_0 - < c >_1|}{\sqrt{\Delta c_0^2 + \Delta c_1^2}}$$

where Δc_n^2 is the variance of the distribution $P_n(c)$. For a coherent probe field, where the local oscillator phase Φ is chosen to make α real,

$$\sigma = 2^{3/2}\, \alpha \sin^2\chi.$$

Thus the SNR improves as the probe field strength is made stronger. The improvements arise because the relative degree of the coherent field "shot noise" with respect to the intensity becomes small.

The output from one port of the interferometer is the field

$$b = e^{i\psi}\, a_1' + i a_2$$

where ψ can be determined by tuning the phase shifter. After many single photon events for each value of ψ, the statistical mean $< b^{\dagger} b >_{\psi}$ is inferred. The procedure is repeated for different values of ψ and the visibility is given by

$$V = \frac{< b^{\dagger}b >_{max} - < b^{\dagger}b >_{min}}{< b^{\dagger}b >_{max} + < b^{\dagger}b >_{min}}$$

which is analogous to (5) with the restriction that $\theta = \pi/4$ (as we employ a 50/50 beamsplitter in the interferometer). The model can be generalized to include arbitrary transmission beamsplitters, of course. The quantity V varies between 0 and 1 and precisely determines the degree of coherence between the two arms of the interferometer (the arms corresponding to levels of the two–level system). For a coherent probe field,

$$V = \exp\left(-2|\alpha|^2 \sin^2\chi\right)$$

which is unity for the probe field in the vacuum state ($\alpha = 0$) and approaches zero exponentially as the mean photon number $|\alpha|^2$ of the probe field increases.

The SNR and visibility are related by the Gaussian expression

$$V = \exp\left[-\frac{1}{4}\frac{\sigma^2}{\sin^2\chi}\right] \tag{6}$$

which is optimized for $\chi = \pi/2$. Expression (6) can be modified by squeezing the quantum fluctuations in the probe field[7], but, for $\chi = \pi/2$, expression (6) is *independent* of the squeezing parameter. In fact, for $\chi = \pi/2$, a SNR of 1.67 can produce a visibility of 50% by fixing the coherent probe field mean photon number at 0.59. Thus, one would obtain a good indication of the photon path and also detect fringes: a dramatic demonstration of complementarity. However, the weak nonlinearities present in $\chi^{(3)}$ media seem to disallow achieving a value near $\chi = \pi/2$ for some time. A test of complementarity can, in principle, be conducted for presently obtainable nonlinearities but V falls off very rapidly with respect to σ. Moreover the phase shift due to a single photon would be very difficult but not necessarily impossible. We must emphasize that this scheme relies on technology which is still in the developing stages.

Atomic Level Measurement and Quantum Beats

Recent work on photon number state preparation in high–Q micromaser cavities offers the possibility of testing complementarity in a quantum beat experiment[10]. The presence of quantum beats in the resonance intensity profile of a coherent collection of initially excited three–level V–type atoms is a well–known and well–studied phenomenon. The two excited levels $|1\rangle$ and $|2\rangle$ decay to the ground state $|0\rangle$ via spontaneous emission with frequencies ω_n and decay constants γ_n for n = 1,2. The detuning $\Delta = \frac{1}{2}(\omega_1 - \omega_2)$ is small compared to the mean frequency $\overline{\omega} = \frac{1}{2}(\omega_1 + \omega_2)$. For $\overline{\gamma} = \frac{1}{2}(\gamma_1 + \gamma_2)$, the time–dependent density matrix for the atom in the level basis is

$$\rho^A_{00}(t) = 1 - e^{-\gamma_1 t}\cos^2\theta - e^{-\gamma_2 t}\sin^2\theta$$

$$\rho^A_{11}(t) = e^{-\gamma_1 t}\cos^2\theta$$

$$\rho^A_{22}(t) = e^{-\gamma_2 t}\sin^2\theta$$

$$\rho^A_{12}(t) = \tfrac{1}{2}\, e^{-(\overline{\gamma} + iA)t - i\phi}\sin 2\theta = \rho^A_{21}(t)^*$$

where each atom is initially prepared in the state (1). For each value of ϕ, the mean intensity can be shown to be[12]

$$< I(t) >_\phi = \lambda_1 \rho^A_{11}(t) + \lambda_2\, \rho^A_{22}(t) + \lambda_3\, \mathrm{Re}(\rho^A_{12}(t)) \qquad (7)$$

where $\{\lambda_n\}$ depends on the dipole moments and the angles between the dipole moments and between each dipole moment and the detector. Over many events, for each value of ϕ, the quantity (7) is inferred and, by varying ϕ, the maximum and minimum can be obtained. The measurement of visibility must proceed long before spontaneous emission is significant and

$$V = \lim_{t\to 0} \frac{< I(t) >_{max} - < I(t) >_{min}}{< I(t) >_{max} + < I(t) >_{min}},$$

analogous to (5). In practice ϕ can be kept constant if $\overline{\gamma} << \Delta$; in this case several beats occur before decay in the intensity profile is significant and $< I(t) >_{max}$ and $< I(t) >_{min}$ can be determined over the characteristic beat period.

Scully and Walther[10] have introduced a scheme for performing perfectly precise measurements of the atomic level occupation number and have shown that the precise measurement of $\mathcal{N}$ destroys the quantum beats in the resonance fluorescence intensity profile. We generalize their scheme to permit imprecise measurements and partial destruction of the coherence between levels[12].

A four–level atom is shown in Fig. 3. The atom is essentially a three–level V–type atom with one level, |1>, coupled to a probe level, |p>, with transition frequency ω_p and

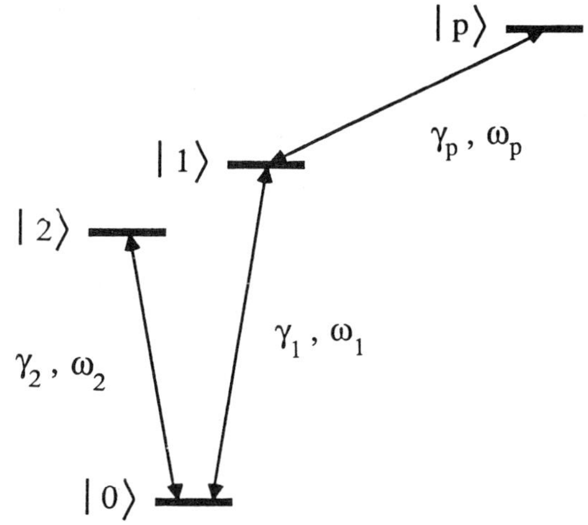

Fig. 3. A four–level atom which can be considered as a three–level V–type atom with one upper level, |1>, coupled to a probe level, |p>. The transition frequencies are ω_i, and γ_i are the decay constants (i = 1,2,p).

decay constant γ_p. The atom is initially excited to a coherent superposition of two levels |1> and |2> with the ground level |0> and the probe level |p> unoccupied. The atom is passed through two micromaser cavities in sequence. The cavities are tuned to the frequency ω_p and the interaction time corresponds to one–half the one–photon Rabi flop time between |p> and |1>. We assume that cavity 1 is prepared in a mixture of the zero and one photon states

$$\rho_1 = P_0 |0\rangle\langle 0| + P_1 |1\rangle\langle 1| \tag{8}$$

such that $P_0 + P_1 = 1$. The second cavity is prepared in the vacuum state. Perfectly precise measurements of the atomic level correspond to $P_0 = 0$. For $P_0 = 0$, an atom prepared in the state |2> does not interact with the cavities and one would detect the photon in cavity 1 with an ideal photon detector. Conversely, the atom prepared in level |1> absorbs the photon in cavity one and emits a photon in cavity 2. Thus, the absence of a photon in cavity 1 allows one to infer that the atom is in level |1>.

For the general cavity 1 preparation (8), we infer from photon counting after the atom–cavity interaction that the atom is in level |2> if a photon is found (correct) and the atom is in level |1> if no photon is found (sometimes incorrect for $P_0 \neq 0$). The probability for being wrong is the conditional measurement $P(0|\rho_n^{(2)}) = P_0$, that is, the conditional probability of detecting no photon given that the atom is prepared in level |2>. The inference quality is

$$\sigma = 1 - P(0|\rho_A^{(2)}) = P_1$$

which, for perfect measurements, is the probability of being correct.

Given the cavity 1 preparation (8), the atomic state after measurement is[12]

$$\begin{bmatrix} \cos^2\theta & \frac{1}{2} P_0 e^{-i\phi} \sin 2\theta \\ \frac{1}{2} P_0 e^{i\phi} \sin 2\theta & \sin^2\theta \end{bmatrix}$$

and the levels |0> and |p> are unoccupied. For $P_0 = 0$, coherences are destroyed and, for $P_0 = 1$, $\rho_A' = \rho_A$ (perfect coherence). Let us introduce the normalized visibility

$$\mathcal{V} = V/V_{max}$$

for $V_{max} = V/P_0 = 0$. Thus $\mathcal{V} = P_0$ and $0 \leq \mathcal{V} \leq 1$. The normalized visibility equals the probability for cavity 1 to be in the vacuum state.

The inference quality and visibility are thus related by

$$\sigma + \mathcal{V} = 1. \qquad (9)$$

An experimental test of complementarity would involve inferring the value of $\cos^2\theta$ and observing the visibility for different values of precision. In this way one can observe the loss of accuracy in determining $\cos^2\theta$ which is necessary to improve visibility. This procedure allows for a test of the complementarity relation (9).

REFERENCES

1. C.M. Caves, K.S. Thorne, R.W.P. Drever, V.D. Sandberg and M. Zimmermann, On the measurement of a weak classical force coupled to a quantum mechanical operator. I. Issues of principle, Rev. Mod. Phys. 52:341 (1980); C.M. Caves, Quantum nondemolition measurements, in: "Quantum Optics, Experimental Gravitation and Measurement Theory", P. Meystre and M.O. Scully, eds., Plenum, New York (1983), p. 567.

2. D. Judge and J.T. Lewis, On the commutator $[L_z, \varphi]_-$, Phys. Lett. 5:190 (1963); P. Carruthers and M.M. Nieto, Phase and angle variables in quantum mechanics, Rev. Mod. Phys. 40:411 (1968).

3. W.K. Wootters and W.H. Zurek, Complementarity in the double–slit experiment: Quantum nonseparability and a quantitative statement of Bohr's principle, Phys. Rev. D. 19:473 (1979).

4. P. Grangier, G. Roger and A. Aspect, Experimental evidence for a photon anticorrelation effect on a beamsplitter: A new light on single–photon interferences, Europhys. Lett. 1:173 (1986).

5. P. Alsing, G.J. Milburn and D.F. Walls, Quantum nondemolition measurements in optical cavities, Phys. Rev. A 37:2970 (1988); H.A. Bachor, M.D. Levenson, D.F. Walls, S.H. Perlmutter and R.M. Shelby, Quantum nondemolition measurements in an optical–fiber ring resonator, Phys. Rev. A 38:180 (1988).

6. B.C. Sanders and G.J. Milburn, Complementarity in a quantum nondemolition measurement, Phys. Rev. A 39:694 (1989).

7. R. Loudon and P.L. Knight, Squeezed Light, J. Mod. Opt. 34:709 (1987), and references therein.

8. M. Xiao, L.–A. Wu and H.J. Kimble, Precision measurement beyond the shot noise limit, Phys. Rev. Lett. 59:278 (1987).

9. J. Krause, M.O. Scully, T. Walther and H. Walther, Preparation of a pure number–state and measurement of the photon statistics in a high–Q micromaser, Phys. Rev. A 39:1915 (1989).

10. M.O. Scully and H. Walther, Quantum optical test of observation and complementarity in quantum mechanics, Phys. Rev. A 39:5229 (1989).

11. S. Haroche, Quantum beats and time–resolved fluorescence spectroscopy, in: "High–Resolution Laser Spectroscopy (Topics in applied physics, V.13)", K. Shimoda, ed., Springer–Verlag, Berlin (1976), p. 256; J.N. Dodd and G.W. Series, Time–resolved fluorescence spectroscopy, in: "Progress in Atomic Spectroscopy: Part A (Physics of atoms and molecules):, W. Hanle and H. Kleinpoppen, eds., Plenum, New York (1979), p. 639.

12. B.C. Sanders and G.J. Milburn, Quantum nondemolition measurement of quantum beats and the enforcement of complementarity, Phys. Rev. A (in press).

MORE COMMENTS ON THE CHOICE OF COUPLINGS $-\underline{\mu}.\underline{E}$ AND $e\underline{p}.\underline{A}/(mc)$

E.A. Power

Department of Mathematics
University College London
Gower Street
London WC1E 6BT
England

ABSTRACT

A summary is given of the minimal coupling and multipolar forms of quantum electrodynamics. The hyperpolarizability and multipolar absorption matrix elements are shown to be identical in the two theories; the explicit proof involves sum-rules over the complete spectrum of the source molecules. A hybrid Hamiltonian allows either interaction to be used for any single mode photon.

I INTRODUCTION

It is over half a century since Goeppert Mayer[1] introduced the transformation $\exp(-ie\underline{A}(t).\underline{q}/\hbar c)$ on the wave function for a charged particle $(-e)$ in a given *external* electromagnetic field which, in dipole approximations, gives the equation for the new wave function $\psi_{\text{Mult}}(\underline{q})$

$$\left(\frac{\underline{p}^2}{2m} + V(\underline{q}) - \underline{\mu}.\underline{E}(t)\right)\psi_{\text{Mult}} = i\hbar\frac{\partial\psi_{\text{Mult}}}{\partial t} . \tag{1}$$

This follows from the Schrödinger equation with minimal coupling

$$\left(\frac{\left(\underline{p}^2 + e\underline{A}/c\right)^2}{2m} + V(\underline{q})\right)\psi_{\text{Min}} = i\hbar\frac{\partial\psi_{\text{Min}}}{\partial t} . \tag{2}$$

To obtain equation (1) from equation (2) it is necessary for the field variables $\underline{A}$ and $\underline{E}$ to commute, which of course they do for given external fields. On the other hand in quantum electrodynamics the electromagnetic fields are part of the dynamical system and are operators which do not commute.

New Frontiers in Quantum Electrodynamics and Quantum Optics
Edited by A. O. Barut, Plenum Press, New York, 1990

It is over 30 years since we (S. Zienau and E.A. Power)[2] examined the change from the minimal coupling to the multipolar for the total system in non-relativistic quantum electrodynamics. The results may be summarized in the two Hamiltonians.

$$H_{Mult} = H_{Molecule} + H_{Field} - \int \underline{P}.\underline{D}^{\perp} + 2\pi \int \underline{P}^{\perp 2} \tag{3}$$

and

$$H_{Min} = H_{Molecule} + H_{Field} + e\underline{p}.\underline{A}(q)/(mc) + e^2\underline{A}^2(\underline{q})/(2mc^2) \tag{4}$$

which in electric-dipole approximation become

$$H_{Mult} = H_{Molecule} + H_{Field} - \underline{\mu}.\underline{D}^{\perp}(\underline{R}) \tag{5}$$

$$H_{Min} = H_{Molecule} + H_{Field} + e\underline{p}.\underline{A}(\underline{R})/(mc) + e^2\underline{A}^2(\underline{R})/(2mc^2) \tag{6}$$

where $\underline{R}$ is the position vector of the centre and a self-energy term in equation (5) is incorporated in $H_{Molecule}$. There have been over a hundred papers since then commenting upon, simplifying, "clarifying" or pointing out differences between the predictions consequent upon using the two Hamiltonians. In this brief talk I have chosen to comment upon a very recent such paper (Hammond 1989)[3] and use the opportunity to show by explicit algebra how the two theories give identical predictions for multipolar absorption rates.

Hammond[3] calculates the second order susceptibility for a molecule with a permanent electric moment. The two theories appear to give different predictions which can be expressed in terms of differing hyperpolarizabilities

$$\beta_{ijk} = \frac{d_i \mu_j \mu_k}{E_{no}^2 - (\hbar\omega)^2} \quad \text{and} \quad \left(\frac{E_{no}}{\hbar\omega}\right)^2 \frac{d_i \mu_j \mu_k}{E_{no}^2 - (\hbar\omega)^2} \tag{7}$$

where $\underline{d} = \underline{\mu}^{nn} - \underline{\mu}^{oo}$ is the difference between the permanent moments of the relevant excited state and the ground state and $\underline{\mu}^{no}$ is the transition moment between these states. The factor $\left(\frac{E_{no}}{\hbar\omega}\right)$ (appearing twice in equation (7)) is precisely that which occurs in comparing matrix elements of $-\underline{\mu}.\underline{E}$ with matrix elements of $e\underline{p}.\underline{A}/(mc)$ for single photon processes. This arises because

$$\underline{E} = -\dot{\underline{A}}/c \quad \text{and} \quad \frac{e}{m}\underline{p}^{nm} = \mp \mu^{nm} E_{nm}/(i\hbar) . \tag{8}$$

II APPARENT DIFFERENCES

A natural question to ask is which process would give the maximum difference between the two theories given the factor $\left(E_{nm}/\hbar\omega\right)$ at each vector? N-photon absorption emphasizes the discrepancy by enormous factors as compared with say the rather weak line-shape difference commented upon by Lamb, Schlicher and Scully[4]. These distinguish between

$$(\underline{\mu}.\underline{\varepsilon})\frac{\gamma}{\left(E_{no}-\hbar\omega\right)^2+(\gamma/2)^2} \quad \text{and} \quad (\underline{\mu}.\underline{\varepsilon})\frac{\gamma}{\left(E_{no}-\hbar\omega\right)^2+(\gamma/2)^2}\left(\frac{E_{no}}{\hbar\omega}\right)^2 \tag{9}$$

[$\underline{\varepsilon}$ is the field intensity and γ the line-width of level n]. The difference near the peak $E_{no} \sim \hbar\omega$ is very small, although measurable. For N-photon absorption with $E_{no} \sim N\hbar\omega$, the matrix element with $e\underline{p}.\underline{A}/(mc)$ at each vertex is

$$M = \frac{\mu_{i_1}^{n\alpha_1}\mu_{i_2}^{\alpha_1\alpha_2}\cdots\mu_{i_N}^{\alpha_{N-1}o}\; e_{i_1}e_{i_1}\cdots e_{i_N}}{\left(E_{n\alpha_1}-\hbar\omega\right)\left(E_{n\alpha_2}-2\hbar\omega\right)\cdots\left(E_{n\alpha_{N-1}}-(N-1)\hbar\omega\right)}\left(\frac{E}{\hbar\omega}\right)^N \tag{10}$$

with E ranging over the molecular energy difference (for a two-level system $E = E_{10}$ for all vertices). Hence

$$M \simeq N^N \text{ x (correct answer)} \tag{11}$$

Since we 'know' the correct answer is that obtained from the multipolar interaction Hamiltonian. Factors of 27 for $N = 3$ have been commented upon in the past[5].

In the case of the hyperpolarizability tensor β_{ijk} the matrix element for two-photon absorption (each with circular frequency ω) with a single-photon emission at double frequency is

$$M = -i\,\beta_{ijk}\,\bar{e}_i(2\omega)e_j(\omega)e_k(\omega)\;\varepsilon_{2\omega}\varepsilon_{\omega}^2\,. \tag{12}$$

This follows from the energy

$$W = -\underline{P}.\underline{\varepsilon}_{-2\omega} \tag{13}$$

with $\underline{P}$ the induced dipole

$$\underline{P} = \underline{\underline{\beta}} : \underline{\varepsilon}_{-\omega}\underline{\varepsilon}_{-\omega}\,. \tag{14}$$

The expression for β_{ijk} is well-known[6]

$$\beta_{ijk} = \frac{1}{2} \sum_{r,s} \left[\frac{\mu_i^{os} \mu_j^{sr} \mu_k^{ro}}{(E_{so} - 2\hbar\omega)(E_{ro} - \hbar\omega)} + \frac{\mu_k^{os} \mu_i^{sr} \mu_j^{ro}}{(E_{so} + \hbar\omega)(E_{ro} - \hbar\omega)} \right.$$

$$\left. + \frac{\mu_j^{os} \mu_k^{sr} \mu_i^{ro}}{(E_{so} + \hbar\omega)(E_{ro} + 2\hbar\omega)} + j \Longleftrightarrow k \text{ (symmetric terms)} \right] . \tag{15}$$

For a two-level system E_n, E_o and for the 2ω-photon emitted having polarization direction parallel to the permanent moment (i.e. $\underline{d}.\underline{e}(2\omega)$) equation (15) reduces to the first alternative in equation (7). The second alternative differs from the first only in the wings similar to the results (9). However it can be shown that a complete calculation of β_{ijk} for the minimal coupling alternative gives the identical expression equation (15) and there is no paradox.

It is of interest to point out that other alternatives to $-\underline{\mu}.\underline{E}$, $e\underline{p}.\underline{A}/(mc)$ exist such as[2] $-e\underline{a}.\underline{Z}/(mc^2)$ [$\underline{Z}$ is the Hertz vector and $\underline{a}$ the acceleration] which can give even larger discrepancies. For example using $-e\underline{a}.\underline{Z}/(mc^2)$ gives $(E/(\hbar\omega))^2$ at each vertex and the apparent matrix element is N^{2N} x (correct answer).

III DERIVATION OF THE TWO HAMILTONIANS

The difference between the two theories, minimal and multipolar, appears in the relationships between the canonical and kinetic momenta. The coordinates are the $\underline{q}$'s for the charges and $\underline{A}(\underline{r})$ for the electromagnetic fields. The Lagrangian can be written

$$L = L_{\text{Molecules}} + \int (\mathcal{L}_{\text{Maxwell}} + \mathcal{L}_{\text{Int}})\, dV . \tag{16}$$

In the minimal coupling theory the Lagrangian density

$$\mathcal{L}_{\text{Min}} = \mathcal{L}_{\text{Maxwell}} + \mathcal{L}_{\text{Int}} = \frac{\dot{\underline{A}}^2/c^2 - (\text{curl } \underline{A})^2}{8\pi} + \underline{j}(\underline{r}).\underline{A}(\underline{r})/c \tag{17}$$

with

$$\underline{j}(\underline{r}) = \sum(-e)\dot{\underline{q}}\delta(\underline{r} - \underline{q}) . \tag{18}$$

Clearly the conjugate momenta are

$$\underline{p} = \frac{\partial L}{\partial \dot{\underline{q}}} = m\dot{\underline{q}} - \frac{e}{c}\underline{A}(\underline{q}) \tag{19}$$

$$\underline{\Pi}(\underline{r}) = \frac{\partial \mathcal{L}}{\partial \dot{\underline{A}}} = \frac{\dot{\underline{A}}}{4\pi c^2} = -\frac{\underline{E}^{\perp}}{4\pi c} \; . \tag{20}$$

Now if a total time derivative is added to the Lagrangian the equations of motion are unchanged. Let[8]

$$\mathcal{L}_{\text{Multi}} = \mathcal{L}_{\text{Min}} - \frac{dF}{c\,dt} \tag{21}$$

with

$$F = \underline{P}(\underline{r}).\underline{A}(\underline{r}) \tag{22}$$

then

$$\mathcal{L}_{\text{Multi}} = \frac{\dot{\underline{A}}^2/c^2 - (\text{curl } \underline{A})^2}{8\pi} + \left(\underline{j}(\underline{r}) - \dot{\underline{P}}(\underline{r})\right).\underline{A}(\underline{r})/c - \underline{P}(\underline{r}).\dot{\underline{A}}(\underline{r})/c \tag{23}$$

In quantum optics it is usually sufficient to keep to the electric-dipole approximation where all currents are polarization currents. In fact

$$\underline{j} = \dot{\underline{P}}(\underline{r}) \quad \text{with} \quad \underline{P} = \sum(-e)\underline{q}\delta(\underline{r} - \underline{R}) \tag{24}$$

and so the multipolar Lagrangian density is

$$\mathcal{L}_{\text{Electric Dipole}} = \frac{\dot{\underline{A}}^2/c^2 - (\text{curl } \underline{A})^2}{8\pi} - \underline{P}(\underline{r}).\dot{\underline{A}}(\underline{r})/c \tag{25}$$

and

$$\underline{p} = \frac{\partial L}{\partial \dot{\underline{q}}} = m\dot{\underline{q}} \tag{26}$$

$$\underline{\Pi}(\underline{r}) = \frac{\partial \mathcal{L}}{\partial \dot{\underline{A}}} = \frac{\dot{\underline{A}}}{4\pi c^2} - \frac{\underline{P}}{c} = -\frac{\underline{D}^{\perp}(\underline{r})}{4\pi c} \; . \tag{27}$$

The corresponding Hamiltonians are precisely equations (6) and (5) respectively. Of course beyond the electric-dipole approximation, say to discuss optical activity, $\underline{P}(\underline{r})$ must be more general. For example to include the $-\underline{m}.\underline{B}$ term in equation (3) we need

$$\underline{j} - \dot{\underline{P}} = \text{curl } \underline{M} \tag{28}$$

with

$$\underline{M} = \frac{-e}{2c}\, \underline{q} \times \dot{\underline{q}}\, \delta(\underline{r} - \underline{R}) \; . \tag{29}$$

The difference between the two Hamiltonians (3) and (4) can be succinctly summarized by two substitutions within the non-interacting Hamiltonian by using the connection derivatives:-

$$\underline{\partial} \longrightarrow \underline{\partial} + ie\underline{A}.c \qquad \text{Minimal} \tag{30}$$

$$\frac{\delta}{\delta \underline{A}} \longrightarrow \frac{\delta}{\delta \underline{A}} + i\underline{P}/c \qquad \text{Multipolar} . \qquad (31)$$

It is also possible to introduce hybrid forms intermediate between minimal and multipolar. One such would be to treat left-handed circularly polarized photons differently from right-handed circularly polarized photons. The Hamiltonian is

$$H_{\text{Molecules}} + H_{\text{Field}} - \underline{\mu}.\underline{E}^{L} - \underline{m}.\underline{B}^{L} + e\underline{p}.\underline{A}^{R}/(mc) + e^{2}\underline{A}^{R^{2}}/(2mc^{2}) \qquad (32)$$

where the generating density function $\underline{P}(\underline{r}).\underline{A}(\underline{r})$ includes a left-handed δ-dyadic

$$\delta^{L}_{ij}(\underline{r}) = \frac{1}{2}\left(\frac{\delta_{ij} - \hat{r}_i\hat{r}_j}{r^3} - i\epsilon_{ijk}\frac{\hat{r}_k}{r^3}\right) . \qquad (33)$$

It is clear that taking this hybridization to its limit we could treat each mode independently; so that one mode could be coupled by $e\underline{p}.\underline{A}/(mc)$ another by $-\underline{\mu}.\underline{E}$.

IV MINIMAL COUPLING GIVES THE IDENTICAL N-PHOTON ABSORPTION MATRIX ELEMENT AS THE MULTIPOLAR COUPLING

In order to simplify the argument we consider the special case where one photon absorbed is from a different mode $\left(\underline{k}',\underline{e}'\right)$ than the remaining $(N - 1)$. The generalization where all N-photons are from different modes is straightforward but algebraically complicated and, of course, the case of all N-photons identical is just a limit of the special case considered. The multipolar theory gives for the matrix element

$$M = \sum_{r=1}^{N} M^{(r)} \qquad (34)$$

where

$$M^{(r)} = \sum_{\alpha,i} \frac{\mu_{i_1}^{n\alpha_1}\mu_{i_2}^{\alpha_1\alpha_2}\cdots\mu_{i_r}^{\alpha_{r-1}\alpha_r}\mu_{i_{r+1}}^{\alpha_r\alpha_{r+1}}\cdots\mu_{i_N}^{\alpha_{r-1}o}\, e_{i_1}e_{i_2}\cdots e'_{i_r}\cdots e_{i_N}}{D_1^{(r)}D_2^{(r)}\cdots D_{r-1}^{(r)}D_r^{(r)}D_{r+1}^{(r)}\cdots D_{N-1}^{(r)}} \qquad (35)$$

where

$$D_s^{(r)} = \left(E_{n\alpha_s} - s\hbar\omega\right) \qquad \text{if } r > s$$

$$= \left(E_{n\alpha_s} - \hbar\omega' - (s-1)\hbar\omega\right) \qquad \text{if } r \le s . \qquad (36)$$

If we now suppose the one photon is coupled by $e\underline{p}.\underline{A}/(mc)$, rather than $-\underline{\mu}.\underline{E}$, then the r'th contribution using equation (8) is

$$\tilde{M}^{(r)} = \left(\frac{E_{\alpha_{r-1}\alpha_r}}{\hbar\omega'} \right) M^{(r)}$$

$$= \frac{\left(E_{n\alpha_r} - \hbar\omega' - (r-1)\hbar\omega\right) - \left(E_{n\alpha_{r-1}} - (r-1)\hbar\omega\right) + \hbar\omega'}{\hbar\omega'} M^{(r)} \qquad (37)$$

We note the special values

$$\tilde{M}^{(1)} = \frac{\left(E_{n\alpha_1} - \hbar\omega'\right) - 0 + \hbar\omega'}{\hbar\omega'} M^{(1)} \qquad (38)$$

and

$$\tilde{M}^{(N)} = \frac{0 - \left(E_{n\alpha_{N-1}} - (N-1)\hbar\omega\right) + \hbar\omega'}{\hbar\omega'} M^{(N)} . \qquad (39)$$

Let us define

$$N^{(r)} = \sum_{\alpha, i} \frac{\mu_{i_1}^{n\alpha_1} \mu_{i_2}^{\alpha_1\alpha_2} \cdots \mu_{i_r}^{\alpha_{r-1}\alpha_r} \mu_{i_{r+1}}^{\alpha_r\alpha_{r+1}} \cdots \mu_{i_N}^{\alpha_{r-1}o} e_{i_1} e_{i_2} \cdots e'_{i_r} \cdots e_{i_N}}{D_1^{(r)} D_2^{(r)} \cdots D_{r-1}^{(r)} . 1 . D_{r+1}^{(r)} \cdots D_{N-1}^{(r)}} \qquad (40)$$

which has the form of $M^{(r)}$ with the denominator $D_r^{(r)}$ removed. In $N^{(r)}$ the sum over α_r can be carried out using closure since no E_{α_r} appears in the denominators. We have

$$\sum_{\alpha_r} \mu_{i_r}^{\alpha_{r-1}\alpha_r} \mu_{i_{r+1}}^{\alpha_r\alpha_{r+1}} = \left\langle \alpha_{r-1} \middle| \mu_{i_r} \mu_{i_{r+1}} \middle| \alpha_{r+1} \right\rangle \qquad (41)$$

which is symmetric in the polarization indices i_r , i_{r+1} . Hence

$$\tilde{M}^{(r)} = \frac{N^{(r)} - N^{(r-1)} + \hbar\omega' M^{(r)}}{\hbar\omega'} . \qquad (42)$$

Since equations (38) and (39) imply the boundary conditions $N^{(0)} \equiv N^{(N)} = 0$ we have

$$\sum_{r=1}^{N} \tilde{M}^{(r)} = \sum_{r=1}^{N} M^{(r)} \tag{43}$$

The essential point within the proof is the closure relation (41). This does not work for a two-level system or a truncated sum over the molecular spectra.

IV CONCLUSION

We have shown by explicit algebraic manipulation that the matrix element for multipolar absorption is the same for minimal and multipolar interactions. Similar arguments can be made if some quanta are emitted such as in frequency doubling. If two photons of the same mode are involved then the minimal coupling formalism requires use of the $e^2\underline{A}^2/(2mc^2)$ term and the calculations are always more complicated then those using the multipolar interaction. This appears to be universal[9] and is especially obvious if higher multipoles are involved in the processes. A clear example is in chiroptic problems involving both $-\underline{\mu}.\underline{E}$ and $-\underline{m}.\underline{B}$ with asymmetric molecules.

REFERENCES

1. M. Goeppert-Mayer, *Am. Phys.* (Leipz.) 9:273 (1931).
2. E.A.Power and S. Zienau, *Philos. Trans. Roy. Soc.* A251:427 (1959).
3. R. Hammond, *Phys. Rev.* A39:3544 (1989).
4. W.E. Lamb, R.R. Schlicher and M.O. Scully, *Phys. Rev.* A36:2763 (1987).
5. E.A.Power, in: "Multipolar Processes," J.H. Eberly and P. Lambropoulos, eds, Wiley, New York (1978).
6. N. Bloembergen, "Nonlinear Optics," Benjamin, New York (1965).
7. W. Henneberger, *Phys. Rev. Lett.* 21:838 (1965).
   See also E.A. Power and T. Thirunamachandran, *Am. J. Phys.* 47:370 (1978).
8. E.A. Power and T. Thirunamachandran, *Proc. Roy. Soc.* A372:265 (1980).
9. J.O. Hirschfelder, "Lasers, Molecules and Methods," Interscience, New York (1989)

TRANSITION FROM CLASSICAL, "MAXWELL-BOLTZMANN" TO QUANTUM, "BOSE-EINSTEIN" PARTITION STATISTICS BY STOCHASTIC SCATTERING OF DEGENERATE LIGHT

F. De Martini and R. Tommasini

Dipartimento di Fisica dell'Universita' "La Sapienza"
Roma, 00185 Italy

INTRODUCTION AND THEORETICAL BACKGROUND

The concept of "quantum statistics for distinguishable particles" has been introduced by a remarkable theoretical work a few years ago (1). There it is shown that the classical, Maxwell-Boltzmann (MB) partition law for an ensemble (or a stream) of n particles to be distributed in M "boxes" (or scattered over M channels) can be transformed formally into the Bose-Einstein (BE) law (or into the Fermi-Dirac (FD) law) by allowing the scattering probability over each channel, W_i, to be a stochastic variable instead of a "stationary" cross-section. Since in this process no use is made of "indistinguishability", i.e., of the fundamental property which is generally attributed to all "quantum" particles, the authors of (1) argue that then the statistics cannot be a criterion for that property. On the other hand, according to the authors of (1), this does not contradict the fundamental prescription according to which "classical" particles obey MB Statistics while "real" particles obey BE or FD quantum-Statistics (2,3). Since the world is actually made of *real* particles, the interesting transformation theorem given by (1) has been generally considered as a mathematical curiosity with no real physical content. In the present paper we give the first experimental demonstration that the statistical behavior of real particles, i.e., the optical photons, *is indeed* described either by the classical-Statistics or by the appropriate quantum-Statistics depending on the statistical character of W_i. Then, in addition to providing the first validation of the mathematical arguments of (1), our work suggests a new and consistent model for all quantum-statistical processes.

Assume that a monochromatic light beam belonging to a field-mode $\vec{k}$, in a chaotic state and with an average photon number per mode $\langle n\rangle$, excites a Beam-Splitter (BS) characterized by "stationary" or "stochastic" scattering cross-sections W_i over the output modes: i=1,2. "StA-BS", "StO-BS" denote the related BS conditions. The field density operator is (4):

$$\rho = \sum_{n(0-\infty)} \underline{P_n} * \sum_{m,h(0-n)} p_{nm} p_{nh} \; |m, n-m\rangle\langle n-h, h|$$

where h,n-h are the photon numbers scattered over the i-modes, $P_n \equiv \langle n\rangle^n/(1+\langle n\rangle)^{(1+n)}$, and $P_{nh} \equiv |p_{nh}|^2$ are the two-way partition probabilities. Consider the StA-BS, QED theory of the process (5). The input state is obtained by successive applications to the vacuum-state of $b_k^+ \equiv \sum_i w_i a_i^+$, the operator expressing the input in terms of output-fields and: $W_i \equiv |w_i|^2$ $\sum_i W_i = 1$. This procedure leads to the Bernoulli partition law:

New Frontiers in Quantum Electrodynamics and Quantum Optics
Edited by A. O. Barut, Plenum Press, New York, 1990

$P_{nh} = W_1^{(n-h)} W_2^h * {}^nC_h$ where: ${}^nC_h \equiv n!/(h!(n-h)!)$ (5). Consider now the StO-BS condition in which the ouput photon numbers (h,n-h) are detected in a series of measurements lasting a time Dt over which $W_1(t)$ shows an appreciable variation. W_1 may be a random function. In this case the observed statistical distribution P_{nh} is the average over all possible probability sets $\{W_1\}$ realized in Dt. According to this prescription and following the path of work (1), the general case is analyzed as follows. Write: $W_1 = X(1+x)$, $W_2 = Y(1-\beta x)$, $\beta \equiv X/Y$, $X+Y=a \leq 1$, where X,Y are stationary cross-sections, x is a stochastic variable and the probability a is introduced to represent the BS loss. $W_1^{(n-h)} W_2^h$ is then replaced by:

$$X^{(n-h)} Y^h * S(x',x") \qquad \text{where,} \quad S(x',x") \equiv \int_{x"}^{x'} [(1+x)^{(n-h)} (1-\beta x)^h] * p(x) * dx$$

is a function of the extreme values taken by x in its range of existence and p(x) is a probability distribution function. By assuming $p(x)=\delta(0)$ the "classical" MB statistics is obtained: $P_{nh} = X^{(n-h)} Y^h * {}^nC_h$. According to quantum theory this law is satisfied by the StA-BS condition, as shown. By assuming the uniform distribution, $p(x)=(x'-x")^{-1}$, and considering the case of lossless, symmetrical-BS ($a=\beta=1$), the partition law is found:

$$P_{nh}(x',x") = [(2^n * (n+1) * (x'-x"))^{-1} \sum_{r(o-h)} {}^{n+1}C_r \ (\Gamma_r' - \Gamma_r")] \tag{1}$$

where $\Gamma_r', \Gamma_r"$ are the values taken by the expression $\Gamma_r \equiv (1+x)^{(n-r+1)} (1-x)^r$ for $x=(x',x")$, respectively. $P_{nh}(x',x")$ expresses the general statistical behavior of a two-channel scattering process within the stated assumptions and under the effect of stochastic perturbations. Interestingly, the size of these ones, viz., the value of $|x'-x"|$, identifies some statistical conditions of extreme physical relevance. In fact, for values (x',x") leading to the corresponding extreme values (1,0) taken by the scattering probabilities $W_{1,2}$ in StO-BS regime, the two-channel Bose-Einstein partition law, $(P_{nh})_{BE} = (1+n)^{-1}$, is obtained (6). In other words, according to our present model, the BE Statistics represents a *fully* stochastic process with *uniform* probability distribution. On the other hand, a zero perturbation level implies MB Statistics, as shown. The intermediate case, $0<|x'-x"|<1$, is found after (1) to correspond to a partition law that is a linear superposition of a hierarchy of modified BE distributions. The double photocount detection method consists of the measurement of the "BS Quantum Noise Function" $G(1/\bar{n}) \equiv \langle \hat{g} \rangle$, and: $\hat{g} \equiv ((n_1/\bar{n}_1) - (n_2/\bar{n}_2))^2$; $(4/\bar{n}) \equiv (1/\bar{n}_1) + (1/\bar{n}_2)$; $\bar{n}_1 \equiv \langle n_1 \rangle \equiv \langle a_1^+ a_1 \rangle$ (2,7). G is related to the "degrees of second-order-coherence", $g_{ij}(0) \equiv \langle a_i^+ a_j^+ a_j \ a_i \rangle / \langle n_i \rangle \langle n_j \rangle$ (i,j=output modes), through a normal-ordering procedure. Define $G' \equiv (g_{11} + g_{22} - 2g_{12})$. We have: $G' = \langle : \hat{g} : \rangle$; $G = \langle \hat{g} \rangle$. Evaluate the ensemble-averages appearing in G by the field density-operators expressed in terms of the classical and of the quantum distributions $(P_{nh})_{MB}$, $(P_{nh})_{BE}$. By making use of the commutation relations for bosons we easily obtain for both cases: $G(1/\bar{n}) = G' + (4/\bar{n})$. In StA-BS condition, when MB law applies, is always $G'=0$ (5,8). To work out the StO-BS case for M=2 we express the output-modes operators in terms of input-mode operators by also using the M=2, BE density-operator:

$$(\varrho)_{BE} = \sum_{n(0-\infty)} P_n * (n+1)^{-1} * \sum_{m,h(0-n)} |m,n-m\rangle \langle n-h,h| .$$

G' is then found to be given in terms of the input operator, $n=(b_k^+ b_k)$, in the form: $G' = 4\langle n(n-1) \rangle / (3 \langle n \rangle^2)$. The mean-square input photon-number average is: $\langle n^2 \rangle = \delta \langle n \rangle^2 + \langle n \rangle$, where $\delta=1$ or $\delta=2$ for input coherent or incoherent fields, respectively (5). At last: $G(1/\bar{n}) = G + 4/\bar{n}$. Relevant values of the "Interstatistics parameter" are $G=0,(4/3),(8/3)$ corresponding respectively to: MB statistics; BE statistics for coherent input light; BE statistics for incoherent light. G expressed in terms of photocounts $\bar{m}_1 \propto \bar{n}_1$ is represented in Fig.1 by the straight lines a,b,c (7,8). We see that the quantum-correlations (BE-condensation) caused by the scrambling effect of StO-BS on the output channels, imply additional contributions of field "wave-fluctuations" on these channels (5).

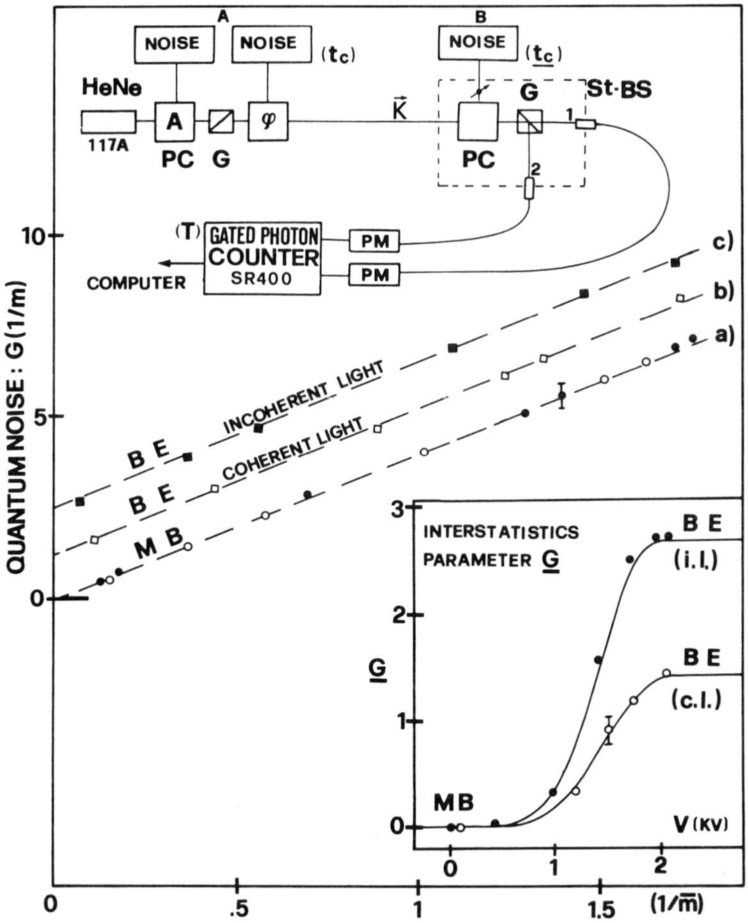

FIG.1-"Beam-Splitter Quantum-Noise Function", G(1/m̄). Plots a,b,c refer to Maxwell-Boltzmann, MB, and to Bose-Einstein, BE, partition-statistics. In the inset, the "interstatistics-parameter" G(V) is plotted as function of the peak-to-peak noise voltage, V, applied to the Pockels-cell of the Stochastic-Beam-Splitter (StO-BS).

As a significant extension of the above theory we consider a set of $M \geq 2$ scattering output states $|\{n_i\}\rangle \equiv |n_m\rangle |n_{m-1}\rangle .. |n_2\rangle |n_1\rangle$ where again photodetection is carried out over states $j=1,2$. Since the relevant operators have the form: $\hat{O}=f(a_j{}^+ a_j, \hat{I})$, where $\hat{I}$ is the identity-operator, we write: $\langle \hat{O} \rangle = \mathrm{Tr}(\varrho \hat{O}) = \sum_{n(0-\infty)} \underline{P_n} \sum_{\{n_i\}} W_i \langle \{n_i\} | \hat{O} | \{n_i\} \rangle$, for: $W_i = |w_i|^2$. Classical absence of inter-particle correlation in the scattering process is expressed by the MB distribution for which the following expression holds (4):

$$\sum_{\{n_i\}} W_i{}^{MB} \equiv n! * \sum_{n_M(0-n)} (W_M{}^{n_M}/n_M!) \ldots \sum_{n_1(0-n_2)} (W_1{}^{n_1}/n_1!) * \delta[n-(n_M+n_{M-1}+..n_1)] = 1$$

while, for quantum-scattering of integer-spin particles, we have:

$\sum_{\{n_i\}} W_i{}^{BE} \equiv W^{BE} * \sum_{\{n_i\}} \langle \hat{I} \rangle = 1$ since the BE statistics is independent of $\{n_i\}$: $W^{BE} \equiv n!(M-1)!/(n+M-1)!$ (4,6). For both distributions we may now evaluate in terms of $\langle n \rangle$ the relevant averages related to 2-channel photodetection: $\langle n_1 \rangle, \langle n_2 \rangle, \langle n_1{}^2 \rangle, \langle n_2{}^2 \rangle, \langle n_1 n_2 \rangle$, by assuming also, with no loss of generality, equal output-channel probabilities: $W_i = M^{-1}$. This is obtained after some algebra by using the properties of the Gamma-function for negative arguments and by a smart re-definition of the limits of the multiple-sums at each step of a repeated elimination of factorials. The results are:

Maxwell-Boltzmann: $\langle n_1 \rangle = \langle n_2 \rangle = \langle n \rangle / M$; $\langle n_1{}^2 \rangle = \langle n_2{}^2 \rangle = [\langle n^2 \rangle + (M-1)\langle n \rangle]/M^2$;
$\langle n_1 n_2 \rangle = (\langle n^2 \rangle - \langle n \rangle)/M^2$;

Bose-Einstein : $\langle n_1 \rangle = \langle n_2 \rangle = \langle n \rangle / M$; $\langle n_1{}^2 \rangle = \langle n_2{}^2 \rangle = [2\langle n^2 \rangle + (M-1)\langle n \rangle]/(M(M+1))$;
$\langle n_1 n_2 \rangle = (\langle n^2 \rangle - \langle n \rangle)/(M(M+1))$.

The "Scattering quantum-noise function" for $M \geq 2$ scattering-channels is then again expressed by: $G(\langle n \rangle) = \underline{G} - 2M/\langle n \rangle$. The "interstatistics-parameter" is now: $\underline{G}=0$ (MB-distribution); $\underline{G}=2\delta M/(M+1)$ (BE-distribution). These formulas, which account correctly for the BS case, lead for $M \approx \infty$ to the asymptotic values: $\underline{G}=2$, $\underline{G}=4$ for coherent and incoherent-light, respectively.

EXPERIMENTAL RESULTS, INTERPRETATION

A TEM00 single k-mode, 1 mW, He-Ne laser beam (Spectra-Physics 117A) was sent through single Amplitude- and Phase-modulators driven by two independent HV noise-generators equipped with RC low-pass filters (9). When activated this set (A, in Fig.1) transformed the coherent beam in a chaotic superposition of $|n\rangle$-states with a coherence-time t_c=1 msec (10). The beam was sent to a StO-BS made of a Pockels-cell (PC), of a Glan-air analyzer (G) and of a z-cut quartz plate inserted in one of the output modes to restore equal polarization for all input/output modes. The StA-BS condition was obtained by deactivating the adjustable noise-generator B. By activating B the beam polarization analyzed by G was randomly driven across the BS-symmetry condition with a BS-coherence-time $\underline{t_c}$=50 μsec. The output light was transferred by two optical-fibers to the cathodes of low-noise 56DVP phototubes (PM) connected to a Stanford-Research SR400 gated counter. The gate-window was T=1 μsec.and each datum of Fig.1 corresponds to computer-evaluated averages made of $\approx 10^5$ samples. The PM noise was about 1 pulse every 100 gate-samples per channel. A continuous transition between curves (a-b) and (a-c) was detected by varying the value of the peak-to-peak noise-voltage V driving StO-BS. The data points on the BE curves b),c), Fig.1, correspond to V=1,96 KV, which is the PC $(\lambda/2)$-voltage at λ=6328 A, the laser wavelength. This corresponds to a variation of the scattering probabilities $W_{1,2}$ over the full randomness range 0-1 in agreement with theory. The plots of $\underline{G}(V)$ (Fig.1, inset) show further details of the transition. The theoretical curves $\underline{G}(V)$ are computer-evaluated by the given interstatistics-theory, Eq.1.

As a significant extension of the above method to a large number of output channels, $M \gg 2$, Fig.2 shows the schematic diagram of a "stochastic"

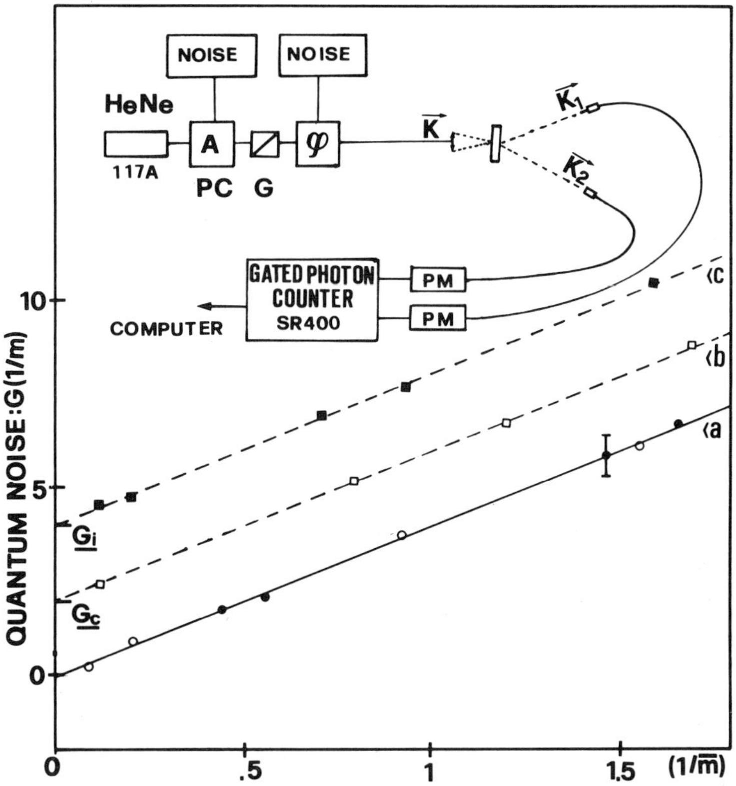

FIG.2-Results of a scattering experiments with microscopic polystyrene-spheres undergoing brownian-motion.

light-scattering apparatus. The device (A) for laser light-modulation is identical to the one already described for the BS case.

The StO-BS of the previous experiment is here replaced by a scattering medium composed by small polystyrene spheres with diameter 9 10^{-6} cm, undergoing brownian-motion in deionized water (11). The laser beam was focused on a 0.4 mm thick scattering-cell by a 5 cm. f.l. lens and the attenuation of the beam never exceeded 15% of the incoming light so that multiple scattering contributions could be neglected. The directions of the output k_1, k_2 wavevectors were taken at 40° from the lens axis. The cell temperature τ, measured by a termocouple, was varied across 0°C by means of a salt-ice refrigerator coupled with an heating resistor. In "stationary" (StA) and "stochastic" (StO) regimes, i.e. for $\tau<0$°C and $\tau>0$°C respectively, the two detectors were excited, over single output spatial-modes, by the random, and time-varying for StO, speckle pattern determined by the scattering dynamics (11,12). The scattering "coherence-time" t_c, evaluated by classical diffusion-theory, was $\approx 10^{-4}$sec at $\tau=+7$°C, while the other characteristic times of the experiment were: $t_c=10^{-3}$, T=1 μsec. The plots of Fig.2 show that, in analogy with the BS case, the MB partition statistics is realized in the StA regime, i.e. when the brownian-motion of the scattering spheres is "frozen" within the ice-crystal. By raising τ, in correspondence with the ice-water structural phase-transition at $\tau\approx 0$ the interstatistics-parameter undergoes a sudden jump from $G=0$ (StA, MB partition) to the values $G\approx 2$ and $G\approx 4$ representing BE partition-statistics for input coherent and incoherent light, respectively.

We have then again demonstrated that the transition from MB to BE statistics may be realized by a, very natural, multi-channel scattering process among microscopic-particles in virtue of *any* W_i-randomization mechanism. This substantiates our model according to which the quantum-statistical behaviour of a gas of colliding integer-spin particles, and then its Planck energy-distribution at equilibrium, is likely due to multiple inter-particle scattering events. The motions of the scatterers, caused either by an "external" mechanism (a brownian process, a StO-BS, a moving mirror) or by inter-particle momentum-exchanges in collisions, are able to determine the overall quantum-statistical behaviour of the gas (13).

CONCLUSIONS

In summary, we have demonstrated that the statistics *is not* a fixed particle property. It is in fact dependent on deterministic scattering conditions leading either to "classical" absence of interparticle correlations, or to intermediate-size or, at last, to full-size BE correlations. The consequences of all that should be profound and far reaching for fundamental physics (14). Furthermore, we have shown on a macroscopic scale what could be the basic mechanism of a most intriguing among physical processes: the interparticle quantum-correlations, or exchange-interactions. We conclude with two quotations by Albert Einstein: "The differences between Boltzmann and BE counting express indirectly a certain hypotesis on a mutual influence of the molecules which for the time being is of quite mysterious nature...", "The statistical method of Herr Bose and myself is by no means beyond doubt but seems only *a posteriori* justified by its success for the case of radiation.." (15, italic by A.E.). In view of our present results the content of these statements should perhaps be reconsidered.

REFERENCES

1. J.Tersoff and D.Bayer, Phys.Rev.Lett. 50, 553, 1983

2. F.De Martini, in: "Squeezed and Nonclassical light", E.Pike, Plenum, N.Y., 1988. F.De Martini, S.Di Fonzo, Europhysics Lett., 10, 123, 1989. F.De Martini and K.H.Strobl, Optics Comm., 75, april 1990.
3. Owing to Bohr's Complementarity our photon-counting experiment reveals physical effects that involve only the photon particle-like aspect. This argument legitimates the present statistical approach.
4. K.Huang, "Statistical Mechanics", Wiley, N.Y., 1963, Ch.9.
5. R.Loudon, "The Quantum Theory of Light", Oxford U. Press, 1987, Ch.10.
6. The M=2, BE partition law given in the text is a particular case of the general BE law: $P_{nh}= n!(M-1)!/(n+M-1)!$ (4).
7. $G(1/\bar{m})$, may be taken as the "quantum-signature" of the electromagnetic field as well as of any Bose-field. An account of this novel method of quantum optics with application to gravity-radiation detection is found in: F. De Martini and K.H. Strobl, Optics Comm., Ref.1.
8. For input $|\alpha\rangle$-states the statistical picture is valid owing to their n-state expansion. Another picture is provided by electrodynamics.
9. The wide-band noise was generated for sets A,B by reverse-biased 2N2369 transistors. The PC were: EOD125 (A), Lasermetrix 1042 (B), Phase-modulator was Lasermetrix 1039B. All electronics was developed in our laboratory. A "uniform" W_1-distribution has been achieved by a suitably clipping via an adjustable saturated amplifier of the wide gaussian-distribution of the noise-voltage V feeding the StO-BS.
10. The state if the input-field coherence was determined by Hanbury Brown-Twiss, StA-BS measurement of $g_{1,2}$ (5).
11. F.T.Arecchi, M.Giglio and U.Tartari, Phys.Rev. 163, 186, 1967.
12. J.C.Dainty (ed.),"Laser Speckle", Springer Verlag, N.Y. 1975; W.Marti enssen and E.Spiller, Am.J.Phys. 32, 919, 1964.
13. Since the Statistics is not a fixed particle property, as demonstrated by our experiments, we should attribute to the *particle-spin* the fundamental quantum property that, among other effects, determines the quantum-statistical behaviour of a gas in equilibrium. In fact, in agreement with quantum-theory, "bosons" and "fermions" are particles characterized by integer and half-integer spins, respectively. The effect of these two classes of spin-values on the dynamics of the inter-particle collisions leading the gas to equilibrium may determine precisely the observed quantum-statistical, BE or FD, equilibrium phenomenology (e.g., the Planck's distribution for bosons). All this establishes, in a very straightforward way, the *Spin-Statistics connection* within our model. For spin-dependent quantum-scattering theory: R.G.Newton, "Scattering Theory of Waves and Particles", Mac Graw-Hill, N.Y. 1966, Ch.8. (B.D'Espagnat, private comm.to F.D.M.).
14. The most obvious one appears to be the need for rejection of the indistinguishability concept (I), which is in contradiction with MB-statistics (4). In other words, there *are no* "classical" and "quantum" particles in physics: as far as I is concerned all particles are the same. Their quantum-statistical behaviour should be determined by scattering events involving nonstationary cross-sections determined by particle motions under momentum transfer in collisions and by the relevance taken in their motion by the dynamical implications of the Heisenberg principle. The classical behaviour of photons interacting with optical instruments has been experimentally verified by: F.De Martini by Michelson-interferometry (2) and by R Lange, J.Brendel, E.Mohler and W.Martienssen, Europhys.Lett. 5, 619, 1988. The effect discovered can be reproduced in first-order-coherence processes and open new trends in interferometry and spectroscopy. The same concepts can be extended to FD-statistics to investigate the Pauli principle. This is being done in our laboratory.
15. A.Einstein, Sitzungs.Preuss.Akad.Wiss. 1924, pag.261 and 1925, pag.18.

A Composite Particle and its Electromagnetic Properties

S. Graf, S. Schramm, B. Müller, W. Greiner

Institut für Theoretische Physik der Johann Wolfgang Goethe - Universität
Robert-Mayer-Straße 8-10, D-6000 Frankfurt am Main, Germany

1 Introduction

Experimental studies of QED of strong fields have been performed at the UNILAC accelerator at GSI by means of heavy-ion collisions at the Coulomb barrier for more than a decade. Besides confirming many theoretical predictions of QED of strong fields [1], these experiments have revealed unexpected line structures in positron spectra [2,3,4,5] and, more recently, in correlated electron-positron sum-energy spectra [6,7,8]. Various properties of these lines, e.g. angular correlations and the narrowness of the sum-energy peak, indicate that the structures could originate from the two-body pair-decay of light neutral objects produced almost at rest in the centre-of-mass frame of the heavy ions. Actually no other mechanism is known which could account for at least a large part of the measured data. In particular, nuclear pair-decay does not seem to present a viable explanation for the peaks.

The observed structures appear to be (almost) invariant against changes in the specific collision system. The same was true for the observed resonance in the positron singles spectra, which led to the first suggestion by Schäfer et. al [9] that the formation of a new object (particle) might be involved. The EPOS collaboration has found correlated structures at kinetic sum energies of $608 \pm 8, 760 \pm 20, 809 \pm 8$ keV in the U+Th system, and at $620 \pm 8, 748 \pm 8$, and 805 ± 8 keV in the U+Ta system. The ORANGE group has detected several significant structures in U+U and U+Pb collisions, at sum energies 809 ± 8 keV (most significant) and 540 ± 16, 640 ± 10, 716 ± 10, 895 ± 10 keV. The 810keV structure is only observed in spectra corresponding to an opening angle of $180^0 \pm 20^0$ between the directions of emission of the two leptons.

Thus the experiments provide indications for the possible existence of a new family of neutral particles in the mass range $1.5 - 2\,\mathrm{MeV/c^2}$. The question then arises whether this hypothesis is compatible with other, well-established experimental facts. This question has been extensively studied during the past three years [9-25], resulting in the conclusion that *new elementary bosons* in this mass range, which could be the origin of the correlated e^+e^--structures, *cannot exist* [10,11,12,17,23,25]. The conclusion is based on new results from Bhabha scattering [26] and beam-dump experiments [27], which unambiguously rule out neutral point-like particles X^0 between 1.5 and 2 MeV/c^2 with lifetimes $\tau_X < 10^{-7}$s against pair decay. Extended particles with internal structure cannot be excluded on this basis, because the beam-dump experiments are not sensitive to particles with a radius much larger than 10^{-11}cm [28]. The model-independent limits from Bhabha scattering, which should still apply in this case, only pose a lower limit $\tau_X > 3.5 \cdot 10^{-13}$s. This bound is not sufficiently strong, because the heavy-ion experiments could accomodate lifetimes up to the range $10^{-9} - 10^{-10}$s. Two further arguments favour extended particles, independent of the exclusion of point-like particles. First, particles with an internal structure would be expected to exhibit a complex excitation spectrum, similar to the one observed in the heavy-ion experiments. Secondly, the momentum distribution of newly produced extended particles is strongly influenced by the form-factor, favouring production at small momenta in the centre-of-mass system.

What could be the nature of these (still hypothetical) new extended particles? Although various attempts have been made to describe such states in the framework given by the standard SU(3)×SU(2)×U(1) gauge model of particle physics , no quantitatively convincing model has been constructed. In particular, no indication of tightly bound e^+e^-, $(e^+e^-)^2$, etc. states or of the existence of a new phase of real-world QED, has been found so far.

We have, therefore, been led to ask the question whether an entirely new sector of particle physics, involving new elementary light fermions as constituents and new gauge interactions, in the relevant range of energies could exist, which has escaped more than 50 years of research in nuclear and particle physics. At first glance, the question appears so unlikely, that one is inclined to dismiss it outright. On more serious thought, however, one finds that the phenomena associated with such a new particle family could be quite different from what may be naively expected, making the presence of new particles and new interactions very difficult to detect. We will discuss two specific aspects of this question in more detail (QED precision experiments and high-energy e^+e^- collisions) and show that the model can be so constructed that there appears to be no (obvious) contradiction to existing data.

We were thus led to conclude that an entirely new family of light extended particles cannot be ruled out on the basis of existing evidence. We find that the constraints derived for any additional contribution to the anomalous moment of the muon and from the total cross section required to explain the GSI data are just compatible. This result came, admittedly, as a surprise, because we had undertaken our investigation with the aim to exclude all extended particles with the properties required to explain the GSI observations. Our model should be regarded as a generic example for the possibility that a combination of a long-range confining force and short-range repulsion could hide a complete sector of low-energy elementary particles from observation. One might even argue that previous particle searches have left an important gap, i.e. light neutral particles with a radius larger than about 100 fm, which should be investigated, independently from the indications furnished by the GSI events. We will show explicitly how future improvements in the accuracy of QED precision experiments could either rule out, or establish the validity of our model.

2 Models of an extended particle structure

In the following we discuss a model [29] of an extended neutral particle X^0 with a mass $M_X \approx 1.8$ MeV which can decay into correlated e^+ and e^-. That particle consists of a pair of hypothetical constituents f^+, f^- which are bound by some new kind of interaction. In order to facilitate copious production of the neutral state in strong electromagnetic fields as they occur in heavy-ion collisions, the constituents are supposed to be electrically charged with unit [1] elementary charge $Q_f = \pm e$. The composite object is then constructed as a bound state of a pair of these hypothetical particles (f^+f^-). As we introduced new electrically charged light particles (f^+,f^-) which could easily be detected in many experiments the interaction of the constituents is assumed to be confining. That implies that solely singlets of the hypothetical interaction exist as physical states of the theory.
In order to get an idea of the possible observational consequences of such a model we will make use of the analogy to the structure of meson states and investigate properties of (f^+f^-) states in the framework of the spherical MIT bag model and a potential model. We start with the bag model [29]:

2.1 Bag model description

The motion of the constituents $f^\pm$ is governed by the free Dirac equation inside the bag

$$(\not{p} - m_f)\psi = 0 \qquad , r < R_X \tag{1}$$

with the linear boundary condition ensuring confinement:

$$-i\gamma_r\psi = \psi \qquad , r = R_X \tag{2}$$

Here R_X is the radius of the bag and m_f is the rest mass of the constituents. The stability of the bag is achieved by imposing the quadratic boundary condition

$$\frac{1}{2}\partial_r\bar{\psi}\psi = B_X \qquad , r = R_X \tag{3}$$

[1]Since the constituents are assumed to be permanently confined, the charge could be any, even non-integer, multiple of the unit charge. Note, however, that a sizeable production cross-section in heavy-ion collisions requires $|Q_f| \tilde{>} e$.

B_X is the unknown vacuum pressure which simulates the confinement property of the hypothetical interaction between the constituents. Solving eq.(1) together with the condition (2) yields the well-known dimensionless momentum and energy eigenvalues $\omega_{\kappa n}$ and $\epsilon_{\kappa n}$

$$\epsilon_{\kappa n} = E_{\kappa n} \cdot R_X = \sqrt{\omega_{\kappa n}^2 + \mu_f^2} \tag{4}$$

depending on the dimensionless mass parameter $\mu_f \equiv m_f \cdot R_X$. $(n-1)$ counts the number of nodes of the wavefunction, κ labels the different angular momentum states. The energy of the spherical bag is given by

$$E_{bag} = \frac{4}{3}\pi R_X^3 B_X + \sum_{i=1}^{N=2} \frac{\epsilon^{(i)}(\mu_f)}{R_X} \tag{5}$$

where the first term describes the volume energy of the bag and the second term sums over the kinetic energy of the (N=2) constituents labeled by the index i. The stability condition (3) can be fulfilled by minimizing the energy (5) with respect to the bag radius R_X yielding

$$4\pi R_X^2 B_X + \sum_{i=1}^{N=2} \frac{d}{dR}\left(\frac{\epsilon^{(i)}(\mu_f)}{R_X}\right)\Bigg|_{R=R_X} = 0 \tag{6}$$

That equation together with demanding that the state has an energy of, e.g., 1.8 MeV

$$E_{bag} \equiv 1.8\text{MeV} \tag{7}$$

can be used to fix two of the initially three undetermined parameters, i.e. bag radius R_X, vacuum pressure B_X, and constituent mass m_f. One of these parameters can be varied independently which is here chosen to be the constituent mass. m_f can take the values

$$0 \leq m_f < \frac{1}{2} M_X = 900 \text{ keV} \tag{8}$$

The vacuum pressure B_X as function of m_f is shown in fig.1. The corresponding radii of the bag are shown in the same figure. The values of the vacuum pressure are very small, i.e. $B_X^{1/4} < 240$ keV, which shows that it is difficult to relate that interaction to one of the known interactions. However, it cannot in principle be ruled out that there exists an unknown phase in the context of the SU(3)×SU(2)×U(1) gauge theory, whose vacuum energy is slightly different from the usual value by the amount B_X giving rise to extended states as discussed here. Such possibility has been discussed by various authors [30-35]

One important argument in favour of an extended particle is the fact that *multiple e^+e^--coincidence structures* can be explained through the decay of various excited states of the particle. As the particle should preferentially decay into e^+e^- from a relative s-state (this point will be discussed further below) one has to consider the excited $\kappa = -1$ states. That can be done by solving eqs.(1) and (2) for $\kappa = -1, n = 2$ and fixing the bag radius by minimizing the bag energy according to eq.(6). The resulting bag energies are shown in fig.2. One can see that a level spacing of 60 keV, comparable to experiment, is obtained for constituent masses $m_f > 830$keV. The corresponding bag radii are enlarged by $\approx 50\%$ with respect to the groundstate value.

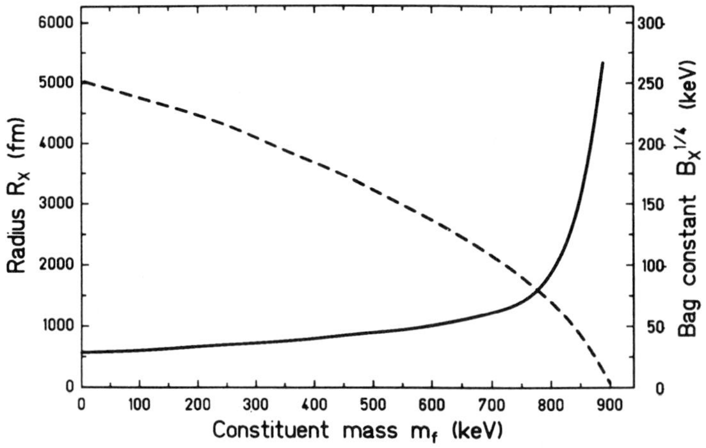

Figure 1. Vacuum pressure $B_X^{1/4}$. (full line) and radius R_X (dashed line) as function of constituent mass m_f

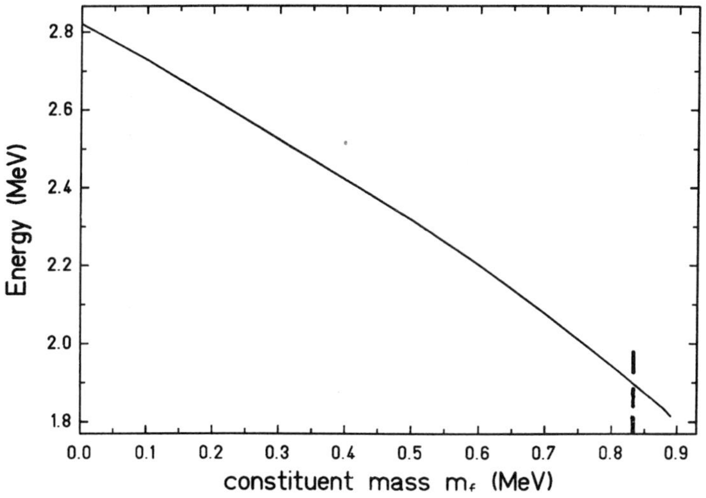

Figure 2. Energy of first excited state ($\kappa = -1, n = 2$) for a bag with 1.8 MeV groundstate energy as function of m_f. The dashed line at $m_f = \bar{m}_f \approx 830$keV indicates that values for m_f which ensures excited states with energies less than 60keV.

2.2 Potential model for the (f^+f^-) state

In the previous section it was shown that the rest masses of the constituents should constitute the major fraction of the total mass of the X^0. Therefore, one may favour a theoretical description of the composite X^0 in terms of a non-relativistic potential model of the constituents. The Schrödinger equation of the relative motion $\phi(\vec{r})$ of the $f^\pm$ particles reads

$$\left(\frac{\hat{\vec{p}}^2}{2m_{red}} + V(r)\right)\phi(\vec{r}) = E\phi(\vec{r}) \quad , \tag{9}$$

$m_{red} = \frac{1}{2}m_f$ being the reduced mass of the $f^\pm$. V(r) is the potential between the constitutents to be specified below. Extending the previous bag model ansatz, where the constituents were treated as free particles inside the bag, one may now consider a more complex structure of the interaction assuming a potential with the general form

$$V(r) = a_x r \quad , \quad r > r_0 \tag{10}$$

$$V(r) = V_0 = const. \quad , \quad r \leq r_0 .$$

Here V_0 is a constant simulating a repulsive interaction between f^+ and f^- ("hard core") at small distances and r_0 describes the range of the repulsion. a_x denotes the string tension of the long-range, confining part of the potential which is the analogue of the vacuum pressure B_X in the bag model. Invoking the flux tube model it is possible to deduce a correspondence between the bag constant B_X and the string tension a_x which has the form [36]

$$a_x = \sqrt{8\pi B_X \alpha_X} \tag{11}$$

with α_X being the coupling constant of the gauge interaction responsible for confinement. Its precise value is of course unknown, but in analogy with the experience from QCD it may be assumed to be of the order of 1, yielding

$$a_x \hat{=} \sqrt{8\pi B_X} \tag{12}$$

The values of a_x derived from a bag model calculation fixing B_X and subsequently applying eq.(12) and compared to a calculation within the potential model shows fair agreement. For masses $m_f \approx 800$ keV the resulting string tension is $a_x \approx 0.1$ keV/fm.

As will be shown later, the particle states within the simple bag model taking free constituents or the potential model with just a linearly rising potential produce additional contributions to the anomalous magnetic moment of leptons and to other QED processes which would spoil the agreement between theory and experiment. This has been the rationale for adding a strongly repulsive interaction at small distances ($r_0 \sim O(1$ fm$)$) to the potential eq.(10).

Like in the bag model, fixing the particle's groundstate at 1.8 MeV one derives a functional dependence $a_x(m_f)$ which is in agreement with the corresponding results of the bag model calculations discussed previously. A level spacing of ≈ 60 keV can be achieved for constituent masses $m_f \approx 0.8$ MeV and a related string tension $a_x \approx 50$ eV/fm, respectively.

3 Production in heavy-ion collisions

3.1 Production of f^+f^- pairs

If the correlated e^+e^--production in heavy-ion collisions is related to the decay of some neutral particle one has to know the properties of the particle in the presence of the strong electromagnetic fields of target and projectile. As the object here discussed has an electromagnetic substructure its energy is considerably affected by the presence of the strong electric fields of the ions. In order to describe this configuration, we first assume that the particle is centered at the center-of-mass of the two ions. The negative constituent f^- is strongly attracted to the charge center by the Coulomb interaction of the nuclei. On the other hand, the positively charged constituent f^+ is repelled to the outer boundary of the confinement region enlarging its radius. *Neglecting* the contribution of the hard core interaction the total energy of the object can be written as

$$E_{X^0}(Z) = E_{f^-}(Z) + E_{f^+}(Z) \tag{13}$$

In order to calculate the energy of the negatively charged constituent E_{f^-} we neglect the influence of the f^+ on the wavefunction of the strongly bound f^- and solves the Dirac equation in the Coulomb field of the two nuclei. The calculations were performed adopting the Coulomb potential $V_c(r)$ of two identical nuclei in the monopole approximation [37], which was very successfully applied for calculating e^+e^- spectra in heavy-ion collisions [38]. Taking that potential the Dirac equation

$$(\hat{p}\cdot\gamma - m_f - \gamma^0 V_c(r))\psi_{f^-} = 0 \quad , r < R_X \tag{14}$$

was numerically solved. The solution yields an energy $E_{f^-}(Z)$ depending on the total charge of the collision system and the internuclear distance R. If the energy decreases below zero, spontaneous supercritical production of the neutral particle could occur.

In the case of the positively charged constituent, the interaction with the strongly localized f^- has been taken into account by solving the Dirac equation for the f^+ in the Coulomb field of the nuclei including the scalar potential $V(r)$ eq.(10) as additional central potential. Thus the equation of motion reads

$$(\hat{p}\cdot\gamma - m_f + \tilde{V}(r))\psi_{f^+} = 0 \quad , r < R_X \tag{15}$$

with the potential

$$\tilde{V}(r) = \gamma^0 V_c(r) + V(r) \tag{16}$$

By solving eqs.(14,15) for the lowest energy eigenvalues one determines the energy of the X^0-particle in the Coulomb field of the two nuclei. The result in the case of two uranium nuclei can be seen in fig. 3. The total energy of the particle in the strong electric field is very small, approaching zero energy for small internuclear distances R. If the energy would decrease below zero, spontaneous supercritical production of the neutral particle could occur.

Anyhow, the small energy gap favours the pair production of f^+f^--pairs in the collision. That point is illustrated in fig.4 where a single particle level diagram of the $f^\pm$ states is schematically shown. The f^--state of lowest energy is strongly bound with a binding energy which - depending on the total charge of the collision system - may exceed $2m_f$. The f^+ states very much resemble positron states, but in contrast the energy levels of the f^+s are discretized due to the confinement in the bag.

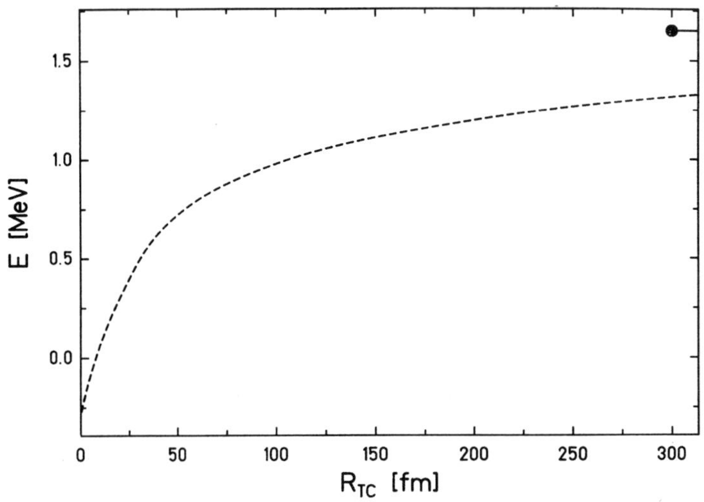

Figure 3. Energy of the extended particle in the Coulomb field of two uranium nuclei for m_f = 850 keV as function of the internuclear distance R. The energy of the particle in the presence of one uranium nuclei is marked to be 1.65 MeV.

As in the case of electrons and positrons the wavefunctions of the constituent particles vary strongly with the nuclear distance due to the rapid change of the Coulomb fields of the nuclei. This yields dynamical transition matrix elements

$$M_{ij} \sim \langle \varphi_{i,f^-} | \partial / \partial t | \varphi_{j,f^+} \rangle \tag{17}$$

which mediate the creation of (f^+f^-)-pairs. These pairs are confined due to the potential eq.(10) giving rise to neutral states. For an exact treatment of the electromagnetic production process in principle one has to take into account the confinement interaction (10) between the particles. However, since the extension of the particle is large, $R_X \sim 1000$ fm, an estimate of the total production probability of the (f^+f^-) states can be obtained by treating the $f^\pm$ as $e^\pm$-like particles with the electron mass replaced by m_f. The production of f^+f^- pairs can be calculated analogously to dynamical electron-positron production [38].

A numerical calculation in a U-U collision for a particle mass m_f = 800 keV resulted in a value of $P_{f^+f^-}$ which is ten times larger than the calculated total e^+-production in the same heavy-ion collision and about three orders of magnitude larger than the corresponding cross section of the observed line structure. The large production probability originates from the fact that in contrast to the case of electrons there are no occupied f^--states in the beginning of the collision and therefore no Pauli suppression for production of f^- particles in bound states occur.

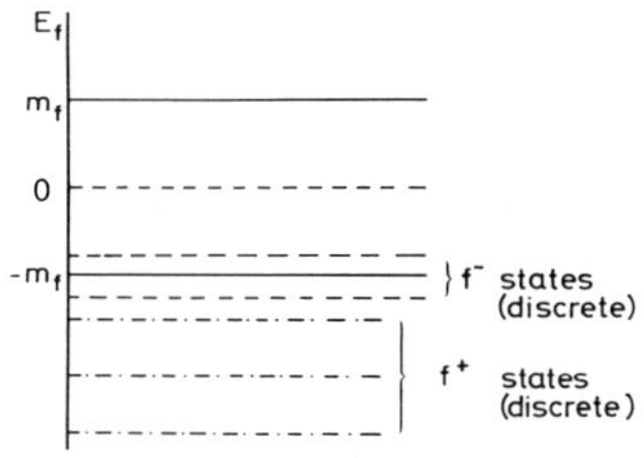

Figure 4. Schematic energy level diagram of f-constituents in the Coulomb field of two heavy ions. The lowermost f^- is supercritically bound whereas the f^+ spectrum is discretized due to the confinement. Three such discrete f^+ states are indicated by the dash-dot-lines; two f^- states by dashed lines.

Therefore, it can be noted that the large cross section of correlated e^+e^--pairs $\sigma_{e^+e^-} \approx 100\mu b$ in heavy-ion collisions can be explained within the model. At this point one should further mention that the similarity of the production process of the bag with dynamical and spontaneous e^+e^- production in the collision suggests that the bag production cross section scales with the total nuclear charge $Z_u = Z_1 + Z_2$ of the collision system like the positron cross section, i.e. roughly

$$\sigma_X \sim (Z_1 + Z_2)^{20} \tag{18}$$

This behaviour would be consistent with the experimental results reported by the ORANGE group stating that the line cross section scales with Z_u like the positron background from pair production in strong electric fields. We will return to that problem, when we discuss the results in the presence of a hard core interaction (chap.(6)).

3.2 Final state effects

When the extended particle has been created as bound state in the center-of-mass frame of the nuclei one has to consider the break-up of the collision system. Although a numerical calculation of the dynamics of the bag in the heavy-ion collision has not yet been performed, one may look at the energy of the particle in the Coulomb field of a single ion in order to get an insight into the strength of binding of the neutral state to target or projectile. The calculation can be done analogously to the case including two nuclei. One solves eq.(14,15) with the Coulomb potential of a single ion. The energy of the total state, given by eq.(14), is about 1.65 MeV and additionaly depicted in fig.3. It can be seen that the X^0 in the Coulomb field of a single nucleus still has a binding energy > 100 keV for constituent masses $m_f > 800$ keV.

Combining these results one cannot definitely answer the question what may happen with the produced particle after the collision. Although adiabatically a bag produced in its lowest energy state should be dragged along with a single target or projectile ion, the influence of the dynamics can change that behaviour, especially for states which are not produced in the groundstate but in higher states. The result shows that one needs a dynamical treatment of the f^+f^- states in the collision. Calculations to that point are in progress. It seems to be plausible to expect that a fraction of the produced particles is getting bound by a single ion and another fraction is set free with small velocity with respect to the CM system. That may explain the experimental finding that the difference energy of some of the correlated e^+e^- lines is not centered at $E_{e^+} - E_{e^-} = 0$ but is shifted to positive values. Sometimes also two 'peaks' in the difference spectrum, one at approximately zero difference energy and one several hundred keV off zero, seem to occur. In addition the decay into e^+e^- pairs from such a bound state would of course not necessarily exhibit a back to back correlation. We will come back to that point in 6.2.

4 Decay of the X^0

The neutral bag state can decay in a similar way as the lowest states of the charmonium system [39]. Since the constituents are electrically charged the object can decay into photons or electrons and positrons depending on the quantum numbers of the specific state. The lowest-order decay diagrams are shown in fig.5. In the case of an 0^{-+}- ('para'-) state with opposite spins of the constituents the f^+f^- may annihilate into two photons.

The 1^{--}- ('ortho'-) state can annihilate into a virtual photon which subsequently decays into a correlated e^+e^--pair. A calculation of the diagrams in fig.6 for non-relativistic bound states yields the decay widths

$$\Gamma_{\gamma\gamma} = \alpha^2 \frac{4\pi}{2\left(\frac{M_X}{2} - m_f - \sqrt{\frac{M_X^2}{4} + m_f^2}\right)^2} \frac{M_X^2}{m_f^2} |\phi_{f\bar{f}}(0)|^2 \tag{19}$$

and

$$\Gamma_{e^+e^-} = \alpha^2 \frac{16\pi}{3}\left(4 + \frac{16}{9}\frac{M_X^2}{m_e^2}\right)\frac{m_e^2}{M_X^5}\left(\frac{M_X^2}{4} - m_e^2\right)^{\frac{1}{2}} |\phi_{f\bar{f}}(0)|^2 \tag{20}$$

The resulting decay widths and corresponding lifetimes for e^+e^- or two γ-decay are shown in table 1 for several parameters of the potential (10). The upper limit for the lifetime of a decaying neutral object set by the heavy-ion collisions is given by the condition that the particle should decay inside of the experimental set-up which yields a value $\tau < 10^{-9}$s which is satisfied by the results shown in table 1.

A lower limit for the lifetime of the particle state can be found from Bhabha scattering experiments [10,40]. In the recent years a number of experiments [41,42,26,43] have been performed searching for a resonance in electron-positron scattering. Apart from two groups claiming the observation of a structure in the excitation function of the Bhabha scattering

[43,44], the results were negative. The most precise measurements were done at Grenoble [26] and showed no structure in the excitation function with a confidence level larger than 3σ in the energy region of the GSI e^+e^- lines. The negative result of the experiment can be used to deduce an upper limit for the width of the resonant state, $\Gamma_R < 1.9$meV, corresponding to a lower limit for the lifetime, $\tau_R > 3.5 \cdot 10^{-13}$s. Comparing this value with the results for the lifetime of the low lying states of our model for X^0 shows that for an appropriate choice of the potential the resulting lifetimes are compatible with the given limits.

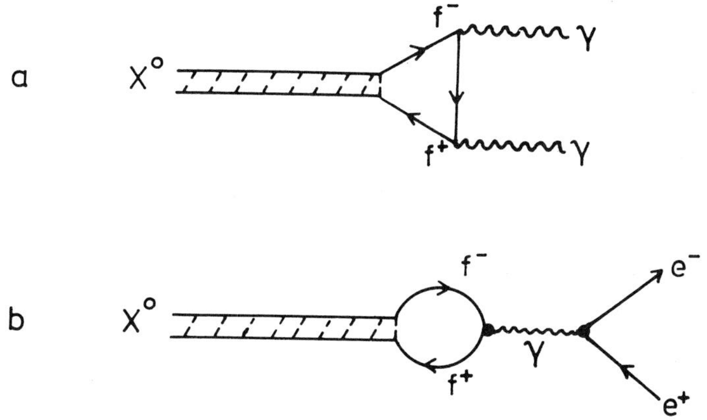

Figure 5. Feynman diagrams for X decay:
(a) 1^{--} state decaying into e^+e^-
(b) 0^{-+} state decaying into two photons

Table 1. The table shows several expectation values and decay widths of f^+f^- states considering different sets of parameters. The obtained values for the width $\Gamma_{e^+e^-}$ are compatible with experimental requirements 10^{-6} eV $< \Gamma_{e^+e^-} < 2 \cdot 10^{-3}$ eV.

m_f	0.85 MeV	0.85 MeV
σ_X	9.05 10^{-3} MeV^2	9.05 10^{-3} MeV^2
r_0	4.64 fm	4.64 10^{-3} fm
V_0	340 MeV	234 10^3 MeV
E_{1S}	1.8 MeV	1.8 MeV
E_{2S}	1.88 MeV	1.88 MeV
E_{2P}	1.847 MeV	1.847 MeV
$< r >_{1S}$	1560 fm	1560 fm
$< r >_{2S}$	2725 fm	2725 fm
$\Gamma_{e^+e^-}$	7.0 10^{-6} eV	2.56 10^{-4} eV
Γ_{rad}	2.4 eV	2.4 eV
$\Gamma_{\gamma\gamma}$	3.1 10^{-5} eV	1.14 10^{-3} eV

The obtained width Γ_{rad} for radiative deexcitation is found to be orders of magnitude larger than the one for pair decay $\Gamma_{e^+e^-}$. An initially excited f^+f^--state, created in high Z collisions, consequently cascades into the lowest S-states before a pair decay happens. Thus the $^3(1S)_1$ ground state acts as a source for sharp e^+e^--lines. If the interaction leads to a level structure such that $E_{2P} \geq E_{2S}$, the $^3(2s)_1$ state will be metastable and show up as a second one. For the presented results, this relation does not hold. This may change, however, if spin-orbit- and spin-spin-interactions are included, thus stabilizing the $^3(2S)_1$ state against radiative decay. Due to the negative outcome of experiments, searching for coincident photons in heavy ion collisions [45], the spin dependent interaction should shift the energies of the lowest $^1S_0$-states such, that $E_{^1(1s)_0}, E_{^1(2s)_0} > E_{^3(1s)_1}, E_{^3(2s)_1}$, in order to suppress two photon decay. We are then left with two sources for two e^+e^--lines separated by ≈ 80 keV. A different concept to explain the occurrence of several lines will be discussed in 6.2.

5 X^0 production in high-energy experiments

If new particle states are introduced in order to explain the GSI-data one has to consider whether these states influence other experiments, either as real particles or through virtual contributions. One experiment where the particles enter as states on mass shell is Bhabha scattering which has already been discussed. Another group of experiments searching especially for new light neutral particles are beam-dump experiments. In these experiments a high-energy beam of electrons hits a target producing a large number of various particles. Whereas electrically charged particles produced in the target will be absorbed or thermalized, neutral particles with a large lifetime may travel through the dump decaying into an electron-positron pair or into photons behind it. These decay products may then be detected as signal of a neutral state.

Several recent beam-dump experiments have shown no signal of a new neutral particle. Therefore one has to consider whether the neutral bag introduced here may show up in these experiments. At that point one should have in mind that the extended state is a large object with electromagnetic substructure. Therefore it should have a small scattering length inside the dump prohibiting its passage through the material. If, the cross section of the particle is roughly given by its geometric size, it should not pass through the dump but decay somewhere inside.

5.1 f^+f^- production in e^+e^- colliders

The next point concerns f^-f^+ production in high-energy electron-positron collisions. As it is known from high-energy experiments, in the collisions jets of hadronic particles may be produced. These jets can be understood in terms of the flux-tube model of QCD. Within that approach a highly excited $q\bar{q}$ state forms a flux tube of color fields, the quarks as sources of the fields being situated at the opposite ends (see fig.6). In the interior of the flux tube $q\bar{q}$

pairs may be spontaneously created. Since these pairs produce an additional color electric field between themselves with opposite orientation but same strength as the original field, the original flux tube breaks into two separate parts. In the case of sufficient excitation energy of the original meson this process may be repeated several times producing two jets of hadrons. The origin of this effect is due to the confinement property of QCD supporting flux tube like excitations.

Since the hypothetical $f^{\pm}$ constituents should also interact via some confining force one has to consider whether a similar effect producing multiparticle events can occur. Here the analogous effect would be the production of a highly excited (f^+f^-) state which could fragment into a large number of neutral X states subsequently decaying into a shower of e^+e^- pairs. As electron-positron showers of that kind have not been observed in experiments, one has to consider the probability of the break-up of an excited X^0 particle. The calculation can be performed analogously as it was done in the case of QCD [46]. There the pair creation

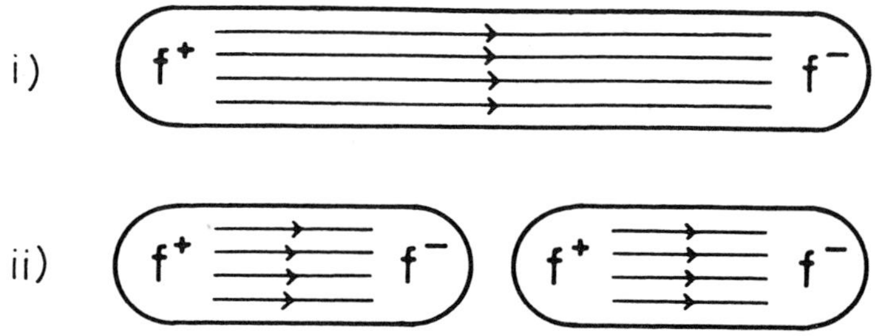

Figure 6. Excited X with the constituents $f^{\pm}$ at the opposite ends.

process in the gluonic fields was treated as a quasiclassical tunneling process, yielding a rate of flux-tube fragmentation per unit volume. For the constituents here discussed the formula reads

$$dP = \frac{a_x^2}{4\pi^3} \sum_{n=1}^{\infty} \frac{1}{n^2} e^{-\pi n m_f^2/a_x} \tag{21}$$

The most important quantity entering the rate is the ratio in the argument of the exponential, $\pi m_f^2/a_x$, with a_x being the string tension of the confining force, which can be understood as the ratio between the rest mass of the created pair $2m_f$ and the field energy stored within a Compton wavelength a_x/m_f. Taking the values for a_x from table 1, one can compute the fragmentation rate (22) per volume to be negligibly small ($\pi m_f^2/a_x \approx 250$).

As the oscillating constituents of a highly excited flux tube are electrically charged the particle may lose its energy by electromagnetic radiation. Treating the state as system of two classical oppositely charged particles interacting via the confinement potential (10) (neglecting the hard core), the power of radiation P is given by the Larmor formula

$$P = -\frac{2}{3}\frac{e^2}{m_f^2}\frac{dp_\mu}{d\tau}\frac{dp^\mu}{d\tau} = -\frac{8}{3}\frac{\alpha}{m_f^2} a_x^2 \tag{22}$$

For a typical value of $a_x = 4.6\ 10^{-2}$keV/fm we get the resulting power P $\approx 1.7\ 10^{13}$MeV/s. Therefore a particle with 100 GeV excitation energy would radiate its energy within 10^{-8}s, and could then decay electromagnetically as described in section 4. The frequency distribution of the radiation is approximately given by the Fourier transform of the velocity of the oscillating particles:

$$v(t) = \frac{p}{E} = \frac{p_0 \pm a_x t}{\sqrt{(p_0 \pm a_x t)^2 + m_f^2}} \tag{23}$$

p_0 denotes the initial momentum imparted on the f^+f^- particles. The $\pm$−sign discriminates the periods of forward and backward motion during the oscillation. In fig. 7 the resulting radiation spectra for different excitation energies are shown. One can see that most of the energy is radiated at low energies < 50 keV where it cannot be easily detected due to the large background of bremsstrahlung radiation in the interaction region.

5.2 Radiative corrections in low-energy QED

A critical problem within the model is given by the contributions of virtual bag states to QED quantities like Lamb shift or leptonic anomalous magnetic moment. These quantities are experimentally measured to a high accuracy and the theoretical values calculated within QED agree with the data very well. New particles which couple in some way to the electron

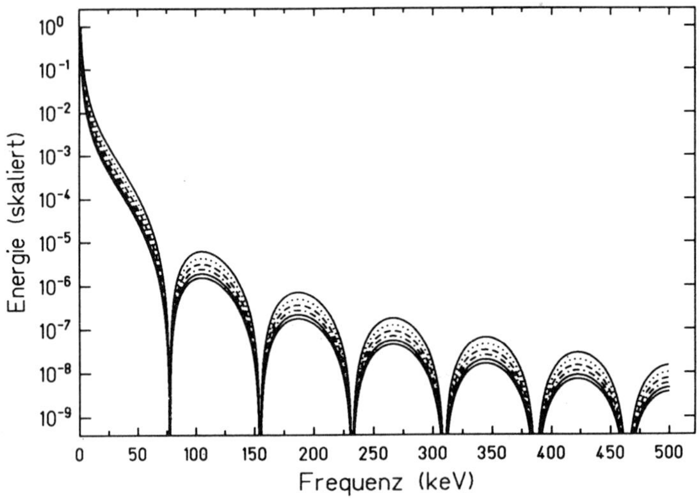

Figure 7. Energy spectrum (scaled units) of the electromagnetic radiation of a highly excited X^0 with excitation energies 50,60,..,100 GeV, resp.

field will naturally yield additional contributions to QED. In the following the contribution of virtual X bags to the anomalous magnetic moment will be discussed. The usual lowest order QED graph which generates the anomalous magnetic moment is given by a virtual photon

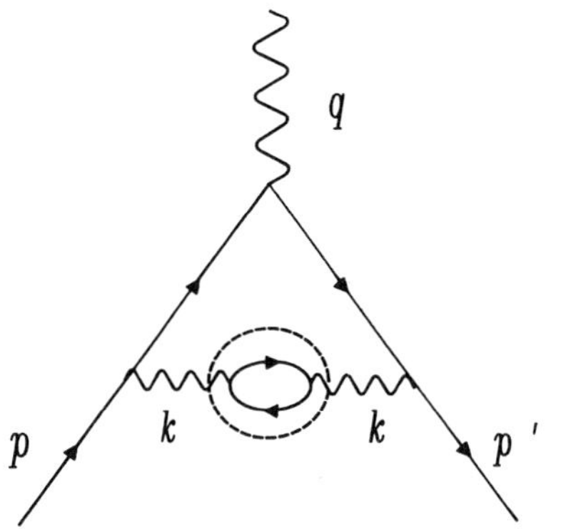

Figure 8. Lowest-order Correction of lepton-photon vertex due to (f^+f^-) exchange

exchange modifying the lepton-photon vertex. A similar process yields the first correction due to some new neutral particle state (fig.8) which couples to the electromagnetic field. The photon creates a virtual X^0 which decays again into a photon. The matrix element of that diagram is given by ($q = p' - p$)

$$\Gamma_\mu(p',p) = -4\pi i e^2 \int \frac{d^4k}{(2\pi)^4} \frac{\Pi^X(k^2)}{k^4 + i\epsilon} \gamma^\nu S_F(p'+k)\gamma_\mu S_F(p+k)\gamma_\nu \tag{24}$$

where $S_F(p)$ denotes the Feynman propagator and m_l is the mass of the lepton ($\mu^\pm, e^\pm$). Here the polarization function Π^X from virtual X states enter. The renormalized polarization function can be written as

$$\Pi^X(k^2) = k^4 \int_0^\infty \frac{\rho(t)}{t^2(k^2 - t + i\epsilon)} dt \tag{25}$$

with the spectral function $\rho(k^2)$ in the Lehmann representation:

$$\rho(k^2) = -\frac{4\pi e^2}{3}(2\pi)^3 \sum_n |\langle 0|j_\mu(0)|n\rangle|^2 \, \delta^{(4)}(k - P_n) \tag{26}$$

The expression (25) then becomes:

$$\Gamma_\mu(p',p) = -4\pi i e^2 \int \frac{dt}{t^2}\rho(t) \int \frac{d^4k}{(2\pi)^4} \frac{1}{t - k^2 + i\epsilon}\gamma^\nu \tag{27}$$

$$S_F(p'+k)\gamma_\mu S_F(p+k)\gamma_\nu$$

The vertex function $\Gamma_\mu(p',p)$ can in general be written in the form

$$\Gamma_\mu(p'-p=q) = f_1(q^2)\gamma^\mu - \frac{1}{2m_l} f_2(q^2)\sigma_{\mu\nu}q^\nu \tag{28}$$

with the electromagnetic form factors $f_1(q^2)$ and $f_2(q^2)$, respectively. The anomalous magnetic moment is given by the value of the form factor $f_2(q^2)$ at q=0. In order to determine $f_2(0)$ one has only to consider these terms in eq.(28) which have the spinor structure of the second term in (29). From that one can derive the expression

$$f_2(0) = \frac{e^2}{\pi^2} \int \frac{dt}{t^2}\rho(t) \int_0^1 dx \frac{x^2(1-x)}{x^2 + (t/m_l^2)(1-x)} \tag{29}$$

In order to evaluate eq.(30) one notes that the cross section for X-boson production in e^+e^- scattering can also be written in terms of the spectral function

$$\sigma_{e^+e^-\to X}(t) \approx 4\pi^2 e^2 \frac{1}{t^2}\rho(t) \tag{30}$$

This yields an expression for $\rho(t)$ which can be inserted into eq.(30) expressing the contribution to the anomalous magnetic moment in terms of the X production cross section. If one assumes the cross section to be a sum of Breit-Wigner resonances located at the mass values M_n and having decay width Γ_n [40]

$$\sigma_{e^+e^-\to X}(t) \sim 12\pi \sum_n \frac{M_n^2}{M_n^2 - 4m_e^2} \frac{\Gamma_n^2}{(t - M_n^2) + M_n^2\Gamma_n^2} \tag{31}$$

one derives the contribution to $f_2(0)$ by X^0-exchange to be

$$f_2(0) = \frac{3}{\pi} \sum_n \frac{M_n^2}{M_n^2 + 2m_e^2} \frac{\Gamma_n}{\sqrt{M_n^2 - 4m_e^2}} \cdot \int_0^1 dx \frac{x^2(1-x)}{x^2 + (M_n^2/m_e^2)(1-x)} \tag{32}$$

The actual calculation of the sum in eq.(33) bears two problems. First. the non-relativistic treatment of the constituents applied for the calculation of groundstate properties of the particle is no longer valid in the case of highly excited states which are included in the sum. Furthermore, it seems to be unpractical to calculate the widths Γ_n of a large number of high-lying intermediate states one by one.

Therefore we proceed as follows. In the first step neglecting the hard core, the calculation of $f_2(0)$ is performed by replacing the cross section of X^0 production as a sum over the individual excitation states eq.(32) by the cross section of non-interacting f^+f^- pairs, i.e. neglecting the confining force. The results of the electronic $f_2(0)$ value is shown in fig.9 in comparison with a calculation taking eq.(33), summing over all excited states up to a cut-off value of the intermediate energy $\sqrt{t_{max}}$. It can be seen that both approaches yield comparable results. Obviously the value of $f_2(0)$ by far exceeds the experimentally allowable value of $\approx 2 \cdot 10^{-10}$. The observed deviations probably originate in the different treatment of the virtual fermion pair: the free $f^\pm$ pair is treated relativistically, whereas the discrete bound states are calculated in the framework of the nonrelativistic equation (9). Therefore, the subsequent calculations were performed by neglecting the confining force between the constituents, using the relativistic $e^+e^-\to f^+f^-$ cross section.

6 Influence of short-range repulsion

6.1 QED precision tests

In order to investigate the influence of a hard core one has to take into account the modification of the free cross section due to the suppression of the f^+f^- wavefunction at small interparticle distances. That has been done by calculating the energy-dependent Gamow factor $g(E)$ describing the suppression from the tunneling of the relativistic wavefunction

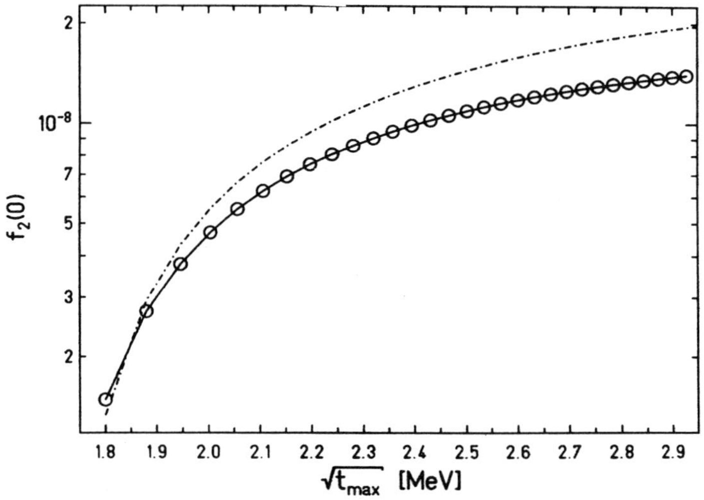

Figure 9. Contributions of X^0 to the anomalous magnetic moment of the electron as function of the cut off energy of the intermediate excited states taken into account. No hard core potential has been taken into account. The dashed-dotted curve shows the result for treating the intermediate states as free f^+f^- pairs whereas the full line denotes the result for summing up the discrete excitation levels of the X^0 explicitly.

through the hard core. The pair production cross section has to be modified in the following manner:

$$\sigma_{f^+f^-} \quad \rightarrow \quad g(E)^2 \sigma_{f^+f^-} \tag{33}$$

$$g(E) = e^{-P_1 r_0} \frac{2P_1}{P_2} \frac{1}{\sqrt{(1 - e^{-2P_1 r_0})^2 + (1 + e^{-2P_1 r_0})^2 \frac{P_1^2}{P_2^2}}} \tag{34}$$

$$P_1 = \sqrt{|V_0 - \epsilon|(\frac{\epsilon + V_0}{2} + m_f)}\ ,\ P_2 = \sqrt{\epsilon(\frac{\epsilon}{2} + m_f)}\ ,$$

The quantity ϵ denotes the binding energy $\epsilon = E - 2m_f$. Within that approach one can systematically vary the parameters of the hardcore (height V_0, width r_0) and study the resulting values of the radiative corrections such as electronic g-2, muonic g-2 and Lamb shift. For more details see [47].

The range of parameters V_0 and r_0, compatible with the corresponding measurements and experimental requirements on the lifetime $\tau_{e^+e^-}$, is depicted in fig.10.

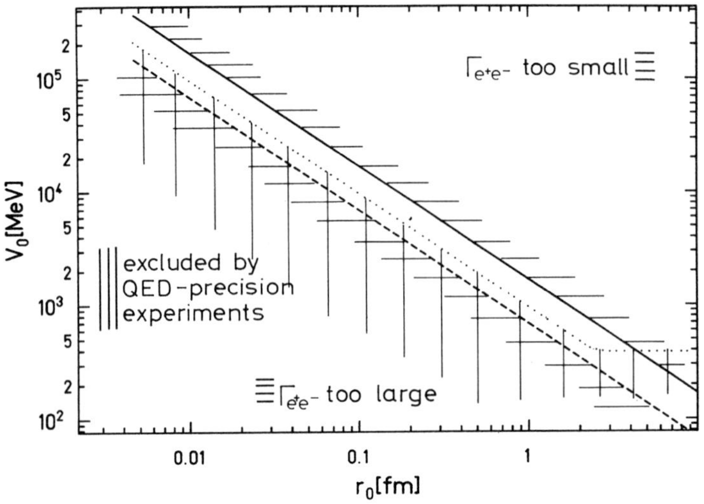

Figure 10. The range of parameters V_0 and r_0, compatible with QED precision measurements and experimental requirements on the decay width $\Gamma_{e^+e^-}$. The region above the solid line is excluded due to values of $\Gamma_{e^+e^-}$ which are too small, whereas the region below the dashed line exhibits the opposite behaviour. The dotted line then excludes the region below, due to the corresponding contributions to QED precision experiments which are too large.

6.2 f^+f^--production in heavy ion collisions

The dynamical calculations presented in section 3.1 , that treated the f^+ f^--particles as non-interacting, have to be modified in the presence of short-range repulsion. Due to the previously investigated systematics we again neglect the confining force and suppose that the total production rate is not seriously altered by it. Since the repulsion acts over a very short range ($r_0 \tilde{<} 1$ fm) it will suppress the coupling matrix elements (17) responsible for dynamic pair production [38]. In the two-body wavefunction of the $f^\pm$ pair the effect of the short-range repulsion can be approximately taken into account by the ansatz [48]:

$$\Psi_{EE'}(\vec{r}, \vec{r}') = G(|\vec{r} - \vec{r}'|)\Psi_E(\vec{r})\Psi_{E'}(\vec{r}') \ , \tag{35}$$

where the Jastrow function $G(|\vec{r} - \vec{r}'|)$ is practically energy-independent for $E, E' << V_0$. $G(|\vec{r} - \vec{r}'|)$ has the limits $G(\infty) = 1$, $G(0) = g(E) \approx g(E')$, where g(E) is the Gamow factor introduced in eq.(35). The matrix elements for dynamical pair creation are thus reduced by an overall factor $g(E)$.

As the total energy of dynamically produced positrons (f^+) and electrons (f^-) lies in the range of $2m_0 \leq E_{tot} \leq 10$ MeV, we may use an overall constant Gamow factor $g(E = 2m_0)$, since $g(E)$ does not vary strongly in this range if the hard core is high enough (fig.11).

We are thus left with the computation of "free" spectra, that have to be multiplied by the corresponding Gamow factor squared. Referring to the previously discussed parameter set, we find the f^+f^--production to be about $4.8 \cdot 10^{-4}$ times the simultaneously produced

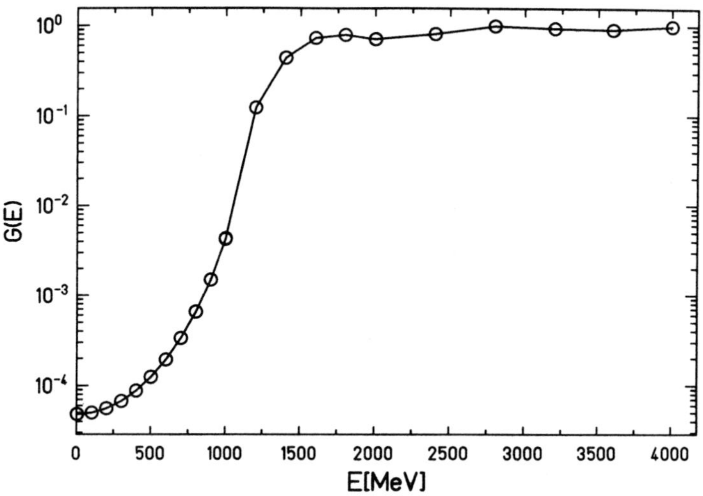

Figure 11. The Gamow factor squared $G(E) = g(E)^2$ as function of the energy E for the parameters $V_0 = 1.13$ GeV and $r_0 = 1.4$ fm.

rate of e^+e^--pairs, which is too small compared to experimental data that require a fraction of the total positron yield of about $\approx 3\%$ generating the line structures in the spectra. Since the parameters of the hard core are mainly restricted by the muonic $(g-2)$, we can gain at low energies by simultaneously increasing its height and localization. It may not be totally unreasonable to push V_0 into the region of electroweak symmetry breaking. As an example we have chosen the following parameters:

$$V_0 = 234 \text{ GeV} \qquad r_0 = 4.64\ 10^{-3}\text{fm} \qquad m_f = 850 \text{ keV} \quad .$$

which yields a low-energy Gamow factor $g(2m_0)^2 = 1.7 \cdot 10^{-3}$. The resulting g-2-contributions, as well as the results for Lamb shift in transitions of muonic lead, are compatible with experimental uncertainties [47]. Fig.12 shows the muonic g-2 contributions that turned out to be most sensitive to the introduction of X^0-states. The total rate of dynamical f^+f^--pair production now yields a fraction of about $\approx 1.7\%$, which compares reasonably well with the experimentally required value.

After the collision, a fraction of the created composite particles may remain bound to one of the separated nuclei. The X^0 ground state and possibly various excited bound states will be populated according to the collision dynamics. Annihilation of a bound X^0 would give rise to e^+e^- coincident lines at discrete energies below M_{X^0}. Our calculations (fig.3) predict that the energy shift is of the order of -150 keV for the ground state (f^- in a 1s orbit around the nucleus) and perhaps -50 keV for a possible excited state (f^- in a 2s orbit). Precise numbers for the mass shifts require three-body calculations: whether the excited states show up in the e^+e^--sum energy spectrum depends on the competition between annihilation and radiative de-excitation. Since the bound X^0 decay proceeds in the vicinity of the nucleus, its two-body characteristics (angular correlation, e^+e^- energy difference) will be disturbed. Precisely this has been observed in recent experiments [49] for all lines except the highest one (809 keV).

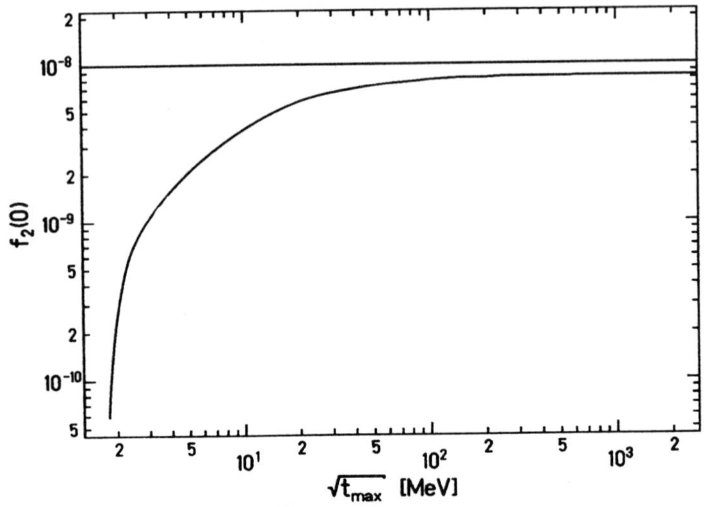

Figure 12. Contributions of X^0 to the anomalous magnetic moment of the muon as function of the cut off energy of the intermediate excited states taken into account. The hard core potential is given by the values $V_0 = 234$ GeV, $r_0 = 0.464 \cdot 10^{-2}$ fm. The full line marks the upper limit supplied by experiment.

6.3 High-energy e^+e^--collisions

Next, we consider the influence of intermediate X^0-production in high-energy e^+e^--collision (E_{Beam} > several GeV) experiments on the value for the ratio R of hadronic production cross section and muon pair creation

$$R = \frac{\sigma_{e^+e^- \to \text{hadrons}}}{\sigma_{e^+e^- \to \mu^+\mu^-}} = 3 \sum_f Q_f^2 \ , \tag{36}$$

with Q_f being the quark charge with flavour f. The sum includes all flavours whose pair creation threshold lies below the available beam energy.

However as discussed before, the preferred values of the hard core are $V_0 > 100$ GeV in order to be consistent with the measured cross section in heavy ion experiments. Then, for energies below $E \leq V_0$, the barrier generates a suppression factor in the f^+f^- creation compared to the production of non-interacting particles as given by the Gamow factor (eq.35). Thus, at presently accessible energies the modifications arising from f^+f^--pair production should be very small compared to the experimental uncertainties ($\Delta R \approx 10\%$). The same argument also holds in other high-energy experiments as long as the energy of possible f^+f^--pair creation lies below the barrier V_0.

7 Conclusions

We discussed a model of an extended neutral particle X^0 with a mass of about ~ 1.8 MeV whose decay into an electron-positron pair may serve as explanation of the e^+e^- coincidence lines as measured in heavy-ion collisions at GSI. The extended object is assumed to be a bound state consisting of new electrically charged constituents (f^+,f^-) interacting via a confining hypothetical force which was implemented by adopting a bag model description or by a potential model exhibiting a linearly rising potential. The most appropriate rest masses of the constituents $f^\pm$ turn out to be relatively large, i.e. $m_f > 800$ keV. For a given value of m_f the vacuum pressure B_X or the coefficient a_x of the confining potential and the corresponding radii R_X of the bag can be determined favouring large bags with $R_X \sim 1000$ fm. The resulting level structure of the bag is comparable to the spacing of the measured coincidence lines. The lifetime of the particle lies within the experimental limits.

Due to its large extension the neutral bag should not show up in beam-dump experiments but will get stuck in the dump until it decays. Since one assumed a confining interaction between the constituents the theoretical possibility of fragmentation of a highly excited X-bag into many X subsequently producing an electron-positron shower was discussed. It could be shown that a process of that kind is suppressed for high rest masses of the constituent particles. However, the state may be deexcited through electromagnetic radiation of low-energy photons within a time of 10^{-8} s and afterwards decays electromagnetically .

Our assumption that the composite object contains electrically charged constituents, has several essential consequences:

i) The virtual contributions of the X^0 to QED high-precision experiments like the leptonic anomalous magnetic moment are potentially large. We have shown that they are smaller than the experimentally allowed values *only* if one assumes an interaction potential between the constituents which exhibits, in addition to long range confinement, a repulsive hard core. Comparison with experimental results requires that the hard core be extremely high and localized, with preferred values $V_0 > 100$ GeV, and $r_0 < 10^{-2}$ fm. In connection with these preferred parameters, the model predicts a muonic $g-2$ contribution of $f_2(0) \approx 9 \cdot 10^{-9}$, which turned out to be the most sensitive bound from high precision QED experiments. An improvement in the accuracy of the $(g-2)_\mu$ measurement, as it is planned at Brookhaven [51], would - according to our present understanding - allow to either confirm or invalidate the constituent model, depending on its result.

ii) Due to electric polarization the energy of the particle may decrease from 1.8 MeV near to zero energy during a heavy-ion collision. That implies large production cross sections of the X^0 in a collision of large-Z nuclei. The most striking success of our model is that it *predicts* a cross section for dynamical production of f^+f^--pairs in heavy ion collisions, that compares reasonably well with the experimental requirements. When target and projectile separate, the bag may stay in the c.m. system of the nuclei or can be dragged along by a single nucleus. In the latter case a decay of the particle in the Coulomb field of one nucleus should result in a) higher energies for the positron compared to the correlated electron; b) an

e^+e^- angular correlation which is not back to back; and c) a strong dependence of the line occurrence on the detailed collision dynamics, that will influence the population of excited bound states. Our model predicts that all e^+e^- sum energy lines, except that with highest energy, should exhibit these particular features (a) and (b). The lower lines should also not show up as resonances in Bhabha scattering. The probabilities of the different final channels for the break-up have to be determined within a dynamical calculation of the heavy-ion scattering, which is currently under study.

iii) Due to the strong suppression of intermediate f^+f^--pair creation arising from the postulated Gamow factor, the modifications of the R-ratio in e^+e^- annihilation at high energy should not violate present experimental results.

One may speculate that the repulsion originates somehow from phenomena at the scale of electroweak symmetry breaking. It should be noted, however, that we have not yet been able to derive a working model for the hard core from an elementary interaction! Another possibility to explain the suppression of the relative f^+f^- wavefunction at zero separation might be a strong coupling to an electrically neutral sector at $r < r_0$, which could be described by an imaginary potential. Such neutral sectors would naturally appear in the context of a gauge theory of the f^+f^--interaction.

One may summarize that the model is adequate to explain a surprising number of the puzzling experimental features of the positron lines found at GSI. If these are in fact caused by neutral particle decays, as many but not all experimental data indicate, one may hope to retain here a working model which could explain the experimental results, without running into contradiction with other well established data.

References

[1] W. Greiner, B. Müller, J. Rafelski, 'Quantum Electrodynamics of Strong Fields', Springer-Verlag Berlin Heidelberg 1985

[2] J. Schweppe, A. Gruppe, K. Bethge, H. Bokemeyer, T. Cowan, H. Folger, J.S. Greenberg, H. Grein, S. Ito, R. Schule, D. Schwalm, K.E. Stiebing, N. Trautmann, P. Vincent, M. Waldschmidt, Phys. Rev. Lett. **51**, 2261 (1983)

[3] M. Clemente, E. Berdermann, P. Kienle, H. Tsertos, W. Wagner, C. Kozhuharov, F. Bosch, W. Koenig, Phys. Lett. **137**, 41 (1984)

[4] T. Cowan, H. Backe, M. Begemann, K. Bethge, H. Bokemeyer, H. Folger, J.S. Greenberg, H. Grein, A. Gruppe, Y. Kido, M. Klüver, D. Schwalm, J. Schweppe, K.E. Stiebing, N. Trautmann, P. Vincent, Phys. Rev. Lett. **54**, 1761 (1985)

[5] W. Koenig, F. Bosch, P. Kienle, C. Kozhuharov, H. Tsertos, E. Berdermann, S. Huchler, W. Wagner, Z. Phys. **A328**, 129 (1987)

[6] T. Cowan, H. Backe, K. Bethge, H. Bokemeyer, H. Folger, J.S. Greenberg, K. Sakaguchi, D. Schwalm, J. Schweppe, K.E. Stiebing, P. Vincent, Phys. Rev. Lett. **56**, 444 (1986)

[7] H. Bokemeyer, H. Folger, T. Cowan, J.S. Greenberg, J. Schweppe, K. Bethge, K. Sakaguchi, P. Salabura, K.E. Stiebing, D. Schwalm, P. Vincent, H. Backe, GSI-87-1, 167 (1987)

[8] E. Berdermann, F. Bosch, P. Kienle, W. Koenig, C. Kozhuharov, H. Tsertos, S. Schuhbeck, S. Huchler, J. Kemmer, A. Schröter, GSI-88-35 preprint (1988)

[9] A. Schäfer, B. Müller, W, Greiner, Phys. Lett. **B149**, 455 (1984)

[10] J. Reinhardt, A. Schäfer, B. Müller, W. Greiner, Phys. Rev. **C33**, 194 (1986)

[11] A. Schäfer, J. Reinhardt, B. Müller, W. Greiner, G. Soff, J. Phys. **G11**, L69 (1985)

[12] A.B. Balantekin, C. Bottcher, M. Strayer, S.J. Lee, Phys. Rev. Lett. **55**, 461 (1985)

[13] L.M. Krauss, M. Zeller, Phys. Rev. **D34**, 3385 (1986)

[14] R. Barbieri, T.E.O. Ericson, Phys. Lett. **B57**, 270 (1975)

[15] U.E. Schröder, Mod. Phys. Lett. **A1**, 157 (1986)

[16] A. Zee, Phys. Lett. **B172**, 377 (1986)

[17] A. Schäfer, J. Reinhardt, W. Greiner, B. Müller, Mod. Phys. Lett. **A1**, 1 (1986)

[18] A. Chodos, L.C.R. Wijewardhana, Phys. Rev. Lett. **56**, 302 (1986)

[19] K. Lane, Phys. Lett. **B169**, 97 (1986)

[20] M. Suzuki, Phys. Lett. **B175**, 364 (1986)

[21] B. Müller, J. Rafelski, Phys. Rev. **D34**, 2896 (1986)

[22] Y. Yamaguchi, H. Sato, Phys. Rev. **C35**, 2156 (1987)

[23] A. Schäfer, B. Müller, J. Reinhardt, Mod. Phys. Lett. **A2**, 159 (1987)

[24] D. Carrier, A. Chodos, L.C.R. Wijewardhana, Phys. Rev. **D34**, 1332 (1986)

[25] A. Schäfer, J. Reinhardt, B. Müller, W. Greiner, Z. Phys. **A324**, 243 (1986)

[26] H. Tsertos, C. Kozhuharov, P. Armbruster, P. Kienle, B. Krusche, K. Schreckenbach, Phys. Lett. **B207**,273 (1988)

[27] A. Konaka, K. Imai, H. Kobayashi, A. Masaike, K. Miyake, T. Nakamura, N. Nagamine, N. Sasao, A. Enomoto, Y. Kukushima, E. Kikutani, H. Koiso, H. Matsumoto, K. Nakahara, S. Ohsawa, T. Taniguchi, I. Sato, J. Urakawa, Phys. Rev. Lett. **57**, 659 (1986)

[28] A. Schäfer, Phys. Lett. **211B**, 207 (1988)

[29] S. Schramm, B. Müller, J. Reinhardt, W. Greiner, Mod. Phys. Lett. **A3**, 783 (1988)

[30] L.S. Celenza, V.K. Mishra, C.M. Shakin, K.F. Liu, Phys. Rev. Lett. 57 (1986)

[31] D.G. Caldi, A. Chodos, Phys. Rev. **D36** 2876 (1987)

[32] Y.J. Ng, Y. Kikuchi, Phys. Rev. **D36** 2880 (1987)

[33] C.W. Wong, Phys. Rev. **D37** 3206 (1988)

[34] A. Schäfer, B. Müller, W. Greiner, Int. J. Mod. Phys. **A3** 1751 (1988)

[35] L.S. Celenza, A. Panziris, C.M. Shakin, H.W. Wang, Nucl. Phys. **A489** 751 (1988)

[36] K. Johnson, C.B. Thorn, Phys. Rev. **D13**, 1934 (1976)

[37] G. Soff, W. Greiner, W. Betz, B. Müller, Phys. Rev. **A20**, 169 (1979)

[38] J. Reinhardt, B. Müller, W. Greiner, Phys. Rev. **A 24**,103 (1981)

[39] V.A. Novikov, L.B. Okun, M.A. Shifman, A.I. Vainshtein, M.B. Voloshin, V.I. Zahkarov, Phys. Rep. **41**, 1 (1978)

[40] J. Reinhardt, A. Scherdin, B. Müller, W. Greiner, Z. Phys. **A327**, 367 (1987)

[41] A.P. Mills, J. Levy, Phys. Rev. **D36**, 707 (1987)

[42] J. van Klinken, W.J. Meiring, F.W.M. de Boer, S.J. Schaafsma, V.A. Wichers, S.Y. van der Werf, G.C.T. Wierda, H.W. Wilschut, H. Bokemeyer, Phys. Lett. **B205**, 223 (1988)

[43] U. von Wimmersperg, S.H. Connell, R.F.A. Hoernle, E. Sideras-Haddad, Phys. Rev. Lett. **59**, 266 (1987)

[44] K. Maier, E. Widmann, W. Bauer, F. Bosch, J. Briggmann, H.-D. Carstanjen, W. Decker, J. Diehl, R. Feldmann, B. Keyerleber, D. Maden, J. Major, H.E. Schaefer, A. Seeger, H. Stoll, Z. Phys. **A330**, 173 (1988)

[45] K. Danzmann, W.E. Meyerhof, E.C.Montenegro, E. Dillard, H.P. Hülskotter, N. Guardala, D.W. Spooner, B. Kotlinski, D. Cline, A. Kavka, C.B. Beausang, J. Burde, M.A. Deleplanque, R.M. Diamond, R.J. Mcdonald, A.O. Macchiavelli, B.S. Rude, F.S. Stephens, J.D. Molitoris, preprint (1989).

[46] N.K. Glendenning, T. Matsui, Phys. Rev. **D28**, 2890 (1983)

[47] S. Graf, S. Schramm, J. Reinhardt, B. Müller, W. Greiner, J. Phys. G **15**, 1467 (1989)

[48] R. Jastrow, Phys. Rev. **98**, 1479 (1955)

[49] H. Bokemeyer, P. Salabura, D. Schwalm, K. E. Stiebing preprint GSI-89-49 (1989)

[50] P.J.E. Peebles, Ap.J. **146**, 542 (1966)

[51] V. Hughes et al, Brookhaven experimental proposal **E821**

This work was supported in part by Bundesministerium für Forschung und Technik (BMFT) under contract no. 060F772 and Gesellschaft für Schwerionenforschung (GSI).

INDEX